# SEMICONDUCTOR DEVICE FUNDAMENTALS

## Robert F. Pierret

School of Electrical and Computer Engineering
Purdue University

**ADDISON-WESLEY PUBLISHING COMPANY**

Reading, Massachusetts · Menlo Park, California · New York
Don Mills, Ontario · Wokingham, England · Amsterdam · Bonn
Sydney · Singapore · Tokyo · Madrid · San Juan · Milan · Paris

Katherine Harutunian   *Associate Editor*
Helen Wythe   *Senior Production Supervisor*
Hugh Crawford   *Manufacturing Supervisor*
Barbara Atkinson   *Associate Cover Design Supervisor*
Peter Blaiwas   *Cover design*
Kenneth J. Wilson   *Text design*
Joyce Grandy   *Copyeditor*
Sandra Rigney   *Production Packaging Services*
G & S Typesetters, Inc.   *Composition*
Publishers' Design and Production Services, Inc.   *Illustrations*


**Library of Congress Cataloging-in-Publication Data**
Pierret, Robert F.
    Semiconductor device fundamentals / Robert F. Pierret.
      p.  cm.
    Includes index.
    ISBN 0-201-54393-1
    1.  Semiconductors.  I.  Title.
  TK7871.85.P484   1996
  621.3815'2–dc20                   95-17387
                                        CIP

MATLAB is a registered trademark of The MathWorks, Inc.,
24 Prime Park Way, Natick, MA 01760-1500.
Phone: (508) 653-1415. Fax: (508) 653-2997
E-mail: info@mathworks.com
WWW:http://www.mathworks.com

Access the latest information about Addison-Wesley books from our Internet gopher site or our World Wide Web page:
gopher aw.com
http://www.aw.com /cseng /

5678910-MA-999897

---

*"The little voice inside never grows any older."*

Frank Pierret (1906–1994)

# PREFACE

Why another text on solid state devices? The author is aware of at least 14 undergraduate texts published on the subject during the past decade. Although several motivating factors could be cited, a very significant factor was the desire to write a book for the next millennium (a Book 2000 so to speak) that successfully incorporates computer-assisted learning. In a recent survey, members of the Undergraduate Curriculum Committee in the School of Electrical and Computer Engineering at Purdue University listed integration of the computer into the learning process as the number one priority. Nationally, university consortiums have been formed which emphasize computer-assisted learning. In January 1992, distribution began of the *Student Edition of MATLAB*, essentially a copy of the original MATLAB manual bundled with a low-cost version of the math-tools software. Over 37,000 copies of the book/software were sold in the first year! Texts and books on a variety of topics from several publishers are now available that make specific use of the MATLAB software. The direction is clear as we proceed into the second millennium: Computer-assisted learning will become more and more prevalent. In dealing with solid state devices, the computer allows one to address more realistic problems, to more readily experiment with "what-if" scenarios, and to conveniently obtain a graphical output. An entire device characteristic can often be computer generated with less time and effort than a small set of manually calculated single-point values.

It should be clarified that the present text is not a totally new entry in the field, but is derived in part from Volumes I–IV of the Addison-Wesley Modular Series on Solid State Devices. Lest there be a misunderstanding, the latest versions of the volumes in the Modular Series were not simply glued together. To the contrary, more than half of the material coverage in the four volumes was completely rewritten. Moreover, several supplemental sections and two additional chapters were added to the Volumes I–IV outline. The new text also contains computer-based text exercises and end-of-chapter problems, plus a number of other special features that are fully described in the General Introduction.

In just about any engineering endeavor there are tradeoffs. Device design is replete with tradeoffs. Tradeoffs also enter into the design of a book. For example, a few topics can be covered in detail (depth) or lesser coverage can be given to several topics (breadth). Similarly one can emphasize the understanding of concepts or optimize the transmission of factual information. Volumes I–IV in the Modular Series are known for their pedantic depth of coverage emphasizing concepts. While retaining the same basic depth of coverage, four "read-only" chapters have been specifically added herein to broaden the coverage and enhance the transmission of factual information. In the read-only chapters the emphasis is more on describing the exciting world of modern-day devices. Compound semiconductor devices likewise receive increased coverage throughout the text. There is also a natural

v

tradeoff between the effort devoted to developing qualitative insight and the implementation of a quantitative analysis. Careful attention has been given to avoid slighting the development of "intuition" in light of the greatly enhanced quantitative capabilities arising from the integrated use of the computer. Lastly, we have not attempted to be all-inclusive in the depth and breadth of coverage—many things are left for later (another course, other books). Hopefully, the proper tradeoffs have been achieved whereby the reader is reasonably knowledgeable about the subject matter and acceptably equipped to perform device analyses after completing the text.

The present text is intended for undergraduate juniors or seniors who have had at least an introductory exposure to electric field theory. Chapters are grouped into three major divisions or "parts," with Part II being further subdivided into IIA and IIB. With some deletions, the material in each of the three parts is covered during a five-week segment of a one-semester, three-credit-hour, junior-senior course in Electrical and Computer Engineering at Purdue University. A day-by-day course outline is supplied on the Instructor's Disk accompanying the Solutions Manual. If necessary to meet time constraints, read-only Chapters 4, 9, 13, and 19 could be deleted from the lecture schedule. (An instructor might preferably assign the chapters as independent readings and reward compliant students by including extra-credit examination questions covering the material.) Standard Chapters 12, 14, and 15, except for the general field-effect introduction in Section 15.1, may also be omitted with little or no loss in continuity.

Although a complete listing of special features is given in the General Introduction, instructors should take special note of the Problem Information Tables inserted prior to the end-of-chapter problems. These tables should prove useful in assigning problems and in dealing with homework graders. When faced with constructing a test, instructors may also be interested in examining the Review Problem Sets found in the mini-chapters (identified by a darkened thumb tab) at the end of the three book parts. The Review Problem Sets are derived from old "open-book" and "closed-book" tests. Concerning the computer-based exercises and problems, the use of either the student or professional version of MATLAB is recommended but not required. The in-text exercise solutions and the problem answers supplied to the instructor, however, do make use of MATLAB. Although it would be helpful, the user need not be familiar with the MATLAB program at the beginning of the book. The MATLAB problems in successive chapters make increasingly sophisticated use of the program. In other words, the early exercises and homework problems provide a learning MATLAB by using MATLAB experience. It is critical, however, that the user complete a large percentage of the computer-based exercises and problems in the first three chapters. The exercises and problems found in later chapters not only assume a reasonably competent use of MATLAB, but also build upon the programs developed in the earlier chapters.

The author gratefully acknowledges the assistance of associates, EE305 students, the respondents to an early marketing survey, the manuscript reviewers, and Addison-Wesley personnel in making Book 2000 a reality. Deserving of special thanks is Ali Keshavarzi for arranging the author's sabbatical at Intel Corporation and for providing photographs of equipment inside the Albuquerque fabrication facility. Prof. Mark Lundstrom at Purdue University was also most helpful in supplying key information and figures for several book sections. Of the undergraduate students asked to examine the manuscript for readability and errors, Eric Bragg stands out as especially perceptive and helpful. The very conscientious manuscript reviewers were Prof. Kenneth A. James, California State University, Long Beach; Prof. Peter Lanyon, Worcester Polytechnic Institute; Prof. Gary S. May, Georgia Institute of Technology; Prof. Dieter K. Schroder, Arizona State University; and Prof. G. W. Stillman, University of Illinois at Urbana-Champaign. In recognition of a fruitful association, a special thanks to Don Fowley, the former editor at Addison-Wesley who enticed the author into writing the book. Last but not least, editor Katherine Harutunian is to be credited with smoothly implementing the project, and executive assistant Anita Devine with cheerfully handling many of the early details.

*Prof. Robert F. Pierret*
School of Electrical and Computer Engineering
Purdue University

# CONTENTS

# GENERAL INTRODUCTION

Coincident with the writing of this book, there has been considerable media discussion about the "Information Superhighway." The envisioned highway itself, the physical link between points supporting the information traffic, is fiber optic cable. Relative to the topic of this book, the on and off ramps, which insert and extract the information from the highway, are semiconductor (solid state) devices. Traffic control, the information processing and the conversion to and from the human interface, is performed by computers. The central processing unit (CPU), memory, and other major components inside the computer are again semiconductor devices. In the modern world, semiconductor devices are incorporated in just about every major system from automobiles to washing machines.

Although roughly a half-century old, the field of study associated with semiconductor devices continues to be dynamic and exciting. New and improved devices are being developed at an almost maddening pace. While the device count in complex integrated circuits increases through the millions and the side-length of chips is measured in centimeters, the individual devices are literally being shrunk to atomic dimensions. Moreover, semiconductor properties desired for a given device structure but not available in nature are being produced artificially; in essence, the semiconductor properties themselves are now being engineered to fit device specifications.

This book should be viewed as a gateway to what the reader will hopefully agree is the fascinating realm of semiconductor devices. It was written for junior- or senior-level students who have at least an introductory exposure to electric field theory. The coverage includes a representative sampling of information about a wide variety of devices. Primary emphasis, however, is placed on developing a fundamental understanding of the internal workings of the more basic device structures. As detailed below, this book contains a number of unique features to assist the reader in learning the material. Alerted at an early stage to their existence, the reader can plan to take full advantage of the cited features.

- *Computer-Based Exercises and End-of Chapter (Homework) Problems.* The majority of chapters contain one or more MATLAB-based exercises requiring the use of a computer. MATLAB is a math-tools software program that has been adapted to run on most computer platforms. A low-cost student edition of MATLAB, which can be used to run all of the files associated with this book, is available in both IBM-compatible and Macintosh versions. The MATLAB program scripts yielding exercise answers are listed in the text and are available in electronic form as detailed below. Computer-based problems, identified by a bullet (●) before the problem number, make up approximately 25% of the problem total. Although other math-tools programs could be employed, the use of MATLAB is recommended in answering computer-based problems. Because computer-based exercises and problems in the early chapters are specifically designed to progressively enhance MATLAB-use proficiency, the user need not be familiar with the MATLAB program at the beginning of the book. It is very important, however, to complete a large percentage of the computer-based exercises and problems in the first three chapters. The exercises and problems found in later chapters not only assume a reasonably competent use of MATLAB, but also build upon the programs developed in the earlier chapters.

- *Computer Program Files.* Program files of the MATLAB scripts associated with computer-based exercises are available via the Internet (`ftp.mathworks.com` in the directory `pub/books/pierret`) or on a floppy disk distributed free of charge by MathWorks, Inc. A pull-out card is provided herein for obtaining the free program disk which is formatted for use with either an IBM-compatible or Macintosh computer. Each floppy disk contains two sets of "m-files" to be used respectively with the pre-4.0 (student 1st edition) or post-4.0 (student 2nd edition) versions of MATLAB. The listings in the text are specifically derived from the Macintosh post-4.0 version, but they are identical to the corresponding IBM-compatible version except for the occasional appearance of a Greek letter.

- *Supplement and Review Mini-Chapters.* The book is divided into three parts. At the end of each part is a Supplement and Review mini-chapter. The mini-chapters, identified by a darkened thumb tab, contain an alternative/supplemental reading list and information table, reference citations for the preceding chapters, an extensive review-list of terms, and review problem sets with answers. The review problem sets are derived from "closed-book" and "open-book" examinations.

- *Read-Only Chapters.* Chapters 4, 9, 13, and 19 have been classified as "read-only." Chapters with the read-only designation contain mostly qualitative information of a supplemental nature. Two of the chapters survey some of the latest device structures. Intended to be fun-reading, the read-only chapters are strategically placed to provide a change of pace. The chapters contain only a small number of equations, no exercises, and few, if any, end-of-chapter problems. In a course format, the chapters could be skipped with little loss in continuity or preferably assigned as independent readings.

- *Problem Information Tables.* A compact table containing information about the end-of-chapter problems in a given chapter is inserted just before the problems. The information provided is (i) the text section or subsection after which the problem can be completed, (ii) the estimated problem difficulty on a scale of 1 (easy or straightforward) to 5 (very difficult or extremely time consuming), (iii) suggested credit or point weighting, and (iv) a short problem description. A bullet before the problem number identifies a computer-based problem. An asterisk indicates computer usage for part of the problem.

- *Equation Summaries.* The very basic carrier modeling equations in Chapter 2 and the carrier action equations in Chapter 3, equations referenced throughout the text, are organized and repeated in Tables 2.4 and 3.3, respectively. These tables would be ideal as "crib sheets" for closed-book examinations covering the material in Part I of the text.

- *Measurements and Data.* Contrary to the impression sometimes left by the sketches and idealized plots often found in introductory texts, device characteristics are real, seldom perfect, and are routinely recorded in measurement laboratories. Herein a sampling of measurement details and results, derived from an undergraduate EE laboratory administered by the author, is included in an attempt to convey the proper sense of reality. For added details on the described measurements, and for a description of additional measurements, the reader is referred to R. F. Pierret, *Semiconductor Measurements Laboratory Operations Manual*, Supplement A in the Modular Series on Solid State Devices, Addison-Wesley Publishing Company, Reading MA, © 1991.

- *Alternative Treatment.* Section 2.1 provides the minimum required treatment on the topic of energy quantization in atomic systems. Appendix A, which contains a more in-depth introduction to the quantization concept and related topics, has been included for those desiring supplemental information. Section 2.1 may be totally replaced by Appendix A with no loss in continuity.

# PART I

# SEMICONDUCTOR FUNDAMENTALS

# 1 Semiconductors: A General Introduction

## 1.1 GENERAL MATERIAL PROPERTIES

The vast majority of all solid state devices on the market today are fabricated from a class of materials known as semiconductors. It is therefore appropriate that we begin the discussion by examining the general nature of semiconducting materials.

### 1.1.1 Composition

Table 1.1 lists the atomic compositions of semiconductors that are likely to be encountered in the device literature. As noted, the semiconductor family of materials includes the elemental semiconductors Si and Ge, compound semiconductors such as GaAs and ZnSe, and alloys like $Al_x Ga_{1-x} As$.[†] Due in large part to the advanced state of its fabrication technology, Si is far and away the most important of the semiconductors, totally dominating the present commercial market. The vast majority of discrete devices and integrated circuits (ICs), including the central processing unit (CPU) in microcomputers and the ignition module in modern automobiles, are made from this material. GaAs, exhibiting superior electron transport properties and special optical properties, is employed in a significant number of applications ranging from laser diodes to high-speed ICs. The remaining semiconductors are utilized in "niche" applications that are invariably of a high-speed, high-temperature, or optoelectronic nature. Given its present position of dominance, we will tend to focus our attention on Si in the text development. Where feasible, however, GaAs will be given comparable consideration and other semiconductors will be featured as the discussion warrants.

Although the number of semiconducting materials is reasonably large, the list is actually quite limited considering the total number of elements and possible combinations of elements. Note that, referring to the abbreviated periodic chart of the elements in Table 1.2, only a certain group of elements and elemental combinations typically give rise to semiconducting materials. Except for the IV-VI compounds, all of the semiconductors listed in Table 1.1 are composed of elements appearing in Column IV of the Periodic Table or are a combination of elements in Periodic Table columns an equal distance to either side of

---

[†] The $x$ (or $y$) in alloy formulas is a fraction lying between 0 and 1. $Al_{0.3}Ga_{0.7}As$ would indicate a material with 3 Al and 7 Ga atoms per every 10 As atoms.

**Table 1.1**  Semiconductor Materials.

| General Classification | Symbol | Semiconductor Name |
|---|---|---|
| (1) Elemental | Si | Silicon |
|  | Ge | Germanium |
| (2) Compounds |  |  |
| (a) IV-IV . . . . . . . . . . . . . . | SiC | Silicon carbide |
| (b) III-V . . . . . . . . . . . . . . | AlP | Aluminum phosphide |
|  | AlAs | Aluminum arsenide |
|  | AlSb | Aluminum antimonide |
|  | GaN | Gallium nitride |
|  | GaP | Gallium phosphide |
|  | GaAs | Gallium arsenide |
|  | GaSb | Gallium antimonide |
|  | InP | Indium phosphide |
|  | InAs | Indium arsenide |
|  | InSb | Indium antimonide |
| (c) II-VI . . . . . . . . . . . . . . | ZnO | Zinc oxide |
|  | ZnS | Zinc sulfide |
|  | ZnSe | Zinc selenide |
|  | ZnTe | Zinc telluride |
|  | CdS | Cadmium sulfide |
|  | CdSe | Cadmium selenide |
|  | CdTe | Cadmium telluride |
|  | HgS | Mercury sulfide |
| (d) IV-VI . . . . . . . . . . . . . . | PbS | Lead sulfide |
|  | PbSe | Lead selenide |
|  | PbTe | Lead telluride |
| (3) Alloys |  |  |
| (a) Binary . . . . . . . . . . . . | $Si_{1-x}Ge_x$ |  |
| (b) Ternary . . . . . . . . . . . | $Al_xGa_{1-x}As$ | (or $Ga_{1-x}Al_xAs$) |
|  | $Al_xIn_{1-x}As$ | (or $In_{1-x}Al_xAs$) |
|  | $Cd_{1-x}Mn_xTe$ |  |
|  | $GaAs_{1-x}P_x$ |  |
|  | $Ga_xIn_{1-x}As$ | (or $In_{1-x}Ga_xAs$) |
|  | $Ga_xIn_{1-x}P$ | (or $In_{1-x}Ga_xP$) |
|  | $Hg_{1-x}Cd_xTe$ |  |
| (c) Quaternary . . . . . . . . | $Al_xGa_{1-x}As_ySb_{1-y}$ |  |
|  | $Ga_xIn_{1-x}As_{1-y}P_y$ |  |

**Table 1.2**    Abbreviated Periodic Chart of the Elements.

| II | III | IV | V | VI |
|---|---|---|---|---|
| 4<br>Be | 5<br>B | 6<br>C | 7<br>N | 8<br>O |
| 12<br>Mg | 13<br>Al | 14<br>Si | 15<br>P | 16<br>S |
| 30<br>**Zn** | 31<br>**Ga** | 32<br>Ge | 33<br>**As** | 34<br>**Se** |
| 48<br>Cd | 49<br>In | 50<br>Sn | 51<br>Sb | 52<br>Te |
| 80<br>Hg | 81<br>Tl | 82<br>Pb | 83<br>Bi | 84<br>Po |

Column IV. The Column III element Ga plus the Column V element As yields the III-V compound semiconductor GaAs; the Column II element Zn plus the Column VI element Se yields the II-VI compound semiconductor ZnSe; the fractional combination of the Column III elements Al and Ga plus the Column V element As yields the alloy semiconductor $Al_x Ga_{1-x} As$. This very general property is related to the chemical bonding in semiconductors, where, on the average, there are four valence electrons per atom.

## 1.1.2 Purity

As will be explained in Chapter 2, extremely minute traces of impurity atoms called "dopants" can have a drastic effect on the electrical properties of semiconductors. For this reason, the compositional purity of semiconductors must be very carefully controlled and, in fact, modern semiconductors are some of the purest solid materials in existence. In Si, for example, the unintentional content of dopant atoms is routinely less than one atom per $10^9$ Si atoms. To assist the reader in attempting to comprehend this incredible level of purity, let us suppose a forest of maple trees was planted from coast to coast, border to border, at 50-ft centers across the United States (including Alaska). An impurity level of one part per $10^9$ would correspond to finding about 25 crabapple trees in the maple tree forest covering the United States! It should be emphasized that the cited material purity refers to *unintentional* undesired impurities. Typically, dopant atoms at levels ranging from one part per $10^8$ to one impurity atom per $10^3$ semiconductor atoms will be *purposely* added to the semiconductor to control its electrical properties.

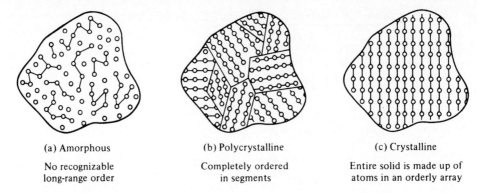

**(a) Amorphous**

No recognizable
long-range order

**(b) Polycrystalline**

Completely ordered
in segments

**(c) Crystalline**

Entire solid is made up of
atoms in an orderly array

**Figure 1.1** General classification of solids based on the degree of atomic order: (a) amorphous, (b) polycrystalline, and (c) crystalline.

### 1.1.3 Structure

The spatial arrangement of atoms within a material plays an important role in determining the precise properties of the material. As shown schematically in Fig. 1.1, the atomic arrangement within a solid causes it to be placed into one of three broad classifications; namely, amorphous, polycrystalline, or crystalline. An amorphous solid is a material in which there is no recognizable long-range order in the positioning of atoms within the material. The atomic arrangement in any given section of an amorphous material will look different from the atomic arrangement in any other section of the material. Crystalline solids lie at the opposite end of the "order" spectrum; in a crystalline material the atoms are arranged in an orderly three-dimensional array. Given any section of a crystalline material, one can readily reproduce the atomic arrangement in any other section of the material. Polycrystalline solids comprise an intermediate case in which the solid is composed of crystalline subsections that are disjointed or misoriented relative to one another.

Upon examining the many solid state devices in existence, one readily finds examples of all three structural forms. An amorphous-Si thin-film transistor is used as the switching element in liquid crystal displays (LCDs); polycrystalline Si gates are employed in Metal-Oxide-Semiconductor Field-Effect Transistors (MOSFETs). In the vast majority of devices, however, the active region of the device is situated within a *crystalline* semiconductor. The overwhelming number of devices fabricated today employ *crystalline* semiconductors.

## 1.2 CRYSTAL STRUCTURE

The discussion at the end of the preceding section leads nicely into the topic of this section. Since in-use semiconductors are typically crystalline in form, it seems reasonable to seek out additional information about the crystalline state. Our major goal here is to present a more detailed picture of the atomic arrangement within the principal semiconductors. To achieve this goal, we first examine how one goes about describing the spatial positioning of atoms within crystals. Next, a bit of "visualization" practice with simple three-dimensional lattices (atomic arrangements) is presented prior to examining semiconductor lattices themselves. The section concludes with an introduction to the related topic of Miller indices. Miller indices are a convenient shorthand notation widely employed for identifying specific planes and directions within crystals.

## 1.2.1 The Unit Cell Concept

Simply stated, a unit cell is a small portion of any given crystal that could be used to reproduce the crystal. To help establish the unit cell (or building-block) concept, let us consider the two-dimensional lattice shown in Fig. 1.2(a). To describe this lattice or to totally specify the physical characteristics of the lattice, one need only provide the unit cell shown in Fig. 1.2(b). As indicated in Fig. 1.2(c), the original lattice can be readily reproduced by merely duplicating the unit cell and stacking the duplicates next to each other in an orderly fashion.

In dealing with unit cells there often arises a misunderstanding, and hence confusion, relative to two points. First, unit cells are not necessarily unique. The unit cell shown in Fig. 1.2(d) is as acceptable as the Fig. 1.2(b) unit cell for specifying the original lattice of

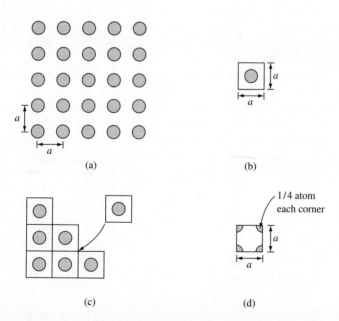

**Figure 1.2**   Introduction to the unit cell method of describing atomic arrangements within crystals. (a) Sample two-dimensional lattice. (b) Unit cell corresponding to the part (a) lattice. (c) Reproduction of the original lattice. (d) An alternative unit cell.

Fig. 1.2(a). Second, a unit cell need not be primitive (the smallest unit cell possible). In fact, it is usually advantageous to employ a slightly larger unit cell with orthogonal sides instead of a primitive cell with nonorthogonal sides. This is especially true in three dimensions where noncubic unit cells are quite difficult to describe and visualize.

## 1.2.2  Simple 3-D Unit Cells

Semiconductor crystals are three-dimensional and are therefore described in terms of three-dimensional (3-D) unit cells. In Fig. 1.3(a) we have pictured the simplest of all three-dimensional unit cells, the simple cubic unit cell. The simple cubic cell is an equal-sided box or cube with an atom positioned at each corner of the cube. The simple cubic lattice associated with this cell is constructed in a manner paralleling the two-dimensional case. In doing so, however, it should be noted that only 1/8 of each corner atom is actually *inside* the cell, as pictured in Fig. 1.3(b). Duplicating the Fig. 1.3(b) cell and stacking the duplicates like blocks in a nursery yields the simple cubic lattice. Specifically, the procedure generates planes of atoms like that previously shown in Fig. 1.2(a). Planes of atoms parallel to the base plane are separated from one another by a unit cell side length or *lattice con-*

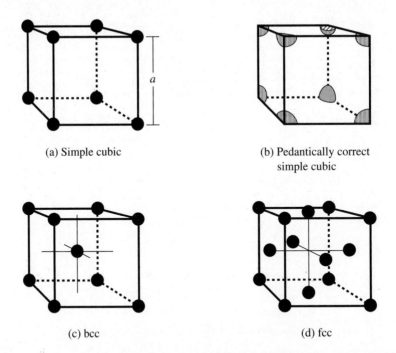

(a) Simple cubic

(b) Pedantically correct
simple cubic

(c) bcc

(d) fcc

**Figure 1.3**  Simple three-dimensional unit cells. (a) Simple cubic unit cell. (b) Pedantically correct simple cubic unit cell including only the fractional portion (1/8) of each corner atom actually within the cell cube. (c) Body centered cubic unit cell. (d) Face centered cubic unit cell.

*stant, a.* When viewed normal to the base plane, atoms in a given plane are positioned directly over atoms in a lower-lying plane.

Figures 1.3(c) and 1.3(d) display two common 3-D cells that are somewhat more complex but still closely related to the simple cubic cell. The unit cell of Fig. 1.3(c) has an atom added at the center of the cube; this configuration is appropriately called the body centered cubic (bcc) unit cell. The face centered cubic (fcc) unit cell of Fig. 1.3(d) contains an atom at each face of the cube in addition to the atoms at each corner. (Note, however, that only one-half of each face atom actually lies inside the fcc unit cell.) Whereas the simple cubic cell contains one atom (1/8 of an atom at each of the eight cube corners), the somewhat more complex bcc and fcc cells contain two and four atoms, respectively. The reader should verify these facts and visualize the lattices associated with the bcc and fcc cells.

---

## (C) Exercise 1.1

The text-associated software distributed via the Internet or on a floppy disk contains a directory or folder named MacMolecule. Files stored inside this directory or folder are ASCII input files generated by the author for use with a computer program by the same name. The MacMolecule program, copyrighted by the University of Arizona, is distributed free of charge to academic users and is available at many academic sites. As the name implies, the program runs only on Macintosh personal computers. The input files supplied to IBM-compatible users, however, can be readily converted for use on a public-access Macintosh.

MacMolecule generates and displays a "ball-and-stick" 3-D color rendering of molecules, unit cells, and lattices. The input files supplied by the author can be used to help visualize the simple cubic, bcc, and fcc unit cells, plus the diamond and zinc-blende unit cells described in the next subsection. The About-MacMolecule file that is distributed with the MacMolecule program provides detailed information about the use of the program and the generation/modification of input files.

The reader is urged to investigate and play with the MacMolecule software. The initial display is always a *z*-direction view and is typically pseudo-two-dimensional in nature. A more informative view is obtained by rotating the model. Rotation is best accomplished by "grabbing and dragging" the edges of the model. The more adventuresome may wish to try their hand at modifying the existing input files or generating new input files.

---

## 1.2.3  Semiconductor Lattices

We are finally in a position to supply details relative to the positioning of atoms within the principal semiconductors. In Si (and Ge) the lattice structure is described by the unit cell pictured in Fig. 1.4(a). The Fig. 1.4(a) arrangement is known as the *diamond lattice* unit cell because it also characterizes diamond, a form of the Column IV element carbon. Examining the diamond lattice unit cell, we note that the cell is cubic, with atoms at each

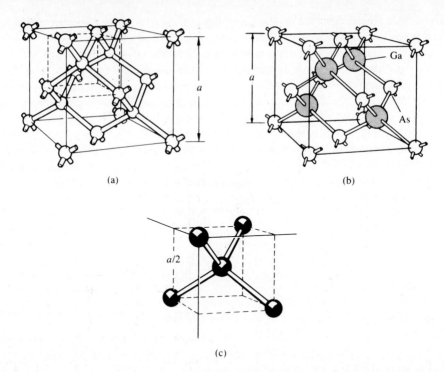

(a)    (b)

(c)

**Figure 1.4**    (a) Diamond lattice unit cell. (b) Zincblende lattice unit cell (GaAs used for illustration). (c) Enlarged top corner of the part (a) diamond lattice emphasizing the four-nearest-neighbor bonding within the structure. The cube side length, *a,* is 5.43 Å and 5.65 Å at $T = 300$ K for Si and GaAs, respectively. [(a) Adapted from Shockley[1]. (b) From Sze[2], ©1981 by John Wiley & Sons, Inc. Reprinted with permission.]

corner and at each face of the cube similar to the fcc cell. The interior of the Fig. 1.4(a) cell, however, contains four additional atoms. One of the interior atoms is located along a cube body diagonal exactly one-quarter of the way down the diagonal from the top front left-hand corner of the cube. The other three interior atoms are displaced one-quarter of the body diagonal length along the previously noted body diagonal direction from the front, top, and left-side face atoms, respectively. Although it may be difficult to visualize from Fig. 1.4(a), the diamond lattice can also be described as nothing more than two interpenetrating fcc lattices. The corner and face atoms of the unit cell can be viewed as belonging to one fcc lattice, while the atoms totally contained within the cell belong to the second fcc lattice. The second lattice is displaced one-quarter of a body diagonal along a body diagonal direction relative to the first fcc lattice.

Most of the III-V semiconductors, including GaAs, crystallize in the *zincblende* structure. The zincblende lattice, typified by the GaAs unit cell shown in Fig. 1.4(b), is essentially identical to the diamond lattice, except that lattice sites are apportioned equally between two different atoms. Ga occupies sites on one of the two interpenetrating fcc sublattices; arsenic (As) populates the other fcc sublattice.

Now that the positioning of atoms within the principal semiconductors has been established, the question may arise as to the practical utilization of such information. Although several applications could be cited, geometrical-type calculations constitute a very common and readily explained use of the unit cell formalism. For example, in Si at room temperature the unit cell side length ($a$) is 5.43 Å (where 1 Å = $10^{-8}$ cm). Since there are eight Si atoms per unit cell and the volume of the unit cell is $a^3$, it follows that there are $8/a^3$ or almost exactly $5 \times 10^{22}$ atoms/cm$^3$ in the Si lattice. Similar calculations could be performed to determine atomic radii, the distance between atomic planes, and so forth. For the purposes of the development herein, however, a major reason for the discussion of semiconductor lattices was to establish that, as emphasized in Fig. 1.4(c), *atoms in the diamond and zincblende lattices have four nearest neighbors.* The chemical bonding (or crystalline glue) within the major semiconductors is therefore dominated by the attraction between any given atom and its four closest neighbors. This is an important fact that should be filed away for future reference.

---

### Exercise 1.2

**P:** If the lattice constant or unit cell side-length in Si is $a = 5.43 \times 10^{-8}$ cm, what is the distance ($d$) from the center of one Si atom to the center of its nearest neighbor?

**S:** As noted in the Fig. 1.4 caption, the atom in the upper front corner of the Si unit cell and the atom along the cube diagonal one-fourth of the way down the diagonal are nearest neighbors. Since the diagonal of a cube is equal to $\sqrt{3}$ times a side length, one concludes $d = (1/4)\sqrt{3}a = (\sqrt{3}/4)(5.43 \times 10^{-8}) = \boxed{2.35 \times 10^{-8} \text{ cm}}$.

---

### (C) Exercise 1.3

**P:** Construct a MATLAB program that computes the number of atoms/cm$^3$ in cubic crystals. Use the MATLAB input function to enter the number of atoms/unit-cell and the unit cell side length ($a$) for a specific crystal. Make a listing of your program and record the program result when applied to silicon.

**S:** MATLAB program script . . .

```
%Exercise 1.3
%Computation of the number of atoms/cm3 in a cubic lattice
N=input('input number of atoms/unit cell, N = ');
a=input('lattice constant in angstrom, a = ');
atmden=N*(1.0e24)/(a^3)   %number of atoms/cm3
```

Program output for Si . . .
    input number of atoms/unit cell, N = 8
    lattice constant in angstrom, a = 5.43
    atmden =
        4.9968e+22

## 1.2.4 Miller Indices

Single crystals of silicon used in device processing normally assume the thin, round form exhibited in Fig. 1.5. The pictured plate-like single crystals, better known as Si wafers, are typical of the starting substrates presently employed by major manufacturers. Of particular interest here is the fact that the surface of a wafer is carefully preoriented to lie along a specific crystallographic plane. Moreover, a "flat" or "notch" is ground along the periphery of the wafer to identify a reference direction within the surface plane. Precise surface orientation is critical in certain device processing steps and directly affects the characteristics exhibited by a number of devices. The flat or notch is routinely employed, for example, to orient device arrays on the wafer so as to achieve high yields during device

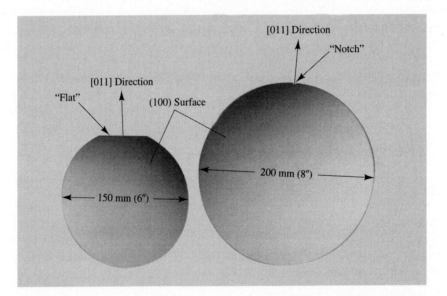

**Figure 1.5** Single-crystal silicon wafers typical of the starting substrates presently employed by major device manufacturers. The 150 mm (6 inch) and 200 mm (8 inch) wafers are nominally 0.625 mm and 0.725 mm thick, respectively. The facing surface is polished and etched yielding a damage-free, mirror-like finish. The figure dramatizes the utility of Miller indices exemplified by the (100) plane and [011] direction designations. (Photograph courtesy of Intel Corporation.)

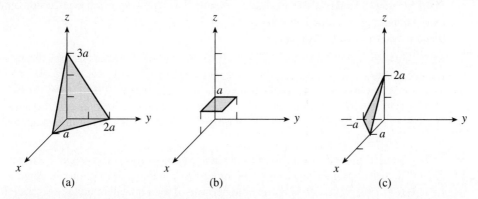

**Figure 1.6**   Sample cubic crystal planes. (a) The (632) plane used in explaining the Miller indexing procedure. (b) The (001) plane. (c) The ($2\overline{2}1$) plane.

separation. The point we wish to make is that the specification of crystallographic planes and directions is of practical importance. Miller indices, exemplified by the (100) plane and [011] direction designations used in Fig. 1.5, constitute the conventional means of identifying planes and directions within a crystal.

The Miller indices for any given plane of atoms within a crystal are obtained by following a straightforward four-step procedure. The procedure is detailed below, along with the simultaneous sample indexing of the plane shown in Fig. 1.6(a).

| Indexing Procedure | Sample Implementation |
|---|---|
| After setting up coordinate axes along the edges of the unit cell, note where the plane to be indexed intercepts the axes. Divide each intercept value by the unit cell length along the respective coordinate axis. Record the resulting normalized (pure-number) intercept set in the order *x, y, z.* | 1, 2, 3 |
| Invert the intercept values; that is, form [1/intercept]s. | 1, ½, ⅓ |
| Using an appropriate multiplier, convert the 1/intercept set to the smallest possible set of whole numbers. | 6, 3, 2 |
| Enclose the whole-number set in curvilinear brackets. | (632) |

To complete the description of the plane-indexing procedure, the user should also be aware of the following special facts:

(i) If the plane to be indexed is parallel to a coordinate axis, the "intercept" along the axis is taken to be at infinity. Thus, for example, the plane shown in Fig. 1.6(b) intercepts the coordinate axes at ∞, ∞, 1, and is therefore a (001) plane.

(ii) If the plane to be indexed has an intercept along the negative portion of a coordinate axis, a minus sign is placed *over* the corresponding index number. Thus the Fig. 1.6(c) plane is designated a $(2\bar{2}1)$ plane.

(iii) Referring to the diamond lattice of Fig. 1.4(a), note that the six planes passing through the cube faces contain identical atom arrangements; that is, because of crystal symmetry, it is impossible to distinguish between the "equivalent" (100), (010), (001), $(\bar{1}00)$, $(0\bar{1}0)$, and $(00\bar{1})$ planes; or, it is impossible to distinguish between {100} planes. A group of equivalent planes is concisely referenced in the Miller notation through the use of { } braces.

(iv) Miller indices cannot be established for a plane passing through the origin of coordinates. The origin of coordinates must be moved to a lattice point not on the plane to be indexed. This procedure is acceptable because of the equivalent nature of parallel planes.

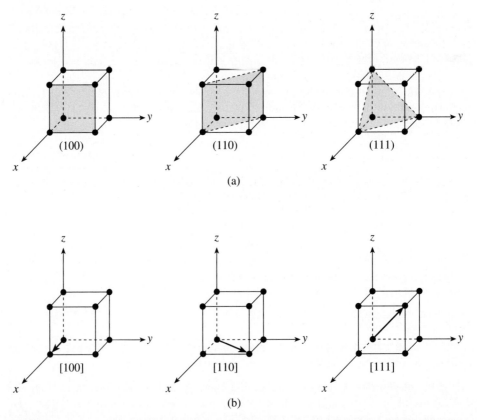

**Figure 1.7**  Visualization and Miller indices of commonly encountered (a) crystalline planes and (b) direction vectors.

**Table 1.3**   Miller Convention Summary.

| Convention | Interpretation |
|------------|----------------|
| (*hkl*) | Crystal plane |
| {*hkl*} | Equivalent planes |
| [*hkl*] | Crystal direction |
| ⟨*hkl*⟩ | Equivalent directions |

Using the simple cubic cell as a reference, the three most commonly encountered crystalline planes are visualized in Fig. 1.7(a).

The Miller indices for *directions* are established in a manner analogous to the well-known procedure for finding the components of a vector. First, set up a vector of arbitrary length in the direction of interest. Next, decompose the vector into its components by noting the projections of the vector along the coordinate axes. Using an appropriate multiplier, convert the component values into the smallest possible whole-number set. This of course changes the length of the original vector but not its direction. Finally, enclose the whole-number set in brackets. Square brackets, [ ], are used to designate specific directions within a crystal; triangular brackets, ⟨ ⟩, designate an equivalent set of directions. Common direction vectors and their corresponding Miller indices are displayed in Fig. 1.7(b). A summary of the Miller bracketing convention for planes and directions is given in Table 1.3.

In the foregoing discussion we presented the procedure for progressing from a given plane or direction in a crystal to the corresponding Miller indices. More often than not, one is faced with the inverse process—visualizing the crystalline plane or direction corresponding to a given set of indices. Fortunately, one seldom encounters other than low-index planes and directions such as (111), (110), [001]. Thus it is possible to become fairly adept at the inverse process by simply memorizing the orientations for planes and directions associated with small-number indices. It is also helpful to note that, *for cubic crystals, a plane and the direction normal to the plane have precisely the same indices*—e.g., the [110] direction is normal to the (110) plane. Of course, any plane or direction can always be deduced by reversing the indexing procedure.

---

**Exercise 1.4**

**P:** For a cubic crystal lattice:

(a) Determine the Miller indices for the plane and direction vector shown in Fig. E1.4(a).

(b) Sketch the plane and direction vector characterized by (011) and [011], respectively.

**S:** (a) The intercepts of the plane are $-1, 1,$ and 2 along the $x, y,$ and $z$ coordinate axes, respectively. Taking [1/intercept]s yields $-1, 1, 1/2$. Multiplying by 2 to obtain the lowest whole-number set, and enclosing in parentheses, gives $\boxed{(\bar{2}21)}$ for the Miller indices of the plane. The direction vector has projections of $2a, a,$ and 0 along the $x, y,$ and $z$ coordinate axes, respectively. The Miller indices for the direction are then $\boxed{[210]}$.

(b) For the (011) plane, [1/intercept]s $= 0, 1, 1$. The normalized $x, y, z$ intercepts of the plane are therefore $\infty, 1, 1$; the plane intersects the $y$-axis at $a$, intersects the $z$-axis at $a$, and is parallel to the $x$-axis. In a cubic crystal the [011] direction is normal to the (011) plane. The deduced plane and direction are as sketched in Fig. E1.4(b).

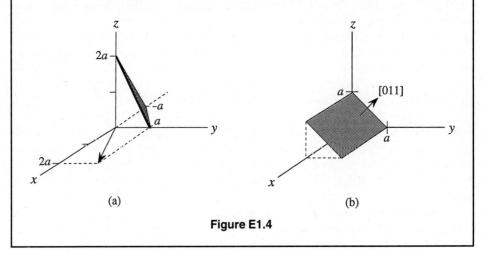

(a)                      (b)

**Figure E1.4**

## 1.3  CRYSTAL GROWTH

### 1.3.1 Obtaining Ultrapure Si

Given the obvious general availability and widespread use of semiconductor materials, particularly Si, it is reasonable to wonder about the origin of the single-crystal Si used in modern-day device production. Could it be that Si is readily available in sandstone deposits?—No. Perhaps, as a sort of by-product, Si single crystals come from South African diamond mines?—Wrong again. As suggested recently in a low-budget science fiction movie, maybe Si is scooped from the ocean bottom by special submarines?—Sorry, no. Although Si is the second most abundant element in the earth's crust and a component in numerous compounds, chief of which are silica (impure $SiO_2$) and the silicates (Si + O + another element), silicon never occurs alone in nature as an element. Single-crystal Si used in device production, it turns out, is a man-made material.

Given the preceding introduction, it should be clear that the initial step in producing device-quality silicon must involve separating Si from its compounds and purifying the

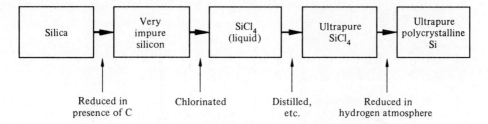

**Figure 1.8**   Summary of the process employed to produce ultrapure silicon.

separated material. The ingenious separation and purification process that has evolved is summarized in Fig. 1.8. Low-grade silicon or ferrosilicon is first produced by heating silica with carbon in an electric furnace. The carbon essentially pulls the oxygen away from the impure $SiO_2$ (i.e., *reduces* the $SiO_2$), leaving behind impure Si. Next, the ferrosilicon is chlorinated to yield $SiCl_4$ or $SiHCl_3$, both of which are liquids at room temperature. Although appearing odd at first glance, the liquefaction process is actually a clever maneuver. Whereas solids are very difficult to purify, a number of standard procedures are available for purifying liquids. An ultrapure $SiCl_4$ (or $SiHCl_3$) is the result after multiple distillation and other liquid purification procedures. Lastly, the high-purity halide is chemically reduced, yielding the desired ultrapure elemental silicon. This is accomplished, for example, by heating $SiCl_4$ in a hydrogen atmosphere [$SiCl_4 + 2H_2 \rightarrow 4\,HCl + Si$].

## 1.3.2  Single-Crystal Formation

Although ultrapure, the silicon derived from the separation and purification process as just described is not a single crystal, but is instead polycrystalline. Additional processing is therefore required to form the large single crystals used in device fabrication. The most commonly employed method yielding large single crystals of Si is known as the Czochralski method. In this method the ultrapure polycrystalline silicon is placed in a quartz crucible and heated in an inert atmosphere to form a melt, as shown in Fig. 1.9. A small single crystal, or Si "seed" crystal, with the normal to its bottom face carefully aligned along a predetermined direction (typically a $\langle 111 \rangle$ or $\langle 100 \rangle$ direction), is then clamped to a metal rod and dipped into the melt. Once thermal equilibrium is established, the temperature of the melt in the vicinity of the seed crystal is reduced, and silicon from the melt begins to freeze out onto the seed crystal, the added material being a structurally perfect extension of the seed crystal. Subsequently, the seed crystal is slowly rotated and withdrawn from the melt; this permits more and more silicon to freeze out on the bottom of the growing crystal. A photograph showing a crystal after being pulled from the melt is displayed in Fig. 1.9(c). The large, cylindrically shaped single crystal of silicon, also known as an ingot, is routinely up to 200 mm (8 inches) in diameter and 1 to 2 meters in length. The Si wafers used in device processing, such as those previously pictured in Fig. 1.5, are ultimately produced by cutting the ingot into thin sections using a diamond-edged saw.

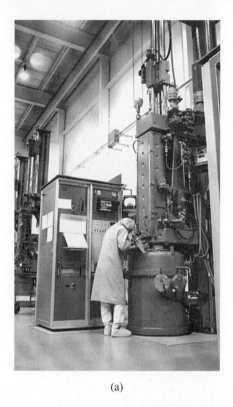

(a)

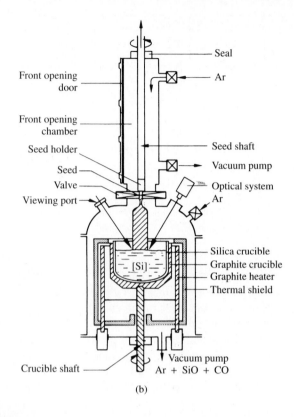

(b)

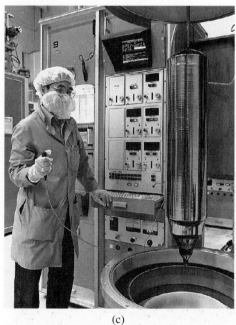

(c)

**Figure 1.9** Silicon crystal-pulling apparatus and resulting single crystal. (a) Photograph of a computer-controlled Czochralski crystal puller. (b) Simplified schematic drawing of the puller. (c) Silicon ingot. [(a) and (c) Photographs courtesy of Wacker Siltronic. (b) From Zuhlehner and Huber[3]; reprinted with permission.]

## 1.4 SUMMARY

This chapter provides basic information about semiconductors in general and silicon in particular. Examining the elemental compositions of semiconductors, one finds that there are, on average, four valence electrons per atom. Moreover, device-quality semiconductors are typically single crystals of high compositional purity. Si crystallizes in the diamond lattice, GaAs crystallizes in the zincblende lattice, and the lattice atoms in both cases have four nearest neighbors. Miller indices, introduced herein as an adjunct to the crystal structure discussion, are the accepted means for identifying planes and directions within a crystal. Finally, large single crystals of device-grade Si are commonly produced by the Czochralski method.

## PROBLEMS

| CHAPTER 1    PROBLEM INFORMATION TABLE | | | | |
|---|---|---|---|---|
| Problem | Complete After | Difficulty Level | Suggested Point Weighting | Short Description |
| 1.1 | 1.4 | 1 | 10 (a::h-1, i-2) | Short answer review |
| 1.2 | 1.2.3 | 1 | 8 | Unit cell for alloy |
| ● 1.3 | " | 1 | 10 (5 each part) | Ge atomic density |
| 1.4 | " | 1 | 8 (4 each part) | Nearest-neighbor distances |
| ● 1.5 | 1.2.4 | 2–3 | 15 (a-2, b-2, c-3, d-3, e-5) | Atoms/cm$^2$ on planes |
| 1.6 | " | 1 | 8 (2 each subpart) | Determine Miller indices |
| 1.7 | " | 1 | 16 (2 each part) | Given (*hkl*), sketch plane |
| 1.8 | " | 1 | 16 (2 each part) | Given [*hkl*], sketch direction |
| 1.9 | " | 1 | 8 (4 each part) | Perpendicular directions |
| 1.10 | " | 2 | 10 | Direction on wafer surface |
| 1.11 | " | 2 | 10 (a-3, b-2, c-3, d-2) | Combination question |
| 1.12 | " | 2 | 12 (2 each part) | Equivalent planes/directions |
| 1.13 | 1.2.3 | 3 | 20 (5 each part) | % of volume occupied |

**1.1** Quick Quiz. Answer the following questions as concisely as possible.

(a) Name (i) one elemental semiconductor and (ii) one III-V compound semiconductor.

(b) What is the difference between a crystalline and a polycrystalline material?

(c) Give a word definition of *unit cell*.

(d) How many atoms are there in a simple cubic unit cell?   in a bcc unit cell?   in a fcc unit cell?   in the unit cell characterizing the diamond lattice?

(e)  $1\text{Å} = ? \, \text{cm}$.

(f)  In terms of the lattice constant $a$, what is the distance between nearest-neighbor atoms in a simple cubic lattice?

(g)  How many nearest-neighbor atoms are there in the diamond and the zincblende lattices?

(h)  What is being indicated by the bracket sets (), [], { }, and ⟨⟩ as employed in the Miller indexing scheme?

(i)  Describe the Czochralski method for obtaining large single crystals of silicon.

**1.2**  The GaAs unit cell is pictured in Fig. 1.4(b). Describe (or sketch) the unit cell for $Al_{0.5}Ga_{0.5}As$.

**1.3**  (a)  The lattice constant of Ge at room temperature is $a = 5.65 \times 10^{-8} \, \text{cm}$. Determine the number of Ge atoms/$cm^3$.

● (b)  Copies of computer-required exercises employing MATLAB are included on disk. Run the Exercise 1.3 program to verify your result from part (a).

**1.4**  In terms of the lattice constant $a$, what is the distance between nearest-neighbor atoms in

(a)  a bcc lattice?

(b)  an fcc lattice?

**1.5**  The surface of a Si wafer is a (100) plane.

(a)  Sketch the placement of Si atoms on the surface of the wafer.

(b)  Determine the number of atoms per $cm^2$ at the surface of the wafer.

(c)/(d)  Repeat parts (a) and (b), this time taking the surface of the Si wafer to be a (110) plane.

● (e)  Primarily for practice in utilizing MATLAB, construct a MATLAB program that computes the surface density of atoms on (100) planes of cubic crystals. The number of atoms with centers on the (100) face of the unit cell, and the lattice constant in Å, are to be the input variables. Confirm the result obtained manually in part (b).

**1.6**  Record all intermediate steps in answering the following questions.

(a)  As shown in Fig. P1.6(a), a crystalline plane has intercepts of $1a$, $3a$, and $1a$ on the $x$, $y$, and $z$ axes, respectively. $a$ is the cubic cell side length.

  (i)  What is the Miller index notation for the plane?

  (ii)  What is the Miller index notation for the direction normal to the plane?

(b) Assuming the crystal structure to be cubic, determine the Miller indices for (i) the plane and (ii) the vector pictured in Fig. P1.6(b).

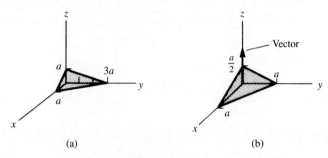

(a)                                                    (b)

**Figure P1.6**

**1.7** Assuming a cubic crystal system, make a sketch of the following planes.

(a) (001),   (b) (111),   (c) (123),   (d) ($\bar{1}$10),

(e) (010),   (f) ($\bar{1}\bar{1}\bar{1}$),   (g) (221),   (h) (0$\bar{1}$0).

**1.8** Assuming a cubic crystal system, use an appropriately directed arrow to identify each of the following directions:

(a) [010],   (b) [101],   (c) [00$\bar{1}$],   (d) [111],

(e) [001],   (f) [110],   (g) [0$\bar{1}$0],   (h) [123].

**1.9** Identify two crystalline directions in a cubic crystal which are perpendicular to

(a) the [100] direction,

(b) the [111] direction.

NOTE: The cosine of the angle $\theta$ between two arbitrary directions, $[h_1 k_1 l_1]$ and $[h_2 k_2 l_2]$, in a cubic crystal is

$$\cos(\theta) = \frac{h_1 h_2 + k_1 k_2 + l_1 l_2}{[(h_1^2 + k_1^2 + l_1^2)(h_2^2 + k_2^2 + l_2^2)]^{1/2}}$$

Consequently, for two directions to be perpendicular, $\cos(\theta) = 0$ and one must have $h_1 h_2 + k_1 k_2 + l_1 l_2 = 0$.

**1.10** As pictured in Fig. 1.5, [011] is the direction in the surface plane normal to the major flat when the surface of the Si wafer is a (100) plane. To construct a particular device structure, a parallel set of grooves must be etched in the (100) surface plane along the [010] direction. Make a sketch indicating how the grooves are to be oriented on the wafer's surface. Explain how you arrived at your sketch.

**1.11** A crystalline lattice is characterized by the cubic unit cell pictured in Fig. P1.11. The cell has a single atom positioned at the center of the cube.

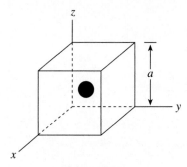

**Figure P1.11**

(a) What is the name of the lattice generated by the given unit cell?

(b) Determine the number of atoms per unit volume in the crystal. (Your answer should be in terms of the lattice constant, $a$.)

(c) Suppose the crystal has a (110) surface plane. Determine the number of atoms per unit area whose centers lie on the (110) plane.

(d) A direction vector is drawn through the center of the atom in the unit cell. Specify the Miller indices of the direction vector.

**1.12** Given a cubic lattice, indicate how many equivalent planes [parts (a) to (c)] or equivalent directions [parts (d) to (f)] are associated with each of the following designations:

$$\text{(a) } \{100\}, \quad \text{(b) } \{110\}, \quad \text{(c) } \{111\},$$
$$\text{(d) } \langle 100 \rangle, \quad \text{(e) } \langle 110 \rangle, \quad \text{(f) } \langle 111 \rangle.$$

**1.13** Treating atoms as rigid spheres with radii equal to one-half the distance between nearest neighbors, show that the ratio of the volume occupied by the atoms to the total available volume in the various crystal structures is

(a) $\pi/6$ or 52% for the simple cubic lattice.

(b) $\sqrt{3}\pi/8$ or 68% for the body centered cubic lattice.

(c) $\sqrt{2}\pi/6$ or 74% for the face centered cubic lattice.

(d) $\sqrt{3}\pi/16$ or 34% for the diamond lattice.

# 2  Carrier Modeling

Carriers are the entities that transport charge from place to place inside a material and hence give rise to electrical currents. In everyday life the most commonly encountered type of carrier is the electron, the subatomic particle responsible for charge transport in metallic wires. Within semiconductors one again encounters the familiar electron, but there is also a second equally important type of carrier—the hole. Electrons and holes are the focal point of this chapter, wherein we examine carrier related concepts, models, properties, and terminology.

Although reminders will be periodically interjected, it should be emphasized from the very start that the development throughout this chapter assumes *equilibrium* conditions exist within the semiconductor. "Equilibrium" is the term used to describe the unperturbed state of a system. Under equilibrium conditions there are no external voltages, magnetic fields, stresses, or other perturbing forces acting on the semiconductor. All observables are invariant with time. This "rest" condition, as we will see, provides an excellent *frame of reference*. Being able to characterize the semiconductor under equilibrium conditions permits one to extrapolate and ascertain the semiconductor's condition when a perturbation has been applied.

The reader should be forewarned that a number of formulas and facts will be presented without justification in this chapter. We would like to provide a complete discussion, properly developing every concept and deriving every formula. Unfortunately, this is not possible. It is our underlying philosophy, moreover, that being able to derive a result is secondary to knowing how to *interpret* and *use* a result. One can, of course, fill in any information gap through supplemental readings. For suggested references see the supplemental reading list at the end of Part I.

> **Lastly, please note that Appendix A provides a more in-depth introduction to the quantization concept and related topics considered in Section 2.1. Section 2.1 may be replaced by Appendix A with no loss in continuity.**

## 2.1 THE QUANTIZATION CONCEPT

Instead of attempting to deal immediately with electrons in crystalline Si, where there are 14 electrons per atom and $5 \times 10^{22}$ atoms/cm$^3$, we will take a more realistic approach and first establish certain ground rules by examining much simpler atomic systems. We begin with the simplest of all atomic systems, the isolated hydrogen atom. This atom, as the reader may recall from a course in modern physics, was under intense scrutiny at the start of the twentieth century. Scientists of that time knew the hydrogen atom contained a negatively charged particle in orbit about a more massive positively charged nucleus. What they could not explain was the nature of the light emitted from the system when the hydrogen atom was heated to an elevated temperature. Specifically, the emitted light was observed at only certain discrete wavelengths; according to the prevailing theory of the time, scientists expected a continuum of wavelengths.

   In 1913 Niels Bohr proposed a solution to the dilemma. Bohr hypothesized that the H-atom electron was restricted to certain well-defined orbits; or equivalently, Bohr assumed that the orbiting electron could take on only certain values of angular momentum. This "quantization" of the electron's angular momentum was, in turn, coupled directly to energy quantization. As can be readily established, if the electron's angular momentum is assumed to be $\mathbf{n}\hbar$, then

$$E_\mathrm{H} = -\frac{m_0 q^4}{2(4\pi\varepsilon_0 \hbar \mathbf{n})^2} = -\frac{13.6}{\mathbf{n}^2}\ \mathrm{eV}, \qquad \mathbf{n} = 1,\ 2,\ 3,\ \ldots \qquad (2.1)$$

where (also see Fig. 2.1) $E_\mathrm{H}$ is the electron binding energy within the hydrogen atom, $m_0$ is the mass of a free electron, $q$ is the magnitude of the electronic charge, $\varepsilon_0$ is the permittivity of free space, $h$ is Planck's constant, $\hbar = h/2\pi$, and $\mathbf{n}$ is the energy quantum number or orbit identifier. The *electron volt* (eV) is a unit of energy equal to $1.6 \times 10^{-19}$ joules. Now, with the electron limited to certain energies inside the hydrogen atom, it follows from the Bohr model that the transition from a higher $\mathbf{n}$ to a lower $\mathbf{n}$ orbit will release quantized energies of light; this explains the observation of emitted light at only certain discrete wavelengths.

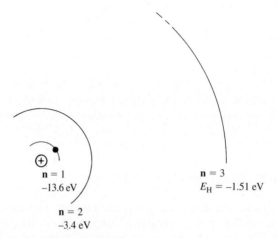

**Figure 2.1**   The hydrogen atom—idealized representation showing the first three allowed electron orbits and the associated energy quantization.

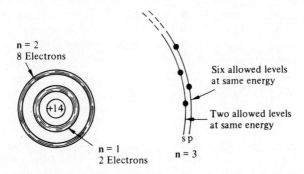

**Figure 2.2**  Schematic representation of an isolated Si atom.

For our purposes, the most important idea to be obtained from the Bohr model is that the energy of electrons in atomic systems is restricted to a limited set of values.[†] Relative to the hydrogen atom, the energy level scheme in a multi-electron atom like silicon is, as one might intuitively expect, decidedly more complex. It is still a relatively easy task, however, to describe the salient energy-related features of an isolated silicon atom. As pictured in Fig. 2.2, ten of the 14 Si-atom electrons occupy very deep-lying energy levels and are tightly bound to the nucleus of the atom. The binding is so strong, in fact, that these ten electrons remain essentially unperturbed during chemical reactions or normal atom–atom interactions, with the ten-electron-plus-nucleus combination often being referred to as the "core" of the atom. The remaining four Si-atom electrons, on the other hand, are rather weakly bound and are collectively called the *valence electrons* because of their strong participation in chemical reactions. As emphasized in Fig. 2.2, the four valence electrons, if unperturbed, occupy four of the eight allowed slots (or states) having the next highest energy above the deep-lying core levels. Finally, we should note, for completeness, that the electronic configuration in the 32-electron Ge-atom (germanium being the other elemental semiconductor) is essentially identical to the Si-atom configuration except that the Ge-core contains 28 electrons.

## 2.2  SEMICONDUCTOR MODELS

Building on the information presented in previous sections, we introduce and describe in this section two very important models or visualization aids that are used extensively in the analysis of semiconductor devices. The inclusion of semiconductor models in a chapter devoted to carrier modeling may appear odd at first glance but is actually quite appropriate. We are, in effect, modeling the carrier "container," the semiconductor crystal.

---

[†] Actually, not only energy but many other observables relating to atomic-sized particles are quantized. An entire field of study, Quantum Mechanics, has been developed to describe the properties and actions of atomic-sized particles and systems.

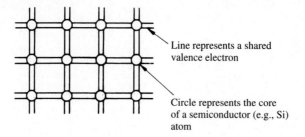

**Figure 2.3**   The bonding model.

## 2.2.1 Bonding Model

The isolated Si atom, or a Si atom not interacting with other atoms, was found to contain four valence electrons. Si atoms incorporated in the diamond lattice, on the other hand, exhibit a bonding that involves an attraction between each atom and its four nearest neighbors (refer to Fig. 1.4c). The implication here is that, in going from isolated atoms to the collective crystalline state, Si atoms come to share one of their valence electrons with each of the four nearest neighbors. This covalent bonding, or equal sharing of valence electrons with nearest neighbors, and the mere fact that atoms in the diamond lattice have four nearest neighbors, give rise to the idealized semiconductor representation, the bonding model, shown in Fig. 2.3. Each circle in the Fig. 2.3 bonding model represents the core of a semiconductor atom, while each line represents a shared valence electron. (There are eight lines connected to each atom because any given atom not only *contributes* four shared electrons but must also *accept* four shared electrons from adjacent atoms.) The two-dimensional nature of the model is, of course, an idealization that facilitates mental visualizations.

Although considerable use will be made of the bonding model in subsequent discussions, it is nevertheless worthwhile at this point to examine sample applications of the model to provide some indication of the model's utility. Two sample applications are presented in Fig. 2.4. In Fig. 2.4(a) we use the bonding model to picture a point defect, a missing atom, in the lattice structure. In Fig. 2.4(b) we visualize the breaking of an atom-to-atom bond and the associated release or freeing of an electron. [Bond breaking (at $T >$ 0 K) and defects occur naturally in all semiconductors, and hence (if we may be somewhat overcritical) the basic model of Fig. 2.3 is strictly valid for an entire semiconductor only at $T \simeq 0$ K when the semiconductor in question contains no defects and no impurity atoms.]

## 2.2.2 Energy Band Model

If our interests were restricted to describing the spatial aspects of events taking place within a semiconductor, the bonding model alone would probably be adequate. Quite often, however, one is more interested in the energy-related aspects of an event. In such instances the bonding model, which says nothing about electron energies, is of little value and the energy band model becomes the primary visualization aid.

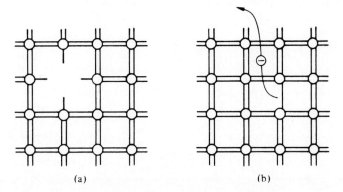

**Figure 2.4**  Sample utilization of the bonding model. (a) Visualization of a missing atom or point defect. (b) Breaking of an atom-to-atom bond and freeing of an electron.

Let us begin the conceptual path leading to the energy band model by recalling the situation inside an isolated Si atom. Reviewing the Section 2.1 discussion, ten of the 14 electrons inside an isolated Si atom are tightly bound to the nucleus and are unlikely to be significantly perturbed by normal atom–atom interactions. The remaining four electrons are rather weakly bound and, if unperturbed, occupy four of the eight allowed energy states immediately above the last core level. Moreover, it is implicitly understood that the electronic energy states within a group of Si atoms, say $N$ Si atoms, are all identical—as long as the atoms are isolated, that is, far enough apart so that they are noninteracting.

Given the foregoing knowledge of the isolated atom situation, the question next arises as to whether we can use the knowledge to deduce information about the crystalline state. Assuredly we can omit any further mention of the core electrons because these electrons are not significantly perturbed by normal interatomic forces. The opposite, however, is true of the valence electrons. If $N$ atoms are brought into close proximity (the case in crystalline Si), it is quite reasonable to expect a modification in the energy states of the valence electrons.

The modification in the valence-electron energy states actually known to take place is summarized in Fig. 2.5. Starting with $N$-isolated Si atoms, and conceptually bringing the atoms closer and closer together, one finds the interatomic forces lead to a progressive spread in the allowed energies. The spread in energies gives rise to closely spaced sets of allowed states known as *energy bands*. At the interatomic distance corresponding to the Si lattice spacing, the distribution of allowed states consists of two bands separated by an intervening energy gap. The upper band of allowed states is called the *conduction band;* the lower band of allowed states, the *valence band;* and the intervening energy gap, the *forbidden gap* or *band gap.* In filling the allowed energy band states, electrons tend to gravitate to the lowest possible energies. Noting that electrons are restricted to single occupancy in allowed states (the Pauli exclusion principle) and remembering that the $4N$ valence band states can just accommodate what were formerly $4N$ valence electrons, we typically find that the valence band is almost completely filled with electrons and the

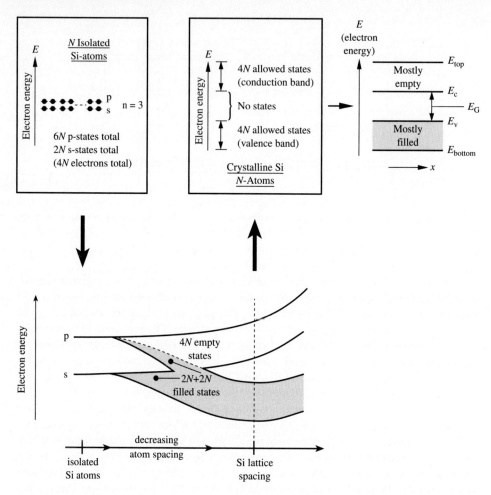

**Figure 2.5** Conceptual development of the energy band model starting with $N$ isolated Si atoms on the top left and concluding with a "dressed-up" version of the energy band model on the top right.

conduction band is all but devoid of electrons. Indeed, the valence band is completely filled and the conduction band completely empty at temperatures approaching $T = 0$ K.

To complete our plausibilization of the energy band model, we need to introduce and utilize one additional fact: Unlike the valence electrons in the isolated-atom case, the band electrons in crystalline silicon are not tied to or associated with any one particular atom. True, on the average, one will typically find four electrons being shared between any given Si atom and its four nearest neighbors (as in the bonding model). However, the identity of the shared electrons changes as a function of time, with the electrons moving around from point to point in the crystal. In other words, the allowed electronic states are no longer atomic states but are associated with the crystal as a whole; independent of the point ex-

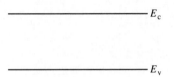

**Figure 2.6**   The energy band diagram—widely employed simplified version of the energy band model.

amined in a perfect crystal, one sees the same allowed-state configuration. We therefore conclude that for a perfect crystal under equilibrium conditions a plot of the allowed electron energies versus distance along any preselected crystalline direction (always called the $x$-direction) is as shown on the right-hand side of Fig. 2.5. The cited plot, a plot of allowed electron energy states as a function of position, is the basic energy band model. $E_c$ introduced in the Fig. 2.5 plot is the lowest possible conduction band energy, $E_v$ is the highest possible valence band energy, and $E_G = E_c - E_v$ is the band gap energy.

Finally, Fig. 2.6 displays the form of the basic energy band model encountered in practice. In this widely employed "shorthand" version, the line to indicate the top energy in the conduction band, the line to indicate the bottom energy in the valence band, the fill pattern drawing attention to mostly filled states in the valence band, the labeling of the $y$- or electron-energy axis, and the labeling of the $x$- or position axis are all understood to exist implicitly, but are not shown explicitly.

### 2.2.3  Carriers

With the semiconductor models having been firmly established, we are now in a position to introduce and to visualize the current-carrying entities within semiconductors. Referring to Fig. 2.7, we note first from part (a) that there are no carriers or possible current flow if the bonding model has no broken bonds. Equivalently, in the energy band model, if the valence band is completely filled with electrons and the conduction band is devoid of electrons, there are no carriers or possible current flow. This lack of carriers and associated current flow is easy to understand in terms of the bonding model where the shared electrons are viewed as being tied to the atomic cores. As accurately portrayed in the energy band model, however, the valence band electrons actually move about in the crystal. How is it then that no current can arise from this group of electrons? As it turns out, the momentum of the electrons is quantized in addition to their energy. Moreover, for each and every possible momentum state in a band, there is another state with an oppositely directed momentum of equal magnitude. Thus, if a band is completely filled with electrons, the *net* momentum of the electrons in the band is always identically zero. It follows that no current can arise from the electrons in a completely filled energy band.

The electrons that do give rise to charge transport are visualized in Fig. 2.7(b). When a Si–Si bond is broken and the associated electron is free to wander about the lattice, the released electron is a carrier. Equivalently, in terms of the energy band model, excitation of valence band electrons into the conduction band creates carriers; that is, *electrons in the*

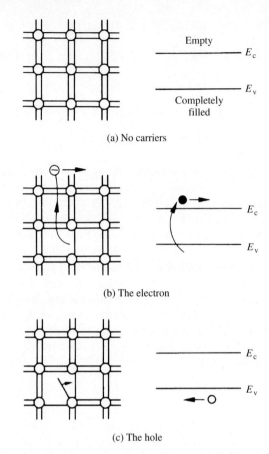

**Figure 2.7**   Visualization of carriers using the bonding model (left) and the energy band model (right). (a) No-carrier situation; (b) visualization of an electron; (c) visualization of a hole.

*conduction band are carriers.* Note that the energy required to break a bond in the bonding model and the band gap energy, $E_G$, are one and the same thing. Likewise, freed bonding-model electrons and conduction band electrons are just different names for the same electrons; in subsequent discussions the word "electrons," when used without a modifier, will be understood to refer to these conduction band electrons.

In addition to releasing an electron, the breaking of a Si–Si bond also creates a missing bond or void in the bonding structure. Thinking in terms of the bonding model, one can visualize the movement of this missing bond from place to place in the lattice as a result of nearby bound electrons jumping into the void (see Fig. 2.7c). Alternatively, one can think in terms of the energy band model where the removal of an electron from the valence band creates an empty state in an otherwise vast sea of filled states. The empty state, like a bubble in a liquid, moves about rather freely in the lattice because of the cooperative motion of the valence band electrons. What we have been describing, the missing bond in the bonding scheme, the empty state in the valence band, is the second type of carrier found in semi-conductors—the hole.

Although perhaps not obvious from the preceding introductory description, the valence band hole is an entity on an equal footing with the conduction band electron. Both electrons and holes participate in the operation of most semiconductor devices. Holes are even the primary carrier in some devices. As one's familiarity with carrier modeling grows, the comparable status of electrons and holes becomes more and more apparent. Ultimately, it is commonplace to conceive of the hole as a type of subatomic particle.

### 2.2.4 Band Gap and Material Classification

We conclude this section by citing a very important tie between the band gap of a material, the number of carriers available for transport in a material, and the overall nature of a material. As it turns out, although specifically established for semiconductors, the energy band model of Fig. 2.6 can be applied with only slight modification to all materials. The major difference between materials lies not in the nature of the energy bands, but rather in the magnitude of the energy gap between the bands.

Insulators, as illustrated in Fig. 2.8(a), are characterized by very wide band gaps, with $E_G$ for the insulating materials diamond and $SiO_2$ being $\simeq 5$ eV and $\simeq 8$ eV, respectively. In these wide band gap materials the thermal energy available at room temperature excites very few electrons from the valence band into the conduction band; thus very few carriers exist inside the material and the material is therefore a poor conductor of current. The band

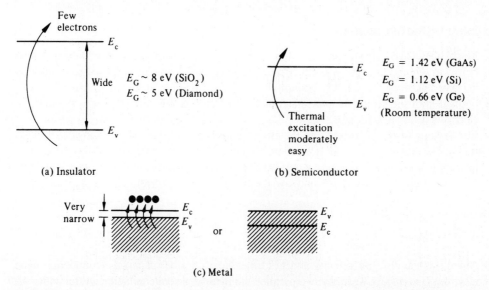

**Figure 2.8** Explanation of the distinction between (a) insulators, (b) semiconductors, and (c) metals using the energy band model.

gap in metals, by way of comparison, is either very small, or no band gap exists at all due to an overlap of the valence and conduction bands (see Fig. 2.8c). There is always an abundance of carriers in metals, and hence metals are excellent conductors. Semiconductors present an intermediate case between insulators and metals. At room temperature $(T = 300 \text{ K})$, $E_G = 1.42$ eV in GaAs, $E_G = 1.12$ eV in Si, and $E_G = 0.66$ eV in Ge. Thermal energy, by exciting electrons from the valence band into the conduction band, creates a moderate number of carriers in these materials, giving rise in turn to a current-carrying capability intermediate between poor and excellent.

## 2.3  CARRIER PROPERTIES

Having formally introduced the electron and the hole, we next seek to learn as much as possible about the nature of these carriers. In this particular section we examine a collage of carrier-related information, information of a general nature including general facts, properties, and terminology.

### 2.3.1 Charge

Both electrons and holes are charged entities. Electrons are negatively charged, holes are positively charged, and the *magnitude* of the carrier charge, $q$, is the same for the two types of carriers. To three-place accuracy in MKS units, $q = 1.60 \times 10^{-19}$ coul. Please note that, under the convention adopted herein, the electron and hole charges are $-q$ and $+q$, respectively; i.e., the sign of the charge is displayed explicitly.

### 2.3.2 Effective Mass

Mass, like charge, is another very basic property of electrons and holes. Unlike charge, however, the carrier mass is not a simple property and cannot be disposed of simply by quoting a number. Indeed, the apparent or "effective" mass of electrons within a crystal is a function of the semiconductor material (Si, Ge, etc.) and is different from the mass of electrons within a vacuum.

Seeking to obtain insight into the effective mass concept, let us first consider the motion of electrons in a vacuum. If, as illustrated in Fig. 2.9(a), an electron of rest mass $m_0$ is moving in a vacuum between two parallel plates under the influence of an electric field $\mathscr{E}$, then, according to Newton's second law, the force **F** on the electron will be

$$\mathbf{F} = -q\mathscr{E} = m_0 \frac{d\boldsymbol{v}}{dt} \tag{2.2}$$

where $\boldsymbol{v}$ is the electron velocity and $t$ is time. Next consider electrons (conduction band electrons) moving between the two parallel end faces of a semiconductor crystal under the influence of an applied electric field, as envisioned in Fig. 2.9(b). Does Eq. (2.2) also describe the overall motion of electrons within the semiconductor crystal? The answer is *no*.

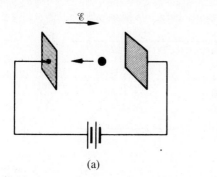 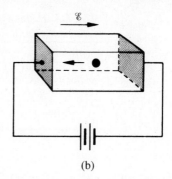

(a)                                    (b)

**Figure 2.9**  An electron moving in response to an applied electric field (a) within a vacuum, and (b) within a semiconductor crystal.

Electrons moving inside a semiconductor crystal will collide with semiconductor atoms, thereby causing a periodic deceleration of the carriers. However, should not Eq. (2.2) apply to the portion of the electronic motion occurring *between* collisions? The answer is again *no*. In addition to the applied electric field, electrons in a crystal are also subject to complex crystalline fields not specifically included in Eq. (2.2).

The foregoing discussion delineated certain important differences between electrons in a crystal and electrons in a vacuum, but it left unresolved the equally important question as to how one properly describes the motion of carriers in a crystal. Strictly speaking, the motion of carriers in a crystal can be described only by using Quantum Mechanics, the formalism appropriate for atomic-sized systems. Fortunately, if the dimensions of the crystal are large compared to atomic dimensions, the complex quantum mechanical formulation for the carrier motion between collisions simplifies to yield an equation of motion identical to Eq. (2.2), except that $m_0$ is replaced by an effective carrier mass. In other words, for the Fig. 2.9(b) electrons one can write

$$\mathbf{F} = -q\mathcal{E} = m_n^* \frac{d\boldsymbol{v}}{dt} \tag{2.3}$$

where $m_n^*$ is the electron effective mass. A similar equation can be written for holes with $-q \rightarrow q$ and $m_n^* \rightarrow m_p^*$. In each case the internal crystalline fields and quantum-mechanical effects are all suitably lumped into the effective mass factor multiplying $d\boldsymbol{v}/dt$. This is a very significant result. *It allows us to conceive of electrons and holes as quasi-classical particles and to employ classical particle relationships in most device analyses.*

Although the effective mass formulation is a significant simplification, it should be mentioned that the carrier acceleration can vary with the direction of travel in a crystal; i.e., the effective masses can have multiple components. Moreover, depending on how a macroscopic observable is related to the carrier motion, a different grouping of mass components can lead to a different $m^*$ being utilized in a particular relationship. There are, for example, cyclotron resonance effective masses, conductivity effective masses, density of

**Table 2.1**    Density of States Effective Masses at 300 K.

| Material | $m_n^*/m_0$ | $m_p^*/m_0$ |
|----------|-------------|-------------|
| Si       | 1.18        | 0.81        |
| Ge       | 0.55        | 0.36        |
| GaAs     | 0.066       | 0.52        |

states effective masses, among others. It is also probably not too surprising that the effective masses vary somewhat with temperature. Herein we make direct use of only the density of states effective masses (see Subsection 2.4.1). The density of states effective masses for electrons and holes in Si, Ge, and GaAs at 300 K are listed in Table 2.1.

### 2.3.3 Carrier Numbers in Intrinsic Material

The term *intrinsic semiconductor* in common usage refers to an extremely pure semiconductor sample containing an insignificant amount of impurity atoms. More precisely, an intrinsic semiconductor is a semiconductor whose properties are native to the material (that is, not caused by external additives). The number of carriers in an intrinsic semiconductor fits into the scheme of things in that it is an identifiable intrinsic property of the material.

Defining, quite generally,

$$n = \text{number of electrons/cm}^3$$

$$p = \text{number of holes/cm}^3$$

existing inside a semiconductor, then, given an intrinsic semiconductor under equilibrium conditions, one finds

$$n = p = n_i \tag{2.4}$$

and

$$
\left.
\begin{aligned}
n_i &\simeq 2 \times 10^6/\text{cm}^3 && \text{in GaAs} \\
&\simeq 1 \times 10^{10}/\text{cm}^3 && \text{in Si} \\
&\simeq 2 \times 10^{13}/\text{cm}^3 && \text{in Ge}
\end{aligned}
\right\} \text{at room temperature}
$$

The electron and hole concentrations in an intrinsic semiconductor are equal because carriers within a very pure material can be created only in pairs. Referring to Fig. 2.7, if a semiconductor bond is broken, a free electron and a missing bond or hole are created simultaneously. Likewise, the excitation of an electron from the valence band into the conduction band automatically creates a valence band hole along with the conduction band electron. Also note that the intrinsic carrier concentration, although large in an absolute sense, is relatively small compared with the number of bonds that could be broken. For example, in Si there are $5 \times 10^{22}$ atoms/cm$^3$ and four bonds per atom, making a grand total of $2 \times 10^{23}$ bonds or valence band electrons per cm$^3$. Since $n_i \simeq 10^{10}$/cm$^3$, one finds less than one bond in $10^{13}$ broken in Si at room temperature. To accurately represent the situation inside intrinsic Si at room temperature, we could cover all the university chalkboards in the world with the bonding model and possibly show only *one* broken bond.

### 2.3.4  Manipulation of Carrier Numbers—Doping

Doping, in semiconductor terminology, is the addition of controlled amounts of specific impurity atoms with the express purpose of increasing either the electron or the hole concentration. The addition of dopants in controlled amounts to semiconductor materials occurs routinely in the fabrication of almost all semiconductor devices. Common Si dopants are listed in Table 2.2. To increase the electron concentration, one can add either phosphorus, arsenic, or antimony atoms to the Si crystal, with phosphorus followed closely by arsenic being the most commonly employed donor (electron-increasing) dopants. To increase the hole concentration, one adds either boron, gallium, indium, or aluminum atoms to the Si crystal, with boron being the most commonly employed acceptor (hole-increasing) dopant.

To understand how the addition of impurity atoms can lead to a manipulation of carrier numbers, it is important to note that the Table 2.2 donors are all from Column V in the Periodic Table of Elements, while all of the cited acceptors are from Column III in the Periodic Table of Elements. As visualized in Fig. 2.10(a) using the bonding model, when a Column V element with five valence electrons is substituted for a Si atom in the semiconductor lattice, four of the five valence electrons fit snugly into the bonding structure. The fifth donor electron, however, does not fit into the bonding structure and is weakly bound to the donor site. At room temperature this electron is readily freed to wander about the lattice and hence becomes a carrier. Please note that this donation (hence the name "donor") of carrier electrons does not increase the hole concentration. The donor ion left behind when the fifth electron is released cannot move, and there are no broken atom–atom bonds associated with the release of the fifth electron.

**Table 2.2**  Common Silicon Dopants. Arrows indicate the most widely employed dopants.

| *Donors (Electron-increasing dopants)* | *Acceptors (Hole-increasing dopants)* |
|---|---|
| P ← ⎫ <br> As ← ⎬ Column V <br> Sb ⎭ elements | B ← ⎫ <br> Ga ⎪ Column III <br> In ⎬ elements <br> Al ⎭ |

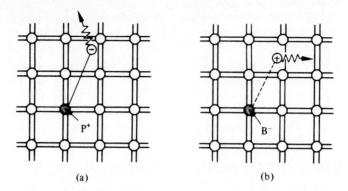

(a)                              (b)

**Figure 2.10**   Visualization of (a) donor and (b) acceptor action using the bonding model. In (a) the Column V element P is substituted for a Si atom; in (b) the Column III element B is substituted for a Si atom.

The explanation of acceptor action follows a similar line of reasoning. The Column III acceptors have three valence electrons and cannot complete one of the semiconductor bonds when substituted for Si atoms in the semiconductor lattice (see Fig. 2.10b). The Column III atom, however, readily accepts (hence the name "acceptor") an electron from a nearby Si–Si bond, thereby completing its own bonding scheme and in the process creating a hole that can wander about the lattice. Here again there is an increase in only one type of carrier. The negatively charged acceptor ion (acceptor atom plus accepted electron) cannot move, and no electrons are released in the hole-creation process.

The foregoing bonding-model-based explanation of dopant action is reasonably understandable. There are, nevertheless, a few loose ends. For one, we noted that the fifth donor electron was rather weakly bound and readily freed at room temperature. How does one interpret the relative term "weakly bound"? It takes ≈1 eV to break Si–Si bonds and very few of the Si–Si bonds are broken at room temperature. Perhaps "weakly bound" means a binding energy ≈0.1 eV or less? The question also arises as to how one visualizes dopant action in terms of the energy band model. Both of the cited questions, as it turns out, involve energy considerations and are actually interrelated.

Let us concentrate first on the binding energy of the fifth donor electron. Crudely speaking, the positively charged donor-core-plus-fifth-electron may be likened to a hydrogen atom (see Fig. 2.11). Conceptually, the donor core replaces the hydrogen-atom nucleus and the fifth donor electron replaces the hydrogen-atom electron. In the real hydrogen atom the electron moves of course in a vacuum, can be characterized by the mass of a free electron, and, referring to Eq. (2.1), has a ground-state binding energy of $-13.6$ eV. In the pseudo-hydrogen atom, on the other hand, the orbiting electron moves through a sea of Si atoms and is characterized by an effective mass. Hence, in the donor or pseudo-atom case, the permittivity of free space must be replaced by the permittivity of Si and $m_0$ must be replaced by $m_n^*$. We therefore conclude that the binding energy ($E_B$) of the fifth donor electron is approximately

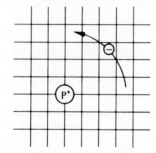

**Figure 2.11**  Pseudo-hydrogen atom model for the donor-site bond.

$$E_B \simeq -\frac{m_n^* q^4}{2(4\pi K_S \varepsilon_0 \hbar)^2} = \frac{m_n^*}{m_0} \frac{1}{K_S^2} E_{H|n=1} \simeq -0.1 \text{ eV} \qquad (2.5)$$

where $K_S$ is the Si dielectric constant ($K_S = 11.8$). Actual donor-site binding energies in Si are listed in Table 2.3. The observed binding energies are seen to be in good agreement with the Eq. (2.5) estimate and, confirming the earlier speculation, are roughly $\simeq 1/20$ the Si band gap energy.

Having established the strength of dopant-site bonds, we are now in a position to tackle the problem of how one visualizes dopant action using the energy band model. Working on the problem, we note first that when an electron is released from a donor it becomes a conduction band electron. If the energy absorbed at the donor is precisely equal to the electron binding energy, the released electron will moreover have the lowest possible energy in the conducting band—namely, $E_c$. Adding an energy $|E_B|$ to the bound electron, in other words, *raises* the electron's energy to $E_c$. Hence, we are led to conclude that the bound electron occupies an allowed electronic level an energy $|E_B|$ below the conduction band edge, or, as visualized in Fig. 2.12, donor sites can be incorporated into the energy band scheme by adding allowed electronic levels at an energy $E_D = E_c - |E_B|$. Note that the donor energy level is represented by a set of dashes, instead of a continuous line, because an electron bound to a donor site is localized in space; i.e., a bound electron does not leave the general $\Delta x$ vicinity of the donor. The relative closeness of $E_D$ to $E_c$ of course reflects the fact that $E_c - E_D = |E_B| \simeq (1/20)E_G(\text{Si})$.

**Table 2.3**  Dopant-Site Binding Energies.

| *Donors* | $|E_B|$ | *Acceptors* | $|E_B|$ |
|----------|---------|-------------|---------|
| Sb | 0.039 eV | B | 0.045 eV |
| P | 0.045 eV | Al | 0.067 eV |
| As | 0.054 eV | Ga | 0.072 eV |
|  |  | In | 0.16  eV |

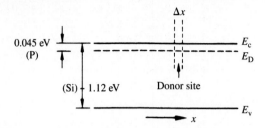

**Figure 2.12**  Addition of the $E = E_D$ donor levels to the energy band diagram. Dashes of width $\Delta x$ emphasize the localized nature of the bound donor-site states.

The actual visualization of dopant action using the energy band model is pictured in Fig. 2.13. Examining the left-hand side of Fig. 2.13(a), one finds all the donor sites filled with bound electrons at temperatures $T \rightarrow 0$ K. This is true because very little thermal energy is available to excite electrons from the donor sites into the conduction band at these very low temperatures. The situation changes, of course, as the temperature is increased, with more and more of the weakly bound electrons being donated to the conduction band. At room temperature the ionization of the donors is all but total, giving rise to the situation pictured at the extreme right of Fig. 2.13(a). Although we have concentrated on donors, the situation for acceptors is completely analogous. As visualized in Fig. 2.13(b), acceptors introduce allowed electronic levels into the forbidden gap at an energy slightly above the valence band edge. At low temperatures, all of these sites will be empty—there is insuffi-

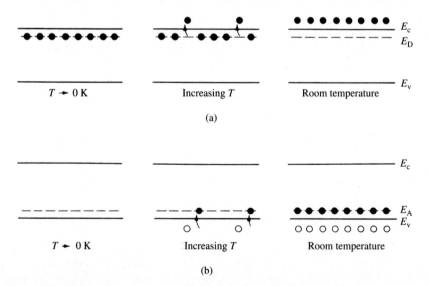

**Figure 2.13**  Visualization of (a) donor and (b) acceptor action using the energy band model.

cient energy at temperatures $T \rightarrow 0$ K for a valence band electron to make the transition to an acceptor site. Increasing temperature, implying an increased store of thermal energy, facilitates electrons jumping from the valence band onto the acceptor levels. The removal of electrons from the valence band of course creates holes. At room temperature, essentially all of the acceptor sites are filled with an excess electron and an increased hole concentration is effected in the material.

To complete this subsection, a few words are in order concerning the doping of non-elemental semiconductors such as GaAs. Dopant action in GaAs follows the same general principles but is slightly more involved because of the existence of two different lattice-site atoms. Completely analogous to the doping of a Si crystal, the Column VI elements S, Se, and Te act as donors when replacing the Column V element As in the GaAs lattice. Similarly, the Column II elements Be, Mg, and Zn act as acceptors when replacing the Column III element Ga. A new situation arises when Column IV elements such as Si and Ge are incorporated into GaAs. Si typically replaces Ga in the GaAs lattice and is a popular $n$-type dopant. However, under certain conditions Si can be made to replace As in the GaAs lattice, thereby functioning as an acceptor. In fact, GaAs $pn$ junctions have been fabricated where Si was both the $p$-side and $n$-side dopant. An impurity that can act as either a donor or an acceptor is referred to as an *amphoteric* dopant.

---

**Exercise 2.1**

*Energy Quiz*

**P:** (a) 1 eV is equal to how many joules of energy?

(b) $kT$ is equal to how many eV at 300 K?

(c) The ionization energy of acceptors and donors in Si is roughly equal to _____?

(d) $E_G(\text{Si}) = $ _____?

(e) $E_G(\text{SiO}_2) = $ _____?

(f) The energy required to ionize a hydrogen atom initially in the $\mathbf{n} = 1$ state is _____?

**S:** (a) 1 eV $= 1.60 \times 10^{-19}$ joules.

(b) $kT = (8.617 \times 10^{-5})(300) = 0.0259$ eV at 300 K.

(c) The ionization energy of dopants is equal to $|E_B|$ of the dopant sites. As discussed in Subsection 2.3.4, $|E_B| \cong 0.1$ eV for dopants in Si.

(d) $E_G(\text{Si}) = 1.12$ eV at 300 K (see Subsection 2.2.4).

(e) $E_G(\text{SiO}_2) \cong 8$ eV (see Subsection 2.2.4).

(f) Making use of Eq. (2.1) in Section 2.1, the hydrogen atom ionization energy is $|E_{H|\mathbf{n}=1}| = 13.6$ eV.

### 2.3.5 Carrier-Related Terminology

Since terminology is often a stumbling block to understanding and since this particular section is replete with specialized terms, it seems appropriate to conclude the section with an overview of carrier-related terminology. Approximately one half of the carrier-related terms that follow were introduced and defined earlier in this section; the remainder of the terms are being listed here for the first time. All of the terms are widely employed and their definitions should be committed to memory.

*Dopants*—specific impurity atoms that are added to semiconductors in controlled amounts for the express purpose of increasing either the electron or the hole concentration.

*Intrinsic semiconductor*—undoped semiconductor; extremely pure semiconductor sample containing an insignificant amount of impurity atoms; a semiconductor whose properties are native to the material.

*Extrinsic semiconductor*—doped semiconductor; a semiconductor whose properties are controlled by added impurity atoms.

*Donor*—impurity atom that increases the electron concentration; *n*-type dopant.

*Acceptor*—impurity atom that increases the hole concentration; *p*-type dopant.

*n-type material*—a donor-doped material; a semiconductor containing more electrons than holes.

*p-type material*—an acceptor-doped material; a semiconductor containing more holes than electrons.

*Majority carrier*—the most abundant carrier in a given semiconductor sample; electrons in an *n*-type material, holes in a *p*-type material.

*Minority carrier*—the least abundant carrier in a given semiconductor sample; holes in an *n*-type material, electrons in a *p*-type material.

## 2.4 STATE AND CARRIER DISTRIBUTIONS

Up to this point in the modeling process we have concentrated on carrier properties and information of a conceptual, qualitative, or, at most, semiquantitative nature. Practically speaking, there is often a need for more detailed information. For example, most semiconductors are doped, and the precise numerical value of the carrier concentrations inside doped semiconductors is of routine interest. Another property of interest is the distribution of carriers as a function of energy in the respective energy bands. In this section we begin the process of developing a more detailed description of the carrier populations. The development will eventually lead to relationships for the carrier distributions and concentrations within semiconductors under equilibrium conditions.

## 2.4.1 Density of States

When the energy band model was first introduced in Section 2.2 we indicated that the total number of allowed states in each band was four times the number of atoms in the crystal. Not mentioned at the time was how the allowed states were distributed in energy; i.e., how many states were to be found at any given energy in the conduction and valence bands. We are now interested in this energy distribution of states, or *density of states,* as it is more commonly known, because the state distribution is an essential component in determining carrier distributions and concentrations.

To determine the desired density of states, it is necessary to perform an analysis based on quantum-mechanical considerations. Herein we will merely summarize the results of the analysis; namely, for energies not too far removed from the band edges, one finds

$$g_c(E) = \frac{m_n^* \sqrt{2m_n^*(E - E_c)}}{\pi^2 \hbar^3}, \qquad E \geq E_c \tag{2.6a}$$

$$g_v(E) = \frac{m_p^* \sqrt{2m_p^*(E_v - E)}}{\pi^2 \hbar^3}, \qquad E \leq E_v \tag{2.6b}$$

where $g_c(E)$ and $g_v(E)$ are the density of states at an energy $E$ in the conduction and valence bands, respectively.

What exactly should be known and remembered about the cited density of states? For one, it is important to grasp the general density of states concept. The density of states can be likened to the description of the seating in a football stadium, with the number of seats in the stadium a given distance from the playing field corresponding to the number of states a specified energy interval from $E_c$ or $E_v$. Second, the general form of the relationships should be noted. As illustrated in Fig. 2.14, $g_c(E)$ is zero at $E_c$ and increases as the square root of energy when one proceeds upward into the conduction band. Similarly, $g_v(E)$ is precisely zero at $E_v$ and increases with the square root of energy as one proceeds downward from $E_v$ into the valence band. Also note that differences between $g_c(E)$ and $g_v(E)$ stem from differences in the carrier effective masses. If $m_n^*$ were equal to $m_p^*$, the seating (states) on both sides of the football field (the band gap) would be mirror images of each other. Finally, considering closely spaced energies $E$ and $E + dE$ in the respective bands, one can state

$g_c(E)dE$ represents the number of conduction band states/cm$^3$ lying in the energy range between $E$ and $E + dE$ (if $E \geq E_c$),

$g_v(E)dE$ represents the number of valence band states/cm$^3$ lying in the energy range between $E$ and $E + dE$ (if $E \leq E_v$).

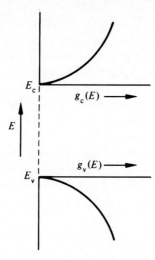

**Figure 2.14**   General energy dependence of $g_c(E)$ and $g_v(E)$ near the band edges. $g_c(E)$ and $g_v(E)$ are the density of states in the conduction and valence bands, respectively.

It therefore follows that $g_c(E)$ and $g_v(E)$ themselves are numbers/unit volume-unit energy, or typically, numbers/cm$^3$-eV.

## 2.4.2  The Fermi Function

Whereas the density of states tells one how many states exist at a given energy $E$, the Fermi function $f(E)$ specifies how many of the existing states at the energy $E$ will be filled with an electron, or equivalently,

> $f(E)$   specifies, under equilibrium conditions, the probability that an available state at an energy $E$ will be occupied by an electron.

Mathematically speaking, the Fermi function is simply a probability distribution function. In mathematical symbols,

$$f(E) = \frac{1}{1 + e^{(E-E_F)/kT}} \tag{2.7}$$

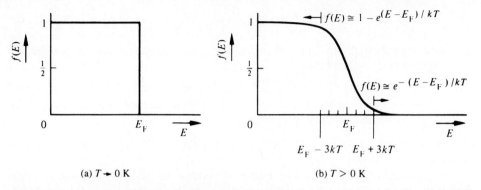

**Figure 2.15** Energy dependence of the Fermi function. (a) $T \rightarrow 0$ K; (b) generalized $T > 0$ K plot with the energy coordinate expressed in $kT$ units.

where

$$E_F = \text{Fermi energy or Fermi level}$$

$$k = \text{Boltzmann constant } (k = 8.617 \times 10^{-5} \text{ eV/K})$$

$$T = \text{temperature in Kelvin (K)}$$

Seeking insight into the nature of the newly introduced function, let us begin by investigating the Fermi function's energy dependence. Consider first temperatures where $T \rightarrow 0$ K. As $T \rightarrow 0$ K, $(E - E_F)/kT \rightarrow -\infty$ for all energies $E < E_F$ and $(E - E_F)/kT \rightarrow +\infty$ for all energies $E > E_F$. Hence $f(E < E_F) \rightarrow 1/[1 + \exp(-\infty)] = 1$ and $f(E > E_F) \rightarrow 1/[1 + \exp(+\infty)] = 0$. This result is plotted in Fig. 2.15(a) and is simply interpreted as meaning that all states at energies below $E_F$ will be filled and all states at energies above $E_F$ will be empty for temperatures $T \rightarrow 0$ K. In other words, there is a sharp cutoff in the filling of allowed energy states at the Fermi energy $E_F$ when the system temperature approaches absolute zero.

Let us next consider temperatures $T > 0$ K. Examining the Fermi function, we make the following pertinent observations.

(i) If $E = E_F$, $f(E_F) = 1/2$.

(ii) If $E \geq E_F + 3kT$, $\exp[(E - E_F)/kT] \gg 1$ and $f(E) \simeq \exp[-(E - E_F)/kT]$. Consequently, above $E_F + 3kT$ the Fermi function or filled-state probability decays exponentially to zero with increasing energy. Moreover, most states at energies $3kT$ or more above $E_F$ will be empty.

(iii) If $E \leq E_F - 3kT$, $\exp[(E - E_F)/kT] \ll 1$ and $f(E) \simeq 1 - \exp[(E - E_F)/kT]$. Below $E_F - 3kT$, therefore, $[1 - f(E)]$, the probability that a given state will be *empty*, decays exponentially to zero with decreasing energy. Most states at energies $3kT$ or more below $E_F$ will be filled.

(iv) At room temperature ($T = 300$ K), $kT = 0.0259$ eV and $3kT = 0.0777$ eV $\ll E_G$(Si). Compared to the Si band gap, the $3kT$ energy interval that appears prominently in the $T > 0$ K formalism is typically quite small.

The properties just cited are reflected and summarized in the $T > 0$ K Fermi function plot displayed in Fig. 2.15(b).

Before concluding the discussion here, it is perhaps worthwhile to reemphasize that the Fermi function applies only under equilibrium conditions. The Fermi function, however, is universal in the sense that it applies with equal validity to all materials—insulators, semiconductors, and metals. Although introduced in relationship to semiconductors, the Fermi function is not dependent in any way on the special nature of semiconductors but is simply a statistical function associated with electrons in general. Finally, the relative positioning of the Fermi energy $E_F$ compared to $E_c$ (or $E_v$), an item of obvious concern, is treated in subsequent subsections.

---

### Exercise 2.2

**P:** The probability that a state is filled at the conduction band edge ($E_c$) is precisely equal to the probability that a state is *empty* at the valence band edge ($E_v$). Where is the Fermi level located?

**S:** The Fermi function, $f(E)$, specifies the probability of electrons occupying states at a given energy $E$. The probability that a state is empty (not filled) at a given energy $E$ is equal to $1 - f(E)$. Here we are told

$$f(E_c) = 1 - f(E_v)$$

Since

$$f(E_c) = \frac{1}{1 + e^{(E_c - E_F)/kT}}$$

and

$$1 - f(E_v) = 1 - \frac{1}{1 + e^{(E_v - E_F)/kT}} = \frac{1}{1 + e^{(E_F - E_v)/kT}}$$

we conclude

$$\frac{E_c - E_F}{kT} = \frac{E_F - E_v}{kT}$$

or

$$E_F = \frac{E_c + E_v}{2}$$

The Fermi level is positioned at midgap.

## (C) Exercise 2.3

In this exercise we wish to examine in more detail how the Fermi function varies with temperature.

**P:** Successively setting $T = 100, 200, 300$, and then 400 K, compute and plot $f(E)$ versus $\Delta E = E - E_F$ for $-0.2 \text{ eV} \leq \Delta E \leq 0.2 \text{ eV}$. All $f(E)$ versus $\Delta E$ curves should be superimposed on a single set of coordinates.

**S:** MATLAB program script . . .

```
%Fermi Function Calculation, f(ΔE,T)

% Constant
k=8.617e-5;

%Computation proper
for ii=1:4;
   T=100*ii;
   kT=k*T;
   dE(ii,1)=-5*kT;
   for jj=1:101
        f(ii,jj)=1/(1+exp(dE(ii,jj)/kT));
        dE(ii,jj+1)=dE(ii,jj)+0.1*kT;
   end
end
dE=dE(:,1:jj);    %This step strips the extra dE value
```

```
%Plotting result
close
plot(dE',f'); grid;    %Note the transpose (') to form data columns
xlabel('E-EF(eV)');    ylabel('f(E)');
text(.05,.2,'T=400K');    text(-.03,.1,'T=100K');
```

Program output . . .

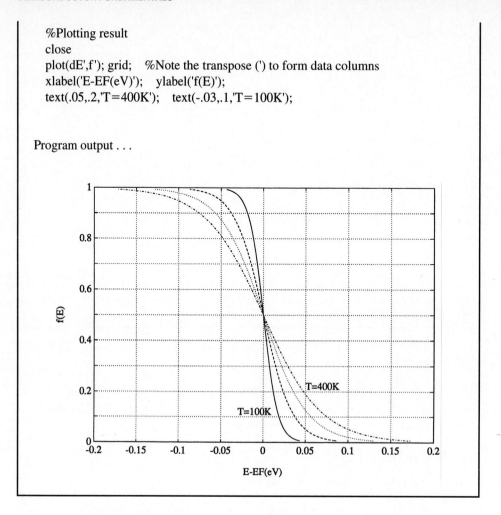

## 2.4.3  Equilibrium Distribution of Carriers

Having established the distribution of available band states and the probability of filling those states under equilibrium conditions, we can now easily deduce the distribution of carriers in the respective energy bands. To be specific, the desired distribution is obtained by simply multiplying the appropriate density of states by the appropriate occupancy factor—$g_c(E)f(E)$ yields the distribution of electrons in the conduction band and $g_v(E)[1 - f(E)]$ yields the distribution of holes (unfilled states) in the valence band. Sample carrier distributions for three different assumed positions of the Fermi energy (along with associated energy band diagram, Fermi function, and density of states plots) are pictured in Fig. 2.16.

Examining Fig. 2.16, we note in general that all carrier distributions are zero at the

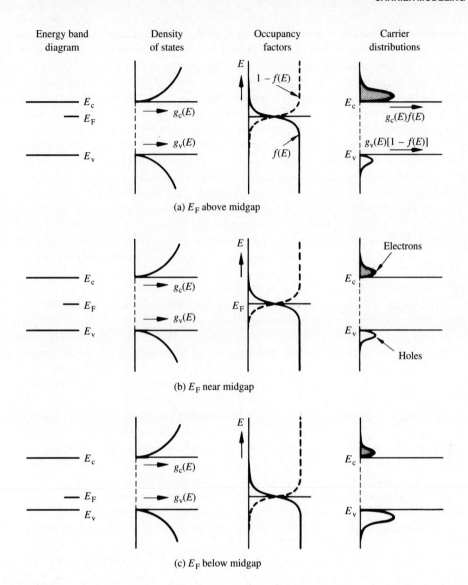

**Figure 2.16**  Carrier distributions (not drawn to scale) in the respective bands when the Fermi level is positioned (a) above midgap, (b) near midgap, and (c) below midgap. Also shown in each case are coordinated sketches of the energy band diagram, density of states, and the occupancy factors (the Fermi function and one minus the Fermi function).

band edges, reach a peak value very close to $E_c$ or $E_v$, and then decay very rapidly toward zero as one moves upward into the conduction band or downward into the valence band. In other words, most of the carriers are grouped energetically in the near vicinity of the band edges. Another general point to note is the effect of the Fermi level positioning on the relative magnitude of the carrier distributions. When $E_F$ is positioned in the upper half of the band gap (or higher), the electron distribution greatly outweighs the hole distribution. Although both the filled-state occupancy factor $[f(E)]$ and the empty-state occupancy factor $[1 - f(E)]$ fall off exponentially as one proceeds from the band edges deeper into the conduction and valence bands, respectively, $[1 - f(E)]$ is clearly much smaller than $f(E)$ at corresponding energies if $E_F$ lies in the upper half of the band gap. Lowering $E_F$ effectively slides the occupancy plots downward, giving rise to a nearly equal number of carriers when $E_F$ is at the middle of the gap. Likewise, a predominance of holes results when $E_F$ lies below the middle of the gap. The argument here assumes, of course, that $g_c(E)$ and $g_v(E)$ are of the same order of magnitude at corresponding energies (as is the case for Si and Ge). Also, referring back to the previous subsection, the statements concerning the occupancy factors falling off exponentially in the respective bands are valid provided $E_c - 3kT \geq E_F \geq E_v + 3kT$.

The information just presented concerning the carrier distributions and the relative magnitudes of the carrier numbers finds widespread usage. The information, however, is often conveyed in an abbreviated or shorthand fashion. Figure 2.17, for example, shows a common way of representing the carrier energy distributions. The greatest number of circles or dots are drawn close to $E_c$ and $E_v$, reflecting the peak in the carrier concentrations near the band edges. The smaller number of dots as one progresses upward into the conduction band crudely models the rapid falloff in the electron density with increasing energy. An extensively utilized means of conveying the relative magnitude of the carrier numbers is displayed in Fig. 2.18. To represent an intrinsic material, a dashed line is drawn in approximately the middle of the band gap and labeled $E_i$. The near-midgap positioning of $E_i$, the intrinsic Fermi level, is of course consistent with the previously cited fact that the electron and hole numbers are about equal when $E_F$ is near the center of the band gap. Similarly, a solid line labeled $E_F$ appearing above midgap tells one at a glance that the semiconductor in question is *n*-type; a solid line labeled $E_F$ appearing below midgap signifies that the semiconductor is *p*-type. Note finally that the dashed $E_i$ line also typically appears on the

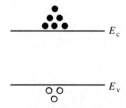

**Figure 2.17**   Schematic representation of carrier energy distributions.

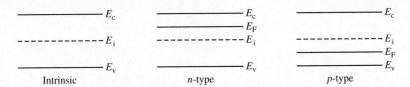

**Figure 2.18** "At a glance" representation of intrinsic (left), *n*-type (middle), and *p*-type (right) semiconductor materials using the energy band diagram.

energy band diagrams characterizing extrinsic semiconductors. The $E_i$ line in such cases represents the expected positioning of the Fermi level *if* the material were intrinsic, and it serves as a reference energy level dividing the upper and lower halves of the band gap.

## 2.5  EQUILIBRIUM CARRIER CONCENTRATIONS

We have arrived at an important point in the carrier modeling process. For the most part, this section simply embodies the culmination of our modeling efforts, with working relationships for the equilibrium carrier concentrations being established to complement the qualitative carrier information presented in previous sections. Unfortunately, the emphasis on the development of mathematical relationships makes the final assault on the carrier modeling summit a little more arduous (and perhaps just a bit boring). Hopefully, the reader can stay focused. A comment is also in order concerning the presentation herein of alternative forms for the carrier relationships. The alternative forms can be likened to the different kinds of wrenches used, for example, in home and automobile repairs. The open-end wrench, the box wrench, and the ratchet wrench all serve the same general purpose. In some applications one can use any of the wrenches. In other applications, however, a special situation restricts the type of wrench employed or favors the use of one wrench over another. The same is true of the alternative carrier relationships. Finally, boxes are drawn around expressions that find widespread usage. A single-walled box signifies a moderately important result; a double-walled box, a very important result.

### 2.5.1 Formulas for *n* and *p*

Since $g_c(E)\,dE$ represents the number of conduction band states/cm$^3$ lying in the $E$ to $E + dE$ energy range, and $f(E)$ specifies the probability that an available state at an energy $E$ will be occupied by an electron, it then follows that $g_c(E)f(E)dE$ gives the number of conduction band electrons/cm$^3$ lying in the $E$ to $E + dE$ energy range, and $g_c(E)f(E)dE$ integrated over all conduction band energies must yield the total number of electrons in the conduction band. In other words, integration over the equilibrium distribution of electrons in the conduction band yields the equilibrium electron concentration. A similar statement can be made relative to the hole concentration. We therefore conclude

$$n = \int_{E_c}^{E_{top}} g_c(E)f(E)dE \tag{2.8a}$$

$$p = \int_{E_{bottom}}^{E_v} g_v(E)[1 - f(E)]dE \tag{2.8b}$$

Seeking explicit expressions for the carrier concentrations, let us focus our efforts on the *n*-integral. (The analogous *p*-integral manipulations are left to the reader as an exercise.) Substituting the Eq. (2.6a) expression for $g_c(E)$ and the Eq. (2.7) expression for $f(E)$ into Eq. (2.8a), one obtains

$$n = \frac{m_n^* \sqrt{2m_n^*}}{\pi^2 \hbar^3} \int_{E_c}^{E_{top}} \frac{\sqrt{E - E_c}\, dE}{1 + e^{(E - E_F)/kT}} \tag{2.9}$$

Now letting

$$\eta = \frac{(E - E_c)}{kT} \tag{2.10a}$$

$$\eta_c = \frac{(E_F - E_c)}{kT} \tag{2.10b}$$

$$E_{top} \to \infty \tag{2.10c}$$

yields

$$n = \frac{m_n^* \sqrt{2m_n^*}\, (kT)^{3/2}}{\pi^2 \hbar^3} \int_0^\infty \frac{\eta^{1/2} d\eta}{1 + e^{\eta - \eta_c}} \tag{2.11}$$

The (2.10c) simplification on the upper integration limit makes use of the fact that the integrand in question falls off rapidly with increasing energy and is essentially zero for energies only a few $kT$ above $E_c$. Hence, extending the upper limit to $\infty$ has a totally negligible effect on the value of the integral.

Even with the cited simplification, the Eq. (2.11) integral cannot be expressed in a closed form containing simple functions. The integral itself is in fact a tabulated function that can be found in a variety of mathematical references. Identifying

$$F_{1/2}(\eta_c) \equiv \int_0^\infty \frac{\eta^{1/2} d\eta}{1 + e^{\eta - \eta_c}}, \quad \text{the Fermi-Dirac integral of order 1/2} \tag{2.12}$$

and also defining

$$N_C = 2 \left[ \frac{m_n^* kT}{2\pi\hbar^2} \right]^{3/2}, \quad \text{the ``effective'' density of conduction band states} \qquad (2.13a)$$

$$N_V = 2 \left[ \frac{m_p^* kT}{2\pi\hbar^2} \right]^{3/2}, \quad \text{the ``effective'' density of valence band states} \qquad (2.13b)$$

one obtains

$$n = N_C \frac{2}{\sqrt{\pi}} F_{1/2}(\eta_c) \qquad (2.14a)$$

and, by analogy,

$$p = N_V \frac{2}{\sqrt{\pi}} F_{1/2}(\eta_v) \qquad (2.14b)$$

where $\eta_v \equiv (E_v - E_F)/kT$.

The Eq. (2.14) relationships are a very general result, valid for any conceivable positioning of the Fermi level. The constants $N_C$ and $N_V$ are readily calculated; at 300 K, $N_{C,V} = (2.510 \times 10^{19}/\text{cm}^3)(m_{n,p}^*/m_0^*)^{3/2}$. The value of the Fermi integral can be obtained from available tables, from plots, or by direct computation. The general-form relationships, nonetheless, are admittedly cumbersome and inconvenient to use in routine analyses. Fortunately, simplified closed-form expressions do exist that can be employed in the vast majority of practical problems. To be specific, if $E_F$ is restricted to values $E_F \leq E_c - 3kT$, then $1/[1 + \exp(\eta - \eta_c)] \simeq \exp[-(\eta - \eta_c)]$ for all $E \geq E_c (\eta \geq 0)$, and

$$F_{1/2}(\eta_c) = \frac{\sqrt{\pi}}{2} e^{(E_F - E_c)/kT} \qquad (2.15a)$$

Likewise, if $E_F \geq E_v + 3kT$, then

$$F_{1/2}(\eta_v) = \frac{\sqrt{\pi}}{2} e^{(E_v - E_F)/kT} \qquad (2.15b)$$

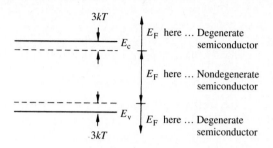

**Figure 2.19**   Definition of degenerate/nondegenerate semiconductors.

It therefore follows that, if $E_v + 3kT \leq E_F \leq E_c - 3kT$,

$$n = N_C e^{(E_F - E_c)/kT} \qquad (2.16a)$$

$$p = N_V e^{(E_v - E_F)/kT} \qquad (2.16b)$$

The mathematical simplification leading to Eqs. (2.16) is equivalent to approximating the occupancy factors, $f(E)$ and $1 - f(E)$, by simple exponential functions—an approximation earlier shown to be valid provided $E_F$ was somewhere in the band gap no closer than $3kT$ to either band edge. Whenever $E_F$ is confined, as noted, to $E_v + 3kT \leq E_F \leq E_c - 3kT$, instead of continually repeating the $E_F$ restriction, the semiconductor is simply said to be *nondegenerate*. Whenever $E_F$ lies in the band gap closer than $3kT$ to either band edge or actually penetrates one of the bands, the semiconductor is said to be *degenerate*. These very important terms are also defined pictorially in Fig. 2.19.

### 2.5.2 Alternative Expressions for *n* and *p*

Although in closed form, the Eq. (2.16) relationships are not in the simplest form possible, and, more often than not, it is the simpler alternative form of these relationships that one encounters in device analyses. The alternative-form relationships can be obtained by recalling that $E_i$, the Fermi level for an intrinsic semiconductor, lies close to midgap, and hence Eqs. (2.16) most assuredly apply to an intrinsic semiconductor. If this be the case, then specializing Eqs. (2.16) to an intrinsic semiconductor, i.e., setting $n = p = n_i$ and $E_i = E_F$, one obtains

$$n_i = N_C e^{(E_i - E_c)/kT} \qquad (2.17a)$$

and

$$n_i = N_V e^{(E_v - E_i)/kT} \qquad (2.17b)$$

Solving Eqs. (2.17) for $N_C$ and $N_V$ yields

$$N_C = n_i e^{(E_c - E_i)/kT} \qquad (2.18a)$$

$$N_V = n_i e^{(E_i - E_v)/kT} \qquad (2.18b)$$

Finally, eliminating $N_C$ and $N_V$ in the original Eq. (2.16) relationships using Eqs. (2.18) gives

$$n = n_i e^{(E_F - E_i)/kT} \qquad (2.19a)$$

$$p = n_i e^{(E_i - E_F)/kT} \qquad (2.19b)$$

Like Eqs. (2.16), the Eq. (2.19) expressions are valid for any semiconductor in equilibrium whose doping is such as to give rise to a nondegenerate positioning of the Fermi level. Whereas two constants and three energy levels appear in the original relationships, however, only one constant and two energy levels appear in the alternative relationships. Because of their symmetrical nature, the alternative expressions are also easier to remember, requiring merely an interchange of $E_F$ and $E_i$ in going from the $n$-formula to the $p$-formula.

### 2.5.3 $n_i$ and the $np$ Product

As can be inferred from its appearance in Eqs. (2.19), the intrinsic carrier concentration can figure prominently in the quantitative calculation of the carrier concentrations. Continuing to establish pertinent carrier concentration relationships, we next interject considerations specifically involving this important material parameter.

First, if the corresponding sides of Eqs. (2.17a) and (2.17b) are multiplied together, one obtains

$$n_i^2 = N_C N_V e^{-(E_c - E_v)/kT} = N_C N_V e^{-E_G/kT} \qquad (2.20)$$

or

$$n_i = \sqrt{N_C N_V} e^{-E_G/2kT} \qquad (2.21)$$

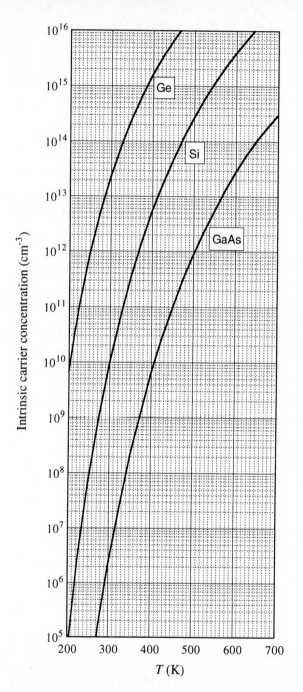

| Si | |
|---|---|
| $T(°C)$ | $n_i(\text{cm}^{-3})$ |
| 0 | $8.86 \times 10^8$ |
| 5 | $1.44 \times 10^9$ |
| 10 | $2.30 \times 10^9$ |
| 15 | $3.62 \times 10^9$ |
| 20 | $5.62 \times 10^9$ |
| 25 | $8.60 \times 10^9$ |
| 30 | $1.30 \times 10^{10}$ |
| 35 | $1.93 \times 10^{10}$ |
| 40 | $2.85 \times 10^{10}$ |
| 45 | $4.15 \times 10^{10}$ |
| 50 | $5.97 \times 10^{10}$ |
| 300 K | $1.00 \times 10^{10}$ |

| GaAs | |
|---|---|
| $T(°C)$ | $n_i(\text{cm}^{-3})$ |
| 0 | $1.02 \times 10^5$ |
| 5 | $1.89 \times 10^5$ |
| 10 | $3.45 \times 10^5$ |
| 15 | $6.15 \times 10^5$ |
| 20 | $1.08 \times 10^6$ |
| 25 | $1.85 \times 10^6$ |
| 30 | $3.13 \times 10^6$ |
| 35 | $5.20 \times 10^6$ |
| 40 | $8.51 \times 10^6$ |
| 45 | $1.37 \times 10^7$ |
| 50 | $2.18 \times 10^7$ |
| 300 K | $2.25 \times 10^6$ |

**Figure 2.20**   Intrinsic carrier concentrations in Ge, Si, and GaAs as a function of temperature.

Equation (2.21) expresses $n_i$ as a function of known quantities and can be used to compute $n_i$ at a specified temperature or as a function of temperature. Numerical values for the intrinsic carrier concentration in Si and Ge at room temperature were cited previously; the best available plots of $n_i$ as functions of temperature in Si, Ge, and GaAs are displayed in Fig. 2.20.

A second very important $n_i$-based relationship follows directly from Eqs. (2.19). Multiplying the corresponding sides of Eqs. (2.19a) and (2.19b) together yields

$$np = n_i^2 \qquad\qquad (2.22)$$

Although appearing trivial, the $np$ product relationship (Eq. 2.22) often proves to be extremely useful in practical computations. If one of the carrier concentrations is known, the remaining carrier concentration is readily determined using Eq. (2.22)—provided of course that the semiconductor is in equilibrium and nondegenerate.

---

### (C) Exercise 2.4

Substituting the Eq. (2.13) definitions of $N_C$ and $N_V$ into Eq. (2.21), introducing normalization factors, and replacing known constants by their numerical values, one obtains

$$n_i = (2.510 \times 10^{19})\left(\frac{m_n^*}{m_0}\frac{m_p^*}{m_0}\right)^{3/4}\left(\frac{T}{300}\right)^{3/2} e^{-E_G/2kT}$$

$E_G$ and the effective masses exhibit a weak but non-negligible temperature dependence. The $E_G$ versus $T$ variation can be modeled to four-place accuracy by the fit relationship noted in Problem 2.1(a). As deduced from the analysis by Barber [Solid-State Electronics, **10**, 1039 (1967)], the temperature dependence of the effective masses over the range $200\ \text{K} \le T \le 700\ \text{K}$ can be approximated by

$$\frac{m_n^*}{m_0} = 1.028 + (6.11 \times 10^{-4})T - (3.09 \times 10^{-7})T^2$$

$$\frac{m_p^*}{m_0} = 0.610 + (7.83 \times 10^{-4})T - (4.46 \times 10^{-7})T^2$$

**P:** (a) Confirm that the $n_i$ versus $T$ curve for Si graphed in Fig. 2.20 is generated employing the relationships just cited, *provided* $E_G$ in the $n_i$ expression is replaced by $E_G - E_{ex}$, where $E_{ex} = 0.0074$ eV. (In the previously cited article, Barber suggested using an exciton correction factor of $E_{ex} = 0.007$ eV. The slightly larger value employed in the Fig. 2.20 computation was specifically chosen to give $n_i = 10^{10}/$ cm$^3$ at 300 K.)

(b) The accepted value of $n_i$ in Si at 300 K has been revised recently to be in agreement with the experimental $n_i$ versus $T$ data acquired by Sproul and Green [Journal of Applied Physics, **70,** 846 (July 1991)]. The authors concluded $n_i = 1.00 \pm 0.03 \times 10^{10}/\text{cm}^3$ in Si at 300 K. Over the probed temperature range of 275 K $\leq T \leq$ 375 K, their experimental data could be fitted by the relationship

$$n_i = (9.15 \times 10^{19})\left(\frac{T}{300}\right)^2 e^{-0.5928/kT}$$

Compare the $n_i$ computed from this experimental-fit relationship to the part (a) output over the temperature range of mutual validity.

**S:** (a) MATLAB program script . . .

```
%ni vs. T calculation for Si (200K - 700K) used in Fig. 2.20

%Initialization
format short e

%Constants and T-range
k=8.617e-5;
A=2.510e19;
Eex=0.0074;    %Value was adjusted to match S&G ni(300K) value
T=200:25:700;

%Band Gap vs. T
EG0=1.17;
a=4.730e-4;
b=636;
EG=EG0-a.*(T.^2)./(T+b);

%Effective mass ratio (mnr=mn*/m0, mpr=mp*/m0)
mnr=1.028 + (6.11e-4).*T - (3.09e-7).*T.^2;
mpr=0.610 + (7.83e-4).*T - (4.46e-7).*T.^2;

%Computation of ni
ni=A.*((T./300).^(1.5)).*((mnr.*mpr).^(0.75)).*exp(-(EG-Eex)./(2 .*k.*T));

%Display output on screen
j=length(T);
fprintf('\n \n T      ni\n');    %There are ten spaces between T and ni.
for ii=1:j,
fprintf('%-10.f%-10.3e\n',T(ii),ni(ii));
end
```

(b) MATLAB program script . . .

%Experimental fit of Sproul-Green ni data (275K - 375K)

```
%ni calculation
T=275:25:375;
k=8.617e-5;
ni=(9.15e19).*(T./300).^2 .*exp(-0.5928./(k*T));
```

```
%Display result on screen
j=length(T);
fprintf('\n\n T       ni\n');   %There are ten spaces between T and ni.
for ii=1:j,
fprintf('%-10.f%-10.3e\n',T(ii),ni(ii));
end
```

Output from the two programs is reproduced below. Note the excellent agreement of the computational results (to within 2%) over the temperature range of mutual validity.

| $T$ (K) | (a) $n_i$ (cm$^{-3}$) | (b) $n_i$ (cm$^{-3}$) |
|---|---|---|
| 200 | $5.246 \times 10^4$ | — |
| 275 | $1.059 \times 10^9$ | $1.051 \times 10^9$ |
| 300 | $1.000 \times 10^{10}$ | $1.006 \times 10^{10}$ |
| 325 | $6.798 \times 10^{10}$ | $6.887 \times 10^{10}$ |
| 350 | $3.565 \times 10^{11}$ | $3.623 \times 10^{11}$ |
| 375 | $1.518 \times 10^{12}$ | $1.542 \times 10^{12}$ |
| 400 | $5.449 \times 10^{12}$ | — |
| 500 | $2.716 \times 10^{14}$ | — |
| 600 | $3.988 \times 10^{15}$ | — |
| 700 | $2.865 \times 10^{16}$ | — |

## 2.5.4 Charge Neutrality Relationship

The relationships established to this point are devoid of an explicit dependence on the dopant concentrations introduced into a semiconductor. It is the charge neutrality relationship that provides the general tie between the carrier and dopant concentrations.

To establish the charge neutrality relationship, let us consider a *uniformly doped* semiconductor, a semiconductor where the number of dopant atoms/cm$^3$ is the same everywhere. Systematically examining little sections of the semiconductor far from any surfaces, and assuming equilibrium conditions prevail, one must invariably find that each and every section is charge-neutral, i.e., contains no net charge. If this were not the case, electric fields would exist inside the semiconductor. The electric fields in turn would give rise to carrier motion and associated currents—a situation totally inconsistent with the assumed equilibrium conditions. There are, however, charged entities inside all semiconductors. Electrons, holes, ionized donors (donor atoms that have become positively charged by donating an electron to the conduction band) and negatively-charged ionized acceptors can all exist simultaneously inside any given semiconductor. For the uniformly doped material to be everywhere charge-neutral clearly requires

$$\frac{charge}{cm^3} = qp - qn + qN_D^+ - qN_A^- = 0 \tag{2.23}$$

or

$$\boxed{p - n + N_D^+ - N_A^- = 0} \tag{2.24}$$

where, by definition,

$$N_D^+ = \text{number of ionized (positively charged) donors/cm}^3,$$

$$N_A^- = \text{number of ionized (negatively charged) acceptors/cm}^3.$$

As previously discussed, there is sufficient thermal energy available in a semiconductor at room temperature to ionize almost all of the shallow-level donor and acceptor sites. Defining

$$N_D = \text{total number of donors/cm}^3,$$

$$N_A = \text{total number of acceptors/cm}^3,$$

and setting

$$N_D^+ = N_D$$

$$N_A^- = N_A$$

one then obtains

$$\boxed{p - n + N_D - N_A = 0} \qquad \text{assumes total ionization} \atop \text{of dopant atoms} \tag{2.25}$$

Equation (2.25) is the standard form of the charge neutrality relationship.

## 2.5.5  Carrier Concentration Calculations

We are finally in a position to calculate the carrier concentrations in a uniformly doped semiconductor under equilibrium conditions. In the computations to be presented we specifically make the assumptions of NONDEGENERACY (allowing us to use the $np$ product relationship) and TOTAL IONIZATION of the dopant atoms. Note that $n_i$, which appears in the $np$ product expression, has been calculated and plotted and must be considered a known quantity. Likewise, $N_A$ and $N_D$, which appear in the charge neutrality relationship, are typically controlled and determined experimentally and should also be considered known quantities. The only other symbols used in the two equations are $n$ and $p$. Thus, under the cited assumptions of nondegeneracy and total ionization of dopant atoms, we have two equations and two unknowns from which $n$ and $p$ can be deduced.

Starting with the $np$ product expression, one can write

$$p = \frac{n_i^2}{n} \tag{2.26}$$

Eliminating $p$ in Eq. (2.25) using Eq. (2.26) gives

$$\frac{n_i^2}{n} - n + N_D - N_A = 0 \tag{2.27}$$

or

$$n^2 - n(N_D - N_A) - n_i^2 = 0 \tag{2.28}$$

Solving the quadratic equation for $n$ then yields

$$n = \frac{N_D - N_A}{2} + \left[ \left( \frac{N_D - N_A}{2} \right)^2 + n_i^2 \right]^{1/2} \tag{2.29a}$$

and

$$p = \frac{n_i^2}{n} = \frac{N_A - N_D}{2} + \left[ \left( \frac{N_A - N_D}{2} \right)^2 + n_i^2 \right]^{1/2} \tag{2.29b}$$

Only the plus roots have been retained in Eqs. (2.29) because physically the carrier concentrations must be greater than or equal to zero.

Equations (2.29) are general-case solutions. In the vast majority of practical computations it is possible to simplify these equations prior to substituting in numerical values for $N_D$, $N_A$, and $n_i$. Special cases of specific interest are considered next.

(1) *Intrinsic Semiconductor* ($N_A = 0$, $N_D = 0$). With $N_A = 0$ and $N_D = 0$, Eqs. (2.29) simplify to $\boxed{n = n_i}$ and $\boxed{p = n_i}$. $n = p = n_i$ is of course the expected result for the equilibrium carrier concentrations in an intrinsic semiconductor.

(2) *Doped Semiconductor where either* $N_D - N_A \simeq N_D \gg n_i$ *or* $N_A - N_D \simeq N_A \gg n_i$. This is the special case of greatest practical interest. The unintentional doping levels in Si are such that the controlled addition of dopants routinely yields $N_D \gg N_A$ or $N_A \gg N_D$. Moreover, the intrinsic carrier concentration in Si at room temperature is about $10^{10}/cm^3$, while the dominant doping concentration ($N_A$ or $N_D$) is seldom less than $10^{14}/cm^3$. Thus the special case considered here is the usual case encountered in practice. If $N_D - N_A \simeq N_D \gg n_i$, the square root in Eq. (2.29a) reduces to $N_D/2$ and

$$\boxed{\begin{aligned} n &\simeq N_D \\ p &\simeq n_i^2/N_D \end{aligned}} \qquad \begin{aligned} N_D \gg N_A, \quad N_D \gg n_i \\ \text{(nondegenerate, total ionization)} \end{aligned} \qquad \begin{aligned} &(2.30a) \\ &(2.30b) \end{aligned}$$

Similarly

$$\boxed{\begin{aligned} p &\simeq N_A \\ n &\simeq n_i^2/N_A \end{aligned}} \qquad \begin{aligned} N_A \gg N_D, \quad N_A \gg n_i \\ \text{(nondegenerate, total ionization)} \end{aligned} \qquad \begin{aligned} &(2.31a) \\ &(2.31b) \end{aligned}$$

As a numerical example, suppose a Si sample maintained at room temperature is uniformly doped with $N_D = 10^{15}/cm^3$ donors. Using Eqs. (2.30), one rapidly concludes $n \simeq 10^{15}/cm^3$ and $p \simeq 10^5/cm^3$.

(3) *Doped Semiconductor where* $n_i \gg |N_D - N_A|$. Systematically increasing the ambient temperature causes a monotonic rise in the intrinsic carrier concentration (see Fig. 2.20). At sufficiently elevated temperatures, $n_i$ will eventually equal and then exceed the net doping concentration. If $n_i \gg |N_D - N_A|$, the square roots in Eqs. (2.29) reduce to $n_i$ and $\boxed{n \simeq p \simeq n_i}$. In other words, *all semiconductors become intrinsic at sufficiently high temperatures where* $n_i \gg |N_D - N_A|$.

(4) *Compensated Semiconductor.* As is evident from Eqs. (2.29), donors and acceptors tend to negate each other. Indeed, it is possible to produce intrinsic-like material by making $N_D - N_A = 0$. In some materials, such as GaAs, $N_A$ may be comparable to $N_D$ in the as-grown crystal. When $N_A$ and $N_D$ are comparable and nonzero, the material is said to be *compensated*. If the semiconductor is compensated, both $N_A$ and $N_D$ must be retained in all carrier concentration expressions.

In summary, Eqs. (2.29) can always be used to compute the carrier concentrations if the semiconductor is nondegenerate and the dopant atoms are totally ionized. In the vast majority of practical situations, however, it is possible to simplify these equations prior to performing numerical computations. Equations (2.29) must be used to compute the carrier concentrations only in those rare instances when $|N_D - N_A| \sim n_i$. The simplified relationships of greatest practical utility are Eqs. (2.30) and (2.31).

---

### Exercise 2.5

**P:** A Si sample is doped with $10^{14}$ boron atoms per $cm^3$.

(a) What are the carrier concentrations in the Si sample at 300 K?

(b) What are the carrier concentrations at 470 K?

**S:** (a) Boron in Si is an acceptor (see Table 2.2). Thus $N_A = 10^{14}/cm^3$. At 300 K, $n_i = 1.00 \times 10^{10}/cm^3$ and the given $N_A$ is clearly much greater than $n_i$. Moreover, since the $N_D$ doping was omitted from the problem statement, we infer $N_D \ll N_A$. With $N_A \gg n_i$ and $N_A \gg N_D$, Eqs. (2.31) may be used to calculate the carrier concentrations: $p = N_A = 10^{14}/cm^3$; $n = n_i^2/N_A = 10^6/cm^3$.

(b) As deduced from Fig. 2.20, $n_i \simeq 10^{14}/cm^3$ at 470 K. Because $n_i$ is comparable to $N_A$, Eqs. (2.29) must be used to calculate at least one of the carrier concentrations. (Once one of the carrier concentrations is known, the second is more readily computed using the $np$ product expression.) Performing the indicated calculations gives $p = N_A/2 + [(N_A/2)^2 + n_i^2]^{1/2} = 1.62 \times 10^{14}/cm^3$; $n = n_i^2/p = 6.18 \times 10^{13}/cm^3$.

---

### 2.5.6  Determination of $E_F$

Knowledge concerning the exact position of the Fermi level on the energy band diagram is often of interest. For example, when discussing the intrinsic Fermi level, we indicated that $E_i$ was located somewhere near the middle of the band gap. It would be useful to know the *precise* positioning of $E_i$ in the band gap. Moreover, we have developed computational formulas for $n$ and $p$ appropriate for nondegenerate semiconductors. Whether a doped semiconductor is nondegenerate or degenerate depends, of course, on the value or positioning of $E_F$.

Before running though the mechanics of finding the Fermi level in selected cases of interest, it is useful to make a general observation; namely, Eqs. (2.19) or (2.16) [or even more generally, Eqs. (2.14)] provide a one-to-one correspondence between the Fermi energy and the carrier concentrations. Thus, having computed any one of the three variables—$n$, $p$, or $E_F$—one can always determine the remaining two variables under equilibrium conditions.

(1) *Exact positioning of $E_i$.* In an intrinsic material

$$n = p \tag{2.32}$$

Substituting for $n$ and $p$ in Eq. (2.32) using Eqs. (2.16), and setting $E_F = E_i$, yields

$$N_C e^{(E_i - E_c)/kT} = N_V e^{(E_v - E_i)/kT} \tag{2.33}$$

Solving for $E_i$, one obtains

$$E_i = \frac{E_c + E_v}{2} + \frac{kT}{2} \ln\left(\frac{N_V}{N_C}\right) \tag{2.34}$$

But

$$\frac{N_V}{N_C} = \left(\frac{m_p^*}{m_n^*}\right)^{3/2} \tag{2.35}$$

Consequently,

$$\boxed{E_i = \frac{E_c + E_v}{2} + \frac{3}{4} kT \ln\left(\frac{m_p^*}{m_n^*}\right)} \tag{2.36}$$

According to Eq. (2.36), $E_i$ lies precisely at midgap only if $m_p^* = m_n^*$ or if $T = 0$ K. For the more practical case of silicon at room temperature, Table 2.1 gives $m_p^*/m_n^* = 0.69$, $(3/4)kT \ln(m_p^*/m_n^*) = -0.0073$ eV, and $E_i$ therefore lies 0.0073 eV below midgap. Although potentially significant in certain problems, this small deviation from midgap is typically neglected in drawing energy band diagrams, etc.

(2) *Doped semiconductors (nondegenerate, dopants totally ionized).* The general positioning of the Fermi level in donor- and acceptor-doped semiconductors assumed to be nondegenerate, in equilibrium, and maintained at temperatures where the dopants are fully ionized, is readily deduced from Eqs. (2.19). Specifically, solving Eqs. (2.19) for $E_F - E_i$, one obtains

$$E_F - E_i = kT \ln(n/n_i) = -kT \ln(p/n_i) \tag{2.37}$$

Depending on the simplifications inherent in a particular problem, the appropriate carrier concentration solution [Eqs. (2.29), (2.30), (2.31), etc.] is then substituted into Eq. (2.37) to determine the positioning of $E_F$. For example, per Eqs. (2.30a) and (2.31a), $n \simeq N_D$ in typical donor-doped semiconductors and $p \simeq N_A$ in typical acceptor-doped semiconductors maintained at or near room temperature. Substituting into Eq. (2.37), we therefore conclude

$$E_F - E_i = kT \ln(N_D/n_i) \qquad \ldots N_D \gg N_A, \quad N_D \gg n_i \qquad (2.38a)$$

$$E_i - E_F = kT \ln(N_A/n_i) \qquad \ldots N_A \gg N_D, \quad N_A \gg n_i \qquad (2.38b)$$

From Eqs. (2.38) it is obvious that the Fermi level moves systematically upward in energy from $E_i$ with increasing donor doping and systematically downward in energy from $E_i$ with increasing acceptor doping. The exact Fermi level positioning in Si at room temperature, nicely reinforcing the foregoing statement, is displayed in Fig. 2.21. Also note that for a given semiconductor material and ambient temperature, there exist maximum nondegenerate donor and acceptor concentrations, doping concentrations above which the material becomes degenerate. In Si at room temperature the maximum nondegenerate doping concentrations are $N_D \simeq 1.6 \times 10^{18}/\text{cm}^3$ and $N_A \simeq 9.1 \times 10^{17}/\text{cm}^3$. The large Si doping values required for degeneracy, we should interject, have led to the common usage of "highly doped" (or $n^+$-material/$p^+$-material) and "degenerate" as interchangeable terms.

Finally, the question may arise: What procedure should be followed in computing $E_F$ when one is not sure whether a material is nondegenerate or degenerate? Unless a material is known to be degenerate, always assume nondegeneracy and compute $E_F$ employing the appropriate nondegenerate relationship. If $E_F$ derived from the nondegenerate formula lies in the degenerate zone, one must then, of course, recompute $E_F$ using the more complex formalism valid for degenerate materials.

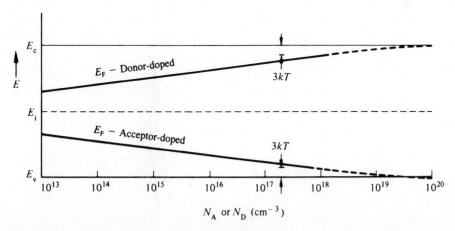

**Figure 2.21**  Fermi level positioning in Si at 300 K as a function of the doping concentration. The solid $E_F$ lines were established using Eq. (2.38a) for donor-doped material and Eq. (2.38b) for acceptor-doped material ($kT = 0.0259$ eV, and $n_i = 10^{10}/\text{cm}^3$).

## Exercise 2.6

**P:** For each of the conditions specified in Exercise 2.5, determine the position of $E_i$, compute $E_F - E_i$, and draw a carefully dimensioned energy band diagram for the Si sample. Note that $E_G(\text{Si}) = 1.08$ eV and $m_p^*/m_n^* = 0.71$ at 470 K.

**S:** (a) In part (a) of Exercise 2.5, the $N_A = 10^{14}/\text{cm}^3$ Si sample is mainted at 300 K. Using Eq. (2.36), we conclude $E_i$ is located 0.0073 eV below midgap. (The positioning of $E_i$ in Si at 300 K is also noted in the text following Eq. 2.36). Next applying Eq. (2.38b), we find

$$
\begin{aligned}
E_i - E_F &= kT \ln(N_A/n_i) \\
&= 0.0259 \ln(10^{14}/10^{10}) \\
&= 0.239 \text{ eV}
\end{aligned}
$$

The energy band diagram constructed from the deduced positioning of $E_i$ and $E_F$ is shown in Fig. E2.6(a).

(b) In part (b) of Exercise 2.5 the Si sample is heated to 470 K. With $m_p^*/m_n^* = 0.71$ and $kT = 0.0405$ eV at 470 K, $(3/4)kT \ln(m_p^*/m_n^*) = -0.0104$ eV and $E_i$ is deduced to be located 0.0104 eV below midgap. Because $N_A$ is comparable to $n_i$ at 470 K, Eq. (2.37) must be used to compute the positioning of $E_F$. Specifically, with $n_i = 10^{14}/\text{cm}^3$ and $p = 1.62 \times 10^{14}/\text{cm}^3$,

$$
\begin{aligned}
E_i - E_F &= kT \ln(p/n_i) \\
&= 0.0405 \ln(1.62 \times 10^{14}/10^{14}) \\
&= 0.0195 \text{ eV}
\end{aligned}
$$

Here $E_F$ is only slightly removed from $E_i$ as pictured in Fig. E2.6(b).

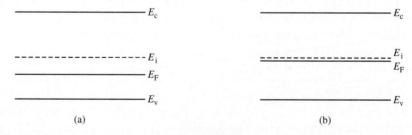

(a)                              (b)

**Figure E2.6**

## 2.5.7 Carrier Concentration Temperature Dependence

A number of isolated facts about the carrier concentration temperature dependence have already been presented at various points in this chapter. The discussion in Section 2.3 concerned with dopant action, for example, described the increased ionization of dopant sites and the associated increase in the majority carrier concentration when the temperature of a semiconductor is raised from near $T = 0$ K toward room temperature. More recent subsections have included a plot of the intrinsic carrier concentration versus temperature (Fig. 2.20) and a calculation indicating that all semiconductors become intrinsic ($n \to n_i$, $p \to n_i$) at sufficiently high temperatures. In this subsection, which concludes the carrier concentration discussion, temperature-related facts are combined and embellished to provide a broader, more complete description of how the carrier concentrations vary with temperature.

Figure 2.22(a), a typical majority-carrier concentration-versus-temperature plot constructed assuming a phosphorus-doped $N_D = 10^{15}/cm^3$ Si sample, nicely illustrates the general features of the concentration-versus-temperature dependence. Examining Fig. 2.22(a), we find that $n$ is fixed at approximately $N_D$ over a broad temperature range extending roughly from 150 K to 450 K for the given Si sample. This $n \simeq N_D$ or "extrinsic temperature region" constitutes the normal operating range for most solid-state devices. Below 100 K or so, in the "freeze-out temperature region," $n$ drops significantly below $N_D$ and approaches zero as $T \to 0$ K. In the "intrinsic temperature region" at the opposite end of the temperature scale, $n$ rises above $N_D$, asymptotically approaching $n_i$ with increasing $T$.

To qualitatively explain the just-described concentration-versus-temperature dependence, it is important to recall that the equilibrium number of carriers within a material is affected by two separate mechanisms. Electrons donated to the conduction band from donor atoms and valence band electrons excited across the band gap into the conduction band (broken Si–Si bonds) both contribute to the majority-carrier electron concentration in a donor-doped material. At temperatures $T \to 0$ K the thermal energy available in the system is insufficient to release the weakly bound fifth electron on donor sites and totally insufficient to excite electrons across the band gap. Hence $n = 0$ at $T = 0$ K, as visualized on the left-hand side of Fig. 2.22(b). Slightly increasing the material temperature above $T = 0$ K "defrosts" or frees some of the electrons weakly bound to donor sites. Band-to-band excitation, however, remains extremely unlikely, and therefore the number of observed electrons in the freeze-out temperature region equals the number of ionized donors—$n = N_D^+$. Continuing to increase the system temperature eventually frees almost all of the weakly bound electrons on donor sites, $n$ approaches $N_D$, and one enters the extrinsic temperature region. In progressing through the extrinsic temperature region, more and more electrons are excited across the band gap, but the number of electrons supplied in this fashion stays comfortably below $N_D$. Ultimately, of course, electrons excited across the band gap equal, then exceed, and, as pictured on the right-hand side of Fig. 2.22(b), finally swamp the fixed number of electrons derived from the donors.

As a practical note, it should be pointed out that the wider the band gap, the greater the energy required to excite electrons from the valence band into the conduction band, and

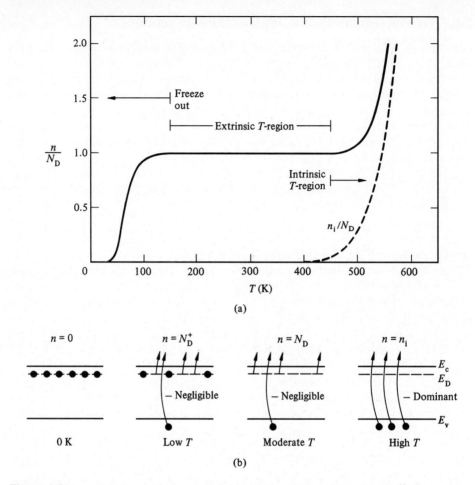

**Figure 2.22** (a) Typical temperature dependence of the majority-carrier concentration in a doped semiconductor. The plot was constructed assuming a phosphorus-doped $N_D = 10^{15}/$ cm³ Si sample. $n_i/N_D$ versus $T$ (dashed line) has been included for comparison purposes. (b) Qualitative explanation of the concentration-versus-temperature dependence displayed in part (a).

the higher the temperature at the onset of the intrinsic temperature region. Since the temperature at the onset of the intrinsic temperature region corresponds to the upper end of the normal operating range for most solid-state devices, GaAs devices can inherently operate at a higher maximum temperature than similarly doped Si devices, which in turn can operate at a higher maximum temperature than similarly doped Ge devices. If we assume, for example, the critical doping concentration is $N_D = 10^{15}/\text{cm}^3$, and the onset of the intrinsic temperature region is approximated as the temperature where $n_i = N_D$, then from Fig. 2.20

the maximum operating temperatures are deduced to be 385 K, 540 K, and $> 700$ K for Ge, Si, and GaAs, respectively. Indeed, GaAs and SiC ($E_G > 2$ eV) devices continue to be under development for use in high temperature environments.

## 2.6 SUMMARY AND CONCLUDING COMMENTS

Under the general heading of carrier modeling we have described, examined, and characterized the carriers within a semiconductor under "rest" or equilibrium conditions. The many important topics addressed in this chapter included the introduction of two "visualization" models: the bonding model and the energy band model. The extremely useful energy band model is actually more than just a model—it is a sophisticated sign language providing a concise means of communicating on a nonverbal level. Relative to the carriers themselves, the reader by now has been successfully prodded into thinking of electrons and holes as classical ball-like "particles," where the charge on an electron is $-q$, the charge on a hole is $+q$, and the effective masses of the particles are $m_n^*$ and $m_p^*$, respectively. The reader should also know that the carrier numbers in an intrinsic material are equal and relatively small; the carrier concentrations, however, can be selectively increased by adding special impurity atoms or dopants to the semiconductor.

In addressing the problem of determining the carrier concentrations in doped semiconductors, we developed or derived a number of useful mathematical relationships. The density of states functions (Eqs. 2.6), the Fermi function (Eq. 2.7), the symmetrical non-degenerate relationships for $n$ and $p$ (Eqs. 2.19), the $np$ product (Eq. 2.22), the charge neutrality relationship (Eq. 2.25), and the simplified $n$ and $p$ expressions appropriate for a typical semiconductor maintained at room temperature (Eqs. 2.30 and 2.31) deserve special mention. The cited equations and a few others are collected in Table 2.4. With regard to the use of these relationships, the reader should be cautioned against "no-think plug and chug." Because semiconductor problems are replete with exceptions, special cases, and nonideal situations, it is imperative that the formula user be aware of derivational assumptions and the validity limits of any and all expressions used in an analysis or computation. In addition to the quantitative carrier relationships, the reader should also have a qualitative "feel" for the carrier distributions in the respective energy bands, the temperature dependence of the intrinsic concentration, and the typical temperature dependence of the majority carrier concentration in a doped semiconductor.

Finally, some attention should be given to the many technical terms and the key parametric values presented in this chapter. The terms extrinsic semiconductor, donor, acceptor, nondegenerate semiconductor, Fermi level, and so on, will be encountered again and again in the discussion of semiconductor devices. Likewise, a knowledge of typical numerical values for key parameters, such as $E_G = 1.12$ eV and $n_i = 10^{10}/cm^3$ in Si at 300 K, will be invaluable in subsequent work when performing both "back-of-the-envelope" and computer-assisted computations. Key parametric values also serve as yardsticks for gauging whether newly encountered quantities are relatively small or relatively large.

**Table 2.4** Carrier Modeling Equation Summary.

---

*Density of States and Fermi Function*

$$g_c(E) = \frac{m_n^*\sqrt{2m_n^*(E - E_c)}}{\pi^2\hbar^3} \;, \quad E \geq E_c$$

$$g_v(E) = \frac{m_p^*\sqrt{2m_p^*(E_v - E)}}{\pi^2\hbar^3} \;, \quad E \leq E_v$$

$$f(E) = \frac{1}{1 + e^{(E - E_F)/kT}}$$

---

*Carrier Concentration Relationships*

$$n = N_C \frac{2}{\sqrt{\pi}} F_{1/2}(\eta_c) \qquad N_C = 2\left[\frac{m_n^*kT}{2\pi\hbar^2}\right]^{3/2}$$

$$p = N_V \frac{2}{\sqrt{\pi}} F_{1/2}(\eta_v) \qquad N_V = 2\left[\frac{m_p^*kT}{2\pi\hbar^2}\right]^{3/2}$$

$$n = N_C e^{(E_F - E_c)/kT}$$
$$p = N_V e^{(E_v - E_F)/kT}$$
$$n = n_i e^{(E_F - E_i)/kT}$$
$$p = n_i e^{(E_i - E_F)/kT}$$

---

*$n_i$, np-Product, and Charge Neutrality*

$$n_i = \sqrt{N_C N_V}\, e^{-E_G/2kT} \qquad np = n_i^2 \qquad p - n + N_D - N_A = 0$$

---

*n, p, and Fermi Level Computational Relationships*

$$n = \frac{N_D - N_A}{2} + \left[\left(\frac{N_D - N_A}{2}\right)^2 + n_i^2\right]^{1/2} \qquad E_i = \frac{E_c + E_v}{2} + \frac{3}{4}kT\ln\left(\frac{m_p^*}{m_n^*}\right)$$

$$n \simeq N_D$$
$$p \simeq n_i^2/N_D$$
$$N_D \gg N_A, N_D \gg n_i$$
$$E_F - E_i = kT\ln(n/n_i) = -kT\ln(p/n_i)$$

$$p \simeq N_A$$
$$n \simeq n_i^2/N_A$$
$$N_A \gg N_D, N_A \gg n_i$$
$$E_F - E_i = kT\ln(N_D/n_i) \quad N_D \gg N_A, N_D \gg n_i$$
$$E_i - E_F = kT\ln(N_A/n_i) \quad N_A \gg N_D, N_A \gg n_i$$

# PROBLEMS

| | CHAPTER 2 | PROBLEM | INFORMATION TABLE | |
|---|---|---|---|---|
| *Problem* | *Complete After* | *Difficulty Level* | *Suggested Point Weighting* | *Short Description* |
| ● 2.1 | 2.2.4 | 2 | 10 (5 each part) | $E_G$ vs. $T$ computation |
| 2.2 | 2.3.4 | 1 | 10 (2 each part) | Bonding model applications |
| 2.3 | 2.3.4 (a–d) 2.5.7 all | 1 | 24 (2 each part) | E-band model applications |
| 2.4 | 2.3.4 | 2 | 10 (2 each part) | Si-doped GaAs |
| 2.5 | 2.4.1 | 2 | 5 | States/cm$^3$ in $\Delta E$ |
| 2.6 | 2.4.2 | 2 | 8 (a-2, b-3, c-3) | Fermi function questions |
| 2.7 | 2.5.1 | 2–3 | 10 | Find distribution peak |
| 2.8 | " | 2 | 5 | Population at $E_c + \Delta E$ |
| ● 2.9 | " | 2 | 15 | Plot distributions in bands |
| ● 2.10 | " | 3 | 20 (a-3, b-17..discuss-2) | $T$ variation of distribution |
| 2.11 | " | 2 | 10 | Derive Eqs. (2.14b), (2.16b) |
| 2.12 | " | 3 | 15 (a-3, b-12) | Hypothetical $g_c$ = constant |
| 2.13 | " | 2 | 10 (a-7, b-3) | Compute $N_C$, $N_V$ |
| 2.14 | 2.5.3 | 1 | 8 (4 each part) | $n_i$ comparison |
| ● 2.15 | " | 1 | 5 | Plot $n_i$ vs. $T$ for Ge |
| 2.16 | 2.5.5 | 2 | 12 (a::d-2, e-4) | Tricky conc. questions |
| 2.17 | " | 1–2 | 10 (2 each part) | Compute $n$ and $p$ |
| 2.18 | 2.5.6 | 2 | 15 (3 each part) | Determine $E_i$, $E_F - E_i$, etc. |
| ● 2.19 | " | 2 | 15 (a-12, b-3) | Computer check P2.17/2.18 |
| 2.20 | " | 2 | 8 | Verify nondegenerate limits |
| ● 2.21 | " | 2 | 10 | Plot $E_F - E_i$ vs. $N_A$, $N_D$ |
| 2.22 | " | 2–3 | 12 (a-3, b-3, c-2, d-4) | GaAs considerations |
| ● 2.23 | 2.5.7 | 4 | 25 (a-8, b-10, c-2, d-3, e-2) | $E_F$ variation with $T$ |

● **2.1** $E_G$ versus $T$ Computation

With increasing temperature an expansion of the crystal lattice usually leads to a weakening of the interatomic bonds and an associated decrease in the band gap energy. For many semiconductors the cited variation of the band gap energy with temperature can be modeled by the empirical relationship

$$E_G(T) = E_G(0) - \frac{\alpha T^2}{(T + \beta)}$$

where $\alpha$ and $\beta$ are constants chosen to obtain the best fit to experimental data and $E_G(0)$ is the limiting value of the band gap at 0 K. As far as Si is concerned, a fit accurate to four places is obtained by employing

$$E_G(0) = 1.170 \text{ eV}$$
$$\alpha = 4.730 \times 10^{-4} \text{ eV/K}$$
$$\beta = 636 \text{ K}$$
$$T \text{ in Kelvin}$$

(a) Make a plot of $E_G$ versus $T$ for Si spanning the temperature range from $T = 0$ K to $T = 600$ K. Specifically note the value of $E_G$ at 300 K.

(b) For $T > 300$ K, the temperature variation is nearly linear. Noting this fact, some authors have employed

$$E_G(T) = 1.205 - 2.8 \times 10^{-4} T \quad \ldots T > 300 \text{ K}$$

How does this simplified relationship compare with the more precise relationship over the temperature region of mutual validity?

**2.2** Using the *bonding* model for a semiconductor, indicate how one visualizes (a) a missing atom, (b) an electron, (c) a hole, (d) a donor, (e) an acceptor.

**2.3** Using the *energy band* model for a semiconductor, indicate how one visualizes (a) an electron, (b) a hole, (c) donor sites, (d) acceptor sites, (e) freeze-out of majority carrier electrons at donor sites as the temperature is lowered toward 0 K, (f) freeze-out of majority carrier holes at acceptor sites as the temperature is lowered toward 0 K, (g) the energy distribution of carriers in the respective bands, (h) an intrinsic semiconductor, (i) an $n$-type semiconductor, (j) a $p$-type semiconductor, (k) a nondegenerate semiconductor, (l) a degenerate semiconductor.

**2.4** The bonding model for GaAs is pictured in Fig. P2.4.

(a) Draw the bonding model for GaAs depicting the removal of the shaded Ga and As atoms in Fig. P2.4. HINT: Ga and As take their bonding electrons with them when they are removed from the lattice. Also see Fig. 2.4(a) showing the results of removing an atom from the Si lattice.

(b) Redraw the bonding model for GaAs showing the insertion of Si atoms into the missing Ga and As atom sites.

(c) Is the GaAs doped $p$- or $n$-type when Si atoms replace Ga atoms? Explain.

(d) Is the GaAs doped $p$- or $n$-type when Si atoms replace As atoms? Explain.

(e) Draw the *energy band diagram* for GaAs when the GaAs is doped with Si on (i) Ga sites and (ii) on As sites.

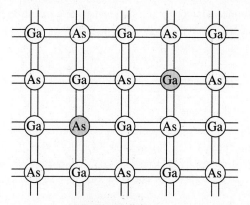

**Figure P2.4**

**2.5** Develop an expression for the total number of available STATES/cm$^3$ in the conduction band between energies $E_c$ and $E_c + \gamma kT$, where $\gamma$ is an arbitrary constant.

**2.6** (a)  Under equilibrium conditions and $T > 0$ K, what is the probability of an electron state being occupied if it is located at the Fermi level?

(b)  If $E_F$ is positioned at $E_c$, determine (numerical answer required) the probability of finding electrons in states at $E_c + kT$.

(c)  The probability a state is filled at $E_c + kT$ is equal to the probability a state is empty at $E_c + kT$. Where is the Fermi level located?

**2.7** The carrier distributions or numbers of carriers as a function of energy in the conduction and valence bands were noted to peak at an energy very close to the band edges. (See the carrier distribution sketches in Fig. 2.16.) Taking the semiconductor to be nondegenerate, show that the energy at which the carrier distributions peak is $E_c + kT/2$ and $E_v - kT/2$ for the conduction and valence bands, respectively.

**2.8** For a nondegenerate semiconductor, the peak in the electron distribution versus energy inside the conduction band noted in Fig. 2.16 occurs at $E_c + kT/2$. Expressed as a fraction of the electron population at the peak energy, what is the electron population in a nondegenerate semiconductor at $E = E_c + 5kT$?

● **2.9** The Fermi level in a Si sample maintained at $T = 300$ K is located at $E_c - E_G/4$. Compute and plot the electron and hole distributions (numbers/cm$^3$–eV) as a function of energy in the conduction and valence bands, respectively.

● **2.10** Let us investigate how the electron energy distribution in the conduction band varies as a function of temperature.

(a) Assuming the semiconductor to be nondegenerate and employing the Eq. (2.16a) expression for $n$, confirm that the electron distribution in the conduction band normalized to the total electron concentration is given by

$$\frac{g_c(E)f(E)}{n} = \frac{2\sqrt{E - E_c}}{\sqrt{\pi}\,(kT)^{3/2}}\,e^{-(E - E_c)/kT}$$

(b) Compute and plot the normalized electron distribution in the conduction band versus $E - E_c$ for temperatures $T = 300$ K, 600 K, and 1200 K. Plot the distribution values along the $x$-axis ($0 \leq g_c(E)f(E)/n \leq 20$ eV$^{-1}$) and $E - E_c$ ($0 \leq E - E_c \leq 0.4$ eV) along the $y$-axis on a single set of coordinates. Discuss your results.

**2.11** Starting with Eq. (2.8b) and following a procedure analogous to that outlined in the text, present the intermediate steps and arguments leading to Eqs. (2.14b) and (2.16b).

**2.12** The density of states in the conduction band of a hypothetical semiconductor is

$$g_c(E) = \text{constant} = N_C/kT \qquad \ldots E \geq E_c$$

(a) Assuming $E_F < E_c - 3kT$, sketch the electron distribution in the conduction band of the hypothetical semiconductor.

(b) Following the procedure outlined in the text, establish relationships for the electron concentration in the hypothetical semiconductor analogous to Eqs. (2.14a) and (2.16a).

**2.13** (a) Verify the statement in Subsection 2.5.1 that, at 300 K,

$$N_{C,V} = (2.510 \times 10^{19}/\text{cm}^3)(m^*/m_0)^{3/2}$$

where one sets $m^* = m_n^*$ in computing $N_C$ and $m^* = m_p^*$ in computing $N_V$. $m_0 = 9.109 \times 10^{-31}$ kg; $h = 6.625 \times 10^{-34}$ joule-sec; and $q = 1.602 \times 10^{-19}$ coul.

(b) Using the effective mass values recorded in Table 2.1, construct a table that lists the numerical values of $N_C$ and $N_V$ for Si, Ge, and GaAs at 300 K.

**2.14** (a) Determine the temperature at which the intrinsic carrier concentration in (i) Si and (ii) GaAs are equal to the room temperature (300 K) intrinsic carrier concentration of Ge.

(b) Semiconductor A has a band gap of 1 eV, while semiconductor B has a band gap of 2 eV. What is the ratio of the intrinsic carrier concentrations in the two materials $(n_{iA}/n_{iB})$ at 300 K. Assume any differences in the carrier effective masses may be neglected.

● **2.15** Confirm that the $n_i$ versus $T$ curve for Ge graphed in Fig. 2.20 can be generated employing the empirical-fit relationship

$$n_i(\text{Ge}) = (1.76 \times 10^{16})T^{3/2}e^{-0.392/kT}$$

**2.16** Concentration questions with a twist.

(a) A silicon wafer is uniformly doped $p$-type with $N_A = 10^{15}/\text{cm}^3$. At $T \approx 0$ K, what are the equilibrium hole and electron concentrations?

(b) A semiconductor is doped with an impurity concentration $N$ such that $N \gg n_i$ and all the impurities are ionized. Also, $n = N$ and $p = n_i^2/N$. Is the impurity a donor or an acceptor? Explain.

(c) The electron concentration in a piece of Si maintained at 300 K under equilibrium conditions is $10^5/\text{cm}^3$. What is the hole concentration?

(d) For a silicon sample maintained at $T = 300$ K, the Fermi level is located 0.259 eV above the intrinsic Fermi level. What are the hole and electron concentrations?

(e) In a nondegenerate germanium sample maintained under equilibrium conditions near room temperature, it is known that $n_i = 10^{13}/\text{cm}^3$, $n = 2p$, and $N_A = 0$. Determine $n$ and $N_D$.

**2.17** Determine the equilibrium electron and hole concentrations inside a uniformly doped sample of Si under the following conditions:

(a) $T = 300$ K, $N_A \ll N_D$, $N_D = 10^{15}/\text{cm}^3$.

(b) $T = 300$ K, $N_A = 10^{16}/\text{cm}^3$, $N_D \ll N_A$.

(c) $T = 300$ K, $N_A = 9 \times 10^{15}/\text{cm}^3$, $N_D = 10^{16}/\text{cm}^3$.

(d) $T = 450$ K, $N_A = 0$, $N_D = 10^{14}/\text{cm}^3$.

(e) $T = 650$ K, $N_A = 0$, $N_D = 10^{14}/\text{cm}^3$.

**2.18** (a to e) For each of the conditions specified in Problem 2.17, determine the position of $E_i$, compute $E_F - E_i$, and draw a carefully dimensioned energy band diagram for the Si sample. NOTE: $E_G(\text{Si}) = 1.08$ eV at 450 K and 1.015 eV at 650 K.

● **2.19** (a) Assuming a nondegenerate Si sample and total impurity atom ionization, construct a MATLAB program that computes $n$, $p$, and $E_F - E_i$ given acceptable input values of $T$ (temperature in Kelvin), $N_D$ (cm$^{-3}$), and $N_A$ (cm$^{-3}$). Incorporate the program presented in part (a) of Exercise 2.4 to compute $n_i$ at a specified $T$. Make use of the MATLAB input function to enter the input variables from the command window.

(b) Use your program to check the relevant answers to Problems 2.17 and 2.18.

**2.20** According to the text, the maximum nondegenerate donor and acceptor doping concentrations in Si at room temperature are $N_D \simeq 1.6 \times 10^{18}/cm^3$ and $N_A \simeq 9.1 \times 10^{17}/cm^3$, respectively. Verify the text statement.

● **2.21** Construct a computer program to produce a plot of $E_F - E_i$ versus $N_A$ or $N_D$ similar to Fig. 2.21. Use the program to verify the accuracy of the cited figure. (You may find it convenient to employ the MATLAB `logspace` function in constructing your program.)

**2.22** GaAs considerations.

(a) Make a sketch similar to Fig. 2.14 that is specifically appropriate for GaAs. Be sure to take into account the fact that $m_n^* \ll m_p^*$ in GaAs.

(b) Based on your answer to part (a), would you expect $E_i$ in GaAs to lie above or below midgap? Explain.

(c) Determine the precise position of the intrinsic Fermi level in GaAs at room temperature (300 K).

(d) Determine the maximum nondegenerate donor and acceptor doping concentrations in GaAs at room temperature.

**2.23** Given an $N_A = 10^{14}/cm^3$ doped Si sample:

(a) Present a *qualitative* argument that leads to the approximate positioning of the Fermi level in the material as $T \rightarrow 0$ K.

● (b) Construct a MATLAB (computer) program to calculate and plot $E_F - E_i$ in the material as a function of temperature for 200 K $\leq T \leq$ 500 K.

(c) What do you conclude relative to the general behavior of the Fermi level positioning as a function of temperature?

(d) Run your part (b) program to determine what happens if $N_A$ is progressively increased in decade steps from $N_A = 10^{14}/cm^3$ to $N_A = 10^{18}/cm^3$. Summarize your observations.

(e) How would your foregoing answers be modified if the Si sample was doped with donors instead of acceptors?

# 3 Carrier Action

Carrier modeling under "rest" or equilibrium conditions, considered in Chapter 2, is important because it establishes the proper frame of reference. From a device standpoint, however, the zero current observed under equilibrium conditions is rather uninteresting. Only when a semiconductor system is perturbed, giving rise to carrier action or a net carrier response, can currents flow within and external to the semiconductor system. Action, carrier action, is the general concern of this chapter.

Under normal operating conditions the three primary types of carrier action occurring inside semiconductors are *drift, diffusion,* and *recombination–generation.* In this chapter we first describe each primary type of carrier action qualitatively and then quantitatively relate the action to the current flowing within the semiconductor. Special emphasis is placed on characterizing the "constants of the motion" associated with each type of action, and wherever appropriate the discussion is extended to subsidiary topics of a relevant nature. Although introduced individually, the various types of carrier action are understood to occur simultaneously inside any given semiconductor. Mathematically combining the various carrier activities next leads to the culmination of our carrier-action efforts; we obtain the basic set of starting equations employed in solving device problems of an electrical nature. Finally, simple example problems are considered to illustrate solution approaches and to introduce key supplemental concepts.

## 3.1 DRIFT

### 3.1.1 Definition–Visualization

*Drift,* by definition, *is charged-particle motion in response to an applied electric field.* Within semiconductors the drifting motion of the carriers on a microscopic scale can be described as follows: When an electric field ($\mathscr{E}$) is applied across a semiconductor as visualized in Fig. 3.1(a), the resulting force on the carriers tends to accelerate the $+q$ charged holes in the direction of the electric field and the $-q$ charged electrons in the direction opposite to the electric field. Because of collisions with ionized impurity atoms and thermally agitated lattice atoms, however, the carrier acceleration is frequently interrupted (the carriers are said to be scattered). The net result, pictured in Fig. 3.1(b), is carrier motion generally along the direction of the electric field, but in a disjointed fashion involving repeated periods of acceleration and subsequent decelerating collisions.

The microscopic drifting motion of a single carrier is obviously complex and quite tedious to analyze in any detail. Fortunately, measurable quantities are *macroscopic*

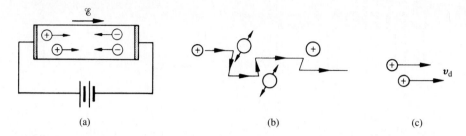

**Figure 3.1**   Visualization of carrier drift: (a) motion of carriers within a biased semiconductor bar; (b) drifting hole on a microscopic or atomic scale; (c) carrier drift on a macroscopic scale.

observables that reflect the average or overall motion of the carriers. Averaging over all electrons or holes at any given time, we find that the resultant motion of each carrier type can be described in terms of a constant drift velocity, $v_d$. In other words, on a macroscopic scale, drift may be visualized (see Fig. 3.1c) as nothing more than all carriers of a given type moving along at a constant velocity in a direction parallel or antiparallel to the applied electric field.

By way of clarification, it is important to point out that the drifting motion of the carriers arising in response to an applied electric field is actually superimposed upon the always-present thermal motion of the carriers. Electrons in the conduction band and holes in the valence band gain and lose energy via collisions with the semiconductor lattice and are nowhere near stationary even under equilibrium conditions. In fact, under equilibrium conditions the thermally related carrier velocities average ~1/1000 the speed of light at room temperature! As pictured in Fig. 3.2, however, the thermal motion of the carriers is completely random. Thermal motion therefore averages out to zero on a macroscopic scale, does not contribute to current transport, and can be conceptually neglected.

### 3.1.2  Drift Current

Let us next turn to the task of developing an analytical expression for the current flowing within a semiconductor as a result of carrier drift. By definition

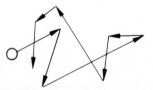

**Figure 3.2**   Thermal motion of a carrier.

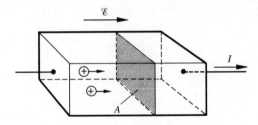

**Figure 3.3** Expanded view of a biased $p$-type semiconductor bar of cross-sectional area $A$.

$I$ (current) = the charge per unit time crossing an arbitrarily chosen plane of observation oriented normal to the direction of current flow.

Considering the $p$-type semiconductor bar of cross-sectional area $A$ shown in Fig. 3.3, and specifically noting the arbitrarily chosen $v_d$-normal plane lying within the bar, we can argue:

$v_d t$   ...   All holes this distance back from the $v_d$-normal plane will cross the plane in a time $t$.

$v_d t A$   ...   All holes in this volume will cross the plane in a time $t$.

$p v_d t A$   ...   Holes crossing the plane in a time $t$.

$q p v_d t A$   ...   Charge crossing the plane in a time $t$.

$q p v_d A$   ...   Charge crossing the plane per unit time.

The word definition of the last quantity is clearly identical to the formal definition of current. Thus

$$I_{\text{P|drift}} = q p v_d A \qquad \text{hole drift current} \qquad (3.1)$$

As a practical matter, the cross-sectional area $A$ appearing in Eq. (3.1) and other current formulas is often excess baggage. Current, moreover, is generally thought of as a scalar quantity, while in reality it is obviously a vector. These deficiencies are overcome by introducing a related parameter known as the current density, **J**. **J** has the same orientation as the direction of current flow and is equal in magnitude to the current per unit area (or $J = I/A$). By inspection, the current density associated with hole drift is simply

$$\mathbf{J}_{\text{P|drift}} = q p \mathbf{v}_d \qquad (3.2)$$

Since the drift current arises in response to an applied electric field, it is reasonable to proceed one step further and seek a form of the current relationship that explicitly relates $\mathbf{J}_{\text{P|drift}}$ to the applied electric field. To this end, we make reference to the representative drift velocity versus electric field data presented in Fig. 3.4. Note that $v_d$ is proportional to $\mathscr{E}$ at

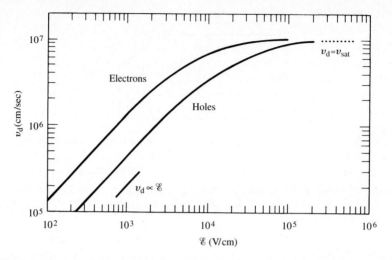

**Figure 3.4** Measured drift velocity of the carriers in ultrapure silicon maintained at room temperature as a function of the applied electric field. Constructed from the data fits and the data respectively in Jacoboni et al.[4] and Smith et al.[5]

low electric fields, while at high electric fields $v_d$ saturates and becomes independent of $\mathscr{E}$. To be more precise,

$$v_d = \frac{\mu_0 \mathscr{E}}{\left[1 + \left(\dfrac{\mu_0 \mathscr{E}}{v_{sat}}\right)^{\beta}\right]^{1/\beta}} = \begin{cases} \mu_0 \mathscr{E} & \dots \mathscr{E} \to 0 \\ v_{sat} & \dots \mathscr{E} \to \infty \end{cases} \tag{3.3}$$

where $\beta \cong 1$ for holes and $\beta \cong 2$ for electrons in silicon, $\mu_0$ is the constant of proportionality between $v_d$ and $\mathscr{E}$ at low to moderate electric fields, and $v_{sat}$ is the limiting or *saturation velocity* approached at very high electric fields. Obviously, in the high-field limit $v_d$ in Eq. (3.2) is simply replaced by $v_{sat}$ and $\mathbf{J}_{P|drift}$ does not exhibit an $\mathscr{E}$-field dependence. In the low-field limit, which is of greatest practical interest and assumed herein unless specified otherwise, $v_d = \mu_p \mathscr{E}$ ($\mu_0 \to \mu_p$ for holes) and substitution into Eq. (3.2) yields

$$\boxed{\mathbf{J}_{P|drift} = q\mu_p p\mathscr{E}} \tag{3.4a}$$

Similarly, for electrons one obtains

$$\boxed{\mathbf{J}_{N|drift} = q\mu_n n\mathscr{E}} \tag{3.4b}$$

Respectively known as the *electron mobility* and *hole mobility*, $\mu_n$ and $\mu_p$ are always taken to be positive quantities. Note that, although electrons drift in the direction opposite to the applied electric field ($v_d = -\mu_n \mathscr{E}$), the current transported by negatively charged particles is in turn counter to the direction of drift ($\mathbf{J}_{N|drift} = -qnv_d$). The net result, as indicated in Eq. (3.4b), is an electron current in the direction of the applied electric field.

### 3.1.3 Mobility

Mobility is obviously a central parameter in characterizing electron and hole transport due to drift. As further readings will reveal, the carrier mobilities also play a key role in characterizing the performance of many devices. It is reasonable therefore to examine $\mu_n$ and $\mu_p$ in some detail to enhance our general familiarity with the parameters and to establish a core of useful information for future reference.

*Standard Units:*  cm²/V-sec.

*Sample Numerical Values:*  $\mu_n \simeq 1360$ cm²/V-sec and $\mu_p \simeq 460$ cm²/V-sec at 300 K in $N_D = 10^{14}$/cm³ and $N_A = 10^{14}$/cm³ doped Si, respectively. In uncompensated high-purity ($N_D$ or $N_A \leq 10^{15}$/cm³) GaAs, the room-temperature mobilities are $\mu_n \cong 8000$ cm²/V-sec and $\mu_p \cong 400$ cm²/V-sec. The quoted values are useful for comparison purposes and when performing order-of-magnitude computations. Also note that $\mu_n > \mu_p$ for both Si and GaAs. In the major semiconductors, $\mu_n$ is consistently greater than $\mu_p$ for a given doping and system temperature.

*Relationship to Scattering:*  The word *mobility* in everyday usage refers to a general freedom of movement. Analogously, in semiconductor work the mobility parameter is a measure of the ease of carrier motion in a crystal. Increasing the motion-impeding collisions within a crystal decreases the mobility of the carriers. In other words, the carrier mobility varies inversely with the amount of scattering taking place within the semiconductor. As visualized in Fig. 3.1(b), the dominant scattering mechanisms in nondegenerately doped materials of device quality are typically (i) lattice scattering involving collisions with thermally agitated lattice atoms, and (ii) ionized impurity (i.e., donor-site and/or acceptor-site) scattering. Relative to lattice scattering, it should be emphasized that it is the thermal vibration, the *displacement* of lattice atoms from their lattice positions, that leads to carrier scattering. The internal field associated with the stationary array of atoms in a crystal is already taken into account in the effective mass formulation.

Although quantitative relationships connecting the mobility and scattering can become quite involved, it is readily established that $\mu = q\langle \tau \rangle / m^*$, where $\langle \tau \rangle$ is the mean free time between collisions and $m^*$ is the conductivity effective mass. Since increasing the number of motion-impeding collisions decreases the mean free time between collisions, we again conclude $\mu$ varies inversely with the amount of scattering. However, $\mu$ is also noted to vary inversely with the carrier effective mass—lighter carriers move more readily. The $m_n^*$ in GaAs is significantly smaller than the $m_n^*$ in Si, thereby explaining the higher mobility of the GaAs electrons.

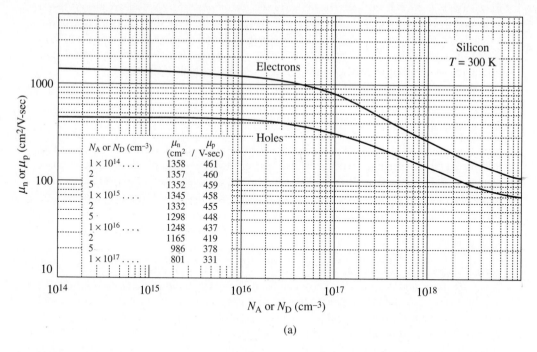

(a)

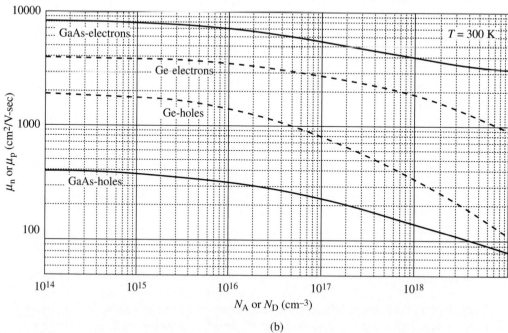

(b)

**Figure 3.5**    Room temperature carrier mobilities as a function of the dopant concentration in (a) Si and (b) Ge and GaAs. $\mu_n$ is the electron mobility; $\mu_p$ is the hole mobility.

*Doping Dependence:* Figure 3.5 exhibits the observed doping dependence of the electron and hole mobilities in Si, Ge, and GaAs. All semiconductors exhibit the same general dependence. At low doping concentrations, below approximately $10^{15}/cm^3$ in Si, the carrier mobilities are essentially independent of the doping concentration. For dopings in excess of $\sim 10^{15}/cm^3$, the mobilities monotonically decrease with increasing $N_A$ or $N_D$.

With the aid of Fig. 3.6 the explanation of the observed doping dependence is relatively straightforward. Invoking an electrical analogy, one can associate a resistance to motion with each of the scattering mechanisms. These resistances are in series. At sufficiently low doping levels, ionized impurity scattering can be neglected compared to lattice scattering, or in the analogy, $R_{TOTAL} = R_L + R_I \cong R_L$. When lattice scattering, which is not a function of $N_A$ or $N_D$, becomes the dominant scattering mechanism, it automatically follows that the carrier mobilities will be likewise independent of $N_A$ or $N_D$. For dopings in excess of $\sim 10^{15}/cm^3$ in Si, ionized impurity scattering and the associated resistance to motion can no longer be neglected. Increasing the number of scattering centers by adding more and more acceptors or donors progressively increases the amount of ionized impurity scattering and systematically decreases the carrier mobilities.

*Temperature Dependence:* The temperature dependence of the electron and hole mobilities in Si with doping as a parameter is displayed in Fig. 3.7. For dopings of $N_A$ or $N_D \leq 10^{14}/cm^3$, the data merge into a single curve and there is a near power-law increase in the carrier mobilities as the temperature is decreased. Roughly, $\mu_n \propto T^{-2.3 \pm 0.1}$ and $\mu_p \propto T^{-2.2 \pm 0.1}$. For progressively higher dopings, the carrier mobilities still increase with decreasing temperature, but at a systematically decreasing rate. In fact, some $N_D \geq 10^{18}/cm^3$ experimental data (not shown) exhibit a *decrease* in $\mu_n$ as $T$ is decreased below 200 K.

The general temperature dependence of the carrier mobilities is relatively easy to explain in the low doping limit. As noted in the doping dependence discussion, lattice scattering is the dominant scattering mechanism ($R_{TOTAL} \cong R_L$) in lightly doped samples. Decreasing the system temperature causes an ever-decreasing thermal agitation of the semiconductor atoms, which in turn decreases the lattice scattering. The decreased scattering

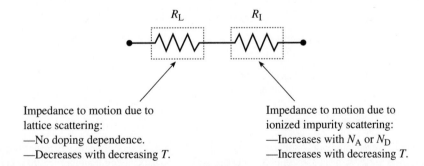

**Figure 3.6**  Electrical analogy for scattering in a semiconductor. $R_L$ and $R_I$ represent the impedance to motion due to lattice scattering and ionized impurity scattering, respectively.

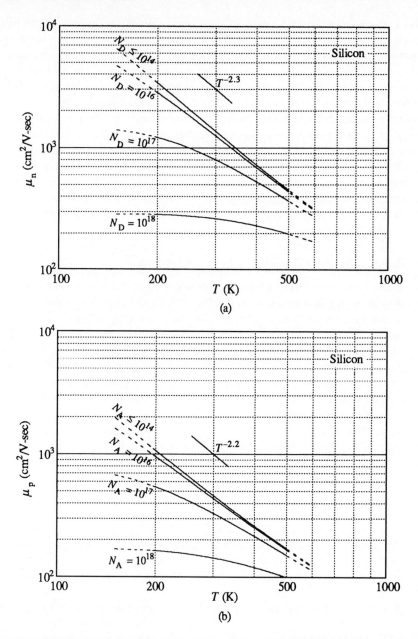

**Figure 3.7** Temperature dependence of (a) electron and (b) hole mobilities in silicon for dopings ranging from $\leq 10^{14}/cm^3$ to $10^{18}/cm^3$. The curves were constructed using the empirical fit relationships and parameters presented in Exercise 3.1. The dashed line portion of the curves correspond to a slight extension of the fit beyond the verified 200 K $\leq T \leq$ 500 K range of validity.

roughly follows a simple power-law dependence. Being inversely proportional to the amount of lattice scattering, the mobility of lightly doped samples is therefore expected to increase with decreasing temperature, varying roughly as the temperature to a negative power.

The more complex dependence of the higher doped samples reflects the added effect of ionized impurity scattering. In the electrical analogy, $R_I$ can no longer be neglected. Moreover, whereas lattice scattering ($R_L$) decreases with decreasing $T$, ionized impurity scattering ($R_I$) increases with decreasing $T$. Ionized impurities become more and more effective in deflecting the charged carriers as the temperature and hence the speed of the carriers decreases. Thus ionized impurity scattering becomes a larger and larger percentage of the overall scattering as the temperature is decreased ($R_{TOTAL} \rightarrow R_I$). Clearly, this explains the decreased slope of the mobility versus temperature dependence exhibited by the higher doped samples.

---

## (C) Exercise 3.1

There exist surprisingly accurate "empirical-fit" relationships that are widely employed to compute the carrier mobilities at a given doping and temperature. Figures 3.5 and 3.7 were constructed using such relationships. The form of a computational relationship is typically established on an empirical basis by noting the general functional dependencies predicted theoretically and observed experimentally. Parameters in the relationship are then adjusted until an acceptable match is obtained to the best available experimental data.

The majority carrier mobility versus doping at room temperature is popularly computed from

$$\mu = \mu_{min} + \frac{\mu_0}{1 + (N/N_{ref})^\alpha}$$

where $\mu$ is the carrier mobility ($\mu_n$ or $\mu_p$), $N$ is the doping concentration ($N_A$ or $N_D$), and all other quantities are fit parameters. To model the temperature dependence, one additionally employs

$$A = A_{300} \left( \frac{T}{300} \right)^\eta$$

$A$ in the above equation represents $\mu_{min}$, $\mu_0$, $N_{ref}$, or $\alpha$; $A_{300}$ is the 300 K value of the parameter, $T$ is temperature in Kelvin, and $\eta$ is the temperature exponent. The fit parameters appropriate for Si are listed in the following table:

| Parameter | Value at 300 K | | Temperature Exponent ($\eta$) |
| --- | --- | --- | --- |
| | Electrons | Holes | |
| $N_{ref}$ (cm$^{-3}$) | $1.3 \times 10^{17}$ | $2.35 \times 10^{17}$ | 2.4 |
| $\mu_{min}$ (cm$^2$/V-sec) | 92 | 54.3 | $-0.57$ |
| $\mu_0$ (cm$^2$/V-sec) | 1268 | 406.9 | $-2.33$ electrons $-2.23$ holes |
| $\alpha$ | 0.91 | 0.88 | $-0.146$ |

**P:** (a) Construct a log-log plot of $\mu_n$ and $\mu_p$ versus $N_A$ or $N_D$ for $10^{14}$/cm$^3 \leq N_A$ or $N_D \leq 10^{19}$/cm$^3$ using the quoted fit relationship and the listed Si fit parameters. Compare your result with Fig. 3.5(a).

(b) Construct log-log plots of $\mu_n$ versus $T$ and $\mu_p$ versus $T$ for 200 K $\leq T \leq$ 500 K and $N_D$ or $N_A$ stepped in decade values from $10^{14}$/cm$^3$ to $10^{18}$/cm$^3$. Compare your results with Figs. 3.7(a) and 3.7(b), respectively.

**S:** (a) MATLAB program script . . .

```
%Mobility versus Dopant Concentration (Si,300K)

%Fit Parameters
NDref=1.3e17; NAref=2.35e17;
μnmin=92; μpmin=54.3;
μn0=1268; μp0=406.9;
an=0.91; ap=0.88

%Mobility Calculation
N=logspace(14,19);
μn=μnmin+μn0./(1+(N/NDref).^an);
μp=μpmin+μp0./(1+(N/NAref).^ap);

%Plotting results
close
loglog(N,μn,N,μp); grid;
axis([1.0e14 1.0e19 1.0e1 1.0e4]);
xlabel('NA or ND (cm-3)');
ylabel('Mobility (cm2/V-sec)');
text(1.0e15,1500,'Electrons');
text(1.0e15,500,'Holes');
text(1.0e18,2000,'Si,300K');
```

The results obtained by running the preceding program should be numerically identical to Fig. 3.5(a).

(b) Part (b) is left for the reader to complete.

## 3.1.4 Resistivity

Resistivity is an important material parameter that is closely related to carrier drift. Qualitatively, *resistivity* is a measure of a material's inherent resistance to current flow—a "normalized" resistance that does not depend on the physical dimensions of the material. Quantitatively, resistivity ($\rho$) is defined as the proportionality constant between the electric field impressed across a homogeneous material and the total particle current per unit area flowing in the material; that is,

$$\mathscr{E} = \rho \mathbf{J} \tag{3.5a}$$

or

$$\mathbf{J} = \sigma \mathscr{E} = \frac{1}{\rho} \mathscr{E} \tag{3.5b}$$

where $\sigma = 1/\rho$ is the material *conductivity*. In a homogeneous material, $\mathbf{J} = \mathbf{J}_{\text{drift}}$ and, as established with the aid of Eqs. (3.4),

$$\mathbf{J}_{\text{drift}} = \mathbf{J}_{\text{N|drift}} + \mathbf{J}_{\text{P|drift}} = q(\mu_n n + \mu_p p)\mathscr{E} \tag{3.6}$$

It therefore follows that

$$\rho = \frac{1}{q(\mu_n n + \mu_p p)} \tag{3.7}$$

In a nondegenerate donor-doped semiconductor maintained in the extrinsic temperature region where $N_D \gg n_i$, $n \simeq N_D$ and $p \simeq n_i^2/N_D \ll n$. This result was established in Subsection 2.5.5. Thus, for typical dopings and mobilities, $\mu_n n + \mu_p p \simeq \mu_n N_D$ in an *n*-type semiconductor. Similar arguments yield $\mu_n n + \mu_p p \simeq \mu_p N_A$ in a *p*-type semiconductor. Consequently, under conditions normally encountered in Si samples maintained at or near room temperature, Eq. (3.7) simplifies to

$$\rho = \frac{1}{q\mu_n N_D} \qquad \text{. . . } n\text{-type semiconductor} \tag{3.8a}$$

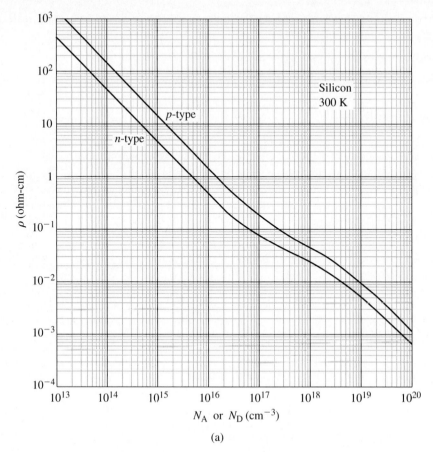

(a)

**Figure 3.8**  Resistivity versus impurity concentration at 300 K in (a) Si and (b) other semiconductors. [(b) From Sze[2], © 1981 by John Wiley & Sons, Inc. Reprinted with permission.]

and

$$\rho = \frac{1}{q\mu_p N_A} \qquad \text{. . . } p\text{-type semiconductor} \tag{3.8b}$$

When combined with mobility-versus-doping data, Eqs. (3.8) provide a one-to-one correspondence between the resistivity, a directly measurable quantity, and the doping inside a semiconductor. In conjunction with plots of $\rho$ versus doping (see Fig. 3.8), the measured resistivity is in fact routinely used to determine $N_A$ or $N_D$.

The measured resistivity required in determining the doping can be obtained in a number of different ways. A seemingly straightforward approach would be to form the semiconductor into a bar, apply a bias $V$ across contacts attached to the ends of the bar as in

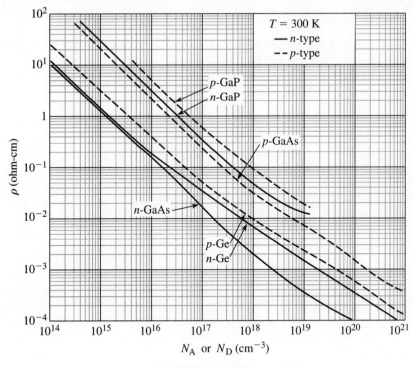

**Figure 3.8 (b)**

Fig. 3.1(a), measure the current $I$ flowing in the circuit, and then deduce $\rho$ from the measured resistance. [$R$(resistance) $= V/I = \rho l/A$, where $l$ is the bar length and $A$ is the cross-sectional area.] Unfortunately, the straightforward approach is deceptively difficult, is destructive (wastes semiconductor material), and is not readily adaptable to the wafers used in device processing.

A measurement method finding widespread usage in practice is the four-point probe technique. In the standard four-point probe technique, four collinear, evenly spaced probes, as shown in Fig. 3.9(a), are brought into contact with the surface of the semiconductor. A known current $I$ is passed through the outer two probes, and the potential $V$ thereby developed is measured across the inner two probes. The semiconductor resistivity is then computed from

$$\rho = 2\pi s \frac{V}{I} \Gamma \tag{3.9}$$

where $s$ is the probe-to-probe spacing and $\Gamma$ is a well-documented "correction" factor. The correction factor typically depends on the thickness of the sample and on whether the bottom of the semiconductor is touching an insulator or a metal. Commercial instruments are

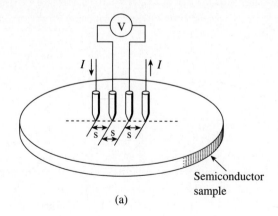

(a)

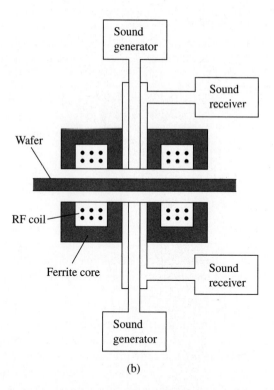

(b)

**Figure 3.9**    Resistivity measurement techniques. (a) Schematic drawing of the probe arrangement, placement, and biasing in the four-point probe measurement. (b) Schematic of a commercial eddy-current apparatus showing the RF coils and sonic components. [(b) From Schroder[6], © 1990 by John Wiley & Sons, Inc. Reprinted with permission.]

available that automatically compute the appropriate correction factor based on the sample thickness entered by the operator. Unlike the semiconductor-bar measurement, the four-point probe technique is obviously easy to implement, causes only slight surface damage in the vicinity of the probe contacts, and is ideally suited for working with wafers. The surface damage, although slight, does exclude the technique from being used on wafers intended for producing high-yield devices and ICs.

A second technique worthy of note utilizes a noncontacting eddy-current approach. The schematic of a commercial eddy-current apparatus is pictured in Fig. 3.9(b). There are fringing fields near a ferrite core excited with an RF coil. If a conducting material such as a semiconductor wafer is placed near the ferrite core, the fringing fields cause localized (eddy) currents to flow in the conducting material. The current flow in turn absorbs some of the RF power. The *sheet* resistivity of the conducting material is deduced by monitoring the power consumption in a calibrated system. The ultrasound generators and receivers built into the commercial apparatus are for the in-situ determination of the wafer thickness. The change in phase of ultrasound bounced off the top and bottom wafer surfaces permits the instrument to compute the intercore spacing occupied by the wafer. The sheet resistivity multiplied by the wafer thickness yields the wafer resistivity.

## 3.1.5 Band Bending

In previous encounters with the energy band diagram, we have consistently drawn $E_c$ and $E_v$ to be energies independent of the position coordinate $x$. When an electric field ($\mathscr{E}$) exists inside a material, the band energies become a function of position. The resulting variation of $E_c$ and $E_v$ with position on the energy band diagram, exemplified by Fig. 3.10(a), is popularly referred to as "band bending."

Seeking to establish the precise relationship between the electric field within a semiconductor and the induced band bending, let us carefully re-examine the energy band diagram. The diagram itself, as emphasized in Fig. 3.10(a), is a plot of the allowed electron energies within the semiconductor as a function of position, where $E$ increasing upward is understood to be the *total* energy of the electrons. Furthermore, we know from previous discussions that if an energy of precisely $E_G$ is added to break an atom-atom bond, the created electron and hole energies would be $E_c$ and $E_v$, respectively, and the created carriers would be effectively motionless. Absorbing an energy in excess of $E_G$, on the other hand, would in all probability give rise to an electron energy greater than $E_c$ and a hole energy less than $E_v$, with both carriers moving around rapidly within the lattice. We are led, therefore, to interpret $E - E_c$ to be the kinetic energy (K.E.) of the electrons and $E_v - E$ to be the kinetic energy of the holes [see Fig. 3.10(b)]. Moreover, since the total energy equals the sum of the kinetic energy and the potential energy (P. E.), $E_c$ minus the energy reference level ($E_{ref}$) must equal the electron potential energy, as illustrated in Fig. 3.10(c). (Potential energy, it should be noted, is arbitrary to within a constant, and the position-independent reference energy, $E_{ref}$, may be chosen to be any convenient value.)

The potential energy is the key in relating the electric field within a semiconductor to positional variations in the energy bands. Specifically, assuming normal operational conditions, where magnetic field, temperature gradient, and stress-induced effects are negligible, only the force associated with an existing electric field can give rise to changes in the

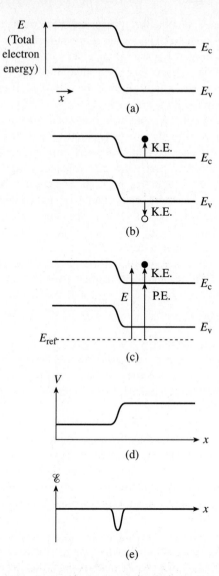

**Figure 3.10**  Relationship between band bending and the electrostatic variables inside a semiconductor: (a) sample energy band diagram exhibiting band bending; (b) identification of the carrier kinetic energies; (c) specification of the electron potential energy; (d) electrostatic potential and (e) electric field versus position dependence deduced from and associated with the part (a) energy band diagram.

potential energy of the carriers. Elementary physics, in fact, tells us that the potential energy of a $-q$ charged particle under such conditions is simply related to the electrostatic potential $V$ at a given point by

$$\text{P.E.} = -qV \tag{3.10}$$

Having previously concluded that

$$\text{P.E.} = E_c - E_{\text{ref}} \tag{3.11}$$

we can state

$$V = -\frac{1}{q}(E_c - E_{\text{ref}}) \tag{3.12}$$

By definition, moreover,

$$\mathscr{E} = -\nabla V \tag{3.13}$$

or, in one dimension,

$$\mathscr{E} = -\frac{dV}{dx} \tag{3.14}$$

Consequently,

$$\mathscr{E} = \frac{1}{q}\frac{dE_c}{dx} = \frac{1}{q}\frac{dE_v}{dx} = \frac{1}{q}\frac{dE_i}{dx} \tag{3.15}$$

The latter forms of Eq. (3.15) follow from the fact that $E_c$, $E_v$, and $E_i$ differ by only an additive constant.

The preceding formulation provides a means of readily deducing the general form of the electrostatic variables associated with the "band bending" in Fig. 3.10(a) and other energy band diagrams. Making use of Eq. (3.12), or simply inverting $E_c(x)$ in Fig. 3.10(a), one obtains the $V$ versus $x$ dependence presented in Fig. 3.10(d). [Please note that the electrostatic potential, like the potential energy, is arbitrary to within a constant—the Fig. 3.10(d) plot can be translated upward or downward along the voltage axis without modifying the physical situation inside the semiconductor.] Finally, recording the slope of $E_c$ versus position, as dictated by Eq. (3.15), produces the $\mathscr{E}$ versus $x$ plot shown in Fig. 3.10(e).

In summary, the reader should be aware of the fact that the energy band diagram contains information relating to the electrostatic potential and electric field within the semiconductor. Moreover, the general form of the $V$ and $\mathscr{E}$ dependencies within the semiconductor can be obtained almost by inspection. To deduce the form of the $V$ versus $x$

relationship, merely sketch the "upside down" of $E_c$ (or $E_v$ or $E_i$) versus $x$; to determine the general $\mathcal{E}$ versus $x$ dependence, simply note the slope of $E_c$ (or $E_v$ or $E_i$) as a function of position.

---

### Exercise 3.2

**P:** Consider the following energy band diagram. Take the semiconductor represented to be Si maintained at 300 K with $E_i - E_F = E_G/4$ at $x = \pm L$ and $E_F - E_i = E_G/4$ at $x = 0$. Note the choice of $E_F$ as the energy reference level and the identification of carriers at various points on the diagram.

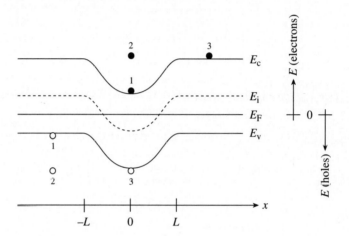

(a) Sketch the electrostatic potential $(V)$ inside the semiconductor as a function of $x$.

(b) Sketch the electric field $(\mathcal{E})$ inside the semiconductor as a function of $x$.

(c) Ascertain the K.E. and P.E. of the electrons and holes pictured on the diagram.

(d) Determine the resistivity of the $x > L$ portion of the semiconductor.

**S:** (a) The $V$ versus $x$ relationship must have the same functional form as the "upside down" of $E_c$ (or $E_i$ or $E_v$). If the arbitrary voltage reference point is taken to be $V = 0$ at $x = L$, then one concludes the $V$ versus $x$ dependence is as sketched here:

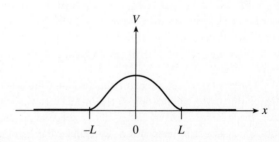

(b) The electric field, $\mathcal{E}$, is proportional to the slope of the bands:

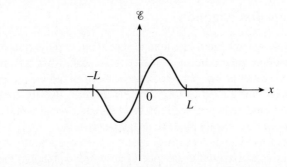

(c) For the electrons, K.E. $= E - E_c$ and P.E. $= E_c - E_{ref} = E_c - E_F$. Since the total energy of the holes increases *downward* on the diagram, K.E. $= E_v - E$ and P.E. $= E_{ref} - E_v = E_F - E_v$ for the holes. The energy differences are of course evaluated at the same $x$. The deduced K.E. and P.E. values are summarized in the following table:

| Carrier | K.E. (eV) | P.E. (eV) |
|---------|-----------|-----------|
| Electron 1 | 0 | 0.28 |
| Electron 2 | 0.56 | 0.28 |
| Electron 3 | 0 | 0.84 |
| Hole 1 | 0 | 0.28 |
| Hole 2 | 0.56 | 0.28 |
| Hole 3 | 0 | 0.84 |

(d) In the $x > L$ region, $E_i - E_F = E_G/4 = 0.28$ eV and

$$N_A = p = n_i e^{(E_i - E_F)/kT} = 10^{10} e^{0.28/0.0259} = 4.96 \times 10^{14}/cm^3$$

From Fig. 3.5(a) one deduces $\mu_p = 459$ cm²/V-sec, and therefore,

$$\rho = \frac{1}{q\mu_p N_A} = \frac{1}{(1.6 \times 10^{-19})(459)(4.96 \times 10^{14})} = \boxed{27.5 \text{ ohm-cm}}$$

Alternatively making use of the Fig. 3.8(a) Si resistivity plot, one similarly concludes $\rho \cong 25$ ohm-cm.

## 3.2  DIFFUSION

### 3.2.1 Definition–Visualization

*Diffusion* is a process whereby particles tend to spread out or redistribute as a result of their random thermal motion, migrating on a macroscopic scale from regions of high particle concentration into regions of low particle concentration. If allowed to progress unabated, the diffusion process operates so as to produce a uniform distribution of particles. The diffusing entity, it should be noted, need not be charged; thermal motion, not interparticle repulsion, is the enabling action behind the diffusion process.

To cite an everyday example of diffusion, suppose an open bottle of perfume is placed in one corner of a room. Even in the absence of air currents, random thermal motion will spread the perfume molecules throughout the room in a relatively short period of time, with intermolecular collisions helping to uniformly redistribute the perfume to every nook and cranny within the room.

Seeking to obtain a more detailed understanding of the diffusion process, let us next conceptually "monitor" the process on a microscopic scale employing a simple hypothetical system. The system we propose to monitor is a one-dimensional box containing four compartments and 1024 mobile particles (see Fig. 3.11). The particles within the box obey certain stringent rules. Specifically, thermal motion causes all particles in a given compartment to "jump" into an adjacent compartment every $\tau$ seconds. In keeping with the random nature of the motion, each and every particle has an equal probability of jumping to the left and to the right. Hitting an "external wall" while attempting to jump to the left or right reflects the particle back to its pre-jump position. Finally, at time $t = 0$ it is assumed that all of the particles are confined in the left-most compartment.

Figure 3.11 records the evolution of our 1024 particle system as a function of time. At time $t = \tau$, 512 of the 1024 particles originally in compartment 1 jump to the right and come to rest in compartment 2. The remaining 512 particles jump to the left and are reflected back into the left-most compartment. The end result is 512 particles each in compartments 1 and 2 after $\tau$ seconds. At time $t = 2\tau$, 256 of the particles in compartment 2 jump into compartment 3 and the remainder jump back into compartment 1. In the meantime, 256 of the particles from compartment 1 jump into compartment 2 and 256 undergo a reflection at the left-hand wall. The net result after $2\tau$ seconds is 512 particles in compartment 1 and 256 particles each in compartments 2 and 3. The state of the system after $3\tau$ and $6\tau$ seconds, also shown in Fig. 3.11, can be deduced in a similar manner. By $t = 6\tau$, the particles, once confined to the left-most compartment, have become almost uniformly distributed throughout the box, and it is unnecessary to consider later states. Indeed, the fundamental nature of the diffusion process is clearly self-evident from an examination of the existing states.

In semiconductors the diffusion process on a microscopic scale is similar to that occurring in the hypothetical system except, of course, the random motion of the diffusing particles is three-dimensional and not "compartmentalized." On a *macroscopic* scale the net effect of diffusion is precisely the same within both the hypothetical system and semi-

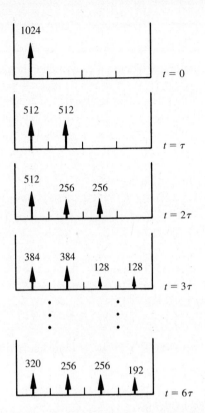

**Figure 3.11** Diffusion on a microscopic scale in a hypothetical one-dimensional system. The numbers over the arrows indicate the number of particles in a given compartment; observation times are listed to the extreme right.

conductors; there is an overall migration of particles from regions of high particle concentration to regions of low particle concentration. Within semiconductors the mobile particles—the electrons and holes—are charged, and diffusion-related carrier transport therefore gives rise to particle currents as pictured in Fig. 3.12.

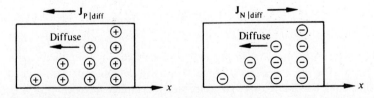

**Figure 3.12** Visualization of electron and hole diffusion on a macroscopic scale.

## (C) Exercise 3.3

The "DiffDemo" MATLAB file listed below was written to help the user visualize the diffusion process. The program provides a pseudo-animation of a one-dimensional particle system similar to that described in the text. The y= [ ] statement in the program controls the initial condition, N specifies the maximum number of monitored "jumps," and the number in the pause statements controls the time between jumps.

```
%Simulation of Diffusion (DiffDemo)
%One-dimensional system, right and left jumps equally probable

%Initialization
close
x=[0.5 1.5 2.5 3.5 4.5];
y=[1.0e6 0 0 0 0];    %NOTE: initial position can be changed
[xp,yp]=bar(x,y);
plot(xp,yp); text (0.5,1.1e6,'t = 0');
axis([0,5,0,1.2e6]);
pause (0.5)
N=15;                 %NOTE: increase N for extended run

%Computations and Plotting
for ii=1:N,
    %Diffusion step calculation
    bin(1)=round(y(1)/2 + y(2)/2);
    bin(2)=round(y(1)/2 + y(3)/2);
    bin(3)=round(y(2)/2 + y(4)/2);
    bin(4)=round(y(3)/2 + y(5)/2);
    bin(5)=round(y(4)/2 + y(5)/2);
    y=bin;
    %Plotting the result
    [xp,yp]=bar(x,y);
    axis(axis);
    plot(xp,yp); text(0.5,1.1e6,['t = ',num2str(ii)]);
    axis([0,5,0,1.2e6]);
    pause (0.5)
end
```

**P:** (a) Enter the program into your computer or locate the copy supplied on disk. Run the program. (The *command-period* on a Macintosh computer or *break* on an IBM-compatible computer can be used to prematurely terminate the program.)

(b) Rerun the program after changing the initial conditions so that all of the carriers are in the middle box at $t = 0$.

(c) Experiment with initially placing particles in more than one box and changing the number in the `pause` statements. You might also try increasing the number of particle boxes.

**S:** Sample output is reproduced in Fig. E3.3.

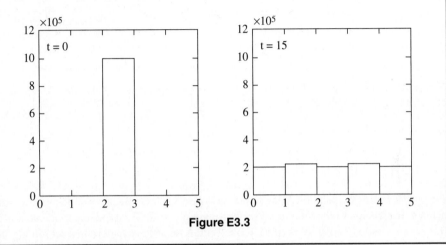

**Figure E3.3**

## 3.2.2 Hot-Point Probe Measurement

Before establishing the current associated with diffusion, let us digress somewhat and briefly consider the hot-point probe measurement. The *hot-point probe measurement* is a common technique for rapidly determining whether a semiconductor is *n*- or *p*-type. From a practical standpoint, knowledge of the semiconductor type is essential in device processing and must be known even before determining the doping concentration from resistivity measurements (refer back to Fig. 3.8). The hot-point probe "typing" experiment is considered here because it simultaneously provides an informative example of the diffusion process.

Examining Fig. 3.13(a), we find the only equipment required for performing the hot-point probe measurement is a hot-point probe, a cold-point probe, and a center-zero milliammeter. The hot-point probe is sometimes a refugee from a wood burning set or simply a soldering iron; the cold-point probe typically assumes the form of an electrical probe like that used with hand-held multimeters. No special requirements are imposed on the center-zero milliammeter connected between the probes. The measurement procedure itself is also extremely simple: After allowing the hot probe to heat up, one brings the two probes into contact with the semiconductor sample, and the ammeter deflects to the right or left, thereby indicating the semiconductor type. Observe that the spacing between the probes is set arbitrarily and can be reduced to enhance the ammeter deflection.

A simplified explanation of how the hot-point probe measurement works is presented in Fig. 3.13(b). In the vicinity of the probe contact the heated probe creates an increased

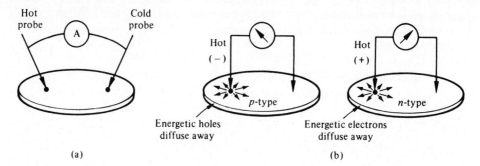

**Figure 3.13**  The hot-point probe measurement: (a) required equipment; (b) simplified explanation of how the measurement works.

number of higher-energy carriers. These energetic carriers will be predominantly holes in the case of a *p*-type material and electrons in an *n*-type material. With more energetic carriers near the heated probe than elsewhere, diffusion acts so as to spread the higher-energy carriers throughout the semiconductor wafer. The net effect is a deficit of holes or a net negative charge in the vicinity of the hot-point probe for a *p*-type material, and a positive charge buildup near the heated probe within an *n*-type material. Accordingly, the center-zero meter deflects in a different direction for *p*- and *n*-type materials.

### 3.2.3  Diffusion and Total Currents

#### Diffusion Currents

In defining the diffusion process and in citing diffusion examples, we have sought to emphasize the direct correlation between diffusion and a spatial variation in particle numbers. For diffusion to occur, more of the diffusing particles must exist at one point than at other points or, in mathematical terms, there must be a nonzero concentration gradient ($\nabla p \neq 0$ for holes, $\nabla n \neq 0$ for electrons). Logically, moreover, the greater the concentration gradient, the larger the expected flux of particles. Quantitative analysis of the diffusion process indeed confirms the foregoing and leads to what is known as Fick's law:

$$\mathscr{F} = -D\nabla\eta \tag{3.16}$$

where $\mathscr{F}$ is the flux or particles/cm²-sec crossing a plane perpendicular to the particle flow, $\eta$ is the particle concentration, and $D$ is the *diffusion coefficient*, a positive proportionality factor. The diffusion current density due to electrons and holes is obtained by simply multiplying the carrier flux by the carrier charge:

$$\mathbf{J}_{P|diff} = -qD_P\nabla p \tag{3.17a}$$

$$\mathbf{J}_{N|diff} = qD_N\nabla n \tag{3.17b}$$

The constants of proportionality, $D_P$ and $D_N$, have units of cm²/sec and are referred to as the hole and electron diffusion coefficients, respectively.

Upon examining Eqs. (3.17), note that the current directions deduced from the equations are consistent with the Fig. 3.12 visualization of the macroscopic diffusion currents. For the positive concentration gradient shown in Fig. 3.12 ($dp/dx > 0$ and $dn/dx > 0$ for the pictured one-dimensional situation), both holes and electrons will diffuse in the $-x$ direction. $J_{P|diff}$ will therefore be negative or directed in the $-x$ direction, while $J_{N|diff}$ will be oriented in the $+x$ direction.

## Total Currents

The total or net carrier currents in a semiconductor arise as the combined result of drift and diffusion. Summing the respective $n$ and $p$ segments of Eqs. (3.4) and Eqs. (3.17), one obtains

$$\mathbf{J}_P = \mathbf{J}_{P|drift} + \mathbf{J}_{P|diff} = q\mu_p p\boldsymbol{\mathscr{E}} - qD_P \nabla p \qquad (3.18a)$$
$$\updownarrow \text{drift} \qquad \updownarrow \text{diffusion}$$
$$\mathbf{J}_N = \mathbf{J}_{N|drift} + \mathbf{J}_{N|diff} = q\mu_n n\boldsymbol{\mathscr{E}} + qD_N \nabla n \qquad (3.18b)$$

The total particle current flowing in a semiconductor is in turn computed from

$$\mathbf{J} = \mathbf{J}_N + \mathbf{J}_P \qquad (3.19)$$

The double boxes emphasize the importance of the total-current relationships; they are used directly or indirectly in the vast majority of device analyses.

## 3.2.4 Relating Diffusion Coefficients/Mobilities

The diffusion coefficients, the constants of the motion associated with diffusion, are obviously central parameters in characterizing carrier transport due to diffusion. Given the importance of the diffusion coefficients, one might anticipate an extended examination of relevant properties paralleling the mobility presentation in Subsection 3.1.3. Fortunately, an extended examination is unnecessary because the $D$'s and the $\mu$'s are all interrelated. It is only necessary to establish the connecting formula known as the Einstein relationship.

In deriving the Einstein relationship, we consider a *nonuniformly* doped semiconductor maintained under equilibrium conditions. Special facts related to the nonuniform doping and the equilibrium state are invoked in the derivation. These facts, important in themselves, are reviewed prior to presenting the derivation proper.

### Constancy of the Fermi Level

Consider a nondegenerate, nonuniformly doped $n$-type semiconductor sample, a sample where the doping concentration varies as a function of position. A concrete example of what we have in mind is shown in Fig. 3.14(a). Assuming equilibrium conditions prevail, the energy band diagram characterizing the sample will have the form sketched in Fig. 3.14(b). In Chapter 2 the Fermi level in *uniformly* doped *n*-type semiconductors was found to move closer and closer to $E_c$ when the donor doping was systematically increased (see Fig. 2.21). Consistent with this fact, $E_c$ was drawn closer to $E_F$ in going from $x = 0$ to $x = L$ on the Fig. 3.14(b) diagram. Construction of the diagram, however, could not have been completed without making use of a critical new piece of information. Specifically,

*under equilibrium conditions, $dE_F/dx = dE_F/dy = dE_F/dz = 0$; i.e., the Fermi level inside a material or a group of materials in intimate contact is invariant as a function of position.*

The constancy of the Fermi level means that $E_F$ is to appear as a horizontal line on equilibrium energy band diagrams.

The position independence of the Fermi energy is established by examining the trans-

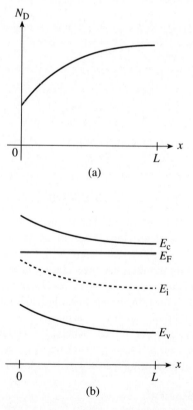

**Figure 3.14** Nonuniformly doped semiconductor: (a) assumed doping variation with position; (b) corresponding equilibrium energy band diagram.

fer of carriers between allowed states with the same energy but at adjacent positions in an energy band. It is concluded the probability of filling the states at a given energy, $f(E)$, must be the same everywhere in the sample under equilibrium conditions. If this were not the case, carriers would preferentially transfer between states and thereby give rise to a net current. The existence of a net current is inconsistent with the specified equilibrium conditions. Referring to the Eq. (2.7) expression for $f(E)$, we find that the constancy of the Fermi function in turn requires the Fermi level to be independent of position.

## Current Flow Under Equilibrium Conditions

Under equilibrium conditions the total current is of course identically zero. Because electron and hole activity is totally decoupled under equilibrium conditions, the electron and hole current densities, $\mathbf{J_N}$ and $\mathbf{J_P}$, must also independently vanish. Setting $\mathbf{J_N} = \mathbf{J_P} = 0$ in Eq. (3.18), however, leads to the conclusion that the drift and diffusion currents associated with a given carrier are merely required to be of equal magnitude and opposite polarity. In fact, the drift and diffusion components will vanish under equilibrium conditions only if $\mathscr{E} = 0$ and $\nabla n = \nabla p = 0$.

To provide a concrete illustration of non-vanishing equilibrium current components, let us return to an examination of the nonuniformly doped semiconductor sample characterized by Fig. 3.14. The dopant variation pictured in Fig. 3.14(a) clearly gives rise to a significant electron concentration gradient. An electron diffusion current flowing in the $+x$ direction must exist inside the sample. The band bending shown in the band diagram of Fig. 3.14(b), moreover, implies the existence of a "built-in" electric field oriented in the $-x$ direction. This gives rise to a drift current also in the $-x$ direction. Note that the drift and diffusion components are of opposite polarity. Since equilibrium conditions are assumed to prevail, the components must also be of equal magnitude. In point of fact, the electric field arises inside the semiconductor to counteract the diffusive tendencies of the carriers caused by the nonuniform doping. The major point here is that, even under equilibrium condition, nonuniform doping will give rise to carrier concentration gradients, a built-in electric field, and non-zero current components.

## Einstein Relationship

Having laid the proper foundation, we can now proceed to derive the connecting formula between the $D$'s and the $\mu$'s known as the Einstein relationship. To simplify the development, we work in only one dimension. The sample under analysis is taken to be a nondegenerate, nonuniformly doped semiconductor maintained under equilibrium conditions. Citing the fact that the net carrier currents must be identically zero under equilibrium conditions, and focusing on the electrons, we can state

$$J_{N|\text{drift}} + J_{N|\text{diff}} = q\mu_n n\mathscr{E} + qD_N\frac{dn}{dx} = 0 \tag{3.20}$$

However,

$$\mathscr{E} = \frac{1}{q}\frac{dE_i}{dx} \tag{3.21}$$

(same as 3.15)

and

$$n = n_i\,e^{(E_F - E_i)/kT} \tag{3.22}$$

Moreover, with $dE_F/dx = 0$ (due to the positional invariance of the Fermi level under equilibrium conditions),

$$\frac{dn}{dx} = -\frac{n_i}{kT}e^{(E_F - E_i)/kT}\frac{dE_i}{dx} = -\frac{q}{kT}n\mathscr{E} \tag{3.23}$$

Substituting $dn/dx$ from Eq. (3.23) into Eq. (3.20), and rearranging the result slightly, one obtains

$$(qn\mathscr{E})\mu_n \quad (qn\mathscr{E})\frac{q}{kT}D_N = 0 \tag{3.24}$$

Since $\mathscr{E} \neq 0$ (a consequence of the nonuniform doping), it follows from Eq. (3.24) that

$$\boxed{\frac{D_N}{\mu_n} = \frac{kT}{q}} \qquad \text{Einstein relationship for electrons} \tag{3.25a}$$

A similar argument for holes yields

$$\boxed{\frac{D_P}{\mu_p} = \frac{kT}{q}} \qquad \text{Einstein relationship for holes} \tag{3.25b}$$

    Although it was established while assuming equilibrium conditions, we can present more elaborate arguments that show the Einstein relationship to be valid even under non-equilibrium conditions. The nondegenerate restriction, however, still applies; slightly modified forms of Eqs. (3.25) result when the argument is extended to degenerate materials. We should emphasize again that the preliminary results, such as the positional invariance of the equilibrium Fermi level and the situation inside a nonuniformly doped semiconductor under equilibrium conditions, are important in themselves. Relative to numerical values, note that $kT/q$ is a voltage, and at room temperature (300 K) is equal to 0.0259V.[†] Hence, for an $N_D = 10^{14}$/cm³ Si sample maintained at room temperature, $D_N = (kT/q)\mu_n = (0.0259)(1358$ cm²/V-sec$) = 35.2$ cm²/sec. Finally, the Einstein relationship is one of the easiest equations to remember because it rhymes internally—$D$ over $\mu$ equals $kT$ over $q$. The rhyme holds even if the equation is inverted—$\mu$ over $D$ equals $q$ over $kT$.

---

### Exercise 3.4

**P:** In this exercise we continue the examination of the semiconductor sample represented by the energy band diagram presented in Exercise 3.2.

(a) The semiconductor is in equilibrium. How does one deduce this fact from the given energy band diagram?

(b) What is the electron current density ($J_N$) and hole current density ($J_P$) at $x = \pm L/2$?

(c) Roughly sketch $n$ and $p$ versus $x$ inside the sample.

(d) Is there an electron diffusion current at $x = \pm L/2$? If there is a diffusion current at a given point, indicate the direction of *current* flow.

(e) Noting the solution for the electric field presented in Exercise 3.2, is there an electron drift current at $x = \pm L/2$? If there is a drift current at a given point, indicate the direction of current flow.

(f) What is the hole diffusion coefficient ($D_P$) in the $x > L$ region of the semiconductor?

**S:** (a) The semiconductor is concluded to be in equilibrium because the Fermi level is invariant as a function of position.

(b) $\boxed{J_N = 0}$ and $\boxed{J_P = 0}$ at both $x = -L/2$ and $x = +L/2$. The net electron and hole currents are always identically zero everywhere under equilibrium conditions.

---

[†] At room temperature $kT = 0.0259$ eV $= (1.6 \times 10^{-19})(0.0259)$ joule. Thus $kT/q = (1.6 \times 10^{-19})(0.0259)/(1.6 \times 10^{-19})$ joule/coul $= 0.0259$ V.

(c) Since $n = n_i \exp[(E_F - E_i)/kT]$ and $p = n_i \exp[(E_i - E_F)/kT]$, we conclude

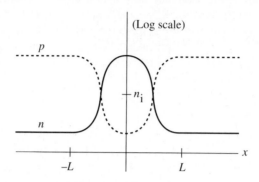

(d) There is an electron diffusion current at both $x = -L/2$ and $x = +L/2$. From the answer to the preceding question we note that $dn/dx \neq 0$ at the cited points. With $dn/dx > 0$, $J_{N|diff}$ at $x = -L/2$ flows in the $+x$ direction. Conversely, $dn/dx < 0$ at $x = L/2$ and $J_{N|diff}$ at this point flows in the $-x$ direction.

(e) There is an electron drift current at both $x = -L/2$ and $x = +L/2$. Since both $n$ and $\mathscr{E}$ are non-zero at $x = \pm L/2$, it follows that $J_{N|drift} = q\mu_n n\mathscr{E} \neq 0$. The drift component of the current always has the same direction as the electric field; $J_{N|drift}$ is in the $-x$ direction at $x = -L/2$ and in the $+x$ direction at $x = l\,L/2$. The drift component of the current must of course cancel the diffusion component of the current so that $J_N = J_{N|drift} + J_{N|diff} = 0$ under equilibrium conditions. The direction-related answers to parts (d) and (e), summarized next, are consistent with this requirement:

|            | $x = -L/2$ | $x = +L/2$ |
|------------|:----------:|:----------:|
| $J_{N|diff}$ | $\rightarrow$ | $\leftarrow$ |
| $J_{N|drift}$ | $\leftarrow$ | $\rightarrow$ |

(f) In working part (d) of Exercise 3.2, we concluded the $x > L$ region of the semiconductor (Si, 300 K) had a doping of $N_A \cong 5 \times 10^{14}/\text{cm}^3$ and a corresponding hole mobility of $\mu_p = 459$ cm²/V-sec. Thus, employing the Einstein relationship, $D_P = (kT/q)\mu_p = (0.0259)(459) = \boxed{11.9 \text{ cm}^2/\text{sec}}$.

# 3.3  RECOMBINATION–GENERATION

## 3.3.1 Definition–Visualization

When a semiconductor is perturbed from the equilibrium state, an excess or deficit in the carrier concentrations relative to their equilibrium values is invariably created inside the semiconductor. Recombination–generation (R–G) is nature's order-restoring mechanism, the means whereby the carrier excess or deficit inside the semiconductor is stabilized (if the perturbation is maintained) or eliminated (if the perturbation is removed). Since non-equilibrium conditions prevail during device operation, recombination–generation more often than not plays a major role in shaping the characteristics exhibited by a device. For-mally, recombination and generation can be defined as follows:

*Recombination*—a process whereby electrons and holes (carriers) are annihilated or destroyed.

*Generation*—a process whereby electrons and holes are created.

Unlike drift and diffusion, the terms *recombination* and *generation* do not refer to a single process. Rather, they are collective names for a group of similar processes; carriers can be created and destroyed within a semiconductor in a number of ways. The most common R–G processes are visualized in Fig. 3.15. The individual processes are described and discussed next.

### Band-to-Band Recombination

Band-to-band recombination is conceptually the simplest of all recombination processes. As pictured in Fig. 3.15(a), it merely involves the direct annihilation of a conduction band electron and a valence band hole. An electron and hole moving in the semiconductor lattice stray into the same spatial vicinity and zap!—the electron and hole annihilate each other. The excess energy released during the process typically goes into the production of a pho-ton (light).

### R–G Center Recombination

The process pictured in Fig. 3.15(b) involves a "third party," or intermediary, and takes place only at special locations within the semiconductor known as R–G centers. Physically, R–G centers are lattice defects or special impurity atoms such as gold in Si. Lattice defects and the special atoms in the form of unintentional impurities are present even in semicon-ductors of the highest available purity. The R–G center concentration, however, is normally very low compared to the acceptor and donor concentrations in device-quality materials. The most important property of the R–G centers is the introduction of allowed electronic levels near the center of the band gap. These levels are identified by the $E_T$ energy label in Fig. 3.15(b). The near midgap positioning of the levels is all-important because it distin-guishes R–G centers from donors and acceptors. The midgap levels introduced by select R–G center impurities in Si are summarized in Fig. 3.16.

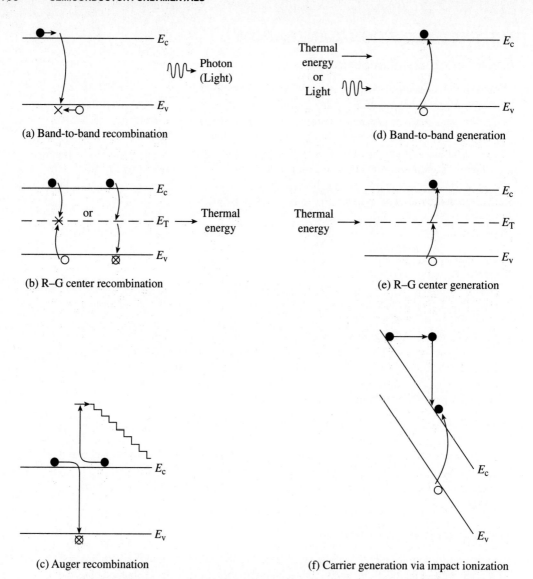

(a) Band-to-band recombination

(d) Band-to-band generation

(b) R–G center recombination

(e) R–G center generation

(c) Auger recombination

(f) Carrier generation via impact ionization

**Figure 3.15** Energy band visualization of recombination and generation processes.

As noted in Fig. 3.15(b), recombination at an R–G center is a two-step process. First, one type of carrier, say an electron, strays into the vicinity of an R–G center, is caught by the potential well associated with the center, loses energy, and is trapped. Subsequently, a hole comes along, is attracted to the trapped electron, loses energy, and annihilates the electron within the center. Alternatively, as also shown in Fig. 3.15(b), one can think of the electron as losing energy a second time and annihilating the hole in the valence band. R–G center recombination, also called *indirect recombination,* typically releases thermal energy (heat) during the process or, equivalently, produces lattice vibrations.

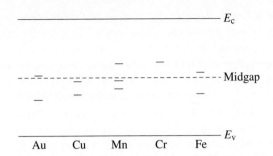

**Figure 3.16** Near-midgap energy levels introduced by some common impurities in Si. When an impurity introduces multiple levels, one of the levels tends to dominate in a given semiconductor sample.

## Auger Recombination

In the Auger (pronounced Oh-jay) process pictured in Fig. 3.15(c), band-to-band recombination occurs simultaneously with the collision between two like carriers. The energy released by the recombination is transferred during the collision to the surviving carrier. This highly energetic carrier subsequently "thermalizes"—loses energy in small steps through heat-producing collisions with the semiconductor lattice. The "staircase" in Fig. 3.15(c) represents the envisioned stepwise loss of energy.

## Generation Processes

Any of the foregoing recombination processes can be reversed to generate carriers. Band-to-band generation, where an electron is excited directly from the valence band into the conduction band, is pictured in Fig. 3.15(d). Note that either thermal energy or light can provide the energy required for the band-to-band transition. If thermal energy is absorbed, the process is alternatively referred to as *direct thermal generation;* if externally introduced light is absorbed, the process is called *photogeneration*. The thermally assisted generation of carriers with R–G centers acting as intermediaries is envisioned in Fig. 3.15(e). Finally, impact ionization, the inverse of Auger recombination, is visualized in Fig. 3.15(f). In this process an electron–hole pair is produced as a result of energy released when a highly energetic carrier collides with the crystal lattice. The generation of carriers through impact ionization routinely occurs in the high $\mathscr{E}$-field regions of devices. We will have more to say about this process when treating breakdown in *pn* junctions.

## 3.3.2 Momentum Considerations

All of the various recombination–generation processes occur at all times in all semiconductors—even under equilibrium conditions. The critical issue is not whether the processes are occurring, but rather, the rates at which the various processes are occurring. Typically, one need be concerned only with the dominant process, the process proceeding at the fastest rate. A number of the processes are important only under special conditions or are highly improbable, occurring at a much slower rate than competing processes. Auger recombination provides a case in point. Because the number of carrier-carrier collisions increases with increased carrier concentration, the frequency of Auger recombination likewise increases with carrier concentration. Auger recombination must be considered, for example, in treating highly doped regions of a device structure, but it is usually negligible otherwise. Similarly, significant impact ionization occurs only in the high $\mathscr{E}$-field regions of a device. In a somewhat different vein, photogeneration at room temperature becomes a significant process only when the semiconductor is exposed to external illumination. What then is the R–G process pair deserving of special attention, the process pair that typically dominates in the low $\mathscr{E}$-field regions of a nondegenerately doped semiconductor maintained at room temperature? From the information provided, the choice is obviously between thermal band-to-band R–G and recombination–generation via R–G centers.

From the energy band explanation alone one might speculate that band-to-band recombination–generation would dominate under the cited "standard" conditions. However, visualization of R–G processes using the energy band diagram can be misleading. The $E$–$x$ plot examines only changes in energy, whereas crystal momentum in addition to energy must be conserved in any R–G process. Momentum conservation requirements, as it turns out, play an important role in setting the process rate.

As first noted in Subsection 2.2.3, the momentum of an electron in an energy band, like the electron energy, can assume only certain quantized values. Momentum-related aspects of R–G processes are conveniently discussed with the aid of plots where the allowed electron energies in the conduction and valence bands are plotted versus the allowed momentum. These are referred to as $E$–$\mathbf{k}$ plots, where $\mathbf{k}$ is a parameter proportional to the electron crystal momentum. The $E$–$\mathbf{k}$ plots associated with semiconductors fall into two general categories. In the first, sketched in Fig. 3.17(a), both the minimum conduction band energy and maximum valence band energy occur at $\mathbf{k} = 0$. In the second, pictured in Fig. 3.17(b), the conduction band minimum is displaced to a $\mathbf{k} \neq 0$. Semiconductors exhibiting the former type of plot are referred to as *direct semiconductors,* and semiconductors exhibiting the latter type of plot, as *indirect semiconductors.* GaAs is a notable member of the direct semiconductor family; Ge and Si are both indirect semiconductors.

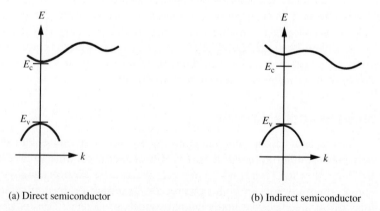

(a) Direct semiconductor                (b) Indirect semiconductor

**Figure 3.17**   General forms of $E$–$\mathbf{k}$ plots for direct and indirect semiconductors.

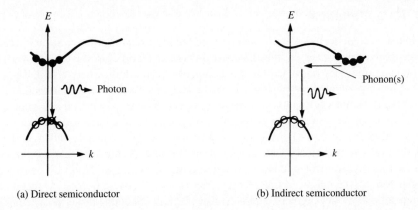

(a) Direct semiconductor                    (b) Indirect semiconductor

**Figure 3.18**    $E$–$\mathbf{k}$ plot visualizations of recombination in direct and indirect semiconductors.

To make use of the $E$–$\mathbf{k}$ plots in visualizing an R–G process, one needs to know the nature of transitions associated with the absorption or emission of photons (light) and lattice vibration quanta called *phonons*. Photons, being massless entities, carry very little momentum, and a photon-assisted transition is essentially vertical on the $E$–$\mathbf{k}$ plot. Conversely, the thermal energy associated with lattice vibrations (phonons) is very small (in the 10–50 meV range), whereas the phonon momentum is comparatively large. Thus on an $E$–$\mathbf{k}$ plot a phonon-assisted transition is essentially horizontal. It is also important to point out that, as pictured in Fig. 3.18, electrons and holes normally only occupy states very close to the $E_c$ minimum and $E_v$ maximum, respectively. This is of course consistent with our earlier discussion of the distribution of carriers in the bands.

Let us now specifically re-examine the band-to-band recombination process. In a direct semiconductor where the $\mathbf{k}$-values of electrons and holes are all bunched near $\mathbf{k} = 0$, little change in momentum is required for the recombination process to proceed. The conservation of both energy and momentum is readily met simply by the emission of a photon (see Fig. 3.18a). In an indirect semiconductor, on the other hand, there is a large change in momentum associated with the recombination process. The emission of a photon will conserve energy but cannot simultaneously conserve momentum. Thus for band-to-band recombination to proceed in an indirect semiconductor, the emission of a photon must be accompanied by the emission or absorption of a phonon (see Fig. 3.18b).

The rather involved nature of the band-to-band process in indirect semiconductors understandably leads to a vastly diminished recombination rate. Band-to-band recombination is in fact totally negligible compared to R–G center recombination in Si and other indirect semiconductors. Band-to-band recombination does proceed at a much faster rate in direct semiconductors and is the process producing the light observed from LEDs and junction lasers. Even in direct materials, however, the R–G center mechanism is often the dominant process. Given the universal importance of the R–G center process, and the fact that most devices are made from Si, which is an indirect material, the quantitative considerations in the next subsection are devoted primarily to the R–G center mechanism. Other processes are considered as the need arises.

### 3.3.3  R–G Statistics

R–G statistics is the technical name given to the mathematical characterization of recombination–generation processes. "Mathematical characterization" in the case of recombination–generation does not mean the development of a current-density relationship. An R–G event occurs at localized positions in the crystal, at a given $x$-value, and therefore the R–G action does not lead *per se* to charge transport. Rather, recombination–generation acts to change the carrier concentrations, thereby indirectly affecting the current flow. It is the time rate of change in the carrier concentrations ($\partial n/\partial t,\ \partial p/\partial t$) that must be specified to achieve the desired mathematical characterization. Because its characterization is relatively simple and its involvement in device operation quite common, we begin by considering photogeneration. The major portion of the subsection is of course devoted to recombination–generation via R–G centers, alternatively referred to as indirect thermal recombination–generation.

**Photogeneration**

Light striking the surface of a semiconductor as pictured in Fig. 3.19 will be partially reflected and partially transmitted into the material. Assume the light to be monochromatic with wavelength $\lambda$ and frequency $\nu$. If the photon energy ($h\nu$) is greater than the band gap energy, then the light will be absorbed and electron–hole pairs will be created as the light passes through the semiconductor. The decrease in intensity as monochromatic light passes through a material is given by

$$I = I_0 e^{-\alpha x} \tag{3.26}$$

where $I_0$ is the light intensity just inside the material at $x = 0^+$ and $\alpha$ is the *absorption coefficient*. Note from Fig. 3.20 that $\alpha$ is material dependent and a strong function of $\lambda$. Since there is a one-to-one correspondence between the absorption of photons and the creation of electron–hole pairs, the carrier creation rate should likewise exhibit an $\exp(-\alpha x)$ dependence. Moreover, referring to Fig. 3.15(d), the photogeneration process always acts so as to create an *equal* number of electrons and holes. We therefore conclude

$$\left.\frac{\partial n}{\partial t}\right|_{\text{light}} = \left.\frac{\partial p}{\partial t}\right|_{\text{light}} = G_{\text{L}}(x,\lambda) \tag{3.27}$$

where, for the situation analyzed,

$$G_{\text{L}}(x,\lambda) = G_{\text{L0}} e^{-\alpha x} \tag{3.28}$$

$G_{\text{L}}$ is the simplified symbol for the photogeneration rate (a number/cm$^3$-sec), while $G_{\text{L0}}$ is the photogeneration rate at $x = 0$. The $|_{\text{light}}$ designation means "due to light" and is necessary because, as we have seen, $n$ and $p$ can be affected by a number of processes.

In working problems involving the illumination of a sample or device, we will often invoke one of two simplifying assumptions: (i) The illumination is assumed to be uniform everywhere inside the sample and (ii) the light is assumed to be absorbed in an

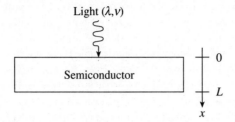

**Figure 3.19**  Semiconductor, light propagation, and coordinate orientation in the photogeneration analysis. The semiconductor is assumed to be sufficiently thick ($L \gg 1/\alpha$) so that bottom surface reflections may be neglected.

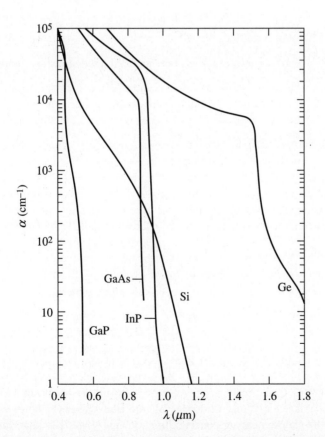

**Figure 3.20**  The absorption coefficient as a function of wavelength in Si and other select semiconductors. (From Schroder[6], © 1990 by John Wiley & Sons, Inc. Reprinted with permission.)

infinitesimally thin layer near the semiconductor surface. Seeking to identify when the assumed conditions are approached in practice, we note first that $1/\alpha$, which has the dimensions of a length, represents the average depth of penetration of the light into a material. At wavelengths where $h\nu \approx E_G$, the absorption coefficient is very small and $1/\alpha$ is very large, about 1 cm in Si. This makes $\exp(-\alpha x) \cong 1$ for all practical depths. To a very good approximation, $G_L \cong$ constant and the light is being uniformly absorbed. On the other hand, if $\lambda = 0.4$ $\mu$m, an $\alpha(Si) \cong 10^5$ cm$^{-1}$ is deduced from Fig. 3.20. At this wavelength the light penetration depth is only a few thousand angstroms and the assumption of surface absorption is closely approximated.

## Indirect Thermal Recombination–Generation

Although the general case development of R–G center statistics is beyond the scope of this text, it is possible to present arguments leading to the special-case expression employed in many practical problems. Let us begin by carefully defining the various carrier and center concentrations involved in the analysis.

$n_0, p_0$     . . .     carrier concentrations in the material under analysis when equilibrium conditions prevail.

$n, p$     . . .     carrier concentrations in the material under arbitrary conditions.

$\Delta n \equiv n - n_0$ . . .     deviations in the carrier concentrations from their equilibrium values.
$\Delta p \equiv p - p_0$     $\Delta n$ and $\Delta p$ can be both positive and negative, where a positive deviation corresponds to a carrier excess and a negative deviation corresponds to a carrier deficit.

$N_T$     . . .     number of R–G centers/cm$^3$.

In the special-case development, perturbation of the carrier concentrations from their equilibrium values must correspond to *low-level injection*. This requirement is just an alternative way of saying the perturbation must be relatively small. To be more precise,

$$
\begin{array}{ll}
\textit{low-level injection implies} & \\
\Delta p \ll n_0, \quad n \simeq n_0 & \text{in an } n\text{-type material} \\
\Delta n \ll p_0, \quad p \simeq p_0 & \text{in a } p\text{-type material}
\end{array}
$$

Consider a specific example of $N_D = 10^{14}$/cm$^3$ doped Si at room temperature subject to a perturbation where $\Delta p = \Delta n = 10^9$/cm$^3$. For the given material, $n_0 \simeq N_D = 10^{14}$/cm$^3$ and $p_0 \simeq n_i^2/N_D \simeq 10^6$/cm$^3$. Consequently, $n = n_0 + \Delta n \simeq n_0$ and $\Delta p = 10^9$/cm$^3 \ll n_0 = 10^{14}$/cm$^3$. The situation is clearly one of low-level injection. Observe, however, that $\Delta p \gg p_0$. Although the majority carrier concentration remains essentially unperturbed under low-level injection, the minority carrier concentration can, and routinely does, increase by many orders of magnitude.

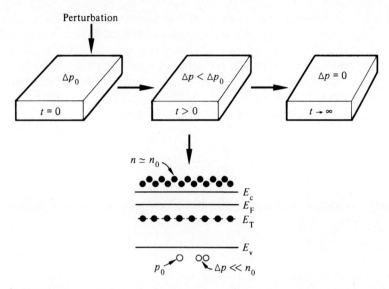

**Figure 3.21**   Situation inside an *n*-type semiconductor after a perturbation causing the low-level injection of excess holes.

We are now ready to analyze the specific situation envisioned in Fig. 3.21. A perturbation has caused a $\Delta p \ll n_0$ excess of holes in an *n*-type semiconductor sample. Assuming the system is in the midst of relaxing back to the equilibrium state from the pictured perturbed state via the R–G center interaction, what factors would you expect to have the greatest effect on $\partial p/\partial t|_R$, the rate of hole recombination? Well, to eliminate a hole, the hole must make the transition from the valence band to an electron-filled R–G center. Logically, the greater the number of filled R–G centers, the greater the probability of a hole annihilating transition and the faster the rate of recombination. Under equilibrium conditions, essentially all of the R–G centers are filled with electrons because the equilibrium Fermi level is comfortably above $E_T$. With $\Delta p \ll n_0$, electrons always vastly outnumber holes and rapidly fill R–G levels that become vacant. This keeps the number of filled R–G centers during the relaxation process $\approx N_T$. Thus, we expect $\partial p/\partial t|_R$ to be approximately proportional to $N_T$. Moreover, it also seems logical that the number of hole-annihilating transitions should increase almost linearly with the number of holes. All other factors being equal, the more holes available for annihilation, the greater the number of holes removed from the valence band per second. Consequently, we also expect $\partial p/\partial t|_R$ to be approximately proportional to $p$. Considering additional factors leads to no additional dependencies. Hence, introducing a positive proportionality constant, $c_p$, and realizing $\partial p/\partial t|_R$ is negative because $p$ is decreasing, we conclude

$$\left.\frac{\partial p}{\partial t}\right|_R = -c_p N_T p \qquad (3.29)$$

If the hole generation process is analyzed in a similar manner, $\partial p/\partial t|_G$ is deduced to depend only on the number of empty R–G centers. This number, which is small for the situation under consideration, is held approximately constant at its equilibrium value over the range of perturbations where hole generation is significant. This means that $\partial p/\partial t|_G$ can be replaced by $\partial p/\partial t|_{G\text{-equilibrium}}$. Moreover, the recombination and generation rates must precisely balance under equilibrium conditions,[†] or $\partial p/\partial t|_G = \partial p/\partial t|_{G\text{-equilibrium}} = -\partial p/\partial t|_{R\text{-equilibrium}}$. Thus, making use of Eq. (3.29), we obtain

$$\left.\frac{\partial p}{\partial t}\right|_G = c_p N_T p_0 \tag{3.30}$$

Finally, the net rate of change in the hole concentration due to the R–G center interaction is given by

$$\left.\frac{\partial p}{\partial t}\right|_{\substack{\text{i-thermal}\\ \text{R–G}}} = \left.\frac{\partial p}{\partial t}\right|_R + \left.\frac{\partial p}{\partial t}\right|_G = -c_p N_T(p - p_0) \tag{3.31}$$

or

$$\left.\frac{\partial p}{\partial t}\right|_{\substack{\text{i-thermal}\\ \text{R–G}}} = -c_p N_T \Delta p \qquad \text{for holes in an } n\text{-type material} \tag{3.32a}$$

An analogous set of arguments yields

$$\left.\frac{\partial n}{\partial t}\right|_{\substack{\text{i-thermal}\\ \text{R–G}}} = -c_n N_T \Delta n \qquad \text{for electrons in a } p\text{-type material} \tag{3.32b}$$

$c_n$ like $c_p$ is a positive proportionality constant. $c_n$ and $c_p$ are referred to as the electron and hole *capture coefficients,* respectively.

Although completely functional as is, it is desirable to manipulate Eqs. (3.32) into a slightly more compact and meaningful form. An examination of the left-hand sides of Eqs. (3.32) reveals the dimensional units are those of a concentration divided by time. Since $\Delta p$ and $\Delta n$ on the right-hand sides of the same equations are also concentrations, the constants $c_p N_T$ and $c_n N_T$ must have units of 1/time. It is therefore reasonable to introduce the time constants

---

[†] Under equilibrium conditions each fundamental process and its inverse must self-balance independent of any other process that may be occurring inside the material. This is known as the *Principle of Detailed Balance.*

$$\tau_p = \frac{1}{c_p N_T} \tag{3.33a}$$

$$\tau_n = \frac{1}{c_n N_T} \tag{3.33b}$$

which, when substituted into Eqs. (3.32), yield

$$\left.\frac{\partial p}{\partial t}\right|_{\substack{\text{i-thermal} \\ \text{R-G}}} = -\frac{\Delta p}{\tau_p} \qquad \text{for holes in an } n\text{-type material} \tag{3.34a}$$

$$\left.\frac{\partial n}{\partial t}\right|_{\substack{\text{i-thermal} \\ \text{R-G}}} = -\frac{\Delta n}{\tau_n} \qquad \text{for electrons in a } p\text{-type material} \tag{3.34b}$$

Equations (3.34) are the desired end result, the special-case characterization of R–G center (indirect thermal) recombination–generation. Although steady state or slowly varying conditions are implicitly assumed in the development of Eqs. (3.34), the relationships can be applied with little error to most transient problems of interest. It should be restated that a $\Delta p < 0$ is possible and will give rise to a $\partial p/\partial t|_{\text{i-thermal R-G}} > 0$. A positive $\partial p/\partial t|_{\text{i-thermal R-G}}$ simply indicates that a carrier deficit exists inside the semiconductor and generation is occurring at a more rapid rate than recombination. $\partial p/\partial t|_{\text{i-thermal R-G}}$ and $\partial n/\partial t|_{\text{i-thermal R-G}}$, it must be remembered, characterize the *net* effect of the thermal recombination and thermal generation processes.

As duly emphasized, the Eq. (3.34) relationships apply only to minority carriers and to situations meeting the low-level injection requirement. The more general steady state result[7], valid for arbitrary injection levels and both carrier types in a nondegenerate semiconductor, is noted for completeness and future reference to be

$$\left.\frac{\partial p}{\partial t}\right|_{\substack{\text{i-thermal} \\ \text{R-G}}} = \left.\frac{\partial n}{\partial t}\right|_{\substack{\text{i-thermal} \\ \text{R-G}}} = \frac{n_i^2 - np}{\tau_p(n + n_1) + \tau_n(p + p_1)} \tag{3.35}$$

where

$$n_1 \equiv n_i e^{(E_T - E_i)/kT} \tag{3.36a}$$

$$p_1 \equiv n_i e^{(E_i - E_T)/kT} \tag{3.36b}$$

It is left as an exercise to show that the general-case relationship reduces to Eqs. (3.34) under the assumed special-case conditions.

### 3.3.4  Minority Carrier Lifetimes

#### General Information

The time constants $\tau_n$ and $\tau_p$, which were introduced without comment in writing down Eqs. (3.34), are obviously the "constants of the action" associated with recombination–generation. Seeking to provide insight relative to the standard interpretation and naming of the $\tau$'s, let us consider once again the situation pictured in Fig. 3.21. Examining the change in the hole concentration with time, it goes almost without saying that the excess holes do not all disappear at the same time. Rather, the hole excess present at $t = 0$ is systematically eliminated, with some of the holes existing for only a short period and others "living" for comparatively long periods of time. If thermal recombination–generation is the sole process acting to relax the semiconductor, the average excess hole lifetime, $\langle t \rangle$, can be computed in a relatively straightforward manner. Without going into details, the computation yields $\langle t \rangle = \tau_n$ (or $\tau_p$). Physically, therefore, $\tau_n$ and $\tau_p$ have come to be interpreted as *the average time an excess minority carrier will live in a sea of majority carriers.* For identification purposes, $\tau_n$ and $\tau_p$ are simply referred to as the *minority carrier lifetimes.*

Like the $\mu$'s and the $D$'s, the $\tau$'s are important material parameters invariably required in the modeling of devices. Unlike the $\mu$'s and the $D$'s, there are few plots indicating the $\tau$-values to be employed in a given device structure. In fact, subsidiary experimental measurements (see the following subsection) must be performed to determine the minority carrier lifetime in a given semiconductor sample. The reason for the lack of catalogued information can be traced to the extreme variability of the $\tau_n$ and $\tau_p$ parameters. Referring to Eqs. (3.33), we see that the carrier lifetimes depend on the often poorly controlled R–G center concentration ($N_T$), *not* on the carefully controlled doping parameters ($N_A$ and $N_D$). Moreover, the physical nature of the dominant R–G center concentration changes even within a given sample during device fabrication. A fabrication procedure called *gettering* can reduce the R–G center concentration to a very low level and give rise to a $\tau_n(\tau_p) \sim$ 1 msec in Si. The intentional introduction of gold into Si, on the other hand, can controllably increase the R–G center concentration and give rise to a $\tau_n(\tau_p) \sim$ 1 nsec. Minority carrier lifetimes in completed Si devices tend to lie about midway between the cited extremes.

#### A Lifetime Measurement

The carrier lifetimes are measured using a variety of different methods. The majority of measurements employ a device structure of some sort and deduce $\tau_n$ and/or $\tau_p$ by matching the theoretical and experimentally observed device characteristics. The measurement to be described employs a bar-like piece of semiconductor with ohmic contacts at its two ends. In a commercial implementation this structure is called a *photoconductor,* a type of photodetector. A situation very similar to that envisioned in Fig. 3.21 is created inside the semi-

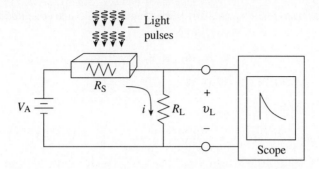

**Figure 3.22**   Schematic illustration of the photoconductivity decay measurement.

conductor during the measurement. The perturbation is in the form of a light pulse that gives rise to a carrier excess inside the semiconductor and a detectable change in conductivity. Once the light is removed the excess carriers are eliminated via recombination and the conductivity decays back to its value in the dark. The carrier lifetime is determined by monitoring the conductivity decay.

The experiment is schematically illustrated in Fig. 3.22. $R_S$ is the sample resistance, $R_L$ a load resistor, $V_A$ the applied d.c. voltage, and $v_L$ the measured load or output voltage. Obviously, $R_S$ and therefore $v_L$ are functions of time, with $v_L$ reflecting the changes in conductivity inside the semiconductor sample.

We are specifically interested in how $v_L$ changes with time after the light is turned off. $v_L$ is related to the conductivity, which in turn is related to the excess carrier concentrations. Given an $n$-type sample and assuming uniform photogeneration throughout the semiconductor, we will verify later in the chapter (Section 3.5) that the decay of the excess minority carrier hole concentration ($\Delta p$) is described by

$$\Delta p(t) = \Delta p_0 e^{-t/\tau_p} \tag{3.37}$$

$\Delta p_0$ is the excess hole concentration at $t = 0$ when the light is turned off. Next considering the conductivity, we note

$$\sigma = 1/\rho = q(\mu_n n + \mu_p p) \tag{3.38a}$$

$$= q[\mu_n(n_0 + \Delta n) + \mu_p(p_0 + \Delta p)] \tag{3.38b}$$

$$= \underbrace{q\mu_n N_D}_{\sigma_{dark}} + \underbrace{q(\mu_n + \mu_p)\Delta p}_{\Delta\sigma(t)} \tag{3.39}$$

In establishing Eq. (3.39), we made use of the fact that $n_0 \cong N_D \gg p_0$ in an $n$-type semiconductor. Also $\Delta n$ was assumed equal to $\Delta p$. $\sigma_{dark}$ is the unperturbed sample conductivity in the dark and $\Delta\sigma(t) \propto \Delta p$ represents the time decaying component of the conductivity.

The last step is to relate $v_L$ and $\Delta\sigma$. Inspecting Fig. 3.22, one concludes

$$v_L = iR_L = V_A \frac{R_L}{R_L + R_S} \tag{3.40}$$

but

$$R_S = \frac{\rho l}{A} = \frac{l}{\sigma A} = \frac{l}{(\sigma_{dark} + \Delta\sigma)A} = R_{Sd}\left(\frac{\sigma_{dark}}{\sigma_{dark} + \Delta\sigma}\right) \tag{3.41}$$

where $l$ is the length, $A$ the cross-sectional area, and $R_{Sd} = l/\sigma_{dark}A$ the dark resistance of the sample, respectively. Substituting Eq. (3.41) into Eq. (3.40) and rearranging yields

$$v_L = \frac{V_A}{1 + \dfrac{R_{Sd}/R_L}{1 + \Delta\sigma/\sigma_{dark}}} \tag{3.42}$$

or, if $R_L$ is matched to $R_{Sd}$ and the light intensity restricted so $\Delta\sigma/\sigma_{dark} \ll 1$ (equivalent to specifying low-level injection conditions), we obtain

$$v_L \cong \frac{V_A}{2}\left(1 + \frac{\Delta\sigma}{\sigma_{dark}}\right) = \underbrace{V_L}_{\text{d.c.}} + \underbrace{v_0 e^{-t/\tau_p}}_{\text{decay transient}} \tag{3.43}$$

An exponential decay is predicted with $\tau_p$ being the decay constant.

An actual implementation of the measurement is shown in Fig. 3.23.[†] The required $V_A$ is supplied by the Tektronix PS5004 power supply (just about any power supply would do), while the Tektronix 11401 Digitizing Oscilloscope is used to capture the photoconductive decay. The bar-shaped Si test structures are housed in a special measurements box along with the $R_L \cong R_{Sd}$ load resistor. The box permits convenient handling of the samples and facilitates switching between samples. The light pulses required in the measurement are derived from the General Radio 1531-AB Stroboscope. The GR Stroboscope puts out high-intensity light flashes of $\sim 1$ $\mu$sec duration spaced at controllable time intervals ranging from 2.5 msec to 0.5 sec. Fairly uniform photogeneration can be achieved in a sample

---

[†] This being the first text reference, it is perhaps worthwhile reiterating the General Introduction statement that measurement particulars, included at various points in the text, are derived from experiments performed in an undergraduate EE laboratory administered by the author. A somewhat expanded discussion of background theory, the measurement system, and measurement procedures can be found in R. F. Pierret, *Semiconductor Measurements Laboratory Operations Manual*, Supplement A, in the Modular Series on Solid State Devices, Addison-Wesley Publishing Co., Reading MA, © 1991.

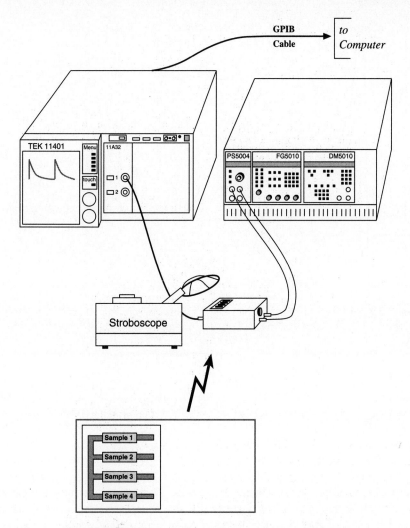

**Figure 3.23**   Photoconductive decay measurement system.

by first passing the stroboscope light through a small-diameter Si wafer. The Si wafer acts as a filter allowing only near-bandgap radiation to reach the sample under test. Although the 11401 Oscilloscope is capable of in situ data analysis, a computer is connected to the oscilloscope to add data manipulation and display flexibility.

An example of the photoconductive decay as displayed on the computer screen is reproduced in Fig. 3.24. The transient component of $v_L$ in millivolts is plotted on the $y$-axis versus time in milliseconds on the $x$-axis. The part (a) linear-scale plot nicely exhibits the generally expected dependence. The part (b) semilog plot verifies the simple exponential nature of the decay (at least for times $0.1$ msec $\leq t \leq 0.6$ msec) and is more useful for

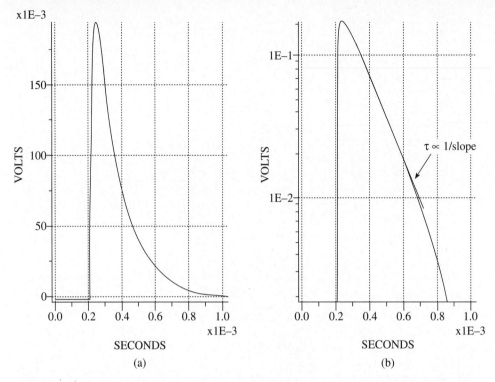

**Figure 3.24** Photoconductivity transient response. The time-varying component of $v_L$ in millivolts is plotted on the $y$-axis versus time in milliseconds on the $x$-axis. (a) is a linear and (b) is a semilog plot.

extracting the minority carrier lifetime. Specifically, from the slope of the straight-line region in the latter plot, the minority carrier lifetime in the probed sample is readily deduced to be $\tau_p \cong 150\ \mu sec$.

## 3.4 EQUATIONS OF STATE

In the first three sections of this chapter we examined and separately modeled the primary types of carrier action taking place inside semiconductors. Within actual semiconductors all the various types of carrier action occur at the same time, and the state of any semiconductor can be determined only by taking into account the combined effect of the individual types of carrier action. "Putting it all together," so to speak, leads to the basic set of starting equations employed in solving device problems, herein referred to as the *equations of state*. The first portion of this section is devoted to developing the equations of state; the latter portion includes a summary of common simplifications, a listing of special-case solutions, and problem-solving examples.

## 3.4.1 Continuity Equations

Each and every type of carrier action—whether it be drift, diffusion, indirect or direct thermal recombination, indirect or direct generation, or some other type of carrier action— gives rise to a change in the carrier concentrations with time. The combined effect of all types of carrier action can therefore be taken into account by equating the overall change in the carrier concentrations per unit time ($\partial n/\partial t$ or $\partial p/\partial t$) to the sum of the $\partial n/\partial t$'s or $\partial p/\partial t$'s due to the individual processes; that is,

$$\frac{\partial n}{\partial t} = \frac{\partial n}{\partial t}\bigg|_{\text{drift}} + \frac{\partial n}{\partial t}\bigg|_{\text{diff}} + \frac{\partial n}{\partial t}\bigg|_{\substack{\text{thermal}\\ \text{R–G}}} + \frac{\partial n}{\partial t}\bigg|_{\substack{\text{other processes}\\ \text{(light, etc.)}}} \tag{3.44a}$$

$$\frac{\partial p}{\partial t} = \frac{\partial p}{\partial t}\bigg|_{\text{drift}} + \frac{\partial p}{\partial t}\bigg|_{\text{diff}} + \frac{\partial p}{\partial t}\bigg|_{\substack{\text{thermal}\\ \text{R–G}}} + \frac{\partial p}{\partial t}\bigg|_{\substack{\text{other processes}\\ \text{(light, etc.)}}} \tag{3.44b}$$

The overall effect of the individual processes is established, in essence, by invoking the requirement of conservation of carriers. Electrons and holes cannot mysteriously appear and disappear at a given point, but must be transported to or created at the given point via some type of ongoing carrier action. There must be a spatial and time continuity in the carrier concentrations. For this reason, Eqs. (3.44) are known as the *continuity equations*.

The continuity equations can be written in a somewhat more compact form by noting

$$\frac{\partial n}{\partial t}\bigg|_{\text{drift}} + \frac{\partial n}{\partial t}\bigg|_{\text{diff}} = \frac{1}{q}\left(\frac{\partial J_{\text{N}x}}{\partial x} + \frac{\partial J_{\text{N}y}}{\partial y} + \frac{\partial J_{\text{N}z}}{\partial z}\right) = \frac{1}{q}\nabla \cdot \mathbf{J}_{\text{N}} \tag{3.45a}$$

$$\frac{\partial p}{\partial t}\bigg|_{\text{drift}} + \frac{\partial p}{\partial t}\bigg|_{\text{diff}} = -\frac{1}{q}\left(\frac{\partial J_{\text{P}x}}{\partial x} + \frac{\partial J_{\text{P}y}}{\partial y} + \frac{\partial J_{\text{P}z}}{\partial z}\right) = -\frac{1}{q}\nabla \cdot \mathbf{J}_{\text{P}} \tag{3.45b}$$

Equations (3.45), which can be established by a straightforward mathematical manipulation, merely state that there will be a change in the carrier concentrations within a given small region of the semiconductor if an imbalance exists between the total carrier currents into and out of the region. Utilizing Eqs. (3.45), we obtain

$$\frac{\partial n}{\partial t} = \frac{1}{q}\nabla \cdot \mathbf{J}_{\text{N}} + \frac{\partial n}{\partial t}\bigg|_{\substack{\text{thermal}\\ \text{R–G}}} + \frac{\partial n}{\partial t}\bigg|_{\substack{\text{other}\\ \text{processes}}} \tag{3.46a}$$

$$\frac{\partial p}{\partial t} = -\frac{1}{q}\nabla \cdot \mathbf{J}_{\text{P}} + \frac{\partial p}{\partial t}\bigg|_{\substack{\text{thermal}\\ \text{R–G}}} + \frac{\partial p}{\partial t}\bigg|_{\substack{\text{other}\\ \text{processes}}} \tag{3.46b}$$

The (3.46) continuity equations are completely general and directly or indirectly constitute the starting point in most device analyses. In computer simulations the continuity equations are often employed directly. The appropriate relationships for $\partial n/\partial t|_{\text{thermal R-G}}$, $\partial p/\partial t|_{\text{thermal R-G}}$ [which may or may not be the special-case relationships given by Eqs. (3.34)], along with the concentration changes due to "other processes," are substituted into Eqs. (3.46), and numerical solutions are sought for $n(x, y, z, t)$ and $p(x, y, z, t)$. In problems where a closed-form type of solution is desired, the continuity equations are typically utilized only in an indirect fashion. The actual starting point in such analyses is a simplified version of the continuity equations to be established in the next subsection.

### 3.4.2 Minority Carrier Diffusion Equations

The "workhorse" minority carrier diffusion equations are derived from the continuity equations by invoking the following set of simplifying assumptions:

(1) The particular system under analysis is *one-dimensional;* i.e., all variables are at most a function of just one coordinate (say the $x$-coordinate).

(2) The analysis is limited or restricted to *minority carriers.*

(3) $\mathcal{E} \simeq 0$ in the semiconductor or regions of the semiconductor subject to analysis.

(4) The equilibrium minority carrier concentrations are not a function of position. In other words, $n_0 \neq n_0(x)$, $p_0 \neq p_0(x)$.

(5) *Low-level injection* conditions prevail.

(6) *Indirect* thermal recombination–generation is the dominant thermal R–G mechanism.

(7) There are *no "other processes,"* except possibly photogeneration, taking place within the system.

Working on the continuity equation for electrons, we note that

$$\frac{1}{q}\, \nabla \cdot \mathbf{J}_{\text{N}} \to \frac{1}{q}\frac{\partial J_{\text{N}}}{\partial x} \tag{3.47}$$

if the system is one-dimensional. Moreover,

$$J_{\text{N}} = q\mu_{\text{n}} n \mathcal{E} + qD_{\text{N}}\frac{\partial n}{\partial x} \simeq qD_{\text{N}}\frac{\partial n}{\partial x} \tag{3.48}$$

when $\mathcal{E} \simeq 0$ and one is concerned only with minority carriers [simplifications (2) and (3)]. By way of explanation, the drift component can be neglected in the current density expression because $\mathcal{E}$ is small by assumption and minority carrier concentrations are also small, making the $n\mathcal{E}$ product extremely small. (Note that the same argument cannot be applied to majority carriers.) Since by assumption $n_0 \neq n_0(x)$, and by definition $n = n_0 + \Delta n$, we can also write

$$\frac{\partial n}{\partial x} = \frac{\partial n_0}{\partial x} + \frac{\partial \Delta n}{\partial x} = \frac{\partial \Delta n}{\partial x} \tag{3.49}$$

Combining Eqs. (3.47) through (3.49) yields

$$\frac{1}{q} \nabla \cdot \mathbf{J}_{\mathrm{N}} \rightarrow D_{\mathrm{N}} \frac{\partial^2 \Delta n}{\partial x^2} \tag{3.50}$$

Turning to the remaining terms in the continuity equation for electrons, the assumed dominance of recombination–generation via R–G centers, combined with the low-level injection and minority carrier restrictions, allows us to replace the thermal R–G term with the Eq. (3.34) special-case expression.

$$\left.\frac{\partial n}{\partial t}\right|_{\substack{\text{thermal} \\ \text{R–G}}} = -\frac{\Delta n}{\tau_{\mathrm{n}}} \tag{3.51}$$

In addition, applying simplification (7) yields

$$\left.\frac{\partial n}{\partial t}\right|_{\substack{\text{other} \\ \text{processes}}} = G_{\mathrm{L}} \tag{3.52}$$

where it is understood that $G_{\mathrm{L}} = 0$ if the semiconductor is not subject to illumination. Finally, the equilibrium electron concentration is never a function of time, $n_0 \neq n_0(t)$, and we can therefore write

$$\frac{\partial n}{\partial t} = \frac{\partial n_0}{\partial t} + \frac{\partial \Delta n}{\partial t} = \frac{\partial \Delta n}{\partial t} \tag{3.53}$$

Substituting Eqs. (3.50) through (3.53) into the (3.46a) continuity equation, and simultaneously recording the analogous result for holes, one obtains

$$\frac{\partial \Delta n_{\mathrm{p}}}{\partial t} = D_{\mathrm{N}} \frac{\partial^2 \Delta n_{\mathrm{p}}}{\partial x^2} - \frac{\Delta n_{\mathrm{p}}}{\tau_{\mathrm{n}}} + G_{\mathrm{L}} \tag{3.54a}$$

$$\frac{\partial \Delta p_{\mathrm{n}}}{\partial t} = D_{\mathrm{P}} \frac{\partial^2 \Delta p_{\mathrm{n}}}{\partial x^2} - \frac{\Delta p_{\mathrm{n}}}{\tau_{\mathrm{p}}} + G_{\mathrm{L}} \tag{3.54b}$$

Minority carrier diffusion equations

Subscripts are added to the carrier concentrations in Eqs. (3.54) to remind the user that the equations are valid only for minority carriers, applying to electrons in *p*-type materials and to holes in *n*-type materials.

**Table 3.1**   Common Diffusion Equation Simplifications.

| *Simplification* | *Effect* |
| --- | --- |
| Steady state | $\dfrac{\partial \Delta n_p}{\partial t} \to 0 \qquad \left( \dfrac{\partial \Delta p_n}{\partial t} \to 0 \right)$ |
| No concentration gradient or no diffusion current | $D_N \dfrac{\partial^2 \Delta n_p}{\partial x^2} \to 0 \qquad \left( D_P \dfrac{\partial^2 \Delta p_n}{\partial x^2} \to 0 \right)$ |
| No drift current or $\mathcal{E} = 0$ | No further simplification. ($\mathcal{E} \simeq 0$ is assumed in the derivation.) |
| No thermal R–G | $\dfrac{\Delta n_p}{\tau_n} \to 0 \qquad \left( \dfrac{\Delta p_n}{\tau_p} \to 0 \right)$ |
| No light | $G_L \to 0$ |

### 3.4.3  Simplifications and Solutions

In the course of performing device analyses, the conditions of a problem often permit additional simplifications that drastically reduce the complexity of the minority carrier diffusion equations. Common simplifications and their effect on the minority carrier diffusion equations are summarized in Table 3.1. Extensively utilized solutions to simplified forms of the minority carrier diffusion equations are collected for future reference in Table 3.2.

### 3.4.4  Problem Solving

Although we have presented all necessary information, it may not be completely obvious how one works with the diffusion equations to obtain the carrier concentration solutions. Device analyses of course provide examples of "problem" solutions, and a number of device analyses are presented later in the text. Nevertheless, it is worthwhile to examine a few well-defined simple problems to illustrate problem-solving procedures. The chosen sample problems also provide a basis for the two supplemental concepts introduced in the next section.

#### Sample Problem No. 1

**P:** A uniformly donor-doped silicon wafer maintained at room temperature is suddenly illuminated with light at time $t = 0$. Assuming $N_D = 10^{15}/\text{cm}^3$, $\tau_p = 10^{-6}$ sec, and a light-induced creation of $10^{17}$ electrons and holes per $\text{cm}^3$-sec throughout the semiconductor, determine $\Delta p_n(t)$ for $t > 0$.

**S:** *Step 1*—Review precisely what information is given or implied in the statement of the problem.

**Table 3.2**    Common Special-Case Diffusion Equation Solutions.

*Solution no. 1*

*GIVEN:*    Steady state, no light.

*SIMPLIFIED*
*DIFF. EQN:*

$$0 = D_N \frac{d^2 \Delta n_p}{dx^2} - \frac{\Delta n_p}{\tau_n}$$

*SOLUTION:*    $\Delta n_p(x) = A e^{-x/L_N} + B e^{x/L_N}$

where    $L_N \equiv \sqrt{D_n \tau_n}$

and *A, B* are solution constants.

*Solution no. 2*

*GIVEN:*    No concentration gradient, no light.

*SIMPLIFIED*
*DIFF. EQN:*

$$\frac{d \Delta n_p}{dt} = - \frac{\Delta n_p}{\tau_n}$$

*SOLUTION:*    $\Delta n_p(t) = \Delta n_p(0) e^{-t/\tau_n}$

*Solution no. 3*

*GIVEN:*    Steady state, no concentration gradient.

*SIMPLIFIED*
*DIFF. EQN:*

$$0 = - \frac{\Delta n_p}{\tau_n} + G_L$$

*SOLUTION:*    $\Delta n_p = G_L \tau_n$

*Solution no. 4*

*GIVEN:*    Steady state, no R–G, no light.

*SIMPLIFIED*
*DIFF. EQN:*

$$0 = D_N \frac{d^2 \Delta n_p}{dx^2} \quad \text{or} \quad 0 = \frac{d^2 \Delta n_p}{dx^2}$$

*SOLUTION:*    $\Delta n_p(x) = A + Bx$

---

The semiconductor is silicon, $T = 300$ K, the donor doping is the same everywhere with $N_D = 10^{15}/\text{cm}^3$, and $G_L = 10^{17}/\text{cm}^3$-sec at all points inside the semiconductor. Also, the statement of the problem *implies* equilibrium conditions exist for $t < 0$.

*Step 2*—Characterize the system under equilibrium conditions.

In Si at room temperature $n_i = 10^{10}/\text{cm}^3$. Since $N_D \gg n_i$, $n_0 = N_D = 10^{15}/\text{cm}^3$ and $p_0 = n_i^2/N_D = 10^5/\text{cm}^3$. With the doping being uniform, the equilibrium $n_0$ and $p_0$ values are the same everywhere throughout the semiconductor.

*Step 3*—Analyze the problem qualitatively.

Prior to $t = 0$, equilibrium conditions prevail and $\Delta p_n = 0$. Starting at $t = 0$ the light creates added electrons and holes and $\Delta p_n$ will begin to increase. The growing excess carrier numbers, however, in turn lead to an increased indirect thermal recombination rate which is proportional to $\Delta p_n$. Consequently, as $\Delta p_n$ grows as a result of photogeneration, more and more of the excess holes are eliminated per second by recombination through R–G centers. Eventually, a point is reached where the carriers annihilated per second by indirect thermal recombination balance the carriers created per second by the light, and a steady state condition is attained.

Summarizing, we expect $\Delta p_n(t)$ to start from zero at $t = 0$, to build up at a decreasing rate, and to ultimately become constant. Since the light generation and thermal recombination rates must balance under steady state conditions, we can even state $G_L = \Delta p_n(t \to \infty)/\tau_p$ or $\Delta p_n(t \to \infty) = \Delta p_{n|max} = G_L \tau_p$ if low-level injection prevails.

*Step 4*—Perform a quantitative analysis.

The minority carrier diffusion equation is the starting point for most first-order quantitative analyses. After examining the problem for obvious conditions that would invalidate the use of the diffusion equation, the appropriate minority carrier diffusion equation is written down, the equation is simplified, and a solution is sought subject to boundary conditions stated or implied in the problem.

For the problem under consideration a cursory inspection reveals that all simplifying assumptions involved in deriving the diffusion equations are readily satisfied. Specifically, only the minority carrier concentration is of interest; the equilibrium carrier concentrations are not a function of position, indirect thermal R–G is dominant in Si, and there are no "other processes" except for photogeneration. Because the photogeneration is uniform throughout the semiconductor, the perturbed carrier concentrations are also position-independent and the electric field $\mathscr{E}$ must clearly be zero in the perturbed system. Finally, a $\Delta p_{n|max} = G_L \tau_p = 10^{11}/cm^3 \ll n_0 = 10^{15}/cm^3$ is consistent with low-level injection prevailing at all times.

With no obstacles to utilizing the diffusion equation, the desired quantitative solution can now be obtained by solving

$$\frac{\partial \Delta p_n}{\partial t} = D_p \frac{\partial^2 \Delta p_n}{\partial x^2} - \frac{\Delta p_n}{\tau_p} + G_L \tag{3.55}$$

subject to the boundary condition

$$\Delta p_n(t)\big|_{t=0} = 0 \tag{3.56}$$

Since $\Delta p_n$ is not a function of position, the diffusion equation becomes an ordinary differential equation and simplifies to

$$\frac{d\Delta p_n}{dt} + \frac{\Delta p_n}{\tau_p} = G_L \tag{3.57}$$

The general solution of Eq. (3.57) is

$$\Delta p_n(t) = G_L\tau_p + Ae^{-t/\tau_p} \tag{3.58}$$

Applying the boundary condition yields

$$A = -G_L\tau_p \tag{3.59}$$

and

$$\Delta p_n(t) = G_L\tau_p(1 - e^{-t/\tau_p}) \quad \Leftarrow \text{solution} \tag{3.60}$$

*Step 5*—Examine the solution.

Failing to examine the mathematical solution to a problem is like growing vegetables and then failing to eat the produce. Relative to the Eq. (3.60) result, $G_L\tau_p$ has the dimensions of a concentration (number/cm$^3$) and the solution is at least dimensionally correct. A plot of the Eq. (3.60) result is shown in Fig. 3.25. Note that, in agreement with qualitative predictions, $\Delta p_n(t)$ starts from zero at $t = 0$ and eventually saturates at $G_L\tau_p$ after a few $\tau_p$.

*Epilogue*—We would be remiss if we did not point out the connection between the hypothetical problem just completed and the photoconductive decay measurement described in Subsection 3.3.4. Light output from the stroboscope used in the measurement can be modeled to first order by the pulse train pictured in Fig. 3.26(a). The Eq. (3.60) solution therefore approximately describes the carrier build-up during a light pulse. It should be noted, however, that the stroboscope light pulses have a duration of $t_{on} \cong 1$ $\mu$sec compared to a minority carrier lifetime of $\tau_p \cong 150$ $\mu$sec. With $t_{on}/\tau_p \ll 1$, one sees only the very

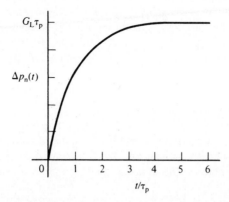

**Figure 3.25**   Solution to Sample Problem No. 1. Photogeneration-induced increase in the excess hole concentration as a function of time.

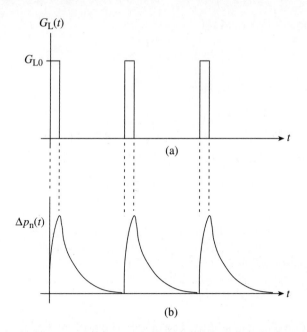

**Figure 3.26**   (a) Approximate model for the light output from the stroboscope in the photoconductivity decay measurement. (b) Sketch of the combined light-on/light-off solution for the excess minority-carrier concentration.

first portion of the Fig. 3.25 transient before the light is turned off and the semiconductor is allowed to decay back to equilibrium.

We are also interested in verifying the correctness of the light-off $\Delta p(t)$ expression used in the photoconductivity decay analysis. Derivation of this expression closely parallels the light-on sample problem solution. Exceptions are $\Delta p_n$ at the beginning of the light-off transient equals $\Delta p_n$ at the end of the light-on transient and $G_L$ is set equal to zero in Eqs. (3.55) and (3.57). The solution to Eq. (3.57) with $G_L = 0$ is the same as Solution No. 2 in Table 3.2; i.e.,

$$\Delta p_n(t) = \Delta p_n(0)e^{-t/\tau_p} \tag{3.61}$$

where $t = 0$ has been reset to correspond to the beginning of the light-off transient. Equation (3.37) in the measurement analysis and Eq. (3.61) are of course identical. For completeness, the combined light-on and light-off $\Delta p_n(t)$ solution is sketched in Fig. 3.26(b).

### Sample Problem No. 2

**P:** As pictured in Fig. 3.27(a), the $x = 0$ end of a uniformly doped semi-infinite bar of silicon with $N_D = 10^{15}/cm^3$ is illuminated so as to create $\Delta p_{n0} = 10^{10}/cm^3$ excess holes at

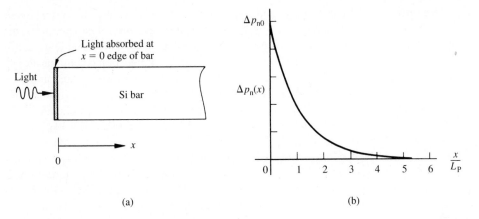

**Figure 3.27** (a) Pictorial definition of Sample Problem No. 2. (b) Solution to Sample Problem No. 2 showing the excess hole concentration inside the Si bar as a function of position.

$x = 0$. The wavelength of the illumination is such that no light penetrates into the interior $(x > 0)$ of the bar. Determine $\Delta p_n(x)$.

**S:** The semiconductor is again silicon uniformly doped with an $N_D = 10^{15}/\text{cm}^3$. Steady state conditions are inferred from the statement of the problem, since we are asked for $\Delta p_n(x)$ and not $\Delta p_n(x, t)$. Moreover, at $x = 0$, $\Delta p_n(0) = \Delta p_{n0} = 10^{10}/\text{cm}^3$, and $\Delta p_n \to 0$ as $x \to \infty$. The latter boundary condition follows from the semi-infinite nature of the bar. The perturbation in $\Delta p_n$ due to the nonpenetrating light can't possibly extend out to $x = \infty$. The nonpenetrating nature of the light allows us to set $G_L = 0$ for $x > 0$. Finally, note that the problem statement fails to mention the temperature of operation. When this happens, it is reasonable to assume an intended $T = 300$ K.

If the light were removed, the silicon bar in Sample Problem 2 maintained at 300 K would revert to an equilibrium condition identical to that described in Sample Problem 1. Under equilibrium conditions, then, $n_0 = 10^{15}/\text{cm}^3$, $p_0 = 10^5/\text{cm}^3$, and the carrier concentrations are uniform throughout the semiconductor bar.

Qualitatively it is a simple matter to predict the expected effect of the nonpenetrating light on the silicon bar. The light first creates excess carriers right at $x = 0$. With more carriers at one point than elsewhere in the bar, the diffusion process next comes into play and the carrier excess spreads into the semiconductor proper. At the same time, however, the appearance of an excess hole concentration inside the bar enhances the thermal recombination rate. Thus, as the diffusing holes move into the bar their numbers are reduced by recombination. In addition, since the minority carrier holes live for only a limited period, a time $\tau_p$ on the average, fewer and fewer excess holes will survive as the depth of penetration into the bar becomes larger and larger. Under steady state conditions it is therefore reasonable to expect an excess distribution of holes near $x = 0$, with $\Delta p_n(x)$ monotonically decreasing from $\Delta p_{n0}$ at $x = 0$ to $\Delta p_n = 0$ as $x \to \infty$.

In preparation for obtaining a quantitative solution, we observe that the system under consideration is one-dimensional, the analysis is restricted to the minority carrier holes, the equilibrium carrier concentrations are position-independent, indirect thermal R–G dominates, there are no "other processes" for $x > 0$, and low-level injection conditions clearly prevail ($\Delta p_{n|max} = \Delta p_{n0} = 10^{10}/cm^3 \ll n_0 = 10^{15}/cm^3$). The only question that might be raised concerning the use of the diffusion equation as the starting point for the quantitative analysis is whether $\mathscr{E} \simeq 0$. With the light on, a nonuniform distribution of holes and associated distribution of positive charge will appear near the $x = 0$ surface. The excess hole pile-up, however, is very small ($\Delta p_{n|max} \simeq n_i$) and the associated electric field is therefore expected to be correspondingly small. Moreover, in problems of this type it is found that the majority carriers, negatively charged electrons in the given problem, redistribute in such a way as to partly cancel the minority carrier charge. Thus experience indicates the $\mathscr{E} \simeq 0$ assumption to be reasonable, and use of the minority carrier diffusion equation to be justified.

Under steady state conditions with $G_L = 0$ for $x > 0$ the hole diffusion equation reduces to the form

$$D_P \frac{d^2 \Delta p_n}{dx^2} - \frac{\Delta p_n}{\tau_p} = 0 \qquad \text{for } x > 0 \tag{3.62}$$

which is to be solved subject to the boundary conditions

$$\Delta p_{n|x=0^+} = \Delta p_{n|x=0} = \Delta p_{n0} \tag{3.63}$$

and

$$\Delta p_{n|x \to \infty} = 0 \tag{3.64}$$

Equation (3.62) should be recognized as one of the simplified diffusion equations cited in Table 3.2, with the general solution

$$\Delta p_n(x) = Ae^{-x/L_P} + Be^{x/L_P} \tag{3.65}$$

where

$$L_P \equiv \sqrt{D_P \tau_p} \tag{3.66}$$

Because $\exp(x/L_P) \to \infty$ as $x \to \infty$, the only way that the Eq. (3.64) boundary condition can be satisfied is for $B$ to be identically zero. With $B = 0$, application of the Eq. (3.63) boundary condition yields

$$A = \Delta p_{n0} \tag{3.67}$$

and

$$\Delta p_n(x) = \Delta p_{n0} e^{-x/L_P} \qquad \Leftarrow \text{solution} \qquad (3.68)$$

The Eq. (3.68) result is plotted in Fig. 3.27(b). In agreement with qualitative arguments, the nonpenetrating light merely gives rise to a monotonically decreasing $\Delta p_n(x)$ starting from $\Delta p_{n0}$ at $x = 0$ and decreasing to $\Delta p_n = 0$ as $x \to \infty$. Note that the precise functional form of the falloff in the excess carrier concentration is exponential with a characteristic decay length equal to $L_P$.

## 3.5 SUPPLEMENTAL CONCEPTS

### 3.5.1 Diffusion Lengths

The situation just encountered in Sample Problem No. 2—the creation (or appearance) of an excess of minority carriers along a given plane in a semiconductor, the subsequent diffusion of the excess from the point of injection, and the exponential falloff in the excess carrier concentration characterized by a decay length ($L_P$)—occurs often enough in semiconductor analyses that the characteristic length has been given a special name. Specifically,

$$L_P \equiv \sqrt{D_P \tau_p} \qquad \begin{array}{l} \text{associated with the minority carrier} \\ \text{holes in an } n\text{-type material} \end{array} \qquad (3.69a)$$

and

$$L_N \equiv \sqrt{D_N \tau_n} \qquad \begin{array}{l} \text{associated with the minority carrier} \\ \text{electrons in a } p\text{-type material} \end{array} \qquad (3.69b)$$

are referred to as *minority carrier diffusion lengths.*

Physically, $L_P$ and $L_N$ represent the average distance minority carriers can diffuse into a sea of majority carriers before being annihilated. This interpretation is clearly consistent with Sample Problem No. 2, where the average position of the excess minority carriers inside the semiconductor bar is

$$\langle x \rangle = \int_0^\infty x \Delta p_n(x) dx \Big/ \int_0^\infty \Delta p_n(x) dx = L_P \qquad (3.70)$$

For memory purposes, minority carrier diffusion into a sea of majority carriers might be likened to a small group of animals attempting to cross a piranha-infested stretch of the Amazon River. In the analogy, $L_P$ and $L_N$ correspond to the average distance the animals advance into the river before being eaten.

Seeking an idea as to the size of diffusion lengths, let us assume $T = 300$ K, the semiconductor is $N_D = 10^{15}/cm^3$ doped Si, and $\tau_p = 10^{-6}$ sec. In this sample

$$L_P = \sqrt{D_p \tau_p} = \sqrt{(kT/q)\mu_p \tau_p} = [(0.0259)(458)(10^{-6})]^{1/2}$$

$$= 3.44 \times 10^{-3} \text{ cm}$$

Although the computed value is fairly representative, it should be understood that diffusion lengths can vary over several orders of magnitude because of wide variations in the carrier lifetimes.

### 3.5.2  Quasi-Fermi Levels

*Quasi-Fermi* levels are energy levels used to specify the carrier concentrations inside a semiconductor under *non*equilibrium conditions.

To understand the need for introducing quasi-Fermi levels, let us first refer to Sample Problem No. 1. In this problem, equilibrium conditions prevailed prior to $t = 0$, with $n_0 = N_D = 10^{15}/cm^3$ and $p_0 = 10^5/cm^3$. The energy band diagram describing the equilibrium situation is shown in Fig. 3.28(a). A simple inspection of the energy band diagram and the Fermi level positioning also conveys the equilibrium carrier concentrations, since

$$n_0 = n_i e^{(E_F - E_i)/kT} \tag{3.71a}$$

$$p_0 = n_i e^{(E_i - E_F)/kT} \tag{3.71b}$$

The point we wish to emphasize is that under equilibrium conditions there is a one-to-one correspondence between the Fermi level and the carrier concentrations. Knowledge of $E_F$ completely specifies $n_0$ and $p_0$ and vice versa.

Let us turn next to the nonequilibrium (steady state) situation inside the Problem 1 semiconductor at times $t \gg \tau_p$. For times $t \gg \tau_p$, $\Delta p_n = G_L \tau_p = 10^{11}/cm^3$, $p = p_0 + \Delta p \cong 10^{11}/cm^3$, and $n \cong n_0 = 10^{15}/cm^3$. Although $n$ remains essentially unperturbed, $p$ has

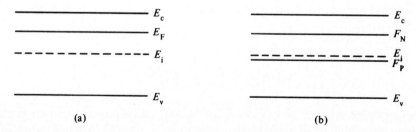

(a)                                                                              (b)

**Figure 3.28**  Sample use of quasi-Fermi levels. Energy band description of the situation inside the semiconductor of Sample Problem No. 1 under (a) equilibrium conditions and (b) nonequilibrium conditions ($t \gg \tau_p$).

increased by many orders of magnitude; and it is clear that the Fig. 3.28(a) diagram no longer describes the state of the system. In fact, the Fermi level is defined only for a system under equilibrium conditions and cannot be used to deduce the carrier concentrations inside a system in a nonequilibrium state.

The convenience of being able to deduce the carrier concentrations by inspection from the energy band diagram is extended to nonequilibrium conditions through the use of quasi-Fermi levels. This is accomplished by introducing two energies, $F_N$ (the quasi-Fermi level for electrons) and $F_P$ (the quasi-Fermi level for holes), which are *by definition* related to the nonequilibrium carrier concentrations in the same way $E_F$ is related to the equilibrium carrier concentrations. To be specific, under nonequilibrium conditions and assuming the semiconductor to be nondegenerate,

$$n \equiv n_i e^{(F_N - E_i)/kT} \quad \text{or} \quad F_N \equiv E_i + kT \ln\left(\frac{n}{n_i}\right) \tag{3.72a}$$

and

$$p \equiv n_i e^{(E_i - F_P)/kT} \quad \text{or} \quad F_P \equiv E_i - kT \ln\left(\frac{p}{n_i}\right) \tag{3.72b}$$

Note that $F_N$ and $F_P$ are conceptual constructs, with the values of $F_N$ and $F_P$ being totally determined from a prior knowledge of $n$ and $p$. Moreover, the quasi-Fermi level definitions have been carefully chosen so that when a perturbed system relaxes back toward equilibrium, $F_N \to E_F$, $F_P \to E_F$, and Eqs. (3.72) $\to$ Eqs. (3.71).

To provide a straightforward application of the quasi-Fermi level formalism, let us again consider the $t \gg \tau_p$ state of the semiconductor in Sample Problem No. 1. First, since $n \simeq n_0$, $F_N \simeq E_F$. Next, substitution of $p = 10^{11}/\text{cm}^3$ ($n_i = 10^{10}/\text{cm}^3$ and $kT = 0.0259$ eV) into Eq. (3.72b) yields $F_P = E_i - 0.06$ eV. By eliminating $E_F$ from the energy band diagram and drawing lines at the appropriate energies to represent $F_N$ and $F_P$, one obtains the diagram displayed in Fig. 3.28(b). Figure 3.28(b) clearly conveys to any observer that the system under analysis is in a nonequilibrium state. When referenced to part (a), part (b) of Fig. 3.28 further indicates at a glance that low-level injection is taking place inside the semiconductor, creating a concentration of minority carrier holes in excess of $n_i$. A second example use of the quasi-Fermi level formalism, a use based on Sample Problem No. 2 and involving position dependent quasi-Fermi levels, is presented in Exercise 3.5 at the end of the section.

As a final point, it should be mentioned that the quasi-Fermi level formalism can be used to recast some of the carrier-action relationships in a more compact form. For example, the standard form of the equation for the total hole current reads

$$\mathbf{J}_P = q\mu_p p \mathscr{E} - qD_P \nabla p \tag{3.73}$$

$$\text{(same as 3.18a)}$$

Differentiating both sides of Eq. (3.72b) with respect to position, one obtains

$$\nabla p = \left(\frac{n_i}{kT}\right) e^{(E_i - F_P)/kT} (\nabla E_i - \nabla F_P) \tag{3.74a}$$

$$= \left(\frac{qp}{kT}\right) \mathscr{E} - \left(\frac{p}{kT}\right) \nabla F_P \tag{3.74b}$$

The identity $\mathscr{E} = \nabla E_i/q$ (the three-dimensional version of Eq. 3.15) is employed in progressing from Eq. (3.74a) to Eq. (3.74b). Next, eliminating $\nabla p$ in Eq. (3.73) using Eq. (3.74b) gives

$$\mathbf{J}_P = q\left(\mu_p - \frac{qD_P}{kT}\right) p\mathscr{E} + \left(\frac{qD_P}{kT}\right) p\nabla F_P \tag{3.75}$$

From the Einstein relationship, however, $qD_P/kT = \mu_p$. We therefore conclude

$$\mathbf{J}_P = \mu_p p \nabla F_P \tag{3.76a}$$

Similarly,

$$\mathbf{J}_N = \mu_n n \nabla F_N \tag{3.76b}$$

Since $\mathbf{J}_P \propto \nabla F_P$ and $\mathbf{J}_N \propto \nabla F_N$ in Eqs. (3.76), one is led to a very interesting general interpretation of energy band diagrams containing quasi-Fermi levels. Namely, a quasi-Fermi level that varies with position ($dF_P/dx \neq 0$ or $dF_N/dx \neq 0$) indicates at a glance that current is flowing inside the semiconductor.

---

### Exercise 3.5

**P:** In Sample Problem No. 2, nonpenetrating illumination of a semiconductor bar was found to cause a steady state, excess-hole concentration of $\Delta p_n(x) = \Delta p_{n0} \exp(-x/L_P)$. Given the prevailing low-level injection conditions, and noting that $p = p_0 + \Delta p$, we can therefore state

$$n \simeq n_0$$

$$p = p_0 + \Delta p_{n0} e^{-x/L_P}$$

for the illuminated sample.

(a) Making use of Eqs. (3.72), establish relationships for $F_N$ and $F_P$ in the illuminated bar.

(b) Show that $F_P$ is a linear function of $x$ at points where $\Delta p_n(x) \gg p_0$.

(c) Using the results of parts (a) and (b), sketch the energy band diagrams describing the semiconductor bar of Sample Problem No. 2 under equilibrium and illuminated steady state conditions. (Assume $\mathscr{E} = 0$ in the illuminated bar.)

(d) Is there a hole current in the illuminated bar under steady state conditions? Explain.

(e) Is there an electron current in the illuminated bar under steady state conditions? Explain.

**S:** (a) Since $n \cong n_0$, it follows from Eq. (3.72a) that $F_N \cong E_F$. Likewise, substituting the preceding $p$-expression into Eq. (3.72b), we conclude

$$F_P = E_i - kT \ln(p/n_i) = E_i - kT \ln[p_0/n_i + (\Delta p_{n0}/n_i)e^{-x/L_P}]$$

(b) If $\Delta p_n(x) \gg p_0$, then $(\Delta p_{n0}/n_i) \exp(-x/L_P) \gg p_0/n_i$ and

$$F_P \simeq E_i - kT \ln[(\Delta p_{n0}/n_i)e^{-x/L_P}]$$

or

$$F_P = E_i - kT \ln(\Delta p_{n0}/n_i) + kT(x/L_P)$$

(c) We know from Sample Problem No. 2 that $\Delta p_{n0} = 10^{10}/\text{cm}^3$, $n_i = 10^{10}/\text{cm}^3$, and $p_0 = n_i^2/N_D = 10^5/\text{cm}^3$. Thus

  (i) Near $x = 0$, $\Delta p_n(x) \gg p_0$ and $F_P$ is a linear function of $x$.

  (ii) At $x = 0$, $\Delta p_{n0} = n_i$ and we deduce from the part (b) result that $F_P = E_i$.

  (iii) For large $x$, $F_P$ eventually approaches $F_N = E_F$.

  (iv) $F_N - E_i \simeq E_F - E_i = kT \ln(N_D/n_i) = 0.30$ eV.

Utilizing the preceding information, one concludes

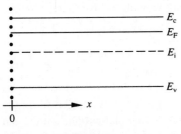

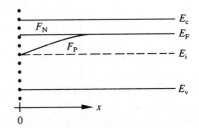

Equilibrium                                Steady state illuminated

(d) Assuming $p \neq 0$, it follows from Eq. (3.76a) that there will be a hole current whenever $dF_P/dx \neq 0$. There is obviously a hole current in the illuminated bar near $x = 0$.

(e) Appearances can sometimes be deceiving; $J_N \neq 0$ near $x = 0$! One might conclude from the part (c) result that $dF_N/dx = 0$ and therefore $J_N = 0$. Under steady state conditions, however, one must have $J$(total current) $= J_N + J_P =$ constant at all points in the bar. Since no current leaves the bar at $x = 0$, $J$ moreover must vanish at $x = 0$. Thus $J_N(x) = -J_P(x)$ and we know $J_P \neq 0$ near $x = 0$. The apparent discrepancy here stems from the fact that $J_N$ is proportional to both $n$ and $\nabla F_N$. Because the majority carrier electron concentration is much larger than the minority carrier hole concentration, $dF_N/dx$ must be correspondingly smaller than $dF_P/dx$. The slope in $F_N$ simply cannot be detected by inspecting the energy band diagram.

## 3.6 SUMMARY AND CONCLUDING COMMENTS

Most of the chapter was devoted to examining the three primary types of carrier action occurring inside of semiconductors: drift, diffusion, and recombination–generation. In each case the carrier action was first defined and visualized. Drift is charged particle motion in response to an applied electric field. Diffusion is migration from regions of high particle concentration to regions of low particle concentration due to the random thermal motion of the carriers. Recombination and generation are respectively the annihilation and creation of carriers. Next the quantitative effect of each type of carrier action was analyzed. Drift and diffusion give rise to particle currents (Eqs. 3.4 and 3.17–3.19); recombination–generation acts to change the carrier concentrations as a function of time (Eqs. 3.34). Associated with the quantitative analysis of each carrier action there arises a "constant of the motion," an important material-based parameter that specifies the magnitude of the carrier action in a given semiconductor sample. The carrier mobilities, the diffusion coefficients, and the minority carrier lifetimes are the material parameters associated with drift, diffusion, and recombination–generation, respectively. In the major semiconductors the carrier mobilities are very well characterized as a function of doping and temperature. Select mobility data are graphed in Figs. (3.5) and (3.7); an empirical fit relationship for the carrier mobilities in Si is presented in Exercise 3.1. The diffusion coefficients in nondegenerate semiconductors are computed from the carrier mobilities using the Einstein relationship (Eqs. 3.25). Conversely, an experimental measurement is normally performed to determine the carrier lifetimes in a given semiconductor sample.

Although introduced and examined individually, the various types of carrier action occur simultaneously inside semiconductors. Mathematically combining the overall effect of the various carrier activities leads to the continuity equations (Eqs. 3.46). The continuity equations in turn simplify to the minority carrier diffusion equations (Eqs. 3.54) under conditions encountered in many practical problems. Additional simplifications and their effect on the minority carrier diffusion equations are noted in Table 3.1; extensively utilized solutions to simplified forms of the equations are collected in Table 3.2. The continuity or minority carrier diffusion equations combined with the other relationships established in the chapter allow one to model the state of a semiconductor subject to an external perturbation. The more important working relationships are organized and repeated in Table 3.3.

**Table 3.3**   Carrier Action Equation Summary.

*Equations of State*

$$\frac{\partial n}{\partial t} = \frac{1}{q}\nabla \cdot \mathbf{J}_N + \frac{\partial n}{\partial t}\bigg|_{\substack{\text{thermal} \\ \text{R–G}}} + \frac{\partial n}{\partial t}\bigg|_{\substack{\text{other} \\ \text{processes}}} \qquad \frac{\partial \Delta n_p}{\partial t} = D_N \frac{\partial^2 \Delta n_p}{\partial x^2} - \frac{\Delta n_p}{\tau_n} + G_L$$

$$\frac{\partial p}{\partial t} = -\frac{1}{q}\nabla \cdot \mathbf{J}_P + \frac{\partial p}{\partial t}\bigg|_{\substack{\text{thermal} \\ \text{R–G}}} + \frac{\partial p}{\partial t}\bigg|_{\substack{\text{other} \\ \text{processes}}} \qquad \frac{\partial \Delta p_n}{\partial t} = D_P \frac{\partial^2 \Delta p_n}{\partial x^2} - \frac{\Delta p_n}{\tau_p} + G_L$$

*Current and R–G Relationships*

$$\mathbf{J}_N = \mathbf{J}_{N|\text{drift}} + \mathbf{J}_{N|\text{diff}} = q\mu_n n\mathscr{E} + qD_N\nabla n \qquad\qquad \frac{\partial n}{\partial t}\bigg|_{\substack{\text{i-thermal} \\ \text{R–G}}} = -\frac{\Delta n}{\tau_n}$$

$$\qquad\qquad\qquad \Updownarrow \text{ drift} \qquad \Updownarrow \text{ diffusion}$$

$$\mathbf{J}_P = \mathbf{J}_{P|\text{drift}} + \mathbf{J}_{P|\text{diff}} = q\mu_p p\mathscr{E} - qD_P\nabla p \qquad\qquad \frac{\partial p}{\partial t}\bigg|_{\substack{\text{i-thermal} \\ \text{R–G}}} = -\frac{\Delta p}{\tau_p}$$

$$\mathbf{J} = \mathbf{J}_N + \mathbf{J}_P$$

*Key Parametric Relationships*

$$L_N \equiv \sqrt{D_N\tau_n} \qquad\qquad \frac{D_N}{\mu_n} = \frac{kT}{q} \qquad\qquad \tau_n = \frac{1}{c_n N_T}$$

$$L_P \equiv \sqrt{D_P\tau_p} \qquad\qquad \frac{D_P}{\mu_p} = \frac{kT}{q} \qquad\qquad \tau_p = \frac{1}{c_p N_T}$$

*Resistivity and Electrostatic Relationships*

$$\rho = \frac{1}{q\mu_n N_D} \qquad \ldots \text{ }n\text{-type semiconductor}$$

$$\rho = \frac{1}{q(\mu_n n + \mu_p p)}$$

$$\rho = \frac{1}{q\mu_p N_A} \qquad \ldots \text{ }p\text{-type semiconductor}$$

$$\mathscr{E} = \frac{1}{q}\frac{dE_c}{dx} = \frac{1}{q}\frac{dE_v}{dx} = \frac{1}{q}\frac{dE_i}{dx} \qquad\qquad V = -\frac{1}{q}(E_c - E_{\text{ref}})$$

*Quasi-Fermi Level Relationships*

$$F_N \equiv E_i + kT \ln\left(\frac{n}{n_i}\right) \qquad\qquad \mathbf{J}_N = \mu_n n\nabla F_N$$

$$F_P \equiv E_i - kT \ln\left(\frac{p}{n_i}\right) \qquad\qquad \mathbf{J}_P = \mu_p p\nabla F_P$$

A number of topics related to carrier action were also addressed in the chapter, including resistivity and resistivity measurements, the hot-point probe measurement, constancy of the equilibrium Fermi level, nonuniform doping and the associated built-in electric field, $E$–$\mathbf{k}$ diagrams, measurement of the minority carrier lifetimes, and diffusion lengths. Moreover, although not identified as such, the energy band diagram was subject to further development. Specifically, it was pointed out that the existence of an electric field inside the semiconductor causes band bending or a variation of the energy bands with position. By simply inspecting an energy band diagram, it is possible to ascertain the general functional dependence of the electrostatic potential and electric field present in the material. In the discussion of recombination–generation, another level was added near midgap. This level, arising from R–G centers, plays a dominant role in the thermal communication between the energy bands. Lastly, quasi-Fermi levels were introduced to describe nonequilibrium conditions.

## PROBLEMS

| CHAPTER 3 PROBLEM INFORMATION TABLE | | | | |
|---|---|---|---|---|
| Problem | Complete After | Difficulty Level | Suggested Point Weighting | Short Description |
| 3.1 | 3.3.1 | 1 | 16 (2 each part) | Energy band visualization |
| 3.2 | 3.1.4 (a-d) 3.3.4 (e-h) | 1–2 | 16 (2 each part) | Short answer |
| ● 3.3 | 3.1.3 | 3 | 25 | $\mu$ vs. $T$ plots |
| ● 3.4 | " | 3 | 18 (b-15, c-3) | $\mu_n$ vs. $T$ student data |
| 3.5 | 3.1.4 | a-1, b-3 | 12 (a-6, b-6) | Intrinsic/maximum $\rho$ |
| 3.6 | " | 2 | 15 (3 each part) | Resistivity questions |
| ● 3.7 | " | 2 | 15 | $\rho$ vs. $N_A$, $N_D$ plot |
| ● 3.8 | " | 3–4 | 20 (10 each part) | $N_D(x)$ variation in resistor |
| ● 3.9 | " | 3 | 18 (a-12, b-6) | $\rho$ vs. $T$ plots |
| 3.10 | " | 4 | 20 (5 each part) | Temperature sensor |
| 3.11 | 3.1.5 | 2 | 5 | Compute thermal speed |
| 3.12 | 3.1.5 (a-d) 3.2.4 (e, f) | 2, e-3 | 16/diagram (a::c-2, d-4, e-2, f-4) | Interpret E-band diagrams |
| 3.13 | 3.2.4 | 3 | 15 (a-10, b-3, c-2) | Built-in electric field |
| ● 3.14 | " | 2 | 15 (a-3, b-10, c-2) | $D$ vs. $N_A$, $N_D$ computation |
| 3.15 | 3.3.3 | 3 | 10 | Eq. (3.35) → Eqs. (3.34) |
| 3.16 | 3.4.2 | 1 | 6 (2 each part) | Simple Diff. Eq. questions |

| 3.17 | 3.4.4 | 1 | 6 | Diff. Eq., no R–G region |
|------|-------|---|---|--------------------------|
| 3.18 | " | 2 | 12 (a::c-2, d-6) | Diff. Eq., mysterious ray |
| 3.19 | " | 2 | 8 | Diff. Eq., $G_L \rightarrow G_L/2$ |
| 3.20 | " | 2–3 | 10 | Diff. Eq., light + edge R |
| 3.21 | " | 3 | 15 (a::d-2, e-7) | Diff. Eq., double-ended |
| 3.22 | " | 3–4 | 15 (a::c-2, d-9) | Diff. Eq., half-illuminated |
| 3.23 | " | 2 | 8 (a-2, b-4, c-2) | CdS Photoconductor |
| 3.24 | 3.5.2 | 2 | 12 (a-2, b-4, c::e-2) | Quasi-Fermi levels |
| 3.25 | " | 3–4 | 15 | Quasi-Fermi levels |
| • 3.26 | " | 3–4 | 15 (a-13, b-2) | Plot quasi-Fermi levels |

**3.1** Using the energy band diagram, indicate how one visualizes

(a) The existence of an electric field inside a semiconductor.

(b) An electron with a K.E. = 0.

(c) A hole with a K.E. = $E_G/4$.

(d) Photogeneration.

(e) Direct thermal generation.

(f) Band-to-band recombination.

(g) Recombination via R–G centers.

(h) Generation via R–G centers.

**3.2** Short Answer

(a) An average hole drift velocity of $10^3$ cm/sec results when 2 V is applied across a 1-cm-long semiconductor bar. What is the hole mobility inside the bar?

(b) Name the two dominant carrier *scattering* mechanisms in nondegenerately doped semiconductors of device quality.

(c) For a given semiconductor the carrier mobilities in intrinsic material are (choose one: higher than, lower than, the same as) those in heavily doped material. Briefly explain why the mobilities in intrinsic material are (chosen answer) those in heavily doped material.

(d) Two GaAs wafers, one *n*-type and one *p*-type, are uniformly doped such that $N_D$(wafer 1) = $N_A$(wafer 2) $\gg n_i$. Which wafer will exhibit the larger resistivity? Explain.

(e) The electron mobility in a silicon sample is determined to be 1300 cm²/V-sec at room temperature. What is the electron diffusion coefficient?

(f) What is the algebraic statement of low-level injection?

(g) Light is used to create excess carriers in silicon. These excess carriers will predominantly recombine via (choose one: band-to-band, R–G center, or photo) recombination.

(h) Prior to processing, a silicon sample contains $N_D = 10^{14}/cm^3$ donors and $N_T = 10^{11}/cm^3$ R–G centers. After processing (say in the fabrication of a device), the sample contains $N_D = 10^{17}/cm^3$ donors and $N_T = 10^{10}/cm^3$ R–G centers. Did processing increase or decrease the minority carrier lifetime? Explain.

● **3.3** Complete part (b) of Exercise 3.1.

● **3.4** (a) (Optional) Read the Experiment No. 7 Introduction and Measurement System description in R. F. Pierret, *Semiconductor Measurement Laboratory Operations Manual,* Supplement A, in the Modular Series on Solid State Devices, Addison-Wesley Publishing Co., Reading, MA, © 1991.

(b) The measurement described in the part (a) reference is used to determine the mobility in an $N_D < 10^{14}/cm^3$ Si sample as a function of temperature from roughly room temperature to $T = 150$ K. Representative $\mu_n$ versus $T$ (K) data derived by undergraduate students is tabulated in the following table. Assuming $\mu_n \propto T^{-b}$ [or $\ln(\mu_n) = $ constant $- $ **b** $\ln(T)$] can be fitted to the data, use the MATLAB polyfit function or an equivalent least squares fit[†] to determine the best fit value of the power factor **b.** Draw the fit line and note the **b**-fit value on a log-log plot of the data.

(c) Compare your part (b) result with Fig. 3.7(a). Is it significant that the experimental data was derived from a lowly doped sample? Is the experimentally derived **b**-value consistent with the **b**-value noted on the text plot? Are the measured mobility values of the proper magnitude?

---

[†] Given $N$ data points, $(x_i, y_i)$ with $i = 1$ to $N$, the straight line $y = a + bx$, which is the "best fit" to the data, will have a $y$-axis intercept and slope respectively equal to

$$a = \frac{\Sigma y_i \Sigma x_i^2 - \Sigma x_i \Sigma x_i y_i}{N\Sigma x_i^2 - (\Sigma x_i)^2} \quad \text{and} \quad b = \frac{N\Sigma x_i y_i - \Sigma x_i \Sigma y_i}{N\Sigma x_i^2 - (\Sigma x_i)^2}$$

All summations are from $i = 1$ to $i = N$. The "best fit" criterion leading to the expressions for $a$ and $b$ is that the square of the difference between the data points and the fitted line is minimized; hence the name "least squares fit."

| $T\,(\text{K})$ | $\mu_n\,(\text{cm}^2/\text{V-sec})$ |
|:---:|:---:|
| 290 | 1501 |
| 280 | 1646 |
| 270 | 1805 |
| 260 | 1985 |
| 250 | 2185 |
| 240 | 2415 |
| 230 | 2675 |
| 220 | 2978 |
| 210 | 3306 |
| 200 | 3743 |
| 190 | 4209 |
| 180 | 4619 |
| 170 | 5216 |
| 160 | 5910 |
| 150 | 6757 |

**3.5** Intrinsic and Maximum Resistivity

(a) Determine the resistivity of intrinsic Ge, Si, and GaAs at 300 K.

(b) Determine the maximum possible resistivity of Ge, Si, and GaAs at 300 K.

**3.6** More Resistivity Questions

(a) A silicon sample maintained at room temperature is uniformly doped with $N_D = 10^{16}/\text{cm}^3$ donors. Calculate the resistivity of the sample using Eq. (3.8a). Compare your calculated result with the $\rho$ deduced from Fig. 3.8(a).

(b) The silicon sample of part (a) is "compensated" by adding $N_A = 10^{16}/\text{cm}^3$ acceptors. Calculate the resistivity of the compensated sample. (Exercise caution in choosing the mobility values to be employed in this part of the problem.)

(c) Compute the resistivity of intrinsic ($N_A = 0$, $N_D = 0$) silicon at room temperature. How does your result here compare with that for part (b)?

(d) A 500-ohm resistor is to be made from a bar-shaped piece of $n$-type Si. The bar has a cross-sectional area of $10^{-2}$ cm$^2$ and a current-carrying length of 1 cm. Determine the doping required.

(e) A lightly doped ($N_D < 10^{14}$/cm$^3$) Si sample is heated up from room temperature to 100°C. $N_D \gg n_i$ at both room temperature and 100°C. Is the resistivity of the sample expected to increase or decrease? Explain.

● **3.7** Making use of the mobility fit relationships and parameters quoted in Exercise 3.1, construct a plot of the silicon resistivity versus impurity concentration at $T = 300$ K. Include curves for both $n$- and $p$-type silicon over the range $10^{13}$/cm$^3 \le N_A$ or $N_D \le 10^{20}$/cm$^3$. Compare your result with Fig. 3.8(a).

**3.8** Resistors in ICs are sometimes thin semiconductor layers near the surface of a wafer. However, formation of the layer by diffusion or ion implantation (discussed in Chapter 4) is likely to give rise to a doping concentration that varies with depth into the layer. Let us examine how the resistance is computed when the doping varies with depth.

(a) Given a bar-shaped layer of width $W$, length $L$, and depth $d$, and assuming an arbitrary $N_D(x)$ variation with the depth $x$ from the wafer surface, show that the resistance of the layer is to be computed from

$$R = \frac{L}{W}\left[\frac{1}{q\displaystyle\int_0^d \mu_n(x)N_D(x)\,dx}\right]$$

● (b) Taking $N_D(x) = N_{D0}\exp(-ax) + N_{DB}$, compute and plot $R$ versus $N_{D0}$ for $10^{14}$/cm$^3 \le N_{D0} \le 10^{18}$/cm$^3$ when $L = W$, $N_{DB} = 10^{14}$/cm$^3$, $d = 5$ $\mu$m, and $1/a = 1$ $\mu$m.

● **3.9** (a) Modify your $\mu$ versus $T$ program from Problem 3.3 to compute and construct *semilog* plots of $\rho$ (log scale) versus $T$ for 200 K $\le T \le$ 500 K and $N_D$ or $N_A$ stepped in decade values from $10^{14}$/cm$^3$ to $10^{18}$/cm$^3$. [NOTE: The lower doping curves are in error at the higher temperatures if Eqs. (3.8) are used to compute $\rho$. Can you explain why? Assuming Eqs. (3.8) were used in the computation, how must the $\rho$ versus $T$ program be modified to correct the error?]

(b) Compare your $n$-type plot with Fig. 7 in Li and Thurber, Solid-State Electronics, **20**, 609 (1977). Compare your $p$-type plot with Fig. 7 in Li, Solid-State Electronics, **21**, 1109 (1978). Comment on the comparisons.

**3.10** Your boss asks you to construct a temperature sensor for measuring the ambient outside temperature in W. Lafayette IN ($-30$°C $\le T \le$ 40°C). You decide to base operation of the temperature sensor on the change in resistance associated with a bar-shaped piece of Si.

(a) Restricting yourself to nondegenerate dopings and reasonable device dimensions, and also requiring the resistance to be readily measured with a hand-held multimeter (say 1 $\Omega \le R \le$ 1000 $\Omega$), indicate the doping and dimensions of your sensor.

(b) Derive an expression for the sensitivity ($dR/dT$ in $\Omega/°C$) of your sensor over the operating temperature range. Relative to sensitivity, is it preferable to use low or high dopings? Explain.

(c) What are the upper and lower temperature limits of operation of your sensor (approximately)? Explain.

● (d) Plot $R$ versus $T$ of your sensor over the required temperature range of operation.

**3.11** Thermal energy alone was noted to contribute to relatively high carrier velocities inside of a semiconductor. Considering a nondegenerate semiconductor maintained at room temperature, compute the thermal velocity of electrons having a kinetic energy corresponding to the peak of the electron distribution in the conduction band. Set $m^* = m_0$ in performing your calculation. (This problem assumes the prior working of either Problem 2.7 or Problem 2.8.)

**3.12** Interpretation of Energy Band Diagrams
Six different silicon samples maintained at 300 K are characterized by the energy band diagrams in Fig. P3.12. Answer the questions that follow after choosing a specific diagram for analysis. Possibly repeat using other energy band diagrams. (Excessive repetitions have been known to lead to the onset of insanity.)

(a) Do equilibrium conditions prevail? How do you know?

(b) Sketch the electrostatic potential ($V$) inside the semiconductor as a function of $x$.

(c) Sketch the electric field ($\mathscr{E}$) inside the semiconductor as a function of $x$.

(d) The carrier pictured on the diagram moves back and forth between $x = 0$ and $x = L$ without changing its total energy. Sketch the K.E. and P.E. of the carrier as a function of position inside the semiconductor. Let $E_F$ be the energy reference level.

(e) Roughly sketch $n$ and $p$ versus $x$.

(f) On the same set of coordinates, make a rough sketch of the electron drift-current density ($J_{N|drift}$) and the electron diffusion-current density ($J_{N|diff}$) inside the Si sample as a function of position. Be sure to graph the proper polarity of the current densities at all points and clearly identify your two current components. Also briefly explain how you arrived at your sketch.

**3.13** The nonuniform doping in the central region of bipolar junction transistors (BJTs) creates a built-in field that assists minority carriers across the region and increases the maximum operating speed of the device. Suppose the BJT is a Si device maintained under equilibrium conditions at room temperature with a central region of length $L$. Moreover, the nonuniform acceptor doping is such that

$$p(x) \cong N_A(x) = n_i e^{(a-x)/b} \quad \ldots 0 \leq x \leq L$$

where $a = 1.8 \ \mu m$, $b = 0.1 \ \mu m$, and $L = 0.8 \ \mu m$.

(a) Draw the energy band diagram for the $0 \leq x \leq L$ region specifically showing $E_c$, $E_F$, $E_i$, and $E_v$ on your diagram. Explain how you arrived at your diagram.

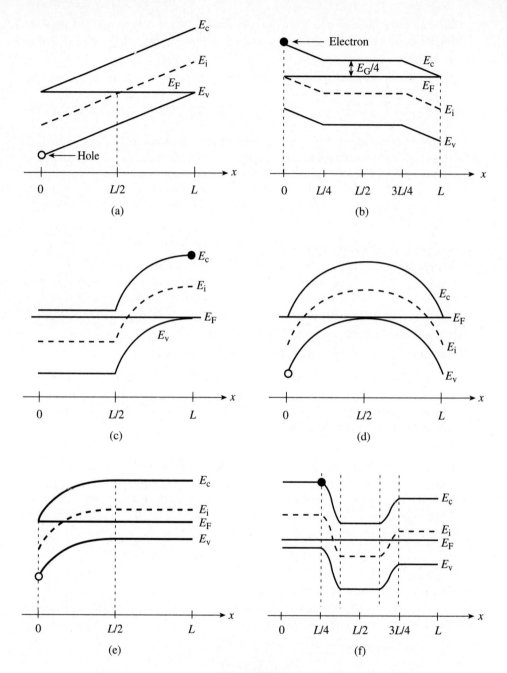

**Figure P3.12**

(b) Make a sketch of the $\mathcal{E}$-field inside the region as a function of position, and compute the value of $\mathcal{E}$ at $x = L/2$.

(c) Is the built-in electric field such as to aid the motion of minority carrier electrons in going from $x = 0$ to $x = L$? Explain.

**3.14** (a) Based on information found in the text proper, *roughly sketch* the expected variation of $D_N$ and $D_P$ versus doping appropriate for $10^{14}/cm^3 \leq N_A$ or $N_D \leq 10^{18}/cm^3$ doped Si maintained at $T = 300$ K. Explain how you arrived at the form of your sketch.

● (b) Making use of the fit relationships and parameters quoted in Exercise 3.1, construct a plot of $D_N$ and $D_P$ versus $N_A$ or $N_D$ for $10^{14}/cm^3 \leq N_A$ or $N_D \leq 10^{18}/cm^3$ doped Si maintained at $T = 300$ K.

(c) Why was the upper doping limit taken to be $10^{18}/cm^3$ in the part (b) computation?

**3.15** Taking $E_T \cong E_i$ so that $n_1 \cong p_1 \cong n_i$, setting $\Delta n = \Delta p$, and assuming $\tau_n$ is comparable to $\tau_p$, show that the general-case R–G relationship of Eq. (3.35) reduces to the special-case relationships of Eqs. (3.34) when one has low-level injection in a specific material type.

**3.16** $\partial \Delta n_p / \partial t = D_N \partial^2 \Delta n_p / \partial x^2 - \Delta n_p / \tau_n + G_L$ is known as the minority carrier diffusion equation for electrons.

(a) Why is it called a *diffusion equation?*

(b) Why is it referred to as a *minority carrier equation?*

(c) The equation is valid only under low-level injection conditions. Why?

**3.17** Show that, under steady state conditions

$$\Delta p_n(x) = \Delta p_{n0}(1 - x/L) \quad \ldots 0 \leq x \leq L$$

is the special-case solution of the minority carrier diffusion equation that will result if (1) one assumes *all* recombination–generation processes are negligible within an *n*-type semiconductor of length $L$, and (2) one employs the boundary conditions $\Delta p_n(0) = \Delta p_{n0}$ and $\Delta p_n(L) = 0$. (Neglecting recombination–generation is an excellent approximation when $L$ is much less than a minority carrier diffusion length. A $\Delta p(x)$ solution of the above type is frequently encountered in practical problems.)

**3.18** The earth is hit by a mysterious ray that momentarily wipes out all minority carriers. Majority carriers are unaffected. Initially in equilibrium and not affected by room light, a uniformly doped silicon wafer sitting on your desk is struck by the ray at time $t = 0$. The wafer doping is $N_A = 10^{16}/cm^3$, $\tau_n = 10^{-6}$ sec, and $T = 300$ K.

(a) What is $\Delta n$ at $t = 0^+$? ($t = 0^+$ is an imperceptible fraction of a second after $t = 0$.)

(b) Does generation or recombination dominate at $t = 0^+$? Explain.

(c) Do low-level injection conditions exist inside the wafer at $t = 0^+$? Explain.

(d) Starting from the appropriate differential equation, derive $\Delta n_p(t)$ for $t > 0$.

**3.19** A silicon wafer ($N_A = 10^{14}/cm^3$, $\tau_n = 1$ $\mu$sec, $T$ = room temperature) is first illuminated for a time $t \gg \tau_n$ with light which generates $G_{L0} = 10^{16}$ electron-hole pairs per cm³-sec uniformly throughout the volume of the silicon. At time $t = 0$ the light intensity is reduced, making $G_L = G_{L0}/2$ for $t \geq 0$. Determine $\Delta n_p(t)$ for $t \geq 0$.

**3.20** A semi-infinite $p$-type bar (see Fig. P3.20) is illuminated with light which generates $G_L$ electron-hole pairs per cm³-sec uniformly throughout the volume of the semiconductor. Simultaneously, carriers are extracted at $x = 0$ making $\Delta n_p = 0$ at $x = 0$. Assuming a steady state condition has been established and $\Delta n_p(x) \ll p_0$ for all $x$, determine $\Delta n_p(x)$.

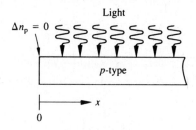

**Figure P3.20**

**3.21** The two ends of a uniformly doped $n$-type Si bar of length $L$ are simultaneously illuminated so as to create $\gamma N_D$ excess holes at both $x = 0$ and $x = L$ (see Fig. P3.21). The wavelength and intensity of the illumination are such that no light penetrates into the interior ($0 < x < L$) of the bar and $\gamma = 10^{-3}$. Also, steady state conditions prevail, $T = 300$ K, and $N_D \gg n_i$.

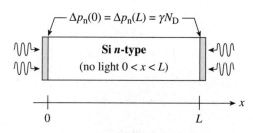

**Figure P3.21**

(a) Based on *qualitative reasoning,* sketch the expected general form of the $\Delta p_n(x)$ solution.

(b) Do low-level injection conditions prevail inside the illuminated bar? Explain.

(c) Write down the differential equation (simplest form possible) you must solve to determine $\Delta p_n(x)$ inside the bar.

(d) Write down the general form of the $\Delta p_n(x)$ solution and the boundary condition(s) appropriate for this particular problem.

(e) Establish an expression for the hole current ($J_p$) flowing in the illuminated bar at $x = 0$. [Your answer may be left in terms of the arbitrary constant(s) appearing in the general form of the $\Delta p_n(x)$ solution.]

**3.22** As pictured in Fig. P3.22, the $x > 0$ portion of an infinite semiconductor is illuminated with light. The light generates $G_L = 10^{15}$ electron-hole pairs/cm³-sec uniformly throughout the $x > 0$ region of the bar. $G_L = 0$ for $x < 0$, steady state conditions prevail, the semiconductor is silicon, the entire bar is uniformly doped with $N_D = 10^{18}$/cm³, $\tau_p = 10^{-6}$ sec, and $T = 300$ K.

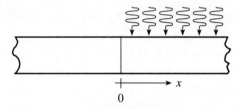

**Figure P3.22**

(a) What is the hole concentration at $x = -\infty$? Explain.

(b) What is the hole concentration at $x = +\infty$? Explain.

(c) Do low-level injection conditions prevail? Explain.

(d) Determine $\Delta p_n(x)$ for all $x$.
   NOTE: (1) Separate $\Delta p_n(x)$ expressions apply for $x > 0$ and $x < 0$.
        (2) Both $\Delta p_n$ and $d\Delta p_n/dx$ must be continous at $x = 0$.

**3.23** CdS is the most widely used material for constructing commercial photoconductors operating in the visible portion of the spectrum. The CdS photoconductor has a high sensitivity and its spectral response closely matches that of the human eye. A model VT333 CdS Photoconductor is pictured in Fig. P3.23.

**Figure P3.23**

(a) Speculate why the conducting film has a snake-like pattern.

(b) The VT333 resistor pattern is approximately 0.3 mm wide and 3 cm long. Estimating the deposited CdS film to be 5 $\mu$m thick, $N_D = 10^{13}$/cm³ $\gg n_i$, and $\mu_n = 100$ cm²/V-sec, compute the dark resistance of the device.

(c) The VT333 exhibited a 250 $\Omega$ resistance when illuminated with a microscope light. Can the usual relationships be used to determine the $G_L$ required to produce the cited resistance? Explain.

**3.24** The equilibrium and steady state conditions before and after illumination of a semiconductor are characterized by the energy band diagrams shown in Fig. P3.24. $T = 300$ K, $n_i = 10^{10}$/cm³, $\mu_n = 1345$ cm²/V-sec, and $\mu_p = 458$ cm²/V-sec. From the information provided, determine

(a) $n_0$ and $p_0$, the equilibrium carrier concentrations.

(b) $n$ and $p$ under steady state conditions.

(c) $N_D$.

(d) Do we have "low-level injection" when the semiconductor is illuminated? Explain.

(e) What is the resistivity of the semiconductor before and after illumination?

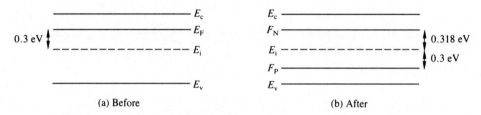

(a) Before          (b) After

**Figure P3.24**

**3.25** A portion ($0 \le x \le L$) of a Si sample, uniformly doped with $N_D = 10^{15}$/cm³ donors and maintained at room temperature, is subject to a *steady state* perturbation such that

$$n \simeq N_D$$
$$p = n_i(1 - x/L) + n_i^2/N_D \qquad \ldots \, 0 \le x \le L$$

Since $n \simeq N_D$, it is reasonable to assume $\mathscr{E} \simeq 0$ in the $0 \le x \le L$ region. Given $\mathscr{E} \simeq 0$, sketch the energy band diagram for the perturbed region specifically including $E_c$, $E_i$, $E_v$, $F_N$, and $F_P$ on your diagram.

● **3.26** (a) Write a MATLAB (computer) program that automatically plots $F_N$ and $F_P$ versus $x$ for the experimental situation first considered in Sample Problem No. 2 and further analyzed in Exercise 3.5. Assume a Si sample maintained at 300 K with $\tau_p = 10^{-6}$ sec. In writing the program, take $N_D$ and $\Delta p_{n0}$ to be input variables. Scale the $x$-axis in $L_P$ units with $(x/L_P)_{max} = \ln[100\Delta p_{n0}/p_0]$. Scale the $y$-axis in $kT$ units plotting $(F_N - E_v)/kT$ and $(F_P - E_v)/kT$. Set $y_{min} = -5$ and $y_{max} = 45$. Identify the positions of $E_v$ and $E_c$ on your plot.

(b) Run your program with $N_D = 10^{15}$/cm³ and $\Delta p_{n0} = 10^{10}$/cm³. Compare the resultant plot with the corresponding diagram sketched in Exercise 3.5.

(c) Remembering the assumed low-level injection requires $\Delta p_{n0} \ll N_D$, try running your program using different values of $N_D$ and $\Delta p_{n0}$.

# 4 Basics of Device Fabrication

This chapter provides a very brief overview of silicon device fabrication. The fabrication overview constitutes the final preparatory step before the consideration and analysis of specific devices. The goal is to develop a general feel for the physical nature of device structures. Moreover, the way a device is made affects key device parameters, which in turn affect the simplifications, assumptions, and so forth possible in the device analysis. Even in purely theoretical endeavors, some familiarity with fabrication processes and procedures is obviously necessary.

In what follows we first examine common "building block" processes. The processes are used in combination and are routinely repeated several times in producing a modern device or IC. Subsequently we describe the sequencing of processes to produce a *pn* junction diode, the first device to be analyzed in Part II of the text. To better illustrate IC fabrication, the chapter concludes with a subsection on the making of a CPU (central processing unit). For additional information the reader is referred to one of the excellent device-fabrication books cited in the Part I reading list.

Finally, it should be noted that this is the first "read only" chapter. As explained in the General Introduction, chapters with the read-only designation are primarily intended to convey qualitative information of general interest. They contain a small number of equations, no exercises, and few, if any, end-of-chapter problems. Also, there is no reference to the contents of read-only chapters in the test-like Review Problem Sets that are included at the end of the three text segments. If the material in this chapter were being transmitted verbally, you would be advised to put down your note-taking pencils and to just sit back and listen.

## 4.1 FABRICATION PROCESSES

### 4.1.1 Oxidation

The existence of a readily produced high-quality oxide helped establish the dominance of Si in the production of commercial devices. $SiO_2$ functions as both an insulator in a number of device structures and as a barrier to diffusion (see Subsection 4.1.2) during device fabrication. Silicon is highly reactive and a thin native oxide layer is rapidly produced by simply exposing virgin Si to the atmosphere at room temperature. $SiO_2$ layers of controllable thickness are produced during device fabrication by reacting Si with either oxygen gas or water vapor at an elevated temperature. In either case the oxidizing species diffuses

**Figure 4.1** Production-line (Intel Fab-9 fabrication facility) bank of furnaces of the type used for oxidation and diffusion. The ends of three horizontal furnace tubes are clearly visible in the center-right of the photograph. An operator in standard cleanroom garb is seated next to the furnace controls. (Photograph courtesy of Intel Corporation.)

through the existing oxide and reacts at the $Si–SiO_2$ interface to form more $SiO_2$. The relevant overall reactions are

$$Si + O_2 \rightarrow SiO_2 \tag{4.1a}$$

$$Si + 2H_2O \rightarrow SiO_2 + 2H_2 \tag{4.1b}$$

Growth of $SiO_2$ using $O_2$ and water vapor are referred to as *dry* and *wet oxidation,* respectively. Dry oxidation is used to form the most critical insulator regions in a device structure, such as the gate oxide in a MOSFET, primarily because it yields superior $Si–SiO_2$ interface properties. Wet oxidation, on the other hand, proceeds at a faster rate and is therefore preferred in forming the thicker barrier oxides.

A photograph of a production-line oxidation/diffusion furnace is shown in Fig. 4.1. A simplified sketch of the furnace and input system is presented in Fig. 4.2 along with a short pictorial description of the oxidation process. Standard oxidation temperatures of 800°C to 1200°C are produced in the furnace by resistance heating. The coil-like heating elements are commonly arranged in three separately controlled zones. The outer zones operate at a higher power to compensate for heat losses out the ends of the furnace. A relatively long center zone can thereby be maintained at a uniform temperature controllable to $\pm 1$°C. The tube down the center of the furnace is usually made of clear fused quartz, although SiC and polycrystalline Si tubes are also used. The Si wafers to be oxidized are loaded vertically onto a slotted quartz boat and pushed into the center of the furnace. During dry oxidation the oxygen gas is simply fed into the back of the tube. Wet oxidation is performed by bubbling a carrier gas (Ar or $N_2$) through water in a heated flask or burning $O_2$ and $H_2$ at the input to the tube. The latter approach is called *pyrogenic wet oxidation*. The required time in the furnace depends on the furnace temperature, the desired oxidation thickness, and the Si surface orientation. (Other factors can enter under certain conditions.) Representative dry and wet oxidation growth curves are graphed in Fig. 4.3. It should be noted that in a commercial system the wafer loading, insertion into the furnace, ramping of the furnace temperature, and gas control are all automated.

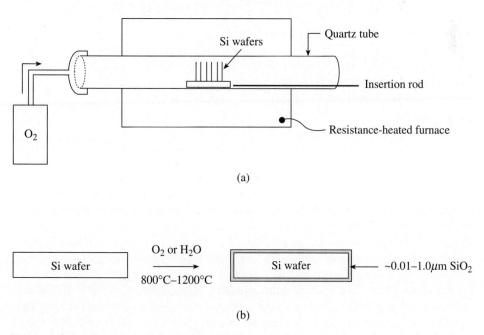

(a)

(b)

**Figure 4.2** (a) Simplified schematic illustration of an oxidation system. (b) Short pictorial description of the oxidation process.

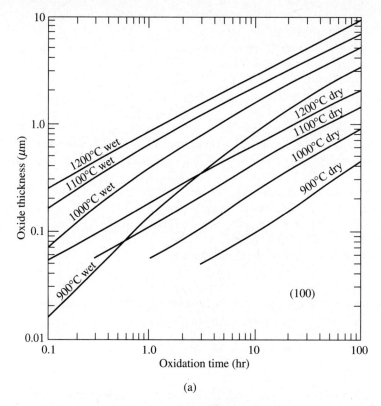

(a)

**Figure 4.3** Dry and wet oxidation growth curves. The $SiO_2$ thickness formed on (a) (100) and (b) (111) silicon surfaces as a function of time. (From Jaeger[8], © 1988 by Addison-Wesley Publishing Co., Inc. Reprinted by permission of the publisher.)

### 4.1.2 Diffusion

Solid state diffusion is a historically important and still widely used method for introducing dopant atoms into a semiconductor lattice. The basic process is sketched in Fig. 4.4. The semiconductor crystal, a Si wafer in Fig. 4.4, is exposed to a solid, liquid, or gaseous source containing the desired impurity. A reaction at the wafer surface establishes a supply of dopant atoms immediately adjacent to the semiconductor crystal. At elevated temperatures the atoms slowly diffuse into regions of the semiconductor not protected by the oxide. Diffusion takes place of course because the concentration of impurity atoms is greater outside the crystal than inside the crystal. The surface doping concentrations produced by this method are very high (up to $10^{21}/cm^3$), and thus the surface region takes on the *n*- or *p*-type character associated with the diffusing atoms. It should be noted that diffusion also takes place into the $SiO_2$, albeit at a relatively slow rate for the common dopants. This means the $SiO_2$ can protect the underlying Si for only a limited period of time. For a given dopant the masking time depends on the oxide thickness, the diffusion temperature, and the background doping.

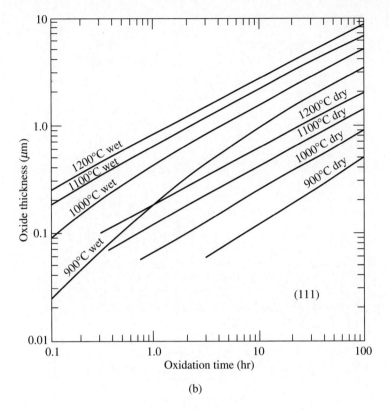

**Figure 4.3** *Continued.*

Diffusion is commonly performed in an open tube system identical in construction to that used for oxidation. Diffusion temperatures are also similar, ranging from roughly 900°C to 1200°C. The only major difference is the introduction of a dopant source in place of the oxidizing ambient. The specific example of phosphorus diffusion employing a liquid source is schematically illustrated in Fig. 4.5. The pictured $N_2$ carrier gas is passed through a bubbler containing phosphorus oxychloride ($POCl_3$) that is a liquid at room temperature. The $N_2$ picks up vapors from the liquid source and carries them into the furnace tube. The small percentage of oxygen simultaneously admitted into the furnace reacts with the $POCl_3$ to deposit $P_2O_5$ on the surface of the Si wafer, and Si in turn reacts with $P_2O_5$ liberating phosphorus atoms that diffuse into the silicon. Because undesirable surface compounds tend to form at the higher diffusion temperatures, the foregoing is often combined with a second step where the dopant source is cut off, the furnace temperature increased, and the already-introduced impurities driven deeper into the semiconductor at a more rapid rate. When a two-step process is used, the portion with the source present is called the *pre-deposition,* the latter portion with the source removed, the *drive-in.*

Neglecting second-order effects, the impurity concentration versus position inside the semiconductor after the predeposition and drive-in steps can be computed from

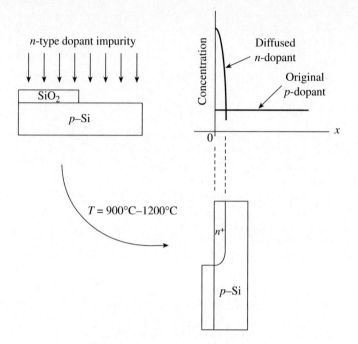

**Figure 4.4**   The basic diffusion process.

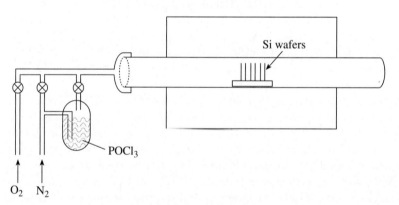

**Figure 4.5**   Schematic illustration of phosphorus diffusion using a liquid source.

$$N_1(x,t_1) = N_0 \text{erfc}(x/2\sqrt{D_1 t_1}) \qquad \qquad \dots \text{after predep} \qquad (4.2a)$$

$$N_2(x,t_2) = N_0\left(\frac{2}{\pi}\sqrt{D_1 t_1/D_2 t_2}\right)e^{-(x/2\sqrt{D_2 t_2})^2} \qquad \dots \begin{array}{l} \text{after drive-in} \\ D_2 t_2 \gg D_1 t_1 \end{array} \qquad (4.2b)$$

The subscripts 1 and 2 refer to the predeposition and drive-in, respectively, $x$ is the distance into the semiconductor as measured from the semiconductor surface, $N(x, t)$ the impurity

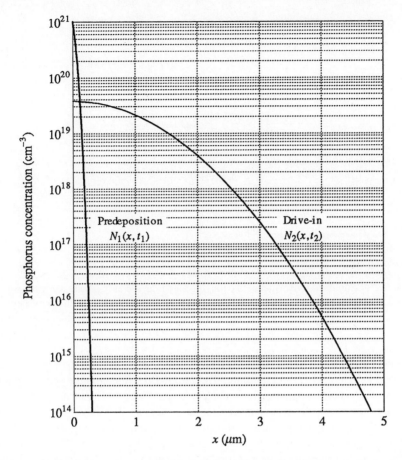

**Figure 4.6**  Computed phosphorus diffusion profiles. ($N_0 = 10^{21}$/cm³, $D_1 = 2.58 \times 10^{-14}$ cm²/sec, $t_1 = 600$ sec, $D_2 = 2.49 \times 10^{-12}$ cm²/sec, $t_2 = 1800$ sec.)

concentration at a depth $x$ after a given time $t$, $N_0$ the concentration at $x = 0$ during the predeposition, $D$ the diffusion coefficient for the given impurity and furnace temperature, and $t$ the time duration of the diffusion step. The complementary error function, erfc, is a function cited in most mathematical handbooks. The results of a sample computation corresponding to a phosphorus predeposition at 1000°C for 10 minutes followed by a drive-in at 1200°C for 30 minutes are presented in Fig. 4.6.

### 4.1.3  Ion Implantation

Ion implantation affords an alternative means of introducing dopants and other atoms into the near-surface region of a semiconductor. With the need for shallower junctions, lower-temperature processing, and more precise control, ion implantation has become a work-horse manufacturing process. In ion implantation an impurity is introduced into the

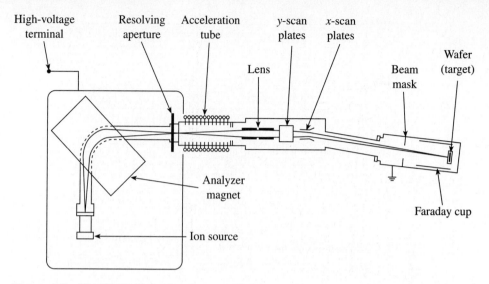

**Figure 4.7**  Simplified schematic of an ion implantation system. (From Runyan and Bean[9], © 1990 by Addison-Wesley Publishing Co., Inc. Reprinted by permission of the publisher.)

semiconductor by creating ions of the impurity, accelerating the ions to high energies ranging from 5 keV to 1 MeV, and then literally shooting the ions into the semiconductor. As one might suspect, the implanted ions displace semiconductor atoms along their path into the crystal. Moreover, the ions themselves do not necessarily come to rest on lattice sites. A follow-up anneal (heating of the semiconductor), an integral part of the overall process, is therefore necessary to remove the crystal damage and "activate" implanted dopants.

A simplified schematic of an ion implantation system and photographs of a commercial implantation end-station are presented in Figs. 4.7 and 4.8, respectively. Ions of the desired impurity are produced in the ion source shown at the extreme left in Fig. 4.7. The ions are next accelerated into the mass analyzer where unwanted ions, also produced in the ion source, are removed. The resulting ion beam is then accelerated to the preset operating potential, focused, and finally scanned over the surface of the wafer. Scanning is accomplished electrostatically, by mechanically moving the wafer, or by a combination of the two methods. An electrical contact to the wafer allows a flow of electrons to neutralize the implanted ions. A very precise determination of the total number of implanted ions/ cm$^2$, called the *dose* and given the symbol $\phi$, is obtained by integrating the target current over the time of the implant.

The concentration profile produced by ion implantation has the general form of a Gaussian distribution function and is mathematically described by

$$N(x) = \frac{\phi}{\sqrt{2\pi}(\Delta R_{\mathrm{p}})}e^{-(1/2)[(x-R_{\mathrm{p}})/\Delta R_{\mathrm{p}}]^2} \tag{4.3}$$

**Figure 4.8** Photographs of an ion implantation end-station at the Intel Fab-9 fabrication facility. The top photograph shows the two ports associated with the station. The left port is open and awaiting the loading of wafers onto the wafer carousel, and ion implantation is taking place in the closed right port. The bottom photograph is a close-up of the wafer carousel and the robot arms that automatically load and unload the carousel. (Photographs courtesy of Intel Corporation.)

For those familiar with statistics, $R_p$ is the mean and $\Delta R_p$ is the standard deviation of the distribution function. In the ion implantation literature, $R_p$ and $\Delta R_p$ are called the *projected range* and *straggle,* respectively. These parameters vary with the implant ion and the substrate material and are roughly proportional to the ion energy. Sample computed distributions for phosphorus implantation into Si at various energies are shown in Fig. 4.9.

Like diffusion, portions of a Si wafer can be masked during ion implantation. Thin films of $SiO_2$, $Si_3N_4$, and, less frequently, photoresist and Al have been used for this pur-

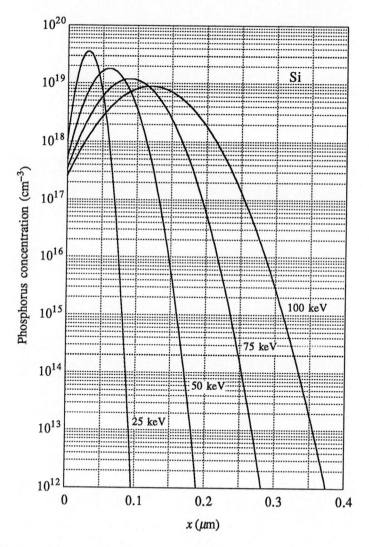

**Figure 4.9**  Computed phosphorus implantation profiles assuming a constant dose of $10^{14}/cm^2$.

pose. Basically the implanted ions are stopped in the masking material before reaching the underlying silicon.

Ion implantation offers a number of advantages over diffusion. It is intrinsically a low-temperature process, with the implantation itself usually being performed at room temperature. The follow-up anneal is often some other high-temperature step in the IC fabrication, such as a subsequent oxidation. If a separate anneal is required, it can be performed at temperatures as low as 600°C under optimum conditions. In any event, undesirable spreading of a concentration profile is minimized using ion implantation. Ion implantation also affords very precise control of the impurity concentration, and virtually any atom can be implanted into any given substrate. An interesting recent development is the implantation of high doses of oxygen into silicon, producing a $SiO_2$ layer below the surface of the wafer and thereby leading to silicon on insulator (SOI) structures. Finally, note from Fig. 4.9 that implantation typically yields very shallow concentration profiles. Implants have thus come to replace the predeposition step in diffusions and are ideally suited for a number of modern device structures requiring extemely shallow junctions.

## 4.1.4 Lithography

In discussing diffusion and ion implantation, we talked about $SiO_2$ or some other masking material covering portions of a wafer. Selective removal of a thin film at prescribed regions across the surface of a wafer, the patterning of insulators and metals required in forming ICs, is achieved through a process called lithography. Photographs of a production-line lithography station are reproduced in Fig. 4.10.

Major steps in the lithography process are depicted in Fig. 4.11 using the patterning of an $SiO_2$ film as an example. The top surface of the $SiO_2$-covered wafer is first coated with an ultraviolet-light sensitive material called *photoresist,* which is supplied as a liquid. A few drops of the liquid are placed on the wafer, and the wafer is spun at high speeds to produce a thin uniform coating. After spinning, a short "prebake" at relatively low temperatures (80°C–100°C) is performed to drive solvent out of the resist and to improve surface adhesion. The hardened photoresist is similar to a photographic emulsion.

The next step is to expose the resist through a "mask" using UV-light as illustrated in Fig. 4.11(b). The mask here is a carefully prepared glass or quartz photoplate containing a copy of the pattern to be transferred to the $SiO_2$ film. Darkened regions in the emulsion on the mask block the UV-light. Regions of the photoresist exposed to the light undergo a chemical reaction which varies with the type of resist being employed. In negative resists the areas where the light strikes become polymerized and more difficult to remove. When placed in a developer the polymerized regions remain, while the unexposed regions dissolve and wash away. The net result after development is pictured on the right-hand side of Fig. 4.11(c). Positive resists contain large amounts of a sensitizer that dramatically slows down the dissolution rate of the resist in an alkaline developer. This sensitizer breaks down when exposed to light leading to the preferential removal of the exposed regions as shown on the left-hand side of Fig. 4.11(c). As an informational aside, negative resist was widely used in early IC processing. Positive resist is now the main type in use because it affords better small-geometry control.

**Figure 4.10** Production-line lithography station consisting of an SSI 150 coat and develop system (center-left in upper photograph) and a Nikon stepper (center-right in the upper photograph). The top of the SSI 150 system is shown more clearly in the lower photograph. The SSI 150 automatically performs the coat, prebake, develop, and harden operations described in the text. The intervening pattern exposure is performed in the Nikon stepper. Without human intervention, the wafer moves along the input track of the SSI 150, enters the Nikon stepper, exits the Nikon stepper, and moves back along one of the two output tracks of the SSI 150. (Photographs courtesy of Intel Corp.)

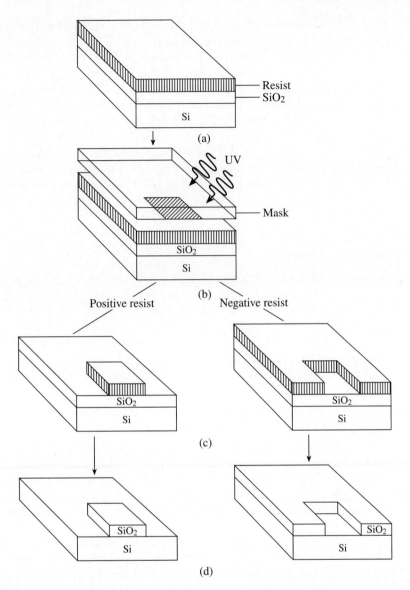

**Figure 4.11**  Major steps in the lithography process: (a) Application of resist; (b) resist exposure through a mask; (c) after development; (d) after oxide etching and resist removal. (From Jaeger[8], © 1988 by Addison-Wesley Publishing Co., Inc. Reprinted by permission of the publisher.)

The final set of steps involves transferring the pattern to the $SiO_2$ film. Immediately following exposure and development the photoresist is given a hard bake, heating for 20 to 30 minutes at 120°C–180°C, to further increase adhesion to the wafer and to improve resistance to the subsequent etch. An acid etch, buffered HF in the case of $SiO_2$, is then used to dissolve unprotected regions of the underlying film. Lastly, the photoresist, having completed its job, is stripped away. This is accomplished by using a chemical solution that swells and lifts the resist or by oxidizing (burning) the resist in an oxygen plasma system called an asher.

The limits of the UV-based lithography process just described are being pushed by the ever shrinking dimensions of modern devices. UV sources with shorter wavelengths and special compensating techniques have expanded the small dimension limit. Nevertheless, lithographic systems employing x-rays are being developed and are likely to come on-line in the near future.

### 4.1.5 Thin-Film Deposition

To connect device structures to the "outside world" requires the deposition and patterning of a metal layer. In fact, complex ICs have three and sometimes four electrically isolated metallization layers. Electrical isolation of the metal layers in turn requires the deposition of intervening dielectric layers. Thin films are also deposited to prevent the interdiffusion of materials and to protect the device or circuit from contamination. In-use methods for depositing the required films are reviewed next.

#### Evaporation

Evaporation is one of the older and more straightforward methods of thin-film deposition. As envisioned in Fig. 4.12, the material to be evaporated is placed in or on a resistance-heated source holder inside a vacuum chamber. To evaporate Al, for example, a short piece of Al wire would be placed on a tungsten filament or boat. The substrate on which the film is to be deposited is also positioned inside the chamber facing the source. The chamber is then evacuated, power supplied to the holder, and the source vaporized. Because of the reduced pressure, the source material travels unimpeded to the substrate and deposits as a thin film.

Generally speaking, hot-filament evaporation is subject to moderately high levels of contamination. Electron-beam evaporation, a variation of the process where the source is heated by an electron beam, eliminates contamination but generates device-degrading x-rays. Consequently, evaporation is seldom used in the production-line fabrication of modern ICs, although it is still extensively used in making simple devices where the cited problems are of minimal concern.

#### Sputtering

Sputtering, like evaporation, is performed in a vacuum chamber. The source material and the substrate (wafer) are placed on opposing parallel plates connected to a high-voltage power supply as pictured in Fig. 4.13. During a deposition the chamber is first evacuated

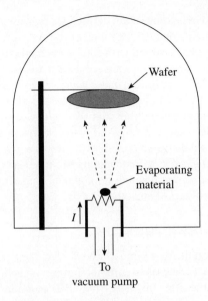

**Figure 4.12**   Hot-filament evaporation.

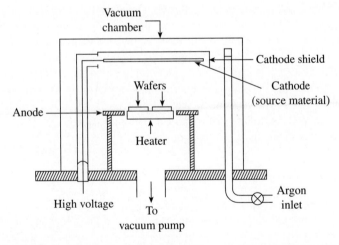

**Figure 4.13**   Schematic of a d.c. sputtering system. The source material covers the cathode while the wafer is mounted on the system anode. (From Jaeger[8], © 1988 by Addison-Wesley Publishing Co., Inc. Reprinted by permission of the publisher.)

of air and then a low-pressure amount of sputtering gas, typically Ar, is admitted into the chamber. Applying an interelectrode voltage ionizes the Ar gas and creates a plasma between the plates. Since the plate covered with source material is maintained at a negative potential relative to the substrate, $Ar^+$ ions are accelerated toward and into the source covered plate. The impacting $Ar^+$ ions in turn cause source atoms or molecules to be ejected from this plate. Being neutral, the ejected atoms or molecules readily travel to the substrate where they deposit to form the desired thin film. A d.c. power supply can be used when depositing metals, and an RF supply is necessary when depositing insulating films. When sputtering compounds it is sometimes necessary to introduce a gas of one of the components to assure the formation of a near-stoichiometric film. Providing low-temperature, low-contamination films with an acceptable throughput, sputtering has become the chief commercial method of depositing Al and other metals.

## Chemical Vapor Deposition (CVD)

In chemical vapor deposition the thin film is formed from one or more gas phase components. Either a compound decomposes to form the film or a reaction between gas components takes place to form the film. Invariably the CVD reactions are surface catalyzed, preferentially taking place on the surface of wafers inserted into the gas stream. In-use CVD processes fall into one of three general categories. They are atmospheric pressure (APCVD or simply CVD), low-pressure (LPCVD), and plasma-enhanced (PECVD) processes. Atmospheric pressure depositions can be performed in relatively simple systems. Low-pressure often offers comparable kinetics with improved uniformity and less gas consumption. In plasma CVD the electrons in the plasma impart energy to the reaction gases, thereby enhancing the reactions and permitting very low deposition temperatures.

   CVD reactors come in a variety of shapes and configurations; an example configuration employed in AP/LPCVD depositions is pictured in Fig. 4.14. CVD processing is routinely used to produce the masking and intermetallic dielectric films required in the formation of complex ICs. Polycrystalline Si, which functions as a pseudo-metal when heavily doped, is also deposited employing atmospheric and low-pressure CVD. APCVD, LPCVD, and PECVD are all typically used at some point in an IC process flow.

## 4.1.6 Epitaxy

Epitaxy is a special type of thin-layer deposition. Whereas depositions described in the preceding subsection yield either amorphous or polycrystalline films, epitaxy produces a crystalline layer that is an extension of the underlying semiconductor lattice. The word *epitaxy* is in fact derived from two Greek words meaning "upon-arranged." Commonly formed from the vapor-phase decomposition of silicon tetrachloride ($SiCl_4$) or a silane compound ($SiH_4$, $SiH_2Cl_2$, $SiHCl_3$) in a reactor very similar to those employed in CVD, additional Si literally grows following the lattice pattern of the pre-existing crystal. The doping of the epi-layer, intentionally distinct from that of the substrate, is controlled by introducing a dopant containing gas such as phosphine ($PH_3$), diborane ($B_2H_6$), or arsine

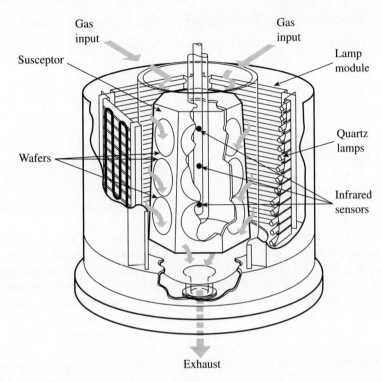

**Figure 4.14**  Cutaway view of a radiantly heated barrel-type reactor employed in both CVD processes and Si epitaxy. Wafers positioned vertically on the central susceptor are heated by the light from the surrounding quartz lamps. Process gases enter at the top, flow across the wafers, and exit coaxially at the bottom. (From Deacon[10], © 1984 by Lake Publishing. All rights reserved. Reprinted by permission of the publisher.)

($AsH_3$) during the deposition. Standard whole-wafer epitaxy must of course be completed early in the process flow, prior to the growth or deposition of any permanent surface films. We should also note that starting Si wafers are sometimes purchased from the wafer supplier with the desired epi-layer already in place.

## 4.2  DEVICE FABRICATION EXAMPLES

The purpose of this section is to illustrate how the individual fabrication processes are combined and sequenced to produce solid state devices. Two examples are presented: The first describes the fabrication of a simple *pn* junction diode. The second outlines the making of a complex IC. Both presentations are intentionally qualitative in nature, including only major processing steps and a minimum of processing details.

### 4.2.1 *pn* Junction Diode Fabrication

Figure 4.15 graphically summarizes the major processing steps in the formation of a *pn* junction diode. The starting point is a flat, damage-free, single-crystal Si wafer. It is assumed that a preclean has removed all particulates, organic films, and adsorbed metal ions from the semiconductor surfaces. For this particular illustration we further assume the wafer is *p*-type, having been uniformly doped with boron during the formation of the crystal.

The initial steps in the process flow are in preparation for a subsequent phosphorus diffusion. First a thermal oxide is grown that will serve as a diffusion barrier. The oxide thickness must be comfortably greater than the projected masking thickness. Step 2 is a lithography process performed to open "diffusion" holes in the oxide that will eventually become the positions of the *pn* junction diodes. Specifically note that the Step 2 illustration in Fig. 4.15 assumes the use of a positive photoresist.

After a proper clean-up the wafer is next inserted into a phosphorus predeposition furnace and subsequently into a phosphorus drive furnace. The net result, as pictured in Step 3 in Fig. 4.15, is the formation of $n^+$-$p$ junctions in surface regions not protected by the oxide. (The $+$ in $n^+$ indicates a very high doping.) By way of clarification, some oxygen is required during the phosphorus predeposition as noted in the diffusion discussion. Also, the drive may be performed in an oxidizing atmosphere to minimize out-diffusion. Thus, to reopen the oxide holes, a subsidiary lithography step, not shown in the simplified process flow of Fig. 4.15, is usually necessary after diffusion.

The final steps facilitate connecting the device to the "outside world." Sputtering or possibly evaporation of Al yields a thin metal film over the entire surface of the wafer as pictured in Step 4 of Fig. 4.15. A lithography process, Step 5 in Fig. 4.15, is then performed to remove excess metal external to the area of the diffused junction. Normally a low-temperature ($\leq 500°C$) anneal would also be performed to promote a low-resistance contact between the metal and Si.

With the completion of the metallization contact, the diodes become functional across the wafer. To produce commercial diodes, a diamond-edged saw would be used to cut the wafer into pieces containing a single device. (A wafer piece containing a single device or IC is called a *die.*) A large area metal contact is then made to the back of the die, a lead attached to the top surface contact, and the device encapsulated in protective plastic or hermetically sealed in a metallic package.

### 4.2.2 Computer CPU Process Flow

In this subsection we examine a fabrication process flow that has been used by Intel Corporation to produce computer CPUs and other ICs. The process flow description and associated figures are a direct excerpt, reprinted with permission, from *Components Quality and Reliability 1991/1992,* Intel Corporation, © 1990. No changes have been made to the excerpted material except for a renumbering of the figures and a modification of the figure captions. Figure 4.16, reproduced from the cited reference, displays a simplified block dia-

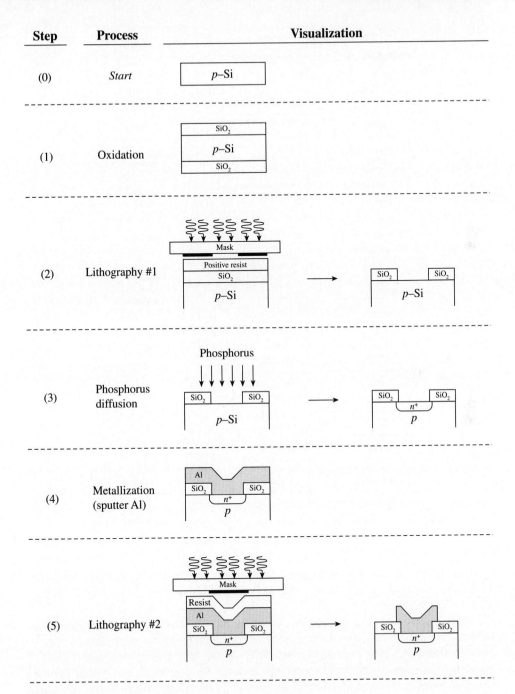

**Figure 4.15** Graphical summary of the major processing steps in the formation of a *pn* junction diode.

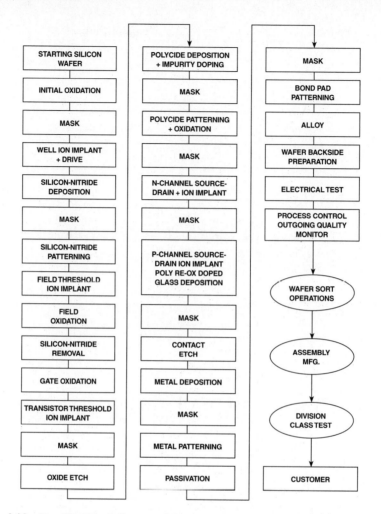

**Figure 4.16** Simplified block diagram of the single poly, single metal, CHMOS process flow.

gram of the overall process flow. Figure 4.17, which is correlated with the text description of the component processes, shows cross-sectional sketches of the IC structure at various stages of construction.

Although most of the terms employed in the process description will be familiar to the reader, a few need to be clarified. For one, the fabrication sequence is referred to as a single poly, single metal, CHMOS process flow. "Poly" is short for polycrystalline Si. CHMOS stands for Complementary High-density Metal-Oxide-Semiconductor—a type of fabrication technology used to produce MOS transistors. "Single poly, single metal" indicates that one level of heavily doped polycrystalline Si and one level of metal are used to contact and interconnect the MOS devices. Text references are also made to "wet" and "dry" etches. Wet etching is the familiar dissolution of a material, such as $SiO_2$, in a liquid chemical bath. Dry etching is the removal of a material using a plasma-enhanced gas-phase reaction.

Finally, the following process flow description is intended to be "looked over," not digested. At this point in the development, the reader is not equipped and not expected to understand the intricacies of either the processing or the IC structure.

1. **Starting Silicon Material and Well Definition** [Fig. 4.17(a)] To start CHMOS device production 150 mm (6″), high-resistivity, ⟨100⟩ orientation, single crystal, p-type (Boron doped) silicon wafers are used. P-type silicon is required to create n-channel transistors. To create p-channel transistors, necessary for CHMOS devices n-type (arsenic or phosphorous doped) silicon regions (n-wells) are implanted. The wafer is masked, then implanted to create p-type and n-type silicon regions on the same wafer. The n-well provides the background doping for the p-channel transistors while the p-type, start material (protected from the implant by the unexposed photoresist) serves as the background doping for n-channel transistors. A high temperature drive cycle completes the formation of the well by thermal dopant transition.

2. **Field Threshold Implant and Field Oxidation** [Fig. 4.17(b)]. Nitride ($Si_3N_4$) is deposited, masked, then etched. The etched nitride regions define the location of the field threshold ion plant [Fig. 4.17(b)] and the locations where oxide $SiO_2$ is permitted to grow during field oxidation [Fig. 4.17(c)]. Areas where nitride remains mask (prevent) oxide growth. These regions are where transistors will be built. The thick (approximately 6000 Å) field oxide isolates adjacent transistors in order to prevent electrical interactions. After field oxidation, the nitride mask is removed.

3. **Gate Oxidation** [Fig. 4.17(d)]. A thin thermal oxide is grown across the wafer. The portions of this oxide remaining after subsequent processing will provide the required gate oxide for the MOS transistors. Device performance is closely related to the growth of a dense, high-quality gate oxide.

4. **Transistor Threshold Ion Implant** [Fig. 4.17(d)]. The boron implant adjusts the threshold voltage (Vt) of p-channel and n-channel devices to the desired level. Thick field oxide prevents the boron from penetrating into the isolation regions.

5. **Polysilicon-to-Diffusion Contacts** [Fig. 4.17(e)]. Openings are defined in the thin gate oxide region where contact between poly conductors and diffused regions in the silicon substrate (buried contacts) is required.

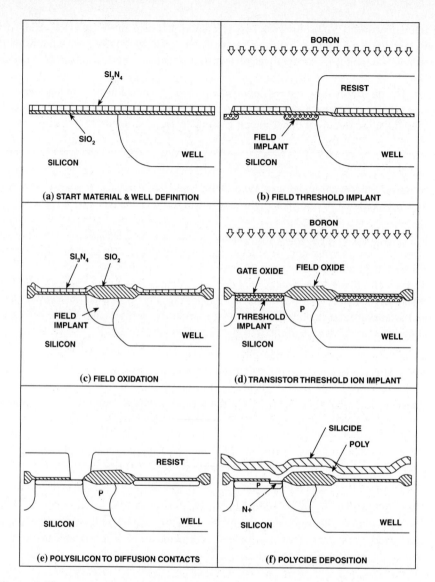

**Figure 4.17** Cross-sections through the CHMOS structure at select points in the process flow.

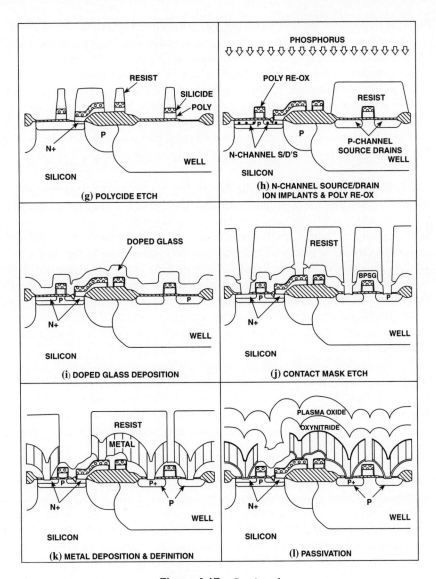

**Figure 4.17** *Continued.*

6. **Polycide Deposition, Impurity Doping, Patterning and Oxidation** [Fig. 4.17(f)]. A polycrystalline silicon layer, about 1500 Å thick, is deposited over the entire wafer by CVD techniques. The polysilicon is subsequently doped with phosphorous via gas-source diffusion techniques for conductivity. The phosphorous is diffused through the poly and into the substrate, minimizing the resistance of the buried contact. About 1500 Å of silicide is conformally deposited on the doped polysilicon. Silicide is used in small geometry transistors to increase device performance speeds. The polycide stack is defined by plasma dry etch techniques with resist acting as the mask [Fig. 4.17(g)]. The polycide stack is oxidized to protect the transistor gates.

7. **Source and Drain Implant and Poly Re-Ox** [Fig. 4.17(h)]. The remaining polycide stack, thin oxide over the poly, field oxide and resist provide a mask for p-channel and n-channel source-drain implants which occurs at this point. (This technique is commonly known as a "self-aligned source/drain" process, since the source and drain are aligned directly to the gate, which defines their location relative to the channel area.) All remaining poly oxide is then removed and the poly re-ox step performed. The re-ox step is a high-temperature thermal oxidation step which drives the source-drain implant into the silicon, provides a high-integrity dielectric on the polysilicon (essential in a double poly process) and grows an oxide on the exposed source-drain regions to prevent dopant out-diffusion during subsequent phosphosilicate glass deposition and processing.

8. **Doped Glass Deposition** [Fig. 4.17(i)]. A doped (boron, phosphorous or both) silicate glass layer, 5000 Å to 10,000 Å thick, is deposited via low-temperature CVD techniques to provide electrical isolation between the subsequent metal conductor lines and the underlying polycide gates and active device structures. The glass is then prepared for subsequent masking steps.

9. **Contact Mask and Etch** [Fig. 4.17(j)]. Windows in the resist are exposed and opened to define the location of metal-to-polycide and metal-to-silicon contact holes. Depending on the process, contact holes are wet, dry, or wet and dry etched. In a wet/dry combination, the wet etch gives a shallow slope near the top of the contact. This improves metal step coverage. The remaining doped silicon glass is then removed by an isotropic dry etch which leaves the contact sidewalls vertical.

10. **Metal Deposition and Definition** [Fig. 4.17(k)]. Metal conductor layer is sputter-deposited into the wafer and defined. The metal electrically connects the transistors to the outside world. The voltage applied by these metal lines turns the transistors "on" and "off."

11. **Passivation** [Fig. 4.17(l)]. Plasma enhanced CVD techniques are used to deposit a dual film passivation layer over the entire wafer surface. The dual layer consist of an

oxynitride underneath a plasma oxide. The oxynitride provides long-term field reliability, by being designed to be a contamination and moisture barrier. The latter is particularly critical if the dice are to be assembled in plastic packages, plastic itself being a poor moisture barrier. The plasma oxide layer provides handling protection for the wafer and individual devices.

---

Most of the remaining production-line and testing steps cited in the Fig. 4.16 block diagram are in preparation for eventual IC die separation and packaging. The photograph of a sample CPU die, along with an identification of its functional units, are displayed in Fig. 4.18. The pictured Pentium™ Processor die, fabricated following an upgraded version of the described processing, is approximately 1.5 cm square and contains 3.1 million transistors.

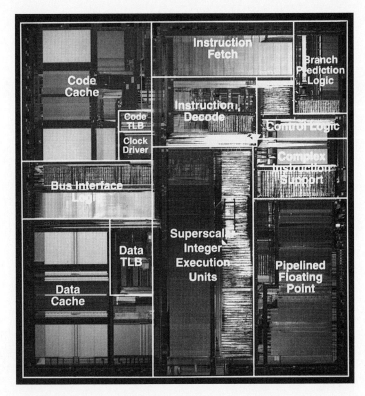

**Figure 4.18** The Intel Pentium™ Processor and its functional units. (Photograph and functional unit identification courtesy of Intel Corporation.)

## 4.3  SUMMARY

The overview of silicon device fabrication hopefully provided insight into the construction and physical nature of device structures. First, common component processes were examined, including oxidation, diffusion, ion implantation, lithography, evaporation, sputtering, chemical vapor deposition, and epitaxy. The fabrication of a simple *pn* junction diode and a moderately complex IC process flow were then considered to illustrate how the individual processes are combined and sequenced in producing actual structures.

# R1  Part I
## SUPPLEMENT AND REVIEW

## ALTERNATIVE / SUPPLEMENTAL READING LIST

### Recommended Readings

| Author(s) | Type (A = Alt., S = Supp.) | Level | Relevant Chapters |
|---|---|---|---|
| *For Semiconductor Physics* | | | |
| Ferendeci | A/S | Advanced Undergraduate/ Introductory Graduate | 1–6 |
| Neamen | A | Undergraduate | 1–6 |
| Pierret | S | Introductory Graduate | 1–6 |
| Streetman | A | Undergraduate | 1–4 |
| Tyagi | A/S | Advanced Undergraduate/ Introductory Graduate | 1–5 |
| *For Device Fabrication* | | | |
| Jaeger | S | Undergraduate | All |
| Runyan and Bean | S | Professional | All |
| Sze (editor) | S | Professional | All |

(1)  A. M. Ferendeci, *Physical Foundations of Solid State and Electron Devices,* McGraw-Hill, New York, © 1991.

(2)  R. C. Jaeger, *Introduction to Microelectronic Fabrication,* Vol. V in the Modular Series on Solid State Devices edited by G. W. Neudeck and R. F. Pierret, Addison-Wesley, Reading, MA, © 1988.

(3)  D. A. Neamen, *Semiconductor Physics and Devices, Basic Principles,* Irwin, Homewood, IL, © 1992.

(4)  R. F. Pierret, *Advanced Semiconductor Fundamentals,* Vol. VI in the Modular Series on Solid State Devices edited by G. W. Neudeck and R. F. Pierret, Addison-Wesley, Reading, MA, © 1987.

**R1**

(5) W. R. Runyan and K. E. Bean, *Semiconductor Integrated Circuit Processing Technology,* Addison-Wesley, Reading, MA, © 1990.

(6) B. G. Streetman, *Solid State Electronic Devices,* 4th edition, Prentice Hall, Englewood Cliffs, NJ, © 1995.

(7) S. M. Sze (editor), *VSLI Technology,* 2nd edition, McGraw Hill, New York, © 1988.

(8) M. S. Tyagi, *Introduction to Semiconductor Materials and Devices,* John Wiley & Sons, New York, © 1991.

## Other Recent Texts

(1) A. Bar-Lev, *Semiconductors and Electronic Devices,* 3rd edition, Prentice Hall, Inc., New York, © 1993.

(2) D. H. Navon, *Semiconductor Microdevices and Materials,* Holt, Rinehart and Winston, New York, © 1986.

(3) D. L. Pulfrey and N. G. Tarr, *Introduction to Microelectronic Devices,* Prentice Hall, Englewood Cliffs, NJ, © 1989.

(4) C. T. Sah, *Fundamentals of Solid-State Electronics,* World Scientific, Singapore, © 1991.

(5) J. Singh, *Semiconductor Devices, an Introduction,* McGraw-Hill, New York, © 1994.

(6) S. M. Sze, *Semiconductor Devices, Physics and Technology,* John Wiley & Sons, New York, © 1985.

(7) E. S. Yang, *Microelectronic Devices,* McGraw Hill, New York, © 1988.

(8) M. Zambuto, *Semiconductor Devices,* McGraw Hill, New York, © 1989.

(↓ Reader-added texts)

# FIGURE SOURCES / CITED REFERENCES

(1) W. Shockley, *Electrons and Holes in Semiconductors,* Litton Educational Publishing, Inc., © 1950.

(2) S. M. Sze, *Physics of Semiconductor Devices,* 2nd edition, John Wiley & Sons, New York, © 1981.

(3) W. Zuhlehner and D. Huber, "Czochralski Grown Silicon," *Crystals* **8,** Springer-Verlag, Berlin, © 1982.

(4) C. Jacoboni, C. Canali, G. Ottaviani, and A. A. Quaranta, "A Review of Some Charge Transport Properties of Silicon," Solid-State Electronics, **20,** 77 (1977).

(5) P. M. Smith, J. Frey, and P. Chatterjee, "High-Field Transport of Holes in Silicon," Applied Physics Letters, **39,** 332 (Aug. 1981).

(6) D. K. Schroder, *Semiconductor Material and Device Characterization,* John Wiley & Sons, New York, © 1990.

(7) R. F. Pierret, *Advanced Semiconductor Fundamentals,* Vol. VI in the Modular Series on Solid State Devices edited by G. W. Neudeck and R. F. Pierret, Addison-Wesley, Reading, MA, © 1987.

(8) R. C. Jaeger, *Introduction to Microelectronic Fabrication,* Vol. V in the Modular Series on Solid State Devices edited by G. W. Neudeck and R. F. Pierret, Addison-Wesley, Reading, MA, © 1988.

(9) W. R. Runyan and K. E. Bean, *Semiconductor Integrated Circuit Processing Technology,* Addison-Wesley, Reading, MA, © 1990.

(10) T. Deacon, "Silicon Epitaxy: An Overview," Microelectronic Manufacturing and Testing, September 1984.

($\downarrow$ Reader-added references)

R1

## REVIEW LIST OF TERMS

Defining the following terms using your own words provides a rapid review of the Part I material.

(1) amorphous
(2) polycrystalline
(3) crystalline
(4) lattice
(5) unit cell
(6) ingot
(7) carrier
(8) electron
(9) hole
(10) conduction band
(11) valence band
(12) band gap
(13) effective mass
(14) intrinsic semiconductor
(15) extrinsic semiconductor
(16) dopant
(17) donor
(18) acceptor
(19) $n$-type material
(20) $p$-type material
(21) $n^+$ (or $p^+$) material
(22) majority carrier
(23) minority carrier
(24) density of states
(25) Fermi function
(26) Fermi energy (or level)
(27) nondegenerate semiconductor
(28) degenerate semiconductor
(29) charge neutrality
(30) extrinsic temperature region
(31) intrinsic temperature region
(32) freeze-out
(33) drift
(34) scattering
(35) thermal motion
(36) drift velocity
(37) saturation velocity
(38) current density
(39) mobility
(40) resistivity

(41) conductivity
(42) band bending
(43) diffusion
(44) diffusion coefficient
(45) recombination
(46) generation
(47) band-to-band R–G
(48) R–G centers
(49) $E$–$\mathbf{k}$ plots
(50) direct semiconductor
(51) indirect semiconductor
(52) photogeneration
(53) absorption coefficient
(54) low-level injection
(55) equilibrium
(56) perturbation
(57) steady state
(58) minority carrier lifetime
(59) minority carrier diffusion length
(60) quasi-Fermi level
(61) dry oxidation
(62) wet oxidation
(63) predeposition
(64) drive-in
(65) diffusion mask
(66) ion implantation
(67) dose
(68) projected range
(69) photoresist
(70) negative resist
(71) positive resist
(72) sputtering
(73) CVD
(74) LPCVD
(75) PECVD
(76) epitaxy
(77) die
(78) poly
(79) wet etch
(80) dry etch

# PART I—REVIEW PROBLEM SETS AND ANSWERS

The following problem sets were designed assuming a knowledge—at times an integrated knowledge—of the subject matter in Chapters 1–3 of Part I. The sets could serve as a review or as a means of evaluating the reader's mastery of the subject. Problem Set A was adapted from a one-hour "open-book" examination; Problem Set B was adapted from a one-hour "closed-book" examination. An answer key is included at the end of the problem sets.

## Problem Set A

*Special Instructions:*

(1) You may employ an OPEN book in answering the Problem Set A questions.

(2) Unless instructed otherwise, employ parameters accurate to three-places in performing numerical computations.

## Problem A1

Answer the questions that follow making use of the unit cell for the Si crystal lattice reproduced here:

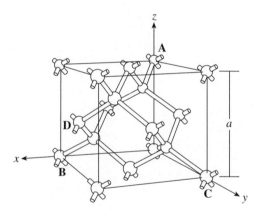

(a) If, as pictured, the origin of coordinates is located at the lower back corner of the unit cell, what are the Miller indices of the plane passing through the points ABC?

(b) What are the Miller indices of the direction vector running from the origin of coordinates to the point D?

(c) In terms of the lattice constant $a$, what is the distance between nearest neighbors in the Si lattice?

(d) Modifying the pictured Si unit cell, indicate how one would visualize a *donor*.

R1

## Problem A2

A Si wafer is uniformly doped with $N_D = 10^{17}/\text{cm}^3$ donors and maintained under equilibrium conditions at $T = 300$ K. $\tau_p = 10^{-6}$ sec. Complete the following table characterizing the wafer.

| Parameter | Value | Units | Explanation |
|---|---|---|---|
| $E_G$ | | | — |
| $n_i$ | $10^{10}$ | $\text{cm}^{-3}$ | — |
| $n_0$ | | | |
| $p_0$ | | | |
| $E_F - E_i$ | | | |
| $\rho$ (charge density) | | | |
| $\mu_n$ | | | |
| $\rho$ (resistivity) | | | |
| $L_P$ | | | |
| $D_P$ | | | |

## Problem A3

(a) On the energy band diagram provided below, indicate the usual positioning of the following energy levels:

 (i)   $E_i$ . . . instrinsic Fermi level
 (ii)  $E_D$ . . . donor energy level
 (iii) $E_A$ . . . acceptor energy level
 (iv)  $E_T$ . . . trap or R–G center energy level
 (v)   $E_F$ for a *degenerately* doped p-type material

Add comments as necessary to forestall any misinterpretation of your graphical answer.

(b) A Si device is maintained under equilibrium conditions at $T = 300$ K. Given the electric field inside the device as is pictured next, $N_A = 10^{16}/\text{cm}^3$ for $0 \le x \le x_a$, and $N_D = 10^{16}/\text{cm}^3$ for $x_b \le x \le x_c$, draw the energy band diagram for the device. Include $E_c$, $E_v$, $E_i$, and a carefully positioned $E_F$ on your diagram. Also indicate how you arrived at your answer.

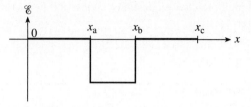

(c) What is the electrostatic potential drop across the device described in part (b); i.e., what is $V(x = x_c) - V(x = 0)$? Record your work.

## Problem A4

An $N_D = 10^{15}$/cm³ Si sample maintained at room temperature is subject to two flashes of light from a stroboscope. Each flash creates $\Delta p_f = 10^{10}$/cm³ holes *uniformly* throughout the Si sample. The first flash occurs at $t = 0$ and the second occurs at $t = t_f$, where $t_f$ is comparable to $\tau_p$. The flashes of light themselves are of infinitesimally short duration compared to $\tau_p$.

(a) Sketch (based on qualitative reasoning) the expected general variation of $\Delta p(t)$ as a function of time.

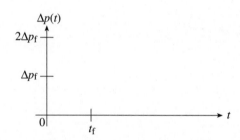

(b) Do low-level injection conditions prevail at all times inside the Si sample? Explain.

(c) Starting from the appropriate differential equation, solve for $\Delta p(t)$ for all $t \geq 0$.

(d) Assuming $\Delta n(t) = \Delta p(t)$ inside the bar, determine the fractional change in the Si sample conductivity immediately after the first flash of light. Give both a symbolic and numerical value for the desired $[\sigma(0^+) - \sigma(0^-)]/\sigma(0^-)$. $0^+$ and $0^-$ refer to times just before and just after the first flash, respectively.

## Problem Set B

*Special Instructions:*

(1) Problem Set B questions are to be answered with the book CLOSED.

**R1**

(2) In answering multiple-choice questions, choose the *best available answer.* Select only *one answer* per question.

(3) Employ $n_i = 10^{10}/\text{cm}^3$ and $kT = 0.0259$ eV in numerical computations involving Si samples maintained at 300 K.

## I. General Information

(1) As shown in the accompanying figure, a cubic unit cell has one atom centered halfway down each of the vertical edges of the cell *and* one atom positioned in the middle of the top and bottom faces of the cell. How many atoms are there within the cell?

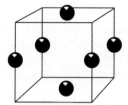

    (a) 4

    (b) 2

    (c) 1

    (d) 1/2

(2) What is the name of the *lattice* characterized by the unit cell in Question (1)?

    (a) Simple cubic

    (b) bcc

    (c) fcc

    (d) Diamond

(3) Determine the Miller indices of the plane pictured below.

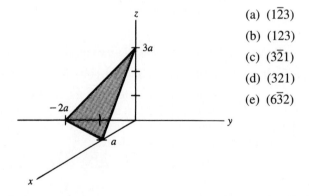

    (a) $(1\bar{2}3)$

    (b) $(123)$

    (c) $(3\bar{2}1)$

    (d) $(321)$

    (e) $(6\bar{3}2)$

R1

(4) Which of the following directions is *not* perpendicular to the [100] direction?

(a) $[0\bar{1}1]$

(b) $[032]$

(c) $[001]$

(d) $[\bar{1}00]$

(5) In a particular semiconductor the probability of electrons occupying states at an energy $kT$ above the bottom of the conduction band is $e^{-10}$. Determine the position of the Fermi level in the given material.

(a) $E_F = E_c$

(b) $E_c - E_F \simeq 9kT$

(c) $E_c - E_F = 10kT$

(d) $E_F = E_c + kT$

(6) Which of the following is the dominant carrier scattering mechanism in lowly doped (high-purity) silicon at room temperature?

(a) Carrier-carrier scattering

(b) Lattice scattering

(c) Impurity ion scattering

(d) Piezoelectric scattering

(7) Which of the following sketches best describes the $D_N$ versus $N_D$ dependence of electrons in silicon at room temperature?

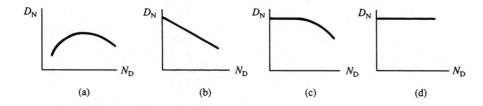

(8) Using the *bonding* model, indicate how one visualizes a donor.

(9) Using the energy band diagram, indicate how one would visualize the freeze-out of electrons on donor sites as $T \to 0$ K.

(10) The resistivity of *n*-type material is typically smaller than the resistivity of comparably doped *p*-type material. Explain why.

(11) Using the energy band diagram, indicate how one visualizes *recombination* via R–G centers.

(12) Define *drift velocity*.

## II.  Interpretation of Energy Band Diagrams

A silicon device maintained at 300 K is characterized by the following energy band diagram. Use the cited energy band diagram in answering Questions (13) to (20).

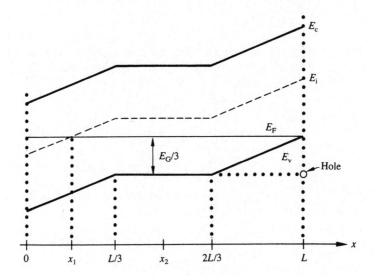

(13) The electrostatic potential ($V$) inside the semiconductor is as sketched below.

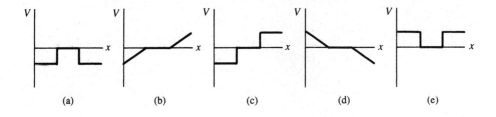

(14) The electric field ($\mathscr{E}$) is as sketched below.

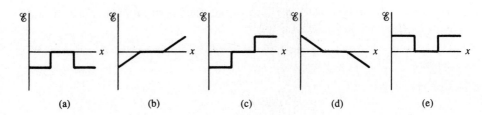

(15) Do equilibrium conditions prevail?

    (a) Yes

    (b) No

    (c) Can't be determined

(16) The semiconductor is degenerate

    (a) Near $x = 0$

    (b) For $L/3 \leq x \leq 2L/3$

    (c) Near $x = L$

    (d) Nowhere

(17) At $x = x_2$, $p = ?$

    (a) $7.63 \times 10^6/\text{cm}^3$

    (b) $1.35 \times 10^{13}/\text{cm}^3$

    (c) $10^{10}/\text{cm}^3$

    (d) $1.72 \times 10^{16}/\text{cm}^3$

(18) The electron current density $(J_N)$ flowing at $x = x_1$ is

    (a) Zero

    (b) $\mu_n n_i E_G / L$

    (c) $-\mu_n n_i E_G / L$

    (d) $D_N[n(x_2) - n(0)]/L$

(19) The hole drift current density $(J_{P|\text{drift}})$ flowing at $x = x_1$ is

    (a) Zero

    (b) $\mu_p n_i E_G / L$

    (c) $-\mu_p n_i E_G / L$

    (d) $q\mu_p N_D(kT/q)/L$

(20) The kinetic energy of the hole shown on the diagram is

    (a) $E_v$

    (b) $-E_G/3$

    (c) $E_G/3$

    (d) Zero

## III. Problem Solving

A uniformly doped $n$-type silicon bar of length $L$ is maintained at room temperature under steady state conditions such that $\Delta p_n(0) = \Delta p_{n0} > 0$ and $\Delta p_n(L) = 0$. $N_D = 10^{15}$ / cm$^3$, $\Delta p_{n0} \ll n_0$, and there are no "other processes" (including photogeneration) occurring in side the bar.

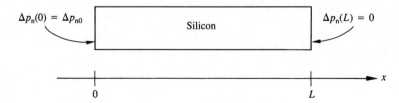

(21) Determine $p_0$, the equilibrium hole concentration.

    (a) $p_0 \approx 10^{10}$/cm$^3$

    (b) $p_0 \approx 10^{15}$/cm$^3$

    (c) $p_0 \approx 10^5$/cm$^3$

    (d) Not given

(22) Does low-level injection exist under steady state conditions?

    (a) Yes

    (b) No

    (c) Can't tell

(23) The equation (simplest form possible) one must solve to determine $\Delta p_n(x)$ is

    (a) $0 = D_P \dfrac{d^2 \Delta p_n}{dx^2}$

    (b) $\dfrac{d \Delta p_n}{dt} = \dfrac{\Delta p_n}{\tau_p}$

    (c) $\dfrac{\partial \Delta p_n}{\partial t} = D_P \dfrac{\partial^2 \Delta p_n}{\partial x^2} - \dfrac{\Delta p_n}{\tau_p}$

    (d) $\dfrac{\partial p_n}{\partial t} = D_P \dfrac{\partial^2 \Delta p_n}{\partial x^2}$

    (e) $0 = D_P \dfrac{d^2 \Delta p_n}{dx^2} - \dfrac{\Delta p_n}{\tau_p}$

(24) The general solution for $\Delta p_n(x)$ will be ($L_P^2 \equiv D_P \tau_p$)

    (a) $A \exp(-x/L_P) + B \exp(x/L_P)$

    (b) $A \exp(-t/\tau_p)$

    (c) $A + Bx$

    (d) $A \exp(-x/L_P)$

(25) The boundary conditions to be employed in solving for $\Delta p_n(x)$ are

    (a) $\Delta p_n(0) = \Delta p_{n0}$ and $\Delta p_n(\infty) = 0$

    (b) $\Delta p_n(0) = 10^{15}/cm^3$ and $\Delta p_n(\infty) = 0$

    (c) $\Delta p_n(0) = \Delta p_{n0}$ and $\Delta p_n(L) = 0$

    (d) $\Delta p_n(0) = 10^{15}/cm^3$ and $\Delta p_n(L) = 0$

## Answers—Set A

### Problem A1

(a) (111)

(b) [211]

(c) $(\sqrt{3}/4)a$

(d) One of the Si atoms in the unit cell is to be replaced with a non-Si atom exhibiting an extra bond (representing a weakly bound 5th valence electron).

### Problem A2

| Parameter | Value | Units | Explanation |
|---|---|---|---|
| $E_G$ | 1.12 | eV | — |
| $n_i$ | $10^{10}$ | $cm^{-3}$ | — |
| $n_0$ | $10^{17}$ | $cm^{-3}$ | $n_0 = N_D$ $(N_D \gg n_i)$ |
| $p_0$ | $10^3$ | $cm^{-3}$ | $p_0 = n_i^2/N_D$ |
| $E_F - E_i$ | 0.417 | eV | $E_F - E_i = kT\ln(N_D/n_i) = (0.0259)\ln(10^{17}/10^{10})$ |
| $\rho$ (charge density) | 0 | coul/$cm^3$ | Charge neutrality prevails under the specified conditions. |
| $\mu_n$ | 801 | $cm^2$/V-sec | From Fig. 3.5 |

| $\rho$ (resistivity) | 0.078 | ohm-cm | $\rho = \dfrac{1}{q\mu_n N_D} = \dfrac{1}{(1.6 \times 10^{-19})(801)(10^{17})}$ |
|---|---|---|---|
| $L_P$ | $2.93 \times 10^{-3}$ | cm | $L_P = \sqrt{D_P \tau_P} = \sqrt{(8.57)(10^{-6})}$ |
| $D_P$ | 8.57 | cm²/sec | $D_P = (kT/q)\mu_p = (0.0259)(331)$ |

## Problem A3

(a) $E_T$ is located near midgap, while $E_F$ must be positioned below $E_v + 3kT$.

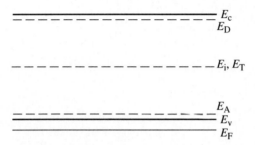

(b) Under equilibrium conditions, we know that $E_F$ is position-independent. Since $\mathscr{E} = (1/q)(dE_c/dx) = (1/q)(dE_i/dx) = (1/q)(dE_v/dx)$, from the $\mathscr{E}$ versus $x$ plot, we deduce that the bands are flat for $0 \leq x \leq x_a$ and $x_b \leq x \leq x_c$. In the $x_a \leq x \leq x_b$ region the bands must exhibit a constant negative slope. Also, for $0 \leq x \leq x_a$, $E_i - E_F = kT\ln(N_A/n_i) = (0.0259)\ln(10^{16}/10^{10}) = 0.358$ eV; for $x_b \leq x \leq x_c$, $E_F - E_i = kT\ln(N_D/n_i) = 0.358$ eV. Combining the cited information, one concludes the energy band diagram is as sketched below.

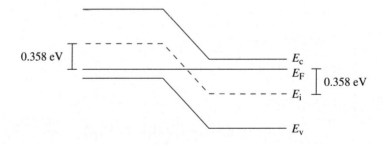

(c) $\Delta V = -\dfrac{1}{q}\Delta E_c = \dfrac{1}{q}[E_c(0) - E_c(x_c)] = \dfrac{1}{q}[E_i(0) - E_i(x_c)] = 0.716$ V

## Problem A4

(a)

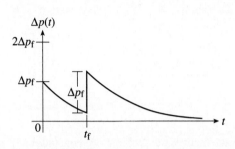

(b) Yes. $\Delta p_{max} \leq 2\Delta p_f = 2 \times 10^{10}/\text{cm}^3 \ll n_0 = N_D = 10^{15}/\text{cm}^3$. Thus $\Delta p \ll n_0$ as required for low-level injection.

(c) In the given problem, the minority carrier diffusion equation simplifies to

$$\frac{d\Delta p_n}{dt} + \frac{\Delta p_n}{\tau_p} = 0$$

For $0 \leq t \leq t_f$ . . .

$\Delta p(0) = \Delta p_f$    . . . boundary condition

$\Delta p(t) = Ae^{-t/\tau_p}$    . . . general solution

$A = \Delta p_f$

$\Delta p(t) = \Delta p_f e^{-t/\tau_p}$    $\Leftarrow$ answer

For $t \geq t_f$ . . .

$\Delta p(t_f) = \Delta p_f + \Delta p_f e^{-t_f/\tau_p} = \Delta p_f(1 + e^{-t_f/\tau_p})$    . . . boundary condition

$\Delta p(t) = Ae^{-(t-t_f)/\tau_p}$    . . . general solution

$A = \Delta p_f(1 + e^{-t_f/\tau_p})$

$\Delta p(t) = \Delta p_f(1 + e^{-t_f/\tau_p})e^{-(t-t_f)/\tau_p}$    $\Leftarrow$ answer

(d)    $\sigma = 1/\rho = q(\mu_n n + \mu_p p)$

$\sigma(0^-) \cong q\mu_n N_D$    . . . $n_0 \cong N_D$ and $n_0 \gg p_0$

$\sigma(0^+) = q[\mu_n(n_0 + \Delta n) + \mu_p(p_0 + \Delta p)] \cong q\mu_n N_D + q(\mu_n + \mu_p)\Delta p_f$

$\dfrac{\Delta\sigma}{\sigma(0^-)} = \dfrac{q(\mu_n + \mu_p)\Delta p_f}{q\mu_n N_D} = \left(1 + \dfrac{\mu_p}{\mu_n}\right)\dfrac{\Delta p_f}{N_D}$    $\Leftarrow$ answer symbolic

$= \left(1 + \dfrac{458}{1345}\right)\dfrac{10^{10}}{10^{15}} = 1.34 \times 10^{-5}$    $\Leftarrow$ answer numeric

## Answers—Set B

(1) b . . . 1/2 × (2 face atoms) + 1/4 × (4 edge atoms) = 2

(2) b

(3) e . . .  1, −2, 3          — intercepts
             1, −1/2, 1/3       — 1/intercepts
             ($6\bar{3}2$)       — indices

(4) d . . . Any direction with a nonzero first integer will *not* be perpendicular to the [100] direction.

(5) b . . . From the statement of the problem, we conclude

$$f(E_c + kT) = \frac{1}{1 + e^{(E_c + kT - E_F)/kT}} = e^{-10}$$

$$E_c - E_F = kT[\ln(e^{10} - 1) - 1] \simeq 9kT$$

(6) b

(7) c . . . For nondegenerate dopings $D_N = (kT/q)\mu_n$ and $\mu_n$ versus $N_D$ has the form of sketch (c). (See Fig. 3.5.)

(8)

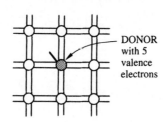

DONOR
with 5
valence
electrons

(9)

$E_c$
$E_D$

$E_v$

(10)

$$\rho = \frac{1}{q\mu_n N_D} \quad \ldots \text{ } n\text{-type}$$

$$\rho = \frac{1}{q\mu_p N_A} \quad \ldots \text{ } p\text{-type}$$

Typically, $\mu_p < \mu_n$ for a given semiconductor and temperature. Thus $\rho(p\text{-type})$ is greater than $\rho(n\text{-type})$ if $N_D = N_A$.

(11)

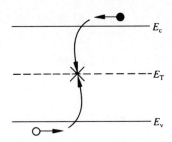

(12) *Drift velocity* is the average electron or hole velocity resulting from an applied electric field. (One might also mention that at low fields $v_d = \mu \mathscr{E}$, where $\mu$ is the carrier mobility.)

(13) d . . . $V$ has the same shape as the "upside-down" of the bands.

(14) e . . . $\mathscr{E}$ is proportional to the slope of the bands.

(15) a . . . $E_F$ is invariant with position.

(16) c . . . $E_F$ approaches $E_v$ near $x = L$.

(17) b . . .

$$E_i - E_F = E_G/2 - E_G/3 = E_G/6$$

$$p = n_i e^{(E_i - E_F)/kT} = 10^{10} e^{1.12/(6 \times 0.0259)} = 1.35 \times 10^{13}/cm^3$$

(18) a . . . Under equilibrium conditions $J_N = 0$.

(19) b . . .

$$J_{P|drift} = q\mu_p p \mathscr{E}$$
$$\text{At } x = x_1$$
$$p = n_i \qquad (E_i = E_F)$$
$$\mathscr{E} = \frac{1}{q} \frac{dE_i}{dx} = \frac{1}{q} \frac{\Delta E_i}{\Delta x} = \frac{1}{q} \frac{(E_G/3)}{(L/3)} = \frac{E_G}{qL}$$
$$J_{P|drift} = \mu_p n_i E_G / L$$

(20) c . . . K. E. = $E_v(L) - E_{hole} = E_G/3$

(21) c . . . $n_0 \simeq N_D = 10^{15}/cm^3$

$$p_0 \simeq n_i^2/n_0 \simeq 10^{20}/10^{15} = 10^5/cm^3$$

(22) a . . . $\Delta p_{n|max} = \Delta p_{n0} \ll n_0$

(23) e . . . Steady state conditions and the absence of photogeneration allow one to eliminate the $\partial/\partial t$ and $G_L$ terms in the minority carrier diffusion equation (Eq. 3.54b).

(24) a . . . This is Solution No. 1 cited in Subsection 3.4.3.

(25) c

# PART IIA

# *pn* JUNCTION DIODES

# 5  *pn* Junction Electrostatics

In Part II we primarily examine devices whose operation is intimately tied to the one or more *pn* junctions built into the structure. Considerable attention is initially given to the charter member of the device family, the *pn* junction diode. The diode analysis in Part IIA is of particular importance because it establishes basic concepts and analytical procedures that are of universal utility. Special *pn* junction diodes are described at the end of Part IIA in the "read-only" chapter on optoelectronic diodes. In Part IIB the concepts and procedures are applied and somewhat expanded in treating another important member of the family, the two-junction, three-terminal, bipolar junction transistor (BJT). Devices with more than two *pn* junctions are briefly considered in the chapter on PNPN devices. The concluding chapter in Part IIB treats metal–semiconductor contacts and the Schottky diode, a "cousin" of the *pn* junction diode.

A complete, systematic device analysis is typically divided into four major segments. Providing the foundation for the entire analysis, the first treats the charge density, the electric field, and the electrostatic potential—collectively referred to as the electrostatics—existing inside the device under equilibrium and steady state conditions. Subsequent segments are in turn devoted to modeling the steady state (d.c.) response, the small-signal (a.c.) response, and the transient (pulsed) response of the device. The *pn* junction diode analysis to be presented follows the noted four-step development. In this chapter we take the first step by exploring the electrostatics associated with a *pn* junction.

## 5.1 PRELIMINARIES

### 5.1.1 Junction Terminology/Idealized Profiles

Suppose for the sake of discussion that a *pn* junction has been formed by diffusing a *p*-type dopant into a uniformly doped *n*-type wafer. The assumed situation is pictured in Fig. 5.1(a). In the near-surface region where the in-diffused $N_A > N_D$, the semiconductor is obviously *p*-type. Deeper in the semiconductor where $N_D > N_A$, the semiconductor is *n*-type. Clearly, the dividing line between the two regions, known formally as the *metallurgical junction,* occurs at the plane in the semiconductor where $N_D - N_A = 0$.

Note that only the *net* doping is relevant in determining the position of the metallurgical junction. Only the net doping concentration is likewise required in establishing the electrostatic variables. Thus, instead of presenting superimposed plots of $N_A$ and $N_D$ versus $x$ as in Fig. 5.1(a), it is more useful to combine $N_A$ and $N_D$ information into a single

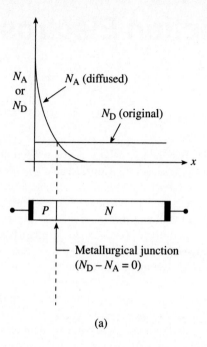

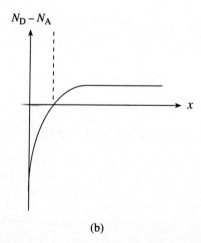

**Figure 5.1** Junction definitions: (a) Location of the metallurgical junction, (b) doping profile—a plot of the net doping versus position.

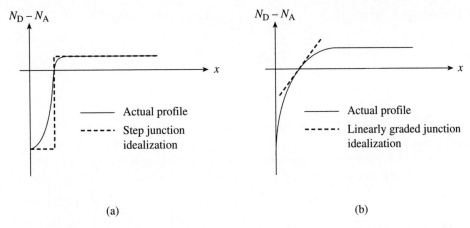

**Figure 5.2**   Idealized doping profiles: (a) Step junction, (b) linearly graded junction.

$N_D - N_A$ versus $x$ plot as illustrated in Fig. 5.1(b). A plot of the net doping concentration as a function of position is referred to as the *doping profile*.

When incorporated into the analysis, the actual doping profiles created by the commonplace diffusion and ion implantation processes drastically complicate the mathematics. It becomes difficult to obtain and interpret results. Fortunately, only the doping variation in the immediate vicinity of the metallurgical junction is of prime importance. Surprisingly accurate results can be obtained using rather idealized doping profiles. The two most common idealizations are the *step junction* and the *linearly graded junction* profiles graphically defined in Fig. 5.2. The more appropriate of the two idealizations depends on the slope of the actual doping profile at the metallurgical junction and the background doping of the starting wafer. The step junction is an acceptable approximation to an ion-implantation or shallow diffusion into a lightly doped starting wafer, whereas the linearly graded profile would be more appropriate for deep diffusions into a moderate to heavily doped starting wafer. In most *pn* junction analyses presented herein, we arbitrarily invoke the step junction idealization to minimize the mathematical complexity of the analysis.

### 5.1.2  Poisson's Equation

Poisson's equation is a well-known relationship from Electricity and Magnetism. In semiconductor work it often constitutes the starting point in obtaining quantitative solutions for the electrostatic variables. The three-dimensional version appropriate for semiconductor analyses is

$$\nabla \cdot \mathscr{E} = \frac{\rho}{K_S \varepsilon_0} \qquad (5.1)$$

In one-dimensional problems where $\mathscr{E} = \mathscr{E}_x$, Poisson's equation simplifies to

$$\frac{d\mathscr{E}}{dx} = \frac{\rho}{K_S \varepsilon_0} \qquad (5.2)$$

$K_S$ is the semiconductor dielectric constant and $\varepsilon_0$ is the permittivity of free space. $\rho$, previously associated with resistivity, is understood to be the charge density (charge/cm$^3$) in analyses involving the electrostatic variables. Assuming the dopants to be totally ionized, the charge density inside a semiconductor is given by

$$\rho = q(p - n + N_D - N_A) \qquad (5.3)$$

Eq. (5.3) was originally presented as the first portion of Eq. (2.23) in deriving the charge neutrality relationship. The charge density is identically zero far from any surfaces inside a uniformly doped semiconductor in equilibrium. However, $\rho$ is often nonzero and a function of position under less restrictive conditions.

Lastly note from Eq. (5.2) that $\rho$ is proportional to $d\mathscr{E}/dx$ in one-dimensional problems. The general functional form of $\rho$ versus $x$ can therefore be deduced from an $\mathscr{E}-x$ plot by simply noting the slope of the plot as a function of position.

### 5.1.3  Qualitative Solution

Prior to performing a quantitative analysis, it is always useful to have a general idea as to the expected form of the solution. Based on the Subsection 3.2.4 discussion leading to the derivation of the Einstein relationship, we already know there should be band bending and an internal electric field associated with the inherently nonuniform doping of a *pn* junction diode. Let us assume a one-dimensional step junction and equilibrium conditions in seeking to determine the general functional form of the potential, electric field, and charge density inside the diode.

Our approach will be to first construct the energy band diagram for a *pn* junction diode under equilibrium conditions and then to utilize previously established procedures in deducing the electrostatic variables. For the assumed step junction it is reasonable to expect regions far removed from the metallurgical junction to be identical in character to an isolated, uniformly doped semiconductor. Thus, the energy band diagrams for the regions far removed from the junction are concluded to be of the simple form shown in Fig. 5.3(a). Under equilibrium conditions we know in addition that the Fermi level is a constant independent of position. This leads to Fig. 5.3(b), where the diagrams of Fig. 5.3(a) are properly aligned to the position-independent Fermi level. The missing near-junction portion of the Fig. 5.3(b) diagram is completed in Fig. 5.3(c) by connecting the $E_c$, $E_i$, and $E_v$ endpoints on the two sides of the junction. Although the exact form of the band bending near the

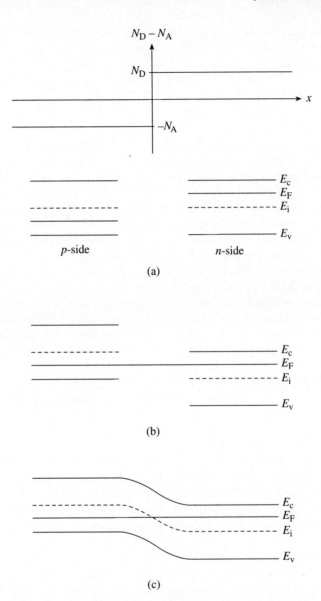

**Figure 5.3**  Step-by-step construction of the equilibrium energy band diagram for a *pn* junction diode. (a) Assumed step junction profile and energy band diagrams for the semiconductor regions far removed from the metallurgical junction. (b) Alignment of the part (a) diagrams to the position-independent Fermi level. (c) The completed energy band diagram.

metallurgical boundary is not known, it is reasonable to assume the variation is monotonic in nature, with a zero slope at the two ends of the central region. Figure 5.3(c) is of course the desired equilibrium energy band diagram for a *pn* junction diode.

It is now a relatively simple matter to deduce the functional form of the electrostatic variables. For one, referring to Subsection 3.1.5 on band bending, the $V$ versus $x$ relationship must have the same functional form as the "upside-down" of $E_c$ (or $E_i$ or $E_v$). This leads to the $V$ versus $x$ dependence sketched in Fig. 5.4(b) with $V$ arbitrarily set equal to zero on the far $p$-side of the junction. Next, the $\mathscr{E}$ versus $x$ dependence shown in Fig. 5.4(c) is obtained by recording the graphical derivative of $E_c$ as a function of position.[†] Finally, the general functional form of $\rho$ versus $x$ sketched in Fig. 5.4(d) is deduced from the slope of the $\mathscr{E} - x$ plot.

Perhaps the most interesting features of the Fig. 5.4 solution are the voltage drop across the junction under equilibrium conditions and the appearance of charge near the metallurgical boundary. The "built-in" voltage ($V_{bi}$) will be given separate consideration in the next subsection. Of immediate interest is the region of charge near the junction pictured in Fig. 5.4(d). The question arises, Where does this charge come from? Answering this question provides considerable physical insight.

Suppose that the $p$- and $n$-regions were initially separated as pictured in Fig. 5.5(a). Charge neutrality is assumed to prevail in the isolated, uniformly doped semiconductors. In the $p$-material the positive hole charges, the $\oplus$'s in Fig. 5.5, balance the immobile acceptor-site charges shown as $\boxminus$'s in Fig. 5.5. Likewise, in the $n$-material the electronic charge ($\ominus$) everywhere balances the immobile charge associated with the ionized donors ($\boxplus$). Next suppose a structurally perfect connection is made between the $p$ and $n$ materials as envisioned in Fig. 5.5(b). Naturally, since there are many more holes on the $p$-side than on the $n$-side, holes begin to diffuse from the $p$-side to the $n$-side an instant after the connection is made. Similarly, electrons diffuse from the $n$-side to the $p$-side of the junction. Although the electrons and holes can move to the opposite side of the junction, the donors and acceptors are fixed in space. Consequently, the diffusing away of the carriers from the near-vicinity of the junction leaves behind an unbalanced dopant site charge as shown in Fig. 5.5(c). This is the source of the charge around the metallurgical junction, nicely correlating with the previously deduced $\rho$ versus $x$ dependence redrawn in Fig. 5.5(d). The near-vicinity of the metallurgical junction where there is a significant nonzero charge is called the *space charge region* or *depletion region*. The latter name follows from the fact that the carrier concentrations in the region are greatly reduced or depleted. We should also mention that the build-up of charge and the associated electric field continues until the diffusion of carriers across the junction is precisely balanced by the carrier drift. The individual carrier diffusion and drift components must of course cancel to make $J_N$ and $J_P$ separately zero under equilibrium conditions.

---

[†] Actually, from the Fig. 5.4(a) energy band diagram, one can conclude only that the magnitude of $\mathscr{E}$ first increases from zero on the $p$-side of the junction, reaches a maximum near the metallurgical boundary, and then decreases again to zero far on the $n$-side of the junction. The pseudo-linear dependence sketched in Fig. 5.4(c) reflects an advance knowledge of the quantitative solution.

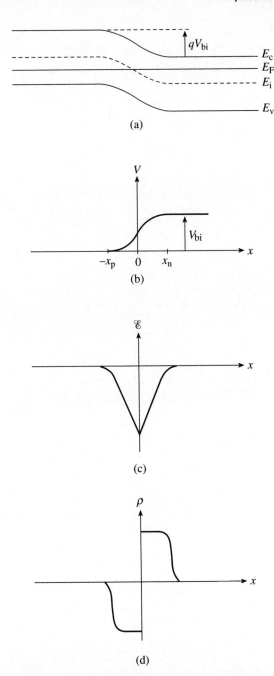

(a)

(b)

(c)

(d)

**Figure 5.4** General functional form of the electrostatic variables in a *pn* junction under equilibrium conditions. (a) Equilibrium energy band diagram. (b) Electrostatic potential, (c) electric field, and (d) charge density as a function of position.

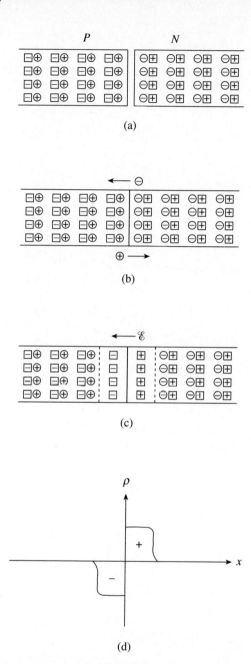

**Figure 5.5** Conceptual *pn* junction formation and associated charge redistribution. (a) Isolated *p* and *n* regions. (b) Electrons and holes diffuse to the opposite side of the junction moments after joining the *p* and *n* regions. (c) Charge redistribution completed and equilibrium conditions re-established. (d) Previously deduced charge density versus position. (⊕–holes, ⊟–ionized acceptors, ⊖–electrons, and ⊞–ionized donors.)

## 5.1.4 The Built-in Potential ($V_{bi}$)

The voltage drop across the depletion region under equilibrium conditions, known as the built-in potential ($V_{bi}$), is a junction parameter of sufficient importance to merit further consideration. We are particularly interested in establishing a computational relationship for $V_{bi}$. Working toward the stated goal, we consider a nondegenerately-doped *pn* junction maintained under equilibrium conditions with $x = 0$ positioned at the metallurgical boundary. The ends of the equilibrium depletion region are taken to occur at $-x_p$ and $x_n$ on the *p*- and *n*-sides of the junction respectively (see Fig. 5.4b).

Proceeding with the derivation, we know

$$\mathscr{E} = -\frac{dV}{dx} \tag{5.4}$$

Integrating across the depletion region gives

$$-\int_{-x_p}^{x_n} \mathscr{E} \, dx = \int_{V(-x_p)}^{V(x_n)} dV = V(x_n) - V(-x_p) = V_{bi} \tag{5.5}$$

Furthermore, under equilibrium conditions,

$$J_N = q\mu_n n\mathscr{E} + qD_N \frac{dn}{dx} = 0 \tag{5.6}$$

Solving for $\mathscr{E}$ in Eq. (5.6) and making use of the Einstein relationship, we obtain

$$\mathscr{E} = -\frac{D_N}{\mu_n} \frac{dn/dx}{n} = -\frac{kT}{q} \frac{dn/dx}{n} \tag{5.7}$$

Substituting Eq. (5.7) into Eq. (5.5), and completing the integration then yields

$$V_{bi} = -\int_{-x_p}^{x_n} \mathscr{E} \, dx = \frac{kT}{q} \int_{n(-x_p)}^{n(x_n)} \frac{dn}{n} = \frac{kT}{q} \ln\left[\frac{n(x_n)}{n(-x_p)}\right] \tag{5.8}$$

For the specific case of a nondegenerately doped step junction where $N_D$ and $N_A$ are the *n*- and *p*-side doping concentrations, one identifies

$$n(x_n) = N_D \tag{5.9a}$$

$$n(-x_p) = \frac{n_i^2}{N_A} \tag{5.9b}$$

and therefore

$$V_{bi} = \frac{kT}{q} \ln\left(\frac{N_A N_D}{n_i^2}\right)$$

(5.10)

It is useful to perform a sample computation to gauge the relative magnitude of the built-in voltage. Choosing $N_A = N_D = 10^{15}/cm^3$ and a Si diode maintained at 300 K, one computes $V_{bi} = (0.0259)\ln(10^{30}/10^{20}) \cong 0.6$ V. This is a typical result. In nondegenerately doped diodes $V_{bi} < E_G/q$, or $V_{bi}$ is less than the band gap energy converted to volts. Nondegenerately doped Ge, Si, and GaAs diodes maintained at room temperature exhibit a $V_{bi}$ less than 0.66 V, 1.12 V, and 1.42 V, respectively.

The relationship between $V_{bi}$ and $E_G$ alluded to in the preceding paragraph is nicely explained by an alternative $V_{bi}$ derivation based on the energy band diagram. Referring to Figs. 5.4(a) and (b), we note

$$V_{bi} = V(x_n) - V(-x_p)$$

(5.11a)

$$= \frac{1}{q}[E_c(-x_p) - E_c(x_n)] = \frac{1}{q}[E_i(-x_p) - E_i(x_n)]$$

(5.11b)

or

$$V_{bi} = \frac{1}{q}[(E_i - E_F)_{p\text{-side}} + (E_F - E_i)_{n\text{-side}}]$$

(5.12)

Clearly, for a nondegenerately doped diode both $(E_i - E_F)_{p\text{-side}}$ and $(E_F - E_i)_{n\text{-side}}$ are less than $E_G/2$, making $V_{bi} < E_G/q$. Moreover, in a nondegenerately doped step junction under equilibrium conditions,

$$(E_i - E_F)_{p\text{-side}} = kT \ln(N_A/n_i)$$

(5.13a)

$$(E_F - E_i)_{n\text{-side}} = kT \ln(N_D/n_i)$$

(5.13b)

Substituting Eqs. (5.13) into Eq. (5.12) and simplifying yields Eq. (5.10). Note that, while Eqs. (5.13) and hence Eq. (5.10) are valid only for nondegenerate dopings, there are no doping-related restrictions on the validity of Eq. (5.12).

## (C) Exercise 5.1

**P:** Most real diodes are very heavily doped on one side of the junction. In computing the built-in voltage of $p^+$-$n$ and $n^+$-$p$ step junctions, it is common practice to assume that the Fermi level on the heavily doped side is positioned at the band edge; i.e., $E_F = E_v$ in a $p^+$ material and $E_F = E_c$ in an $n^+$ material. Making the cited assumption, compute and plot $V_{bi}$ as a function of the doping ($N_A$ or $N_D$) on the lightly doped side of Si $p^+$-$n$ and $n^+$-$p$ step junctions maintained at 300 K. The plot is to cover the range $10^{14}/cm^3 \leq N_A$ or $N_D \leq 10^{17}/cm^3$.

**S:** Specifically considering a $p^+$-$n$ junction, we can write

$$(E_i - E_F)_{p\text{-side}} \text{ assumed} = E_i - E_v = E_G/2$$

$$(E_F - E_i)_{n\text{-side}} = kT \ln(N_D/n_i)$$

Substituting into Eq. (5.12), which is valid for arbitrary doping levels, we rapidly conclude

$$V_{bi} = \frac{E_G}{2q} + \frac{kT}{q} \ln\left(\frac{N_D}{n_i}\right)$$

For $n^+$-$p$ junctions, $N_A$ simply replaces $N_D$, yielding a computationally equivalent relationship. The $V_{bi}$ computational program and resultant plot (Fig. E5.1) follow:

MATLAB program script...

```
%Vbi Computation (p+/n and n+/p junctions)

%Constants
EG=1.12;
kT=0.0259;
ni=1.0e10;

%Computation
ND=logspace(14,17);
Vbi=EG/2+kT.*log(ND./ni);

%Plotting
close
semilogx(ND,Vbi); grid
axis([1.0e14 1.0e17 0.75 1])
xlabel('NA or ND (cm-3)'); ylabel('Vbi (volts)')
text(1e16,0.8,'Si, 300K')
text(1e16,0.78,'p+/n and n+/p diodes')
```

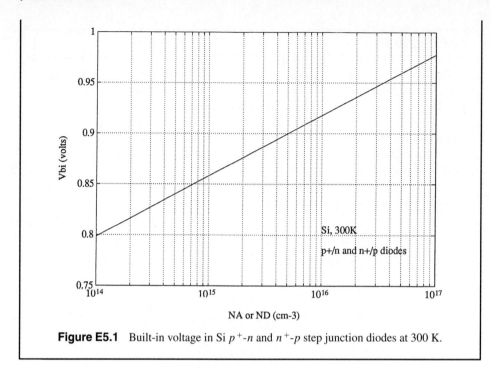

**Figure E5.1** Built-in voltage in Si $p^+$-$n$ and $n^+$-$p$ step junction diodes at 300 K.

## 5.1.5 The Depletion Approximation

To obtain a quantitative solution for the electrostatic variables, it will be necessary to solve Poisson's equation. The depletion approximation facilitates obtaining closed-form solutions to the equation. Appearing in a myriad of device analyses, the depletion approximation is far and away the most important and most widely encountered of the simplifying approximations used in the modeling of devices.

To understand why the depletion approximation is introduced, consider the one-dimensional Poisson equation rewritten as Eq. (5.14). The doping profile, $N_D - N_A$, appearing in the equation

$$\frac{d\mathscr{E}}{dx} = \frac{\rho}{K_S \varepsilon_0} = \frac{q}{K_S \varepsilon_0}(p - n + N_D - N_A) \tag{5.14}$$

is assumed to be known. However, to write down $\rho$ as a function of $x$, solve the differential equation for $\mathscr{E}$ versus $x$, and eventually obtain $V$ versus $x$, one must also have explicit expressions for the carrier concentrations as a function of $x$. Unfortunately, the carrier concentrations in $\mathscr{E} \neq 0$ regions, like the *pn* junction depletion region, are not specified prior to solving Poisson's equation. Rather, $p$ and $n$ in the depletion region are functions of the potential as is obvious from an examination of the Fig. 5.4(a) energy band diagram. Although exact closed-form (usually complex) solutions do exist in certain instances, the depletion approximation provides a simple, nearly universal way of obtaining approximate solutions without prior knowledge of the carrier concentrations.

We have already established the basis for the depletion approximation. A prominent feature of the qualitative solution was the appearance of a nonzero charge density straddling the metallurgical junction. This charge arises because the carrier numbers are reduced by diffusion across the junction. The carrier "depletion" tends to be greatest in the immediate vicinity of the metallurgical boundary and then tails off as one proceeds away from the junction. The depletion approximation introduces an idealization of the actual charge distribution. The approximation has two components that can be stated as follows: (1) The carrier concentrations are assumed to be negligible compared to the net doping concentration in a region $-x_p \leq x \leq x_n$ straddling the metallurgical junction. (2) The charge density outside the depletion region is taken to be identically zero. The depletion approximation is summarized pictorially in Fig. 5.6(a) and is illustrated assuming a step junction in

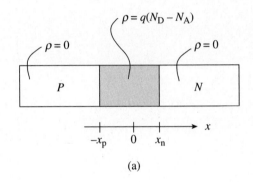

(a)

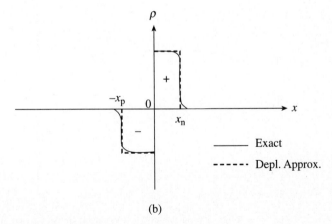

(b)

**Figure 5.6**   (a) Pictorial summary of the depletion approximation. (b) Illustration of the approximation as applied to a step junction.

Fig. 5.6(b). When the depletion approximation is invoked, the one-dimensional Poisson equation simplifies to

$$\frac{d\mathscr{E}}{dx} \cong \begin{cases} \dfrac{q}{K_S \varepsilon_0}(N_D - N_A) & \ldots \; -x_p \leq x \leq x_n \qquad (5.15a) \\[2mm] 0 & \ldots \; x \leq -x_p \text{ and } x \geq x_n \qquad (5.15b) \end{cases}$$

Note that, except for the values of $-x_p$ and $x_n$, the charge density is totally specified by invoking the depletion approximation. Moreover, the charge density will have precisely

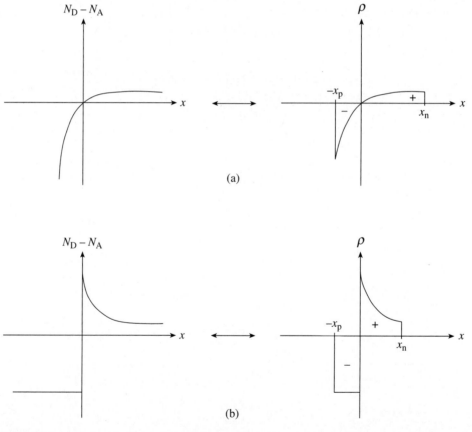

**Figure 5.7** Relationship between the doping profile and the charge density based on the depletion approximation. (a) and (b) are two separate examples.

the same functional form as $N_D - N_A$ within the depletion region. Figure 5.7 is presented to emphasize this last point by showing the $\rho$ versus $x$ plots associated with sample doping profiles of a moderately complex nature.

## 5.2 QUANTITATIVE ELECTROSTATIC RELATIONSHIPS

With the preliminaries out of the way, we proceed to the main event: development of quantitative relationships for the electrostatic variables. The development primarily deals with the step junction and is presented in detail not only to obtain the desired relationships but also to establish the derivational procedures that can be applied to other doping profiles. We begin by carefully specifying the system under analysis, turn to the step junction under equilibrium conditions, and then consider the modifications required when there is an applied bias. The section concludes with a concise derivation and examination of the electrostatic relationships appropriate for a linearly graded junction.

### 5.2.1 Assumptions/Definitions

Major features of the *pn* junction diode subject to analysis are identified in Fig. 5.8. All variables are taken to be functions only of $x$, the coordinate normal to the semiconductor surface. The device is thus said to be "one-dimensional"; obvious two-dimensional effects associated with the lateral ends of a real device are assumed to be negligible. For the electrostatic analysis, $x = 0$ is positioned at the metallurgical boundary. External contacts to the ends of the diode are specified to be "ohmic" in nature. By definition, a negligible portion of an externally applied voltage appears across an ohmic contact. Note that the

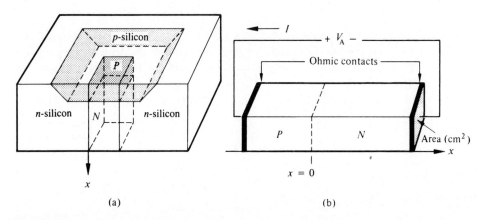

**Figure 5.8** (a) Sketch of a physical diode. (b) One-dimensional diode subject to analysis including the applied voltage, coordinate, and contact specifications.

symbol $V_A$ is used for the applied voltage. The subscript A distinguishes the applied voltage from the internal junction voltage. In the initial development $V_A$ is set equal to zero, or equivalently, the device is assumed to be in equilibrium.

## 5.2.2 Step Junction with $V_A = 0$

### Solution for $\rho$

We consider a step junction under equilibrium conditions. $N_A$ is drawn greater than $N_D$ in the Fig. 5.9(a) sketch of the doping profile for the sake of illustration. As summarized in Fig. 5.9(b), invoking the depletion approximation yields the charge density solution

$$\rho = \begin{cases} -qN_A & \ldots -x_p \leq x \leq 0 & (5.16a) \\ qN_D & \ldots 0 \leq x \leq x_n & (5.16b) \\ 0 & \ldots x \leq -x_p \text{ and } x \geq x_n & (5.16c) \end{cases}$$

The values of $x_n$ and $x_p$ are not known at this point but will be determined later in the analysis.

### Solution for $\mathscr{E}$

Substituting the charge density solution into Poisson's equation gives the equations to be solved for the electric field.

$$\frac{d\mathscr{E}}{dx} = \begin{cases} -qN_A/K_S\varepsilon_0 & \ldots -x_p \leq x \leq 0 & (5.17a) \\ qN_D/K_S\varepsilon_0 & \ldots 0 \leq x \leq x_n & (5.17b) \\ 0 & \ldots x \leq -x_p \text{ and } x \geq x_n & (5.17c) \end{cases}$$

$\mathscr{E} = 0$ far from the metallurgical boundary and therefore $\mathscr{E} = 0$ everywhere outside of the depletion region. Since $\mathscr{E}$ must also vanish right at the edges of the depletion region, $\mathscr{E} = 0$ at $x = -x_p$ and $\mathscr{E} = 0$ at $x = x_n$ respectively become the boundary conditions for the (5.17a) and (5.17b) differential equations. Separating variables and integrating from the depletion region edge to an arbitrary point $x$, one obtains for the $p$-side of the depletion region

$$\int_0^{\mathscr{E}(x)} d\mathscr{E}' = -\int_{-x_p}^x \frac{qN_A}{K_S\varepsilon_0} dx' \tag{5.18}$$

or

$$\mathscr{E}(x) = -\frac{qN_A}{K_S\varepsilon_0}(x_p + x) \quad \ldots -x_p \leq x \leq 0 \tag{5.19}$$

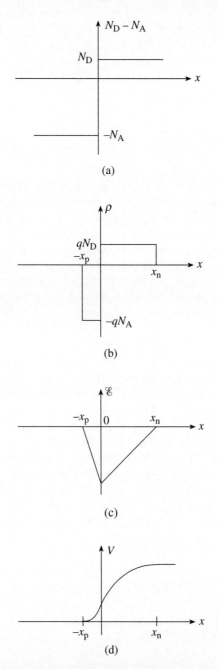

**Figure 5.9** Step junction solution. Depletion approximation based quantitative solution for the electrostatic variables in a *pn* step junction under equilibrium conditions ($V_A = 0$). (a) Step junction profile. (b) Charge density, (c) electric field, and (d) electrostatic potential as a function of position.

Similarly on the *n*-side

$$\int_{\mathscr{E}(x)}^{0} d\mathscr{E}' = \int_{x}^{x_n} \frac{qN_D}{K_S\varepsilon_0} dx' \tag{5.20}$$

or

$$\mathscr{E}(x) = -\frac{qN_D}{K_S\varepsilon_0}(x_n - x) \quad \dots 0 \le x \le x_n \tag{5.21}$$

The electric field solution is plotted in Fig. 5.9(c). This result is of course consistent with the qualitative solution sketched in Fig. 5.4(c). Within the depletion region the field is always negative and exhibits a linear variation with position. It should be noted that in constructing Fig. 5.9(c) the electric field was taken to be continuous at $x = 0$; the *p*- and *n*-side solutions were simply matched at the metallurgical boundary. From Electricity and Magnetism, we know the electric field will be continuous across a boundary as long as a sheet of charge does not lie along the interface between the two regions. If the (5.19) and (5.21) expressions for the electric field are evaluated at $x = 0$ and equated, the continuity of the electric field is found to require

$$N_A x_p = N_D x_n \tag{5.22}$$

For those familiar with Gauss' law, the fact that $\mathscr{E} = 0$ outside the depletion region means the total charge within the depletion region must sum to zero, or the minus charge on the *p*-side of the junction must balance the plus charge on the *n*-side of the junction. Previous charge density plots (Figs. 5.4–5.7) have all been drawn with this fact in mind. Applied to the step junction $\rho$ versus $x$ plot in Fig. 5.9(b), the balance of charge requires the rectangular areas on the *p*- and *n*-sides of the junction to be equal, or $qN_A x_p = qN_D x_n$. Thus Eq. (5.22) may be viewed alternatively as a reflection of the fact that the total charge within the depletion region must sum to zero.

## Solution for *V*

Since $\mathscr{E} = -dV/dx$, the electrostatic potential is obtained by solving

$$\frac{dV}{dx} = \begin{cases} \dfrac{qN_A}{K_S\varepsilon_0}(x_p + x) & \dots -x_p \le x \le 0 \tag{5.23a} \\[2mm] \dfrac{qN_D}{K_S\varepsilon_0}(x_n - x) & \dots 0 \le x \le x_n \tag{5.23b} \end{cases}$$

With the arbitrary reference potential set equal to zero at $x = -x_p$, and remembering the voltage drop is $V_{bi}$ across the depletion region under equilibrium conditions, Eqs. (5.23a) and (5.23b) are respectively subject to the boundary conditions

$$V = 0 \quad \text{at} \quad x = -x_p \tag{5.24a}$$

$$V = V_{bi} \quad \text{at} \quad x = x_n \tag{5.24b}$$

Separating variables and integrating from the depletion region edge to an arbitrary point $x$, one obtains for the $p$-side of the depletion region

$$\int_0^{V(x)} dV' = \int_{-x_p}^x \frac{qN_A}{K_S \varepsilon_0}(x_p + x')\,dx' \tag{5.25}$$

or

$$V(x) = \frac{qN_A}{2K_S \varepsilon_0}(x_p + x)^2 \quad \ldots -x_p \le x \le 0 \tag{5.26}$$

Similarly on the $n$-side of the junction

$$\int_{V(x)}^{V_{bi}} dV' = \int_x^{x_n} \frac{qN_D}{K_S \varepsilon_0}(x_n - x')\,dx' \tag{5.27}$$

or

$$V(x) = V_{bi} - \frac{qN_D}{2K_S \varepsilon_0}(x_n - x)^2 \quad \ldots 0 \le x \le x_n \tag{5.28}$$

The electrostatic potential solution given by Eqs. (5.26) and (5.28) is plotted in Fig. 5.9(d). The $V$ versus $x$ dependence is quadratic in nature, with a concave curvature on the $p$-side of the junction and a convex curvature on the $n$-side of the junction. Paralleling the $\mathscr{E}$-field procedure, Fig. 5.9(d) was constructed simply by matching the $p$- and $n$-side solutions at $x = 0$. The assumed continuity of the electrostatic potential at $x = 0$ is justified because there is no dipole layer (closely spaced sheets of plus and minus charge) along the metallurgical boundary. Note that if the (5.26) and (5.28) expressions for the potential are evaluated at $x = 0$ and equated, one obtains

$$\frac{qN_A}{2K_S \varepsilon_0}x_p^2 = V_{bi} - \frac{qN_D}{2K_S \varepsilon_0}x_n^2 \tag{5.29}$$

## Solution for $x_n$ and $x_p$

The electrostatic solution is not complete until the values of $x_n$ and $x_p$ are determined. In the course of the development, we have already laid the groundwork for obtaining the $n$- and $p$-side depletion widths. Specifically, $x_n$ and $x_p$ are the only unknowns in Eqs. (5.22) and (5.29). Eliminating $x_p$ in Eq. (5.29) using Eq. (5.22) and solving the resulting equation for $x_n$ rapidly yields

$$x_n = \left[ \frac{2K_S \varepsilon_0}{q} \frac{N_A}{N_D(N_A + N_D)} V_{bi} \right]^{1/2} \tag{5.30a}$$

and

$$x_p = \frac{N_D x_n}{N_A} = \left[ \frac{2K_S \varepsilon_0}{q} \frac{N_D}{N_A(N_A + N_D)} V_{bi} \right]^{1/2} \tag{5.30b}$$

It also follows that

$$W \equiv x_n + x_p = \left[ \frac{2K_S \varepsilon_0}{q} \left( \frac{N_A + N_D}{N_A N_D} \right) V_{bi} \right]^{1/2} \tag{5.31}$$

*W*, the total width of the depletion region, better known simply as the *depletion width,* is often encountered in practical device computations.

---

### Exercise 5.2

**P:** Perform a sample computation to gauge the size of $W$ and $|\mathscr{E}|_{max}$ under equilibrium conditions. Specifically assume a Si step junction operated at 300 K with $N_A = 10^{17}/cm^3$ and $N_D = 10^{14}/cm^3$.

**S:** For the given junction,

$$V_{bi} = \frac{kT}{q} \ln\left( \frac{N_A N_D}{n_i^2} \right) = (0.0259) \ln\left[ \frac{(10^{17})(10^{14})}{(10^{20})} \right] = 0.656 \text{ V}$$

Making use of Eqs. (5.30), one computes

$$x_n \cong \left[ \frac{2K_S \varepsilon_0}{q N_D} V_{bi} \right]^{1/2} = \left[ \frac{(2)(11.8)(8.85 \times 10^{-14})(0.656)}{(1.6 \times 10^{-19})(10^{14})} \right]^{1/2}$$

$$= 2.93 \times 10^{-4} \text{ cm} = 2.93 \ \mu m$$

$$x_p = \left( \frac{N_D}{N_A} \right) x_n = (10^{-3}) x_n = 2.93 \times 10^{-7} \text{ cm}$$

and

$$W = x_n + x_p \cong x_n = \boxed{2.93 \ \mu m}$$

Also

$$|\mathcal{E}|_{max} = |\mathcal{E}(0)| = \frac{qN_D}{K_S\varepsilon_0}x_n = \frac{(1.6 \times 10^{-19})(10^{14})(2.93 \times 10^{-4})}{(11.8)(8.85 \times 10^{-14})}$$

$$= \boxed{4.49 \times 10^3 \text{ V/cm}}$$

$W$ could have been computed directly using Eq. (5.31). However, we wished to make the additional point that in an asymmetrically doped junction ($N_A \gg N_D$ or $N_D \gg N_A$) the depletion region lies almost exclusively on the lightly doped side of the metallurgical boundary.

## 5.2.3 Step Junction with $V_A \neq 0$

The solution for the electrostatic variables must be extended to $V_A \neq 0$ operating conditions if it is to be of practical utility. One solution approach would be to pedantically repeat the derivations from the previous subsection with $V_A \neq 0$. Fortunately, there is an easier approach.

Consider the diode in Fig. 5.10 with a voltage $V_A \neq 0$ applied to the diode terminals. This voltage must be dropped somewhere inside the diode. However, in a well-made device a negligible portion of the applied voltage appears across the contacts to the device. More-over, under low-level injection conditions (reasonable current levels) the resistive voltage drop across the quasineutral $p$- and $n$-regions extending from the contacts to the edges of the depletion region will also be negligible. The applied voltage must therefore be dropped across the depletion region. When $V_A > 0$, this externally imposed voltage drop *lowers* the potential on the $n$-side of the junction relative to the $p$-side of the junction. Conversely, when $V_A < 0$, the potential on the $n$-side increases relative to the $p$-side. In other words,

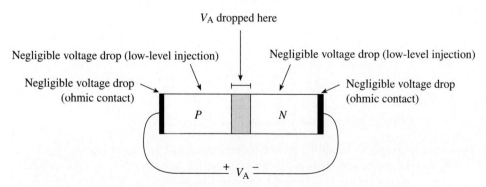

**Figure 5.10**   Voltage drops internal to a diode resulting from an externally applied voltage.

the voltage drop across the depletion region, and hence the boundary condition at $x = x_n$, becomes $V_{bi} - V_A$.

Since the only modification to the formation of the problem is a change in one boundary condition, the $V_A \neq 0$ electrostatic relationships can be extrapolated from the $V_A = 0$ relationships by simply replacing all explicit appearances of $V_{bi}$ by $V_{bi} - V_A$. Making the indicated substitution yields the $V_A \neq 0$ solution for the electrostatic variables given in Eqs. (5.32)–(5.38).

For $-x_p \leq x \leq 0$ . . .

$$\mathscr{E}(x) = -\frac{qN_A}{K_S \varepsilon_0}(x_p + x) \tag{5.32}$$

$$V(x) = \frac{qN_A}{2K_S \varepsilon_0}(x_p + x)^2 \tag{5.33}$$

$$x_p = \left[\frac{2K_S \varepsilon_0}{q} \frac{N_D}{N_A(N_A + N_D)}(V_{bi} - V_A)\right]^{1/2} \tag{5.34}$$

For $0 \leq x \leq x_n$ . . .

$$\mathscr{E}(x) = -\frac{qN_D}{K_S \varepsilon_0}(x_n - x) \tag{5.35}$$

$$V(x) = V_{bi} - V_A - \frac{qN_D}{2K_S \varepsilon_0}(x_n - x)^2 \tag{5.36}$$

$$x_n = \left[\frac{2K_S \varepsilon_0}{q} \frac{N_A}{N_D(N_A + N_D)}(V_{bi} - V_A)\right]^{1/2} \tag{5.37}$$

and

$$W = \left[\frac{2K_S \varepsilon_0}{q}\left(\frac{N_A + N_D}{N_A N_D}\right)(V_{bi} - V_A)\right]^{1/2} \tag{5.38}$$

To prevent an imaginary result, $V_A$ is obviously restricted to $V_A \leq V_{bi}$ in Eqs. (5.34), (5.37), and (5.38). The formulation fails because a large current begins to flow, and the quasineutral region voltage drops cannot be neglected, when $V_A$ approaches $V_{bi}$.

### (C) Exercise 5.3

**P:** Construct a log-log plot of the depletion width ($W$) versus the impurity concentration ($N_A$ or $N_D$) on the lightly doped side of Si $p^+$-$n$ and $n^+$-$p$ step junctions maintained at 300 K. Include curves for $V_A = 0.5$ V, 0 V, and $-10$ V covering the range $10^{14}$/cm$^3 \leq N_A$ or $N_D \leq 10^{17}$/cm$^3$.

**S:** The $V_{bi}$ associated with the $p^+$-$n$ and $n^+$-$p$ step junctions is computed using the relationship established in Exercise 5.1. Also, with the junction asymmetrically doped, the doping factor in the Eq. (5.38) expression for $W$ simplifies to

$$\frac{N_A + N_D}{N_A N_D} \cong \frac{1}{N_B}$$

where $N_B$ is the doping ($N_A$ or $N_D$) on the lightly doped side of the junction. The MATLAB program for the $W$ versus doping computation and the program results (Fig. E5.3) follow:

MATLAB program script...

```
% This program calculates and plots the depletion width vs impurity
% concentration in Silicon p+/n and n+/p step junctions at 300K.
%
% Three plots are generated corresponding to VA = 0.5V, 0.0V, and -10V
%
%    The Vbi relationship employed is Vbi=(EG/2q)+(kT/q)ln(NB/ni)
%    where NB is the impurity concentration on the lightly doped side.

%Constants and Parameters
T=300;          % Temperature in Kelvin
k=8.617e-5;     % Boltzmann constant (eV/K)
e0=8.85e-14;    % permittivity of free space (F/cm)
q=1.602e-19;    % charge on an electron (coul)
KS=11.8;        % dielectric constant of Si at 300K
ni=1e10;        % intrinsic carrier conc. in Silicon at 300K (cm^-3)
EG=1.12;        % band gap of Silicon (eV)

%Choose variable values
NB=logspace(14,17);  % doping ranges from 1e14 to 1e17
VA=[0.5 0 -10];      % VA values set
```

```
%Depletion width calculation
Vbi=EG/2+k*T.*log(NB./ni);
W(1,:)=1.0e4*sqrt(2*KS*e0/q.*(Vbi-VA(1))./NB);
W(2,:)=1.0e4*sqrt(2*KS*e0/q.*(Vbi-VA(2))./NB);
W(3,:)=1.0e4*sqrt(2*KS*e0/q.*(Vbi-VA(3))./NB);

%Plot
close
loglog(NB, W,'-'); grid
axis([1.0e14 1.0e17 1.0e-1 1.0e1])
xlabel('NA or ND (cm^-3)')
ylabel('W (micrometers)')
set(gca,'DefaultTextUnits','normalized')
text(.38,.26,'VA=0.5V')
text(.38,.50,'VA=0')
text(.38,.76,'VA=-10V')
text(.77,.82,'Si,300K')
text(.77,.79,'p+/n and n+/p')
set(gca,'DefaultTextUnits','data')
```

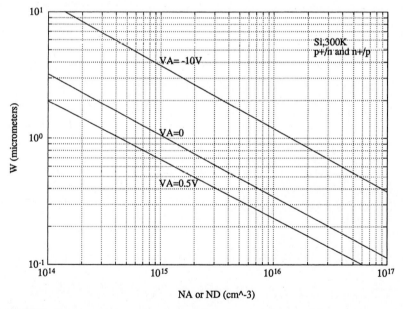

**Figure E5.3**   Depletion width at select applied biases as a function of doping (Si, $p^+$-$n$ and $n^+$-$p$ step junctions, 300 K).

### 5.2.4 Examination/Extrapolation of Results

Having expended considerable effort in establishing the results, it is reasonable to spend a few moments examining the results. We are particularly interested in how the electrostatic variables change as a function of the applied bias. Examining the (5.34) and (5.37) relationships for $x_p$ and $x_n$, we conclude these widths decrease under forward biasing ($V_A > 0$) and increase under reverse biasing ($V_A < 0$). This conclusion is of course consistent with the Exercise 5.3 computation showing a smaller depletion width for $V_A > 0$ and a larger depletion width for $V_A < 0$. The changes in $x_p$ and $x_n$ likewise translate into changes in the electric field. As deduced from Eqs. (5.32) and (5.35), a smaller $x_p$ and $x_n$ under forward biases cause the $\mathscr{E}$-field to decrease everywhere inside the depletion region, while the larger $x_p$ and $x_n$ associated with reverse biases give rise to a larger $\mathscr{E}$-field. This conclusion is also reasonable from a physical standpoint. A decreased depletion width when $V_A > 0$ means less charge around the junction and a correspondingly smaller $\mathscr{E}$-field. $V_A < 0$, on the other hand, creates a larger space charge region and a bigger electric field. Similarly, the potential given by Eqs. (5.33) and (5.36) decreases at all points when $V_A > 0$ and increases at all points when $V_A < 0$. The potential hill shrinks in both size and $x$-extent under forward biasing, whereas reverse biasing gives rise to a wider and higher potential hill. The foregoing discussion is graphically summarized in Fig. 5.11.

In Subsection 3.1.5 we established a procedure for deducing the form of the electrostatic potential from a given energy band diagram. The procedure was applied in Subsection 5.1.3 to obtain the qualitative solution for the electrostatic potential inside a *pn* junction under equilibrium conditions. Having just described and envisioned (Fig. 5.11d) how the potential changes as a function of bias, we should be able to reverse the cited procedure to construct the energy band diagrams appropriate for a *pn* junction under forward and reverse bias. Specifically, we know what the energy band diagram looks like under equilibrium conditions (redrawn in Fig. 5.12a). Conceptually taking the upside-down of the potential plots and appropriately modifying the equilibrium energy band diagram— smaller depletion width and smaller hill for forward bias, larger depletion width and larger hill for reverse bias—yields the diagrams for forward and reverse bias respectively pictured in Figs. 5.12(b) and (c).

Several comments are in order concerning the $V_A \neq 0$ diagrams. For one, the Fermi level is omitted from the depletion region because the device is no longer in equilibrium and a single level cannot be used to describe the carrier concentrations in this region. In fact, the levels labeled $E_{Fn}$ and $E_{Fp}$, occupying the former position of the Fermi level in the quasineutral regions, are actually majority-carrier quasi-Fermi levels. However, the deviation from equilibrium in the nondepleted portions of the diode is normally small, especially far from the junction, and it is therefore acceptable to continue using the $E_F$ designation. Finally, by carefully inspecting the diagrams, it is readily established that

$$E_{Fp} - E_{Fn} = -qV_A \tag{5.39}$$

Equation (5.39) suggests one may conceive of the diode terminals as providing direct access to the *p*- and *n*-ends of the equilibrium Fermi level. Conceptually grabbing onto the

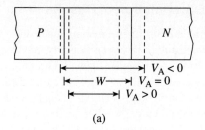

(a)

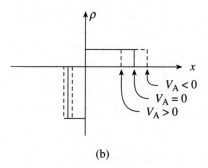

(b)

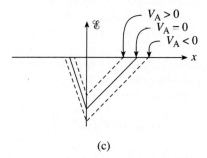

(c)

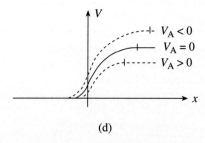

(d)

**Figure 5.11** Effect of forward and reverse biasing on the (a) depletion width, (b) charge density, (c) electric field, and (d) electrostatic potential inside a *pn* junction diode.

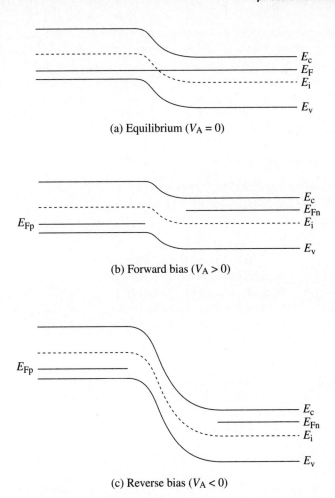

(a) Equilibrium ($V_A = 0$)

(b) Forward bias ($V_A > 0$)

(c) Reverse bias ($V_A < 0$)

**Figure 5.12**  *pn* junction energy band diagrams. (a) Equilibrium ($V_A = 0$), (b) forward bias ($V_A > 0$), and (c) reverse bias ($V_A < 0$).

ends of the equilibrium Fermi level, one progresses from the equilibrium diagram to the forward bias diagram by moving the *n*-side upward by $qV_A$ while holding the *p*-side fixed. Similarly, the reverse bias diagram is obtained from the equilibrium diagram by pulling the *n*-side Fermi level downward.

---

**(C) Exercise 5.4**

Once a quantitative relationship has been established for the electrostatic potential, it becomes possible to construct a fully dimensioned energy band diagram. The "Diagram Generator" program that follows draws the equilibrium energy band diagram for a nondegenerately doped Si step junction maintained at room temperature. The user is prompted to input the *p*- and *n*-side doping concentrations. Run the program trying different $N_A$ and $N_D$ combinations. It is informative to include at least one combination each where $N_A \gg N_D$, $N_A \cong N_D$, and $N_A \ll N_D$. The asymmetrical junctions are of particular interest because the resultant "one-sided" diagrams differ from those normally included in textbooks. The user might also consider modifying the program so that it draws the energy band diagram for an arbitrary applied bias.

MATLAB program script...

```
% Equilibrium Energy Band Diagram Generator
%(Si, 300K, nondegenerately doped step junction)

%Constants
T=300;               % Temperature in Kelvin
k=8.617e-5;          % Boltzmann constant (eV/K)
e0=8.85e-14;         % permittivity of free space (F/cm)
q=1.602e-19;         % charge on an electron (coul)
KS=11.8;             % Dielectric constant of Si
ni=1.0e10;           % intrinsic carrier conc. in Silicon at 300K (cm^-3)
EG=1.12;             % Silicon band gap (eV)

%Control constants
xleft = -3.5e-4;     % Leftmost x position
xright = -xleft;     % Rightmost x position
NA=input ('Please enter p-side doping (cm^-3), NA = ');
ND=input ('Please enter n-side doping (cm^-3), ND = ');

%Computations
Vbi=k*T*log((NA*ND)/ni^2);
xN=sqrt(2*KS*e0/q*NA*Vbi/(ND*(NA+ND)));      % Depletion width n-side
xP=sqrt(2*KS*e0/q*ND*Vbi/(NA*(NA+ND)));      % Depletion width p-side
x = linspace(xleft, xright, 200);
Vx1=(Vbi-q*ND.*(xN-x).^2/(2*KS*e0).*(x<=xN)).*(x>=0);
Vx2=0.5*q*NA.*(xP+x).^2/(KS*e0).*(x>=-xP & x<0);
Vx=Vx1+Vx2;                                  % V as a function of x
VMAX = 3;                                    % Maximum Plot Voltage
EF=Vx(1)+VMAX/2-k*T*log(NA/ni);              % Fermi level

%Plot Diagram
close
plot (x, -Vx+EG/2+VMAX/2);
axis ([xleft xright 0 VMAX]);
axis ('off'); hold on
plot (x, -Vx-EG/2+VMAX/2);
```

```
plot (x, -Vx+VMAX/2,'w:');
plot ([xleft xright], [EF EF], 'w');
plot ([0 0], [0.15 VMAX-0.5], 'w--');
text(xleft*1.08,(-Vx(1)+EG/2+VMAX/2-.05),'Ec');
text(xright*1.02,(-Vx(200)+EG/2+VMAX/2-.05),'Ec');
text(xleft*1.08,(-Vx(1)-EG/2+VMAX/2-.05),'Ev');
text(xright*1.02,(-Vx(200)-EG/2+VMAX/2-.05),'Ev');
text(xleft*1.08,(-Vx(1)+VMAX/2-.05),'Ei');
text(xright*1.02, EF-.05,'EF');
set(gca,'DefaultTextUnits','normalized')
text(.18, 0,'p-side');
text(.47, 0, 'x=0');
text(.75, 0,'n-side');
set(gca,'DefaultTextUnits','data')
hold off
```

## 5.2.5  Linearly Graded Junctions

The linearly graded profile, as noted in the preliminary discussion, is a more realistic approximation for junctions formed by deep diffusions into moderate to heavily doped wafers. Redrawn in Fig. 5.13(a), the linearly graded profile is mathematically modeled by

$$N_D - N_A = ax \tag{5.40}$$

where $a$ has units of $cm^{-4}$ and is called the *grading constant.*

Since we seek a quantitative solution for the electrostatic variables associated with the linearly graded profile, this subsection might be viewed as an unnumbered exercise—an illustration of how the procedures established in the step junction analysis are applied to another profile. However, there are features of the linearly graded analysis that are sufficiently different to merit special consideration. First and foremost, the Eq. (5.40) profile is continuous through $x = 0$. This actually simplifies the mathematical development. It is not necessary to treat the $p$- and $n$-sides of the depletion region separately or to match the solutions at $x = 0$. There is only one $\rho$, $\mathscr{E}$, and $V$ solution for the entire depletion region. Moreover, because the profile is symmetrical about $x = 0$, all of the electrostatic variables likewise exhibit a symmetry about $x = 0$. A somewhat complicating feature is the nonuniform doping *outside* of the depletion region. From previous work we know the nonuniform doping means there is a residual $\rho$, electric field, and potential drop external to the central depletion region. We ignore this fact, taking $\rho = 0$, $\mathscr{E} = 0$, and $V =$ constant outside of the depletion region in the development to be presented. Finally, although $V$ is taken to be a constant outside of the depletion region, a modification of the Eq. (5.10) expression for $V_{bi}$ is still necessary and will be included at the end of the analysis.

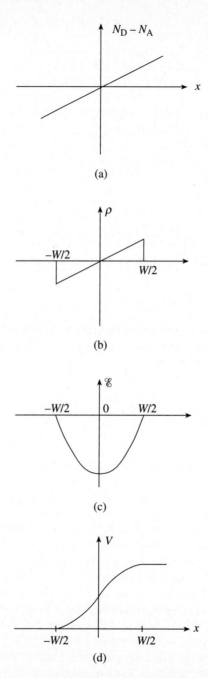

**Figure 5.13** Linearly graded solution. Depletion-approximation-based quantitative solution for the electrostatic variables in a linearly graded junction. (a) Linearly graded profile. (b) Charge density, (c) electric field, and (d) electrostatic potential as a function of position.

Launching into the analysis proper by invoking the depletion approximation yields

$$\rho(x) = \begin{cases} qax & \dots -W/2 \leq x \leq W/2 & (5.41a) \\ 0 & \dots x \leq -W/2 \quad \text{and} \quad x \geq W/2 & (5.41b) \end{cases}$$

Note from Fig. 5.13(b) that the charge density is symmetrical about $x = 0$, and one must have

$$x_p = x_n = \frac{W}{2} \tag{5.42}$$

Substituting into Poisson's equation, separating variables, and integrating from the $p$-edge of the depletion region where $\mathscr{E} = 0$ to an arbitrary point $x$ inside the depletion region then gives

$$\mathscr{E}(x) = \frac{qa}{2K_S \varepsilon_0}\left[x^2 - \left(\frac{W}{2}\right)^2\right] \quad \dots -\frac{W}{2} \leq x \leq \frac{W}{2} \tag{5.43}$$

A linear $\rho$ versus $x$ dependence naturally gives rise to a quadratic $\mathscr{E}$ versus $x$ dependence.

Next setting $\mathscr{E}(x) = -dV/dx$, separating variables, and integrating once again from the $p$-edge of the depletion region where $V = 0$ to an arbitrary point $x$, one obtains the cubic relationship

$$V(x) = \frac{qa}{6K_S \varepsilon_0}\left[2\left(\frac{W}{2}\right)^3 + 3\left(\frac{W}{2}\right)^2 x - x^3\right] \quad \dots -\frac{W}{2} \leq x \leq \frac{W}{2} \tag{5.44}$$

To complete the solution, to determine $W$, we note that the voltage drop across the depletion region must be equal to $V_{bi} - V_A$, or $V(x) = V_{bi} - V_A$ at $x = W/2$. Making the indicated substitutions into Eq. (5.44) and solving for $W$, one finds

$$W = \left[\frac{12K_S \varepsilon_0}{qa}(V_{bi} - V_A)\right]^{1/3} \tag{5.45}$$

For future reference it is important to note that the linearly graded depletion width varies as the cube root of $V_{bi} - V_A$, whereas the step junction depletion width varies as the square root of $V_{bi} - V_A$.

Numerical computations of the linearly graded $\rho$, $\mathscr{E}$, and $V$ versus $x$ would of course require an expression for $V_{bi}$. The Eq. (5.10) expression for $V_{bi}$ cannot be used because it was specifically established assuming a step junction. However, the $V_{bi}$ derivation up to and including Eq. (5.8) places no restrictions on the doping profile other than requiring that the doping be nondegenerate. Consequently, for nonstep junctions one merely needs to re-evaluate $n(x_n)$ and $n(-x_p)$, the electron concentrations at the edges of the equilibrium depletion region. For a linearly graded junction

$$n(x_n)_{\text{equilibrium}} \cong (N_D - N_A)|_{W_0/2} = aW_0/2 \qquad (5.46a)$$

$$n(-x_p)_{\text{equilibrium}} = \frac{n_i^2}{p(-x_p)_{\text{equilibrium}}} \cong \frac{n_i^2}{-(N_D - N_A)|_{-W_0/2}} = \frac{n_i^2}{aW_0/2} \qquad (5.46b)$$

where $W_0 \equiv W|_{V_A=0}$. Substituting Eqs. (5.46) into Eq. (5.8) then gives

$$V_{\text{bi}} = \frac{kT}{q} \ln\left(\frac{aW_0}{2n_i}\right)^2 = \frac{2kT}{q} \ln\left(\frac{aW_0}{2n_i}\right) \qquad (5.47)$$

or, making use of Eq. (5.45),

$$V_{\text{bi}} = \frac{2kT}{q} \ln\left[\frac{a}{2n_i}\left(\frac{12K_S\varepsilon_0}{qa}V_{\text{bi}}\right)^{1/3}\right] \qquad (5.48)$$

Equation (5.48) cannot be solved explicitly for $V_{\text{bi}}$ but must be numerically iterated to determine $V_{\text{bi}}$ for a given grading constant.

## 5.3  SUMMARY

The *pn* junction electrostatics covered in this chapter provides a foundation for the operational modeling of the *pn* junction diode and other devices that incorporate *pn* junctions. Early in the development we defined terms such as *profile* and *metallurgical boundary,* introduced the idealized step junction and linearly graded junction profiles used extensively in analyses, and referenced Poisson's equation, which often constitutes the starting point in obtaining quantitative solutions for the electrostatic variables. Other preliminary considerations included a qualitative solution for the electrostatic variables based on energy band arguments and the derivation of computational relationships for the built-in voltage. The depletion approximation, the most important and widely encountered of the simplifying approximations used in the modeling of devices, was introduced and illustrated.

The established formalism was initially applied to obtain quantitative solutions for the charge density, electric field, and electrostatic potential inside a step junction under equilibrium ($V_A = 0$) conditions. The analysis was subsequently extended to $V_A \neq 0$. The effect of an applied bias on the electrostatic variables was carefully examined and used to deduce the energy band diagrams for *pn* junctions under forward and reverse biasing. Finally, quantitative solutions were obtained for the electrostatic variables inside a linearly graded *pn* junction.

It is hoped the reader has acquired a qualitative feel for the electrostatic situation inside a *pn* junction. Moreover, with the information provided, the reader, if desired, should be able to obtain quantitative solutions for the electrostatic variables associated with other doping profiles.

## PROBLEMS

| Problem | Complete After | Difficulty Level | Suggested Point Weighting | Short Description |
|---------|---------|---------|---------|---------|
| 5.1 | 5.2.5 | 1 | 10 (1 each part) | True-or-false quiz |
| 5.2 | 5.1.4 | 1 | 5 (a-2, b-3) | $V_{bi}$ from energy band |
| 5.3 | 5.1.4 | 2 | 13 (a-2, b-3, c-3, d-2, e-3) | Isotype step junction |
| 5.4 | 5.2.2 | 2 | 12 (a-2, b-3, c-2, d-2, e-3) | Step jct. compute, $N_A \sim N_D$ |
| 5.5 | 5.2.2 | 2 | 12 (a-2, b-3, c-2, d-2, e-3) | Step jct. compute, $N_A \gg N_D$ |
| ● 5.6 | 5.2.3 | 2 | 9 (a-2, b-3, c-2, d-2) | Compute using `diary` |
| ● 5.7 | 5.2.4 | 3 | 25 (a-15, b-5, c-5) | Step junction program |
| 5.8 | 5.2.5 | 2 | 9 (3 each sketch) | Combination profile |
| 5.9 | " | 2 | 9 (a-2, b-2, c-5) | Exponential profile |
| 5.10 | " | 2–3 | 10 (a-3, b-2, c-5) | Modified-step profile |
| 5.11 | " | 3–4 | 18 (a-8, b-2, c-8) | PIN diode |
| 5.12 | " | 2–3 | 12 (4 each part) | Given *V*, find $\mathscr{E}$, $\rho$, $N_D - N_A$ |
| ● 5.13 | " | 3–4 | 25 (a-15, b-5, c-5) | Linearly graded program |
| ● 5.14 | " | 5 | 35 (a-30, b-5) | Exact solution |

**CHAPTER 5   PROBLEM INFORMATION TABLE**

**5.1** True or false:

(a) The step junction is an idealized doping profile used to model $p^+$-*n* and $n^+$-*p* junctions.

(b) The $\rho$ that appears in Poisson's equation is the charge density and has units of couls/cm$^3$.

(c) The space charge region about the metallurgical junction is due to a pile-up of electrons on the *p*-side and holes on the *n*-side.

(d) The built-in potential is typically less than the band gap energy converted to volts.

(e) Invoking the depletion approximation makes the charge density inside the depletion region directly proportional to the net doping concentration.

(f) Ohmic contacts reduce the built-in voltage drop across a junction.

(g) In solutions based on the depletion approximation, the magnitude of the electric field reaches a maximum right at the metallurgical boundary.

(h) If one has a $p^+$-$n$ step junction, a junction where $N_A$($p$-side) $\gg N_D$($n$-side), then it follows that $x_p \ll x_n$.

(i) The potential hill between the $n$-side and the $p$-side of a junction increases with forward biasing.

(j) The depletion width in a linearly graded junction varies as $(V_{bi} - V_A)^{1/3}$.

**5.2** A silicon step junction maintained at room temperature is doped such that $E_F = E_v - 2kT$ on the $p$-side and $E_F = E_c - E_G/4$ on the $n$-side.

(a) Draw the equilibrium energy band diagram for this junction.

(b) Determine the built-in voltage ($V_{bi}$) giving both a symbolic and a numerical result.

**5.3** Consider the $p1$–$p2$ "isotype" step junction shown in Fig. P5.3.

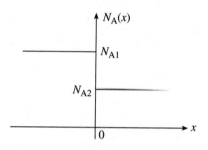

**Figure P5.3**

(a) Draw the equilibrium energy band diagram for the junction, taking the doping to be nondegenerate and $N_{A1} > N_{A2}$.

(b) Derive an expression for the built-in voltage ($V_{bi}$) that exists across the junction under equilibrium conditions.

(c) Make rough sketches of the potential, electric field, and charge density inside the junction.

(d) Briefly describe the depletion approximation.

(e) Can the depletion approximation be invoked in solving for the electrostatic variables inside the pictured $p1$–$p2$ junction? Explain.

**5.4** A Si step junction maintained at room temperature under equilibrium conditions has a $p$-side doping of $N_A = 2 \times 10^{15}$/cm$^3$ and an $n$-side doping of $N_D = 10^{15}$/cm$^3$. Compute

(a) $V_{bi}$.

(b) $x_p$, $x_n$, and $W$.

(c) $\mathcal{E}$ at $x = 0$.

(d) $V$ at $x = 0$.

(e) Make sketches that are roughly to scale of the charge density, electric field, and electrostatic potential as a function of position.

**5.5** Repeat Problem 5.4, taking $N_A = 10^{17}/\text{cm}^3$ to be the *p*-side doping. Briefly compare the results here with those of Problem 5.4.

● **5.6** A Si step junction maintained at room temperature has a *p*-side doping of $N_A =$ (instructor-supplied value) and an *n*-side doping of $N_D =$ (instructor-supplied value). The applied voltage $V_A =$ (instructor-supplied value). Working in the *Command* window and making use of the MATLAB `diary` function to record your work session, compute

(a) $V_{bi}$.

(b) $x_p$, $x_n$, and $W$.

(c) $\mathcal{E}$ at $x = 0$.

(d) $V$ at $x = 0$.

● **5.7** Given a nondegenerately doped silicon *pn* step junction maintained at $T = 300$ K:

(a) Compute and present coordinated plots of the electric field ($\mathcal{E}$) and electrostatic potential ($V$) inside the junction as a function of position ($x$). Assume $N_A = 10^{15}/\text{cm}^3$, $N_D = 2 \times 10^{14}/\text{cm}^3$, and $V_A = -20$ V in performing a sample computation.

*Suggestions*
   (i) Employ the MATLAB function `subplot` to achieve coordinated plots of the electric field and electrostatic potential.
   (ii) Use ($\mathcal{E}_{min}$, 0) and (0, $V_{max}$), where $\mathcal{E}_{min} = 1.1\mathcal{E}|_{x=0}$ and $V_{max} = 1.1(V_{bi} - V_A)$, as the endpoint *y*-values of the electric field and potential plots, respectively. Also, with $x_{max} \equiv 2.5\max(x_n, x_p)$, `max` a MATLAB function, use ($-x_{max}$, $x_{max}$) as the *x*-coordinate limits.

(b) Modify the part (a) program so that results corresponding to multiple $V_A$ values, say $V_A = V_{A0}/2^n$ with n = 0 to 3, are simultaneously displayed. Make a printout of your results.

(c) Modify the part (b) program so that, in addition to the plots, one obtains an output list of relevant parameters and computational constants. The list is to include the following quantities: $N_A$, $N_D$, $V_A$, $V_{bi}$, $x_n$, $x_p$, $W$, $\mathcal{E}$ at $x = 0$, and $V$ at $x = 0$. Print out a sample set of results.

(d) Use your program to examine how $\mathcal{E}$ versus $x$ and $V$ versus $x$ vary with the relative magnitude of the doping on the two sides of the junction. Experiment, for example, with the following combinations of ($N_A$, $N_D$): ($10^{17}/\text{cm}^3$, $10^{15}/\text{cm}^3$), ($10^{16}/\text{cm}^3$, $10^{15}/$

cm$^3$), ($10^{15}$/cm$^3$, $10^{15}$/cm$^3$), ($10^{15}$/cm$^3$, $10^{16}$/cm$^3$), ($10^{15}$/cm$^3$, $10^{17}$/cm$^3$), ($10^{18}$/cm$^3$, $10^{16}$/cm$^3$), and ($10^{16}$/cm$^3$, $10^{14}$/cm$^3$). How would you describe the results when either $N_A \geq 100 N_D$ or $N_D \geq 100 N_A$?

(e)  Use your program to generate answers to Problems 5.4 and 5.5. Check your computer-generated results against those obtained manually.

**5.8** The doping around the metallurgical junction of a special diode is pictured in Fig. P5.8. Sketch the expected charge density, electric field, and electrostatic potential inside the diode based on the depletion approximation. Properly scale and label relevant lengths. Include a few words of explanation as necessary to forestall a misinterpretation of your sketches.

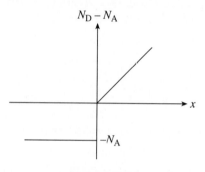

**Figure P5.8**

**5.9** A *pn* junction diode has the doping profile sketched in Fig. P5.9. Mathematically, $N_D - N_A = N_0[1 - \exp(-\alpha x)]$, where $N_0$ and $\alpha$ are constants.

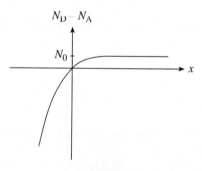

**Figure P5.9**

(a)  Give a concise statement of the depletion approximation.

(b) Invoking the depletion approximation, make a sketch of the charge density inside the diode.

(c) Establish an expression for the electric field, $\mathscr{E}(x)$, inside the depletion region.

NOTE: The interested reader may wish to complete the electrostatic solution by obtaining expressions or computational relationships for $V(x)$, $x_p$, $x_n$, and $V_{bi}$. Be forewarned, however, that a considerable amount of mathematical manipulation is involved.

**5.10** A *pn* junction diode has the doping profile sketched in Fig. P5.10. Make the assumption that $x_n > x_0$ for all applied biases of interest.

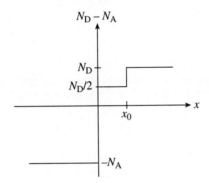

**Figure P5.10**

(a) What is the built-in voltage across the junction? Justify your answer.

(b) Invoking the depletion approximation, sketch the charge density $\rho$ versus $x$ inside the diode.

(c) Obtain an analytical solution for the electric field, $\mathscr{E}(x)$, inside the depletion region.

**5.11** The *p-i-n* diode shown schematically in Fig. P5.11 is a three-region device with a middle region that is intrinsic (actually lightly doped) and relatively narrow. Assuming the *p*- and *n*-regions to be uniformly doped and $N_D - N_A = 0$ in the *i*-region:

(a) Roughly sketch the expected charge density, electric field, and electrostatic potential inside the device. Also draw the energy band diagram for the device under equilibrium conditions.

(b) What is the built-in voltage drop between the *p*- and *n*-regions? Justify your answer.

(c) Establish quantitative relationships for the charge density, electric field, electrostatic potential, and the *p*- and *n*-region depletion widths.

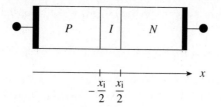

**Figure P5.11**

**5.12** The electrostatic potential in the depletion region of a *pn* junction diode under equilibrium conditions is determined to be

$$V(x) = \frac{V_{bi}}{2}\left[1 + \sin\left(\frac{\pi x}{W}\right)\right] \quad \ldots \quad -W/2 \leq x \leq W/2$$

(a) Establish an expression for the electric field as a function of position in the depletion region $(-W/2 \leq x \leq W/2)$ and sketch $\mathscr{E}(x)$ versus $x$.

(b) Establish an expression for the charge density as a function of position in the depletion region and sketch $\rho(x)$ versus $x$.

(c) Invoking the depletion approximation, determine and sketch $N_D - N_A$ versus $x$ in the depletion region.

● **5.13** Given a nondegenerately doped silicon *linearly graded* junction maintained at $T = 300$ K:

(a) Compute and present coordinated plots of the electric field $(\mathscr{E})$ and electrostatic potential $(V)$ inside the junction as a function of position $(x)$. Employ a dopant gradient constant of $a = 10^{20}/\text{cm}^4$ and an applied voltage $V_A = -20$ V in performing a sample computation. Note that Eq. (5.48) must be iterated to determine $V_{bi}$ for a given grading constant.

(b) Modify the part (a) program so that results corresponding to multiple $V_A$ values, say $V_A = V_{A0}/2^n$ with n = 0 to 3, are simultaneously displayed. Make a printout of your results.

(c) Change the part (b) program so that in addition to the plots one obtains an output list of relevant parameters and computational constants. The list is to include the following quantities: $a$, $V_A$, $V_{bi}$, $W$, $\mathscr{E}$ at $x = 0$, and $V$ at $x = 0$. Print out a sample set of results.

(d) Use your program to examine how $\mathscr{E}$ versus $x$ and $V$ versus $x$ vary with gradient constants over the range $10^{18}/\text{cm}^4 \leq a \leq 10^{23}/\text{cm}^4$.

(e) Compare the results obtained in this problem with the corresponding step junction results obtained in Problem 5.7.

● **5.14** If *equilibrium* conditions prevail, it is possible to obtain a closed-form solution for the electrostatic variables inside a *pn* step junction without invoking the depletion approximation. The "exact" solution valid under equilibrium ($V_A = 0$) conditions is detailed as follows:

*p*-side ($x \leq 0$) solution . . .

$$\int_{U_0}^{U} \frac{dU'}{F(U',U_{FP})} = \frac{x}{L_D}$$

$$\mathcal{E} = -\frac{kT}{q} \frac{1}{L_D} F(U,U_{FP})$$

$$\rho = qn_i(e^{U_{FP}-U} - e^{U-U_{FP}} + e^{-U_{FP}} - e^{U_{FP}})$$

*n*-side ($x \geq 0$) solution . . .

$$\int_{U_0}^{U} \frac{dU'}{F(U'-U_{BI},U_{FN})} = \frac{x}{L_D}$$

$$\mathcal{E} = -\frac{kT}{q} \frac{1}{L_D} F(U-U_{BI},U_{FN})$$

$$\rho = qn_i(e^{U_{FN}-U+U_{BI}} - e^{U-U_{BI}-U_{FN}} + e^{-U_{FN}} - e^{U_{FN}})$$

where . . .

$$L_D = \left[ \frac{K_S \varepsilon_0 kT}{2q^2 n_i} \right]^{1/2}$$

$$F(U1,U2) = [e^{U2}(e^{-U1} + U1 - 1) + e^{-U2}(e^{U1} - U1 - 1)]^{1/2} \quad \ldots\text{``}F\text{''-function}$$

$$U_{FP} = \ln(N_A/n_i) \quad \ldots N_A \text{ is the } p\text{-side doping concentration}$$

$$U_{FN} = -\ln(N_D/n_i) \quad \ldots N_D \text{ is the } n\text{-side doping concentration}$$

$$U_{BI} = U_{FP} - U_{FN} = \frac{V_{bi}}{kT/q}$$

$$U = \frac{V}{kT/q}$$

$U_0$ is the normalized potential ($U$) at $x = 0$. The value of $U_0$ is obtained by solving the transcendental equation: $F(U_0, U_{FP}) = F(U_0 - U_{BI}, U_{FN})$. This relationship results from the fact that the electric field must be continuous at $x = 0$.

(a) Using the preceding relationships, construct coordinated plots of the electric field ($\mathscr{E}$) and the electrostatic potential, $V = (kT/q)U$, inside the *pn* junction as a function of position. Assume $N_A = 10^{15}/\text{cm}^3$ and $N_D = 2 \times 10^{14}/\text{cm}^3$ in performing a sample computation. If available, also run the Problem 5.7(a) program utilizing an identical set of parameters. Compare and discuss the two sets of $\mathscr{E}$, $V$ versus $x$ plots.

*Suggestions*

  (i) First determine $U_0$ employing the MATLAB function `fzero` with an initial guess of $U_0 = U_{BI}/2$.

 (ii) Stepping $U$ from 0.1 to $U_0$ and utilizing the integral relationship, compute $U = V/(kT/q)$ versus $x$ on the *p*-side of the junction. Repeat for the *n*-side, stepping $U$ from $U_0$ to $U_{BI} - 0.1$.

(iii) For each value of $U$ in step (ii), likewise compute $\mathscr{E}$. Knowing $\mathscr{E}$ versus $U$ and $U$ versus $x$ allows one to construct an $\mathscr{E}$ versus $x$ plot.

 (iv) In constructing the $\mathscr{E}$ and $V$ plots, follow the suggestions cited in Problem 5.7(a).

(b) Extend the part (a) computations to obtain the normalized charge density ($\rho/q$) versus $x$. On a single set of coordinates, construct a sample plot of $\rho/q$ versus $x$ as deduced from both the exact and depletion-approximation-based solutions. Discuss your plotted results.

# 6 *pn* Junction Diode: *I–V* Characteristics

This chapter is devoted to modeling the steady state response of the *pn* junction diode. The current flowing through the diode as a function of the applied d.c. voltage, the *I–V* characteristic, is qualitatively and quantitatively correlated to the inside-the-device processes and parameters. The initial development treats the "ideal diode" and works toward the derivation of the ideal diode equation—a simple well-known *I–V* relationship. Although some of the idealizations may not be realized in practice, the ideal diode development permits unobstructed insight into the operation of the device and provides a relatively simple starting point for more exacting analyses. After comparing the ideal theory with experiment, we next focus on adjustments to the theory to correct obvious discrepancies. Several deviations from the ideal are systematically identified and explained, and appropriate modifications are introduced. Finally, we present analytical supplements to the usual analysis that will be particularly useful in subsequent chapters.

## 6.1 THE IDEAL DIODE EQUATION

As noted in the chapter introduction, the ideal diode is very useful for providing insight and for facilitating a base-level analysis. The *I–V* characteristics of the ideal diode are modeled by the ideal diode equation. Derivation of the equation is ostensibly the task of this section, although it should be understood that the analytical procedures, subsidiary results, and insight established in the process are really of prime importance. Making use of the *pn* junction energy band diagrams constructed in Chapter 5, we first pursue a qualitative "derivation" of the ideal diode equation. This exercise illustrates the power and utility of the energy band diagram, yielding the general form of the desired result without writing down a single mathematical relationship. Preparing for the quantitative derivation, we next detail our solution strategy. This is followed by the actual mathematical manipulations leading to the ideal diode equation. A probing examination of the final and subsidiary results concludes the section.

### 6.1.1 Qualitative Derivation

To set the stage, so to speak, consider the equilibrium energy band diagram for a *pn* junction shown in Fig. 6.1(a). The groups of dots (●) and circles (○) added to the figure crudely model the carrier distribution on the two sides of the junction. On the quasineutral

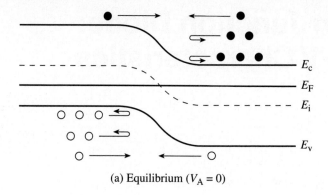

(a) Equilibrium ($V_A = 0$)

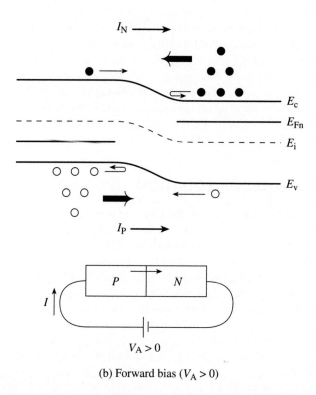

(b) Forward bias ($V_A > 0$)

**Figure 6.1**    *pn* junction energy band diagram, carrier distributions, and carrier activity in the near vicinity of the depletion region under (a) equilibrium ($V_A = 0$), (b) forward bias, and (c) reverse bias conditions. (d) Deduced form of the *I–V* characteristic.

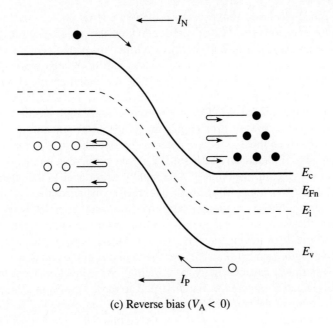

(c) Reverse bias ($V_A < 0$)

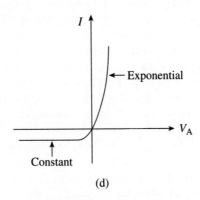

(d)

**Figure 6.1**  *Continued.*

$n$-side of the junction there are a large number of electrons and a few holes. The pyramid-like arrangement of dots (first introduced in Fig. 2.17) schematically represents the roughly exponential decrease in the electron population as one progresses upward into the conduction band. Conversely, on the quasineutral $p$-side of the junction there are a high concentration of holes and a small number of electrons. The hole population drops off in a roughly exponential fashion as one moves downward into the valence band.

The envisioned electrons and holes have thermal energy and are of course moving around inside the semiconductor. Concentrating first on the *n*-side electrons, we see that most of these carriers have insufficient energy to "climb" the potential hill. Excursions into the depletion region merely result in the lower-energy carriers being reflected back toward the *n*-side quasineutral region. However, there will be some high-energy electrons that can surmount the hill and travel over to the *p*-side of the junction. What we have been describing should be recognized as the diffusion of electrons from the high-electron population *n*-side of the junction to the low-electron population *p*-side of the junction.

Whereas electrons on the *n*-side see a potential barrier, electrons on the *p*-side are not restricted in any way. If a member of the small electron population on the *p*-side happens to wander into the depletion region, it will be rapidly swept over to the other side of the junction. Naturally, this *p*- to *n*-side drift current precisely balances the *n*- to *p*-side diffusion current under equilibrium conditions. The hole situation is completely analogous. The few *p*-side holes that have sufficient energy to surmount the potential energy barrier and gain entry to the *n*-side of the junction are precisely balanced by *n*-side holes wandering into the depletion region and being swept over to the *p*-side of the junction.

Aware of the primary carrier activity in the vicinity of the junction, let us now consider the forward bias situation pictured in Fig. 6.1(b). The most significant change relative to zero bias is a lowering of the potential hill between the *p*- and *n*-sides of the junction. The same number of minority carriers are still wandering into the depletion region and being swept over to the other side of the junction. However, with the potential hill decreased in size, more *n*-side electrons and *p*-side holes can now surmount the hill and travel to the opposite side of the junction. This gives rise to both an electron current ($I_N$) and a hole current ($I_P$) directed from the *p*-side to the *n*-side of the junction. Note from the circuit sketched below the Fig. 6.1(b) energy band diagram that the deduced current ($I = I_N + I_P$) flows in the proper direction for a forward biased diode. Moreover, because the potential hill decreases linearly with the applied forward bias and the carrier concentrations vary exponentially as one progresses away from the band edges, the number of carriers that have sufficient energy to surmount the potential barrier goes up exponentially with $V_A$. Thus, as summarized in Fig. 6.1(d), the forward current is expected to be an exponentially increasing function of the applied voltage.

The reverse bias situation is described by the energy band diagram in Fig. 6.1(c). Relative to equilibrium, the major effect of the bias is to increase the potential hill between the *p*- and *n*-sides of the junction. Whereas some *n*-side electrons and *p*-side holes can surmount the hill under equilibrium conditions, even a very small reverse bias, anything greater than a few $kT/q$ in magnitude, reduces the majority carrier diffusion across the junction to a negligible level. The *p*-side electrons and *n*-side holes, on the other hand, can still wander into the depletion region and be swept to the other side of the junction. Reverse biasing thus gives rise to a current flow directed from the *n*-side to the *p*-side of the junction. Being associated with minority carriers, the reverse bias current is expected to be extremely small in magnitude. Note in addition that the minority carrier drift currents are not affected by the height of the potential hill. It is the number of minority carriers wandering into the depletion region per second that determines the current flow. (The situation is similar to a waterfall. The water flowing over the falls is independent of the height of the falls.) Therefore, as sketched in Fig. 6.1(d), the reverse current is expected to saturate—become bias independent—once the majority carrier diffusion currents are reduced to a negligible level at a small reverse bias. If the reverse bias saturation current is taken to be $-I_0$, the overall *I–V* dependence is concluded to be of the general form

$$I = I_0(e^{V_A/V_{ref}} - 1) \tag{6.1}$$

Equation (6.1) is identical to the ideal diode equation if $V_{ref}$ is set equal to $kT/q$.

In addition to essentially yielding the ideal diode equation, the foregoing analysis very nicely explains how a solid state diode manages to rectify a signal; i.e., how the diode passes a large current when forward biased and a very small current when reverse biased. Forward biasing reduces the potential hill between the two sides of the junction, permitting large numbers of majority carriers to be injected across the depletion region. Reverse biasing increases the potential hill, cutting off majority carrier injection and leaving only a residual current supplied by minority carriers.

Once, after completing the qualitative derivation and feeling rather smug about the insight provided, the author was asked, "Yes, but, doesn't the injection of majority carriers under forward bias and the extraction of minority carriers under reverse bias cause a charge build-up inside the device?" The immediate answer is that steady state conditions were assumed in the analysis and a charge build-up, or a change of any type, does not occur under steady state conditions. The question, however, has deeper implications. The author, concentrating solely on the carrier activity in the immediate vicinity of the depletion region, had failed to provide an overall view of carrier activity inside the device. It is the overall view that explains how injected and extracted carriers are resupplied and the status quo maintained.

In presenting the "big picture," we take the diode to be reversed biased for illustrative purposes and refer to the composite energy-band/circuit diagram shown in Fig. 6.2. The capacitor-like plates at the outer ends of the energy band diagram schematically represent the ohmic contacts to the device. The major reverse-bias activity in the immediate vicinity of the depletion region, minority carriers wandering into the depletion region and being swept to the other side of the junction, is again pictured on the diagram. Added to the diagram are $E_T$ levels associated with R–G centers. Whenever an electron on the *p*-side moves to the *n*-side, it is replaced by an electron generated through one of the R–G centers. As pictured just to the left of the depletion region in Fig. 6.2, an electron from the valence band jumps up to the R–G center and then into the conduction band. Similarly, whenever a minority carrier hole is swept from the *n*-side to the *p*-side, the hole is quickly replaced by the carrier generation process. The electrons falling down the hill to the *n*-side, and the electrons simultaneously generated during the *n*-side replacement of lost holes, in turn give rise to an excess of majority carrier electrons on the *n*-side of the junction. (Two extra electrons are pictured adjacent to the depletion region on the *n*-side of the junction in Fig. 6.2.) The extra majority carrier electrons set up a local electric field that pushes adjacent electrons toward the contact. With great rapidity this displacement propagates until

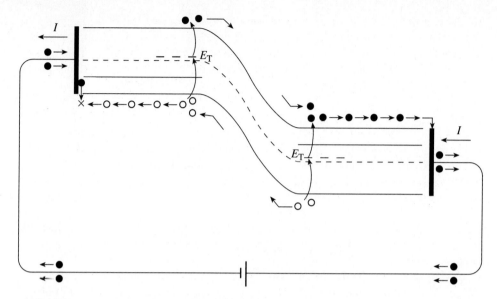

**Figure 6.2** Composite energy-band/circuit diagram providing an overall view of carrier activity inside a reverse-biased *pn* junction diode. The capacitor-like plates at the outer ends of the energy band diagram schematically represent the ohmic contacts to the diode.

the entire in-line group of *n*-side electrons moves slightly toward the contact. Electrons equal in number to the excess and immediately adjacent to the contact are pushed into the contact and out into the external circuit.[†] The hole activity on the *p*-side of the junction is similar. The excess appearing at the depletion region edge causes in-line holes throughout the quasineutral *p*-region to move over slightly. A number of holes equal to the excess are pushed into the contact where they recombine with electrons from the metal. This recombination may be viewed as completing the circle, eliminating the extra electrons pushed into the external circuit on the *n*-side of the diode.

As a somewhat unrelated observation, it is interesting to note from Fig. 6.2 that both electrons and holes contribute to the current through the depletion region, while the hole current dominates far from the junction on the *p*-side of the device and the electron current dominates far from the junction on the *n*-side of the device. The total current through the diode must be constant, but the component electron and hole currents obviously vary with position inside the diode.

---

[†] The phenomenon just described might be likened to the real-life situation where latecomers to a party enter a room absolutely crammed with people. It is assumed there are two doors at opposite ends of the room. By forcing their way into one door, the newcomers cause a like number of people to be pushed out the other door.

## 6.1.2 Quantitative Solution Strategy

Going into a football game, coaches always have a game plan, a strategy for winning the game. Herein we develop and explain the strategy used in the quantitative derivation of the ideal diode equation. Although mathematical steps in the derivation are relatively few and quite straightforward, the strategy underlying the steps is rather involved and needs to be clearly understood in applying the solution approach to other problems and in implementing modifications to the theory.

### General Considerations

We begin by listing the basic assumptions made in the analysis. (Additional assumptions will be necessary as the strategy unfolds.)

(1) The diode is being operated under steady state conditions.

(2) A nondegenerately doped step junction models the doping profile.

(3) The diode is one-dimensional.

(4) Low-level injection prevails in the quasineutral regions.

(5) There are no processes other than drift, diffusion, and thermal recombination–generation taking place inside the diode. Specifically, $G_L = 0$.

The preceding assumptions seem reasonable since they were all explicitly or implicitly invoked in establishing the *pn* junction electrostatics.

Let us next consider the general relationships available for computing the current. They are:

$$I = AJ \qquad (A = \text{cross-sectional area}) \tag{6.2}$$

$$J = J_N(x) + J_P(x) \tag{6.3}$$

$$J_N = q\mu_n n\mathscr{E} + qD_N\frac{dn}{dx} \tag{6.4a}$$

$$J_P = q\mu_p p\mathscr{E} - qD_P\frac{dp}{dx} \tag{6.4b}$$

Equation (6.3) reflects the fact that the total current density is constant throughout the diode, but the electron and hole components vary with position. Equations (6.4) are the one-dimensional versions of Eqs. (3.18). Clearly, if exact analytical solutions for $\mathscr{E}$, $n$, and $p$ versus $x$ were available, we would need to proceed no further. However, only an approximate three-region electrostatic solution is available, as schematically summarized in

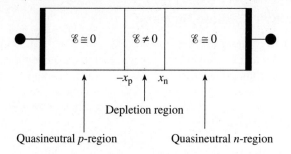

**Figure 6.3** Diode electrostatic regions.

Fig. 6.3. We do note that the conditions required for the use of the minority carrier diffusion equations, including $\mathscr{E} \cong 0$ and low-level injection, are satisfied in the quasineutral regions of the diode.

## Quasineutral Region Considerations

Under the assumed steady state conditions with $G_L = 0$, the minority carrier diffusion equations appropriate for the $p$ and $n$ quasineutral regions are

$$0 = D_N \frac{d^2 \Delta n_p}{dx^2} - \frac{\Delta n_p}{\tau_n} \qquad \ldots x \leq -x_p \tag{6.5a}$$

$$0 = D_P \frac{d^2 \Delta p_n}{dx^2} - \frac{\Delta p_n}{\tau_p} \qquad \ldots x \geq x_n \tag{6.5b}$$

Moreover, since $\mathscr{E} \cong 0$ and $dn_0/dx = dp_0/dx = 0$, Eqs. (6.4) for the carrier current densities in the quasineutral regions simplify to

$$J_N = qD_N \frac{d\Delta n_p}{dx} \qquad \ldots x \leq -x_p \tag{6.6a}$$

$$J_P = -qD_P \frac{d\Delta p_n}{dx} \qquad \ldots x \geq x_n \tag{6.6b}$$

We know the general solution to Eqs. (6.5) and it is a trivial matter to compute the carrier current densities from Eqs. (6.6). Unfortunately, the current density solutions so obtained are limited to non-overlapping segments of the diode. We can only determine $J_N(x)$ in the quasineutral $p$-region and $J_P(x)$ in the quasineutral $n$-region. To solve for $J$ using Eq. (6.3), there must be at least one point inside the diode where one knows both

$J_N(x)$ and $J_P(x)$. Being centrally located, the depletion region is the obvious place to seek overlapping $J_N$ and $J_P$ solutions, solutions possibly extrapolated from the quasineutral regions.

## Depletion Region Considerations

The full-blown continuity equations, Eqs. (3.46), must be used in seeking solutions for the carrier currents within the $\mathscr{E} \neq 0$ depletion region. Under the previously specified assumptions, the continuity equations simplify to

$$0 = \frac{1}{q}\frac{dJ_N}{dx} + \left.\frac{\partial n}{\partial t}\right|_{\substack{\text{thermal} \\ \text{R–G}}} \tag{6.7a}$$

$$0 = -\frac{1}{q}\frac{dJ_P}{dx} + \left.\frac{\partial p}{\partial t}\right|_{\substack{\text{thermal} \\ \text{R–G}}} \tag{6.7b}$$

Suppose the additional assumption is now made that *thermal recombination–generation is negligible throughout the depletion region;* i.e., $\partial n/\partial t|_{\text{thermal R–G}}$ and $\partial p/\partial t|_{\text{thermal R–G}}$ are arbitrarily set equal to zero in Eqs. (6.7). Eliminating the R–G terms in Eqs. (6.7) yields $dJ_N/dx = 0$ and $dJ_P/dx = 0$. $J_N$ and $J_P$ are therefore determined to be constants independent of position inside the depletion region under the stated assumption. The constancy of the carrier currents throughout the depletion region (including the edges) in turn allows one to write

$$J_N(-x_p \leq x \leq x_n) = J_N(-x_p) \tag{6.8a}$$

$$J_P(-x_p \leq x \leq x_n) = J_P(x_n) \tag{6.8b}$$

$J_N(-x_p)$ and $J_P(x_n)$ can be deduced of course from the quasineutral region solutions evaluated at the edges of the depletion region. Summing the $J_N$ and $J_P$ solutions in the depletion region then gives

$$\boxed{J = J_N(-x_p) + J_P(x_n)} \tag{6.9}$$

Clearly a solution strategy has been formulated: to solve for the minority carrier current densities in the quasineutral regions, evaluate the current densities at the depletion region edges, add the edge current densities together, and finally multiply by $A$ to obtain the current.

The rather critical assumption that thermal recombination–generation is negligible in the depletion region can be viewed as a defining property of the ideal diode. There is absolutely no *a priori* justification for the assumption other than it leads to a simple solution for the total current flowing in the diode. The validity of the assumption as far as real diodes are concerned will be considered when comparing theory and experiment.

## Boundary Conditions

There is one more matter to address. Two boundary conditions each are required in solving Eqs. (6.5a) and (6.5b) for $\Delta n_p$ and $\Delta p_n$ in the *p*- and *n*-side quasineutral regions. In particular, as summarized in Fig. 6.4(a), $\Delta n_p$ and $\Delta p_n$ must be specified at the ohmic contacts and at the edges of the depletion region.

### At the Ohmic Contacts

The ideal diode is usually taken to be a "wide-base" diode, or a diode whose contacts are several minority carrier diffusion lengths or more from the edges of the depletion region. In a wide-base diode any perturbation in the carrier concentrations created at the edges of the depletion region will decay to zero before reaching the contacts. The contacts may effectively be viewed as being positioned at $x = \pm\infty$. Thus, in the mathematical derivation

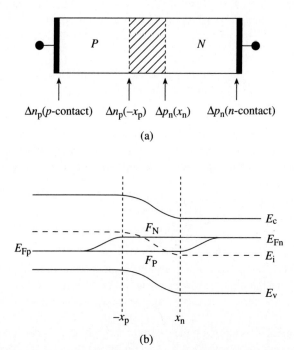

**Figure 6.4** Boundary-condition related considerations. (a) Boundary positions and required values. (b) Approximate variation of the quasi-Fermi levels with position inside a forward-biased diode.

the boundary conditions to be employed are

$$\Delta n_p(x \to -\infty) = 0 \tag{6.10a}$$

$$\Delta p_n(x \to +\infty) = 0 \tag{6.10b}$$

### At the Depletion Region Edges

To establish the boundary conditions at the edges of the depletion region, we make use of the quasi-Fermi level formalism. Equations (3.72) are the defining equations for the electron quasi-Fermi level, $F_N$, and hole quasi-Fermi level, $F_P$. If the left- and right-hand sides of the exponential versions of Eqs. (3.72a) and (3.72b) are multiplied together, one obtains

$$np = n_i^2 e^{(F_N - F_P)/kT} \tag{6.11}$$

The Eq. (6.11) relationship is valid throughout the diode under arbitrary operating conditions. Generally speaking, one does not know the variation of the quasi-Fermi levels as a function of position prior to solving for the carrier concentrations inside the diode. However, as envisioned in Fig. 6.4(b), it is reasonable to assume the $F_N$ and $F_P$ levels will vary monotonically from $E_{Fp}$ far on the *p*-side of the junction to $E_{Fn}$ far on the *n*-side of the junction. Note from Fig. 6.4(b) that the monotonic variation in the levels in turn makes $F_N - F_P \leq E_{Fn} - E_{Fp} = qV_A$ at all points inside the diode. If the equal sign in the preceding expression for $F_N - F_P$ is assumed to hold throughout the depletion region, one concludes

$$np = n_i^2 e^{qV_A/kT} \quad \ldots -x_p \leq x \leq x_n \tag{6.12}$$

Eq. (6.12) has been referred to as the "law of the junction." Evaluating Eq. (6.12) at the depletion region edges very rapidly leads to the desired boundary conditions. Specifically, evaluating Eq. (6.12) at the *p*-edge of the depletion region gives

$$n(-x_p)p(-x_p) = n(-x_p)N_A = n_i^2 e^{qV_A/kT} \tag{6.13}$$

or

$$n(-x_p) = \frac{n_i^2}{N_A} e^{qV_A/kT} \tag{6.14}$$

and

$$\Delta n_p(-x_p) = \frac{n_i^2}{N_A}(e^{qV_A/kT} - 1) \tag{6.15}$$

Similarly,

$$n(x_n)p(x_n) = p(x_n)N_D = n_i^2 e^{qV_A/kT} \tag{6.16}$$

or

$$p(x_n) = \frac{n_i^2}{N_D} e^{qV_A/kT} \tag{6.17}$$

and

$$\Delta p_n(x_n) = \frac{n_i^2}{N_D}(e^{qV_A/kT} - 1) \tag{6.18}$$

The assumption that $F_N - F_P = qV_A$ throughout the depletion region, or equivalently, assuming the quasi-Fermi levels are constant at $F_N = E_{Fn}$ and $F_P = E_{Fp}$ within the depletion region, is obviously central to obtaining the depletion-edge boundary conditions and critical to the overall analysis. Rather lengthy subsidiary analyses and a comparison with experiment indicate the assumption is typically justified.†

**"Game Plan" Summary**

To obtain an analytical solution for the current flowing in an ideal diode as a function of the applied voltage, proceed as follows:

(1) Solve the minority carrier diffusion equations (Eqs. 6.5) employing the (6.10) and (6.15/6.18) boundary conditions to obtain $\Delta n_p$ and $\Delta p_n$ in the quasineutral regions.

(2) Compute the minority carrier current densities in the quasineutral regions using Eqs. (6.6).

(3) Per Eq. (6.9), evaluate the quasineutral region solutions for $J_N(x)$ and $J_P(x)$ at the edges of the depletion region and then sum the two edge-current densities. Finally, multiply the result by the cross-sectional area of the diode.

---

† Assuming $F_N$ and $F_P$ are approximately constant across the depletion region is equivalent to assuming $J_N \cong 0$ and $J_P \cong 0$ within the depletion region (see Eqs. 3.76). The $J_N = 0$ and $J_P = 0$ assumption is used by some authors in an alternative derivation of the depletion-edge boundary conditions that closely parallels the $V_{bi}$ derivation in Subsection 5.1.4. Like the constant quasi-Fermi assumption, and contrary to statements in some texts, there is really no simple a priori justification of the $J_N = 0$ and $J_P = 0$ assumption.

### 6.1.3 Derivation Proper

In implementing the solution procedure, let us first work with holes on the quasineutral *n*-side of the junction. To simplify the mathematics in solving Eq. (6.5b), it is convenient to shift the origin of coordinates to the *n*-edge of the depletion region as shown in Fig. 6.5(a). In terms of the translated $x'$-coordinate, we must solve

$$0 = D_P \frac{d^2 \Delta p_n}{dx'^2} - \frac{\Delta p_n}{\tau_p} \quad \dots x' \geq 0 \tag{6.19}$$

subject to the boundary conditions

$$\Delta p_n(x' \to \infty) = 0 \tag{6.20a}$$

$$\Delta p_n(x' = 0) = \frac{n_i^2}{N_D}(e^{qV_A/kT} - 1) \tag{6.20b}$$

Eq. (6.19) is one of the special-case diffusion equations listed in Table 3.2. The general solution (solution no. 1 in Table 3.2) is

$$\Delta p_n(x') = A_1 e^{-x'/L_P} + A_2 e^{x'/L_P} \quad \dots x' \geq 0 \tag{6.21}$$

where

$$L_P = \sqrt{D_P \tau_p} \tag{6.22}$$

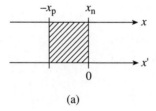

(a)

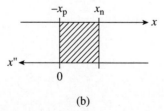

(b)

**Figure 6.5**  Graphical definition of the (a) $x'$- and (b) $x''$-coordinate systems.

Because $\exp(x'/L_P) \to \infty$ as $x' \to \infty$, the only way that the Eq. (6.20a) boundary condition can be satisfied is for $A_2$ to be identically zero. With $A_2 = 0$, application of the Eq. (6.20b) boundary condition yields $A_1 = \Delta p_n(x' = 0)$. We therefore conclude

$$\Delta p_n(x') = \frac{n_i^2}{N_D}(e^{qV_A/kT} - 1)e^{-x'/L_P} \qquad \ldots x' \geq 0 \qquad (6.23)$$

and

$$J_P(x') = -qD_P\frac{d\Delta p_n}{dx'} = q\frac{D_P}{L_P}\frac{n_i^2}{N_D}(e^{qV_A/kT} - 1)e^{-x'/L_P} \qquad \ldots x' \geq 0 \qquad (6.24)$$

On the quasineutral $p$-side of the junction with the $x''$-coordinate as defined in Fig. 6.5(b), one obtains the analogous solutions

$$\Delta n_p(x'') = \frac{n_i^2}{N_A}(e^{qV_A/kT} - 1)e^{-x''/L_N} \qquad \ldots x'' \geq 0 \qquad (6.25)$$

and

$$J_N(x'') = -qD_N\frac{d\Delta n_p}{dx''} = q\frac{D_N}{L_N}\frac{n_i^2}{N_A}(e^{qV_A/kT} - 1)e^{-x''/L_N} \qquad \ldots x'' \geq 0 \qquad (6.26)$$

All that remains is to evaluate Eqs. (6.24) and (6.26) at the depletion region edges, sum the results, and multiply by $A$. We find

$$J_N(x=-x_p) = J_N(x''=0) = q\frac{D_N}{L_N}\frac{n_i^2}{N_A}(e^{qV_A/kT} - 1) \qquad (6.27a)$$

$$J_P(x=x_n) = J_P(x'=0) = q\frac{D_P}{L_P}\frac{n_i^2}{N_D}(e^{qV_A/kT} - 1) \qquad (6.27b)$$

and

$$I = AJ = qA\left(\frac{D_N}{L_N}\frac{n_i^2}{N_A} + \frac{D_P}{L_P}\frac{n_i^2}{N_D}\right)(e^{qV_A/kT} - 1) \tag{6.28}$$

or

$$I = I_0(e^{qV_A/kT} - 1) \tag{6.29}$$

$$I_0 \equiv qA\left(\frac{D_N}{L_N}\frac{n_i^2}{N_A} + \frac{D_P}{L_P}\frac{n_i^2}{N_D}\right) \tag{6.30}$$

Equation (6.29) is the ideal diode equation. It is also sometimes referred to as the Shockley equation.

## 6.1.4  Examination of Results

It is worthwhile at this point to pause and examine the final and intermediate results of the derivation. We hope to become more familiar with the results while simultaneously gaining deeper insight into the operation of the *pn* junction diode. Working backward through the derivation, we first examine the ideal diode equation and the associated saturation current. Subsequently we investigate the carrier currents and carrier concentrations inside the diode. Several exercises are also presented to supplement the discussion.

### Ideal *I–V*

The major features of the predicted *I–V* characteristics are summarized in Fig. 6.6. For reverse biases greater than a few $kT/q$, a few tenths of a volt at room temperature, the exponential voltage term in the ideal diode equation becomes negligible and $I \to -I_0$. According to the ideal diode theory, this saturation current would be observed for reverse voltages of unlimited magnitude. For forward biasing greater than a few $kT/q$, the exponential term dominates and $I \to I_0 \exp(qV_A/kT)$. Reflecting the expected exponential dependence, the forward-bias characteristics are often plotted on a semilog scale as illustrated in Fig. 6.6(b). Since

$$\ln(I) = \ln(I_0) + \frac{q}{kT}V_A \quad \ldots \text{if } V_A > \text{few } \frac{kT}{q} \tag{6.31}$$

the ideal theory predicts a $V_A > 0$ semilog plot that has a linear region slope of $q/kT$ and an extrapolated intercept of $\ln(I_0)$.

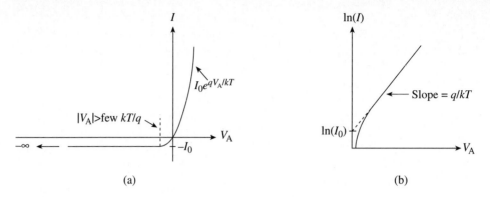

**Figure 6.6**    Ideal diode $I$–$V$ characteristics: (a) linear plot identifying major features; (b) forward-bias semilog plot.

## The Saturation Current

Two rather significant observations can be made concerning the saturation current. First, the size of $I_0$ can vary by many orders of magnitude depending on the semiconductor used to fabricate the diode. This strong material dependence enters through the $n_i^2$ factor in the $I_0$ expression. At room temperature $n_i = 10^{10}/cm^3$ in Si while $n_i \cong 10^{13}/cm^3$ in Ge. Thus, Ge diodes are expected to exhibit a reverse-bias saturation current roughly $10^6$ times larger than that of comparable Si diodes!

The second observation relates to asymmetrically doped junctions. The $I_0$ expression has two terms that vary inversely with the dopings respectively on the $p$- and $n$-side of the junction. Because of the cited doping dependence, the term associated with the heavily doped side of $p^+$-$n$ and $n^+$-$p$ junctions becomes negligible; i.e.,

$$I_0 \cong qA \frac{D_P}{L_P} \frac{n_i^2}{N_D} \quad \text{. . . } p^+\text{-}n \text{ diodes} \quad\quad (6.32a)$$

and

$$I_0 \cong qA \frac{D_N}{L_N} \frac{n_i^2}{N_A} \quad \text{. . . } n^+\text{-}p \text{ diodes} \quad\quad (6.32b)$$

In essence, one has to consider only the lightly doped side of such junctions in working out the diode $I$–$V$ characteristics. We also found one can all but neglect the heavily doped side of asymmetrical junctions in computing the depletion width and other electrostatic variables. These similar conclusions suggest that, *as a general rule, the heavily doped side of an asymmetrical junction can be ignored in determining the electrical characteristics of the junction.* If an asymmetrical junction is specified in a problem statement or at the beginning of an analysis, it should be understood that the heavily doped side is to be ignored in completing the problem or analysis.

We might mention that it is rather fortuitous that the current contribution from the heavily doped side of asymmetrical junctions is negligible. In most real diodes the $p^+$- or $n^+$-side doping is degenerate. If the current contribution from the heavily doped side were significant, comparison with experiment would necessitate a modification of the ideal diode theory to account for the degenerate doping.

---

### Exercise 6.1

**P:** Two ideal $p^+$-$n$ step junction diodes maintained at room temperature are identical except that $N_{D1} = 10^{15}/cm^3$ and $N_{D2} = 10^{16}/cm^3$. Compare the *I–V* characteristics of the two diodes; sketch both characteristics on a single set of axes.

**S:** For $p^+$-$n$ diodes

$$I_0 \cong qA \frac{D_P}{L_P} \frac{n_i^2}{N_D}$$

Also

$$\frac{D_P}{L_P} = \sqrt{\frac{D_P}{\tau_p}} = \sqrt{\frac{(kT/q)\mu_p}{\tau_p}}$$

$\tau_{p1} = \tau_{p2}$, since the R–G center concentrations are taken to be identical in the two diodes. Moreover, although the semiconductor material is not specified in the problem statement, there is only a small difference between the $N_D = 10^{15}/cm^3$ and $N_D = 10^{16}/cm^3$ mobilities in most materials (see Fig. 3.5). Thus

$$\frac{I_{01}}{I_{02}} = \frac{N_{D2}}{N_{D1}} \sqrt{\frac{\mu_{p1}}{\mu_{p2}}} \cong 10$$

The diode 1 current is approximately ten times larger than the diode 2 current for all applied voltages (see Fig. E6.1).

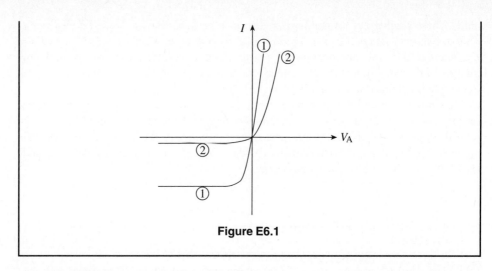

**Figure E6.1**

## (C) Exercise 6.2

**P:** An *ideal* Si $p^+$-$n$ step junction diode is maintained at 300 K. The diode has a cross-sectional area $A = 10^{-4}$ cm$^2$ and $\tau_p = 10^{-6}$ sec. Using the empirical-fit relationship for the hole mobility introduced in Exercise 3.1, write a MATLAB program that computes and plots the ideal $I$–$V$ characteristic of the diode. The $n$-side donor doping is to be considered an input variable. Employ a linear plot with the MATLAB `axis` function set to $[-1, 0.2, -2*I_0, 5*I_0]$, where $I_0$ is the reverse bias saturation current. Generally use your program to explore how the ideal diode characteristic varies as a function of the semiconductor doping.

**S:** The program that follows was written to handle multiple doping inputs. Enclosing your input in square brackets, simply type the desired doping values separated by spaces in response to the "ND= " prompt. The sample plot (Fig. E6.2) exhibits the same general doping dependence noted in Exercise 6.1. However, examining the numerical $I_0$ values sent by the program to the *Command* window, one finds a noticeable mobility dependence because of the higher assumed dopings. The user should also take note of the extremely small size of the computed saturation currents.

MATLAB program script...

```
%Variation of Ideal-Diode I-V with semiconductor doping.
%Si step junction, T = 300K.
%In response to the "ND=" prompt type [ND1 ND2 ...] to input
%multiple doping values.

%Initialization and Universal Constants
clear
k=8.617e-5;
q=1.6e-19;
```

```
%Device, Material, and System Parameters
A=1.0e-4;
ni=1.0e10;
taup=1.0e-6;
ND=input('Input the n-side doping concentration, ND=');
T=300;

%Hole Mobility Calculation
NAref=2.35e17;
μpmin=54.3;
μp0=406.9;
ap=0.88;
μp=μpmin+μp0./(1+(ND./NAref).^ap);
%The mobility calculation here assumes the hole minority carrier
   %mobility is equal to the hole majority carrier mobility.

%I-V Calculation
VA=linspace(-1,0.2);
DP=k.*T.*μp;
LP=sqrt(DP.*taup);
I0=q.*A.*(DP./LP).*(ni^2 ./ND)
I=I0.'*(exp(VA./(k.*T))-1);

%Plotting Result
close
plot(VA,I); grid;
ymin=-2*I0(1); ymax=5*I0(1);
axis([-1,0.2,ymin,ymax]);
xlabel('VA (volts)'); ylabel('I (amps)');

%Adding axes,key
xx=[-1 0.2]; yx=[0 0];
xy=[0 0]; yy=[ymin,ymax];
hold on
plot(xx,yx,'-w',xy,yy,'-w');
j=length(ND);
for i=1:j;
   yput=(0.70-0.06*i)*ymax;
   yk(i,1)=yput; yk(i,2)=yput;
   text(-0.68,(0.69-0.06*i)*ymax,['ND=',num2str(ND(i),'/cm3']);
end
xk=[-0.8 -0.7];
plot(xk,yk);
text(-0.74,0.75*ymax,'Si, 300K');
hold off
```

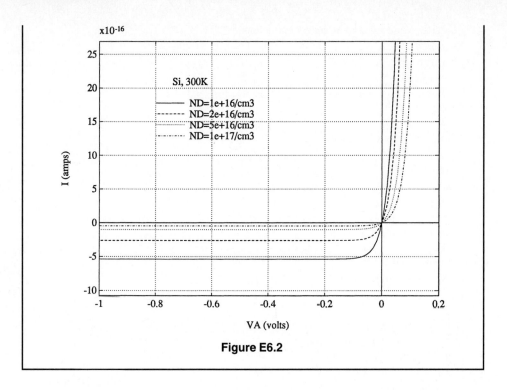

**Figure E6.2**

## Carrier Currents

A sample plot of the carrier and total current densities as a function of position inside a forward-biased diode is displayed in Fig. 6.7. A reverse-bias plot is essentially identical except all current densities are negative. Steps in the construction of the Fig. 6.7 plot are as follows: Examining Eqs. (6.24) and (6.26), one concludes the minority carrier current densities in the quasineutral regions decay exponentially away from the edges of the depletion region. Within the depletion region, $J_N$ and $J_P$ per Eq. (6.8) are next drawn constant at their respective depletion-edge values. The total current density in the depletion region is then just the graphical sum of $J_N$ and $J_P$. Since the total current density is constant everywhere inside the diode, the value deduced for the depletion region can be extended throughout the diode. Finally, the majority-carrier current densities in the quasineutral regions are obtained by graphically subtracting the minority-carrier current densities from the total current density.

The construction of Fig. 6.7 shows that there is sufficient information available to deduce the current densities everywhere inside the diode. Moreover, the plot helps one visualize the nature of the $J_N(x)$ and $J_P(x)$ solutions. The general form of the solutions, we might note, is consistent with the observation made at the end of the qualitative derivation. Whereas electrons and holes both contribute to the current through the depletion region, the hole current dominates far from the junction on the $p$-side of the device and the electron current dominates far from the junction on the $n$-side of the device.

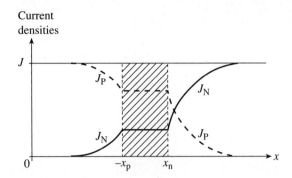

**Figure 6.7**   Carrier and total current densities versus position inside a forward-biased *pn* junction diode.

## Carrier Concentrations

Equations (6.23) and (6.25) specify the deviation from equilibrium of the minority carrier concentrations in the quasineutral regions. From these solutions we conclude forward biasing increases the carrier concentrations over their respective equilibrium values and that reverse biasing lowers the concentrations below the equilibrium values. In either case, the perturbations decay exponentially as one proceeds away from the edges of the depletion region. Moreover, after several diffusion lengths the perturbations effectively die out and the minority carrier concentrations approach their equilibrium values; i.e., $n_p \rightarrow n_{p0} = n_i^2/N_A$ as $x \rightarrow -\infty$ and $p_n \rightarrow p_{n0} = n_i^2/N_D$ as $x \rightarrow +\infty$ independent of the applied bias. Because low-level injection is assumed to prevail in the quasineutral regions, we can also assert that the *majority* carrier concentrations in these regions are everywhere approximately equal to their equilibrium values regardless of the applied bias. Plots of the forward- and reverse-bias carrier concentrations incorporating the foregoing information are shown in Fig. 6.8. Note that the exponential decays of $\Delta n_p$ and $\Delta p_n$ show up as straight lines on the forward-bias plot because the carrier concentrations are being plotted on a logarithmic scale.

Under forward biasing the majority carriers are injected in large numbers over the potential hill to the other side of the junction. Once on the other side of the junction, the injected carriers become minority carriers and are progressively eliminated by recombination as they attempt to diffuse deeper into the region. (The situation and solution are all but identical to that of Sample Problem No. 2 considered in Chapter 3.) The net result, pictured in Fig. 6.8(a), is a build-up of minority carriers in the quasineutral regions immediately adjacent to the edges of the depletion region. The build-up of excess minority carriers adjacent to the depletion region is a consequence of forward biasing, which will prove to be important in subsequent analyses.

Under reverse biasing the depletion region acts like a "sink" for minority carriers, draining the carriers from the adjacent quasineutral regions as pictured in Fig. 6.8(b). A

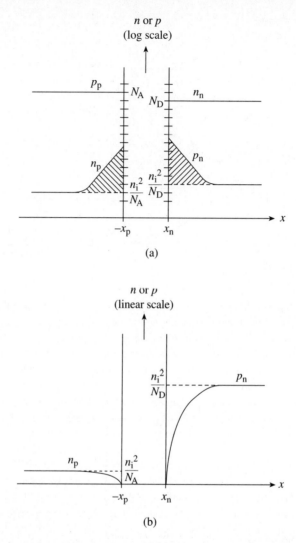

**Figure 6.8** Carrier concentrations inside a *pn* junction diode under (a) forward biasing and (b) reverse biasing. The cross-hatching identifies excess minority carriers. Note that (a) is a semilog plot while (b) is a linear plot. $N_A > N_D$ was assumed in constructing the sample plots.

reverse bias of only a few $kT/q$ effectively reduces to zero the minority carrier concentrations at the edges of the depletion region. Larger reverse biases have little effect on the carrier distributions. (This is consistent with the fact that the current, which is directly related to the slope of the carrier distributions at the depletion region edges, saturates for applied biases greater than a few $kT/q$.) Overall, reverse biasing gives rise to a relatively small deficit of minority carriers in the near vicinity of the depletion region.

---

## Exercise 6.3

**P:** Figure E6.3 is a dimensioned plot of the steady state carrier concentrations inside a *pn* junction diode maintained at room temperature.

(a) Is the diode forward or reverse biased? Explain how you arrived at your answer.

(b) Do low-level injection conditions prevail in the quasineutral regions of the diode? Explain how you arrived at your answer.

(c) Determine the applied voltage, $V_A$.

(d) Determine the hole diffusion length, $L_P$.

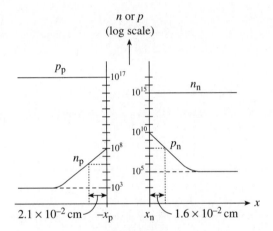

**Figure E6.3**

**S:** (a) The diode is forward biased. There is a pile-up or minority carrier excess ($\Delta n_p > 0$ and $\Delta p_n > 0$) at the edges of the depletion region.

(b) Low-level injection conditions *do* prevail. $\Delta p_n \ll n_n$ and $\Delta n_p \ll p_p$ everywhere inside the quasineutral regions.

(c) We can make use of either the depletion edge boundary conditions or the "law of the junction" to determine $V_A$. Specifically, solving Eq. (6.12) for $V_A$ gives

$$V_A = \frac{kT}{q} \ln\left(\frac{np}{n_i^2}\right) \qquad \ldots -x_p \leq x \leq x_n$$

Evaluating the $V_A$ expression at the $n$-edge of the depletion region and noting $np \to n_i^2$ for $x \to \pm\infty$, we compute

$$V_A = \frac{kT}{q} \ln\left[\frac{n_n(x_n)p_n(x_n)}{n_n(\infty)p_n(\infty)}\right] = (0.0259) \ln\left(\frac{10^{25}}{10^{20}}\right) \cong \mathbf{0.3\ V}$$

(d) Equation (6.23) can be rewritten

$$\Delta p_n(x') = \Delta p_n(x'=0)e^{-x'/L_P}$$

In the near vicinity of the depletion region edge $\Delta p_n \cong p_n$, giving

$$p_n(x') = p_n(0)e^{-x'/L_P}$$

or

$$\ln\left[\frac{p_n(x')}{p_n(0)}\right] = -\frac{x'}{L_P}$$

and

$$L_P = \frac{x'}{\ln\left[\dfrac{p_n(0)}{p_n(x')}\right]} = \frac{1.6 \times 10^{-2}}{\ln\left(\dfrac{10^{10}}{10^8}\right)} = \mathbf{3.47 \times 10^{-3}\ cm}$$

---

## Exercise 6.4

When reverse biased greater than a few $kT/q$, the current flowing in a diode is equal to $q$ times the number of minority carriers per second that wander into the depletion region and are swept to the other side of the junction. Under steady state conditions the number of minority carriers thereby extracted per second from the $p$- and $n$-sides of the junction must be precisely equal to the number of minority carriers generated per second in the quasineutral regions. In other words, the reverse-bias saturation current may be alternatively viewed as arising from minority carrier generation in the quasineutral regions, with $I_0$ being equal to $q$ times the number of minority carriers generated per second.

**P:** Suppose the actual reverse-bias minority carrier distributions in the quasineutral regions are approximated by squared-off distributions as pictured in Fig. E6.4. *All minority carriers are taken to be depleted a minority carrier diffusion length to either* side of the depletion region. Remembering that low-level injection prevails in the quasineutral regions, and referring to the problem introduction, derive an expression for the reverse-bias saturation current based on the approximate distributions.

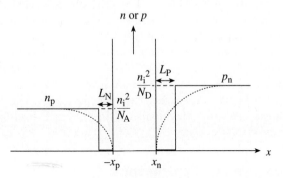

**Figure E6.4**

**S:** The carrier generation per second per unit volume based on the approximate distributions is

$$\frac{\partial n}{\partial t}\bigg|_{\substack{\text{thermal}\\ \text{R–G}}} = -\frac{\Delta n_p}{\tau_n} = \frac{n_i^2/N_A}{\tau_n} \quad \ldots -L_N - x_p \le x \le -x_p$$

and

$$\frac{\partial p}{\partial t}\bigg|_{\substack{\text{thermal}\\ \text{R–G}}} = -\frac{\Delta p_n}{\tau_p} = \frac{n_i^2/N_D}{\tau_p} \quad \ldots x_n \le x \le x_n + L_P$$

Generation is taking place in volumes of $AL_N$ and $AL_P$ on the *p*- and *n*-sides of the junction, respectively. Thus

$$I_0 = q(AL_N)\left(\frac{n_i^2/N_A}{\tau_n}\right) + q(AL_P)\left(\frac{n_i^2/N_D}{\tau_p}\right)$$

$$= qA\left(\frac{D_N}{L_N}\frac{n_i^2}{N_A} + \frac{D_P}{L_P}\frac{n_i^2}{N_D}\right)$$

One obtains the usual $I_0$ expression!

The approximation introduced in this problem, replacing an exponential distribution with a square distribution extending over one decay length (the same area being enclosed by both distributions), is often a useful analytical tool.

## 6.2  DEVIATIONS FROM THE IDEAL

A moderately long list of assumptions was involved in deriving the ideal diode equation. Some of the assumptions were made without any a priori justification. It is therefore reasonable to expect that a careful examination of experimental *I*–*V* characteristics will reveal discrepancies. Herein we first compare experiment and theory to identify the major deviations from the ideal. Subsequently we treat specific deviations, noting underlying causes and providing required modifications to the ideal theory. Hopefully the reader will not be dismayed by the seemingly band-aid approach in developing an acceptable theoretical description of the diode *I*–*V* characteristics. The iterative approach—formulating a simple theory, comparing theory and experiment, revising the theory, comparing the revised theory with experiment, and so on—is common engineering and scientific practice in treating real problems of a complex nature.

### 6.2.1  Ideal Theory Versus Experiment

The *I*–*V* characteristic derived from a Si diode maintained at room temperature and displayed on a 0.1 $\mu$A/division linear-scale plot is reproduced in Fig. 6.9. For the most part the characteristic exhibits a form consistent with theoretical expectations. The forward current is a rapidly rising function of the applied voltage, and the reverse current is vanishingly small over a better part of the probed voltage range. However, there is one obvious feature not modeled by the ideal theory: A large reverse-bias current flows when the reverse voltage exceeds a certain value. This phenomenon is referred to as *breakdown*. Although the voltage where breakdown occurs can vary over several orders of magnitude, the phenomenon is common to all *pn* junction diodes. Breakdown constitutes a major deviation from the ideal and is addressed in Subsection 6.2.2.

To determine whether there are additional deviations, it is necessary to take a closer look at the *I*–*V* data. A semilog plot of the forward-bias data from the Fig. 6.9 device is displayed in Fig. 6.10(a). Reverse-bias data from the device are replotted in Fig. 6.10(b) employing an expanded current scale of 50 pA/division. Data for both plots were obtained using a HP4145B Semiconductor Parameter Analyzer. The noticeable fluctuations in the extremely small reverse bias current are caused by noise inherent in the device and measurement system.

Examining the Fig. 6.10(a) forward-bias data, we note that the curve exhibits the expected $q/kT$ slope for applied voltages between approximately 0.35 V and 0.7 V. However, there are obvious deviations outside the cited voltage range. At forward biases in excess of 0.7 V the slope progressively decreases or the characteristic "slopes over." This deviation, more or less to be expected, is related to the high level of current flowing in the device when $V_A \rightarrow V_{bi}$ and is treated in Subsection 6.2.4.

For voltages below 0.35 V the current levels are quite small and the deviation is clearly of a different origin. In addition, the observed current is far in excess of the expected value and the slope of the characteristic curiously approaches $q/2kT$. These facts suggest we may be observing a current component either overlooked or neglected in the derivation of the ideal diode equation. This suspicion is further enhanced by an examination of the reverse bias data. For one, the device has an estimated $I_0 \cong 10^{-14}$A, some three orders of magni-

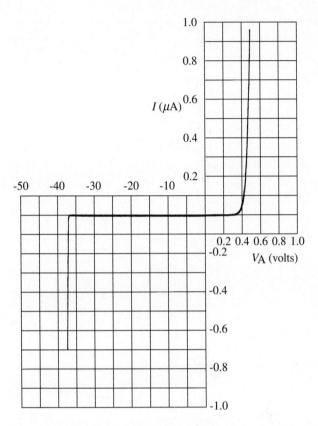

**Figure 6.9**  Linear plot of the measured *I–V* characteristic derived from a commercially available Si *pn* junction diode maintained at room temperature. The plot permits a coarse evaluation of the diode characteristic. Note the change in voltage scale in going from forward to reverse bias.

tude smaller than the reverse current observed at $V_A = -5V$! Moreover, the Fig. 6.10(b) characteristic does not saturate as predicted by the ideal theory. Rather, the current continually increases with increasing reverse bias. As will be verified in Subsection 6.2.3, the cited reverse-bias and small forward-bias deviations are caused by the added current arising from carrier recombination–generation in the depletion region. Carrier R–G was assumed to be negligible throughout the depletion region in deriving the ideal diode equation.

    Although feature details do vary from device to device, and other nonidealities may be encountered, the data displayed in Figs. 6.9 and 6.10 are fairly representative of the *I–V* characteristics derived from commercially available Si *pn* junction diodes maintained at room temperature. Similar characteristics are exhibited by GaAs diodes. Ge diodes maintained at room temperature and Si diodes operated at elevated temperatures, on the other hand, are often found to more closely approach the ideal, specifically exhibiting saturating reverse-bias characteristics. An explanation of this difference in behavior is included in the R–G current discussion.

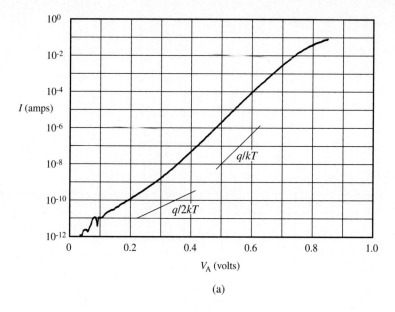

(a)

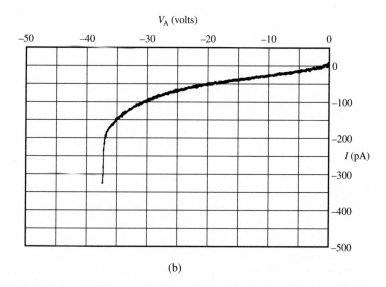

(b)

**Figure 6.10**    Detailed plots of the measured $I$–$V$ characteristic derived from a commercially available Si *pn* junction diode maintained at room temperature. The Fig. 6.9 and Fig. 6.10 characteristics are from the same device. (a) Semilog plot of the forward-bias current versus voltage. (b) Expanded scale plot of the reverse-bias current versus voltage.

## 6.2.2 Reverse-Bias Breakdown

Although referred to as "breakdown," the large reverse current that flows when the reverse voltage exceeds a certain value is a completely reversible process. That is, breakdown does not damage the diode in any way. The current must be limited, of course, to avoid excessive heating. The absolute value of the reverse voltage where the current goes off to infinity is known as the breakdown voltage and is given the symbol $V_{BR}$. Practical $V_{BR}$ measurements typically quote the voltage where the current exceeds a preselected value such as 1 $\mu$A or 1 mA. The expected breakdown voltage of planar (one-dimensional) $p^+$-$n$ and $n^+$-$p$ step-junction diodes as a function of the nondegenerate-side doping is plotted in Fig. 6.11 for select semiconductors. At a given doping, $V_{BR}$ tends to increase with the band gap of the semiconductor used to fabricate the diode. Of greater significance, the doping on the lightly doped side of the junction can be used to vary $V_{BR}$ from a few volts to over a thousand volts. Note that the doping dependence above the dashed line in Fig. 6.11 is roughly described by

$$V_{BR} \propto \frac{1}{N_B^{0.75}} \tag{6.33}$$

where $N_B$ is the doping on the lightly doped side of the junction.

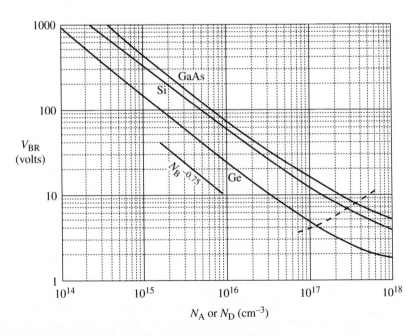

**Figure 6.11**   Breakdown voltage as a function of the nondegenerate-side doping in planar $p^+$-$n$ and $n^+$-$p$ step-junction Ge, Si, and GaAs diodes. Avalanche is the dominant breakdown process for dopings above the dashed line. $T = 300$ K. (After Sze[1], © 1981 by John Wiley & Sons, Inc. Reprinted with permission.)

From a theoretical standpoint, breakdown is directly related to the failure of the "no other processes" assumption in the derivation of the ideal diode equation. In fact, two "other processes"—avalanching and the Zener process—can cause the breakdown current. Avalanching is typically the dominant process, with the Zener process only becoming important when both sides of the junction are heavily doped. In what follows, we delve into the physical nature of the two processes and provide process-specific information.

## Avalanching

Working up to a physical description of the avalanching phenomenon, let us first consider a reverse-biased diode where $V_A$ is relatively small and far below the breakdown voltage. Ideally, the reverse current flowing in the diode is due to minority carriers randomly entering the depletion region and being accelerated by the electric field in the region to the other side of the junction. In crossing the depletion region the carrier acceleration is not continuous but is interrupted by energy-losing collisions with the semiconductor lattice, as envisioned in Fig. 6.12(a). Since the mean free path between collisions is $\sim 10^{-6}$ cm, and a median depletion width is $\sim 10^{-4}$ cm, a carrier can undergo tens to thousands of collisions in crossing the depletion region. Thus at small applied reverse biases the energy lost by the carriers per collision is relatively small. The energy transferred to the lattice simply causes lattice vibrations—there is just localized heating that is readily dissipated.

With increasing reverse bias the amount of energy transferred to the semiconductor lattice per collision systematically increases. Approaching the breakdown voltage, the energy transferred per collision becomes sufficient to ionize a semiconductor atom. By "ionize" we mean the collision frees a valence electron from the atom, or causes an electron from the valence band to jump into the conduction band, thereby creating an electron-hole pair. This phenomenon, called *impact ionization,* was previously visualized in Fig. 3.15(f). The added carriers created by impact ionization are immediately accelerated by the electric field in the depletion region. Consequently, they and the original carriers make additional collisions and create even more carriers as envisioned in Fig. 6.12(b). The result is a snowballing creation of carriers very similar to an avalanche of snow on a mountain side. At the breakdown voltage the carrier creation and reverse current effectively go off to infinity.

A couple of clarifying comments are in order concerning the preceding discussion and avalanching. For one, the Fig. 6.12(b) energy band diagram is not, and realistically could not be, drawn to scale. If the breakdown voltage were 100 V, for example, the distance between the *p*- and *n*-side Fermi levels would have to be approximately 100 times the $E_c$ minus $E_v$ band gap distance. Second, it is important to note that avalanche breakdown does not occur sharply at $V_A = -V_{BR}$. When the reverse-bias characteristics are examined carefully as in Fig. 6.10(b), they are found to exhibit a sloping approach to breakdown. There is considerable carrier multiplication a few volts before reaching breakdown and even some carrier multiplication at voltages far below breakdown. The reason is that the distance between collisions is a random variable statistically distributed about the mean value. Thus carriers can occasionally gain sufficient energy to have an ionizing collision at voltages far below breakdown. At a few volts below breakdown the

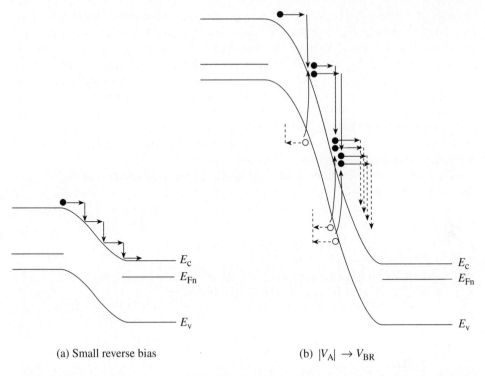

(a) Small reverse bias          (b) $|V_A| \rightarrow V_{BR}$

**Figure 6.12**   Carrier activity inside the depletion region of a reverse-biased *pn* junction diode when (a) $|V_A| \ll V_{BR}$ and (b) $|V_A| \rightarrow V_{BR}$. Carrier multiplication due to impact ionization and the resultant avalanche is pictured in (b).

number of carriers gaining sufficient energy to have an ionizing collision becomes quite large. The increase in current associated with the carrier multiplication is modeled by introducing a *multiplication factor, M*. If $I_0$ is taken to be the current without any carrier multiplication, then

$$M \equiv \frac{|I|}{I_0} \qquad (6.34)$$

An empirical fit to experimental data gives

$$M = \frac{1}{1 - \left[\dfrac{|V_A|}{V_{BR}}\right]^m} \qquad (6.35)$$

where $m$ takes on a value between 3 and 6, depending on the semiconductor used to fabricate the diode. Please note that the multiplication factor can be used to correct the ideal diode equation to account for carrier multiplication and avalanching.

We next seek to explain the $V_{BR}$ dependence noted in the breakdown introduction. From the qualitative description of avalanching, it was concluded that breakdown occurs when the carriers gain an ionizing amount of energy in traveling a lattice-scattering mean free path. This should be true independent of the junction doping. A specific energy gain over a given distance, however, corresponds to a specific electric field. In other words, breakdown occurs when the electric field in the depletion region reaches some critical value, $\mathscr{E}_{CR}$, essentially independent of the junction doping.

Considering a step junction, employing Eqs. (5.35) and (5.37), and evaluating the electric field at $x = 0$, we find

$$\mathscr{E}(0) = -\frac{qN_D}{K_S\varepsilon_0}x_n = -\left[\frac{2q}{K_S\varepsilon_0}\left(\frac{N_AN_D}{N_A + N_D}\right)(V_{bi} - V_A)\right]^{1/2} \tag{6.36}$$

Next, squaring the preceding expression and making use of the fact that $\mathscr{E}(0) \to \mathscr{E}_{CR}$ when $V_{bi} - V_A \to V_{bi} + V_{BR} \cong V_{BR}$, we obtain

$$\mathscr{E}_{CR}^2 = \frac{2q}{K_S\varepsilon_0}\left(\frac{N_AN_D}{N_A + N_D}\right)V_{BR} \tag{6.37}$$

Since $\mathscr{E}_{CR}$ is independent of doping, the right-hand side of Eq. (6.37) must likewise be independent of doping. The right-hand side of Eq. (6.37) will be independent of doping if

$$V_{BR} \propto \frac{N_A + N_D}{N_AN_D} \tag{6.38}$$

or for asymmetrically doped junctions

$$V_{BR} \propto \frac{1}{N_B} \tag{6.39}$$

The Eq. (6.39) result is not precisely as observed experimentally, but it is acceptably close given the simplicity of the argument.

Concerning the $V_{BR}$ band gap (or semiconductor) dependence, we again note that breakdown occurs when the carriers gain an ionizing amount of energy in traveling a lattice-scattering mean free path. The required ionization energy obviously increases with increasing $E_G$, but the mean free path is found to vary only slightly for the semiconductors cited in Fig. 6.11. Thus, consistent with Fig. 6.11, the breakdown voltage is expected to progressively increase in going from Ge to Si to GaAs.

One more dependence is worth mentioning. The $V_{BR}$ due to avalanching is found to increase as the temperature goes up. As noted in the mobility discussion in Subsection 3.1.3, lattice scattering increases as the temperature increases. Increasing lattice scattering means a smaller mean free path, a larger critical electric field for avalanching, and hence a higher breakdown voltage.

Finally, the discussion has focused on planar step-junction diodes. Separate breakdown curves apply to linearly graded diodes. Moreover, step-junction diodes formed by diffusion or ion-implantation through a mask always have curved, nonplanar lateral edges. For a given applied voltage the electric field is greater in the nonplanar region than in the planar part of the device. Hence breakdown occurs sooner in the nonplanar regions, lowering the $V_{BR}$ of the diode. This "curvature" effect becomes more pronounced as the depth of the junction decreases. Special edge modifications have been developed to minimize the effect.

---

### (C) Exercise 6.5

**P:** Compute and make a plot of the multiplication factor ($M$) versus $|V_A|/V_{BR}$ when $m = 3$ and $m = 6$. Orient the axes so a reverse-current-like plot is obtained; i.e., plot $M$ increasing downward and $|V_A|/V_{BR}$ increasing to the left. Discuss the plotted results.

**S:** The computer program script and resultant plot (Fig. E6.5) follow. Using MATLAB, the author was only able to obtain the desired axes orientation by plotting $-M$ versus $-|V_A|/V_{BR}$. The plot is noted to be quite similar to the left-hand side of Fig. 6.10(b). A perhaps surprising $2\times$ increase in current is predicted for applied voltages less than 80% and 90% of the way to $V_{BR}$ for $m = 3$ and $m = 6$, respectively. There is already a carrier enhancement of ~10% ($M = 1.1$) at $0.45V_{BR}$ when $m = 3$ and at $0.65V_{BR}$ when $m = 6$.

MATLAB program script...

```
%Exercise 6.5...Multiplication factor

%Initialization
close
clear

%M calculation
x=linspace(0,.99);      %x=|VA|/VBR
M3=1 ./(1-x.^3);        %M when m=3
M6=1 ./(1-x.^6);        %M when m=6

%Plotting result
plot(-x,-M3,-x,-M6); grid
axis([-1 0 -10 0])
xlabel('-|VA|/VBR')
ylabel('-Multiplication factor')
text(-0.8,-2.5,'m = 3')
text(-0.95,-1.5,'m = 6')
```

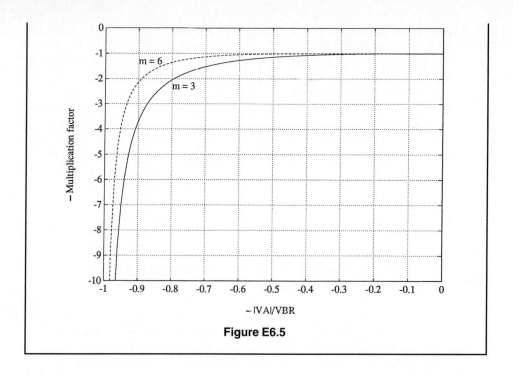

**Figure E6.5**

## Zener Process

*Zener process* is the name given to the occurrence of "tunneling" in a reverse-biased diode. *Tunneling* is the first phenomenon we have encountered that is of a purely quantum-mechanical nature. It has no classical analog. The general nature and basic features of tunneling can be understood with the aid of Fig. 6.13. The particle pictured in the figure is taken to be positioned on the left-hand side of a potential energy barrier. The height of the barrier is assumed to be greater than the kinetic energy of the particle. Classically, the only way a particle such as an electron could move to the other side of the barrier would be to gain additional energy and go over the top of the barrier. Quantum mechanically, there is another way the particle can get to the other side of the barrier: It can go through the barrier. Tunneling is going "through" a potential energy barrier. It should be emphasized that the particle energy remains constant during the process. Also (thinking classically) the particle and the barrier are not damaged in any way.

There are two major requirements for tunneling to occur and be significant:

(1) There must be filled states on one side of the barrier and empty states on the other side of the barrier at the same energy. Tunneling cannot take place into a region void of allowed states.

(2) The width of the potential energy barrier, $d$ in Fig. 6.13, must be very thin. Quantum-mechanical tunneling becomes significant only if $d < 100 \text{ Å} = 10^{-6}$ cm.

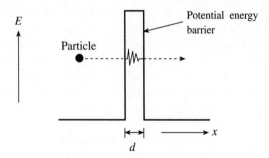

**Figure 6.13**   General visualization of tunneling.

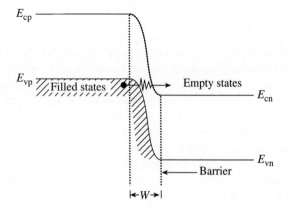

**Figure 6.14**   Visualization of tunneling in a reverse-biased *pn* junction diode.

Visualization of the Zener process, tunneling in a reverse-biased *pn* junction diode, is presented in Fig. 6.14. The particles doing the tunneling are valence band electrons on the *p*-side of the junction. The potential energy barrier classically restricting these electrons to the *p*-side of the junction is outlined in the figure. Tunneling takes place when the electrons pass through the barrier to empty states at the same energy in the conduction band on the *n*-side of the junction. The greater the reverse bias, the larger the number of filled valence-electron states on the *p*-side placed opposite empty conduction-band states on the *n*-side, and hence the greater the reverse-bias tunneling current.

For tunneling to be significant, the barrier thickness, roughly the depletion width in the case of the *pn* junction diode, must be $<10^{-6}$ cm. Referring to Fig. E5.3, we find Si diode depletion widths $<10^{-6}$ cm necessitate dopings in excess of $10^{17}$/cm$^3$ on the "lightly" doped side of the junction. Thus the Zener process is important only in diodes that are heavily doped on both sides of the junction. The breakdown voltage of the diodes is correspondingly small. The Zener process makes a significant contribution to the breakdown current in diodes where $V_{BR} < 6E_G/q$ (~6.7 V in Si at 300 K) and dominates in diodes where $V_{BR} < 4E_G/q$ (~4.5 V in Si at 300 K).

Two experimental observations can be used to distinguish between avalanching and Zener process breakdown. First, whereas the $V_{BR}$ associated with avalanching increases with increasing *T*, $V_{BR}$ *decreases* with increasing *T* if the Zener process is dominant. Second, the breakdown characteristics associated with the Zener process are very "soft." The current exhibits a very slow approach to infinity even when observed on a relatively coarse scale.

Historically, the Zener process was the first proposed to explain reverse-bias breakdown. All diodes specifically fabricated to make use of the breakdown characteristic came to be known as Zener diodes. The name continues to be applied to all diodes making use of the breakdown characteristic, even though those with breakdown voltages in excess of $6E_G/q$ are functionally avalanche diodes.

## 6.2.3 The R–G Current

In comparing theory and experiment, a current far in excess of that predicted by the ideal diode theory was found to exist at small forward biases and all reverse biases in Si diodes maintained at room temperature. The observed "extra" current arises from thermal carrier recombination–generation in the depletion region that was assumed to be negligible in the derivation of the ideal diode equation. How thermal R–G in the depletion region gives rise to an added current component can be understood with the aid of Fig. 6.15. First consider the reverse-bias case modeled in Fig. 6.15(a). Heretofore in all quantitative analyses, and even in the qualitative derivation of the ideal diode equation, we associated the reverse current with the minority carriers wandering into the depletion region from the two sides of the junction. When the diode is reverse biased, however, the carrier concentrations in the depletion region are reduced below their equilibrium values, leading to the thermal generation of electrons and holes throughout the region.[†] The large electric field in the depletion region rapidly sweeps the generated carriers into the quasineutral regions, thereby adding to the reverse current. Forward biasing increases the carrier concentrations in the depletion region above their equilibrium values giving rise to carrier recombination in the region. As envisioned in Fig. 6.15(b), one can effectively view the resulting added forward current as arising from the carriers that cannot make it over the potential hill being partially eliminated via recombination at R–G centers in the depletion region.

Seeking to establish a quantitative expression for the added current, $I_{R-G}$, arising from thermal recombination–generation in the depletion region, we note that the net R–G rate is the same for electrons and holes under steady state conditions. Moreover, for every electron-hole *pair* created or destroyed in the depletion region per second, *one* electron per second flows into or out of the diode contacts. Summing either the electrons or the holes created/destroyed throughout the depletion region per second and multiplying by *q* should

---

[†] Use of the term *depletion region* in the present context sometimes leads to confusion. It must be remembered that the carrier concentrations in the depletion region are not zero or automatically less than their equilibrium values, but are merely small compared to the background doping concentrations.

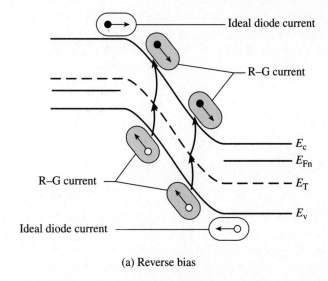

(a) Reverse bias

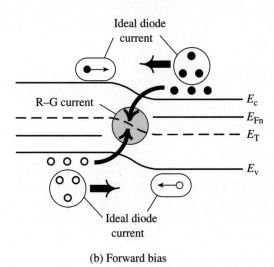

(b) Forward bias

**Figure 6.15**  The R–G current. Visualization of the additional current resulting from (a) reverse-bias generation and (b) forward-bias recombination in the depletion region.

therefore give the magnitude of the added current flowing in the device. Proceeding as indicated and accounting for the polarity of the current yields

$$I_{\text{R-G}} = -qA \int_{-x_p}^{x_n} \left. \frac{\partial n}{\partial t} \right|_{\substack{\text{thermal} \\ \text{R-G}}} dx \tag{6.40}$$

Note that, because of the conditions prevalent in the depletion region, the familiar special-case R–G relationship, $\partial n/\partial t|_{\text{thermal R–G}} = -\Delta n/\tau_n$, is not applicable. Rather, the general-case result initially presented as Eq. (3.35) must be employed:

$$\left.\frac{\partial n}{\partial t}\right|_{\substack{\text{thermal}\\ \text{R–G}}} = -\frac{np - n_i^2}{\tau_p(n + n_1) + \tau_n(p + p_1)} \tag{6.41}$$

and

$$I_{\text{R–G}} = qA\int_{-x_p}^{x_n} \frac{np - n_i^2}{\tau_p(n + n_1) + \tau_n(p + p_1)}\,dx \tag{6.42}$$

For reverse biases greater than a few $kT/q$, the carrier concentrations become quite small throughout most of the depletion region. With the carrier concentrations being negligible ($n \to 0$, $p \to 0$), the integral in Eq. (6.42) is readily evaluated to obtain

$$\boxed{I_{\text{R–G}} = -\frac{qAn_i}{2\tau_0}W} \qquad \ldots \text{ reverse biases} > \text{few } kT/q \tag{6.43}$$

where

$$\tau_0 \equiv \frac{1}{2}\left(\tau_p\frac{n_1}{n_i} + \tau_n\frac{p_1}{n_i}\right) = \frac{1}{2}\left(\tau_p e^{(E_T - E_i)/kT} + \tau_n e^{(E_i - E_T)/kT}\right) \tag{6.44}$$

For forward biases the carrier concentrations cannot be neglected and even an approximate evaluation of the Eq. (6.42) integral becomes rather involved. We merely note that $I_{\text{R–G}}$ is expected to vary roughly as $\exp(qV_A/\eta kT)$, $1 < \eta \leq 2$, for forward biases greater than a few $kT/q$.[2] Typically, the expected $\eta$ is close to 2, and the combined forward and reverse bias dependence is approximately described by

$$\boxed{I_{\text{R–G}} = \frac{qAn_i}{2\tau_0}W\frac{(e^{qV_A/kT} - 1)}{\left(1 + \dfrac{V_{\text{bi}} - V_A}{kT/q}\dfrac{\sqrt{\tau_n\tau_p}}{2\tau_0}e^{qV_A/2kT}\right)}} \tag{6.45}$$

With the introduction of a second current component, it is common practice to refer to the current described by the ideal diode equation as the *diffusion current, $I_{\text{DIFF}}$*. The diffusion current expression is rewritten here to facilitate a comparison of the two components:

$$I_{\text{DIFF}} = qA\left(\frac{D_{\text{N}}}{L_{\text{N}}}\frac{n_i^2}{N_{\text{A}}} + \frac{D_{\text{P}}}{L_{\text{P}}}\frac{n_i^2}{N_{\text{D}}}\right)(e^{qV_{\text{A}}/kT} - 1)$$

(6.46)

The total current flowing in the diode is of course just the sum of the diffusion current and the R–G current:

$$I = I_{\text{DIFF}} + I_{\text{R–G}}$$

(6.47)

We are now in a position to explain earlier experimental observations. In Si diodes at room temperature $qAn_iW/2\tau_0 \gg I_0$ and the $I_{\text{R–G}}$ current dominates at reverse biases and small forward biases. Since the reverse bias $I_{\text{R–G}}$ is proportional to $W$, the reverse current never saturates, but continually increases with increasing reverse bias. The forward bias $I_{\text{R–G}}$ varies as $\exp(qV_{\text{A}}/2kT)$ for $V_{\text{A}} >$ few $kT/q$, consistent with experimental observations at small forward biases. With increasing forward biases the $I_{\text{DIFF}}$ component, which increases more rapidly with voltage, eventually overtakes the $I_{\text{R–G}}$ component, leading to the $q/kT$ region on a semilog plot of the forward-bias characteristics. Because $I_{\text{DIFF}} \propto n_i^2$ while $I_{\text{R–G}} \propto n_i$, the relative weight of the two components varies significantly from semiconductor to semiconductor. Whereas $qAn_iW/2\tau_0 \gg I_0$ in Si and GaAs diodes at room temperature, the larger $n_i$ of Ge typically makes $I_0 > qAn_iW/2\tau_0$ in Ge diodes at room temperature. Also, since $I_0 \propto n_i^2$ while the reverse bias $I_{\text{R–G}} \propto n_i$, the reverse bias diffusion component of the current will increase at a faster rate with increasing temperature. The diffusion component eventually dominates at a sufficiently elevated temperature. The $n_i$ dependence explains the experimental observations that Ge diodes maintained at room temperature and Si diodes operated at elevated temperatures are often found to more closely approach the ideal, specifically exhibiting saturating reverse-bias characteristics.

---

### (C) Exercise 6.6

If the weak voltage dependence of the factors multiplying the exponential in the $I_{\text{R–G}}$ expression is ignored, the total *pn* junction diode current for forward biases greater than a few $kT/q$ may be approximately modeled by

$$I = I_{01}e^{qV_{\text{A}}/kT} + I_{02}e^{qV_{\text{A}}/2kT}$$

$$\underset{I_{\text{DIFF}}}{\uparrow} \qquad \underset{I_{\text{R–G}}}{\uparrow}$$

where *both* $I_{01}$ and $I_{02}$ are taken to be constants independent of voltage.

**P:** (a) Compute $I_{01}$ and $I_{02}$ for a Si $p^+$-$n$ step junction diode maintained at 300 K with $A = 10^{-4}$ cm$^2$, $N_D = 10^{16}$/cm$^3$, and $\tau_n = \tau_p = \tau_0 = 10^{-6}$ sec. In the $I_{02}$ computation, assume that the $\exp(qV_A/2kT)$ term in the denominator of the $I_{R-G}$ expression is much greater than unity and set $V_A = V_{bi}/4$ in evaluating voltage-dependent terms. As in previous exercises compute $V_{bi}$ using $V_{bi} = (E_G/2q) + (kT/q)\ln(N_D/n_i)$.

(b) Repeat part (a) for a Ge $p^+$-$n$ step junction diode maintained at 300 K. Assume a set of parameters identical to the part (a) Si diode except that $\mu_p = 1500$ cm$^2$/V-sec, $n_i = 2.5 \times 10^{13}$/cm$^3$, and $K_S = 16$.

(c) Making use of the simplified current relationship introduced in this problem, construct a semilog plot of $I$ versus $V_A$ that simultaneously displays the forward-bias characteristics of the previously described Si and Ge diodes. Limit the plot axes to $0 \text{ V} \le V_A \le 1 \text{ V}$ and $10^{-12} \text{ A} \le I \le 10^{-3} \text{ A}$. Begin the calculation at $V_A = 0.1$V. Discuss your results.

**S:** The computed $I_{01}$ and $I_{02}$ values for the two diodes are summarized in the table presented below. The part (c) computer program and plot follow the table. The $I_{01}$ and $I_{02}$ values are stored in memory after running the listed computer program and were accessed from the MATLAB *Command* window.

As expected, $I_{02} \gg I_{01}$ for the Si diode while $I_{01} > I_{02}$ for thc Ge diode. Because of the small band gap and corresponding large $n_i$, large forward currents flow in the Ge diode at relatively small voltages, making it all but impossible to observe the $I_{R-G}$ component. Conversely, both the $I_{DIFF}$ and $I_{R-G}$ components are clearly evident on the silicon diode plot.

|  | *Silicon* | *Germanium* |
|---|---|---|
| $I_{01}$ | $5.38 \times 10^{-16}$ A | $6.23 \times 10^{-9}$ A |
| $I_{02}$ | $2.08 \times 10^{-13}$ A | $8.34 \times 10^{-10}$ A |

MATLAB program script...

```
% Comparison of forward bias I-VA for Si and Ge diodes at 300K.
%    This program uses a simplified formula for the current:
%    I = I01*exp(qVA/(kT)) + I02*exp(qVA/(2kT)).
%    Also, Vbi=EG/2q+(kT/q)ln(ND/ni).
```

```
%Initialization
close
clear

% Constants
T=300;                    % Temperature in Kelvin
k=8.617e-5;               % Boltzmann constant eV/K
e0=8.85e-14;              % permittivity of free space (F/cm)
q=1.602e-19;              % charge on an electron (coul)
KS=[11.8 16];             % Dielectric constant [Si Ge]
ni=[1.0e10 2.5e13];       % intrinsic carrier conc. at 300K [Si Ge]
µp=[437 1500];            % hole mobility [Si Ge]
EG=[1.12 0.66];           % band gap [Si Ge]

% Given Constants
A=1.0e-4;      % cm^2
ND=1.0e16;     % cm^(-3)
taun=1.0e-6;   % seconds
taup=1.0e-6;   % seconds

% I01
DP=k*T.*µp;
LP=sqrt(DP.*taup);
I01=q*A.*(DP./LP.*ni.^2./ND);

% I02
Vbi=EG/2+k*T.*log(ND./ni);
W=sqrt(2.*KS*e0/(q*ND).*Vbi);
I02=q*A.*ni/sqrt(taun*taup).*W.*(k*T)./(3 .*Vbi./4);

% Currents for both Silicon (ISi) and Germanium (IGe)
VA=linspace(0.1,1);
ISi=I01(1).*exp(VA./(k*T))+I02(1).*exp(VA./(2*k*T));
IGe=I01(2).*exp(VA./(k*T))+I02(2).*exp(VA./(2*k*T));

% Plot
semilogy(VA,ISi,VA,IGe,'-'); grid
axis([0 1 1.0e-12 1.0e-3]);
xlabel('VA(volts)');
ylabel('I(A)');
text(.7, 1.4e-9,'T = 300K');
text(.7, 4.0e-10,'ND = 1.0e16 /cm^3');
text(.25, 1.4e-5,'Ge');
text(.48, 1.4e-8,'Si');
```

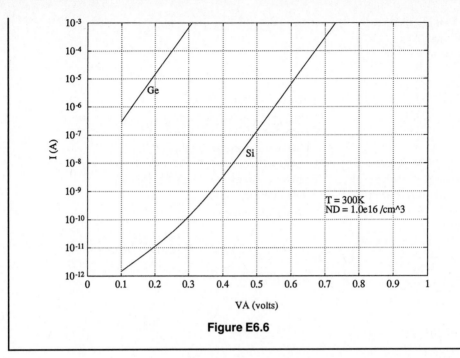

**Figure E6.6**

## Exercise 6.7

A generalization of the forward-bias modeling expression in Exercise 6.6 is commonly introduced in analyzing experimental data. Specifically, one employs

$$I = I_{01}e^{qV_A/n_1kT} + I_{02}e^{qV_A/n_2kT}$$
$$\quad\quad\uparrow \quad\quad\quad\quad\quad \uparrow$$
$$\quad\quad I_{DIFF} \quad\quad\quad\quad I_{R-G}$$

where $n_1$ and $n_2$ are arbitrary constants determined by a match to the near-ideal and R–G current related regions of the forward-bias characteristics, respectively.

**P:** Determine the $I_{01}$, $I_{02}$, $n_1$, and $n_2$ that provide an optimal match to the forward bias *I–V* data presented in Fig. 6.10(a).

**S:** Figure E6.7 is a reproduction of Fig. 6.10(a) with straight lines drawn through the two linear regions on the plot. Based on the generalized modeling expression (and similar to Eq. 6.31), the mathematical description of the straight lines is

$$\ln(I) = \ln(I_{0j}) + \frac{q}{n_jkT}V_A \quad \dots j = 1, 2$$

The $V_A = 0$ intercept therefore yields the $I_{0j}$ values, and the $n_j$-values can be computed from

$$n_j = \frac{V_{A2} - V_{A1}}{(kT/q)\ \ln(I_2/I_1)}$$

where 1 and 2 are any two points along the straight lines. We conclude

| | |
|---|---|
| $I_{01} \cong 10^{-14}$ A | $n_1 = 1.01$ |
| $I_{02} \cong 10^{-12}$ A | $n_2 = 1.56$ |

The $I_{\text{DIFF}}$ component is almost uncharacteristically ideal. The observed $I_{\text{R–G}}$ *n*-factor is on the low side but within the normally observed range. $I_{02} \gg I_{01}$, as expected.

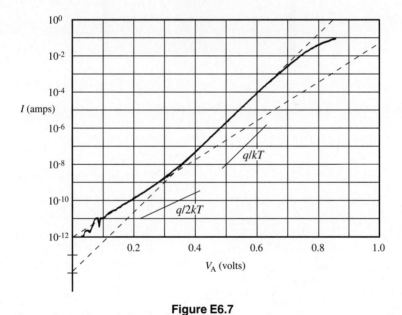

**Figure E6.7**

## 6.2.4  $V_A \rightarrow V_{bi}$ High-Current Phenomena

A significant current begins to flow in the diode when the applied forward bias approaches $V_{bi}$. A large current in turn creates conditions inconsistent with a number of ideal diode assumptions and approximations. For one, the assumption that all of the applied voltage is dropped across the depletion region becomes questionable. If $I = 0.1$ A, for example, a resistance outside the depletion region of only 1 ohm would give rise to a significant voltage drop of 0.1 V. A large current may also produce a high level of injection. The effects and modeling of these deviations are considered next.

## Series Resistance

The quasineutral regions have an inherent resistance determined by the doping and dimensions of the regions. Although quite small in a well-made device, there is also a residual resistance associated with the diode contacts. These combine to form a resistance, $R_S$, in series with the current flow across the junction as envisioned in Fig. 6.16(a). At low current levels the voltage drop across the series resistance, $IR_S$, is totally negligible compared to the applied voltage drop across the depletion region better known as the "junction" voltage, $V_J$. Under the cited condition, $V_J = V_A$, as assumed in the electrostatic and ideal diode

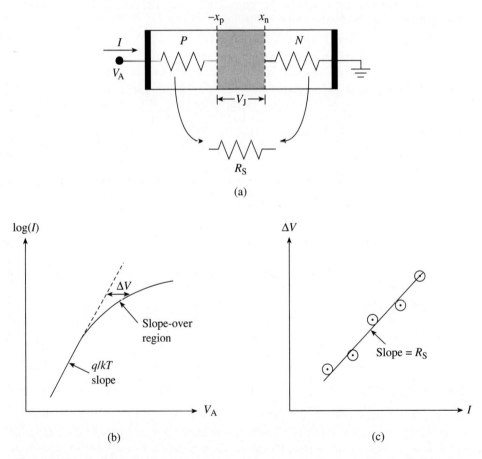

**Figure 6.16**  Identification and determination of the series resistance. (a) Physical origin of $R_S$. (b) Forward-bias semilog plot used to deduce $\Delta V$ versus $I$. (c) $\Delta V$ versus $I$ plot used to deduce $R_S$.

derivations. At current levels where $IR_S$ becomes comparable to $V_A$, however, the applied voltage drop appearing across the depletion region is reduced to

$$V_J = V_A - IR_S \qquad (6.48)$$

Effectively, part of the applied voltage is wasted, a larger applied voltage is necessary to achieve the same level of current compared to the ideal, and the characteristics slope over as illustrated in Fig. 6.10(a).

To correct for the series resistance, one merely replaces $V_A$ by $V_J = V_A - IR_S$ in previously derived $I–V_A$ relationships. Since the diffusion current typically dominates at the current levels where $IR_S$ becomes important, we can write

$$I = I_0 e^{qV_J/kT} = I_0 e^{q(V_A - IR_S)/kT} \qquad \ldots \, V_A \rightarrow V_{bi} \qquad (6.49)$$

Technically Eq. (6.49) is a transcendental equation that cannot be solved for $I$ as a function of $V_A$. However, $I$ versus $V_A$ is readily established by choosing a $V_J$ value, computing $I$ from Eq. (6.49), and then computing $V_A$ from Eq. (6.48). The computation does require a knowledge of $R_S$. $R_S$ can be determined from experimental data as outlined in parts (b) and (c) of Fig. 6.16. Working with a forward-bias semilog plot similar to Fig. 6.10(a), one extends the ideal-diode part of the plot into the slope-over region and notes the $\Delta V$ voltage displacement between the two curves as a function of $I$. Since $\Delta V = V_A - V_J = IR_S$, the slope of the line through a plot of the $\Delta V$ versus $I$ data yields $R_S$.

---

### Exercise 6.8

**P:** Crudely estimate the series resistance of the diode exhibiting the $I–V$ character-istics presented in Fig. 6.10(a).

**S:** A blow-up of the slope-over region is needed to produce a decent $\Delta V$ versus $I$ plot. We can, however, estimate $R_S$ using the $\Delta V$ at a given $I$. The Fig. 6.10(a) $I–V$ curve terminates at a current of about 80 mA when $V_A = 0.85$ V. Referring to Fig. E6.7, we find that an extension of the $q/kT$ region on the plot passes through 80 mA at roughly 0.78 V. Thus $\Delta V \cong 0.07$ V and $R_S = \Delta V/I \approx 0.07/0.08 \sim 1$ ohm.

---

### High-Level Injection

The low-level injection assumption made in the derivation of the ideal diode equation be-gins to fail when the minority carrier concentration at the depletion region edge on the lightly doped side of the junction approaches the doping concentration. In Si at room tem-perature this typically occurs at applied voltages a few tenths of a volt below $V_{bi}$. A further increase in the applied voltage gives rise to high-level injection. Under high-level injection both the minority carrier and the majority carrier concentrations adjacent to the depletion region are perturbed, as pictured in Fig. 6.17(a). The majority carrier concentration must increase to maintain approximate charge neutrality in the quasineutral regions. An analysis of high-level injection leads to a predicted current varying roughly as $\exp(q/2kT)$. In other words, one expects a high-current $q/2kT$ region on a semilog plot of the forward bias $I$–$V$ characteristics as sketched in Fig. 6.17(b). The predicted high-current $q/2kT$ region is seldom observed, however, because it is obscured by the slope-over associated with series resistance. We should note, nonetheless, that the enhanced carrier concentrations as-

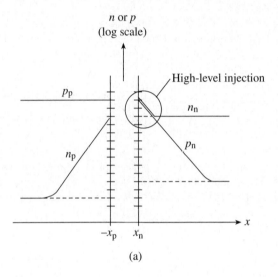

(a)

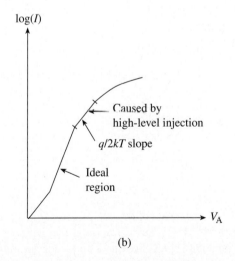

(b)

**Figure 6.17**   High-level injection. (a) Carrier concentrations under high-level injection conditions. (b) Predicted effect on the observed characteristic.

sociated with high-level injection can reduce the observed series resistance. The reduction in resistivity resulting from high levels of carrier injection is referred to as *conductivity modulation.*

---

**Exercise 6.9**

**P:** The measured *I–V* characteristic of a Si diode maintained at room temperature is crudely sketched in Fig. E6.9. Note that the current scale is logarithmic for forward bias and linear for reverse bias. Nonidealities exhibited by the characteristic are identified by capital letters. Various possible sources for the deviations from the ideal are listed to the right of the sketch. Identify the cause of each nonideal *I–V* feature; place the proper source number(s) adjacent to the letters on the sketch.

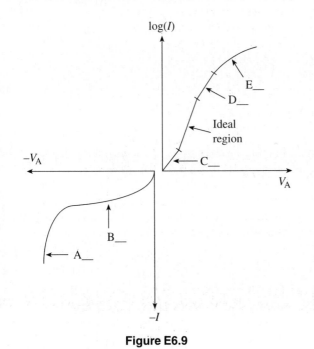

1. Photogeneration
2. Thermal recombination in the depletion region
3. Avalanching and/or Zener process
4. Low-level injection
5. Depletion approximation
6. Thermal generation in the depletion region
7. Band bending
8. Series resistance
9. $V_A > V_{bi}$
10. High-level injection

**Figure E6.9**

**S:**  A3, B6, C2, D10, E8

---

## 6.3 SPECIAL CONSIDERATIONS

The two topics addressed in this section are of a supplemental nature as far as the basic d.c. response of the *pn* junction diode is concerned. Considerations associated with the topics could be delayed until absolutely required later in the text, but that would tend to de-emphasize their importance by submerging them in a larger context. Also, a linkage of ideas and concepts critical to a greater depth of understanding becomes more difficult to establish. The charge control approach, addressed first, is a sort of "big picture" method of analysis that often allows a reasonably accurate approximate solution with a minimum of mathematics. The narrow-base diode, considered in Subsection 6.3.2, is both a special deviation from the ideal and a conceptual link to the bipolar junction transistor.

### 6.3.1 Charge Control Approach

In the charge control approach, the basic carrier variable is the charge associated with the minority carrier excess (or deficit) within an entire quasineutral region. To be specific, consider a forward-biased $p^+$-$n$ junction diode. Let $\Delta p_n(x, t)$ be the minority carrier excess in the $n$-side quasineutral region at a given time $t$ and at a point $x$, $x_n \leq x \leq \infty$. The total excess hole charge, $Q_P$, within the region is then

$$Q_P = qA \int_{x_n}^{\infty} \Delta p_n(x, t)\, dx \qquad (6.50)$$

Equation (6.50) also applies to reverse bias with a negative $Q_P$ being interpreted as a carrier deficit. $Q_P$ is of course independent of $x$, but is potentially a function of time.

Still considering a $p^+$-$n$ diode, we next seek a relationship that describes the overall behavior of this combined minority carrier charge in the $n$-side quasineutral region. A logical place to start is the minority carrier diffusion equation, Eq. (3.54b), which describes the overall behavior of the minority carrier hole concentration at a given point. Specifically, with $G_L = 0$,

$$\frac{\partial \Delta p_n}{\partial t} = D_P \frac{\partial^2 \Delta p_n}{\partial x^2} - \frac{\Delta p_n}{\tau_p} \qquad (6.51)$$

Since in a $\mathscr{E} \cong 0$ region

$$J_P = -qD_P \frac{\partial \Delta p_n}{\partial x} \qquad (6.52)$$

we can alternatively write

$$\frac{\partial(q\Delta p_n)}{\partial t} = -\frac{\partial J_P}{\partial x} - \frac{q\Delta p_n}{\tau_p} \qquad (6.53)$$

Being interested in the combined behavior of the $q\Delta p_n(x, t)$, let us integrate all terms in Eq. (6.53) over the volume ($= A \int dx$) of the *n*-side quasineutral region. After some rearrangement, we arrive at

$$\frac{d}{dt}\left[ qA \int_{x_n}^{\infty} \Delta p_n \, dx \right] = -A \int_{J_P(x_n)}^{J_P(\infty)} dJ_P - \frac{1}{\tau_p}\left[ qA \int_{x_n}^{\infty} \Delta p_n \, dx \right] \quad (6.54)$$

The bracketed quantities are recognized to be $Q_P$. Moreover,

$$-A \int_{J_P(x_n)}^{J_P(\infty)} dJ_P = -AJ_P(\infty) + AJ_P(x_n) = AJ_P(x_n) \cong AJ_{\text{DIFF}} = i_{\text{DIFF}} \quad (6.55)$$

The hole current goes to zero far on the *n*-side of the junction, making $J_P(\infty) = 0$. Since we are treating a $p^+$-*n* junction, $J_{\text{DIFF}} = J_N(-x_p) + J_P(x_n) \cong J_P(x_n)$. A lowercase *i* is used for the final result in Eq. (6.55) because the current is permitted in general to be a function of time. Making the indicated simplifications to Eq. (6.54) yields

$$\boxed{\frac{dQ_P}{dt} = i_{\text{DIFF}} - \frac{Q_P}{\tau_p}} \quad (6.56)$$

The result of our manipulations, Eq. (6.56), has a very simple interpretation. It says there are two ways to change the excess hole charge within a region: Holes can flow into or out of the region ($i_{\text{DIFF}}$), or the excess charge can be modified by recombination–generation within the region ($-Q_P/\tau_p$). Eq. (6.56) is effectively a continuity equation for the excess hole charge.

The charge control approach finds application in both steady state and transient analyses. To illustrate its use, consider a $p^+$-*n* junction diode under steady state conditions. In the steady state $dQ_p/dt = 0$, $i_{\text{DIFF}} = I_{\text{DIFF}}$, and Eq. (6.56) reduces to

$$I_{\text{DIFF}} = \frac{Q_P}{\tau_p} \quad (6.57)$$

Now suppose a solution was not available for $\Delta p_n(x)$. We do know the value of $\Delta p_n$ at $x = x_n$ ($x' = 0$) from the Eq. (6.18) boundary condition and we suspect the solution, if obtained, would be exponential with a decay constant of a diffusion length. These facts suggest an approximation similar to that introduced in Exercise 6.4. We equate $Q_P$ to the excess hole charge associated with a squared-off distribution extending a diffusion length into the *n*-side and equal to the $\Delta p_n$ value at the depletion region edge. Therefore, without actually solving for $\Delta p_n(x)$, we estimate

$$Q_P \cong q(AL_P)\Delta p_n(x'=0) = qAL_P\frac{n_i^2}{N_D}(e^{qV_A/kT} - 1) \tag{6.58}$$

giving

$$I_{DIFF} = \frac{Q_P}{\tau_p} = qA\frac{L_P}{\tau_p}\frac{n_i^2}{N_D}(e^{qV_A/kT} - 1) \tag{6.59}$$

Since $L_P/\tau_p = D_P/L_P$, Eq. (6.59) is recognized to be the usual result for the diffusion current in the $p^+$-*n* diode.

### 6.3.2 Narrow-Base Diode

#### Current Derivation

In the ideal diode derivation the diode contacts were assumed to be several minority carrier diffusion lengths or more from the edges of the depletion region. This gives rise to the $\Delta n_p(-\infty) = 0$ and $\Delta p_n(\infty) = 0$ boundary conditions. For diodes whose width on the lightly doped side is equal to the substrate or wafer thickness, the wide-base diode assumption is typically justified. The assumption comes into question, however, if the diode is fabricated, for example, in a thin epitaxial layer. A diode where the width of the quasineutral region on the lightly doped side of the junction is on-the-order-of or less than a diffusion length is known as a *narrow-base* diode.

To determine the diffusion current flowing in a narrow base diode, it is necessary to revise the ideal diode derivation. We take the diode under analysis to be a $p^+$-*n* step junction diode identical in every way to an ideal diode except that the width of the quasineutral *n*-region is possibly comparable to or less than a minority carrier diffusion length. As graphically defined in Fig. 6.18, $x_c$ is the distance from the metallurgical junction to the *n*-side contact, and $x_c' = x_c - x_n$ the distance from the *n*-edge of the depletion region to the *n*-side contact. The *n*-side contact is assumed to be ohmic, making $\Delta p_n = 0$ at $x' = x_c'$. In general, the minority carrier concentration at a contact a finite distance from the depletion region edge depends on the rate of carrier recombination–generation at the contact. At a well-made or ohmic contact, the recombination–generation rate is high and the minority carrier concentration is maintained near its equilibrium value.

Paralleling the derivation of the ideal diode equation, we must solve

$$0 = D_P\frac{d^2\Delta p_n}{dx'^2} - \frac{\Delta p_n}{\tau_p} \quad \ldots 0 \le x' \le x_c' \tag{6.60}$$

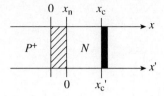

**Figure 6.18**  Specification of the *n*-side contact position in a narrow-base diode. $x'_c$ is assumed to be comparable to or less than $L_P$.

subject to the boundary conditions

$$\Delta p_n(x'=0) = \frac{n_i^2}{N_D}(e^{qV_A/kT} - 1) \qquad (6.61a)$$

$$\Delta p_n(x'=x'_c) = 0 \qquad (6.61b)$$

As in the ideal diode derivation, the general solution is

$$\Delta p_n(x') = A_1 e^{-x'/L_P} + A_2 e^{x'/L_P} \quad \dots 0 \le x' \le x'_c \qquad (6.62)$$

Here, however, $A_2$ is not required to vanish. Rather, applying the boundary conditions gives

$$\Delta p_n(0) = A_1 + A_2 \qquad (6.63a)$$

$$0 = A_1 e^{-x'_c/L_P} + A_2 e^{x'_c/L_P} \qquad (6.63b)$$

Solving Eqs. (6.63) for $A_1$ and $A_2$, and substituting back into the general solution, we conclude

$$\Delta p_n(x') = \Delta p_n(0)\left(\frac{e^{(x'_c-x')/L_P} - e^{-(x'_c-x')/L_P}}{e^{x'_c/L_P} - e^{-x'_c/L_P}}\right) \quad \dots 0 \le x' \le x'_c \quad (6.64)$$

or more compactly in terms of the sinh function

$$\boxed{\Delta p_n(x') = \Delta p_n(0)\frac{\sinh[(x'_c - x')/L_P]}{\sinh[x'_c/L_P]}} \quad \dots 0 \le x' \le x'_c \qquad (6.65)$$

where

$$\sinh(\xi) \equiv \frac{e^{\xi} - e^{-\xi}}{2} \tag{6.66}$$

Finally,

$$I_{\text{DIFF}} \cong AJ_{\text{P}}(x' = 0) = -qAD_{\text{P}}\frac{d\Delta p_{\text{n}}}{dx'}\bigg|_{x'=0} \tag{6.67}$$

leading to

$$I_{\text{DIFF}} = I_0'(e^{qV_{\text{A}}/kT} - 1) \tag{6.68}$$

$$I_0' \equiv qA\frac{D_{\text{P}}}{L_{\text{P}}}\frac{n_{\text{i}}^2}{N_{\text{D}}}\frac{\cosh(x_{\text{c}}'/L_{\text{P}})}{\sinh(x_{\text{c}}'/L_{\text{P}})} \tag{6.69}$$

where

$$\cosh(\xi) \equiv \frac{e^{\xi} + e^{-\xi}}{2} \tag{6.70}$$

## Limiting Cases/Punch-Through

To assist in the examination and interpretation of results, it is useful to note

$$\sinh(\xi) \rightarrow \begin{cases} \xi & \ldots \xi \rightarrow 0 \tag{6.71a} \\ \dfrac{e^{\xi}}{2} & \ldots \xi \rightarrow \infty \tag{6.71b} \end{cases}$$

$$\cosh(\xi) \rightarrow \begin{cases} 1 + \dfrac{\xi^2}{2} & \ldots \xi \rightarrow 0 \tag{6.72a} \\ \dfrac{e^{\xi}}{2} & \ldots \xi \rightarrow \infty \tag{6.72b} \end{cases}$$

Consider first the limit where $x_{\text{c}}' \rightarrow \infty$. If $x_{\text{c}}' \rightarrow \infty$ or $x_{\text{c}}'/L_{\text{P}} \gg 1$, the ratio of the sinh terms in Eq. (6.65) simplifies to $\exp(-x'/L_{\text{P}})$ and Eq. (6.65) reduces to Eq. (6.23) in the ideal diode derivation. Similarly, $\cosh(x_{\text{c}}'/L_{\text{P}})/\sinh(x_{\text{c}}'/L_{\text{P}}) \rightarrow 1$, and Eq. (6.69) reduces to the standard ideal diode result for a $p^{+}$-$n$ junction. In essence, the narrow-base diode analysis may be viewed as a generalization of the ideal diode formulation valid for quasineutral regions of arbitrary thickness.

Although mathematically reassuring, the wide-base limit leads to nothing new. It is the opposite limit where $x_c' \to 0$ that proves to be more interesting. For one, if $x_c'/L_P \ll 1$, the sinh terms in Eq. (6.65) can be replaced by their arguments, and $\Delta p_n(x')$ dramatically simplifies to

$$\Delta p_n(x') = \Delta p_n(0)\left(1 - \frac{x'}{x_c'}\right) \tag{6.73}$$

The perturbed carrier concentration becomes a linear function of position, just a straight line connecting the two endpoint values set by the boundary conditions. The linear dependence is a direct consequence of negligible thermal recombination–generation in a region much shorter than a diffusion length. In fact, the Eq. (6.73) result could have been obtained much more readily by neglecting the R–G term ($-\Delta p_n/\tau_p$) in the original minority carrier diffusion equation. This observation is justification for henceforth *neglecting the thermal R–G term in the minority carrier diffusion equation when the quasineutral width is small compared to a diffusion length.*

A second small-width observation involves the reverse-bias current. In the limit where $x_c'/L_P \ll 1$, $\cosh(x_c'/L_P)/\sinh(x_c'/L_P) \to L_P/x_c'$ and $I_0' \to qA(n_i^2/N_D)(D_P/x_c')$. The width of the quasineutral region, $x_c' = x_c - x_n$, decreases with increasing reverse bias because of the growing depletion width. It therefore follows that $I_{DIFF}(V_A < 0) \cong -I_0' \propto 1/x_c'$ does not saturate, but systematically increases with the applied reverse bias. If $x_c$ is sufficiently small, there is also the possibility of $x_c' \to 0$. The situation where an entire device region becomes depleted is referred to as *punch-through*. Based on the Eq. (6.68)/(6.69) result, $I_{DIFF}(V_A < 0) \to -\infty$ if $x_c' \to 0$. However, because of the depletion approximation used in defining the width and nature of the quasineutral region, the theoretical formulation becomes invalid in the extreme $x_c' \to 0$ limit. Experimentally and in more precise theoretical formulations, the diffusion current remains finite at the punch-through voltage in a narrow-base diode, provided the electric field inside the device is insufficient to produce avalanche breakdown.

---

### Exercise 6.10

**P:** A planar Si $p^+$-$n$ step junction diode maintained at room temperature has an $n$-side doping of $N_D = 10^{16}/cm^3$ and an $n$-side thickness of $x_c = 2~\mu m$. Invoking the depletion approximation, determine the punch-through voltage; i.e., determine the voltage required to completely deplete the $n$-side of the diode.

**S:** At the punch-through voltage $x_n = x_c$, where from Eq. (5.37),

$$x_n \cong \left[\frac{2K_S\varepsilon_0}{qN_D}(V_{bi} - V_A)\right]^{1/2}$$

Thus

$$x_n^2 = x_c^2 = \frac{2K_S\varepsilon_0}{qN_D}(V_{bi} - V_A)$$

$$V_A = V_{bi} - \frac{qN_D}{2K_S\varepsilon_0}x_c^2$$

$$= 0.92 - \frac{(1.6 \times 10^{-19})(10^{16})(2 \times 10^{-4})^2}{(2)(11.8)(8.85 \times 10^{-14})} = \mathbf{-29.7\ V}$$

$V_{bi}$ was read from Fig. E5.1 in Subsection 5.1.4.

Inspecting Fig. 6.11, we note that avalanche breakdown occurs at $V_{BR} \cong 55$ V in a comparable wide-base diode. Thus, punch-through will take place prior to break-down in the given narrow-base diode.

## 6.4  SUMMARY AND CONCLUDING COMMENTS

The chapter was devoted to modeling the steady state response of the *pn* junction diode. We began with a qualitative description of diode operation using the energy band diagram as a visualization tool. The reverse-bias current in an ideal diode is associated with minority carriers wandering into the depletion region and being accelerated to the opposite side of the junction. Forward biasing lowers the potential hill between the two sides of the diode and enhances majority carrier injection across the junction and into the opposite-side quasineutral region. There is of course no carrier build-up anywhere inside the device under steady state conditions. Recombination–generation acts to stabilize the minority carrier concentrations, while a very rapid rearrangement and interchange with the contacts maintain the constancy of the majority carrier concentrations.

Considerable attention was next given to the formulation, derivation, and examination of the ideal diode theory. Although the predicted *I–V* characteristic exhibits serious deficiencies when carefully compared with experiment, the ideal diode development provides valuable insight into the internal operation of the diode and an analytical base for more accurate formulations. Even with its limitations, the theory and extensions of the theory are widely used in diode and other device analyses to obtain manageable first order predictions. The ideal diode equation, it should be cautioned, is only a small part of the ideal diode theory. The theory encompasses a clear understanding of the solution strategy and approximations, the mathematical steps to follow in establishing a quantitative result, and predictions relative to carrier concentrations, carrier currents, material dependencies, doping dependencies, and so on.

As already acknowledged, a careful comparison between the ideal diode and experimental *I–V* characteristics revealed several deviations from the ideal. For a Si diode maintained at room temperature, the deviations included reverse-bias breakdown, excess current levels under small forward bias and all reverse biases, and a reduced *I–V* slope or slope-over at forward biases approaching $V_{bi}$. Breakdown was associated with avalanching or the Zener process, the excess current with thermal carrier recombination–generation in the depletion region, and the reduced *I–V* slope at high forward currents with the voltage drop across internal series resistances or possibly high-level injection. Each of the deviation-causing phenomena was described physically, and the more important phenomena were modeled analytically. Information of special note included $V_{BR}$ due to avalanching varies roughly as the inverse of the doping concentration on the lightly doped side of the junction, carrier multiplication leads to an enhanced current far below the breakdown voltage, the R–G current exhibits an expected dependence approaching $\exp(qV_A/2kT)$ under forward biasing and is proportional to $W$ under reverse biasing, the R–G current varies as $n_i$ while the diffusion current varies as $n_i^2$, the voltage drop across the internal series resistance reduces the voltage drop across the junction to $V_J = V_A - IR_S$, and high-level injection gives rise to an expected high-current $\exp(qV_A/2kT)$ operational region that is usually obscured by the slope-over associated with series resistance. The development, it should be noted, concentrated on the step junction. Analytical modifications may be required in treating other profiles.

    In the special considerations section we introduced an analytical approach known as charge control and generalized the ideal diode analysis to accommodate base widths of arbitrary thickness. When the quasineutral width is small compared to a diffusion length, the perturbed carrier distribution was noted to become linear. For extremely small base widths, there is also the possibility of punch-through or complete depletion of the narrow base at a reverse-bias voltage less than $V_{BR}$.

    Although we have covered the highlights in this summary, there are other pieces of information—terms, techniques, approximations, equations—that are part of the overall information package. The Part II Supplement and Review does contain a Review List of Terms grouped by chapter. As far as equations are concerned, the ideal diode solution (Eqs. 6.29/6.30), the law of the junction (Eq. 6.12), the depletion-edge boundary conditions (Eqs. 6.15/6.18), and the reverse-bias R–G current expression (Eq. 6.43) might be committed to memory. The book should be retained for looking up the remainder of the equations.

## PROBLEMS

| CHAPTER 6   PROBLEM INFORMATION TABLE | | | | |
|---|---|---|---|---|
| Problem | Complete After | Difficulty Level | Suggested Point Weighting | Short Description |
| 6.1 | 6.1.4 | 1 | 10 (1 each part) | Ideal diode quiz |
| 6.2 | 6.2.4 | 1 | 10 | Compare actual-ideal *I–V* |

| 6.3 | 6.1.1 | 1 | 9 (3 each part) | Band diagram sketches |
|---|---|---|---|---|
| 6.4 | " | 2 | 8 | Forward-bias big picture |
| 6.5 | 6.1.2 | 2–3 | 10 | Derive Eq. (6.15) using $J_N=0$ |
| ● 6.6 | 6.1.3 | 2 | 10 (5 each part) | Typical $I$ values using `diary` |
| ● 6.7 | " | 2 | 16 (a-10, b-5, c-1) | Ideal $I$–$V$ variation with $T$ |
| 6.8 | " | 3 | 16 (a-3, b-10, c-3) | Photodiode $I$–$V$ derivation |
| ● 6.9 | " | 1 | 5 | Photodiode $I$–$V$ plot |
| 6.10 | 6.1.4 | 2 | 10 (a::c-2, d-4) | Deduce info from conc. plot |
| 6.11 | " | 3 | 12 | Deduce, plot carrier conc. |
| ● 6.12 | " | 3 | 12 (a-9, b-3) | *pn* diode $T$-sensor |
| 6.13 | 6.2.2 | 2 | 6 (2 each part) | $V_{BR}$-related calculations |
| ● 6.14 | " | 2 | 10 (a-8, b-2) | Breakdown $I$–$V$ ($MI_0$) |
| ● 6.15 | 6.2.3 | 3 | 12 | Breakdown $I$–$V$ ($MI_{R-G}$) |
| 6.16 | " | 3 | 8 | Determine elevated $T$-point |
| 6.17 | " | 3 | 10 | Modified $I_{R-G}$ |
| 6.18 | 6.2.4 | 3 | 9 (a-6, b-3) | $R_S$-related computations |
| 6.19 | " | 3 | 20 (10-line fit, 8-least sq. fit, 2-discuss) | Deduce $I_{0j}$, $n_j$ from $I$–$V$ |
| 6.20 | " | 2–3 | 15 | Deduce $R_S$ from $I$–$V$ |
| ● 6.21 | " | 4 | 40 (a-2, b-10, c::e-8, f-4) | Multiple nonideality $I$–$V$ |
| 6.22 | 6.3.2 | 3 | 10 | Pseudo-narrow-base diode |
| 6.23 | " | 3 | 15 | Narrow/wide diode combo |
| 6.24 | " | 4 | 34 (a-2, b-2, c-4, d-2, e-12, f-12) | Multiple considerations |

### 6.1 Ideal Diode Quiz

Answer the following questions as concisely as possible. Assume the questions refer to an ideal diode.

(a) The forward-bias current is associated with what type of carrier activity?

(b) The reverse-bias current is associated with what type of carrier activity?

(c) Why is the reverse-bias current expected to be small in magnitude and to saturate at a small reverse voltage?

(d) Under reverse biasing, what processes occur in the quasineutral regions adjacent to the depletion region edges? Is it drift and diffusion, diffusion and recombination, generation and diffusion, or generation and drift?

(e) Why can't the minority carrier diffusion equations be used to determine the minority carrier concentrations and currents in the depletion region of a diode?

(f) What a priori justification is there for assuming recombination–generation is negligible throughout the depletion region?

(g) What exactly is a "wide-base" diode?

(h) What is the "law of the junction"?

(i) Given a semilog plot of the forward bias *I–V* characteristic, how does one determine $I_0$?

(j) True or false: The reverse-bias saturation current may be alternatively viewed as arising from minority carrier generation in the quasineutral regions.

**6.2** Compare (preferably with the aid of sketches) the *I–V* characteristic derived from an actual Si *pn* junction diode maintained at room temperature and the theoretical ideal-diode characteristic. In effecting your comparison, carefully note and (as necessary) clarify all deviations from the ideal. Finally, briefly identify the cause of each deviation from the ideal theory. Devote no more than a sentence or two to the cause of each deviation.

**6.3** Sketch the energy band diagram for an ideal $p^+$-$n$ step junction diode showing the carrier activity in and near the depletion region when

(a) $V_A = 0$.

(b) $V_A > 0$.

(c) $V_A < 0$.

**6.4** Construct a composite energy-band/circuit diagram analogous to Fig. 6.2 that provides an overall view (the "big picture" view) of carrier activity inside a *forward-biased pn* junction diode.

**6.5** Referring to the $V_{bi}$ derivation in Subsection 5.1.4, show that the Eq. (6.15) boundary condition can be derived by continuing to assume $J_N = 0$ when $V_A \neq 0$. Use the Eq. (5.10) result to eliminate $V_{bi}$ in your expression for $n(-x_p)$.

● **6.6** Considering an $n^+$-$p$ silicon step junction ($N_A = 10^{15}/\text{cm}^3$, $\tau_n = 10^{-6}$ sec, $A = 10^{-3}$ cm²) to be ideal, perform calculations to determine typical current levels expected from the device under different temperature and biasing conditions. Invoke the MATLAB diary function to record your work.

(a) Compute the ideal diode current at $T = 300$ K when (i) $V_A = -50$ V, (ii) $V_A = -0.1$ V, (iii) $V_A = 0.1$ V, and (iv) $V_A = 0.5$ V.

(b) Assuming $\tau_n$ does not vary significantly with temperature, repeat part (a) for $T = 500$ K.

● **6.7** In this problem we wish to explore how the ideal diode $I$–$V$ characteristic varies as a function of temperature. Consider an ideal Si $p^+$-$n$ step junction diode with an area $A = 10^{-4}$ cm², $N_D = 1.0 \times 10^{16}$/cm³, and a room temperature $\tau_p = 10^{-6}$ sec.

(a) Using the empirical-fit relationship for the hole mobility specified in Exercise 3.1, the $n_i$ fit relationship cited in Exercise 2.4(b), and assuming $\tau_p$ to be temperature-independent, construct a MATLAB program that computes and plots the ideal $I$–$V$ characteristic of the diode with temperature $T$ in Kelvin being considered an input variable. Print out a sample plot simultaneously displaying the characteristics associated with $T = 295$ K, 300 K, and 305 K. (To clearly display both the reverse and forward characteristics, limit the plot axes to $-1 \text{ V} \leq V_A \leq 0.2$ V and $-2I_0 \leq I \leq 5I_0$. Use the largest $I_0$ to set the current limits when displaying several $I$–$V$ curves on the same set of coordinates.)

(b) Appropriately modify your part (a) program to obtain a semilog plot of $I_0$ versus $T$ for 300 K $\leq T \leq$ 400 K.

(c) Comment on the part (a) and (b) results.

## 6.8 Photodiode/Solar Cell

A *pn* junction photodiode is just a *pn* junction diode that has been specially fabricated and encapsulated to permit light penetration into the vicinity of the metallurgical junction. Commercially available solar cells are in essence large-area *pn* junction photodiodes designed to minimize energy losses. The general form of the similar $I$–$V$ characteristics exhibited by photodiodes and solar cells is readily established by a straightforward modification of the ideal diode equation.

Consider an ideal $p^+$-$n$ step junction diode where incident light is uniformly absorbed throughout the device producing a photogeneration rate of $G_L$ electron-hole pairs per cm³-sec. Assume that low-level injection prevails.

(a) What is the excess minority carrier concentration on the $n$-side a large distance $(x \to \infty)$ from the metallurgical junction. [NOTE: $\Delta p_n(x \to \infty) \neq 0$.]

(b) The usual ideal diode boundary conditions (Eqs. 6.15 and 6.18) still hold at the edges of the depletion region. Using the revised boundary condition established in part (a) and Eq. (6.18), derive an expression for the $I$–$V$ characteristic of the $p^+$-$n$ diode under the stated conditions of illumination. As in the derivation of the ideal diode equation, ignore all recombination–generation, including photogeneration, occurring in the depletion region.

(c) Sketch the general form of the photodiode $I$–$V$ characteristics taking in turn $G_L = 0$, $G_L = G_{L0}$, $G_L = 2G_{L0}$, and $G_L = 4G_{L0}$. Assume the light intensity is high enough to significantly perturb the characteristics when $G_L = G_{L0}$. (NOTE: Problem 6.9 can be substituted for this part of the problem.)

● **6.9** *I–V* characteristics of a solar cell are approximately modeled by the relationship,

$$I = I_0(e^{qV_A/kT} - 1) + I_L$$

where $I_L$ is the current due to light. $I_L$ is always negative and a voltage-independent constant for a given level of illumination. Construct a plot illustrating the general nature of the solar cell characteristics. Setting $T = 300$ K, simultaneously plot $I/I_0$ versus $V_A$ for the assumed values of $I_L/I_0 = 0, -1, -2,$ and $-4$. Limit $V_A$ to $-0.5$ V $\leq V_A \leq 0.1$ V.

**6.10** Figure P6.10 is a dimensioned plot of the steady state carrier concentrations inside a *pn* step junction diode maintained at room temperature.

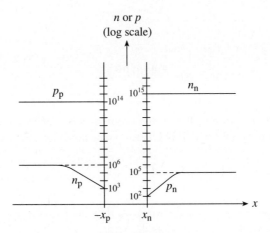

**Figure P6.10**

(a) Is the diode forward or reverse biased? Explain how you arrived at your answer.

(b) Do low-level injection conditions prevail in the quasineutral regions of the diode? Explain how you arrived at your answer.

(c) What are the *p*-side and *n*-side doping concentrations?

(d) Determine the applied voltage, $V_A$.

**6.11** A voltage $V_A = 23.03(kT/q)$ is being applied to a step junction diode with *n*- and *p*-side dopings of $N_A = 10^{17}/cm^3$ and $N_D = 10^{16}/cm^3$, respectively. $n_i = 10^{10}/cm^3$. Make a dimensioned $\log(p$ and $n)$ versus $x$ sketch of both the majority and minority carrier concentrations in the quasineutral regions of the device. Identify points 10 and 20 diffusion lengths from the depletion region edges on your sketch.

● **6.12** As was confirmed graphically in Problem 6.7, the *I–V* characteristics of the *pn* junction diode are very sensitive to temperature. It should not be too surprising then that there are silicon diodes sold commercially as temperature sensors. To use the diode to measure temperature, it is typically forward-biased with a constant current source and $V_A$ is monitored as a function of *T* as shown in Fig. P6.12. Make the assumption the diode is being operated in the forward bias range where $I \cong I_0 \exp(qV_A/kT)$. Also take the intended range of operation to be $0 \leq T(°C) \leq 100°C$.

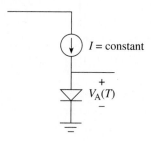

$I = \text{constant}$

$V_A(T)$

**Figure P6.12**

(a) With *I* as an input variable, modify the Problem 6.7(b) program to compute and plot $V_A$ versus *T* for $0 \leq T(°C) \leq 100°C$. Print out a sample plot using $I = 10^{-4}$ A.

(b) Considering the sensitivity ($dV_A/dT$ in mV/°C) of the sensor, would it be preferable to monitor the temperature using $I = 10^{-4}$ A or $I = 10^{-3}$ A? Support your answer.

**6.13** Given a planar $p^+$-$n$ Si step junction diode with an *n*-side doping of $N_D = 10^{15}$/cm³ and $T = 300$ K, determine

(a) The approximate $V_{BR}$ of the diode.

(b) The depletion width at the breakdown voltage.

(c) The maximum magnitude of the electric field in the depletion region at the breakdown voltage.

● **6.14** Let us pursue the modeling of carrier multiplication and avalanche breakdown in an otherwise ideal diode. As noted in the text, Eq. (6.35) can be used to correct the ideal diode equation and approximately account for carrier multiplication. $I_0$ is simply replaced by $MI_0$. Moreover, the expected $V_{BR}$ due to avalanching varies roughly as $N_B^{-0.75}$. A reasonable fit to the Fig. 6.11 dependence for Si diodes is

$$V_{BR} \cong 60(N_B/10^{16})^{-0.75}$$

where $N_B$ is in cm$^{-3}$ and $V_{BR}$ is in volts. Take the Si diode under analysis to be a nearly ideal $p^+$-$n$ step junction maintained at 300 K.

(a) Construct a MATLAB program to perform $I$–$V$ computations that include carrier multiplication and avalanche breakdown. Specifically, compute and plot the modified reverse bias $I/I_0$ versus $V_A$ characteristic covering the range from $-V_{BR}$ to 0. (Note the $I_0$ normalization.) Employ $m = 6$ in your computations and set the `axis` function parameters to $[-1.1*V_{BR}, 0, -5, 0]$. Print out a sample characteristic taking $N_D = 2 \times 10^{16}$/cm$^3$. Is your plot consistent with the curving approach to breakdown shown in Fig. 6.10(b) and Exercise 6.5?

(b) Use your program to explore how the diode characteristic varies as a function of $m$ and the impurity concentration on the lightly doped side of the junction.

● **6.15** Repeat Problem 6.14 replacing $I_0$ by the reverse bias $I_{R-G}$. Normalize the current axis to the magnitude of the current at $V_A = -V_{BR}/2$. Run your program setting $N_D = 2 \times 10^{16}$/cm$^3$. Compare your plot with Fig. 6.10(b).

**6.16** The diffusion component of the current is expected to dominate in Si diodes at sufficiently elevated temperatures. What is a sufficiently elevated temperature? To answer this question, suppose one has a Si $p^+$-$n$ step junction diode where $N_D = 10^{16}$/cm$^3$, $\tau_0 = \tau_p$, and $L_p \cong 10^{-2}$ cm for 300 K $\leq T \leq$ 500 K. Determine the temperature where $I_{DIFF} = I_{R-G}$ at a reverse bias of $V_{bi} - V_A = V_{BR}/2$.

**6.17** Inside a Si diode at 300 K a region of width $d$ contains *three times as many R–G centers* as adjoining regions. This special region, as envisioned in Fig. P6.17, lies totally inside the depletion region when the diode is zero biased. Derive an expression for the R–G current ($I_{R-G}$) to be expected from the diode when reverse-biased greater than a few $kT/q$ volts. Record all derivational steps.

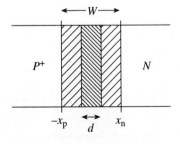

**Figure P6.17**

**6.18** Suppose $I_0 = 10^{-14}$ A and $R_S = 2\ \Omega$ in a diode where high-level injection is negligible. $T = 300$ K.

(a) Determine the forward-bias current at which the applied voltage differs from the ideal by 10%.

(b) Repeat part (a) with $R_S = 20\ \Omega$.

**6.19** Point-by-point forward-bias $I$–$V$ data from a 1N757 Si *pn* junction diode maintained at room temperature is listed in Table P6.19 and plotted in Fig. P6.19. Following the procedure outlined in Exercise 6.7, determine $I_{01}$, $I_{02}$, $n_1$, and $n_2$. Because the line-drawing approach is somewhat subjective, also apply a least squares fit (available in MATLAB) to appropriate segments of the data to determine the fit parameters. Discuss your results.

**Table P6.19**

| $V_A$ (volts) | $I$ (amps) | $V_A$ (volts) | $I$ (amps) |
|---|---|---|---|
| 0.02 | $6.070 \times 10^{-10}$ | 0.48 | $3.615 \times 10^{-6}$ |
| 0.04 | $1.203 \times 10^{-9}$ | 0.50 | $5.200 \times 10^{-6}$ |
| 0.06 | $2.165 \times 10^{-9}$ | 0.52 | $7.576 \times 10^{-6}$ |
| 0.08 | $3.508 \times 10^{-9}$ | 0.54 | $1.122 \times 10^{-5}$ |
| 0.10 | $5.417 \times 10^{-9}$ | 0.56 | $1.711 \times 10^{-5}$ |
| 0.12 | $8.210 \times 10^{-9}$ | 0.58 | $2.694 \times 10^{-5}$ |
| 0.14 | $1.183 \times 10^{-8}$ | 0.60 | $4.426 \times 10^{-5}$ |
| 0.16 | $1.730 \times 10^{-8}$ | 0.62 | $7.587 \times 10^{-5}$ |
| 0.18 | $2.449 \times 10^{-8}$ | 0.64 | $1.359 \times 10^{-4}$ |
| 0.20 | $3.416 \times 10^{-8}$ | 0.66 | $2.531 \times 10^{-4}$ |
| 0.22 | $4.764 \times 10^{-8}$ | 0.68 | $4.852 \times 10^{-4}$ |
| 0.24 | $6.501 \times 10^{-8}$ | 0.70 | $9.444 \times 10^{-4}$ |
| 0.26 | $8.866 \times 10^{-8}$ | 0.72 | $1.841 \times 10^{-3}$ |
| 0.28 | $1.209 \times 10^{-7}$ | 0.74 | $3.518 \times 10^{-3}$ |
| 0.30 | $1.666 \times 10^{-7}$ | 0.76 | $6.433 \times 10^{-3}$ |
| 0.32 | $2.305 \times 10^{-7}$ | 0.78 | $1.103 \times 10^{-2}$ |
| 0.34 | $3.201 \times 10^{-7}$ | 0.80 | $1.752 \times 10^{-2}$ |
| 0.36 | $4.462 \times 10^{-7}$ | 0.82 | $2.585 \times 10^{-2}$ |
| 0.38 | $6.285 \times 10^{-7}$ | 0.84 | $3.579 \times 10^{-2}$ |
| 0.40 | $8.845 \times 10^{-7}$ | 0.86 | $4.706 \times 10^{-2}$ |
| 0.42 | $1.249 \times 10^{-6}$ | 0.88 | $5.941 \times 10^{-2}$ |
| 0.44 | $1.776 \times 10^{-6}$ | 0.90 | $7.264 \times 10^{-2}$ |
| 0.46 | $2.527 \times 10^{-6}$ | | |

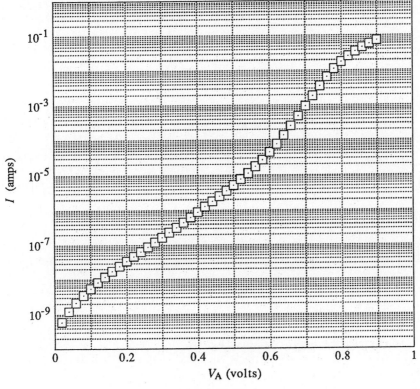

**Figure P6.19**

**6.20** Following the procedure outlined in Fig. 6.16, determine the $R_S$ value of the 1N757 diode characterized by the room temperature data listed in Table P6.19 and plotted in Fig. P6.19.

● **6.21** The model shown in Fig. P6.21 is sometimes employed as a generalized representation of real *pn* junction diodes. The diode symbol is understood to represent the standard model of the device including the diffusion and R–G components of the junction current. The series resistor ($R_S$) accounts for non-negligible voltage drops in the quasineutral regions and contacts of the diode under high-current conditions. The shunt resistor ($R_{SH}$) accounts for a possible leakage current in the semiconductor external to the *pn* junction. Note that the current through the *pn* junction and the voltage drop across the junction are redefined to be $I_J$ and $V_J$, respectively.

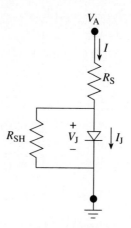

**Figure P6.21**

(a) If the current through the junction proper is described by the relationship,

$$I_J = I_{01}(e^{qV_J/n_1kT} - 1) + I_{02}(e^{qV_J/n_2kT} - 1)$$

where $I_{01}$ and $I_{02}$ are constants and typically $n_1 \cong 1$ and $n_2 \cong 2$, confirm that the total current through the diode is described by

$$I = I_{01}(e^{qV_J/n_1kT} - 1) + I_{02}(e^{qV_J/n_2kT} - 1) + \frac{V_J}{R_{SH}}$$

with $V_A = V_J + IR_S$

(b) Construct a computer program that can be used to investigate the effect of $R_S$, $R_{SH}$, and the size of the current prefactors ($I_{01}$ and $I_{02}$) on the shape of the observed *forward-bias* $I$–$V_A$ characteristic. In performing the computation, it is best to obtain $I$–$V_A$ data pairs associated with evenly stepped values of $V_J$. Limit your plotted $\log(I)$ vs. $V_A$ output to $0 \le V_A \le 1$ V and $10^{-10}$ A $\le I \le 10^{-2}$ A. Employ $I_{01} = 10^{-13}$ A, $I_{02} = 10^{-9}$ A, $n_1 = 1$, $n_2 = 2$, $R_S = 1\ \Omega$, $R_{SH} = 10^{12}\ \Omega$, and $T = 300$ K in your initial calculation.

(c) Employing a manually controlled computational switch, modify your program so that it computes and *simultaneously* displays four $I$–$V_A$ characteristics associated with four different values of *either* $R_{SH}$, $R_S$, $I_{01}$, or $I_{02}$. Using the same basic set of parameters as specified in part (b), compute and record the $I$–$V_A$ characteristics resulting from the following variation of parameters:

  (i) $R_{SH} = 10^{12}$, $10^9$, $10^6$, and $10^3\ \Omega$ ($I_{01} = 10^{-13}$ A, $I_{02} = 10^{-9}$ A, $R_S = 1\ \Omega$);

  (ii) $R_S = 1$, $10$, $10^2$, and $10^3\ \Omega$ ($I_{01} = 10^{-13}$ A, $I_{02} = 10^{-9}$ A, $R_{SH} = 10^{12}\ \Omega$);

(iii) $I_{02} = 10^{-9}$, $10^{-10}$, $10^{-11}$, and $10^{-12}$ A ($I_{01} = 10^{-13}$ A, $R_S = 1\,\Omega$, $R_{SH} = 10^{12}\,\Omega$);

(iv) $I_{01} = 10^{-13}$, $10^{-12}$, $10^{-11}$, and $10^{-10}$ A ($I_{02} = 10^{-9}$ A, $R_S = 1\,\Omega$, $R_{SH} = 10^{12}\,\Omega$).

(d) Following the procedure established in Exercise 6.6, construct a subsidiary program that can be used to compute $I_{01}$ and $I_{02}$ from first principle relationships. Specifically assume a $n^+$-$p$ step junction. Join the subsidiary program to your main program and obtain sample results employing $A = 10^{-2}$ cm$^2$, $N_A = 10^{15}/$cm$^3$, and $\tau_n = \tau_p = \tau_0 = 10^{-6}$ sec. The remainder of the parameters should of course be appropriate for Si at $T = 300$ K.

(e) Construct a modified version of the part (c) program that displays the computed results on a *linear I–V*$_A$ plot. Limit the linear current axis to $0 \leq I \leq 10^{-2}$ A. Compute and record linear plots of the $I$–$V_A$ characteristics resulting from the variation of parameters specified in part (c). How do the results here compare with those obtained in part (c)?

(f) Generally comment on the results obtained in this problem.

**6.22** Consider the special silicon $p^+$-$n$ step junction diode pictured in Fig. P6.22. Note that $\tau_p = \infty$ for $0 \leq x \leq x_b$ and $\tau_p = 0$ for $x_b \leq x \leq x_c$. Excluding biases that would cause high-level injection or breakdown, develop an expression for the room-temperature $I$–$V$ characteristic of the diode. Assume the depletion width ($W$) never exceeds $x_b$ for all biases of interest.

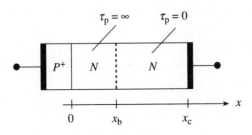

**Figure P6.22**

**6.23** Reconsider the special silicon $p^+$-$n$ junction diode pictured in Fig. P6.22. Instead of $\tau_p = 0$ in the $x_b \leq x \leq x_c$ region, take $\tau_p$ to be nonzero but still sufficiently small so that $L_P \ll x_c - x_b$. Assuming the depletion width ($W$) never exceeds $x_b$ for all biases of interest, and excluding biases that would cause high-level injection or breakdown, derive an expression for the room temperature $I$–$V$ characteristic of the diode.

**6.24** In modern device processing a procedure called *denuding* is used to reduce the R–G center concentration in the near-surface region of devices. A planar Si $p^+$-$n$ step junction diode maintained at 300 K has an $n$-side doping of $N_D = 10^{16}/$cm$^3$. As shown in Fig. P6.24, denuding has created a reduced R–G center concentration of $N_{T1}$ in the $n$-side region from $x = 0$ to $x = x_b = 2 \ \mu$m. The R–G center concentration in the $x > x_b$ $n$-side region is $N_{T2} = 100 N_{T1}$.

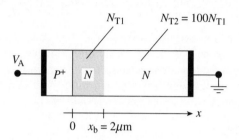

Figure P6.24

(a) What is the junction built-in voltage ($V_{bi}$)?

(b) What is the diode breakdown voltage ($V_{BR}$)?

(c) What applied voltage is necessary to expand the $n$-side depletion region edge to $x = x_b$?

(d) If the minority carrier lifetimes are $\tau_{p1}$ and $\tau_{p2}$ in the near-junction and $x > x_b$ regions, respectively, what is the $\tau_{p1}/\tau_{p2}$ ratio?

(e) Establish an expression or expressions for the diffusion current ($I_{DIFF}$) flowing in the diode as a function of the applied voltage. To simplify the development, assume $L_{P1} \gg x_b$ and take the $n$-side contact to be many $L_{P2}$ diffusion lengths from the depletion region edge. Make a sketch of the expected $I_{DIFF}$ versus $V_A$ characteristic when the diode is reverse-biased.

(f) Assuming $\tau_{01} \cong \tau_{p1}$ for $0 \leq x \leq x_b$ and $\tau_{02} \cong \tau_{p2}$ for $x > x_b$, establish an expression or expressions for the R–G current ($I_{R-G}$) as a function of the applied voltage when the diode is reverse-biased greater than a few $kT/q$ volts. Make a sketch of the expected reverse-bias $I_{R-G}$ versus $V_A$ characteristic.

# 7 *pn* Junction Diode: Small-Signal Admittance

## 7.1 INTRODUCTION

In this chapter we examine and model the small signal response of the *pn* junction diode. A small sinusoidal voltage ($v_a$) is taken to be superimposed on the applied d.c. bias giving rise to an a.c. current ($i$) flowing through the diode, as pictured in Fig. 7.1. The a.c. response of a passive device like the diode is characterized by specifying the small-signal admittance, $Y = i/v_a$. Response information can be alternatively conveyed through the use of a small-signal equivalent circuit. The junction region response of the *pn* diode has capacitive ($C$) and conductive ($G$) components and in general exhibits an admittance of the form

$$Y = G + j\omega C \qquad (7.1)$$

where $j = \sqrt{-1}$ and $\omega$ is the angular frequency of the a.c. signal in radians/sec. The corresponding equivalent circuit, modeling the entire diode and valid for arbitrary d.c. biasing conditions, is shown in Fig. 7.2. The arrows through the capacitance and conductance symbols in the figure indicate that $C$ and $G$ are functions of the applied d.c. bias. $R_S$, the series resistance introduced in Chapter 6, models the portion of the diode outside the junction region. $R_S$ can limit the performance of the diode in certain applications, but it is normally very small compared to the junction impedance except at large forward biases. Unless stated otherwise, we will henceforth assume $R_S$ to be negligible.

In establishing explicit expressions for the admittance of the junction region, it is convenient to divide the development into two parts. In the first we take the diode to be reverse-biased. When reverse-biased, the diode conductance is small and $Y \cong j\omega C$. Moreover, the reverse-bias capacitance is linked solely to majority carrier oscillations inside the device. The second part treats forward bias where the conductance cannot be neglected and minority carriers contribute to the overall response.

## 7.2 REVERSE-BIAS JUNCTION CAPACITANCE

### 7.2.1 General Information

Reiterating the statement in the introduction, when reverse-biased, the *pn* junction diode becomes functionally equivalent to a capacitor. Many "capacitors" in ICs and other circuits are in fact reverse-biased *pn* junction diodes. The *pn* junction diode does differ from

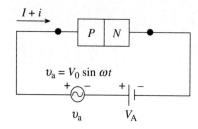

**Figure 7.1** Diode biasing circuit. $v_a$ is the applied small-signal voltage; $i$ is the resultant a.c. current.

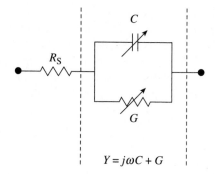

**Figure 7.2** General-case small-signal equivalent circuit for a *pn* junction diode. $Y = j\omega C + G$ is the admittance of the junction region; $R_S$ models the series resistance of the contacts and quasineutral regions.

a standard capacitor in that the diode capacitance monotonically decreases with increasing reverse bias. The sample $C$–$V$ data presented in Fig. 7.3 illustrate the observed dependence.

The general a.c. behavior of the *pn* junction diode under reverse biasing can be explained by examining what happens inside the diode as the a.c. signal runs through a bias cycle. With the a.c. signal superimposed on the d.c. bias, the total voltage drop across the junction becomes $V_A + v_a$. During the positive portion of the $v_a$ cycle, the a.c. signal slightly reduces the reverse bias across the junction, and the depletion width shrinks by a small amount, as envisioned in Fig. 7.4(a). The $\rho$-plot in Fig. 7.4(b), which was drawn assuming a step junction, emphasizes that there is a corresponding decrease in the charge on the two sides of the depletion region. When the a.c. signal reverses and goes negative, $v_a$ now increases the total reverse bias across the junction to greater than $V_A$, and the depletion width increases slightly above its steady state value. A depletion width larger than the steady state value in turn gives rise to an increase in the depletion region charge on the two sides of the junction. The overall effect of the a.c. signal may thus be viewed as a small oscillation of the depletion width about its steady state value and an associated $\Delta\rho$ charge density oscillation, as pictured in Figs. 7.4(c) and (d). The $\Delta\rho$-plots were obtained

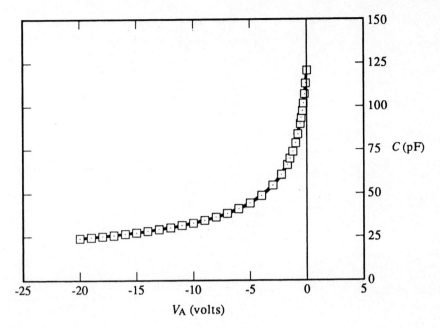

,**Figure 7.3** Monotonic decrease in the diode capacitance with increasing reverse bias. Sample *C–V* data derived from a 1N5472A abrupt junction diode.

by subtracting the d.c. charge density distribution in Fig. 7.4(b) from the $V_A \pm v_a$ charge density distributions.

The small size of $v_a$, typically a few tens of millivolts or less, dictates that the maximum displacement of the $\Delta\rho$ charge from the edges of the steady state depletion region will be extremely small. Effectively, it looks like plus and minus charges are being alternately added and subtracted from two planes inside the diode separated by a width $W$. The described a.c. situation is physically identical to what takes place inside a parallel plate capacitor. It is well known that the parallel plate capacitor exhibits a capacitance equal to the dielectric permittivity of the material between the plates, times the area of the plates, divided by the distance between the plates. The diode capacitance is concluded by analogy to be

$$ C_J = \frac{K_S \varepsilon_0 A}{W} \tag{7.2} $$

The capacitance associated with the depletion width oscillation is known as the *junction* or *depletion-layer capacitance* and is identified by the subscript J. $C = C_J$ if the diode is reverse-biased, since there are no other significant charge oscillations. Also, we know $W$

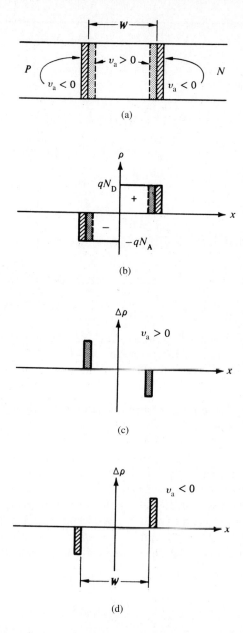

**Figure 7.4**   Depletion-layer charge considerations. (a) Depletion width and (b) total charge density oscillations in response to an applied a.c. signal. (c) $v_a > 0$ and (d) $v_a < 0$ a.c. charge densities.

increases with increasing reverse bias. $C_J$ being proportional to $1/W$ is therefore expected to decrease with increased reverse biasing, in agreement with experimental (Fig. 7.3) observations.

One point must be clarified concerning the oscillation of the depletion width about its steady state value: To achieve this oscillation, majority carriers must move rapidly into and out of the affected region in sync with the a.c. signal. In other words, the carriers are assumed to respond to the a.c. signal as if it were a d.c. bias. When this happens, the device is said to *follow the a.c. signal quasistatically.* In the discussion of the d.c. response, we noted the majority carriers in the quasineutral regions were capable of a rapid rearrangement, with majority carriers being supplied or eliminated at the contacts as required. The majority carrier response time in Si is typically $\sim 10^{-10}$ sec or less, and the quasistatic assumption made here is routinely valid up to very high signal frequencies.

## 7.2.2  *C–V* Relationships

The precise capacitance versus voltage dependence expected from a given diode is established by replacing $W$ in Eq. (7.2) by the appropriate expression relating $W$ to the applied voltage. In treating the junction electrostatics in Chapter 5, the $W$ versus $V_A$ relationship was found to vary with the doping profile. The Chapter 5 analysis specifically gave

$$W = \left[\frac{2K_S\varepsilon_0}{qN_B}(V_{bi} - V_A)\right]^{1/2} \quad \ldots \text{ asymmetrical step junction} \qquad (7.3)$$

and

$$W = \left[\frac{12K_S\varepsilon_0}{qa}(V_{bi} - V_A)\right]^{1/3} \quad \ldots \text{ linearly graded junction} \qquad (7.4)$$

where $N_B$ is the doping ($N_A$ or $N_D$) on the lightly doped side of the asymmetrical step junction and $a$ is the linear grading constant.

Although Eqs. (7.3) and (7.4) could be separately substituted into Eq. (7.2), it is more convenient to deal with a single generalized relationship valid for a wide range of profiles. Working to develop a generalized $W$ versus $V_A$ relationship, we consider a profile with one side ($x < 0$) heavily doped and the concentration on the lightly doped side described by the power law

$$N_B(x) = bx^m \quad \ldots x > 0 \qquad (7.5)$$

where $b > 0$ and $m$ are constants for a given profile. Examples of one-sided power-law profiles for select $m$-values are displayed in Fig. 7.5. Note that $m = 0$ and $m = 1$ correspond to the asymmetrical step junction and the one-sided linearly graded junction, respectively. Also note that negative $m$-values are permitted. A profile where $m < 0$ and the doping decreases as one moves away from the junction is said to be *hyperabrupt.* Hyperabrupt junctions can be formed by ion implantation or epitaxy.

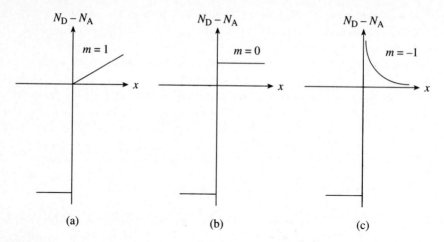

**Figure 7.5** Example one-sided power-law profiles: (a) linearly graded, (b) step, and (c) $m = -1$ hyperabrupt.

Inside a *pn* junction with a one-sided power-law profile, the depletion width dependence for $m > -2$ is

$$W = \left[ \frac{(m + 2)K_S\varepsilon_0}{qb}(V_{bi} - V_A) \right]^{1/(m+2)} \tag{7.6}$$

The derivation of Eq. (7.6) is left as an exercise. For $m = 0$, $b = N_B$—where $N_B$ without a trailing $(x)$ is understood to be a position-independent constant—and Eq. (7.6) reduces to Eq. (7.3). Setting $m = 1$ and $b = a/4$ in Eq. (7.6) yields Eq. (7.4).[†] Satisfied Eq. (7.6) yields acceptable results, we conclude upon substituting Eq. (7.6) into Eq. (7.2) that

$$C_J = \frac{K_S\varepsilon_0 A}{\left[ \frac{(m + 2)K_S\varepsilon_0}{qb}(V_{bi} - V_A) \right]^{1/(m+2)}} \tag{7.7}$$

Alternatively, it is sometimes convenient to introduce

$$C_{J0} \equiv C_J|_{V_A=0} = \frac{K_S\varepsilon_0 A}{\left[ \frac{(m + 2)K_S\varepsilon_0}{qb}V_{bi} \right]^{1/(m+2)}} \tag{7.8}$$

---

[†]Equation (7.4) is appropriate for a *two-sided* linearly graded junction. $b$ must be set equal to $a/4$ instead of $a$ to account for the difference in profiles.

where, in terms of the capacitance at $V_A = 0$,

$$C_J = \frac{C_{J0}}{\left(1 - \dfrac{V_A}{V_{bi}}\right)^{1/(m+2)}} \tag{7.9}$$

A *pn* junction diode that is manufactured to take advantage of the capacitance-voltage variation described by Eqs. (7.7)/(7.9) is called a *varactor*. *Varactor* is a combination of the words *var*iable and re*actor*, where reactor alludes to the reactance $= 1/j\omega C$ of the device. Varactor diodes are widely used in parametric amplification, harmonic generation, mixing, detection, and voltage-variable tuning. In such applications it is often desirable to employ a diode exhibiting the maximum capacitance ratio over a given voltage range. This figure of merit is called the *tuning ratio* (*TR*). Examining Eq. (7.9) for reverse biases such that $-V_A/V_{bi} \gg 1$, we find

$$TR \equiv \frac{C_J(V_{A1})}{C_J(V_{A2})} \cong \left(\frac{V_{A2}}{V_{A1}}\right)^{1/(m+2)} \tag{7.10}$$

As is obvious from Eq. (7.10), the largest tuning ratios are derived from devices with the smallest *m*-values—*TR* progressively increases in going from linearly graded ($m = 1$), to step ($m = 0$), to hyperabrupt ($m < 0$) junctions. This should explain the special interest in hyperabrupt profiles and the commercial availability of hyperabrupt varactor diodes.

---

### (C) Exercise 7.1

**P:** Equation (7.7) can be used to compute and plot fully dimensioned capacitance-voltage characteristics; "universal" normalized *C–V* curves are readily constructed using Eq. (7.9). To examine the general nature of the predicted *C–V* characteristics, compute and plot normalized $C_J/C_{J0}$ versus $V_A/V_{bi}$ curves appropriate for linearly graded, step, and $m = -1$ hyperabrupt junction diodes. Limit the voltage axis to $-25 \le V_A/V_{bi} \le 0$. Comment on the results.

**S:** The normalized characteristics were computed using Eq. (7.9). The MATLAB program script and output plot (Fig. E7.1) follow. The step junction curve is noted to provide a fairly good match to the experimental data presented in Fig. 7.3 if one assumes $V_{bi} \sim 1$ V. Also, if $V_A = 0$ is used for $V_{A1}$, and assuming similar $V_{bi}$ values, the tuning ratio $[C_J(V_{A1})/C_J(V_{A2})]$ employing any $V_{A2}$ is obviously greatest for the hyperabrupt diode and least for the linearly graded diode, in agreement with the text discussion.

MATLAB program script...

%Exercise 7.1...Normalized C-V curves

```
%Computation
clear
m=[1 0 -1];
s=1 ./(m+2);
x=linspace(-25,0);      %x=VA/Vbi
y=[];                   %y=CJ/CJ0
for i=1:3,
y=[y;1 ./(1-x).^s(i)];
end

%Plot
close
plot(x,y,'-'); grid
axis([-25 0 0 1])
xlabel('VA/Vbi'); ylabel('CJ/CJ0')
text(-20,.42,'linear (m=1)')
text(-20,.27,'step (m=0)')
text(-20,.10,'hyperabrupt (m=-1)')
```

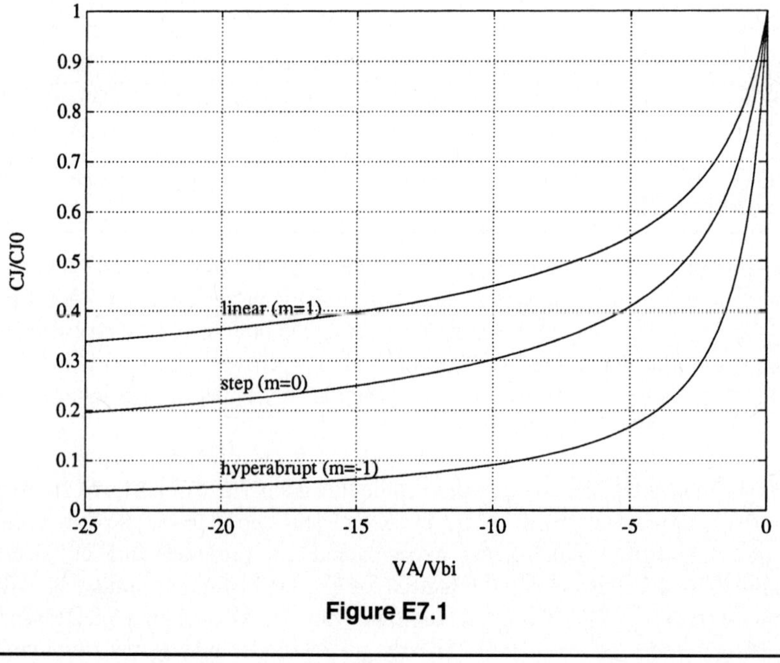

Figure E7.1

## 7.2.3 Parameter Extraction/Profiling

*C–V* data from *pn* junction diodes and other devices are routinely used to determine device parameters, notably the average doping or doping profile on the lightly doped side of a junction. *C–V* measurements have become so commonplace in the characterization and testing of devices that automated systems are available for acquiring and analyzing the *C–V* data. Since the measurement is an integral component of parameter extraction, we begin here with a brief description of an automated *C–V* system.

The *C–V* system in the measurements laboratory administered by the author is pictured schematically in Fig. 7.6. The MSI *C–V* meter is the heart of the system, employing a 15 mV rms a.c. signal at a probing frequency of 1 MHz. Four capacitance ranges with a maximum value of 2 pF, 20 pF, 200 pF, and 2000 pF are available for selection from the front panel. A d.c. bias supply inside the meter has two full-scale ranges of $\pm 9.999$ V, programmable in 0.001 V increments, and $\pm 99.99$ V, programmable in 0.01 V increments. Operator control of the biasing, the data display, data manipulations, and hard-copy output to a printer or plotter are accomplished by software on the personal computer shown to the left of the *C–V* meter in Fig. 7.6. The probe box to the right of the meter is used if the device or devices under test are situated on a wafer. The circular chuck in the probe box provides electrical contact to the back of the wafer and a wire probe contacts the device structure on the top of the wafer. The chuck can be resistance-heated under the local control of the 832 T-Controller. After probe contact, the top of the probe box is typically lowered to keep room light from perturbing the capacitance measurement. If the device under test is encapsulated, the leads to the probe box are removed and the test device is inserted into an adapter connected to the input terminals of the meter. We should mention the meter automatically measures and compensates for the stray capacitance associated with cabling, probe box components, and the encapsulated-device adapter.

Turning to the interpretation of the *C–V* data, suppose the device under test is known to be an asymmetrically doped abrupt junction. (The term *abrupt* is normally used to describe *actual* doping profiles approximately modeled by the idealized step junction.) For the assumed junction profile, $m \to 0$ and $b \to N_B$ in Eq. (7.7). Additionally, if both sides of Eq. (7.7) are inverted and then squared, one obtains

$$\frac{1}{C_J^2} = \frac{2}{qN_B K_S \varepsilon_0 A^2}(V_{bi} - V_A) \tag{7.11}$$

Equation (7.11) indicates that a plot of $1/C_J^2$ versus $V_A$ should be a straight line, with a slope inversely proportional to $N_B$ and an extrapolated $1/C_J^2 = 0$ intercept equal to $V_{bi}$. Thus, assuming the area $A$ of the diode is known, $N_B$ is readily deduced from the slope of the plot. Although perhaps stating the obvious, a straight line $1/C_J^2$ versus $V_A$ plot is also confirmation that the diode can be modeled as a step junction. A sample analysis of *C–V* data based on the $1/C_J^2$ versus $V_A$ plot is presented in Exercise 7.2.

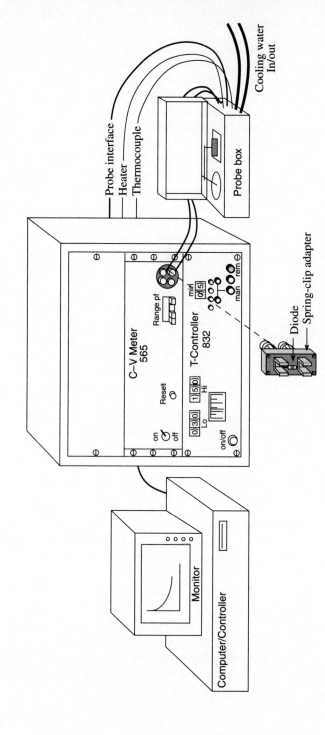

**Figure 7.6** Sketch of a capacitance–voltage ($C$–$V$) measurement system.

## Exercise 7.2

**P:** The manufacturer indicates the 1N5472A is an $n^+$-$p$ abrupt junction diode with an area $A = 3.72 \times 10^{-3}$ cm$^2$. Encapsulation is noted to typically introduce a 2 pF stray capacitance shunting the diode junction. Utilizing the measured 1N5472A *C–V* data presented in Fig. 7.3, apply the $1/C_J^2$ versus $V_A$ plot technique to confirm the abrupt nature of the junction and to determine the *p*-side doping concentration. Also quote the deduced value of $V_{bi}$.

**S:** A $1/C_J^2$ versus $V_A$ plot of the 1N5472A *C–V* data is displayed in Fig. E7.2. Prior to constructing the plot, 2 pF was subtracted from all measured capacitance values to correct for the encapsulation-related shunt capacitance ($C_J = C - 2$ pF). The Fig. E7.2 data points are seen to lie in an almost perfect straight line. The junction is definitely abrupt. Performing a least squares fit to the data gives

$$\frac{1}{C_J^2} = (6.89 \times 10^{19}) - (9.78 \times 10^{19})V_A$$

where $C_J$ is in farads and $V_A$ is in volts. Referring to Eq. (7.11), we conclude

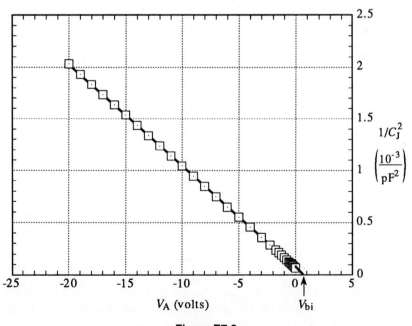

**Figure E7.2**

$$N_A = \frac{2}{qK_S\varepsilon_0 A^2|\text{slope}|}$$

$$= \frac{2}{(1.6 \times 10^{-19})(11.8)(8.85 \times 10^{-14})(3.72 \times 10^{-3})^2(9.78 \times 10^{19})}$$

$$= \mathbf{8.84 \times 10^{15}/cm^3}$$

and

$$V_{bi} = V_A\big|_{1/C_J^2=0} = \frac{6.89 \times 10^{19}}{9.78 \times 10^{19}} = \mathbf{0.70 \ V}$$

It should be noted that the deduced $V_{bi}$ is lower than one would expect from the $N_A \cong 9 \times 10^{15}/cm^3$ *p*-side doping. The $V_{bi}$ value deduced from the *C–V* data is subject to serious extrapolation errors and is sensitive to doping variations in the immediate vicinity of the metallurgical junction.

The foregoing plot approach could obviously be extended to the linearly graded and other profiles. However, this is seldom done. As it turns out, the doping variation with position on the lightly doped side of a junction can be deduced directly from the *C–V* data without prior knowledge about the nature of the doping profile. Omitting the derivational details, we merely note the doping concentration versus position is computed using[3]

$$N_B(x) = \frac{2}{qK_S\varepsilon_0 A^2|d(1/C_J^2)/dV_A|} \tag{7.12}$$

$$x = \frac{K_S\varepsilon_0 A}{C_J} \tag{7.13}$$

where $x$ is the distance into the lightly doped side of the diode as measured from the metallurgical junction. Note that substituting the Eq. (7.11) step junction relationship into Eq. (7.12) yields the required position-independent result.

The process of determining the doping as a function of position is called *profiling*. The profile determined using Eqs. (7.12)/(7.13) becomes inaccurate if the doping is a rapidly varying function of position, and only a limited portion of the junction can be scanned. Moreover, the result tends to be "noisy" because the slope or derivative of the *C–V* data is required. Nevertheless, *C–V* profiling is relatively simple to implement, typically yields useful results, and finds widespread utilization. Software provided with automated *C–V* systems even performs the (7.12)/(7.13) computations and displays the result graphically. A sample profile, that of a hyperabrupt junction diode automatically processed by the Fig. 7.6 *C–V* system, is reproduced in Fig. 7.7.

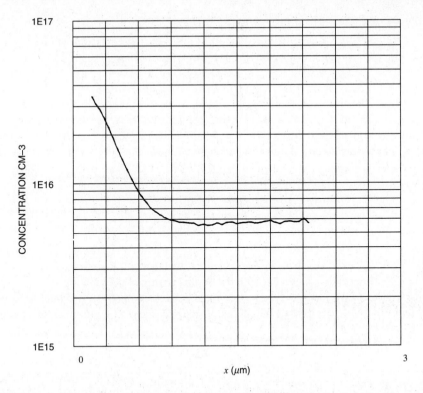

**Figure 7.7**  Doping profile of a hyperabrupt tuning diode (ZC809). Output from the Fig. 7.6 *C–V* measurement system. *x* is the distance from the metallurgical junction into the lightly doped side.

## 7.2.4 Reverse-Bias Conductance

All standard capacitors exhibit a certain amount of conductance. The same is true of the *pn* junction diode. Although predominantly capacitive, the reverse bias admittance does have a small conductive component. A few words are in order concerning the reverse-bias conductance.

By definition, the differential d.c. conductance of a diode is just the slope of the *I–V* characteristic, *dI/dV*, at the d.c. operating point. If the diode is assumed to respond to an a.c. signal quasistatically, then the a.c. conductance = $\Delta I/\Delta V$ = *dI/dV* = differential d.c. conductance. Limiting our considerations to frequencies where the diode can follow the a.c. signal quasistatically, and introducing the symbol $G_0$ for the associated low-frequency conductance, we can write

$$G_0 = \frac{dI}{dV_A}$$

(7.14)

For an ideal diode where $I = I_0[\exp(qV_A/kT) - 1]$,

$$G_0 = \frac{q}{kT} I_0 e^{qV_A/kT} = \frac{q}{kT}(I + I_0) \qquad \ldots \text{ ideal diode} \qquad (7.15)$$

When the reverse bias exceeds a few $kT/q$ volts in an ideal diode, $I \to -I_0$ and we see from Eq. (7.15) that $G_0 \to 0$. This is consistent with the fact that the reverse-bias $I-V$ characteristic saturates and the slope of the ideal $I-V$ curve goes to zero. If the d.c. recombination–generation current dominates in the given diode then, for reverse biases greater than a few $kT/q$ volts,

$$G_0 = \frac{d}{dV_A}\left(-\frac{qAn_i}{2\tau_0}W\right) = \frac{qAn_iW/2\tau_0}{(m + 2)(V_{bi} - V_A)} \qquad \ldots I_{R-G} \text{ dominant} \quad (7.16)$$

where use has been made of the Eq. (7.6) relationship for $W$. When the R–G current dominates, Eq. (7.16) indicates there is a residual conductance at all reverse biases, with the precise voltage dependence varying with the doping profile of the junction.

Regardless of the junction type or dominant current component, it should be emphasized that Eq. (7.14) can always be applied to the measured $I-V$ characteristic to determine $G_0$.

---

### Exercise 7.3

**P:** The diode exhibiting the reverse-bias $I-V$ characteristic displayed in Fig. 6.10(b) has a measured junction capacitance of $C_J = 63$ pF at $V_A = -10$ V. The series resistance of the diode was estimated to be 1 ohm in Exercise 6.8. Taking the d.c. operating point to be $V_A = -10$ V and the a.c. frequency to be $f = 100$ kHz, confirm that $R_S$ and $G = G_0$ may be totally neglected in modeling the admittance of the device.

**S:** Figure 6.10(b) can be used to estimate the low-frequency conductance. We find

$$G_0 = \left.\frac{dI}{dV_A}\right|_{V_A=-10V} \cong \frac{40 \text{ pA}}{18 \text{ V}} = 2.22 \times 10^{-12} \text{ S}$$

By way of comparison, at the operational bias and frequency,

$$\omega C_J = (2\pi)(10^5)(6.3 \times 10^{-11}) = 3.96 \times 10^{-5} \text{ S}$$

The capacitive component is clearly much greater than the conductive component and $Y \cong j\omega C_J$.

Next, consider the series resistance. Since $R_S$ is in series with $Y$, we need to compare $|Z| = 1/|Y|$ with $R_S$. Here we find

$$R_S \cong 1 \ \Omega$$

and

$$\frac{1}{\omega C_J} = 2.52 \ \times \ 10^4 \ \Omega$$

The reactance of the junction is much greater than the series resistance. For the given operational conditions, the diode may be essentially viewed as a pure capacitor.

## 7.3 FORWARD-BIAS DIFFUSION ADMITTANCE

### 7.3.1 General Information

Capacitance arises from charge oscillations inside of a device structure. The junction capacitance introduced in the reverse-bias analysis is caused by the in-and-out movement of majority carriers about the steady state depletion width. The minority carrier concentrations also oscillate about the edges of the depletion width in response to the a.c. signal. The minority carrier numbers are so minuscule under reverse-bias conditions, however, that the contribution to the admittance is negligible. Forward biasing gives rise to nothing new as far as the majority carriers are concerned. These carriers still move back and forth about the edges of the depletion region giving rise to a junction capacitance. In fact, the $C_J$ relationships developed in the previous section can be applied without modification to forward bias. The something new under forward bias is a significant contribution from the minority carrier charge oscillation in response to the a.c. signal.

As noted during the examination of ideal diode results in Subsection 6.1.4, forward biasing of the diode causes a build-up of minority carriers in the quasineutral regions immediately adjacent to the depletion region. The build-up becomes larger and larger with increasing forward bias. In response to an a.c. signal, the voltage drop across the junction is changed to $V_A + v_a$ and the excess minority carrier distributions oscillate about their d.c. values as pictured in Fig. 7.8(a). This results in an additional capacitance. If the minority carriers can follow the signal quasistatically, the carriers move back and forth in unison between the two straight lines in the figure. However, the supply and removal of minority carriers is not as rapid as that of the majority carriers. At angular frequencies approaching the inverse of the minority carrier lifetimes, the minority carrier charge oscillation has difficulty staying in sync with the a.c. signal. The result is an out-of-phase spatial variation of the charge something like the undulating distributions sketched in Fig. 7.8(a). An out-of-phase charge oscillation enhances the observed conductance and reduces the observed

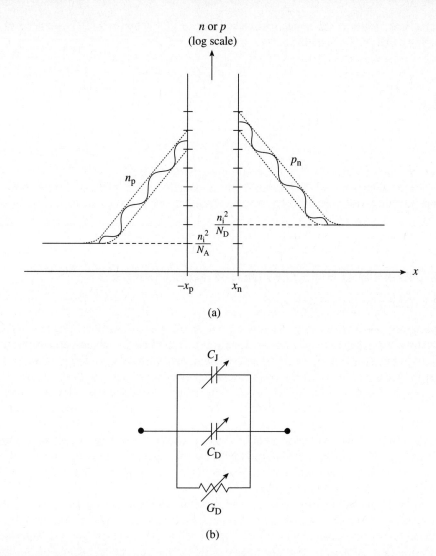

**Figure 7.8** The diffusion admittance. (a) Minority carrier charge fluctuation (greatly exaggerated), giving rise to the diffusion admittance. (b) Forward-bias small-signal equivalent circuit for the *pn* junction diode (series resistance assumed to be negligible).

capacitance. In other words, the capacitance and conductance associated with the minority carrier oscillations are expected to be frequency-dependent.

Because the minority carrier build-up about the edges of the depletion region is caused by the diffusion current, the admittance associated with the minority carrier charge oscillation is called the diffusion admittance, $Y_D$. In general, as we have indicated,

$$Y_D = G_D + j\omega C_D \tag{7.17}$$

where $C_D$ and $G_D$ are the diffusion capacitance and diffusion conductance, respectively. The overall forward-bias admittance is of course the parallel combination of the junction capacitance and the diffusion admittance as noted in Fig. 7.8(b).

Measurement of the forward-bias admittance poses more of a challenge than the reverse-bias capacitance measurement. The larger d.c. current that flows when the diode is forward-biased tends to load down the detection and biasing circuitry in almost all commercially available $C-V$ systems. In measurements on a Si diode at room temperature, a forward bias of only a few tenths of a volt often leads to a fallacious result or automatic termination of the measurement. One instrument that can be used for forward-bias measurements is the HP4284A LCR Meter pictured schematically in Fig. 7.9. The HP4284A with the 001 option installed contains a special isolation circuit that minimizes the loading problem for d.c. currents up to 0.1 A. A very versatile piece of equipment, the HP4284A permits simultaneous capacitance and conductance measurements at frequencies ranging from 20 Hz to 1 MHz, variation of the a.c. signal level from 5 mV to 20 V rms, setting of the d.c. bias between $\pm 40$ V, and capacitance detectability claimed to range from 0.01

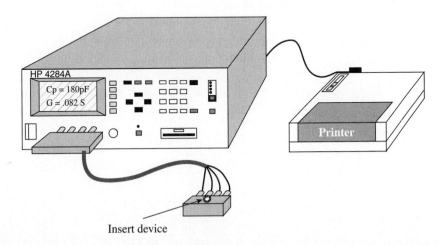

Insert device

**Figure 7.9** The Hewlett-Packard 4284A LCR Meter used for forward-bias measurements. The attached printer conveniently provides a hard-copy record of the display screen data.

femtofarad to 10 farads. The instrument basically performs single-point measurements, but will automatically cycle through ten operator-set values of frequency, d.c. bias, or a.c. signal level.

## 7.3.2 Admittance Relationships

Obtaining explicit relationships for the diffusion admittance is not difficult, but the mathematical manipulations can become rather tedious. In the straightforward brute-force approach, the ideal diode derivation is simply repeated with the d.c. quantities all replaced by d.c. plus a.c. quantities. Separate solutions are then sought for the a.c. quantities. Once a solution is obtained for the a.c. current as a function of the a.c. voltage, the admittance is computed from $Y = i/v_a$. Alternatively, the a.c. current solution can be obtained from the d.c. current solution by noting equivalencies in the two formulations. We will pursue the latter approach.

The device under analysis is taken to be a $p^+$-$n$ junction diode. With an a.c. signal superimposed on the d.c. bias, the $n$-side minority carrier diffusion equation to be solved is

$$\frac{\partial \Delta p_n(x,t)}{\partial t} = D_P \frac{\partial^2 \Delta p_n(x,t)}{\partial x^2} - \frac{\Delta p_n(x,t)}{\tau_p} \tag{7.18}$$

Assuming the a.c. signal is a sine or cosine function, we can write

$$\Delta p_n(x,t) = \overline{\Delta p_n}(x) + \tilde{p}_n(x,\omega)e^{j\omega t} \tag{7.19}$$

where $\overline{\Delta p_n}$ is the time-invariant (d.c.) portion of $\Delta p_n(x, t)$ and $\tilde{p}_n$ is the amplitude of the a.c. component. Substituting the Eq. (7.19) expression for $\Delta p_n(x, t)$ into Eq. (7.18) and explicitly working out the time derivative yields

$$j\omega\tilde{p}_n e^{j\omega t} = D_P \frac{d^2\overline{\Delta p_n}}{dx^2} + D_P \frac{d^2\tilde{p}_n}{dx^2}e^{j\omega t} - \frac{\overline{\Delta p_n}}{\tau_p} - \frac{\tilde{p}_n}{\tau_p}e^{j\omega t} \tag{7.20}$$

The d.c. and a.c. terms in Eq. (7.20) must separately balance. Thus, after collecting like terms and simplifying the a.c. result, one obtains

$$0 = D_P \frac{d^2\overline{\Delta p_n}}{dx^2} - \frac{\overline{\Delta p_n}}{\tau_p} \tag{7.21a}$$

$$0 = D_P \frac{d^2\tilde{p}_n}{dx^2} - \frac{\tilde{p}_n}{\tau_p/(1 + j\omega\tau_p)} \tag{7.21b}$$

Equation (7.21a) is recognized as the usual steady state minority carrier diffusion equation. Note that the a.c. version of the diffusion equation, Eq. (7.21b), has exactly the same form except that $\tau_p$ is replaced by $\tau_p/(1 + j\omega\tau_p)$.

In solving Eqs. (7.21) the usual boundary condition applies at $x = \infty$; i.e., $\overline{\Delta p_n}(\infty) = \tilde{p}_n(\infty) = 0$. The boundary condition at $x = x_n$, however, becomes

$$\Delta p_n(x = x_n) = \overline{\Delta p_n}(x_n) + \tilde{p}_n(x_n) = \frac{n_i^2}{N_D}(e^{q(V_A + v_a)/kT} - 1) \tag{7.22}$$

or

$$\overline{\Delta p_n}(x_n) = \frac{n_i^2}{N_D}(e^{qV_A/kT} - 1) \tag{7.23}$$

$$\tilde{p}_n(x_n) = \frac{n_i^2}{N_D}e^{qV_A/kT}(e^{qv_a/kT} - 1) \tag{7.24a}$$

$$\cong \frac{n_i^2}{N_D}\left(\frac{qv_a}{kT}e^{qV_A/kT}\right) \quad \ldots v_a \ll kT/q \tag{7.24b}$$

$v_a$ in the foregoing is understood to be the amplitude of the a.c. signal.

At this point we have yet to solve for anything. We have merely established the equation and boundary conditions to be employed in solving for the a.c. variables and current. It is now a simple matter, however, to progress to the desired solution. Since the a.c. minority carrier diffusion equation is identical in form to the d.c. equation except $\tau_p \rightarrow \tau_p/(1 + j\omega\tau_p)$, and since the boundary conditions are identical except $[\exp(qV_A/kT) - 1] \rightarrow [(qv_a/kT)\exp(qV_A/kT)]$, the a.c. and d.c. current solution must likewise be identical except for the cited modifications to $\tau_p$ and the voltage factor. For a $p^+$-$n$ diode we know

$$I_{DIFF} = qA \frac{D_P}{L_P} \frac{n_i^2}{N_D}(e^{qV_A/kT} - 1) = qA \sqrt{\frac{D_P}{\tau_p}} \frac{n_i^2}{N_D}(e^{qV_A/kT} - 1) \tag{7.25}$$

Thus, making the noted $\tau_p$ and voltage factor substitutions, we conclude

$$i_{diff} = qA \sqrt{\frac{D_P}{\tau_p}} \sqrt{1 + j\omega\tau_p} \frac{n_i^2}{N_D}\left(\frac{qv_a}{kT}e^{qV_A/kT}\right) = \left(\frac{qv_a}{kT}I_0 e^{qV_A/kT}\right)\sqrt{1 + j\omega\tau_p} \tag{7.26}$$

or in terms of the ideal-diode low-frequency conductance (Eq. 7.15),

$$i_{diff} = G_0\sqrt{1 + j\omega\tau_p}\, v_a \tag{7.27}$$

and

$$\boxed{Y_D = \frac{i_{diff}}{v_a} = G_0\sqrt{1 + j\omega\tau_p}} \quad \ldots p^+\text{-}n \text{ diode} \tag{7.28}$$

For an $n^+$-$p$ diode $\tau_p \rightarrow \tau_n$ in Eq. (7.28). Given a two-sided junction, $G_0$ must be separated into its $n$ and $p$ components and each multiplied by the appropriate $\sqrt{1 + j\omega\tau}$ factor.

If the Eq. (7.28) diffusion admittance for a $p^+$-$n$ diode is separated into real and imaginary parts and the result compared with Eq. (7.17), one finds

$$G_D = \frac{G_0}{\sqrt{2}}\left(\sqrt{1 + \omega^2\tau_p^2} + 1\right)^{1/2} \tag{7.29a}$$

$$C_D = \frac{G_0}{\omega\sqrt{2}}\left(\sqrt{1 + \omega^2\tau_p^2} - 1\right)^{1/2} \tag{7.29b}$$

$G_D$ and $C_D$ are noted to be functions of both the d.c. bias (through $G_0$) and the signal frequency. With $G_0$ varying as $\exp(qV_A/kT)$, the diffusion components increase very rapidly with increasing forward bias. Whereas $C_J$ dominates the a.c. response at small forward biases, the diffusion capacitance surpasses and eventually overshadows the junction capacitance as the forward bias is progressively increased. Relative to the frequency dependence, at low frequencies where $\omega\tau_p \ll 1$, $\sqrt{1 + \omega^2\tau_p^2} \cong 1 + \omega^2\tau_p^2/2$ and

$$G_D \Rightarrow G_0 \qquad \ldots \omega\tau_p \ll 1 \tag{7.30a}$$

$$C_D \Rightarrow G_0\frac{\tau_p}{2} \qquad \ldots \omega\tau_p \ll 1 \tag{7.30b}$$

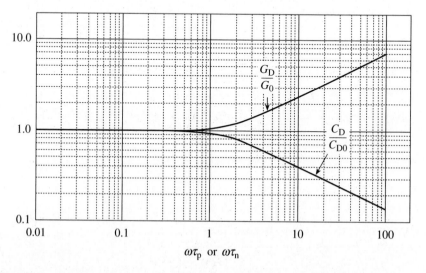

**Figure 7.10** Diffusion capacitance and diffusion conductance normalized to their low-frequency values as a function of $\omega\tau_p$ ($p^+$-$n$ diode) or $\omega\tau_n$ ($n^+$-$p$ diode). ($C_{D0} = G_0\tau/2$)

If $\tau_p = 10^{-6}$ sec, for example, a frequency independent response described by Eqs. (7.30) is expected for $f \lesssim 1/(20\pi\tau_p) \cong 16$ kHz. For signal frequencies where $\omega\tau_p \gtrsim 1$, the conductance increases while the capacitance decreases with increasing frequency. The relative change in $G_D$ and $C_D$ compared to their low-frequency values is graphed as a function of $\omega\tau_p$ in Fig. 7.10.

---

### (C) Exercise 7.4

**P:** The 1N5472A $n^+$-$p$ abrupt junction diode yielding the reverse-bias $C$–$V$ data plotted in Fig. 7.3 and analyzed in Exercise 7.2 has a zero-bias junction capacitance of $C_{J0} = 120$ pF. The $V_{bi}$ value affording the best fit to junction capacitance data was determined to be 0.7 V in Exercise 7.2. From $I$–$V$ data following the procedure outlined in Exercise 6.7, one obtains $I_0 = 8 \times 10^{-13}$ A and $n_1 = 1.22$. The $p$-side minority carrier lifetime is estimated to be $\tau_n = 5 \times 10^{-7}$ sec.

(a) Assuming $\omega\tau_n = 0.01$, but employing relationships valid for arbitrary values of $\omega\tau_n$, compute and plot the expected $C_J$, $C_D$, and $C_J + C_D$ versus $V_A$ for the given diode. Restrict $V_A$ to $0 \leq V_A \leq 0.65$ V. Specifically note the approximate voltage where $C_D = C_J$.

(b) Repeat part (a) setting $\omega\tau_n = 0.1, 1, 10,$ and $100$. Comment on the results.

**S:** The primary relationships employed in the computation are:

$$C_J = \frac{C_{J0}}{\left(1 - \dfrac{V_A}{V_{bi}}\right)^{1/2}}$$

$$C_D = \frac{\tau_n G_0}{\omega\tau_n\sqrt{2}}\left(\sqrt{1 + \omega^2\tau_n^2} - 1\right)^{1/2}$$

and

$$G_0 = \frac{q}{kT}I_0 e^{qV_A/n_1 kT}$$

The $C_J$ expression is Eq. (7.9) with $m = 0$. The $C_D$ relationship is Eq. (7.29b) with $\tau_p$ replaced by $\tau_n$ and a $\tau_n$ associated with the $\omega$ external to the square root. $G_0$ is the Eq. (7.15) low-frequency conductance with $n_1$ inserted into the exponent to account for the nonideality of the diode.

The MATLAB program script and the part (a) computational results are reproduced below. Actual forward-bias $C$–$V$ data derived from the 1N5472A using the

HP4284A LCR Meter (Fig. 7.9) are included on the plot. The junction capacitance is noted to dominate at low forward biases, with the diffusion capacitance rising to $C_D = C_J$ at $V_A \cong 0.545$ V. The same computational results are obtained, as expected, for all $\omega\tau_n < 1$. With increasing $\omega\tau_n$ greater than $\omega\tau_n = 1$, $C_D$ progressively decreases at all biases and the $C_D = C_J$ point shifts to higher and higher voltages. $C_D = C_J$ at $V_A \cong 0.575$ V and 0.62 V when $\omega\tau_n = 10$ and 100, respectively.

MATLAB program script...

```
%Exercise 7.4...Forward-Bias Capacitance

%Computational constants
clear
CJ0=120e-12;        %farads
Vbi=0.7;            %volts
Vth=0.0259;         %Vth=kT/q in volts
taun=5.0e-7;        %seconds
I0=8.0e-13;         %amps
n1=1.22;            %ideality factor
wt=input('input the angular-frequency*lifetime product--');
VA=linspace(0,0.65);

%CJ Computation
CJ=CJ0./sqrt(1-VA./Vbi);

%CD Computation
G0=I0/Vth*exp(VA./(n1*Vth));
CD=taun.*G0./(sqrt(2)*wt)*sqrt(sqrt(1+(wt)^2)-1);

%Measured CD Data
VAm=[0.1 0.2 0.3 0.4 0.42 0.44 0.46 0.48 0.50 0.52 0.54 0.56 0.58];
CDm=[1.31e-10 1.43e-10 1.61e-10 1.88e-10 1.97e-10 2.08e-10 ...
2.23e-10 2.46e-10 2.76e-10 3.46e-10 4.40e-10 6.54e-10 9.38e-10];

%Plot
close
semilogy(VA,CJ,'--r'); axis([0 0.7 1.0e-10 1.0e-9]); grid
hold on; semilogy(VA,CD,'--g'); semilogy(VA,CJ+CD)
semilogy(VAm,CDm-2e-12,'o')
xlabel('VA (volts)'); ylabel('C (farads)')

%Key
semilogy(0.12,7e-10,'o'); text(0.125,7e-10,'...C-V Data')
x=[0.1 0.2];
y1=[6.1e-10 6.1e-10]; semilogy(x,y1,'-y'); text(0.21,6.1e-10,'CJ+CD')
y2=[5.2e-10 5.2e-10]; semilogy(x,y2,'--r'); text(0.21,5.2e-10,'CJ')
y3=[4.3e-10 4.3e-10]; semilogy(x,y3,'-.g'); text (0.21,4.3e-10,'CD')
hold off
```

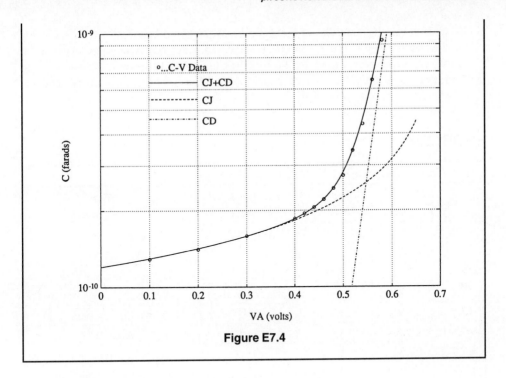

**Figure E7.4**

## 7.4  SUMMARY

The chapter was devoted to examining and modeling the small-signal response of the *pn* junction diode. The discussion was divided into two parts corresponding to reverse-biasing and forward-biasing of the diode. When reverse-biased, the *pn* junction diode is functionally equivalent to a capacitor. The capacitance of the diode can in fact be computed using the well-known parallel plate capacitor formula (Eq. 7.2). The *pn* junction diode differs from a standard capacitor in that the diode capacitance monotonically decreases with increasing reverse bias. The reverse-bias junction capacitance arises physically from the in-and-out movement of the majority carriers about the steady state depletion width in response to the a.c. signal. Reverse-bias diodes are employed as capacitors and variable capacitors (varactors) in numerous circuit applications. Capacitance measurements are used extensively in the characterization and testing of devices, particularly in determining the average doping or doping profile on the lightly doped side of a junction. The relevance of the profile to the use of the diode as a varactor, and parameter extraction/profiling procedures, were noted during the course of the discussion.

When the diode is forward-biased, there is a significant build-up of minority carriers in the quasineutral regions immediately adjacent to the depletion region. The oscillation of the minority carrier charge in response to the a.c. signal gives rise to an additional admittance component, the diffusion admittance. An expression for the diffusion admittance was established by appropriately modifying the ideal diode equation. The diffusion admittance components are strong functions of the d.c. bias, eventually dominating the observed admittance as the forward bias is progressively increased. At signal frequencies where $\omega\tau \gtrsim 1$, the minority carriers have trouble following the a.c. signal and the resulting out-of-phase oscillations enhance the diffusion conductance at the expense of the diffusion capacitance.

## PROBLEMS

| CHAPTER 7 PROBLEM INFORMATION TABLE | | | | |
|---|---|---|---|---|
| Problem | Complete After | Difficulty Level | Suggested Point Weighting | Short Description |
| 7.1 | 7.4 | 1 | 16 (2 each part) | Quick quiz |
| 7.2 | 7.2.2 | 2 | 10 | Derive Eq. (7.6) |
| ● 7.3 | " | 2 | 12 (a/b-10, c-2) | Dimensioned $C$–$V$ plot |
| * 7.4 | 7.2.3 | 2 | 15 (plot-5, i-5, ii-5) | Deduce $N_B$, $V_{bi}$ from $C$–$V$ |
| 7.5 | " | 3 | 10 | Derive Eqs. (7.12)/(7.13) |
| ● 7.6 | " | 3 | 15 | $N_B(x)$ versus $x$ |
| ● 7.7 | 7.3.2 | 1 | 5 | Verify accuracy of Fig. 7.10 |
| ● 7.8 | " | 2 | 10 (prog-8, questions-2) | Relative size of $\omega C_D$, $G_D$ |
| 7.9 | " | 2 | 10 | $\tau$ measurement |

**7.1** Quick Quiz.

Answer the following questions as concisely as possible.

(a) What is the physical origin of the junction capacitance?

(b) Sketch the $m = -1$ hyperabrupt profile of an $n^+$-$p$ junction.

(c) Define *quasistatically*.

(d) Define *varactor*.

(e) Define *profiling*.

(f) Make a sketch of the low-frequency conductance of an ideal diode showing both forward and reverse bias. Comment as necessary to forestall a misinterpretation of your sketch.

(g) What is the physical origin of the diffusion admittance?

(h) Why does the diffusion conductance increase with increasing $\omega\tau_p$ when $\omega\tau_p \gtrsim 1$?

**7.2** Given the one-sided power-law profile described by Eq. (7.5), and generally following the procedures outlined in Chapter 5, derive Eq. (7.6). Assume a $p^+$-$n$ junction where $N_B(x) = N_D(x)$. Why is it necessary to specify $m > -2$?

● **7.3** (a) Construct a computer program that yields fully dimensioned reverse-bias $C$–$V$ curves that can be compared directly with experimental data. The program is to be specifically designed for comparison with the data from Si $p^+$-$n$ abrupt junction diodes maintained at 300 K. The diode area ($A$), the lightly doped side concentration ($N_B$), and the largest reverse-bias voltage of interest ($|V_A|_{max}$) are to be input variables.

(b) Employing $A = 3.72 \times 10^{-3}$ cm$^2$ and $N_B = 8.84 \times 10^{15}$/cm$^3$, compare your program output with the experimental data presented in Fig. 7.3. Does the $V_{bi}$ result cited in Exercise 7.2 have any bearing on the agreement between experiment and theory?— Explain.

(c) How does the lightly doped side concentration affect the junction capacitance? Substantiate your answer.

NOTE: Those seeking a greater challenge might consider generalizing the computer program in this problem to handle any one-sided power-law profile.

* **7.4** The 1N4002 is one of the popular 4000-series general-purpose diodes used in automotive and other applications. $C$–$V$ data from a 1N4002 $p^+$-$n$ junction diode is listed in Table P7.4. Before analyzing the data, subtract 3pF from each capacitance value to account for the stray capacitance shunting the encapsulated diode. Assuming the diode profile to be abrupt and $A = 6 \times 10^{-3}$ cm$^2$, apply the plot approach described in the text to determine the lightly doped side concentration and the "best-fit" $V_{bi}$. Quote the results obtained by (i) "eyeballing" a straight line through the data and (ii) by performing a least squares fit to the data. (NOTE: The 1N4002 is not as ideally abrupt as the 1N5472A of Exercise 7.2. Do not be surprised if your plot points deviate somewhat from a straight line.)

**Table P7.4**    1N4002 Reverse-Bias $C$–$V$ Data.

| $V_A$(V) | $C$(pF) | $V_A$(V) | $C$(pF) |
|---|---|---|---|
| 0.0 | 38.709 | −5.0 | 15.548 |
| −0.2 | 33.717 | −6.0 | 14.599 |
| −0.4 | 30.567 | −7.0 | 13.834 |
| −0.6 | 28.319 | −8.0 | 13.189 |
| −0.8 | 26.598 | −9.0 | 12.639 |
| −1.0 | 25.170 | −10.0 | 12.163 |
| −1.4 | 23.060 | −11.0 | 11.746 |
| −1.8 | 21.490 | −12.0 | 11.373 |
| −2.2 | 20.254 | −13.0 | 11.037 |
| −2.6 | 19.248 | −14.0 | 10.734 |
| −3.0 | 18.405 | −15.0 | 10.458 |
| −4.0 | 16.762 | | |

**7.5** Derive Eqs. (7.12) and (7.13). HINT: See, for example, p. 43 in reference [3].

● **7.6** Construct a computer program that accepts $C$–$V$ data input and outputs a plot of the $N_B(x)$ versus $x$ profile based on Eqs. (7.12) and (7.13). Test run your program using the 1N4002 data in Table P7.4.

● **7.7** Verify the accuracy of Fig. 7.10. Compute and simultaneously plot the diffusion capacitance and diffusion conductance normalized to their low-frequency values as a function of $\omega\tau_p$. Limit the computation to $0.01 \leq \omega\tau_p \leq 100$.

● **7.8** What is the relative size of the capacitive and conductive components of the diffusion admittance at a given d.c. bias and signal frequency? To answer this question, first examine Eqs. (7.29) and (7.30) to determine the limiting values of $\omega C_D/G_D$ when $\omega\tau_p \ll 1$ and $\omega\tau_p \gg 1$. Next, compute and plot $\omega C_D/G_D$ versus $\omega\tau_p$ for $0.01 \leq \omega\tau_p \leq 100$. Does your plot approach the correct limiting values? In words, what is the answer to the original question?

**7.9** Forward bias admittance measurements have been used to determine the minority carrier lifetime on the lightly doped side of a junction. Note from Eqs. (7.30) that, given a $p^+$-$n$ diode, $C_D/G_D = \tau_p/2$ when $\omega\tau_p \ll 1$. Forward bias $C$–$V$ data from a 1N5472A diode was presented in Exercise 7.4. The corresponding $G_D$–$V_A$ data for forward biases between 0.50 V and 0.58 V are listed in the following table. Determine the apparent $\tau_n$ of the 1N5472A $n^+$-$p$ diode at each of the listed voltages assuming $\omega\tau_n \ll 1$. Also quote the average $\tau_n$ deduced from the data.

| $V_A$(V) | $G_D$(S) |
|----------|----------|
| 0.50 | $2.00 \times 10^{-4}$ |
| 0.52 | $3.90 \times 10^{-4}$ |
| 0.54 | $7.15 \times 10^{-4}$ |
| 0.56 | $1.33 \times 10^{-3}$ |
| 0.58 | $2.28 \times 10^{-3}$ |

# 8 *pn* Junction Diode: Transient Response

At the beginning of the *pn* junction diode discussion, we noted that a complete, systematic device analysis was normally divided into four major segments: modeling of the internal electrostatics, steady state response, small-signal response, and transient response. In this chapter we address the final major segment of the *pn* junction diode analysis, the transient or switching response. In a number of applications a *pn* junction diode is used as an electrical switch. A pulse of current or voltage is typically used to switch the diode from forward bias, called the "on" state, to reverse bias, called the "off" state, and vice-versa. Of prime concern to circuit and device engineers is the speed at which the *pn* junction diode can be made to switch states. Generally speaking, it is during the turn-off transient, going from the on to the off state, where speed limitations are most significant. The subsequent development therefore concentrates on the turn-off transient. Moreover, the diode under analysis is assumed to be ideal. This allows us to convey the basic concepts and principles of transient operation with a minimum of mathematical complexity.

## 8.1 TURN-OFF TRANSIENT

### 8.1.1 Introduction

Consider the idealized representation of a switching circuit shown in Fig. 8.1(a). Prior to $t = 0$ the diode is taken to be forward-biased with a steady state forward current, $I_F$, flowing through the diode. At $t = 0$ the switch in the circuit is rapidly moved to the right-hand position. For use in switching applications, one would like the corresponding diode current to decrease instantaneously from $I_F$ to the small steady-state reverse current consistent with the applied reverse voltage. What one actually observes is sketched in Fig. 8.1(b). Instead of a vanishingly small current, the reverse current immediately after switching is comparable in magnitude to the forward current if $V_R/R_R \sim V_F/R_F$. Subsequently, the current through the diode remains essentially constant at the large $-I_R$ for a limited period of time before eventually decaying to the steady state value. The period of time during which the reverse current remains constant is known as the *storage time* or *storage delay time* ($t_s$). The total time required for the reverse current to decay to 10% of its maximum magnitude is defined to be the reverse recovery time ($t_{rr}$), while the recovery time ($t_r$) is the difference between $t_{rr}$ and $t_s$. The cited times characterizing the transient are also defined graphically in Fig. 8.1(b).

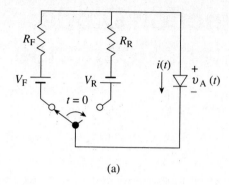

(a)

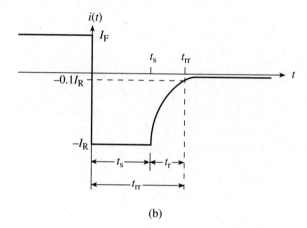

(b)

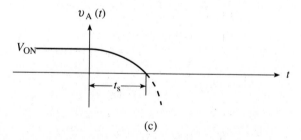

(c)

**Figure 8.1**   The turn-off transient. (a) Idealized representation of the switching circuit. (b) Sketch and characterization of the current-time transient. (c) Voltage-time transient.

The variation of the instantaneous diode voltage ($v_A$) corresponding to the $i$-$t$ transient is shown in Fig. 8.1(c). Specifically note from the figure that (i) the junction remains forward biased for $0 < t < t_s$ even though the externally applied voltage is such as to reverse-bias the diode, and (ii) the $t = t_s$ point correlates with $v_A = 0$.

In analyzing the transient response, we make the assumption that the battery voltages ($V_F$ and $V_R$) are large compared to the maximum forward voltage drop ($V_{ON}$) across the diode. Under the stated assumption

$$I_F = \frac{V_F - V_{ON}}{R_F} \cong \frac{V_F}{R_F} \tag{8.1a}$$

and

$$I_R = \frac{V_R + v_A|_{0<t\leq ts}}{R_R} \cong \frac{V_R}{R_R} \tag{8.1b}$$

Additionally, the qualitative and quantitative analyses to follow focus on the storage delay portion of the transient. Because the decaying $t_r$ portion is readily distorted by stray capacitance in the measurement circuit, it is $t_s$ that has come to be quoted as the primary figure of merit in characterizing the turn-off transient.

## 8.1.2 Qualitative Analysis

In looking at the turn-off transient for the first time, a number of questions undoubtedly come to mind. Why is there a delay in going from the on-state to the off-state? Or perhaps better stated, what is the physical cause of the delay? What goes on inside the diode during the transient? How is it the diode remains forward biased for $0 < t < t_s$ even though the applied voltage is such as to reverse bias the diode?

The root cause of the delay in switching between the on and off states is easy to identify. Forward biasing of the diode, as we have noted previously several times, causes a build-up or storage of excess minority carriers in the quasineutral regions immediately adjacent to the depletion region. When the diode is reverse biased, on the other hand, there is a deficit of minority carriers in the near-vicinity of the depletion region. Simply stated, to progress from the on-state to the off-state, the excess minority carriers pictured in Fig. 8.2 must be removed from the two sides of the junction. The storage delay time derives its name from the fact that the majority of the stored charge is being removed from the diode during the $t_s$ portion of the transient.

As the charge control analysis of Subsection 6.3.1 indicated, removal of the excess minority carrier charge in the quasineutral regions can be achieved in two ways. For one, the carriers can be eliminated in place via recombination. Recombination is of course not instantaneous; several minority carrier lifetimes would be required to go from the on-state to the off-state if recombination were the sole means of carrier removal. The other method

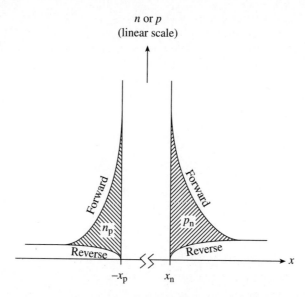

**Figure 8.2**   Stored minority carrier charge leading to the delay in switching between the on and off states. The reverse and forward minority carrier concentrations are plotted simultaneously on a linear scale with the $x$ coordinates matched at the depletion region edges. The break in the $x$-axis inside the depletion region acknowledges a difference in the forward-bias and reverse-bias depletion widths. The cross-hatched areas identify minority carriers that must be removed for switching to be complete.

of reducing the carrier excess is by a net carrier flow out of the region. Once the sustaining external bias is removed, the minority carriers can simply flow back to the other side of the junction where they become majority carriers. This reverse injection could conceivably occur at a very rapid rate. The time required to drift back across the depletion region is only $W/\bar{v}_d \sim 10^{-10}$ sec, where $\bar{v}_d$ is the average drift velocity in the depletion region. However, the number of minority carriers removed per second is limited by the switching circuitry. The maximum reverse current that can flow though the diode is approximately $V_R/R_R = I_R$. The smaller $I_R$, the slower the carrier removal rate. A very rapid transient could be obtained by replacing $R_R$ with a short, but such a procedure would likely lead to a current flow exceeding device specifications and damage to the diode. Summarizing, there are two mechanisms, recombination and reverse current flow, that operate to remove the excess stored charge. Neither mechanism can safely remove the charge at a sufficiently rapid rate to be considered instantaneous. Hence one observes a delay in going from the on to the off state.

We have yet to answer the question how it is the diode remains forward biased during the $0 < t < t_s$ portion of the transient. To answer the question, consider the progressive removal of the hole excess on the $n$-side of a $p^+$-$n$ junction as envisioned in Fig. 8.3. Note from the figure that during the $t < t_s$ stages of the decay the minority carrier concentration

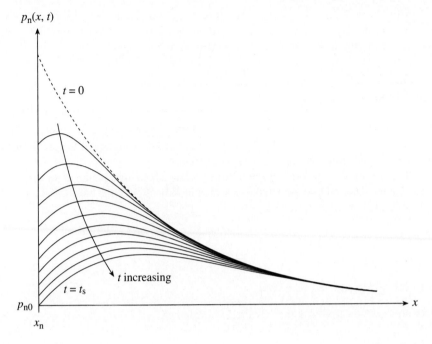

**Figure 8.3** Decay of the stored hole charge inside a $p^+$-$n$ diode as a function of time for $0 \le t \le t_s$.

at the edges of the depletion region ($x = x_n$) is greater than the equilibrium value. In Chapter 6 we established depletion edge boundary conditions that tied the minority carrier concentrations at the depletion region edges to the applied voltage. These same boundary conditions apply under transient conditions with $V_A \rightarrow v_A$. Thus a $v_A > 0$ indicates there is an excess of minority carriers adjacent to the edges of the depletion region. Pertinent to the present discussion, the reverse is also true—a minority carrier excess above the equilibrium value at the edges of the depletion region implies the junction is forward biased. Or stated another way, it is the residual carrier excess at the edges and inside the depletion region that maintains the forward bias across the junction. It is only when the hole concentration at $x = x_n$ drops below the equilibrium value that the diode becomes reverse biased.

Finally, a comment is in order concerning the slope of the Fig. 8.3 curves at $x = x_n$. In an ideal $p^+$-$n$ diode, $i = AJ_P(x_n) = -qAD_P d\Delta p_n/dx|_{x=x_n}$ or

$$\left.\frac{d\Delta p_n}{dx}\right|_{x=x_n} = \left\{\begin{array}{l}\text{slope of } \Delta p_n(x) \text{ or } p_n(x) \\ \text{versus } x \text{ plot at } x = x_n\end{array}\right\} = -\frac{i}{qAD_P} \tag{8.2}$$

All $t > 0$ concentration curves must therefore slope upward at $x = x_n$ because $i < 0$. Moreover, the slopes at $x = x_n$ must be the same for all $t > 0$ curves in Fig. 8.3 because $i = -I_R =$ constant during the $0 < t \leq t_s$ portion of the transient.

---

### Exercise 8.1

**P:** Use the qualitative insight gained into the diode response to predict how key factors are expected to affect the observed $i$–$t$ transient. The accuracy of the predictions will be checked after working out the quantitative theory.

The figures after the problem statement contain a base-line sketch of an $i$–$t$ transient. Using a dashed line, sketch the expected modification to the base-line transient if as indicated on the figures:

(a) $I_F$ is increased to $I_F'$.

(b) $I_R$ is increased to $I_R'$.

(c) $\tau_p$ is decreased (made shorter).

Explain how you arrived at the modified $i$–$t$ sketches.

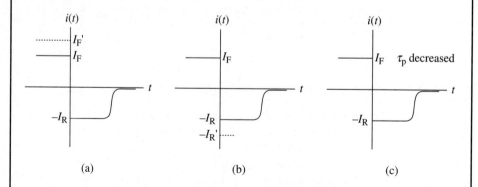

(a)                         (b)                         (c)

**S:** The reasoning leading to the graphical answers presented in Fig. E8.1 are as follows:

(a) Increasing $I_F$ increases the stored charge inside the diode. Since the stored charge is increased and the removal rate is unchanged, it will take longer to remove the stored charge. $t_s$ is expected to increase.

(b) Increasing $I_R$ increases the rate at which the stored charge is removed by the reverse current flow. Thus in this case the storage delay time is reduced.

(c) A shorter minority carrier lifetime increases the carrier recombination rate and will therefore decrease $t_s$.

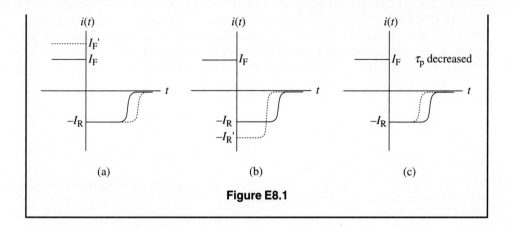

**Figure E8.1**

## 8.1.3 The Storage Delay Time

### Quantitative Analysis

We seek a quantitative relationship that can be used to predict and compute the storage delay time, $t_s$. To simplify the analysis, we treat an ideal $p^+$-$n$ step junction diode and make use of the charge control approach. The electron charge stored on the $p$-side of a $p^+$-$n$ junction is of course negligible compared to the hole charge ($Q_P$) stored on the $n$-side. We also know that $i = i_{DIFF}$ in an ideal diode and, referring to Eq. (6.56) in Subsection 6.3.1,

$$\frac{dQ_P}{dt} = i - \frac{Q_P}{\tau_p} \qquad (8.3)$$

Working toward a solution, we note that $i = -I_R =$ constant for times $0^+ \le t \le t_s$, where $t = 0^+$ is an instant after switching. Thus Eq. (8.3) simplifies to

$$\frac{dQ_P}{dt} = -\left(I_R + \frac{Q_P}{\tau_p}\right) \qquad \ldots 0^+ \le t \le t_s \qquad (8.4)$$

The $Q_P$ and $t$ variables in Eq. (8.4) can be separated and an integration performed over time from $t = 0^+$ to $t = t_s$. We obtain

$$\int_{Q_P(0^+)}^{Q_P(t_s)} \frac{dQ_P}{I_R + Q_P/\tau_p} = -\int_{0^+}^{t_s} dt = -t_s \qquad (8.5)$$

giving

$$t_s = -\tau_p \ln\left(I_R + \frac{Q_P}{\tau_p}\right)\Bigg|_{Q_P(0^+)}^{Q_P(t_s)} = \tau_p \ln\left[\frac{I_R + Q_P(0^+)/\tau_p}{I_R + Q_P(t_s)/\tau_p}\right] \qquad (8.6)$$

The $Q_P(0^+)$ and $Q_P(t_s)$ appearing in Eq. (8.6) must be dealt with to complete the derivation. $Q_P(0^+)$ proves to be readily expressed in terms of known parameters. Because charge cannot be eliminated instantaneously, $Q_P(0^+) = Q_P(0^-)$. However, prior to switching, $dQ_P/dt = 0$ and $i = I_F$. It therefore follows from Eq. (8.3) that

$$I_F = \frac{Q_P(0^-)}{\tau_p} = \frac{Q_P(0^+)}{\tau_p} \tag{8.7}$$

$Q_P(t_s)$, the stored charge remaining at $t = t_s$, poses more of a problem. Wishing to err on the conservative side (i.e., obtain an estimate of $t_s$ that is too large), we take $Q_P(t_s)$ to be approximately zero. Eliminating $Q_P(0^+)$ in Eq. (8.6) using Eq. (8.7) and setting $Q_P(t_s) = 0$, we conclude

$$t_s = \tau_p \ln\left(1 + \frac{I_F}{I_R}\right) \tag{8.8}$$

Equation (8.8) is noted to be in total agreement with the qualitative predictions of Exercise 8.1. $t_s$ increases with increasing $I_F$, decreases with increasing $I_R$, and is directly proportional to $\tau_p$. As a point of information, a more precise analysis, based on a complete $\Delta p_n(x, t)$ solution and properly accounting for the residual stored charge at $t = t_s$, gives[4]

$$\mathrm{erf}\left(\sqrt{\frac{t_s}{\tau_p}}\right) = \frac{1}{1 + \dfrac{I_R}{I_F}} \tag{8.9}$$

Although considerably different in appearance, the Eq. (8.9) solution is likewise noted to be in total agreement with the qualitative predictions of Exercise 8.1.

## Measurement

In both Eqs. (8.8) and (8.9) $t_s$ is directly proportional to $\tau_p$. Moreover, the only other parameters affecting $t_s$ are the currents $I_F$ and $I_R$ controlled by the switching circuitry. This suggests it should be a relatively simple matter to determine the minority carrier lifetime on the lightly doped side of an asymmetrical junction by measuring the turn-off $i$–$t$ transient, noting the storage delay time, and computing $\tau_p$ from Eq. (8.8) or Eq. (8.9).

A measurement system that can be used to observe the transient response of diodes with lifetimes in the $\mu$sec range is pictured in Fig. 8.4. The measurement circuit in combination with the Tektronix PS5004 d.c. power supply and FG5010 function generator simulates the switching circuit of Fig. 8.1(a). To first order, the d.c. bias applied at the power supply node determines $I_F$. Rapid switching of the voltage applied across the test diode is

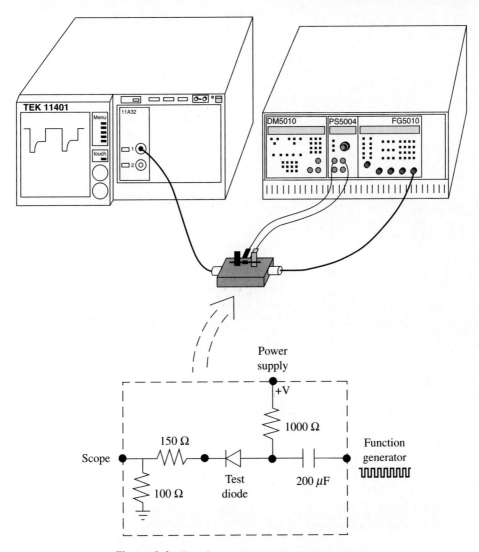

**Figure 8.4**   Transient response measurement system.

provided by the negative-going square-wave pulse derived from the function generator. Because the charge on the plates of the capacitor cannot change instantaneously, the voltage drop across the capacitor must remain constant as the output of the function generator goes from 0 V to the preselected negative value. This forces the voltage on the diode side of the capacitor to decrease by an amount equal to the peak-to-peak value of the square-wave pulse. Subsequent current flow will tend to discharge the capacitor, but the circuit RC time constants are $10^{-2}$ seconds or greater, making the discharge negligible during a typical pulsing period. A voltage proportional to the instantaneous current through the test diode,

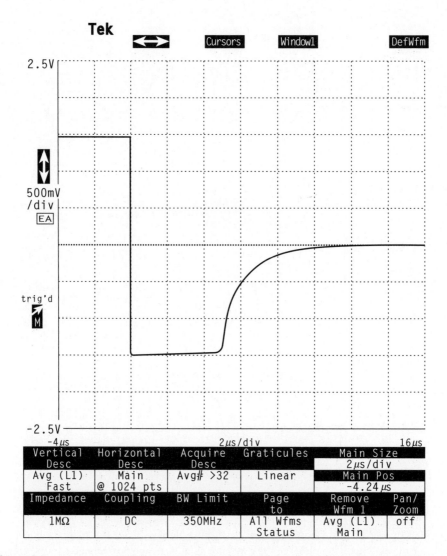

**Figure 8.5** Sample current-time turn-off transient. Output derived from the Fig. 8.4 measurement system. (The *y*-axis display voltage is directly proportional to the instantaneous current through the diode.)

the *i–t* transient, is monitored using the Tektronix 11401 Digitizing Oscilloscope. Waveforms displayed on the 11401 can be analyzed in place, or the screen data can be sent to a printer for subsequent examination.

A sample measured *i–t* transient is reproduced in Fig. 8.5. As extracted from the response curve, $I_R/I_F \cong 1.0$ and $t_s \cong 5.0$ $\mu$sec. One deduces a $\tau_p = t_s/\ln(1 + I_F/I_R) = 7.2$ $\mu$sec employing Eq. (8.8) and a $\tau_p = 22$ $\mu$sec utilizing Eq. (8.9). A more accurate determination of the lifetime, and a check as to whether the theory properly models the diode under test, is obtained by varying the $I_R/I_F$ ratio and choosing the $\tau_p$ yielding the best fit to the normalized $t_s/\tau_p$ versus $I_R/I_F$ plot shown in Fig. 8.6. The dashed and solid lines in Fig. 8.6 were computed employing Eqs. (8.8) and (8.9), respectively. Experimental

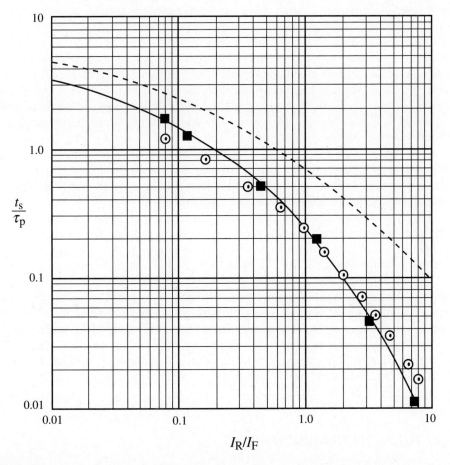

**Figure 8.6**  Theoretical and measured storage delay times normalized to $\tau_p$ versus the reverse to forward current ratio. The dashed line was computed using Eq. (8.8) and the solid line using Eq. (8.9). Experimental data are from a 1N91 Ge diode (■) and a 1N4002 Si diode (⊙).

data from a 1N91 Ge diode and a 1N4002 Si diode have also been added to Fig. 8.6. The Ge diode data can be very closely matched to the more exacting theory if one employs $\tau_p = 14.5$ $\mu$sec. The Si diode data, however, deviates from the predicted dependence. No value of $\tau_p$ can be chosen to fit both the upper and lower ends of the curve. Si diodes, it must be remembered, are seldom ideal diodes as assumed in the derivation of Eqs. (8.8) and (8.9).

### 8.1.4 General Information

We conclude the discussion of the turn-off transient with a few observations of a practical nature. First, as a general rule $t_s \sim \tau_p$ (or $\tau_n$). Increasing the $I_R/I_F$ ratio decreases $t_s$ below $\tau_p$ as is obvious from Fig. 8.6, but more often than not there are constraints that limit the size of the $I_R/I_F$ ratio. Another approach to achieve a rapid switching response is to build diodes with short minority carrier lifetimes. Since $\tau_n$ and $\tau_p$ are proportional to $1/N_T$, where $N_T$ is the R–G center concentration, the minority carrier lifetime can be decreased by the intentional introduction of R–G centers during the fabrication of the diode. The reduction of the minority carrier lifetime in Si devices is typically achieved by diffusing gold into the Si. There is a limit, however, to the R–G center concentration that can be added to a diode. While a shorter lifetime makes for more rapid switching, it also proportionally increases the R–G current ($I_{R-G} \propto 1/\tau_0$)—a high R–G center concentration may increase the off-state current to unacceptable levels. R–G center concentrations approaching the donor or acceptor concentrations also affect the diode electrostatics. In any event, there is no need for *pn* junction diodes with extremely large R–G center concentrations. Other devices with fewer stored carriers, such as the bipolar junction transistor and the metal-semiconductor diode (both addressed in later chapters), are available for use when the application requires subnanosecond switching times.

Finally, mention should be made of the *step-recovery* or *snap-back* diode. The response of the step-recovery diode is special in that the $t_r$ portion of the transient is very short, $\sim 1$ nsec. With a storage delay time $\sim 1$ $\mu$sec, the reverse current part of the *i–t* transient looks like a step, rapidly "snapping back" to the steady state value after reaching $t = t_s$. Step-recovery diodes are used as pulse generators and high-order, single-stage harmonic generators. In fabricating the diodes, a narrow, lowly doped region is sandwiched between heavily doped $p$ and $n$ regions. Formed by employing epitaxial techniques, the junctions in this *p-i-n* type structure are required to be very abrupt. The special doping profile causes the minority carrier charge to be stored very close to the edges of the depletion region. This facilitates almost complete removal of the charge by the end of the storage delay time. With little additional charge to be removed after reaching $t = t_s$, the current drops abruptly to the steady state value.

## 8.2 TURN-ON TRANSIENT

The turn-on transient occurs when the diode is switched from the reverse-bias off-state to the forward-bias on-state. The transition can be accomplished with a current pulse, a voltage pulse, or a mixture of the two pulses. Because of its simplicity and utilization in prac-

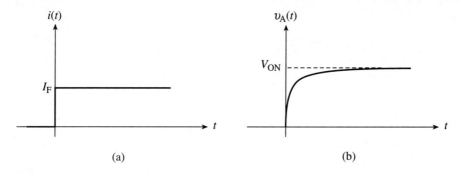

**Figure 8.7**  Turn-on transient assumed to start from $i = 0$: (a) current pulse; (b) voltage-time response.

tical circuits, we consider herein the case where a current pulse is used to switch the diode into the on state.

When the diode current is changed instantaneously from the prevailing reverse bias value to a constant forward current $I_F$, the voltage drop across the diode, $v_A(t)$, monotonically increases from the $V_{OFF}$ at $t = 0$ to $V_{ON}$ at $t = \infty$. The first stage of the response from $t = 0$ to the time when $v_A = 0$ is extremely short in duration. The few minority carriers needed to raise the junction voltage to zero are rapidly injected across the depletion region. Majority carrier rearrangement also acts quickly to shrink the depletion width to its zero-bias value. The short duration of the first portion of the transient allows us to act as if the diode were being pulsed from $i = 0$ to $i = I_F$ at $t = 0$ as depicted in Fig. 8.7(a). Figure 8.7(b) shows the corresponding voltage response assumed to start at $v_A = 0$.

In seeking a quantitative solution for $v_A(t) \geq 0$, we again take the device under analysis to be an ideal $p^+$-$n$ step junction diode and make use of the charge control approach. The envisioned growth of the stored $n$-side hole charge with time is pictured in Fig. 8.8. Note from the figure that $Q_P = 0$ at $t = 0$ consistent with initiating the transient at $v_A = 0$. Since $i = I_F$ throughout the turn-on transient, the Eq. (8.3) relationship for the stored hole charge reduces to

$$\frac{dQ_P}{dt} = I_F - \frac{Q_P}{\tau_p} \tag{8.10}$$

Separating variables and integrating from $t = 0$ when $Q_P = 0$ to an arbitrary time $t$ yields

$$\int_0^{Q_P(t)} \frac{dQ_P}{I_F - Q_P/\tau_p} = \int_0^t dt' = t \tag{8.11}$$

or upon evaluating the $Q_P$ integral

$$t = -\tau_p \ln\left(I_F - \frac{Q_P}{\tau_p}\right)\Bigg|_0^{Q_P(t)} = -\tau_p \ln\left[1 - \frac{Q_P(t)}{I_F \tau_p}\right] \tag{8.12}$$

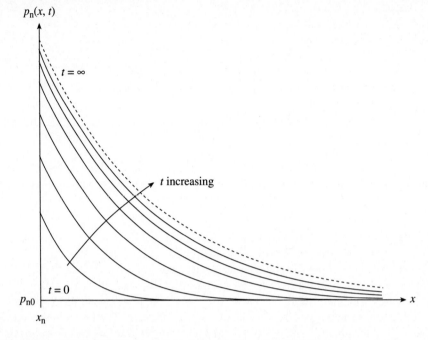

**Figure 8.8**  Build-up of the stored hole charge inside a $p^+$-$n$ diode during the turn-on transient.

Solving Eq. (8.12) for $Q_P(t)$ gives

$$Q_P(t) = I_F \tau_p(1 - e^{-t/\tau_p}) \tag{8.13}$$

Under steady state conditions the stored hole charge in an ideal diode is

$$Q_P = I_{DIFF} \tau_p = I_0 \tau_p(e^{qV_A/kT} - 1) \quad \ldots \text{ steady state} \tag{8.14}$$

As a first-order approximation, let us make the assumption $Q_P(t)$ during the turn-on transient is described by Eq. (8.14) with $V_A \rightarrow v_A$. This is equivalent to assuming the build-up of stored charge occurs quasistatically. One can then write

$$Q_P(t) = I_0 \tau_p(e^{qv_A(t)/kT} - 1) \tag{8.15}$$

Equating the (8.13) and (8.15) expressions for $Q_P(t)$, and solving for $v_A(t)$, we arrive at the solution

$$v_A(t) = \frac{kT}{q} \ln\left[1 + \frac{I_F}{I_0}(1 - e^{-t/\tau_p})\right] \tag{8.16}$$

The turn-on response modeled by Eq. (8.16) is similar to the turn-off response in that the overall length of the transient increases with increasing $I_F$ and $\tau_p$. This is to be more or less expected because the charge required to reach the steady state is directly proportional to $I_F$ and $\tau_p$; i.e., $Q_P(\infty) = I_F\tau_p$ as deduced from either Eq. (8.10) or Eq. (8.13). Perhaps the most interesting feature of the turn-on transient is an initial rapid rise in $v_A(t)$, with $v_A(t)$ increasing to a large fraction of $V_{ON}$ in a very short period of time. If, for example, $T = 300$ K and $V_{ON} = 0.75$ V or $I_F/I_0 = 3.77 \times 10^{12}$, $v_A(t)$ increases to $(0.84)V_{ON}$ after only $t = 0.01\tau_p$. By way of contrast, the final approach to the steady state is considerably slower, requiring a period of time equal to several $\tau_p$.

---

### (C) Exercise 8.2

The CPG (Concentration Plot Generation) program that follows is intended as a visualization and learning aid. The program plots out curves of $\Delta p_n(x', t)/\Delta p_{nmax}$ versus $x'/L_P = (x - x_n)/L_P$ at select $t/\tau_p$ for both the turn-off and turn-on transients. The computations are based on the direct solution of the time-dependent minority carrier diffusion equation for an ideal *pn* step junction[4]. The user chooses the type of plot to be displayed from an opening menu. Menu choices are linear or semilog plots of the turn-off transient concentrations and linear or semilog plots of the turn-on transient concentrations. In the turn-off plots $t$ is stepped from $0.1t_s$ to $t_s$ in $0.1t_s$ increments. In the turn-on plots $t$ is stepped from $0.1\tau_p$ to $2\tau_p$ in $0.1\tau_p$ increments. The user must specify the $I_R/I_F$ ratio when a turn-off plot is desired. We should mention that Figs. 8.3 and 8.8 were drawn based on the CPG program output.

Possible uses of the program are:

(1) Visualize the stored-charge decay during turn-off.

(2) Visualize the stored-charge build-up during turn-on.

(3) Examine and compare corresponding linear and semilog plots.

(4) Confirm that the $x = x_n$ slope of the linear turn-off curves are all the same for $0 < t \le t_s$. (Are the slopes the same on the corresponding semilog plot? Are the $x = x_n$ slopes or curves on a semilog plot related to the current?)

(5) Examine the effect of $I_R/I_F$ on the decay of the stored charge.

(6) Ascertain why the approximate $t_s$ result of Eq. (8.8) becomes less and less accurate as $I_R/I_F$ is increased. (Look at the stored charge remaining at $t = t_s$ as a function of $I_R/I_F$.)

(7) Check the accuracy of the quasistatic approximation employed in the derivation of Eq. (8.16). (On a semilog plot the turn-on curves should all be parallel to the $t = \infty$ curve if the build-up proceeds quasistatically.)

(8) Compare the turn-on $v_A(t)$ computed employing Eq. (8.16) and the $v_A(t)$ values deduced from the turn-on plot.

Hopefully the user will not be limited by the cited suggestions, but will feel free to experiment on his/her own. The user might also consider modifying the program to better display a given output, to extend the computations, or to obtain a specific output such as $v_A(t)$ versus $t$.

MATLAB program script...

```
%Exercise 8.2--Turn-off/Turn-on Concentration Plot Generator

%Determine type of desired plot
clear
close
s=menu('Choose the desired plot','OFF-Linear','OFF-Semilog',...
   'ON-Linear','ON-Semilog');

%Compute ts/taup if turn-off plot is desired
if s<=2,
   %Let Iratio=IR/IF and TS=ts/taup
   Iratio=input('Please input the IR/IF ratio: IR/IF= ');
   if Iratio==0, %Catch if IR=0
      TS=1;
   else
      TS=(erfinv(1./(1+Iratio)))^2;
   end
else
end

%Set values of X and T to be computed for desired plot
%X=x'/LP and T=t/taup
if s==1 | s==2,
   X=0:0.03:3;
   T=TS/10:TS/10:TS;
else
   X=0:0.03:3;
   T=[0.1:0.1:2];
end

%Plot steady-state curve, set axes-labels
y0=exp(-X);
if s==1 | s==3,
   plot(X,y0,'g')
   axis([0 3 0 1])
else
   semilogy(X,y0,'g')
   axis([0 3 1.0e-3 1])
```

```
end
xlabel('x`/LP'); ylabel('Δpn(x`,t)/Δpnmax')
grid; hold on

%Primary computations and time-dependent plots
j=length(T);
for i=1:j,
   A=exp(-X).*(1-erf(X./(2*sqrt(T(i)))-sqrt(T(i))));
   B=exp(X).*(1-erf(X./(2*sqrt(T(i)))+sqrt(T(i))));
   yon=(A-B)/2;   %yon=Δpn(x',t)/Δpnmax during turn-on
   if s==3,
      plot(X,yon);
   elseif s==1,
      yoff=exp(-X)-(1+Iratio).*yon;   %yoff=Δpn(x',t)/Δpnmax during turn-off
      plot(X,yoff);
   else
   end
   if s==4,
      semilogy(X,yon);
   elseif s==2,
      yoff=exp(-X)-(1+Iratio).*yon;
      semilogy(X,yoff);
   else
   end
end; hold off
```

## 8.3  SUMMARY

In this chapter we examined the electrical response and internal carrier response of *pn* junction diodes subjected to a large rapid change in the applied voltage or impressed current, a change intended to switch the diode from the forward-bias on-state to the reverse-bias off-state, or vice-versa. During the turn-off transient the excess minority carriers stored in the quasineutral regions must be removed before steady state conditions can be re-established. The diode initially remains forward biased and a large, constant, reverse current flows through the diode until the carrier concentrations decrease to their equilibrium values at the edges of the depletion region. This takes place in a period of time known as the storage delay time, $t_s$. $t_s$ is the primary figure of merit used to characterize the transient response of *pn* junction diodes. The storage delay time increases in relation to the initial store of carriers, decreases with the rate of carrier removal by the reverse current, and is directly proportional to the minority carrier lifetime. The storage time is decreased by adding R–G centers to the semiconductor during device fabrication, and step-recovery diodes are produced by properly tailoring the doping profile. Measurement of the turn-off transient can also be used to determine the minority carrier lifetime on the lightly doped side of a junction. Overall, the focus of the chapter has been on establishing a basic understanding of transient operation and providing physical insight, knowledge that will prove very useful in treating one of the pre-eminent switching devices, the bipolar junction transistor.

## PROBLEMS

| CHAPTER 8    PROBLEM INFORMATION TABLE | | | | |
|---|---|---|---|---|
| Problem | Complete After | Difficulty Level | Suggested Point Weighting | Short Description |
| 8.1 | 8.3 | 1–2 | 15 (a::c-2, d::h-1, i::j-2) | Quick quiz |
| 8.2 | 8.1.2 | 1 | 6 (2 each part) | Interpret $p_n(x, t)$ plot |
| ● 8.3 | 8.1.3 | 2 | 8 | Improved $Q_P(t_s)$ approx. |
| 8.4 | 8.2 | 2–3 | 10 (a::c-3, d-1) | Open circuit voltage decay |
| 8.5 | " | 2 | 6 | Compute turn-on times |
| 8.6 | " | 2 | 10 (a-3, b-7) | Pulse $I_{F1}$ to $I_{F2} > I_{F1}$ |
| 8.7 | " | 2–3 | 10 (a-6, b-4) | Combined turn-off/turn-on |
| ● 8.8 | " | 3 | 15 (a-10, b-5) | Compare turn-on $v_A(t)$ |

**8.1** Quick Quiz.

Answer the following questions as concisely as possible.

(a) Define *storage delay time.*

(b) Define *recovery time* $(t_r)$.

(c) Is it possible for the *pn* junction to support a reverse current even though $v_A > 0$? Explain.

(d) What is the root cause of the delay in switching from the on-state to the off-state?

(e) Name the two mechanisms that act to remove the excess stored charge during the turn-off transient.

(f) True or false: If $\Delta p_n(x, t) > 0$, $v_A > 0$.

(g) True or false: If $i > 0$, the slope of a linear $p_n(x, t)$ versus $x$ plot must be positive ($p_n$ increases with $x$) at $x = x_n$.

(h) What is special about the electrical and physical properties of a step-recovery diode?

(i) True or false: Increasing both $I_F$ and $I_R$ by a factor of 2 will have no effect on the storage delay time. Indicate how you arrived at your answer.

(j) True or false: Recombination actually acts to *retard* the build-up of stored carriers during the turn-on transient.—Indicate how you arrived at your answer.

**8.2** The hole concentration on the $n$-side of a $pn$ step junction diode at a given instant of time is as pictured in Fig. P8.2.

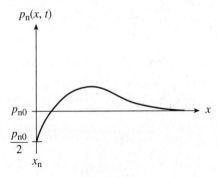

**Figure P8.2**

(a)  Is the junction forward or reverse biased? Explain how you arrived at your answer.

(b)  If $p_{n0} = 10^4/\text{cm}^3$ and $T = 300$ K, determine $v_A$.

(c)  Is there a forward or reverse current flowing through the diode? Explain how you arrived at your answer.

● **8.3** The approximation $Q_P(t_s) = 0$ was used in deriving Eq. (8.8). Researchers have suggested the alternative approximation,[5]

$$Q_P(t_s) = \frac{I_F \tau_p}{1 + I_F/I_R}$$

which leads to the revised charge control expression

$$t_s = \tau_p \ln\left[\frac{(1 + I_F/I_R)^2}{1 + 2I_F/I_R}\right]$$

Determine whether the revised $t_s$ expression is a significant improvement. Construct a plot of $t_s/\tau_p$ versus $I_R/I_F$ similar to Fig. 8.6 using the revised charge control expression in place of Eq. (8.8). Briefly comment on the result.

**8.4** An ideal $p^+$-$n$ step junction diode carrying a forward current $I_F$ is suddenly open-circuited at $t = 0$.

(a)  Sketch the expected variation of $p_n(x, t)$ versus $x$ at progressively increasing times after open-circuiting the diode. (Check your answer using the CPG program in Exercise 8.2.)

(b)  Derive an expression for the stored hole charge, $Q_P(t)$, inside the diode at times $t > 0$. Be sure to express $Q_P(0)$ in terms of known parameters.

(c) Assuming a quasistatic decay of the hole charge, derive an expression for $v_A(t)$. Take the $v_A$ voltages of interest to be greater than a few $kT/q$; i.e., $\exp(qv_A/kT) \gg 1$. Also use the fact that $I_F/I_0 = \exp(qV_{ON}/kT) - 1 \cong \exp(qV_{ON}/kT)$ to simply your result.

(d) Does the part (c) result suggest anything? Explain. (For further information see Subsection 8.5.2 in Schroder[3].)

**8.5** An ideal $p^+$-$n$ step junction diode is switched with a current pulse from $I = 0$ to $I_F = 1$ mA at $t = 0$. Calculate the time necessary for the diode voltage to reach 90% and 95% of its final value. Let $\tau_p = 1$ $\mu$sec and $I_0 = 10^{-15}$ A.

**8.6** An ideal $p^+$-$n$ step junction diode initially forward biased at $I_{F1}$ is pulsed to a constant current of $I_{F2} > I_{F1}$ at $t = 0$.

(a) Sketch the expected variation of $p_n(x, t)$ versus $x$ at progressively increasing times after $t = 0$.

(b) Assuming a quasistatic build-up of the stored charge, derive an expression for $v_A(t)$.

**8.7** At $t = 0$ the current through a $p^+$-$n$ diode is switched from $I_F = 1$ mA to $i = -I_R = -1$ mA. After 1 $\mu$sec a current pulse is applied to switch the diode back to an $I_F = 1$ mA. Assume the diode to be ideal with $\tau_p = 1$ $\mu$sec.

(a) Sketch the $i(t)$ through the diode as a function of time.

(b) Establish an expression for $v_A(t)$ at times $t > 1$ $\mu$sec.

● **8.8** (a) The turn-on curves drawn by the CPG program in Exercise 8.2 correspond to $t/\tau_p$ values stepped from 0.1 to 2 in 0.1 increments. Noting

$$\frac{\Delta p_n(0,t)}{\Delta p_{nmax}} = \frac{e^{qv_A(t)/kT} - 1}{e^{qV_{ON}/kT} - 1}$$

appropriately modify the program to obtain $v_A/V_{ON}$ at the stepped values of $t/\tau_p$. Let $V_{ON} = 0.5$ V and $T = 300$ K in performing a sample computation.

(b) Compare the $v_A(t)$ derived from the part (a) exact solution with the Eq. (8.16) solution that was obtained by assuming turn-on proceeds quasistatically. The comparison assuming $V_{ON} = 0.5$ V and $T = 300$ K may be presented in either a plot or point-by-point format. Note that $I_F/I_0 = \exp(qV_{ON}/kT) - 1$.

# 9 Optoelectronic Diodes

## 9.1 INTRODUCTION

To round out the discussion of *pn* junction diodes, we consider in this chapter special diode structures that are specifically designed and built for optical applications. Many of the diodes to be discussed involve a semiconductor or semiconductors other than Si. The reader is thereby given a glimpse of the larger semiconductor picture where more and more materials are being used in the fabrication of sophisticated special-purpose devices. Semiconductor photodevices quite generally fall into one of three functional categories. Two groups of photodevices convert photo-energy into electrical energy. If the purpose of the photo-to-electrical conversion is to detect or determine information about the photo-energy, the device is called a *photodetector*. If the purpose of the photo-to-electrical energy conversion is to produce electrical power, the device is called a *solar cell*. The third type of photodevice converts electrical energy into photo-energy and includes light emitting diodes (LEDs) and laser diodes. Herein we survey one set of devices from each of the three categories: *pn* junction and related photodiodes belonging to the photodetector family, *pn* junction solar cells, and LEDs.

The commercial marketplace has experienced a virtual explosion of optoelectronic diode applications in recent years. The bar code reader now common at check-out counters, the digital disk reader in audio systems, and the laser printer in the office all make use of LEDs or laser diodes as a photosource. Photodetectors in combination with LEDs or laser diodes are used in circuit isolators, intruder alarms, and remote controls. In addition, the optical signal is generated using a laser diode and detected using a photodiode in modernized telecommunication networks employing optical fibers. Solar cells are used to power hand-held calculators, battery chargers, and communication satellites. Although slow to find significant use in large-scale power generation, the U.S. National Photovoltaics Program calls for an installed solar-cell capacity of 200–1000 megawatts by the year 2000 and 10,000–50,000 megawatt capacity by 2010–2030. Clearly, the variety and scale of optoelectronic diode applications can only be expected to increase with time.

As an aid in the discussion, the visible and adjacent wavelength regions of the optical spectrum are identified in Fig. 9.1. The relative response of the human eye plotted in the upper part of the figure is correlated with the major color bands shown spanning the roughly 0.4 $\mu$m to 0.7 $\mu$m visible portion of the spectrum. Note that the relationship between the optical wavelength ($\lambda$) and the associated photon energy ($E_{ph} = h\nu$) is

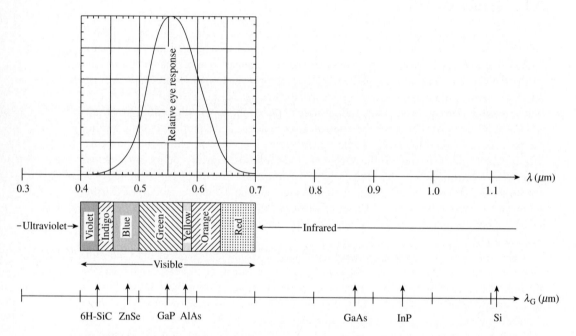

**Figure 9.1** Visible and adjacent regions of the optical spectrum (middle) correlated with the relative eye sensitivity for normal photopic vision (top) and wavelengths where the photon energy is equal to the 300 K band gap energy of select semiconductors (bottom).

given by

$$\lambda = \frac{c}{\nu} = \frac{hc}{h\nu} = \frac{hc}{E_{ph}} \qquad \ldots c = \text{speed of light} \qquad (9.1)$$

or, if $\lambda$ is expressed in $\mu$m and $E_{ph}$ in eV,

$$\boxed{\lambda = \frac{1.24}{E_{ph}}} \qquad (9.2)$$

The wavelength at which the photon energy is equal to the semiconductor band gap energy, $\lambda_G = 1.24/E_G$, is also cited in Fig. 9.1 for some of the more important optoelectronic materials.

## 9.2 PHOTODIODES

### 9.2.1 *pn* Junction Photodiodes

A *pn* junction photodiode is just a *pn* junction diode that has been specifically fabricated and encapsulated to permit light penetration into the vicinity of the metallurgical junction. The absorption of light inside the diode creates electron-hole pairs, as pictured in Fig. 9.2.

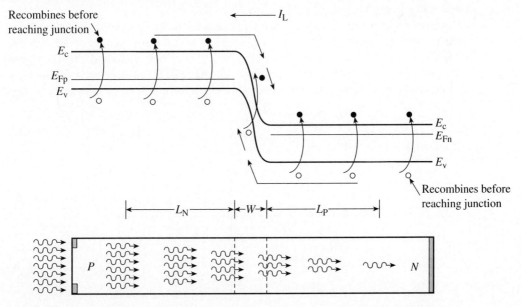

**Figure 9.2** Visualization of light absorption, electron-hole creation, and the light-induced current in a *pn* junction photodiode.

On average, minority carriers created in the quasineutral regions within a diffusion length of $x = -x_p$ (*p*-side) or $x = x_n$ (*n*-side) live long enough to diffuse to the depletion region. These carriers, and carriers photogenerated within the depletion region, are subsequently swept by the $\mathscr{E}$-field to the opposite side of the junction, thereby contributing an added *reverse*-going component to the current through the diode. If the photogeneration rate ($G_L$) is assumed to be uniform throughout the diode, the added component due to light ($I_L$) should be equal to $-q$ times the electron-hole pairs photogenerated per second in the volume $A(L_N + W + L_P)$, or

$$I = I_{dark} + I_L \tag{9.3}$$

with

$$I_L = -qA(L_N + W + L_P)G_L \tag{9.4}$$

A modified derivation of the ideal diode equation and the R–G current relationship with $G_L \neq 0$ confirms the foregoing result (see Problem 6.8).

Examining Eq. (9.4), we note that the depletion width $W$ in a *pn* junction diode is typically small compared to $L_N + L_P$. If $W$ is negligible, $I_L$ becomes independent of the applied bias. The light-on *I–V* characteristics are therefore expected to be essentially identical to the dark *I–V* characteristic, except the light-on curves are translated downward, moved in the $-I$ direction, along the current axis. Moreover, because $I_L \propto G_L$, the downward translation of the characteristics should increase in proportion to the intensity of the incident illumination. The described form of the photodiode *I–V* characteristics are illustrated in Fig. 9.3.

An important characteristic of any photodetector is its spectral or wavelength response—how the observed $I_L$, for example, varies with the wavelength of the incident light. A spectral response curve representative of Si *pn* junction photodiodes is reproduced in Fig. 9.4. The pictured photodiode response, like those of all photodetectors, spans only a limited range of wavelengths. The upper wavelength limit in most photodetectors is tied

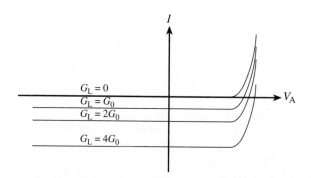

**Figure 9.3**  Photodiode *I–V* characteristics.

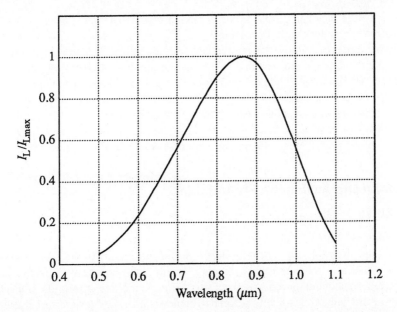

**Figure 9.4**  Spectral response of a Si *pn* junction photodiode. The photo-power incident on the photodiode was the same for all wavelengths. Representative characteristic: The response varies somewhat with diode construction.

directly to the semiconductor band gap. Photons are absorbed and electron-hole pairs are photogenerated in a semiconductor if $E_{ph} > E_G$. When $E_{ph} < E_G$, on the other hand, the semiconductor is all but transparent to the light. The semiconductor spectral response therefore essentially cuts off at $\lambda_G = 1.24/E_G$. For Si, $E_G = 1.12$ eV at 300 K and, consistent with Fig. 9.4, one expects a minimal response at wavelengths greater than $\lambda_G \cong 1.1$ $\mu$m.

Two reasons can be cited for the decrease in the spectral response at shorter wavelengths. First, as is common practice, the photo-power was held constant in accumulating the Fig. 9.4 data. Since the photon energy increases with decreasing wavelength, the flux of photons striking the semiconductor correspondingly decreases. Thus part of the reduced response at shorter wavelengths is simply due to the fact that there are fewer photons to be absorbed.

The remaining decrease in the spectral response is associated with the nonuniform absorption of light inside the diode. During the discussion of photogeneration in Subsection 3.3.3, the light intensity was noted to fall off exponentially with distance from the surface of the semiconductor. The decay constant corresponding to the inverse of the absorption coefficient $(1/\alpha)$ represents the average depth of penetration of light into a material. As $\lambda$ is decreased below $\lambda_G$, $\alpha$ increases rapidly (see Fig. 3.20) and the light is absorbed closer and closer to the semiconductor surface. Eventually the majority of generation takes place on the side of the junction adjacent to the surface. Carrier recombination is greater near the surface and an increased number of the photogenerated carriers recombine before they can diffuse to the depletion region. The net result is a progressive reduction in the response with decreasing $\lambda$.

Another property employed in photodetector characterization is frequency response—how rapidly the detector can respond to a time-varying optical signal. In this regard the standard *pn* junction photodiode exhibits limited capability. Photogenerated minority carriers must diffuse to the depletion region before an electrical current is observed externally. Diffusion being a relatively slow process, the maximum frequency response of *pn* junction photodiodes is at best in the tens of MHz. This response is quite low compared to the frequency response attainable with photodiodes described in the next subsection.

### 9.2.2 *p-i-n* and Avalanche Photodiodes

#### *p-i-n* Photodiodes

A *p-i-n* diode is a three-region structure in which an "intrinsic" (actually lightly doped) *i*-region is sandwiched between heavily doped *p*- and *n*-regions. In the *p-i-n* photodiode, shown schematically in Fig. 9.5(a), an opening is made in the surface metallization to admit light, the top semiconductor region is kept very thin to minimize absorption in the region, and the *i*-layer width is specifically tailored to achieve the desired response characteristics.

Because of the low doping, the *i*-layer is totally depleted under zero bias or becomes depleted at small reverse biases. Furthermore, the heavy doping of the outer *p*- and *n*-regions causes the depletion widths in these regions to be very narrow. Thus as pictured in Fig. 9.5(b), the depletion width inside the device is effectively equal to the *i*-layer width independent of the applied reverse bias. The energy bands in Fig. 9.5(b) are linear functions of position and the $\mathscr{E}$-field is approximately constant in the *i*-region because of the low semiconductor doping. It should also be noted that the heavy doping of the outer *p*- and *n*-regions means the minority carrier diffusion lengths in these regions will be relatively small. As a result, the greater part of the photocurrent flowing in a *p-i-n* photodiode arises from carriers generated in the central depletion region.

Operational advantages of the *p-i-n* photodiode that have made it one of the most widely employed photodetectors stem from the existence and tailorability of the *i*-region. For one, the diode can be optimized for response at a given wavelength by making the *i*-layer width equal to the inverse of the absorption coefficient $(1/\alpha)$ at the specified wavelength. Second, with most of the photocurrent arising from light absorption in the *i*-region, the frequency response is greatly enhanced over that of a *pn* junction photodiode. The large $\mathscr{E}$-field in the depleted *i*-region leads to the rapid collection of photogenerated carriers and a maximum frequency response,

$$f_{\max} \cong \left( \frac{1}{\substack{\text{carrier transit} \\ \text{time across } W_{\mathrm{I}}}} \right) \cong \frac{1}{W_{\mathrm{I}}/v_{\mathrm{sat}}} \qquad (9.5)$$

where $W_{\mathrm{I}}$ is the width of the *i*-region and $v_{\mathrm{sat}}$ is the saturation drift velocity (see Subsection 3.1.2). Typically $v_{\mathrm{sat}} \cong 10^7$ cm/sec. If $W_{\mathrm{I}} = 5$ $\mu$m, for example, then $f_{\max} \cong$ 20 GHz. We should note that $W_{\mathrm{I}}$ cannot be made arbitrarily small to improve the frequency

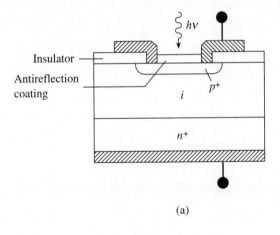

(a)

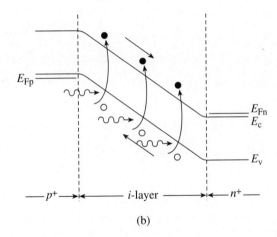

(b)

**Figure 9.5**   *p-i-n* photodiode. (a) Cross section. (b) Reverse-bias energy band diagram emphasizing the *i*-region and picturing photogeneration.

response. The *RC* time constant associated with the internal series resistance ($R_S$) and the junction capacitance ($C_J = K_S \varepsilon_0 A / W_I$) increases with decreasing $W_I$ and eventually limits the response time of the diode.

The excellent frequency response of the *p-i-n* photodiode makes it a prime candidate for use as the photodetector in optical fiber telecommunications. By and large, silica-based optical fiber systems installed prior to 1990 operated at the 1.3 $\mu$m wavelength where chromatic dispersion[†] in the fibers is at a minimum. More advanced high bit rate systems, on

[†] Chromatic dispersion is a spreading out of a light pulse caused by different wavelengths of light traveling at slightly different velocities. The use of multiwavelength photosources requires operation at the chromatic dispersion minimum.

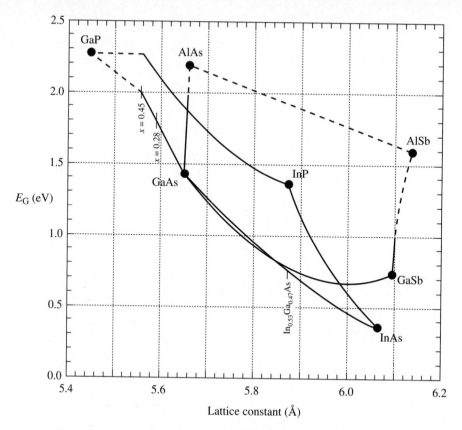

**Figure 9.6** Band gap energy versus lattice constant of select III-V compounds and alloys. The line connecting two compounds specifies the $E_G$ versus lattice constant for alloys made of the two compounds. In most cases the lattice constant is a linear function of the x-value in going from compound-A to compound-B. Thus, for example, $In_{0.8}Ga_{0.2}P$ would have a lattice constant of 5.87 Å − 0.2(5.87 Å − 5.45 Å) = 5.79 Å, and an $E_G \cong 1.5$ eV is deduced from the line connecting InP and GaP. Solid and dashed lines identify alloys that are direct and indirect semiconductors, respectively.

the other hand, employ single-wavelength photosources and operate at 1.55 $\mu$m, where fiber loss is at a minimum. In either case the wavelength of operation is beyond the 1.1 $\mu$m cutoff of Si, thereby necessitating the use of some other semiconductor. The material system of choice for fiber optic applications is the alloy $In_{0.53}Ga_{0.47}As$ deposited on an InP substrate. As shown in the Fig. 9.6 band gap plot of common III-V alloy systems, $In_{0.53}Ga_{0.47}As$ has an $E_G \cong 0.75$ eV ($\lambda_G = 1.65$ $\mu$m) and is lattice matched to InP. Lattice matching, having exactly the same lattice constant, facilitates the deposition of quality $In_{0.53}Ga_{0.47}As$ layers on commercially available InP substrates.

Figure 9.7 shows the cross section of an InGaAs *p-i-n* photodiode. The wide band gap InP with a $\lambda_G = 0.95$ $\mu$m acts as a "window" through which 1.3 $\mu$m or 1.55 $\mu$m light is readily transmitted to the absorbing $In_{0.53}Ga_{0.47}As$ "*i*"-layer. Defects in the InGaAs layer

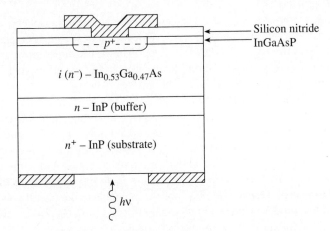

**Figure 9.7**  Cross section of an InGaAs *p-i-n* photodiode.

are minimized by the prior deposition of an InP buffer layer. The wide band gap InGaAsP cap-layer is added to reduce surface-related dark currents. Finally, the insulating silicon nitride layer generally protects the surface and minimizes surface recombination.

## Avalanche Photodiodes

Avalanche photodiodes are specially constructed *p-i-n, pn,* or even metal-semiconductor (see Chapter 14) photodiodes that are operated near the avalanche breakdown point. A standard Si avalanche photodiode configuration is displayed in Fig. 9.8. One obvious

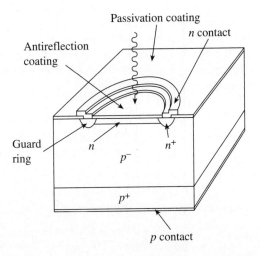

**Figure 9.8**  Si avalanche photodiode. (From Yang[6], ©1988 by McGraw-Hill, Inc. Reprinted by permission of the publisher.)

special structural feature is the guard ring around the junction periphery. As noted in Chapter 6, junction curvature leads to early breakdown about the junction periphery. The guard ring minimizes the edge breakdown problem. Uniformly doped low-defect material and minimization of defect creation during device processing are also required to achieve uniform breakdown across the face of the junction.

The primary advantage of the avalanche photodiode is a photo-signal gain leading to improvement in the signal-to-noise (S/N) ratio. As a general rule, amplification of a signal is accompanied by amplification of the noise and added noise from the amplifier. Signal gain therefore typically leads to a *reduction* in the S/N ratio. Inside an avalanche photodiode, however, avalanche multiplication amplifies the photo-signal without amplifying the typically dominant receiver circuit noise. Thus there is an improvement in the S/N ratio until the added avalanche-related noise becomes comparable to the circuit noise.

Avalanche photodiodes made from InGaAs on InP and from Ge provide alternatives to the *p-i-n* photodiode for use in fiber optic telecommunications.

## 9.3  SOLAR CELLS

### 9.3.1 Solar Cell Basics

A variety of device structures can and have been employed in constructing solar cells. By far and away the most common cells are in essence just large-area *pn* junction photodiodes. Solar cells are designed of course to minimize energy losses, whereas photodiodes are routinely designed to achieve a specific spectral response or a rapid time response. Design differences notwithstanding, the solar cell *I–V* characteristics are of the same general form as the photodiode characteristics of Fig. 9.3. Note that power is derived from the illuminated device if the d.c. operating point lies in the fourth quadrant where $I$ is negative and $V$ is positive. Fourth-quadrant operation can be achieved, for example, by simply placing a resistor in series with the illuminated solar cell.

Understandably, the fourth-quadrant characteristic is of prime interest in evaluating and applying solar cells. It is therefore common practice to show only the fourth-quadrant portion of the characteristics and to orient the $-I$ axis upward on the plot as illustrated in Fig. 9.9. Figure 9.9 also graphically defines the following solar cell parameters of interest:

$V_{oc}$ ... the open circuit voltage
$I_{sc}$ ... the short circuit current
$V_m$, $I_m$ ... the operating point voltage and current yielding the maximum power output.

$V_{oc}$ is obviously the maximum voltage that can be supplied by the cell for a given photoinput, and $I_{sc}$ is the maximum current that can be derived from the cell. It follows that $P_{max} = I_m V_m < I_{sc} V_{oc}$. In assessments of solar cell performance, one often encounters

$$FF \equiv \frac{P_{max}}{I_{sc} V_{oc}} = \frac{I_m V_m}{I_{sc} V_{oc}} \tag{9.6}$$

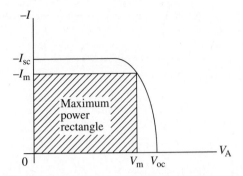

**Figure 9.9** Inverted fourth-quadrant *I–V* characteristic of a solar cell under illumination with key solar cell parameters identified along the coordinate axes.

*FF* is known as the *fill factor* and is of course always less than unity. The ultimate measure of solar cell performance, the power conversion efficiency ($\eta$), is determined from the current–voltage parameters employing

$$\eta \equiv \frac{P_{max}}{P_{in}} = \frac{I_m V_m}{P_{in}} = \frac{FF I_{sc} V_{oc}}{P_{in}} \tag{9.7}$$

where $P_{in}$ is the photo-energy incident per second or input power.

### 9.3.2 Efficiency Considerations

A key issue with solar cells is conversion efficiency—converting the maximum amount of available solar energy into electrical energy. The higher the efficiency of the cells, the lower the cost and collection area required to achieve a desired electrical output. A number of factors enter into the overall efficiency exhibited by a cell and involve both material and design considerations.

Any discussion of conversion efficiency logically begins with the output from the sun. The spectral distribution of the sun's energy reaching the earth is plotted in Fig. 9.10. The AM0 (air mass zero) curve is the measured radiant energy just outside the earth's atmosphere and is of interest in orbiting satellite applications. The AM1.5 curve, normalized to yield a total spectral power density of 100 mW/cm², is representative of average terrestrial conditions in the United States. In either case, most of the spectral power is in the visible, with a long tail extending into the infrared. When considering materials for use in solar cells, the area under the spectral curve beyond the $\lambda_G$ of a semiconductor yields the incident power that is lost or wasted because it cannot be absorbed. Si, with a slightly larger cutoff wavelength, has an advantage in this regard compared to GaAs: ~20% and ~35% of the incident energy is not absorbed in Si and GaAs, respectively. However, it would be incorrect to conclude the overall conversion efficiency increases with decreasing band gap. If photon absorption at $\lambda < \lambda_G$ wavelengths is examined, one finds that only the $E_G$ portion

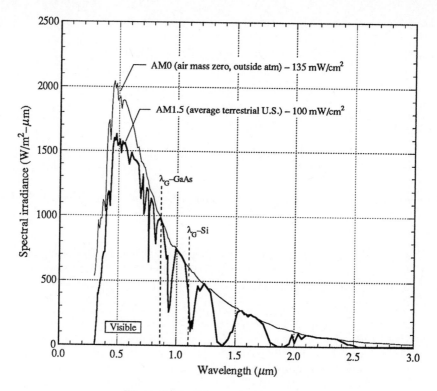

**Figure 9.10**   Solar spectral irradiance.

of the photon energy is used profitably in producing electron-hole pairs. The $E_{ph} > E_G$ portion of the photon energy adds to the kinetic energy of the photogenerated carriers and is eventually dissipated as heat. Calculations indicate that ~40% of the absorbed photon energy is unavoidably wasted, dissipated as heat, in Si. The corresponding loss is only ~30% in GaAs because of its larger band gap. Clearly, a trade-off exists between the two cited loss mechanisms, giving rise to an optimum band gap where the energy conversion is at a maximum. Rather fortuitously, the band gaps of Si and GaAs, the semiconductors with the most advanced technologies, both lie very close to the theoretical maximum.

Given a specific semiconductor material, the next task is to design and fabricate the solar cell to minimize further energy losses. In discussing device-related loss mechanisms, we will refer to the high-efficiency Si solar cell pictured in Fig. 9.11. Observe first of all that the contact to the top (light-incident) side of the cell is made through narrow "fingers." The fingers, all connected together along one edge of the cell, are a design compromise. Zero-width fingers or fingers only along the cell edges would allow maximum light penetration to the underlying silicon. However, a series resistance of only a few ohms can seriously degrade the efficiency of a solar cell. The farther apart the fingers, the longer the current path through the narrow *n*-region at the top of the cell, and the greater the series resistance. Metallization and contact resistances can become important if the fingers are too narrow. The chosen finger size and spacing are calculated to provide an optimum trade-

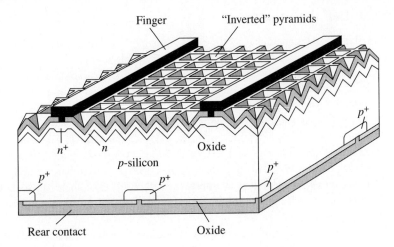

**Figure 9.11**  Schematic diagram of a high-efficiency Si single-crystal solar cell. (From Green et al.[7], ©1990 IEEE.)

off between residual "shadowing," blocking of some of the light that strikes the cell, and cell series resistance.

Another potentially significant loss mechanism is the reflection of light at the Si surface. Approximately 30% of the light striking a bare planar Si surface at normal incidence will be reflected. To minimize losses due to reflection, the top surface of solar cells are typically "textured" and covered with an antireflection coating. Texturing of the surface, the formation of the inverted pyramids in Fig. 9.11, decreases the reflected light by forcing the light to strike the Si surface two or more times before escaping. Texturing is achieved by placing the Si in an anisotropic etching solution, an etch that preferentially removes Si atoms along certain crystalline planes. Having an index of refraction intermediate between air and Si, the top $SiO_2$ layer in the pictured cell, or preferably a deposited antireflection coating with optimized parameters, further reduces the reflection. Solar cells constructed in the described manner have attained a net reflection of less than 1%.

Once the light has entered the semiconductor, the focus shifts to maximizing the light absorption. In the Fig. 9.11 solar cell the bottom surface oxide and metallization effectively form a mirror that reflects light back into the silicon. Long wavelength light literally bounces back and forth between the top and bottom surfaces of the cell. This "light-trapping" dramatically enhances the long ($\lambda \sim \lambda_G$) wavelength absorption.

Finally, the cell must be designed and built to collect as many of the photogenerated minority carriers as possible. This necessitates minimizing carrier recombination throughout the device structure. Very long minority carrier lifetimes are the rule in modern single-crystal Si cells, yielding diffusion lengths greater than the width of the cell. With the top and bottom surfaces also carefully oxidized to minimize surface recombination, carriers generated almost anywhere in the cell volume have a high probability of diffusing to the depletion region and being swept to the opposite side of the junction before they recombine.

### 9.3.3 Solar Cell Technology

Solar cells fall into three general categories: thin-film, single-crystal, and concentrator. The light-absorbing semiconductor in thin-film cells is a deposited amorphous or polycrystalline film. Thin-film amorphous Si cells are commercially available, while CdTe and $CuInSe_2$ cells are under development. Almost all single-crystal cells are made of either Si or GaAs. The cell described in the preceding section is an example of a single-crystal Si cell. Single-crystal GaAs cells often contain AlGaAs or other alloy layers, with the active GaAs layer being epitaxially deposited on GaAs, Ge, or Si substrates. Concentrator cells, designed to operate under light intensities equal to 100 suns or more, again primarily utilize Si and GaAs, although highly efficient cells have also been constructed using InGaAsP and InP/InGaAs. The great strides made at improving solar cell efficiencies in recent years, including the best solar efficiencies attained to date in each category, are summarized in Fig. 9.12.

Deposited thin-film solar cells are of interest because they can be made cheaply and in large area configurations. Commercial amorphous-Si (a-Si) modules have been produced with areas up to 1.2 m². Amorphous-Si cells are produced by vapor depositing Si onto tin-oxide coated glass substrates. The tin oxide is a transparent conducting material that functions as the front or illuminated-side contact. An aluminum or silver layer is deposited over the a-Si to form the back contact. The ~1 $\mu$m thick amorphous Si is appropriately doped during deposition to achieve a *p-i-n*-type structure. The maximum efficiency of a-Si cells in the laboratory is ~13%; the efficiency of commercial units is often a factor of two lower. In addition, the output from a-Si cells decreases by 10%–15% during the first year of operation; exposure to light apparently breaks passivating bonds, giving rise to additional traps and enhanced recombination. Nevertheless, amorphous-Si accounts for a significant fraction of worldwide solar-cell shipments, most of the a-Si cells being destined for use in consumer products. Thin-film CdTe cells, having recently attained an efficiency of 16% in the laboratory and 11% in the field, are likely to provide effective competition in the near future.

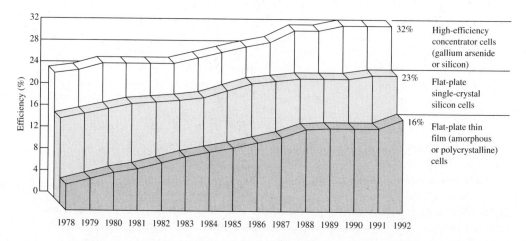

**Figure 9.12**   Maximum efficiencies of laboratory solar cells as a function of time.[8]

Concentrator cells are obviously intended for high-power applications. The basic premise leading to the development of the cells was that a mirror or concentrating lens system would be less expensive than an equivalent area of solar cells. Providing somewhat of a bonus, the higher current and voltage levels in concentrator cells also leads to more efficient cell operation as is evident from Fig. 9.12. The stacking of concentrator cells— placing for example a Si cell or a GaSb ($\lambda_G = 1.7 \ \mu$m) cell under a GaAs cell to absorb wavelengths not processed by the GaAs—is expected to eventually yield efficiencies in excess of 35%. Special concerns with the design and operation of concentrator cells include heat dissipation, elevated device temperatures, and high current densities.

## 9.4  LEDs

### 9.4.1 General Overview

Basic operation of the LED is relatively easy to explain utilizing information about semiconductors and diode operation presented in earlier chapters. We know that forward biasing a *pn* junction diode causes large numbers of majority carrier electrons on the *n*-side to be injected over the reduced potential hill into the *p*-side quasineutral region. Holes on the *p*-side are similarly injected into the *n*-side quasineutral region. These injected carriers subsequently recombine. As explained in Subsection 3.3.2, using *E*-**k** diagrams, the crystal momenta of the electrons and holes are decidedly different in indirect semiconductors like Si. This makes it difficult to conserve momentum in band-to-band transitions. The recombination in indirect semiconductors therefore takes place predominantly through R–G centers, and the energy released during the recombination process is dissipated as heat. In direct semiconductors like GaAs, on the other hand, the crystal momenta of the electrons and holes are about the same, and a significant portion of the injected carriers is eliminated via band-to-band recombination. As visualized in Fig. 9.13, the energy released in the band-to-band recombination process is in the form of photons and, upon escaping from the diode, becomes the light produced by the LED.

In the discussion of the carrier distributions in the conduction and valence bands

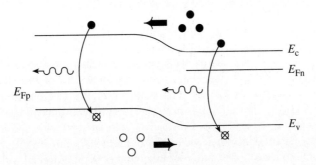

**Figure 9.13**  Production of light in an LED resulting from carrier injection in a forward-biased *pn* junction diode and subsequent band-to-band recombination.

(Subsection 2.4.3), it was noted that the distributions peaked very close to the band edges. The distribution maxima are in fact readily shown to be $kT/2$ from the band edges in a nondegenerate semiconductor. The recombination of conduction band electrons and valence band holes is therefore expected to produce a photon distribution with a spread in energies comparable to $kT$ and a peak energy slightly greater than $E_G$. In some cases the photons are actually created by an electron falling from a band gap center slightly below the conduction band edge and/or excitons are formed prior to recombination. (An *exciton* is an electron and a hole that become electrostatically coupled and function as a unit.) When this happens, the peak photon energy can be slightly less than $E_G$. In any event, the peak photon energy is typically close to $E_G$, and the peak output wavelength is at roughly $\lambda_G = 1.24/E_G$. For the light produced to be visible, the output light wavelength and thus $\lambda_G$ must lie in the range 0.4 $\mu$m $< \lambda_G <$ 0.7 $\mu$m, which implies 1.77 eV $< E_G <$ 3.10 eV.

From the operational description and the energy/wavelength considerations, we conclude a semiconductor used to produce visible LEDs is subject to three requirements. The semiconductor should be direct, have a band gap energy intermediate between 1.77 eV and 3.10 eV, and be amenable to the formation of *pn* junction diodes. Surprisingly, very few semiconductors meet all three requirements. Si and Ge are excluded immediately. GaAs would be ideal except that its band gap is too small. Examining either Fig. 9.1 or Fig. 9.6, we conclude the III–V compounds GaP and AlAs have band gaps in the desired range, but both GaP and AlAs are indirect. Figure 9.1 also indicates the IV–IV compound SiC has a band gap of the correct size. Unfortunately, SiC is also indirect. Several II–VI compound semiconductors including ZnSe cited in Fig. 9.1 are both direct and have band gaps in the desired range. However, until recently it has been impossible to form *pn* junctions in most of the II–VI compounds. Native defects in the II–VI compounds tend to compensate either the *n*- or *p*-type dopants introduced to form a junction. With no elemental or compound semiconductors meeting the three requirements, it is understandable why semiconductor alloys and special "light-enhancing" centers have come to be employed in producing commercially available LEDs, as detailed in the next subsection.

### 9.4.2  Commercial LEDs

A summary of relevant information about commercially available LEDs is presented in Table 9.1. Individual table entries are discussed briefly below.

### GaAs$_{0.6}$P$_{0.4}$

GaAs is a direct semiconductor, but its band gap is too small to produce visible light emission. GaP has a $\lambda_G$ in the visible range but is indirect. Individually neither compound meets the LED material requirements. However, referring to Fig. 9.6, a combination of these two compounds—the GaAs$_{1-x}$P$_x$ alloy—is both direct and has a light-producing $E_G$ for $x$-values in the range $0.28 \leq x \leq 0.45$. GaAs$_{0.6}$P$_{0.4}$ with an $x = 0.4$ is specifically used as the light-emitting material because it yields an output that appears brightest to the human eye. The external efficiency ($\eta \equiv$ photo power out/electrical power in) of a GaAs$_{1-x}$P$_x$ LED decreases by about a factor of 10 as $x$ is increased from $x = 0.28$ to the direct-indirect transition at $x = 0.45$. However, as $x$ is increased, the output wavelength decreases and the

**Table 9.1**   Characteristics of Commercial LEDs. (Adapted from Craford[9].)

| Semiconductor | Color | Peak $\lambda(\mu m)$ | External Efficiency $\eta$ (%) | Performance (lumens/watt)[†] |
|---|---|---|---|---|
| *Established Materials* | | | | |
| $GaAs_{0.6}P_{0.4}$ | Red | 0.650 | 0.2 | 0.15 |
| $GaAs_{0.35}P_{0.65}$:N | Orange-Red | 0.630 | 0.7 | 1 |
| $GaAs_{0.14}P_{0.86}$:N | Yellow | 0.585 | 0.2 | 1 |
| GaP:N | Green | 0.565 | 0.4 | 2.5 |
| GaP:Zn-O | Red | 0.700 | 2 | 0.40 |
| *Recent Additions* | | | | |
| AlGaAs | Red | 0.650 | 4–16 | 2–8 |
| AlInGaP | Orange | 0.620 | 6 | 20 |
| AlInGaP | Yellow | 0.585 | 5 | 20 |
| AlInGaP | Green | 0.570 | 1 | 6 |
| SiC | Blue | 0.470 | 0.02 | 0.04 |
| GaN | Blue | 0.450 | 2 | 0.6 |

[†]Luminous performance of a visible light source is measured in terms of the lumens derived from the device per watt of electrical power input. Lumens are calculated by multiplying the radiant output (in watts) times the relative response of the human eye (Fig. 9.1).

human eye response increases by a factor of ~50. The product of the LED efficiency and the eye response exhibits a maximum at $x = 0.4$, the value used in producing commercial devices.

The low-cost red LEDs made of $GaAs_{0.6}P_{0.4}$ were the first solid state lamps to be successfully mass marketed. They were used in the displays on watches and hand-held calculators manufactured in the early 1970s. Subsequently, of course, they were replaced by liquid crystal displays that permitted longer battery life. Exhibiting rather low brightness, the $GaAs_{0.6}P_{0.4}$ diodes primarily function as indicator lamps in indoor applications.

Figure 9.14(a) shows the cross section of a $GaAs_{0.6}P_{0.4}$ LED. Starting with a GaAs substrate, a layer with $x$ systematically increased or "graded" from $x = 0$ to $x = 0.4$ is first deposited to minimize lattice mismatch problems. Because of the large difference in the lattice constants of GaAs and $GaAs_{0.6}P_{0.4}$, direct deposition of $GaAs_{0.6}P_{0.4}$ on GaAs would yield a material with a large number of defects. Next the $GaAs_{0.6}P_{0.4}$ layer and then another graded layer with $x$ increased from $x = 0.4$ to $x = 0.6$ is added to the structure. Finally, Zn is diffused into the upper layers to form the $p$-side of the junction. The increased $x$ of the top layer produces a wider band gap, thus allowing light generated in the $GaAs_{0.6}P_{0.4}$ material to pass through the top layer with minimum reabsorption.

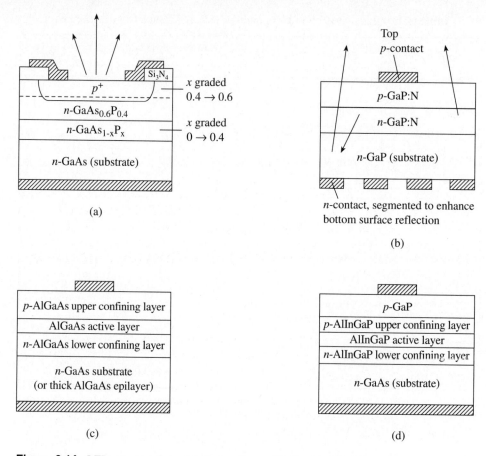

**Figure 9.14**   LED cross sections: (a) $GaAs_{0.6}P_{0.4}$ red LED, (b) GaP:N green LED, (c) AlGaAs "high-brightness" red LED, and (d) AlInGaP LED. (Adapted from Craford[9,10].)

## $GaAs_{0.35}P_{0.65}$:N, $GaAs_{0.14}P_{0.86}$:N, GaP:N

As recorded in Table 9.1, orange-red, yellow, and green LEDs are made from $GaAs_{1-x}P_x$ materials with $x$-values ranging from $x = 0.65$ to $x = 1$. Technically, all of these materials are indirect. However, doping with nitrogen as indicated by the ":N" designation makes these materials pseudo-direct. Nitrogen, it should be noted, is a Column V element like As and P. Because nitrogen replaces another Column V element when it is added to the $GaAs_{1-x}P_x$ lattice, it does not act like a standard dopant. No additional electrons or holes are produced. Rather, the nitrogen introduces an electronic level, an electron trapping level, approximately 0.1 eV below the conduction band edge. This special center is called an *isoelectronic* trap because it is formed from an element with the same (iso = same) valence structure as the atom it replaces. Of importance to LED operation, an electron attracted to the trap becomes highly localized in space, which in turn causes a large spread in the allowed momentum of the trapped electron. Matching the momentum of holes in the valence band, the trapped electron has a dramatically enhanced probability of making a hole-annihilating transition that results in the production of a photon. In essence, the isoelectronic trap acts as a conduit for the relatively efficient radiative recombination of electrons and holes.

The $GaAs_{0.35}P_{0.65}$:N and $GaAs_{0.14}P_{0.86}$:N LEDs are similar in construction to the $GaAs_{0.6}P_{0.4}$ LED except the devices are formed on GaP substrates. The GaP:N structure pictured in Fig. 9.14(b) is different in that a substrate grading layer is not required and both the active $n$- and $p$-regions of the diode are formed epitaxially. Because the light resulting from the nitrogen center transition in any of the structures has a $\lambda_{peak} > \lambda_G$ ($E_{ph} < E_G$), there is very little reabsorption in the epitaxial layers. Moreover, the GaP substrate is also transparent to the LED light. Thus even downwardly directed light is reflected and may eventually escape without being reabsorbed. This yields higher external efficiencies and brighter devices. The orange-red, yellow, and green LEDs described here, along with the GaP:Zn-O red LED described next, presently command the largest share of the LED market.

## GaP:Zn-O

Zn alone acts as an acceptor and O alone acts like a donor in GaP. When added in approximately equal numbers and after proper annealing, however, the Zn replacing Ga and the O replacing P tend to form on adjacent lattice cites and function together as an isoelectronic trap. The Zn-O trap complex differs from N in that the trap level lies deeper in the GaP band gap, approximately 0.3 eV below the conduction band edge. Light produced by transitions through the complex is red in color with $\lambda_{peak} = 0.700$ $\mu$m. Fabrication of GaP:Zn-O LEDs parallels that of the GaP:N LED shown in Fig. 9.14(b).

## AlGaAs

Introduced in the latter half of the 1980s, the AlGaAs LED is a relative newcomer to the commercial marketplace. It is sometimes sold as the "high brightness" red LED, exhibiting a luminance up to 20 times greater than the more common GaAsP:N and GaP:Zn-O red LEDs. Notable applications of the AlGaAs red LED, which has captured roughly 10% of the market, includes use in the spoiler or third automobile tail light and as the heel light on athletic tennis shoes. The cross section of an AlGaAs LED is shown in Fig. 9.14(c). Light is derived from recombination in the narrow active region which has an $E_G \cong 1.9$ eV. The adjacent confining layers have a slightly wider band gap to minimize reabsorption of the generated light and to maximize extraction of randomly directed light from the structure. In the brightest AlGaAs devices, a thick (100–200 $\mu$m) epitaxial layer is grown under the bottom confining layer and the $E_G < E_{ph}$ GaAs substrate is removed. Note from Fig. 9.6 that an AlGaAs alloy with $E_G < 2$ eV is a direct semiconductor. Moreover, all AlGaAs alloys have almost exactly the same lattice constant, are lattice matched, to GaAs. The combination of minimal nonradiative recombination due to the perfection of the lattice-matched layers and the production of light via direct recombination (which is about an order of magnitude more efficient than recombination through isoelectronic traps) leads to the high-efficiency/high-brightness of the structure.

### AlInGaP

AlInGaP LEDs were first offered for sale in 1992. Involving a rather unusual alloy consisting of three Column III elements and one Column V element, AlInGaP technology is still in its infancy. Nevertheless, AlInGaP LEDs already significantly outperform established orange, yellow, and green devices. In the long run, AlInGaP may eventually become the dominant high-performance LED technology for all colors from red through green. The cross section of an AlInGaP LED is shown in Fig. 9.14(d). Like the AlGaAs LED, the confining and active layers are lattice matched to the GaAs substrate and recombination in the active region is direct. Different colors are derived of course from structures with different active layer compositions. GaP is presently used as the top contacting/window layer because it provides a much higher electrical conductivity than that attainable with *p*-type AlInGaP.

### SiC

The SiC story is a rather interesting one. Electroluminescence was first reported from a SiC sample in 1907 that undoubtedly contained built-in *pn* junctions. SiC LEDs were intensely researched in the 1960s and briefly offered for sale in the early 1970s. These initial offerings were very inefficient ($\eta \sim 0.001\%$), and the light output was unacceptably low. Following a breakthrough in the formation of high-quality SiC substrates, and given the established need for a blue LED, SiC LEDs were again offered for sale in 1990. Although significantly improved to 0.02%, the external efficiency remains low in the new offering because LED operation still relies on low-probability radiant recombination in an indirect material.

### GaN

GaN is a direct III–V semiconductor with an $E_G = 3.36$ eV. The GaN blue LED was first offered for sale in April 1994. The somewhat surprising development of the GaN LED by a chemical company with no expertise in manufacturing semiconductor devices was made possible by a breakthrough in forming *pn* junctions in GaN and GaN alloys. The light-producing radiative recombination actually takes place in an $E_G = 2.75$ eV InGaN film sandwiched between wider-band-gap AlGaN layers. It is projected that the more efficient GaN LED will rapidly surpass the SiC LED in sales and become the blue LED of choice. Providing the "missing" color component, the GaN blue LED makes possible full-color LED-based outdoor video displays that operate at reasonable power levels.

## 9.4.3  LED Packaging and Photon Extraction

A cross-sectional sketch of the standard LED package is presented in Fig. 9.15. The LED chip, typically measuring approximately 250 $\mu$m on a side, is placed in a reflective cavity.

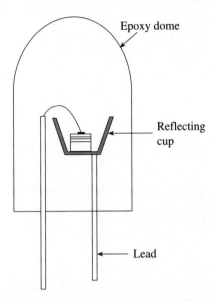

Epoxy dome

Reflecting
cup

Lead

**Figure 9.15**  Cross section of the standard LED package.

The reflective cavity is used because a portion, in some cases a large portion, of the light actually escapes from the sides of the chip. One of the diode leads is connected to the cavity, while a wire bond is made from the top contact on the chip to the second diode lead. The chip and lead frame are embedded in an epoxy that holds the lead frame together and (as will be explained) facilitates photon extraction from the chip. The dome-like shape of the epoxy encapsulant optimizes light transmission through the top of the package.

Getting generated light out of the LED chip poses more of a problem than one might suspect. We have already noted that wider band gap material overlays the light-emitting layers in the $GaAs_{0.6}P_{0.4}$, AlGaAs, and AlInGaP LEDs. The wider band gap material acts like a window, allowing light to reach the surface of the diode without being reabsorbed. A window material is not necessary in the nitrogen or Zn-O doped LEDs because the emitted photons have an energy at least 0.1 eV below the band gap energy of the host material. Ideally it is also desirable to have $E_G > E_{ph}$ material below the light-emitting layer. Half of the generated light is initially directed downward and some light reflects off the top surface. With $E_G > E_{ph}$ material below the light-emitting layer, the light can bounce around inside the entire structure with a minimal probability of reabsorption, thereby dramatically increasing the percentage of the light striking the semiconductor surfaces.

Just because light makes it to the semiconductor surface, however, doesn't mean the light will exit the LED chip. Let us review some basic optics. When light traveling in a dielectric media of refractive index $n_1$ impinges upon the interface with a second dielectric media of refractive index $n_2$, the light in general will be partially reflected and partially

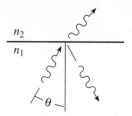

**Figure 9.16**   Light reflection and transmission at an interface between two dielectric materials. The light is shown incident at an angle $\theta$ from the normal in material number 1.

transmitted as pictured in Fig. 9.16. For normal incidence ($\theta = 0$), the fraction of the light transmitted, $T_N$, is given by

$$T_N = \frac{4n_r}{(1 + n_r)^2} \tag{9.8}$$

where $n_r \equiv n_1/n_2$ is the index of refraction ratio. $n_r$ is invariably greater than unity ($n_1 > n_2$) for light being transmitted from a semiconductor into another media. Whenever $n_1 > n_2$, light incident at angles greater than a critical angle, $\theta_c$, will experience total internal reflection ($T = 0$). The critical angle is computed from

$$\sin\theta_c = \frac{n_2}{n_1} \tag{9.9}$$

Accounting for total internal reflection, assuming the incident light is uniformly distributed in all angles and randomly polarized, the usual case at the top surface in LEDs, and summing over all angles, the overall fraction of the light transmitted, $\overline{T}$, is determined to be

$$\overline{T} \cong \frac{(\sin\theta_c)^2}{2} T_N = \frac{2}{n_r(1 + n_r)^2} \tag{9.10}$$

   In performing a sample computation, let us consider light transmission at the GaP–air interface. GaP with a refractive index of 3.4 is fairly representative of the semiconductors used in LEDs. Setting $n_1$(GaP) $= 3.4$ and $n_2$(air) $= 1$, one computes $T_N = 70\%$. However, $\theta_c = 17.1°$ and $\overline{T} = 3.0\%$!! A huge percentage of the light is subject to total internal reflection. The epoxy, with an index of refraction intermediate between air and the semiconductor, allows a much larger fraction of the light to escape from the LED chip. The conically shaped dome of the LED package in turn makes it possible for much of the light to arrive at the epoxy-air interface within the critical angle.

# PART IIB

# BJTs AND OTHER
# JUNCTION DEVICES

# 10  BJT Fundamentals

Chapters 5–8 were devoted to the detailed analysis of the *pn* junction diode. As a logical extension of the diode analysis, the development next progresses from the one-junction/ two-terminal diode to the two-junction/three-terminal transistor. In this chapter we initiate the discussion of the bipolar junction transistor (BJT). The chapter contains a collage of introductory BJT information—definition of terms and symbols, qualitative operational concepts, key relationships, and so on. In combination, the information forms the required knowledge base preparatory to a detailed device analysis.

## 10.1  TERMINOLOGY

The BJT is a semiconductor device containing three adjoining, alternately doped regions, with the middle region being very narrow compared to the minority carrier diffusion length in that region. *pnp* and *npn* junction transistors are pictured schematically in Fig. 10.1. As indicated in the figures, the narrow central region is known as the *base,* and the outer two regions are referred to as the *emitter* and *collector.* It might appear from a cursory inspection of Fig. 10.1 that the outer two regions are interchangeable. However, in practical devices the emitter has a different geometry and is typically more heavily doped than the collector. Interchanging the two terminals therefore significantly modifies the device characteristics.

Standard circuit symbols for the *pnp* and *npn* versions of the bipolar junction transistor are presented in Fig. 10.2. Symbols for the d.c. terminal currents and voltages, plus current and voltage polarities, are also noted in the figure. The "+" and "−" signs used to visually specify the voltage polarities in Fig. 10.2 are actually redundant; the double subscript on the voltage symbol likewise denotes the voltage polarity. The first letter in the double subscript identifies the (+) terminal and the second letter identifies the (−) terminal. $V_{EB}$, for example, is the d.c. voltage drop between the emitter (+) and base (−). Observe that the positive current flow directions specified in Fig. 10.2 are in some cases contrary to IEEE convention, which always takes the current flowing into a terminal to be positive. Given the chosen polarities, however, all terminal currents are positive quantities when the transistor is operated in the standard amplifying mode. The chosen polarities thereby avoid unnecessary complications and are much more convenient in treating the physical operation of the transistor.

Although all three currents and voltages shown in Fig. 10.2 are used in specifying the transistor characteristics, only two of the currents and two of the voltages are independent. The current flowing into a device must be equal to the current flowing out of a device, and

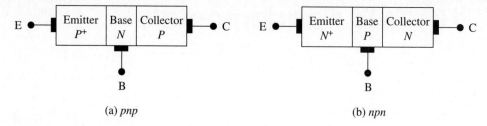

**Figure 10.1** Schematic representation of the (a) *pnp* and (b) *npn* BJT showing device regions and the terminal designations.

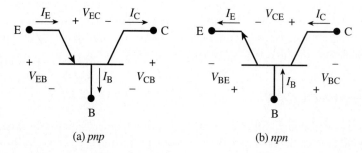

**Figure 10.2** (a) *pnp* and (b) *npn* BJT circuit symbols. The d.c. terminal currents, voltages, and reference polarities are also noted in the figure.

the voltage drop around a closed loop must be equal to zero. Thus, by inspection from Fig. 10.2(a) or (b),

$$I_E = I_B + I_C \tag{10.1}$$

and

$$V_{EB} + V_{BC} + V_{CE} = 0 \qquad (V_{CE} = -V_{EC}) \tag{10.2}$$

If two of the transistor currents or voltages are known, Eq. (10.1) or (10.2) can always be used to determine the third terminal current or voltage.

Because more than one current and one voltage are involved in the operation of the BJT, the device characteristics are inherently multidimensional. For the description of the characteristics to be tractable, it is necessary to focus on the currents, voltages, and polarities of primary interest in a particular application. This is accomplished by specifying the basic *circuit configuration* in which the device is connected and the *biasing mode*.

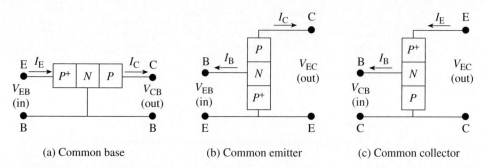

(a) Common base          (b) Common emitter          (c) Common collector

**Figure 10.3**  Circuit configurations: (a) common base; (b) common emitter; and (c) common collector.

In most applications a signal is input across two of the BJT leads, and an output signal is extracted from a second pair of leads. Since the BJT has only three leads, one of the leads must obviously be part of both the input and output circuitry. The terms *common base, common emitter,* and *common collector* are used to identify the lead common to the input and output and to specify the corresponding circuit configurations illustrated in Fig. 10.3. The common emitter is the most widely employed configuration, with the common base finding occasional utilization. The common collector is seldom if ever used and will be henceforth neglected. Inspecting Fig. 10.3, we see the circuit configuration succinctly identifies the currents and voltages of interest. In the common emitter arrangement, for example, $I_C$ and $V_{EC}$ are clearly the relevant output variables. Idealized sketches of the common base and common emitter output characteristics are presented in Fig. 10.4 for future reference.

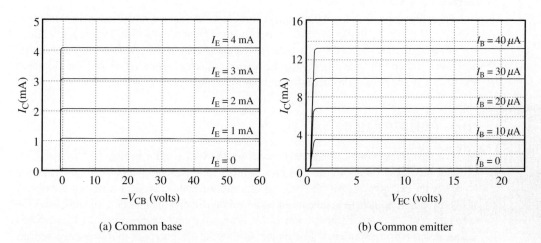

(a) Common base          (b) Common emitter

**Figure 10.4**  Idealized sample sketches of (a) the common base and (b) the common emitter output characteristics of a *pnp* BJT.

**Table 10.1** Biasing Modes.

| Biasing Mode | Biasing Polarity E–B Junction | Biasing Polarity C–B Junction |
|---|---|---|
| Saturation | Forward | Forward |
| **Active** | **Forward** | **Reverse** |
| Inverted | Reverse | Forward |
| Cutoff | Reverse | Reverse |

The biasing mode helps to further specify transistor operation in a given application by identifying the voltage polarities of primary interest. Specifically, the biasing mode tells one the polarity of the bias (forward or reverse) being applied to the two transistor junctions. In all, there are four polarity combinations as summarized in Table 10.1 and Fig. 10.5. *Active* or *forward active* biasing, where the E–B junction is forward biased and the C–B junction reverse biased, is the most widely encountered of the operational modes. Almost all linear signal amplifiers, such as operational amplifiers, are active mode biased. Under active mode biasing the transistor exhibits its largest signal gain and smallest signal distortion. *Saturation,* when both junctions are forward biased, and *cutoff,* when both junctions are reverse biased, give rise respectively to the on-state (high current flow, low voltage) and off-state (low current flow, high voltage) associated with operation of the transistor as a switch. In digital circuits these low-voltage and high-voltage states correspond to the "zero" and "one" logic levels, respectively. Finally, in the *inverted* or *inverted active* mode, the C–B junction is forward biased and the E–B junction is reverse biased. Effectively, the roles of the emitter and collector are thereby interchanged or inverted relative to the forward active mode.

Throughout most of the development to follow, the device under analysis is taken to be a *pnp* BJT. Although the *npn* BJT is used in a far greater number of circuit applications and IC designs, the *pnp* BJT is a more convenient vehicle for establishing operational principles and concepts. It is assumed the reader can readily modify the *pnp* development and results so they are appropriate for an *npn* BJT.

## 10.2 FABRICATION

An idealized pseudo-one-dimensional model for the BJT similar to Fig. 10.1 is used both in qualitative discussions and to obtain first-order quantitative results. BJT fabrication is briefly described here to provide some insight into the actual physical nature of the device structure and to indicate the correlation with the one-dimensional model. In many ways the fabrication of the bipolar transistor is just a straightforward extension of the diode fabrication described in Chapter 4. The major difference, of course, is that two *pn* junctions must be formed in close proximity.

Cross sections of a typical discrete, double-diffused *pnp* transistor and an integrated

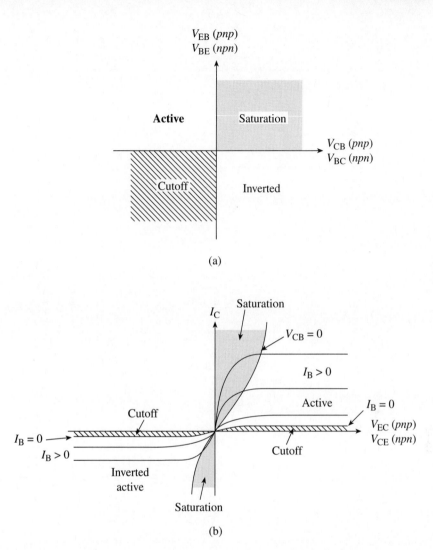

**Figure 10.5** (a) Combinations of the BJT input and output voltages resulting in the four biasing modes. (b) Regions of the BJT common emitter output characteristics associated with the four biasing modes.

circuit *npn* transistor are shown respectively in Figs. 10.6(a) and (b). Starting with a heavily doped $p^+$-Si wafer, the first step in forming the discrete *pnp* is the deposition of a high-resistivity (lightly doped) *p*-type Si epilayer. This layer, typically 5 to 10 $\mu$m thick, ultimately contains the transistor proper. The starting wafer merely provides mechanical support and a low-resistance path to the collector contact on the bottom of the wafer. After growing an oxide and opening oxide windows, a phosphorus or an arsenic diffusion is

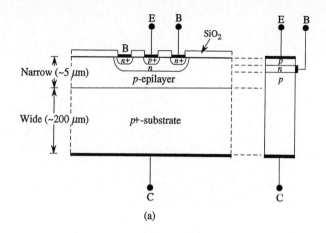

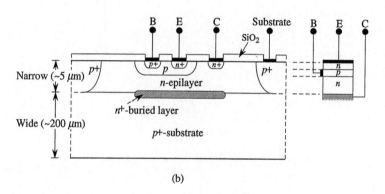

**Figure 10.6** Cross sections and simplified models for (a) a typical discrete, double-diffused *pnp* BJT and (b) an integrated circuit *npn* BJT.

performed yielding the *n*-type base. With the *p*-type dopant boron replacing phosphorus or arsenic, the procedure is next repeated to form the $p^+$-emitter. Finally, aluminum is deposited on top of the wafer and patterned to create the emitter and base contacts. It should be noted that the geometrical features in Fig. 10.6 are not drawn to scale. A better idea as to the true relative size of the structure is obtained by expanding the figure by about a factor of 50 in the horizontal direction.

    The *npn* integrated circuit transistor pictured in Fig. 10.6(b) is similarly fabricated by performing a *p*-type diffusion and then an *n*-type diffusion into a high-resistivity *n*-type epilayer. In the IC case, however, the epilayer has a doping type opposite to that of the substrate and diode isolation of the *npn* transistor is completed by a deep *p*-type diffusion around the transistor periphery. The most negative potential in the circuit is applied to the *p*-type substrate. Being at a higher potential, the *n*-epitaxial region is therefore electrically isolated by a reverse-biased *pn* junction from other devices on the IC chip. In more ad-

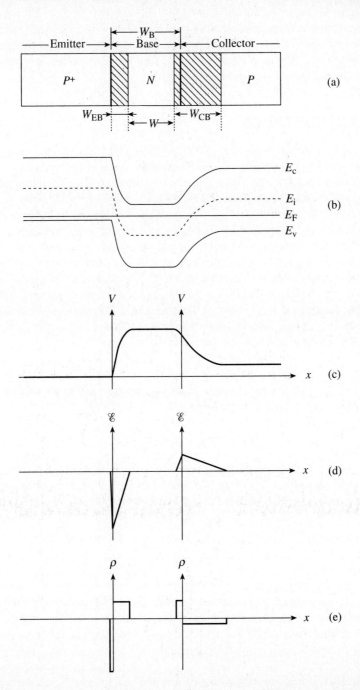

**Figure 10.7** Sketches of the electrostatic variables in a *pnp* BJT under equilibrium conditions. (a) Depletion regions, (b) energy band diagram, (c) electrostatic potential, (d) electric field, and (e) charge density. The transistor regions are assumed to be uniformly doped with $N_{AE} \gg N_{DB} > N_{AC}$.

vanced designs the lateral *pn* junctions have been replaced by silicon dioxide isolation. The $n^+$ buried layer shown in Fig. 10.6(b) is formed by diffusion prior to depositing the epilayer. The buried layer serves as a low-resistance path between the active collector region of the transistor and the top-side collector contact.

## 10.3  ELECTROSTATICS

Paralleling the *pn* junction diode analysis, we begin the BJT analysis by looking into the electrostatic situation inside the transistor. Fortunately, under equilibrium and normal operating conditions, the BJT may be viewed electrostatically as nothing more than two independent *pn* junctions. The *pn* junction electrostatics established in Chapter 5 —relationships for the built-in potential, charge density, electric field, electrostatic potential, and depletion width—may be applied separately, without modification, to the E–B and C–B junctions. Thus, for example, assuming the transistor regions to be uniformly doped and taking $N_{AE}$(emitter doping) $\gg N_{DB}$ (base doping) $> N_{AC}$(collector doping), the usual case in a standard *pnp* transistor, the equilibrium energy band diagram and electrostatic variables based on the depletion approximation are concluded to be as pictured in Fig. 10.7.

Upon examining Fig. 10.7, note that the depletion widths are consistent with the assumed $N_{AE} \gg N_{DB} > N_{AC}$ doping profile. Specifically, almost all of the E–B depletion width ($W_{EB}$) lies in the base while most of the C–B depletion width ($W_{CB}$) lies in the collector. $W_{CB} > W_{EB}$ since the lightly doped side of the C–B junction has a lower doping than the lightly doped side of the E–B junction. Also note that $W_B$ is the total width of the base and $W$ is the standard symbol for the portion of the base that is *not* depleted; i.e., for a *pnp* transistor,

$$W = W_B - x_{nEB} - x_{nCB} \tag{10.3}$$

where $x_{nEB}$ and $x_{nCB}$ are respectively the portions of the E–B and C–B depletion widths lying inside the *n*-type base. The $W$ in BJT analyses, referred to as the *quasineutral base width*, should not be confused with the diode depletion width that was associated with the symbol $W$ in earlier chapters.

---

### Exercise 10.1

**P:** Draw the energy band diagram characterizing: (a) an *npn* transistor under equilibrium conditions; (b) a *pnp* transistor under active mode biasing. Take the transistor regions to be uniformly doped and assume a standard doping profile where the emitter doping $\gg$ base doping $>$ collector doping.

**S:** (a) Given the complementary nature of *npn* and *pnp* transistors, the energy band diagram for an *npn* transistor can be obtained from the equivalent *pnp* diagram by merely flipping the *pnp* diagram upside down. Thus, either by flipping Fig. 10.7(b) upside down or by appropriately combining the equilibrium band diagrams of an *np* and a *pn* junction, the equilibrium energy band diagram for an *npn* transistor is concluded to have the form shown in Fig. E10.1(a).

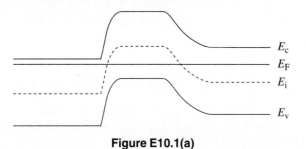

**Figure E10.1(a)**

(b) Under active mode biasing the E–B junction is forward biased and the C–B junction is reverse biased. Thus, referring to Fig. 10.7(b), the Fermi level on the base side of the E–B junction is raised relative to the emitter-side Fermi level and both the E–B depletion width and potential hill are decreased. Conversely, reverse biasing the C–B junction lowers the Fermi level on the base side of the junction relative to the collector-side Fermi level and both the C–B depletion width and potential hill are increased. The deduced energy band diagram is sketched in Fig. E10.1(b).

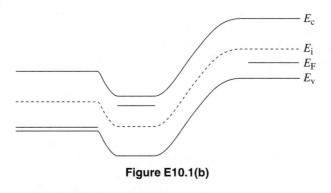

**Figure E10.1(b)**

## (C) Exercise 10.2

**P:** Paralleling Exercise 5.4, construct a MATLAB (computer) program that draws the equilibrium energy band diagram for a nondegenerately doped Si *pnp* or *npn* BJT. Assume that the transistor regions are uniformly doped and the BJT is maintained at room temperature.

**S:** A BJT "diagram generator" is included on disk as file BJT_Eband. The student-produced code for the program is reproduced in Appendix M. As supplied, the program draws the equilibrium energy band diagram for a *pnp* BJT with $N_{AE} = 10^{18}/$ cm$^3$, $N_{DB} = 10^{16}/$cm$^3$, $N_{AC} = 10^{15}/$cm$^3$, and $W_B = 1$ $\mu$m. The doping parameters and total base width can be modified by changing the DOPING and WIDTH entries on program lines 5 and 6. *n*-type dopings are input as negative values.

## 10.4  INTRODUCTORY OPERATIONAL CONSIDERATIONS

To provide insight into the operation of the bipolar transistor, let us consider a *pnp* transistor under active mode biasing and initially focus on the activity of the holes in and adjacent to the base region. As pictured in Fig. 10.8, the primary carrier activity in the vicinity of the forward-biased E–B junction is majority carrier injection across the junction into the opposite-side quasineutral regions. Naturally, the $p^+$-$n$ nature of the junction leads to many more holes being injected from the emitter into the base than electrons being injected from the base into the emitter. The key to transistor action is what subsequently happens to the carriers that are injected into the base. If the quasineutral base width were much larger than a minority carrier diffusion length, the injected holes would simply recombine in the *n*-type base and there would be no interaction between the two junctions. The structure would be nothing more than two back-to-back *pn* junctions. However, by definition, the BJT is a structure where the base is *narrow* compared to a minority carrier diffusion length. Thus, the vast majority of injected holes diffuse completely through the quasineutral base and enter the C–B depletion region. The accelerating electric field in the C–B depletion region then rapidly sweeps these carriers into the collector. The narrow width of the base thereby leads to a coupling of the E–B and C–B junction currents, with carrier activity being decidedly different from two back-to-back *pn* junctions.

In addition to citing the carrier activity underlying transistor action, the foregoing helps to explain the naming of the transistor regions. When active mode biased, the emitter functions as a source of carriers, emitting carriers into the base. Conversely, the collector portion of the reverse-biased C–B junction acts like a sink, collecting the emitted carriers after they pass through the control or base region.

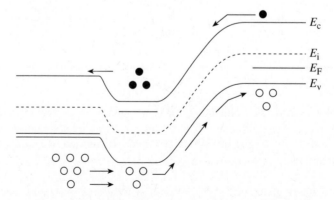

**Figure 10.8**  Carrier activity in a *pnp* BJT under active mode biasing.

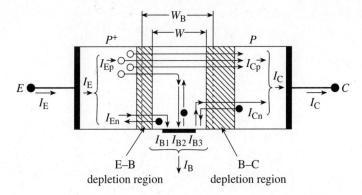

**Figure 10.9** Spatial visualization of the diffusion currents flowing in a *pnp* BJT under active mode biasing.

For a more complete picture, let us next expand the discussion with the aid of Figs. 10.8 and 10.9 to include all of the diffusion currents flowing in the BJT. (Paralleling the diode analysis, the R–G currents associated with the depletion regions are initially neglected.) In Fig. 10.9 the hole diffusion current flowing across the E–B junction, the current associated with the holes injected into the base, is identified as $I_{Ep}$. Similarly, the hole diffusion current flowing across the C–B junction, a current almost exclusively resulting from the injected holes that successfully cross the base, is labeled $I_{Cp}$. As indicated previously, very few of the injected holes are lost by recombination in the base of a well-made transistor. Thus, $I_{Cp} \cong I_{Ep}$. The total emitter and collector currents are of course

$$I_E = I_{Ep} + I_{En} \tag{10.4}$$

and

$$I_C = I_{Cp} + I_{Cn} \tag{10.5}$$

$I_{En}$ is the current associated with electron injection from the base into the emitter. Corresponding to the diffusion current on the heavily doped side of a *pn* junction, $I_{En} \ll I_{Ep}$. $I_{Cn}$ arises from the minority carrier electrons in the collector that wander into the C–B depletion region and are swept into the base. Being a reverse bias current, $I_{Cn} \ll I_{Cp}$. Since both the emitter and collector electron components are small compared to the respective hole components, and since $I_{Cp} \cong I_{Ep}$, it follows that $I_C \cong I_E$. This is a well-known property of the transistor terminal currents. $I_B$, on the other hand, is popularly known to be small compared to $I_E$ and $I_C$ under active mode biasing. This is consistent with the fact $I_B = I_E - I_C$. A direct inspection of Fig. 10.9 also leads to the conclusion that $I_B$ is expected to be relatively small. The three components of the base current, $I_{B1} = I_{En}$, $I_{B3} = I_{Cn}$, and $I_{B2}$ equal to the current flowing into the base to replace electrons lost by recombination with holes injected from the emitter, are all small.

Finally, a few words are in order on how the BJT manages to amplify a signal. When connected in the common emitter configuration, the output current is $I_C$, the input current

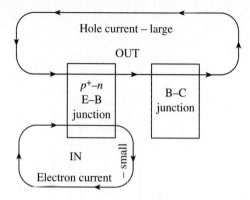

**Figure 10.10**   Schematic visualization of amplification in a *pnp* BJT under active mode biasing.

is $I_B$, and the d.c. current gain is $I_C/I_B$. $I_B$, an electron current in a *pnp* BJT, and $I_C$, predominantly a hole current, are tied together through the operation of the E–B junction; that is, increasing $I_B$ proportionally increases $I_C$. Control of the larger $I_C$ by the smaller $I_B$ is made possible by the coupled junction arrangement that physically divides the small electron and large hole currents crossing the E–B junction into two separate current loops (see Fig. 10.10).

## 10.5  PERFORMANCE PARAMETERS

Several parameters are commonly used to characterize the performance of a BJT as an amplifier. They include the emitter efficiency and the base transport factor, which relate to the internal operation of the device, and the d.c. current gains, which relate to the external operation. In this section we survey the definitions and the interrelationships of these important performance parameters in preparation for the quantitative analysis to be presented in the next chapter.

### Emitter Efficiency

Increasing the hole current injected across the E–B junction in Fig. 10.10 while holding the total emitter current constant would clearly decrease the electron current and increase the overall current gain. Thus, thinking of $I_E$ as the emitter input and $I_{Ep}$ as the useful emitter output, the emitter efficiency is defined to be

$$\gamma = \frac{I_{Ep}}{I_E} = \frac{I_{Ep}}{I_{Ep} + I_{En}} \qquad \ldots pnp \text{ BJT} \qquad (10.6)$$

Obviously $0 \leq \gamma \leq 1$ and the current gain in a BJT is maximized by making $\gamma$ as close as possible to unity.

## Base Transport Factor

The fraction of the minority carriers injected into the base that successfully diffuse across the quasineutral width of the base and enter the collector is known as the base transport factor ($\alpha_T$). In a *pnp* BJT the number of carriers injected into the base from the emitter is proportional to $I_{Ep}$; the residual number of carriers entering the collector is proportional to $I_{Cp}$. The fraction making it across the base is therefore

$$\boxed{\alpha_T = \frac{I_{Cp}}{I_{Ep}}} \quad \ldots \; pnp \;\; \text{BJT} \tag{10.7}$$

Note that $0 \leq \alpha_T \leq 1$. The smaller the loss of injected carriers via recombination in the quasineutral base, the smaller the degradation of BJT performance and the larger $\alpha_T$. Here again maximum amplification occurs when the performance parameter, $\alpha_T$, is as close as possible to unity.

## Common Base d.c. Current Gain

When connected in the common base configuration, the active mode ($-V_{CB} > 0$) portion of the output characteristics sketched in Fig. 10.4(a) is accurately modeled by the relationship

$$I_C = \alpha_{dc} I_E + I_{CB0} \tag{10.8}$$

where $\alpha_{dc}$ is the common base d.c. current gain and $I_{CB0}$ is the collector current that flows when $I_E = 0$. Making use of Eqs. (10.6) and (10.7), we can alternatively write

$$I_{Cp} = \alpha_T I_{Ep} = \gamma \alpha_T I_E \tag{10.9}$$

and

$$I_C = I_{Cp} + I_{Cn} = \gamma \alpha_T I_E + I_{Cn} \tag{10.10}$$

Comparing Eqs. (10.8) and (10.10), we conclude

$$\boxed{\alpha_{dc} = \gamma \alpha_T} \tag{10.11}$$

and

$$I_{CB0} = I_{Cn} \tag{10.12}$$

Eq. (10.11) is significant in that it relates the external gain of the BJT to the internal performance parameters. Also note from Eq. (10.11) that $0 \leq \alpha_{dc} \leq 1$.

## Common Emitter d.c. Current Gain

When connected in the common emitter configuration, the active mode portion of the output characteristics (see Figs. 10.4b and 10.5b) is approximately modeled by the relationship

$$I_C = \beta_{dc} I_B + I_{CE0} \tag{10.13}$$

where $\beta_{dc}$ is the common emitter d.c. current gain and $I_{CE0}$ is the collector current that flows when $I_B = 0$. A second relationship between $I_C$ and $I_B$ can be established by substituting $I_E = I_C + I_B$ into Eq. (10.8). One obtains

$$I_C = \alpha_{dc}(I_C + I_B) + I_{CB0} \tag{10.14}$$

or, after rearranging and solving for $I_C$,

$$I_C = \frac{\alpha_{dc}}{1 - \alpha_{dc}} I_B + \frac{I_{CB0}}{1 - \alpha_{dc}} \tag{10.15}$$

Comparing Eqs. (10.13) and (10.15), we conclude

$$\boxed{\beta_{dc} = \frac{\alpha_{dc}}{1 - \alpha_{dc}}} \tag{10.16}$$

and

$$I_{CE0} = \frac{I_{CB0}}{1 - \alpha_{dc}} \tag{10.17}$$

Also note from Eq. (10.13) that

$$\boxed{\beta_{dc} = \frac{I_C}{I_B}} \tag{10.18}$$

if, as is the usual case, $I_{CE0}$ is negligible compared to $I_C$ at the specified operating point.

Equation (10.16) is significant because it indicates that $\beta_{dc}$ can always be deduced once $\alpha_{dc}$ is known. Since $\alpha_{dc}$ is typically close to unity and $I_C \gg I_B$, a $\beta_{dc} \gg 1$ is to be expected based on either Eq. (10.16) or Eq. (10.18).

## 10.6  SUMMARY

In this chapter we covered very basic information about the bipolar junction transistor. Much of the circuit-related information was probably already familiar to the reader. We noted terminal current and voltage symbols, polarities, and relationships, reviewed the BJT circuit configurations and biasing modes, and cited the general form of the output characteristics. Transistor fabrication was briefly described to obtain insight into the actual physical nature of the device structure and to indicate the correlation with the one-dimensional model appearing in subsequent analyses. Only a minimal effort was also required to establish a rather complete picture of device electrostatics. Under equilibrium and normal operating conditions, the BJT can be viewed electrostatically as nothing more than two independent *pn* junctions. Some insight into how the device works was provided by examining the carrier activity and currents inside the structure under active mode biasing. Finally, parameters used to characterize the performance of the device as an amplifier—the emitter efficiency, the base transport factor, and the external current gains—were defined and interrelated.

## PROBLEMS

| CHAPTER 10   PROBLEM INFORMATION TABLE | | | | |
|---|---|---|---|---|
| *Problem* | *Complete After* | *Difficulty Level* | *Suggested Point Weighting* | *Short Description* |
| 10.1 | 10.6 | 1 | 10 (1 each part) | Quick quiz |
| 10.2 | 10.1 | 1 | 4 (2 each part) | Mode tables |
| 10.3 | 10.3 | 2–3 | 22 (a-4, b-6, c::e-4) | *pnp* equilibrium electrostatics |
| 10.4 | " | 2–3 | 22 (a-4, b-6, c::e-4) | *npn* equilibrium electrostatics |
| 10.5 | " | 2–3 | 10 (a-4, b-6) | *pnp* active mode electrostatics |
| 10.6 | " | 2 | 8 (2 each sketch) | *npn* active mode electrostatics |
| 10.7 | 10.4 | 3 | 12 (6 each figure) | *pnp* carrier activity/currents |
| 10.8 | " | 3 | 12 (6 each figure) | *npn* carrier activity/currents |
| 10.9 | 10.5 | 2 | 11 (a-1, b-1, c-3, d-2, e-2, f-1, g-1) | *pnp* parameter calculations |
| 10.10 | " | 2 | 11 (a-1, b-1, c-3, d-2, e-2, f-1, g-1) | *npn* parameter calculations |
| 10.11 | " | 3 | 10 | $I_{CB0}$ thought problem |

**10.1** Quick Quiz.

Answer the following questions as concisely as possible.

(a) What is the name of the circuit configuration where $I_E$ is the input current and $I_C$ is the output current?

(b) What is the name of the circuit configuration where $I_C$ is the output current and $V_{EC}$ is the output voltage?

(c) Name the four biasing modes.

(d) What is the purpose of the "buried layer" found in integrated circuit BJTs?

(e) In a standard *pnp* BJT, what are the relative sizes of $N_{AE}$, $N_{DB}$, and $N_{AC}$?

(f) What is the standard symbol for the width of the depletion region in a *pn* junction diode? What is the standard symbol for the quasineutral width of the base in a BJT?

(g) The base region in a BJT is narrow. What is the precise definition of "narrow"?

(h) Why is it necessary for the base region in a BJT to be narrow?

(i) Define in words (no equations) what is meant by "emitter efficiency."

(j) Define in words what is meant by "base transport factor."

**10.2** Complete the tables below by indicating the polarity ($+$ or $-$) of the input and output voltages associated with each of the four biasing modes.

(a) *pnp*

| Mode | $V_{EB}$ | $V_{CB}$ |
|---|---|---|
| Active | | |
| Inverted | | |
| Saturation | | |
| Cutoff | | |

(b) *npn*

| Mode | $V_{BE}$ | $V_{BC}$ |
|---|---|---|
| Active | | |
| Inverted | | |
| Saturation | | |
| Cutoff | | |

**10.3** A Si *pnp* BJT with $N_{AE} = 5 \times 10^{17}/\text{cm}^3$, $N_{DB} = 10^{15}/\text{cm}^3$, $N_{AC} = 10^{14}/\text{cm}^3$, and $W_B = 3~\mu\text{m}$ is maintained under equilibrium conditions at room temperature.

(a) Sketch the energy band diagram for the device, properly positioning the Fermi level in the three device regions. (The Exercise 10.2 BJT_Eband program might be used to check your answer.)

(b) Sketch (i) the electrostatic potential, setting $V = 0$ in the emitter region, (ii) the electric field, and (iii) the charge density as a function of position inside the BJT.

(c) Calculate the net potential difference between the collector and emitter.

(d) Determine the quasineutral width of the base.

(e) Calculate the maximum magnitude of the electric fields in the E–B and C–B depletion regions.

**10.4** A Si *npn* with $N_{DE} = 10^{18}/cm^3$, $N_{AB} = 10^{16}/cm^3$, $N_{DC} = 10^{15}/cm^3$, and $W_B = 2 \mu m$ is maintained under equilibrium conditions at room temperature.

(a) Sketch the energy band diagram for the device properly positioning the Fermi level in the three device regions. (The Exercise 10.2 BJT_Eband program might be used to check your answer.)

(b) Sketch (i) the electrostatic potential, setting $V = 0$ in the emitter region, (ii) the electric field, and (iii) the charge density as a function of position inside the BJT.

(c) Calculate the net potential difference between the collector and emitter.

(d) Determine the quasineutral width of the base.

(e) Calculate the maximum magnitude of the electric fields in the E–B and C–B depletion regions.

**10.5** Biases of $V_{EB} = 0.5$ V and $V_{CB} = -2$ V are applied to the Problem 10.3 BJT.

(a) Sketch the energy band diagram for the device, properly positioning the Fermi level in the three device regions.

(b) Superimposed on the respective sketches completed in response to Problem 10.3, sketch the electrostatic potential, electric field, and charge density as a function of position inside the biased BJT.

**10.6** For a typically doped Si *npn* transistor, sketch the energy band diagram, electrostatic potential, electric field, and charge density inside the device as a function of position under active mode biasing.

**10.7** A *pnp* BJT is saturation biased with $I_C > 0$. Construct figures similar to Figs. 10.8 and 10.9 showing the carrier activity and diffusion currents inside the transistor.

**10.8** An *npn* BJT is biased into cutoff. Construct figures similar to Figs. 10.8 and 10.9 that show the carrier activity and diffusion currents inside the transistor.

**10.9** Given a *pnp* BJT where $I_{Ep} = 1$ mA, $I_{En} = 0.01$ mA, $I_{Cp} = 0.98$ mA, and $I_{Cn} = 0.1 \mu A$, calculate:

(a) $\alpha_T$

(b) $\gamma$

(c) $I_E, I_C, I_B$

(d) $\alpha_{dc}$ and $\beta_{dc}$

(e) $I_{CB0}$ and $I_{CE0}$

(f) $I_{Cp}$ is increased to a value closer to 1 mA while all other current components remain fixed. What effect does the $I_{Cp}$ increase have on $\beta_{dc}$? Explain.

(g) $I_{En}$ is increased while all other current components remain fixed. What effect does the $I_{En}$ increase have on $\beta_{dc}$? Explain.

**10.10** Given an *npn* BJT where $I_{En} = 100 \ \mu A$, $I_{Ep} = 1 \ \mu A$, $I_{Cn} = 99 \ \mu A$, and $I_{Cp} = 0.1 \ \mu A$, calculate:

(a) $\alpha_T$

(b) $\gamma$

(c) $I_E$, $I_C$, $I_B$

(d) $\alpha_{dc}$ and $\beta_{dc}$

(e) $I_{CB0}$ and $I_{CE0}$

(f) $I_{Cn}$ is increased to a value closer to 100 $\mu A$ while all other current components remain fixed. What effect does the $I_{Cn}$ increase have on $\beta_{dc}$? Explain.

(g) $I_{Ep}$ is increased while all other current components remain fixed. What effect docs the $I_{Ep}$ increase have on $\beta_{dc}$? Explain.

**10.11** Explain why the $I_{CB0}$ given by Eq. (10.12) is devoid of a hole component. Specifically, shouldn't there be a component associated with minority carrier holes in the base that wander into the C–B depletion region and end up being swept into the collector?

# 11 BJT Static Characteristics

This chapter is primarily devoted to modeling the steady-state response of the bipolar junction transistor. The development is divided into three major segments. Building on the introduction in the preceding chapter, we first perform an ideal transistor analysis that closely parallels the ideal diode analysis. Baseline relationships are established for computing the BJT performance parameters and for constructing the BJT static characteristics. The second portion of the chapter compares the ideal theory with experiment and systematically examines the noted deviations from the ideal. Some of the deviations are expected from the diode analysis while others are unique to the transistor. The third and final segment provides information about the polysilicon emitter BJT and the heterojunction bipolar transistor (HBT). These are recently implemented special transistors that offer performance improvements over the standard BJT.

## 11.1 IDEAL TRANSISTOR ANALYSIS

### 11.1.1 Solution Strategy

The derivation of first-order relationships for the BJT performance parameters and terminal currents involves assumptions and solution procedures very similar to those employed in the derivation of the ideal diode equation. There is of course an increase in complexity associated with the existence of a third quasineutral region of finite extent and the simultaneous handling of three (E, B, C) current-voltage expressions. We summarize here the general solution strategy. Included are listings of the basic assumptions, the material parameters employed in the derivation, the differential equations to be solved along with the associated regional boundary conditions, and computational relationships for the currents and parameters of interest.

#### Basic Assumptions

(1) The device under analysis is taken to be a *pnp* BJT with nondegenerate uniformly doped emitter, base, and collector regions—the E–B and C–B junctions are modeled as step junctions.

(2) The transistor is being operated under steady-state conditions.

(3) The transistor is one-dimensional.

(4) Low-level injection prevails in the quasineutral regions.

(5) There are no processes other than drift, diffusion, and thermal recombination–generation taking place inside the transistor. Specifically, $G_L = 0$.

(6) Thermal recombination–generation is negligible throughout the E–B and C–B depletion regions.

(7) The quasineutral widths of the emitter and collector are much greater than the minority carrier diffusion lengths in these regions. (The emitter and collector are effectively taken to be semi-infinite in extent.)

Note that thermal recombination–generation is *not* ignored in the quasineutral base region and that initially no restrictions are placed on the width of the base.

## Notation

The symbols for material parameters to be employed in the analysis are listed by region in Fig. 11.1. Note that regional subscripts (E, B, C) are used to identify the *minority carrier parameters* in a region as opposed to the previously employed carrier type ($n$, $p$) subscripts. This is necessitated by the existence of two transistor regions with the same doping type. In addition, for consistency in notation, a single regional subscript is used instead of a double subscript to identify the doping concentrations. $n_{E0} = n_i^2/N_E$, $p_{B0} = n_i^2/N_B$, and $n_{C0} = n_i^2/N_C$ are compact symbols for the equilibrium minority carrier concentrations in the emitter, base, and collector, respectively.

## Diffusion Equations / Boundary Conditions

Under the stated assumptions, the minority carrier diffusion equations can be used to solve for the minority carrier concentrations in the quasineutral regions of the transistor. The boundary conditions needed to complete the solutions are analogous in form to those employed in the ideal diode derivation. Specifically, since the quasineutral widths of the emit-

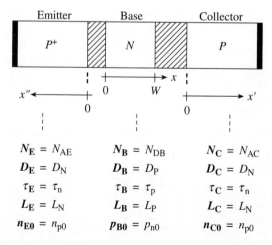

**Figure 11.1** Coordinate systems and material parameter symbols (in bold) employed in the ideal transistor analysis.

ter and collector are taken to be much greater than minority carrier diffusion lengths in these regions, the perturbed emitter and collector carrier concentrations ($\Delta n_E$ and $\Delta n_C$) must vanish as one moves away from the E–B and C–B junctions, respectively. Likewise, the "law of the junction" (Eq. 6.12) can be invoked to give the boundary conditions at the E–B and C–B depletion region edges. Appropriately expressed in terms of the coordinate systems defined in Fig. 11.1, the region-by-region equations to be solved and the corresponding boundary conditions are summarized below.

## Emitter Region

The diffusion equation to be solved is

$$0 = D_E \frac{d^2 \Delta n_E}{dx''^2} - \frac{\Delta n_E}{\tau_E} \tag{11.1}$$

subject to the boundary conditions

$$\Delta n_E(x'' \to \infty) = 0 \tag{11.2a}$$

$$\Delta n_E(x'' = 0) = n_{E0}(e^{qV_{EB}/kT} - 1) \tag{11.2b}$$

## Base Region

The diffusion equation to be solved is

$$0 = D_B \frac{d^2 \Delta p_B}{dx^2} - \frac{\Delta p_B}{\tau_B} \tag{11.3}$$

subject to the boundary conditions

$$\Delta p_B(0) = p_{B0}(e^{qV_{EB}/kT} - 1) \tag{11.4a}$$

$$\Delta p_B(W) = p_{B0}(e^{qV_{CB}/kT} - 1) \tag{11.4b}$$

## Collector Region

The diffusion equation to be solved is

$$0 = D_C \frac{d^2 \Delta n_C}{dx'^2} - \frac{\Delta n_C}{\tau_C} \tag{11.5}$$

subject to the boundary conditions

$$\Delta n_C(x' \to \infty) = 0 \tag{11.6a}$$

$$\Delta n_C(x' = 0) = n_{C0}(e^{qV_{CB}/kT} - 1) \tag{11.6b}$$

### Computational Relationships

Once solutions have been obtained for $\Delta n_E$, $\Delta p_B$, and $\Delta n_C$, the electron and hole current components flowing across the E–B and C–B junctions—the $I_{En}$, $I_{Ep}$, $I_{Cn}$, and $I_{Cp}$ introduced in the previous chapter—are readily computed. For example, with recombination–generation assumed to be negligible in the E–B depletion region, the $I_{En}$ in a *pnp* transistor can be equated to the minority carrier diffusion current in the emitter evaluated at $x'' = 0$, or

$$I_{En} = -qAD_E \left. \frac{d\Delta n_E}{dx''} \right|_{x''=0} \tag{11.7}$$

Similarly,

$$I_{Ep} = -qAD_B \left. \frac{d\Delta p_B}{dx} \right|_{x=0} \tag{11.8}$$

$$I_{Cp} = -qAD_B \left. \frac{d\Delta p_B}{dx} \right|_{x=W} \tag{11.9}$$

and

$$I_{Cn} = qAD_C \left. \frac{d\Delta n_C}{dx'} \right|_{x'=0} \tag{11.10}$$

Finally, expressions for the performance factors and the terminal currents are established employing the foregoing solutions for the internal current components and the following relationships:

$$\gamma = \frac{I_{Ep}}{I_{Ep} + I_{En}} \qquad \begin{array}{c} (11.11) \\ \text{(same as 10.6)} \end{array}$$

$$\alpha_T = \frac{I_{Cp}}{I_{Ep}} \qquad \begin{array}{c} (11.12) \\ \text{(same as 10.7)} \end{array}$$

$$\alpha_{dc} = \gamma \alpha_T \qquad \begin{array}{c} (11.13) \\ \text{(same as 10.11)} \end{array}$$

$$\beta_{dc} = \frac{\alpha_{dc}}{1 - \alpha_{dc}} \qquad \begin{array}{c} (11.14) \\ \text{(same as 10.16)} \end{array}$$

$$I_E = I_{Ep} + I_{En} \qquad \begin{array}{c} (11.15) \\ \text{(same as 10.4)} \end{array}$$

$$I_C = I_{Cp} + I_{Cn} \qquad\qquad (11.16)$$
$$\text{(same as 10.5)}$$

$$I_B = I_E - I_C \qquad\qquad (11.17)$$
$$\text{(same as 10.1)}$$

### 11.1.2  General Solution (W Arbitrary)

Implementing the solution procedure outlined in the preceding subsection, we perform here the mathematical manipulations leading to expressions for the performance parameters and the terminal currents. Before proceeding, the reader may find it helpful to review the ideal diode derivation in Subsection 6.1.3 and the treatment of the narrow-base diode in Subsection 6.3.2.

#### Emitter/Collector Region Solutions

In the emitter quasineutral region the general solution of the minority carrier diffusion equation (Eq. 11.1) is

$$\Delta n_E(x'') = A_1 e^{-x''/L_E} + A_2 e^{x''/L_E} \qquad\qquad (11.18)$$

Because $\exp(x''/L_E) \to \infty$ as $x'' \to \infty$, the Eq. (11.2a) boundary condition can be satisfied only if $A_2$ is identically zero. With $A_2 = 0$, application of the Eq. (11.2b) boundary condition yields $A_1 = \Delta n_E(x''=0)$. Thus

$$\Delta n_E(x'') = n_{E0}(e^{qV_{EB}/kT} - 1)e^{-x''/L_E} \qquad\qquad (11.19)$$

and applying Eq. (11.7),

$$\boxed{I_{En} = qA\,\frac{D_E}{L_E}\,n_{E0}(e^{qV_{EB}/kT} - 1)} \qquad\qquad (11.20)$$

Similarly, in the collector quasineutral region the solution of Eq. (11.5) subject to the Eq. (11.6) boundary conditions yields

$$\Delta n_C(x') = n_{C0}(e^{qV_{CB}/kT} - 1)e^{-x'/L_C} \qquad\qquad (11.21)$$

Utilizing Eq. (11.10), one then concludes

$$\boxed{I_{Cn} = -\,qA\,\frac{D_C}{L_C}\,n_{C0}(e^{qV_{CB}/kT} - 1)} \qquad\qquad (11.22)$$

As should have been anticipated, except for notational differences and the minus sign in Eq. (11.22), the current solutions here are identical to the $p$-side contribution to the ideal diode equation. The minus sign arises in Eq. (11.22) because the collector current is defined to be positive for current flowing from the $n$-base to the $p$-collector; i.e., the positive flow direction is defined opposite to that in the diode.

## Base Region Solution

Whereas the emitter and collector solutions are nothing more than one-sided ideal diode solutions, the base region solution and the associated currents are expected to differ from previous results. Mathematically, this expectation stems from the observations that the base is finite in extent and the perturbed carrier concentration does not necessarily vanish at either $x = 0$ or $x = W$.

Proceeding with the derivation, we note that the general solution to the base region diffusion equation (Eq. 11.3) is of the usual form,

$$\Delta p_B(x) = A_1 e^{-x/L_B} + A_2 e^{x/L_B} \tag{11.23}$$

However, applying the Eq. (11.4) boundary conditions gives

$$\Delta p_B(0) = p_{B0}(e^{qV_{EB}/kT} - 1) = A_1 + A_2 \tag{11.24a}$$

$$\Delta p_B(W) = p_{B0}(e^{qV_{CB}/kT} - 1) = A_1 e^{-W/L_B} + A_2 e^{W/L_B} \tag{11.24b}$$

If Eqs. (11.24) are solved for $A_1$ and $A_2$, and the resulting expressions substituted back into the general solution, we obtain

$$\Delta p_B(x) = \Delta p_B(0) \left( \frac{e^{(W-x)/L_B} - e^{-(W-x)/L_B}}{e^{W/L_B} - e^{-W/L_B}} \right)$$

$$+ \Delta p_B(W) \left( \frac{e^{x/L_B} - e^{-x/L_B}}{e^{W/L_B} - e^{-W/L_B}} \right) \tag{11.25}$$

or more compactly in terms of the sinh function

$$\boxed{\Delta p_B(x) = \Delta p_B(0) \frac{\sinh[(W - x)/L_B]}{\sinh(W/L_B)} + \Delta p_B(W) \frac{\sinh(x/L_B)}{\sinh(W/L_B)}} \tag{11.26}$$

where

$$\sinh(\xi) \equiv \frac{e^{\xi} - e^{-\xi}}{2} \tag{11.27}$$

Please note that if $V_{CB} = 0$, $\Delta p_B(W) = 0$ and Eq. (11.26) reduces to the narrow-base diode solution, Eq. (6.65). This must be the case, of course, since the two problem specifications are identical when $\Delta p_B(W) = 0$.

With the $\Delta p_B(x)$ solution in hand, we can now compute the hole currents flowing across the E–B and C–B junctions. Applying Eqs. (11.8) and (11.9), and explicitly displaying the voltage dependencies of $\Delta p_B(0)$ and $\Delta p_B(W)$, we conclude

$$I_{Ep} = qA \frac{D_B}{L_B} p_{B0} \left[ \frac{\cosh(W/L_B)}{\sinh(W/L_B)} (e^{qV_{EB}/kT} - 1) - \frac{1}{\sinh(W/L_B)} (e^{qV_{CB}/kT} - 1) \right]$$

(11.28)

$$I_{Cp} = qA \frac{D_B}{L_B} p_{B0} \left[ \frac{1}{\sinh(W/L_B)} (e^{qV_{EB}/kT} - 1) - \frac{\cosh(W/L_B)}{\sinh(W/L_B)} (e^{qV_{CB}/kT} - 1) \right]$$

(11.29)

where

$$\cosh(\xi) \equiv \frac{e^\xi + e^{-\xi}}{2}$$

(11.30)

### Performance Parameters/Terminal Currents

In establishing relationships for the performance parameters, the transistor was implicitly assumed to be active mode biased. Under active mode biasing, $V_{EB} > 0$ and $V_{CB} < 0$ in a *pnp* transistor. This biasing combination makes the $[\exp(qV_{CB}/kT) - 1]$ term in Eqs. (11.28) and (11.29) small compared to the $[\exp(qV_{EB}/kT) - 1]$ term. Thus, neglecting the $[\exp(qV_{CB}/kT) - 1]$ term in the $I_{Ep}$ and $I_{Cp}$ expressions, substituting the current expressions into Eqs. (11.11) and (11.12), and noting $n_{E0}/p_{B0} = N_B/N_E$, we obtain

$$\gamma = \frac{1}{1 + \left( \dfrac{D_E}{D_B} \dfrac{L_B}{L_E} \dfrac{N_B}{N_E} \right) \dfrac{\sinh(W/L_B)}{\cosh(W/L_B)}}$$

(11.31)

$$\alpha_T = \frac{1}{\cosh(W/L_B)}$$

(11.32)

Also

$$\alpha_{dc} = \gamma\alpha_T = \dfrac{1}{\cosh(W/L_B) + \left(\dfrac{D_E}{D_B}\dfrac{L_B}{L_E}\dfrac{N_B}{N_E}\right)\sinh(W/L_B)} \tag{11.33}$$

and

$$\beta_{dc} = \dfrac{1}{\dfrac{1}{\alpha_{dc}} - 1} = \dfrac{1}{\cosh(W/L_B) + \left(\dfrac{D_E}{D_B}\dfrac{L_B}{L_E}\dfrac{N_B}{N_E}\right)\sinh(W/L_B) - 1} \tag{11.34}$$

Expressions for the total emitter and collector currents are next obtained by simply adding their respective $n$ and $p$ components.

$$
\begin{aligned}
I_E = qA\Bigg[ &\left(\dfrac{D_E}{L_E}n_{E0} + \dfrac{D_B}{L_B}p_{B0}\dfrac{\cosh(W/L_B)}{\sinh(W/L_B)}\right)(e^{qV_{EB}/kT} - 1) \\
&- \left(\dfrac{D_B}{L_B}p_{B0}\dfrac{1}{\sinh(W/L_B)}\right)(e^{qV_{CB}/kT} - 1)\Bigg]
\end{aligned}
\tag{11.35}
$$

$$
\begin{aligned}
I_C = qA\Bigg[ &\left(\dfrac{D_B}{L_B}p_{B0}\dfrac{1}{\sinh(W/L_B)}\right)(e^{qV_{EB}/kT} - 1) \\
&- \left(\dfrac{D_C}{L_C}n_{C0} + \dfrac{D_B}{L_B}p_{B0}\dfrac{\cosh(W/L_B)}{\sinh(W/L_B)}\right)(e^{qV_{CB}/kT} - 1)\Bigg]
\end{aligned}
\tag{11.36}
$$

Finally, if desired, an explicit expression for the base current could be established employing $I_B = I_E - I_C$.

---

### Exercise 11.1

**P:** (a) Show that the expressions for the emitter and collector terminal currents reduce to the ideal diode equation in the limit where $W \gg L_B$.
(b) Show that setting $V_{CB} = 0$ in the Eq. (11.35) expression for the emitter current leads to the $I$–$V$ relationship established for the narrow-base diode.

**S:** (a) In the limit where $W \gg L_B$ or $W/L_B \to \infty$,

$$\dfrac{\cosh(W/L_B)}{\sinh(W/L_B)} = \dfrac{e^{W/L_B} + e^{-W/L_B}}{e^{W/L_B} - e^{-W/L_B}} \to 1$$

and

$$\frac{1}{\sinh(W/L_B)} = \frac{2}{e^{W/L_B} - e^{-W/L_B}} \to 0$$

When the above limiting values for the hyperbolic functions are substituted into Eqs. (11.35) and (11.36), the result is

$$I_E = qA\left(\frac{D_E}{L_E} n_{E0} + \frac{D_B}{L_B} p_{B0}\right)(e^{qV_{EB}/kT} - 1)$$

$$I_C = -qA\left(\frac{D_C}{L_C} n_{C0} + \frac{D_B}{L_B} p_{B0}\right)(e^{qV_{CB}/kT} - 1)$$

Except for differences in notation and the minus sign in the $I_C$ relationship, the foregoing are clearly ideal diode equations. As previously explained, the minus sign arises from the $n$-to-$p$ definition of positive current flow in the C–B junction. The manipulations here confirm an earlier statement that a structure where $W \gg L_B$ is nothing more than two back-to-back diodes.

(b) If $V_{CB} = 0$, the $[\exp(qV_{CB}/kT) - 1]$ term vanishes and Eq. (11.35) reduces to

$$I_E = qA\left(\frac{D_E}{L_E} n_{E0} + \frac{D_B}{L_B} p_{B0} \frac{\cosh(W/L_B)}{\sinh(W/L_B)}\right)(e^{qV_{EB}/kT} - 1)$$

Given the $p^+$-$n$ nature of the E–B junction, the term involving $n_{E0}$ (the term arising from $I_{En}$) may be neglected. Except for notational differences, the above equation is then identical to the narrow-base result [Eqs. (6.68) and (6.69)]. In general, with the collector and base tied together or $V_{CB} = 0$, the ideal transistor is functionally equivalent to the previously analyzed narrow-base diode.

## 11.1.3 Simplified Relationships ($W \ll L_B$)

The relationships developed in the last subsection constitute a complete solution for the transistor parameters and currents. They could be used directly in performing calculations and constructing characteristics. However, simplified or modified versions of the solution help to provide insight and allow one to more readily manipulate the solution. In this subsection we examine the simplifications resulting from the fact that $W \ll L_B$ in a typical transistor.

The cited narrow width requirement could have been imposed much earlier in the development to simplify the solution for the minority carrier distribution in the quasineutral base. This would have automatically simplified all subsequent expressions. Unfortunately, the early application of the narrow width requirement precludes examining other limiting case solutions, does not permit investigating special case ($W \sim L_B$) solutions, and, most importantly, leads to the loss of significant terms in some expressions. We should also mention that more compact relationships could have been developed through the use of additional hyperbolic functions such as $\coth(\xi) = \cosh(\xi)/\sinh(\xi)$. It was felt preferable, however, to deal exclusively with the sinh and cosh functions. In what follows, we make use of the sinh and cosh small-argument limits first noted in the narrow-base diode discussion and repeated here for the reader's convenience.

$$\sinh(\xi) \rightarrow \xi \qquad \dots \xi \ll 1 \qquad (11.37)$$

and

$$\cosh(\xi) \rightarrow 1 + \frac{\xi^2}{2} \quad \dots \xi \ll 1 \qquad (11.38)$$

### $\Delta p_B(x)$ in the Base

Since $0 \le x \le W$, the $(W-x)/L_B$ and $x/L_B$ arguments in Eq. (11.26) will be much less than unity for all values of $x$ if $W/L_B \ll 1$. In the specified limit all the $\sinh(\xi)$ functions in Eq. (11.26) can therefore be replaced by their arguments and one obtains

$$\Delta p_B(x) \cong \Delta p_B(0) \left( 1 - \frac{x}{W} \right) + \Delta p_B(W) \frac{x}{W} \qquad (11.39)$$

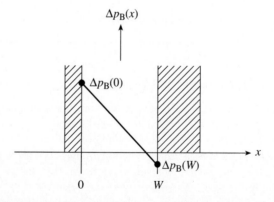

**Figure 11.2**  Perturbed carrier concentration ($p_B - p_{B0}$) in the base of a *pnp* BJT. The pictured distribution corresponds to active mode biasing ($V_{EB} > 0$, $V_{CB} < 0$).

or

$$\Delta p_B(x) = \Delta p_B(0) + [\Delta p_B(W) - \Delta p_B(0)] \frac{x}{W} \tag{11.40}$$

Eq. (11.40) indicates that *the perturbed carrier concentration in the quasineutral base becomes a simple linear function of position in the $W \ll L_B$ limit*. The straight line distribution runs from $\Delta p_B(0) = p_{B0}[\exp(qV_{EB}/kT) - 1]$ at the base edge of the E–B depletion region to $\Delta p_B(W) = p_{B0}[\exp(qV_{CB}/kT) - 1]$ at the base edge of the C–B depletion region as pictured in Fig. 11.2. It is left to the reader as a problem to verify that the distribution remains fairly close to linear even for $W/L_B$ ratios as large as unity.

## Performance Parameters

If we keep at most terms to second order in $W/L_B$ when imposing the $W/L_B \ll 1$ limit, the general solutions for the performance parameters (Eqs. 11.31–11.34) simplify to

$$\gamma = \frac{1}{1 + \dfrac{D_E}{D_B} \dfrac{N_B}{N_E} \dfrac{W}{L_E}} \tag{11.41}$$

$$\alpha_T = \frac{1}{1 + \dfrac{1}{2}\left(\dfrac{W}{L_B}\right)^2} \tag{11.42}$$

$$\alpha_{dc} = \frac{1}{1 + \dfrac{D_E}{D_B} \dfrac{N_B}{N_E} \dfrac{W}{L_E} + \dfrac{1}{2}\left(\dfrac{W}{L_B}\right)^2} \tag{11.43}$$

$$\beta_{dc} = \frac{1}{\dfrac{D_E}{D_B} \dfrac{N_B}{N_E} \dfrac{W}{L_E} + \dfrac{1}{2}\left(\dfrac{W}{L_B}\right)^2} \tag{11.44}$$

Examining Eq. (11.41), we note that, consistent with previous qualitative discussions, an $N_E \gg N_B$ leads to an emitter efficiency near unity. Likewise in agreement with our understanding of transistor operation, Eq. (11.42) indicates a $W \ll L_B$ is required to achieve a base transport factor close to unity.

The Eq. (11.41)–Eq. (11.44) results are sufficiently simple that sample calculations can be readily performed. Suppose a *pnp* BJT operated at 300 K is characterized by the

**Table 11.1**    Sample Set of Assumed and Computed Material Parameters for a *pnp* BJT Operated at 300 K.

| Emitter | Base | Collector |
|---|---|---|
| $N_E = 10^{18}/cm^3$ | $N_B = 10^{16}/cm^3$ | $N_C = 10^{15}/cm^3$ |
| $\mu_E = 263$ cm$^2$/V-sec | $\mu_B = 437$ cm$^2$/V-sec | $\mu_C = 1345$ cm$^2$/V-sec |
| $D_E = 6.81$ cm$^2$/sec | $D_B = 11.3$ cm$^2$/sec | $D_C = 34.8$ cm$^2$/sec |
| $\tau_E = 10^{-7}$ sec | $\tau_B = 10^{-6}$ sec | $\tau_C = 10^{-6}$ sec |
| $L_E = 8.25 \times 10^{-4}$ cm | $L_B = 3.36 \times 10^{-3}$ cm | $L_C = 5.90 \times 10^{-3}$ cm |
| | $W_B = 2 \times 10^{-4}$ cm | |

material parameters recorded in Table 11.1 and a $W \cong W_B$. Substituting the listed material parameters into the performance relationships gives

$$\frac{D_E}{D_B} \frac{N_B}{N_E} \frac{W}{L_E} = 1.46 \times 10^{-3}$$

$$\frac{1}{2}\left(\frac{W}{L_B}\right)^2 = 1.77 \times 10^{-3}$$

and

$$\gamma = 0.9985$$

$$\alpha_T = 0.9982$$

$$\alpha_{dc} = 0.9968$$

$$\beta_{dc} = 310$$

Even though emitter dopings are typically greater than the assumed $10^{18}/cm^3$ (roughly the upper limit imposed by the nondegenerate nature of the theory), the computed results are fairly representative of actual transistors. In modern high-performance transistors, however, $\alpha_T$ is likely to be closer to unity and the gain primarily limited by the emitter efficiency.

---

**Exercise 11.2**

**P:** The Eq. (11.40) linear dependence of $\Delta p_B(x)$ in the $W \ll L_B$ limit should not have come as a surprise. As previously encountered in Problem 3.17 and the narrow-base diode analysis, the negligible thermal recombination–generation in regions much shorter than a diffusion length automatically leads to a linear dependence. Specifically show that Eq. (11.40) is the solution to the minority carrier diffusion equation if thermal recombination–generation is taken to be negligible in the quasineutral base of a *pnp* transistor.

**S:** With thermal recombination–generation assumed to be negligible, the Eq. (11.3) base-region diffusion equation reduces to

$$0 = \frac{d^2 \Delta p_B}{dx^2}$$

which has the general solution

$$\Delta p_B(x) = A_1 + A_2 x$$

Applying the Eq. (11.4) boundary conditions then yields

$$\Delta p_B(0) = A_1$$

and

$$\Delta p_B(W) = A_1 + A_2 W = \Delta p_B(0) + A_2 W$$

or

$$A_2 = [\Delta p_B(W) - \Delta p_B(0)]/W$$

Finally, substituting the deduced $A_1$ and $A_2$ back into the general solution gives the expected result,

$$\Delta p_B(x) = \Delta p_B(0) + [\Delta p_B(W) - \Delta p_B(0)] \frac{x}{W}$$

---

### Exercise 11.3

**P:** Consider a *pnp* transistor ($W \ll L_B$) biased into saturation. Employing a linear scale and using the same set of coordinates, sketch the perturbed minority carrier distribution normalized to the equilibrium carrier concentration ($\Delta n/n_0$ or $\Delta p/p_0$) in the quasineutral regions of the emitter, base, and collector.

**S:** Equations (11.19) and (11.21) allow us to write

$$\Delta n_E(x'')/n_{E0} = (e^{qV_{EB}/kT} - 1)\, e^{-x''/L_E}$$

$$\Delta n_C(x')/n_{C0} = (e^{qV_{CB}/kT} - 1)\, e^{-x'/L_C}$$

When saturation biased, $V_{EB} > 0$ and $V_{CB} > 0$. Thus the distributions in the emitter and collector are similar. Moreover, the distributions have the same general form as the carrier excess in a forward-biased $pn$ diode. There is a pileup of the carriers immediately adjacent to the depletion region edge and an exponential decay of the carrier concentration as one proceeds deeper into the quasineutral region.

At the edges of the quasineutral base, we infer from the Eq. (11.4) boundary conditions that

$$\Delta p_B(0)/p_{B0} = (e^{qV_{EB}/kT} - 1)$$

$$\Delta p_B(W)/p_{B0} = (e^{qV_{CB}/kT} - 1)$$

With $V_{EB} > 0$ and $V_{CB} > 0$, there is a carrier excess at both edges of the quasineutral base. Moreover, note that the normalized carrier excess at the base edge and the emitter edge of the E B depletion region are precisely equal. Similarly, $\Delta p_B(W)/p_{B0} = \Delta n_C(x'=0)/n_{C0}$. Finally, we know the distribution in the base must be approximately linear because $W \ll L_B$.

The desired sketch is concluded to have the general form displayed below.

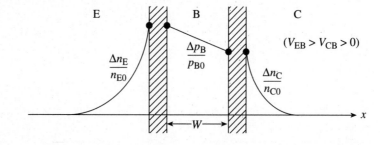

**Exercise 11.4**

**P:** Work out an expression for the base transport factor using the Eq. (11.40) approximation for $\Delta p_B(x)$. Comment on the result.

**S:** Since the approximate $\Delta p_B(x)$ solution is a linear function of $x$, $d\Delta p_B(x)/dx$ is the same for all $x$ in the quasineutral base. It therefore follows from Eqs. (11.8) and (11.9) that $I_{Cp} = I_{Ep}$ and

$$\alpha_T = \frac{I_{Cp}}{I_{Ep}} = 1$$

An $\alpha_T = 1$ implies none of the injected carriers are lost by recombination in crossing the quasineutral base. The result is consistent with the assumption of negligible recombination–generation in the base region, the key assumption invoked in the Exercise 11.2 derivation of Eq. (11.40). Clearly, terms of order $(W/L_B)^2$ are lost in computing the performance parameters using the linear $\Delta p_B(x)$ relationship.

## 11.1.4  Ebers–Moll Equations and Model

In this subsection we shift our interest from the performance parameters to the terminal currents. We begin by re-examining Eqs. (11.35) and (11.36). If $V_{CB}$ is set equal to zero in Eq. (11.35), the remaining part of the equation has the general form of an ideal diode equation for the E–B junction. Likewise, if $V_{EB}$ is set equal to zero in Eq. (11.36), the remaining part of the equation has the general form of an ideal diode equation for the C–B junction. It is reasonable therefore to think of the prefactor to $[\exp(qV_{EB}/kT) - 1]$ in Eq. (11.35) and the prefactor to $[\exp(qV_{CB}/kT) - 1]$ in Eq. (11.36) as effective diode saturation currents and to introduce

$$I_{F0} \equiv qA\left(\frac{D_E}{L_E}\,n_{E0} + \frac{D_B}{L_B}\,p_{B0}\,\frac{\cosh(W/L_B)}{\sinh(W/L_B)}\right) \tag{11.45a}$$

$$I_{R0} \equiv qA\left(\frac{D_C}{L_C}\,n_{C0} + \frac{D_B}{L_B}\,p_{B0}\,\frac{\cosh(W/L_B)}{\sinh(W/L_B)}\right) \tag{11.45b}$$

Next compare the $V_{CB}$ terms in the $I_E$ and $I_C$ equations. Since the prefactor of $[\exp(qV_{CB}/kT) - 1]$ in Eq. (11.35) is smaller than $I_{R0}$, the prefactor of $[\exp(qV_{CB}/kT) - 1]$ in Eq. (11.36), one is led to view the $V_{CB}$ term in Eq. (11.35) as the fraction of the $I_{R0}[\exp(qV_{CB}/kT) - 1]$ current from the C–B junction that makes it to the E–B junction. Analogously, the $V_{EB}$ term in Eq. (11.36) may be viewed as the fraction of the $I_{F0}[\exp(qV_{EB}/kT) - 1]$ current from the E–B junction that makes it to the C–B junction. This suggests the introduction of

$$\alpha_F I_{F0} = \alpha_R I_{R0} \equiv qA\,\frac{D_B}{L_B}\,\frac{p_{B0}}{\sinh(W/L_B)} \tag{11.46}$$

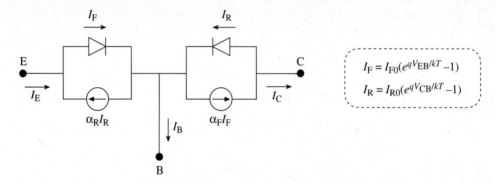

**Figure 11.3**   Large signal equivalent circuit for a *pnp* BJT based on the Ebers–Moll equations.

where $\alpha_F$ and $\alpha_R$ are the forward and reverse "gains," respectively. $\alpha_R I_{R0}$ is to be used in Eq. (11.35) and $\alpha_F I_{F0}$ in Eq. (11.36). Rewriting Eqs. (11.35) and (11.36) explicitly incorporating the newly introduced parameters yields

$$I_E = I_{F0}(e^{qV_{EB}/kT} - 1) - \alpha_R I_{R0}(e^{qV_{CB}/kT} - 1) \qquad (11.47a)$$

$$I_C = \alpha_F I_{F0}(e^{qV_{EB}/kT} - 1) - I_{R0}(e^{qV_{CB}/kT} - 1) \qquad (11.47b)$$

Equations (11.47) are known as the *Ebers–Moll equations*. Computer-aided circuit analysis programs, such as SPICE, often use these equations, or a modified form of the equations, to solve for d.c. operating point variables and to construct the BJT device characteristics. Although established here from first principles, the Ebers–Moll equations may alternatively be viewed as empirical relationships where the four parameters are deduced from a match to experimental data. Actually, only three of the parameters need to be determined, since the fourth can always be computed from the $\alpha_F I_{F0} = \alpha_R I_{R0}$ requirement of Eq. (11.46).

From a circuit viewpoint the Ebers–Moll equations and $I_B = I_E - I_C$ are nothing more than node equations for the three BJT terminals. Associating circuit symbols with the individual terms in the Ebers–Moll equations leads to the large signal equivalent circuit in Fig. 11.3. (The reader should convince himself/herself that the equations and the Fig. 11.3 circuit are indeed equivalent.) The equivalent circuit is sometimes useful in circuit analyses and in treating more complex device structures such as the PNPN device family.

---

### Exercise 11.5

**P:** Show that the forward current gain in the Ebers–Moll formulation is equal to the common base current gain; i.e., $\alpha_F = \alpha_{dc}$.

**S:** Dividing the left- and right-hand sides of Eq. (11.46) by the corresponding sides of Eq. (11.45a) gives

$$\alpha_F = \frac{qA\, \dfrac{D_B}{L_B}\, \dfrac{p_{B0}}{\sinh(W/L_B)}}{qA\left(\dfrac{D_E}{L_E}\, n_{E0} + \dfrac{D_B}{L_B}\, p_{B0}\, \dfrac{\cosh(W/L_B)}{\sinh(W/L_B)}\right)}$$

$$= \frac{1}{\cosh(W/L_B) + \left(\dfrac{D_E}{D_B}\, \dfrac{L_B}{L_E}\, \dfrac{n_{E0}}{p_{B0}}\right)\sinh(W/L_B)}$$

Since $n_{E0}/p_{B0} = N_B/N_E$, the second form of the above $\alpha_F$ equation and the (11.33) $\alpha_{dc}$ equation are seen to be identical.

---

### Exercise 11.6

**P:** In Chapter 10 it was pointed out that the active mode portion of the common-base output characteristics is accurately modeled by the relationship

$$I_C = \alpha_{dc} I_E + I_{CB0} \qquad\qquad \text{(same as 10.8)}$$

Show that the Ebers–Moll equations can be manipulated to obtain an equivalent expression.

**S:** Under active mode operation ($V_{EB} > 0$, $V_{CB} < 0$, $\exp(qV_{CB}/kT) \ll 1$) the Ebers–Moll equations reduce to

$$I_E = I_{F0}(e^{qV_{EB}/kT} - 1) + \alpha_R I_{R0}$$

$$I_C = \alpha_F I_{F0}(e^{qV_{EB}/kT} - 1) + I_{R0}$$

Solving the $I_E$ equation for the voltage factor yields

$$(e^{qV_{EB}/kT} - 1) = \frac{I_E - \alpha_R I_{R0}}{I_{F0}}$$

which when substituted into the $I_C$ equation gives

$$I_C = \alpha_F I_E + (1 - \alpha_F \alpha_R)I_{R0}$$

Since $\alpha_F = \alpha_{dc}$, the above and the originally cited equation are seen to be identical if one identifies $I_{CB0} = (1 - \alpha_F \alpha_R)I_{R0}$.

---

## (C) Exercise 11.7

**P:** Utilizing the Ebers–Moll equations and subsidiary relationships, construct a MATLAB (computer) program that calculates and plots the input and output characteristics of a *pnp* BJT operated in both the common base and the common emitter configurations. In performing a sample computation, assume room temperature operation, a device area $A = 10^{-4}$ cm$^2$, $W = W_B$, and material parameters similar to those listed in Table 11.1.

**S:** The common base input characteristic is a plot of $I_E$ versus $V_{EB}$ with the output voltage $V_{CB}$ taking on select stepped values. Convenient construction of the characteristic requires an equation where $I_E$ is expressed as a function of $V_{EB}$ and $V_{CB}$, or one requires an equation of the form $I_E = I_E(V_{EB}, V_{CB})$. The $I_E$ Ebers–Moll equation, Eq. (11.47a), is precisely of this form and may be used directly to compute the common base input characteristic. To construct any of the other characteristics, however, it is first necessary to combine and/or rearrange the Ebers–Moll equations to obtain a computational equation of the proper form. The algebraic manipulations, similar to those in Exercise 11.6 but somewhat more involved, yield the set of computational relationships:

(1) Common base input $[I_E = I_E(V_{EB}, V_{CB})]$

$$I_E = I_{F0}(e^{qV_{EB}/kT} - 1) - \alpha_R I_{R0}(e^{qV_{CB}/kT} - 1)$$

(2) Common base output $[I_C = I_C(V_{CB}, I_E)]$

$$I_C = \alpha_F I_E - (1 - \alpha_F \alpha_R)I_{R0}(e^{qV_{CB}/kT} - 1)$$

(3) Common emitter input $[I_B = I_B(V_{EB}, V_{EC})]$

$$I_B = [(1 - \alpha_F)I_{F0} + (1 - \alpha_R)I_{R0}e^{-qV_{EC}/kT}]\, e^{qV_{EB}/kT}$$
$$- [(1 - \alpha_F)I_{F0} + (1 - \alpha_R)I_{R0}]$$

(4) Common emitter output $[I_C = I_C(V_{EC}, I_B)]$

$$I_C = \frac{(\alpha_F I_{F0} - I_{R0}e^{-qV_{EC}/kT})[I_B + (1 - \alpha_F)I_{F0} + (1 - \alpha_R)I_{R0}]}{(1 - \alpha_F)I_{F0} + (1 - \alpha_R)I_{R0}e^{-qV_{EC}/kT}} + I_{R0} - \alpha_F I_{F0}$$

The MATLAB program BJT, included on disk and listed in Appendix M, computes the BJT characteristics employing the foregoing relationships. The desired characteristic is chosen from an opening menu. After running the BJT program with the sample set of parameters, the user may wish to investigate the effect of changing the device parameters. Constants, material parameters, and the $W = W_B$ Ebers–Moll parameters are specified or computed in a subsidiary program, BJT0. Modify the appropriate BJT0 program line to change the dopings, lifetimes, base width, and/or device area. The BJT0 program automatically computes the mobilities, diffusion coefficients, and diffusion lengths based on the input dopings and lifetimes. A $W_B = 2.5~\mu m$, $N_B = 1.5 \times 10^{16}/cm^3$, and $N_C = 1.5 \times 10^{15}/cm^3$, values slightly different from those in Table 11.1, were incorporated in the as-provided program to afford a closer match to the experimental characteristics presented in the next section. Axis specifications, it should be noted, are optimized for the sample set of parameters and may have to be revised if the parameter set is modified.

## 11.2  DEVIATIONS FROM THE IDEAL

As demonstrated in Exercise 11.7, the ideal transistor relationships established in the preceding section can be used to generate the expected form of the BJT statistic characteristics. In this section we compare the ideal-device characteristics with measured characteristics and explore reasons for the noted deviations from the ideal. We first concentrate on the underlying phenomena leading to differences in the general form of the predicted and observed characteristics. Subsequent consideration is given to phenomena, such as base resistance and depletion-region recombination–generation, which affect the precise degree of numerical agreement between the computed and observed values.

### 11.2.1 Ideal Theory/Experiment Comparison

A side-by-side comparison of experimental BJT characteristics and the corresponding ideal-theory characteristics is presented in Figs. 11.4 and 11.5. Figure 11.4 displays common-base input and output characteristics; Fig. 11.5 displays common emitter input and output characteristics. The experimental characteristics were derived from a 2N2605 *pnp* BJT maintained at room temperature. The theoretical curves were generated assuming $W = W_B$ and employing the as-provided BJT program described in Exercise 11.7.

The predicted and observed common-base input characteristics of Figs. 11.4(a) and (b) are seen to be of the same general character; namely, the input current ($I_E$) exponentially increases with increasing input voltage ($V_{EB}$). The theoretically characteristics are essentially independent of the applied output voltage ($V_{CB}$). Experimentally, however, the input current at a given input voltage noticeably increases with increasingly negative values of $V_{CB}$.

Turning to the common base output characteristics displayed in Figs. 11.4(c) and (d), we find generally excellent agreement between theory and experiment. In both cases there

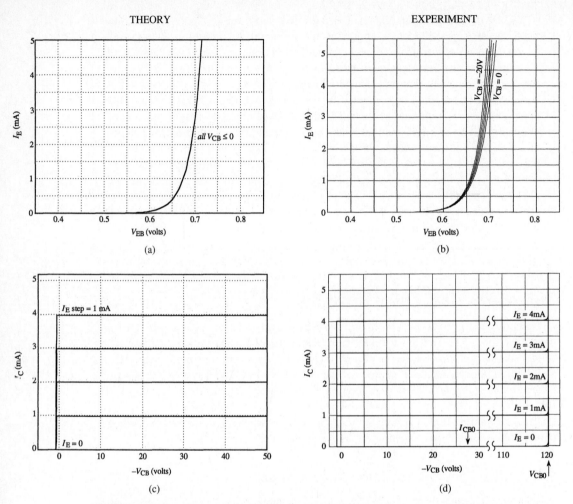

THEORY                                         EXPERIMENT

**Figure 11.4**  Common base *pnp* BJT characteristics: (a) ideal theory and (b) experimental input characteristics; (c) ideal theory and (d) experimental output characteristics. Room temperature data derived from a 2N2605 *pnp* BJT and primarily recorded employing a Hewlett-Packard 4145B Semiconductor Parameter Analyzer. The $-V_{CB} \geq 110$ V portion of (d) is an idealized sketch of the characteristics observed on a curve tracer. The theoretical curves were generated assuming $W = W_B$ and employing the computer program constructed for Exercise 11.7.

is a rapid increase in the output current ($I_C$) starting at sightly negative $-V_{CB}$ values, and a nearly perfect leveling-off or constancy of the output current for $-V_{CB} > 0$. Theoretically there is no limit to the $-V_{CB}$ voltage that can be applied to the transistor. On the other hand, as might be expected from the diode analysis, the real-device output voltage is limited to a maximum value of $V_{CB0}$ by some sort of breakdown phenomenon.

The common emitter input characteristics shown in Figs. 11.5(a) and (b) are, for the most part, in excellent agreement. The $V_{EC} = 0$ characteristic appears by itself in both the

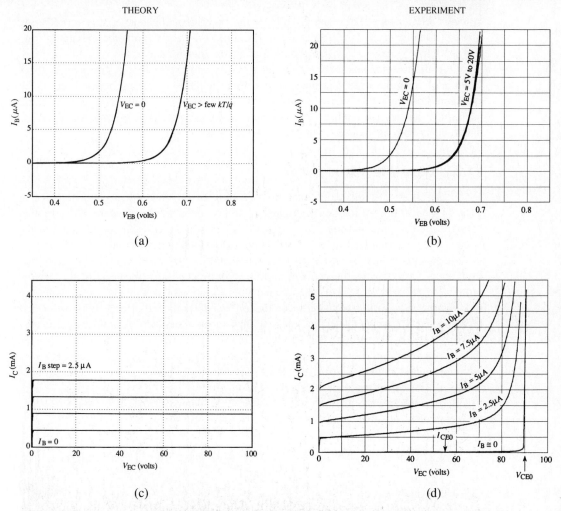

**Figure 11.5** Common emitter *pnp* BJT characteristics: (a) ideal theory and (b) experimental input characteristics; (c) ideal theory and (d) experimental output characteristics. Room temperature data derived from a 2N2605 *pnp* BJT and recorded employing a Hewlett-Packard 4145B Semiconductor Parameter Analyzer. The theoretical curves were generated assuming $W = W_B$ and utilizing the computer program constructed for Exercise 11.7.

theoretical and experimental plots, while characteristics associated with an output voltage of $V_{EC} >$ few $kT/q$ are grouped together and displaced positively along the input voltage ($V_{EB}$) axis.

The greatest deviation between theory and experiment can be found in the common emitter output characteristics displayed in Figs. 11.5(c) and (d). First, whereas the theoretical characteristics are perfectly flat ($V_{EC}$ independent) under active mode biasing, there is a very perceptible upward slope to the experimental curves even at relatively low output voltages. Moreover, at large output voltages the experimental characteristics turn sharply upward. Finally, some sort of breakdown phenomenon clearly limits the maximum low-current output voltage, $V_{CE0}$, that can be applied to a transistor.

Overall the agreement between the ideal theory and experiment is surprisingly good. Deviations from the ideal that must be explained include the strong dependence of the common base input characteristics on the output voltage, the initial quasilinear slope and subsequent upward curvature of the common emitter output characteristics, and the $V_{CB0}$ and $V_{CE0}$ limitations imposed by breakdown phenomena.[†]

### 11.2.2 Base Width Modulation

By setting $W = W_B$ in the computations leading to the theoretical characteristics in Figs. 11.4 and 11.5, we implicitly assumed $W$ to be a constant independent of the applied voltages. In truth, the quasineutral width of the base is not a constant independent of the applied biases. Changing $V_{EB}$ and/or $V_{CB}$ changes the depletion widths about the E–B and/or C–B junctions, thereby shrinking or expanding $W$ as illustrated in Fig. 11.6. Note that even small variations in the depletion widths can be significant because of the narrow physical extent of the base. This phenomenon, the variation of $W$ with the applied voltages, is known as *base width modulation* or the *Early effect* (named after J. Early who first recognized the phenomenon). The practical importance of base width modulation can best be appreciated by referring to Eqs. (11.41)–(11.46). Inspection of these equations reveals the performance and Ebers–Moll parameters are all functions of $W$. Thus, they also vary with the applied voltages. Two of the deviations cited in the previous subsection are direct results of this bias dependence.

Re-examining the equation used to generate the common base input characteristics, we can write

$$I_E = qA\left[\left(\frac{D_E}{L_E}\,n_{E0} + \frac{D_B}{L_B}\,p_{B0}\,\frac{\cosh(W/L_B)}{\sinh(W/L_B)}\right)(e^{qV_{EB}/kT} - 1) + \left(\frac{D_B}{L_B}\,\frac{p_{B0}}{\sinh(W/L_B)}\right)\right] \tag{11.48}$$

assuming $-V_{CB} >$ few $kT/q$. Moreover, if one keeps only lowest-order terms in the $W/L_B \ll 1$ expansion of the hyperbolic functions and neglects the typically small $D_E n_{E0}/L_E$ term, the defining equation for the common base input characteristics reduces to

$$I_E \cong qA\,\frac{D_B}{W}\,p_{B0}\,e^{qV_{EB}/kT} \tag{11.49}$$

When the reverse bias applied to the C–B junction increases, the C–B depletion width increases and $W$ decreases. Thus, according to Eq. (11.49), $I_E$ at a given $V_{EB}$ is expected to

---

[†] $V_{CB0}$ and $V_{CE0}$ are abbreviated forms of the $V_{(BR)CB0}$ and $V_{(BR)CE0}$ symbols often employed in transistor specification sheets to identify the zero-input-current breakdown voltages. $BV_{CB0}$ and $BV_{CE0}$ found in the early device literature have subsequently been supplanted by the noted symbols.

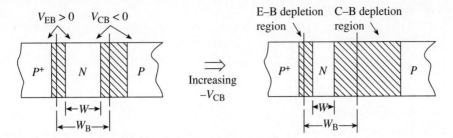

**Figure 11.6** Base width modulation. Decrease in $W$ resulting from an increase in the reverse bias applied across the C–B junction.

increase with increasing $-V_{CB}$. This is precisely the behavior exhibited by the experimental characteristics presented in Fig. 11.4(b).

Base width modulation is also responsible for the quasilinear increase in the common emitter output current recorded in Fig. 11.5(d). Under active mode biasing the Exercise 11.7 equation used to generate the theoretical common-emitter output characteristics reduces to

$$I_C = \beta_{dc} I_B + I_{CE0} \tag{11.50}$$

(same as 10.13)

where, as previously established (Eq. 11.44),

$$\beta_{dc} = \cfrac{1}{\cfrac{D_E}{D_B} \cfrac{N_B}{N_E} \cfrac{W}{L_E} + \cfrac{1}{2}\left(\cfrac{W}{L_B}\right)^2}$$

Since $W$ decreases with increasing reverse bias across the C–B junction, $\beta_{dc}$ systematically increases with increasing $V_{EC}$. Thus, with base width modulation taken into account, the $I_C$ at a given $I_B$ under active mode biasing is likewise expected to increase with increasing $V_{EC}$. It should be noted that base width modulation does *not* explain the sharp upturn in $I_C$ as the output voltage approaches $V_{CE0}$.

The remaining experimental characteristics—the common base output characteristics and the common emitter input characteristics—were found to be in fairly good agreement with the theoretical characteristics constructed assuming $W = W_B$. The question may arise: How is it that the common base output characteristics and the common emitter input characteristics are relatively insensitive to base width modulation? To answer this question, we note that the active mode portion of the common base output characteristics is accurately modeled by (see Exercise 11.6),

$$I_C = \alpha_{dc} I_E + I_{CB0} \tag{11.51}$$

(same as 10.8)

where, as previously established (Eq. 11.43),

$$\alpha_{dc} = \cfrac{1}{1 + \cfrac{D_E}{D_B}\cfrac{N_B}{N_E}\cfrac{W}{L_E} + \cfrac{1}{2}\left(\cfrac{W}{L_B}\right)^2}$$

In a well-made transistor the $W$-dependent terms in the denominator of the $\alpha_{dc}$ expression are small compared to unity. As a result, even large changes in $W$ give rise to an almost imperceptible change in $\alpha_{dc}$. Likewise, the $I_{CB0}$ appearing in Eq. (11.51) can be shown to be essentially $W$-independent (and is moreover negligible in most instances). Base width modulation is therefore concluded to have little effect on the common base output characteristics. It is left as an exercise to show that a similar argument can be applied to the common emitter input characteristics.

A comment is in order concerning how the transistor dopings and the biasing mode affect the sensitivity to base width modulation. In a standard transistor $N_E \gg N_B > N_C$. Thus almost all of the E–B depletion region lies in the base and most of the C–B depletion region lies in the collector. Under active mode biasing the E–B junction experiences only a small forward bias and plays a negligible role in modulating the quasineutral width of the base. The C–B junction, on the other hand, is routinely subjected to a large reverse bias. The effect of base width modulation is minimized, however, because most of the C–B depletion region lies in the collector. The opposite is true if the same transistor is operated in the inverted mode. In the inverted mode a reverse bias is applied across the E–B junction, the associated depletion width extends almost exclusively into the base, and the standard transistor is inherently quite sensitive to base width modulation.

### 11.2.3 Punch-Through

*Punch-through* might be viewed as base width modulation carried to the extreme. Specifically, it refers to the physical situation where base width modulation has resulted in $W \to 0$. That is, the base is said to be punched through when the E–B and C–B depletion regions touch inside the base as illustrated in Fig. 11.7. Once punch-through has occurred, the E–B and C–B junctions become electrostatically coupled. As pictured in Fig. 11.8, an increase in $-V_{CB}$ beyond the punch-through point lowers the E–B potential hill and allows

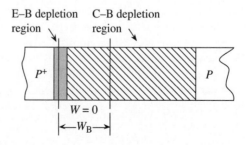

**Figure 11.7**   Visualization of punch-through ($W \to 0$).

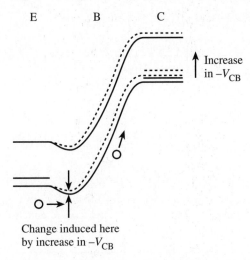

E    B    C

Increase
in $-V_{CB}$

Change induced here
by increase in $-V_{CB}$

**Figure 11.8**  Energy band diagram explanation of why a post-punch-through increase in the output voltage gives rise to a large increase in the output (collector) current.

a large (exponentially increasing) injection of carriers from the emitter directly into the collector. Only a slight increase in $-V_{CB}$ gives rise to a large increase in $I_C$. If it occurs, punch-through limits the maximum voltage ($V_{CB0}$ or $V_{CE0}$) that can be applied to a transistor. We say "if it occurs" because a rapid increase in output current at output voltages approaching $V_{CB0}$ or $V_{CE0}$ could alternatively be caused by avalanche multiplication. Simply stated, the phenomenon that occurs first determines the "breakdown" voltage.

---

### Exercise 11.8

**P:** Consider a *pnp* BJT with the material parameters listed in Table 11.1. If $V_{EB} = 0$ and $T = 300$ K, determine the C–B junction voltage that will lead to punch-through.

**S:** In general, restating Eq. (10.3),

$$W = W_B - x_{nEB} - x_{nCB}$$

where for step junctions as inferred from Eq. (5.37)

$$x_{nEB} = \left[ \frac{2K_S \varepsilon_0}{q} \frac{N_E}{N_B(N_E + N_B)} (V_{bi(EB)} - V_{EB}) \right]^{1/2} ; \qquad V_{bi(EB)} = \frac{kT}{q} \ln\left( \frac{N_E N_B}{n_i^2} \right)$$

$$x_{nCB} = \left[ \frac{2K_S \varepsilon_0}{q} \frac{N_C}{N_B(N_C + N_B)} (V_{bi(CB)} - V_{CB}) \right]^{1/2} ; \qquad V_{bi(CB)} = \frac{kT}{q} \ln\left( \frac{N_C N_B}{n_i^2} \right)$$

At the punch-through voltage, $W = 0$ and one can write

$$0 = W_B - x_{nEB} - \left[ \frac{2K_S\varepsilon_0}{q} \frac{N_C}{N_B(N_C + N_B)} (V_{bi(CB)} - V_{CB}) \right]^{1/2}$$

or on solving for $V_{CB}$

$$V_{CB} = V_{bi(CB)} - \frac{(W_B - x_{nEB})^2}{\frac{2K_S\varepsilon_0}{q} \frac{N_C}{N_B(N_C + N_B)}} \quad \ldots \text{at punch-through}$$

Computations employing the parameters in Table 11.1 give $V_{bi(EB)} = 0.835$ V, $V_{bi(CB)} = 0.656$ V, $x_{nEB} = 3.29 \times 10^{-5}$ cm, and a punch-through voltage of $\boxed{V_{CB} \cong -235 \text{ V}}$.

## 11.2.4 Avalanche Multiplication and Breakdown

### Common Base

If a BJT is connected in the common base configuration and active mode biased to a larger and larger $-V_{CB}$, the carrier activity within the C–B depletion region of the BJT is very similar to that occurring in a simple *pn* junction diode. A growing number of the carriers crossing the C–B depletion region gain enough energy to create additional carriers by the impact ionization of semiconductor atoms. A point is eventually reached where a carrier avalanche occurs and the collector current increases rapidly toward infinity. In terms of carrier multiplication, $M \to \infty$ at the breakdown point, where $M$ is the multiplication factor introduced in Subsection 6.2.2. Assuming avalanche breakdown takes place before punch-through, the maximum magnitude of the voltage that can be applied to the output of the BJT under common base operation, $V_{CB0}$, is clearly just the breakdown voltage of the C–B junction.

### Common Emitter

The common emitter case is far more interesting. The output voltage applied to a transistor connected in the common emitter configuration is $V_{EC} = V_{EB} - V_{CB}$. Under active mode biasing, the E–B junction is forward biased and $V_{EB}$ is typically quite small, less than a volt in Si BJTs. Thus for output voltages exceeding a few volts, $V_{EC} \cong -V_{CB}$. With $V_{EC} \cong -V_{CB}$, it is reasonable to expect a $V_{CE0} \cong V_{CB0}$. However, examining the experimental characteristics from the same device displayed in Figs. 11.4(d) and 11.5(d), one finds a $V_{CE0}$ considerably smaller than $V_{CB0}$; $V_{CE0} \cong 90$ V, whereas $V_{CB0} \cong 120$ V!

This unexpected result can be explained qualitatively with the aid of Fig. 11.9. The initial injection of holes into the base, labeled $\boxed{0}$ in the figure, leads to the holes labeled $\boxed{1}$ entering the base-side of the C–B depletion region. Although the C–B junction is biased far below breakdown, a few of these holes gain a sufficient amount of energy to impact ionize semiconductor atoms and create extra holes and electrons $\boxed{2}$. The added holes drift

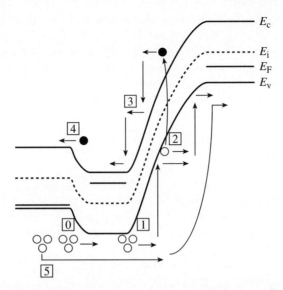

**Figure 11.9**  Step-by-step explanation of the carrier multiplication and feedback mechanism; $\boxed{0}$ initial hole injection; $\boxed{1}$ injected holes entering the C–B depletion region; $\boxed{2}$ electron-hole pair generation by impact ionization; $\boxed{3}$ generated electrons being swept into the base; $\boxed{4}$ excess base electrons injected into the emitter; $\boxed{5}$ holes injected from the emitter into the base in response to the step 4 electron injection.

along with the injected holes into the collector; the added electrons are swept into the base $\boxed{3}$. The extra electrons in the base create a majority carrier imbalance that must be eliminated. The most efficient method of removing the excess would be for electrons to simply flow out of the base lead. This is precisely what happens when the BJT is connected in the common base configuration. When the transistor is connected in the common emitter configuration, however, *the base current is held constant* during the measurement of the common emitter output characteristic—the extra electrons are not allowed to flow out the base lead. Consequently, the carrier imbalance in the base can only be alleviated by injecting electrons from the base into the emitter $\boxed{4}$. As was emphasized in the early operational discussion of the transistor, electron and hole injection across the E–B junction are intimately tied together. Every electron injected from the base into the emitter is accompanied by the injection of $I_{\mathrm{Ep}}/I_{\mathrm{En}}$ added holes from the emitter into the base $\boxed{5}$. Thus for each additional electron-hole pair created in the C–B depletion region by impact ionization, $I_{\mathrm{Ep}}/I_{\mathrm{En}} + 1 = 1/(1 - \gamma) \ge \beta_{\mathrm{dc}} + 1$ additional holes flow into the collector.

Effectively, the carrier multiplication in the C–B depletion region is internally amplified under common emitter operation. Moreover, the process we have just described is regenerative; i.e., it feeds back on itself. With added injection there is added carrier multiplication in the C–B depletion region and an even larger enhancement of the collector current. (Alternatively, the $\beta_{\mathrm{dc}}$ of the transistor may be viewed as being enhanced by the process.) This all leads to an instability point where $I_{\mathrm{C}} \to \infty$ at a $V_{\mathrm{EC}}$ far below the avalanche breakdown voltage of the C–B junction.

To model the process quantitatively, we note that the injection-related component of the collector current in a *pnp* BJT increases to $I_{Cp} = M\alpha_T I_{Ep}$, or $\alpha_T \to M\alpha_T$ and $\alpha_{dc} \to M\alpha_{dc}$, when carrier multiplication in the C–B depletion region is taken into account. In other words, previous relationships can be adjusted to account for carrier multiplication by replacing $\alpha_{dc}$ and $\alpha_F (= \alpha_{dc})$ by $M\alpha_{dc}$ and $M\alpha_F$.

Referring to Eq. (10.15), the simplified expression for the common emitter output characteristics under active mode biasing transforms to

$$I_C = \frac{M\alpha_{dc}}{1 - M\alpha_{dc}} I_B + \frac{I_{CB0}}{1 - M\alpha_{dc}} \tag{11.52}$$

Note from Eq. (11.52) that, for any and all input currents (even $I_B = 0$), $\boxed{I_C \to \infty \text{ when } M \to 1/\alpha_{dc}}$. Typically $\alpha_{dc}$ is only slightly less than unity. $M$ need therefore increase to only slightly greater than unity before reaching the $M = 1/\alpha_{dc}$ point. This is expected to occur at an output voltage far below the $M \to \infty$ breakdown voltage of the C–B junction. Specifically, making use of Eq. (6.35) with $V_{BR} = V_{CB0}$,

$$M\big|_{V_{CE0}} = \frac{1}{1 - \left(\dfrac{V_{CE0}}{V_{CB0}}\right)^m} = \frac{1}{\alpha_{dc}} \tag{11.53}$$

or

$$\boxed{V_{CE0} = V_{CB0}(1 - \alpha_{dc})^{1/m} = \frac{V_{CB0}}{(\beta_{dc} + 1)^{1/m}}} \tag{11.54}$$

where as previously defined, $3 \le m \le 6$. From the latter form of Eq. (11.54), it is immediately obvious that $V_{CE0} < V_{CB0}$.

In the context of BJT operation the mechanism we have just described and modeled must be viewed as detrimental. The maximum voltage one can apply to the transistor is reduced because of the carrier multiplication and feedback. However, in other device contexts the mechanism plays a positive, if not crucial, role. For one, as explained in Chapter 13, the increase in carrier multiplication and feedback with bias helps to trigger switching from the off-state to the on-state in the Silicon Controlled Rectified (SCR) and other PNPN devices. The mechanism is also used to amplify the photosignal in a phototransistor. Phototransistors (or photo-BJTs) are BJTs constructed so that incident light can penetrate to the C–B junction. When operated as a photodetector, the photo-BJT is typically active mode biased with either the base floating or the base current held constant. (Some phototransistors are even produced without a base lead.) As depicted in Fig. 11.10, roughly $\beta_{dc} + 1$ carriers enter the collector for each electron-hole pair created by the absorption of a photon in the C–B depletion region.

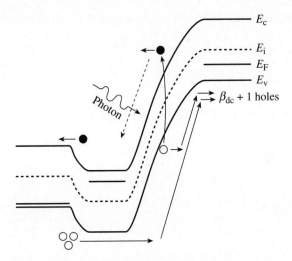

$E_c$

$E_i$

$E_F$

$E_v$

$\beta_{dc} + 1$ holes

Photon

**Figure 11.10** Photosignal amplification in a phototransistor.

**Exercise 11.9**

**P:** Two *pnp* BJTs are similar except $N_B \gg N_C$ in Transistor A while $N_B \ll N_C$ in Transistor B. The doping profiles in the two transistors are graphed below.

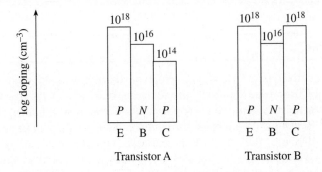

(a) Which transistor is expected to have the larger punch-through voltage under active mode biasing? Explain.

(b) If limited by avalanche breakdown of the C–B junction, which transistor will exhibit the larger $V_{CBO}$? Explain.

**S:**(a) [Transistor A] will have the larger punch-through voltage under active mode biasing. For a given $V_{EB} > 0$ the portion of the base depleted adjacent to the E–B junction ($x_{nEB}$) will be precisely the same in both transistors. On the other hand, the depletion width associated with the reverse biased C–B junction (referring to Eq. 5.37) is

$$x_{nCB} \cong \left[ \frac{2K_S \varepsilon_0}{qN_B} \left( \frac{N_C}{N_B} \right) (V_{bi(CB)} - V_{CB}) \right]^{1/2} \qquad \ldots \text{ in Transistor A}$$

and

$$x_{nCB} \cong \left[ \frac{2K_S \varepsilon_0}{qN_B} (V_{bi(CB)} - V_{CB}) \right]^{1/2} \qquad \ldots \text{ in Transistor B}$$

Since $(N_C/N_B)|_{\text{Transistor A}} = 1/100$ and $V_{bi(CB)|A} < V_{bi(CB)|B}$, $x_{nCB}$ in Transistor A $\leq$ $x_{nCB}/10$ in Transistor B for the same applied $-V_{CB}$. Assuming the base widths ($W_B$) are the same in the two transistors, a larger $-V_{CB}$ is clearly required to completely deplete the base of Transistor A.

(b) $\boxed{\text{Transistor A}}$ will exhibit the larger $V_{CB0}$ if avalanche breakdown of the C–B junction is the limiting phenomenon. The avalanche breakdown voltage of an asymmetrically doped *pn* junction is determined by the doping on the lightly doped side of the junction. This means that $N_C = 10^{14}/cm^3$ and $N_B = 10^{16}/cm^3$ are the dopings of relevance in Transistor A and Transistor B, respectively. Further noting that $V_{BR}$ varies roughly as the inverse of the doping on the lightly doped side of a *pn* junction (see Fig. 6.11 or Eq. 6.33), Transistor A is concluded to exhibit the larger $V_{CB0}$.

---

## (C) Exercise 11.10

**P:** Modify the Exercise 11.7 computer program so that base width modulation and carrier multiplication due to impact ionization can be included in the computation of the BJT characteristics. As a sample application and employing the set of parameters specified in Exercise 11.7, compare the common emitter output characteristics obtained when (i) base width modulation and carrier multiplication are ignored, (ii) only base width modulation is included, and (iii) both base width modulation and carrier multiplication are taken into account.

**S:** The MATLAB program BJTplus, included on disk and listed in Appendix M, permits the optional inclusion of base width modulation and carrier multiplication due to impact ionization when calculating the BJT characteristics. Subprogram BJT0, which inputs constants and material parameters, and subprogram BJTmod, which performs computations related to base width modulation, are run-time requirements. The requested computational results, the common emitter output characteristics progressively showing the effects of base width modulation and carrier multiplication, are displayed on the facing page. Note that base width modulation leads to a moderate quasilinear rise in the output current with increasing output voltage. Also consistent with the text discussion, the addition of carrier multiplication gives rise to a sharp upturn in the output current at output voltage approaching $V_{CE0}$. The computed characteristics taking into account both effects compare very favorably with the observed characteristics presented in Fig. 11.5(d).

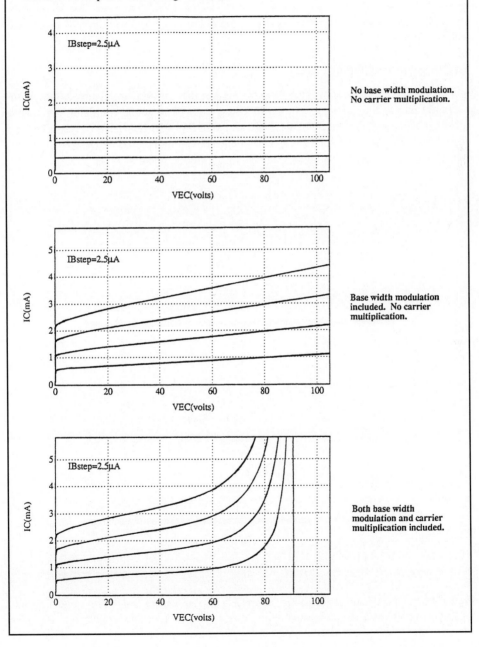

## 11.2.5 Geometrical Effects

Up to this point in the analysis we have assumed the bipolar transistor to be "one-dimensional," with all currents primarily restricted to flowing in one direction. From the cross sections of actual BJTs presented in Fig. 10.6, it is obvious the current flow patterns must be decidedly more complex. This leads to several deviations from the ideal of a geometrical nature. Included under "geometrical effects" are internal voltage drops directly associated with the three-dimensional character of the current.

### Emitter Area ≠ Collector Area

In the ideal model, carriers injected from the emitter into the base are assumed to travel in a straight line to the collector. The current flow path in an actual transistor can have a considerable lateral component. This is illustrated in Fig. 11.11(a) using the cross section of a discrete planar transistor. The spread in current can be approximately taken into account by employing different effective areas for the emitter and collector.

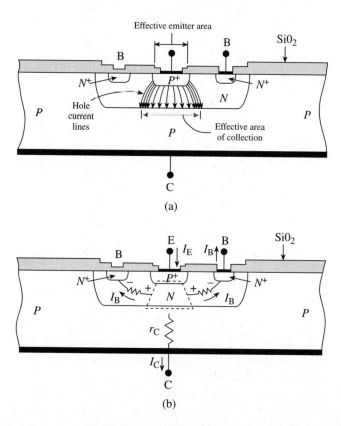

**Figure 11.11**  Illustration of geometrical effects in a discrete planar transistor. (a) Current spreading and crowding. (b) Base and collector series resistances. The dashed line region in (b) identifies the "intrinsic transistor," the "heart" of the device.

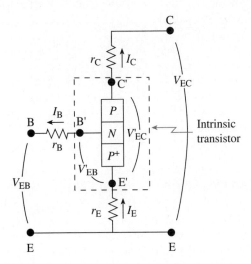

**Figure 11.12**  BJT equivalent circuit including the emitter ($r_E$), base ($r_B$), and the collector ($r_C$) series resistances.

## Series Resistances

The base current flowing between the base contact and the "heart" of the transistor (the dashed line region in Fig. 11.11b) must pass through a resistive bulk region. There is also a small resistance associated with the base contact itself. This means the voltage drop across the E–B junction is somewhat less than the $V_{EB}$ terminal voltage. Although typically quite small in magnitude, the voltage difference is often significant because the E–B junction is forward biased under active mode operation and the emitter current is an exponential function of the junction voltage. To account for the voltage difference, one introduces a base series resistance, $r_B$, equal to the sum of the bulk and contact resistances. If $r_B$ is the only series resistance of importance, then the $V_{EB}$ in previous theoretical relationships is simply replaced by $V'_{EB}$, where $V'_{EB}$ is computed from $V'_{EB} = V_{EB} - I_B r_B$.

In general, one may view previously derived relationships as only applying to the "heart" or central portion of the transistor, often called the "intrinsic transistor." As pictured in Fig. 11.12, individual series resistances connect the conceptual internal terminals of the intrinsic transistor to the external terminals of the physical transistor. In each case the series resistance is the sum of bulk and contact resistances. The voltages at the terminals of the intrinsic transistor are deduced from the applied voltages by accounting for the $Ir$ drop across the relevant series resistance(s). The values of the series resistances are usually deduced from subsidiary measurements.

## Current Crowding

We have already noted the base current flows laterally from the heart of the transistor to the base contact. It also flows laterally across the face of the emitter, giving rise to a voltage drop in the base from the center to the edges of the emitter. The voltage applied across the

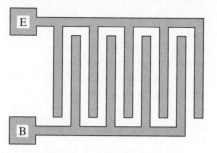

**Figure 11.13**   Interdigitated comb-like contact structure employed in power transistors to minimize current crowding. (Top view of the transistor.)

E–B junction and the associated injection current from the emitter to the collector are therefore greatest at the edges of the emitter. A higher density of current lines was placed near the emitter edges in Fig. 11.11(a) to illustrate the effect. This phenomenon, a larger current flow around the emitter periphery, is called *current crowding*. Current crowding and the onset of localized heating can become particularly detrimental at high current levels. To minimize the problem, power BJTs are constructed with emitters that have a large periphery-to-area ratio. This is achieved by connecting several narrow, long, emitter contacts into a comb-like structure and interdigitating the emitter contacts with base contacts as sketched in Fig. 11.13.

### 11.2.6 Recombination–Generation Current

Thermal recombination–generation was assumed to be negligible throughout the E–B and C–B depletion regions in the ideal transistor analysis. Contrary to the assumption and by analogy with the diode, one might expect to find significant R–G current components in a Si BJT maintained at room temperature. Indeed, under active mode operation with a *small* forward bias applied to the E–B junction, the R–G current from the E–B junction is likely to be the dominant component of the base current. This leads to a lower emitter efficiency[†] and a reduced gain at low current levels (small $V_{EB}$). Also, when the input is open circuited, the reverse-bias R–G current from the C–B junction adds to and is likely to dominate the observed output leakage current ($I_{CB0}$ or $I_{CE0}$). Unlike the diode, however, the ideal transistor characteristics are only nominally affected by the inclusion of the R–G current. Under most active-mode operating conditions, the input bias and current ($I_B$ or $I_E$) are such that the diffusion component dominates the input current. Moreover, at other than low input current levels, the output current extracted from the reverse-biased C–B junction is comprised primarily of the injected (diffusion dominated) current from the emitter.

---

[†] The emitter efficiency in a *pnp* BJT becomes $\gamma = I_{Ep}/(I_{Ep} + I_{En} + I_{R-G})$ when the R–G current is included.

## 11.2.7 Graded Base

A short description of transistor fabrication employing successive diffusions to form the base and emitter was presented in Section 10.2. Each of the diffusions gives rise to a dopant distribution that falls off more or less exponentially with distance into the semiconductor. A rough sketch of the resultant emitter and base doping profiles following the double diffusion process is shown in Fig. 11.14. The point we wish to make is that the doping within the base is not a constant as assumed in the ideal transistor analysis. Rather, the base doping is typically a strong function of position, decreasing from a maximum at the E–B junction to a minimum at the C–B junction. The base is said to be "graded" or to possess a graded (non-constant) doping profile.

When treating semiconductor fundamentals, we established the fact that a doping variation with position will give rise to a built-in electric field. It follows there must be a built-in electric field throughout the quasineutral width of the graded base in a double-diffused transistor. Seeking additional information about the built-in electric field, let $N_B(x)$ be the donor doping in the base of a *pnp* BJT. By definition $\rho = q(p - n + N_B) \cong$ 0, and therefore $n(x) \cong N_B(x)$, within the quasineutral portion ($0 \leq x \leq W$) of the base. Moreover, under equilibrium conditions

$$J_N = q\mu_n n \mathscr{E} + qD_N \frac{dn}{dx} = 0 \tag{11.55}$$

which when solved for $\mathscr{E}$ yields

$$\mathscr{E}(x) = -\frac{D_N}{\mu_n} \frac{dn/dx}{n} \cong -\frac{kT}{q} \frac{dN_B(x)/dx}{N_B(x)} \quad \ldots 0 \leq x \leq W \tag{11.56}$$

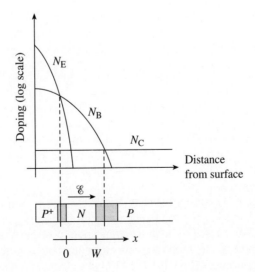

**Figure 11.14**   Sketch of the doping profiles inside a double-diffused transistor.

Assuming the doping profile can be approximated by a simple exponential, we can write

$$N_B(x) = N_B(0)e^{-x/x_{diff}} \tag{11.57}$$

and

$$\frac{dN_B(x)}{dx} = -\frac{N_B(0)}{x_{diff}}e^{-x/x_{diff}} = -\frac{N_B(x)}{x_{diff}} \tag{11.58}$$

Finally, substituting into Eq. (11.56), we arrive at the result

$$\mathscr{E} = \frac{kT/q}{x_{diff}} \quad \ldots 0 \le x \le W \tag{11.59}$$

The exponential doping variation gives rise to a constant electric field in the quasineutral region. The magnitude of the built-in field can be quite substantial. If, for example, the doping decreases by two orders of magnitude in spanning a base width of $W_B = 2\ \mu m$, then $x_{diff} = 4.34 \times 10^{-5}$ cm and $\mathscr{E} \cong 600$ V/cm at room temperature. More importantly, however, the field is directed so as to *assist* the transport of minority carriers across the quasineutral width of the base. This holds for both the emitter-injected holes in a *pnp* transistor and the emitter-injected electrons in an *npn* transistor. ($\mathscr{E} < 0$ in an *npn* transistor.) The drift-enhanced transport of carriers across the base decreases the transit time and thereby reduces recombination in the base. The reduced recombination in turn increases the transport factor and the current gains above the values computed from the ideal theory. The built-in electric field and attendant decrease in the carrier transit time also lead to an improved high-frequency response. The graded doping in the base is one deviation from the ideal that is clearly beneficial.

## 11.2.8  Figures of Merit

We conclude the section with a brief examination of special "figures of merit." These are plots that can be quickly interpreted to determine the extent of deviations from the ideal and other information about the transistor under test. Research papers on new or revised BJT structures invariably contain one or both of the figures to be examined.

The first, called a *Gummel plot* and illustrated in Fig. 11.15, is a simultaneous semilog plot of $I_B$ and $I_C$ as a function of the input voltage $V_{EB}$. The output voltage is set at some convenient value, typically a few volts, during the measurement. Ideally, both the $I_B$ and $I_C$ plots should be straight lines with the same $q/kT$ slope. In reality, as discussed in Subsection 11.2.6, the R–G current typically contributes to the observed base current at small $V_{EB}$ biases. At large $V_{EB}$ biases the collector current plot begins to slope over. The slope-over is most likely caused by high-level injection, possibly aggravated by current crowding. Paralleling the diode, series resistance is a probable contributing cause if a simultaneous slope-over occurs in both the $I_B$ and $I_C$ plots. Note that, assuming $I_{CE0}$ is negligible, the ratio of $I_C$ to $I_B$ at a given $V_{EB}$ yields $\beta_{dc}$ at the chosen operating point. The

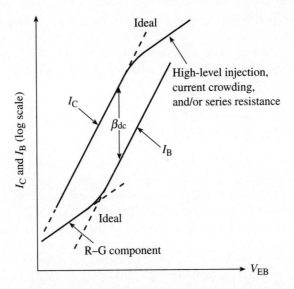

**Figure 11.15** Sketch illustrating the usual form and interpretation of the Gummel plot.

deduced $\beta_{dc}$ is ideally a constant independent of $V_{EB}$. Due to nonidealities, however, the $I_C$ to $I_B$ ratio clearly decreases at both low and high values of $V_{EB}$.

The second figure of merit is a log-log plot of $\beta_{dc}$ versus $I_C$ with the output voltage held constant. A sample plot derived from experimental data is presented in Fig. 11.16.

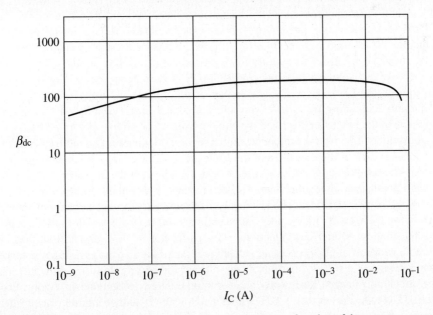

**Figure 11.16** Measured common emitter d.c. current gain as a function of the output current. Data derived from a 2N2605 *pnp* BJT. $V_{EC} = 5$ V.

Truthfully, this plot is merely a continuous record of the same $\beta_{dc}$ information that can be deduced in a point-by-point fashion from the Gummel plot. Again $\beta_{dc}$ is seen to fall off at low $I_C$ (small $V_{EB}$) values because of the increasing importance of $I_{R-G}$. At high $I_C$ (large $V_{EB}$) values the decrease in gain is caused by high-level injection, current crowding, series resistance, or a combination of the listed effects.

## 11.3  MODERN BJT STRUCTURES

This section is primarily intended to expose the reader to recent and ongoing developments related to the continued evolution of the bipolar junction transistor. The discussion of "modern" BJT structures also provides an opportunity to make use of the previously established concepts and information, particularly those associated with nonidealities. The structures to be examined are the polysilicon emitter BJT and the heterojunction bipolar transistor (HBT). The former structure is presently employed in a number of integrated circuits, including the most recent versions of personal computer CPUs.[11] The compound semiconductor HBT is projected for initial use in high-frequency/high-speed applications.[12]

### 11.3.1 Polysilicon Emitter BJT

Complementary MOS or CMOS is presently the preferred large-scale IC technology because of its low power dissipation and high packing density. [Chapters 16–18 are devoted to MOS (metal-oxide-semiconductor) fundamentals and devices.] A BJT, however, can supply several times the drive current of an equal size MOS transistor. A larger drive current translates into the faster switching of large capacitive loads. To take advantage of this fact, modified IC designs have been introduced where bipolar transistors are incorporated into the high-load speed-sensitive portions of predominantly CMOS circuitry. The hybrid bipolar-CMOS technology is referred to as BiCMOS. An idealized cross section of the *npn* BJT formed by the Intel 0.8 $\mu$m BiCMOS process is reproduced in Fig. 11.17.

Carefully examining Fig. 11.17, you will note that a polycrystalline Si film lies between the metallic emitter contact and the silicon proper. The emitter is formed by depositing a polysilicon film over the intended emitter area, $n^+$ doping the film with arsenic, and then diffusing some of the dopant from the film a short distance into the underlying crystalline Si. The resulting emitter configuration, a shallow layer in the Si proper contacted by a similarly doped polysilicon thin film, has been named a polysilicon emitter.

Polysilicon emitters offer several advantages over emitters formed in the usual manner. From a fabrication standpoint the polysilicon emitter contacts are especially suited for producing the shallow emitter/base junctions required in modern ICs. In addition, they are compatible with self-aligned processing techniques that minimize the parasitic resistances and capacitances of a BJT. Of particular relevance to the present discussion, there is also an operational advantage. Specifically, the common emitter current gain of a polysilicon emitter BJT is much larger than a BJT with a shallow all-crystalline emitter of equivalent thickness.

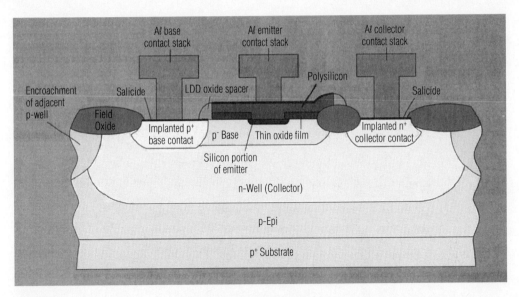

**Figure 11.17**   Idealized cross section of the triple-diffused *npn* polyemitter BJT produced by the Intel 0.8 $\mu$m BiCMOS process. (From Gupta and Bohr[11]. Reprinted with permission.)

Before pursuing the origin of the cited polysilicon emitter gain enhancement, we need to explain that the emitters in modern BJTs are routinely made very thin to achieve high operating speeds. Unfortunately, reducing the emitter thickness ultimately leads to a reduction in gain. In high-performance BJTs the quasineutral width of the emitter ($W_E$) can be less than a minority carrier diffusion length ($L_E$). If $W_E/L_E \ll 1$ in a standard all-crystalline emitter, recombination may be neglected in the emitter and, like in the base, the minority carrier distribution becomes a linear function of position as pictured in Fig. 11.18(a). Associated with the modified minority carrier distribution in the emitter, there is a significant increase in the base-to-emitter injection current under active mode biasing. The emitter efficiency therefore decreases and the gain likewise decreases. Quantitatively, as verified in Problem 11.11, $W_E \ll L_E$ simply replaces $L_E$ in the Eq. (11.41)–(11.44) performance parameter relationships.

The use of a polysilicon emitter compensates in part for the decrease in gain due to the shallow nature of the emitter. Although the actual situation is rather complex, and different theories apply depending on the precise nature of the polysilicon/Si interface[13], to first order the polysilicon portion of the emitter may be modeled as merely low-mobility silicon. A lower mobility in the polysilicon is reasonable, since one would expect additional carrier scattering from the grain boundaries, the disordered regions between the small crystals or "grains" in the polycrystalline film. Assuming the quasineutral widths of both the polycrystalline and crystalline portions of the emitter are much smaller than the respective minority carrier diffusion lengths, recombination will be negligible throughout the emitter, and the minority carrier distributions in both the polycrystalline and crystalline portions of

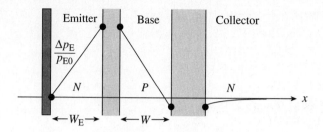

(a) All-crystalline shallow emitter

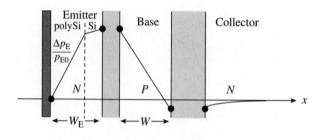

(b) Polysilicon emitter

**Figure 11.18**  Normalized minority carrier distribution under active mode biasing in (a) an all-crystalline emitter BJT and (b) a polysilicon emitter BJT. Quasineutral widths in the base and emitter are assumed to be much smaller than the respective minority carrier diffusion lengths.

the emitter will be linear functions of position. At the polysilicon/Si interface in an *npn* BJT, the minority carrier hole concentration (or equivalently $\Delta p_E$) and the hole current must be continuous. With different mobilities in the two regions, the continuity of the hole current in turn requires

$$qD_{E1}\frac{d\Delta p_{E1}}{dx} = qD_{E2}\frac{d\Delta p_{E2}}{dx} \qquad \ldots \text{ 1–polysilicon; 2–Si} \qquad (11.60)$$

or

$$\frac{d\Delta p_{E1}}{dx} = \frac{D_{E2}}{D_{E1}}\frac{d\Delta p_{E2}}{dx} = \frac{\mu_{E2}}{\mu_{E1}}\frac{d\Delta p_{E2}}{dx} \qquad (11.61)$$

Equation (11.61) indicates the linear slope of the minority carrier concentration must be greater in the polysilicon portion of the emitter as illustrated in Fig. 11.18(b). Comparing Figs. 11.18(a) and (b), we note that the net result of employing a polysilicon emitter is a reduced $\Delta p_E$ slope at the emitter edge of the E–B depletion region, and thus a smaller current back-injected from the base into the emitter. A smaller base current of course signifies improved emitter efficiency and enhanced gain.

## 11.3.2 Heterojunction Bipolar Transistor (HBT)

In Section 11.2 we noted several deviations from the ideal that were associated one way or another with the base doping. Specifically, base-width modulation, the extrinsic base resistance ($r_B$), and the voltage drop across the face of the emitter leading to current crowding are all decreased by increasing the base doping. Unfortunately, increasing the base doping also decreases the emitter efficiency, which in turn decreases the gain of a standard transistor. There is an unavoidable trade-off between gain and the noted deviations from the ideal in designing a standard bipolar transistor.

The situation is different in a heterojunction bipolar transistor where, as a general rule, the base is the most heavily doped region. A *heterojunction* is, in general, a junction between two dissimilar materials. However, when applied to the HBT, it is understood to be a junction between two dissimilar semiconductors. The emitter in an HBT is made from a wider band gap semiconductor than the base. In most semiconductor combinations this sets up a built-in barrier impeding forward-bias injection from the base to the emitter and giving rise to a high injection efficiency even when $N_E \ll N_B$. Although being touted here as a "modern" BJT structure, the heterojunction idea was actually described by W. Shockley in the initial patent on the bipolar junction transistor filed in 1948! The successful fabrication of HBT structures only became possible, however, with the more recent development of sophisticated semiconductor film growth techniques such as molecular beam epitaxy (MBE)[14].

A key issue in dealing with heterojunctions is the band alignment. *Band alignment* is how the respective conduction band edges and valence band edges of the two semiconductors line up relative to each other at the interface between the two materials. Although reasonable estimates are possible from theoretical considerations, highly accurate band alignments for systems of interest are usually determined from experimental observations. Example alignments for some of the most extensively researched systems are pictured in Fig. 11.19. It should be emphasized that the Fig. 11.19 diagrams are not true heterojunction band diagrams, but merely informational diagrams drawn assuming charge neutrality prevails throughout the semiconductors. Band bending in the component materials exists adjacent to an actual heterojunction. We should also mention that in each of the pictured cases the component semiconductors have essentially identical lattice constants. Employing two semiconductors with the same lattice dimensions allows the heterojunction to be formed with a minimum of interfacial and bulk defects. Almost all HBTs have been made utilizing lattice-matched systems.†

---

† An important exception is the $Si/Si_{1-x}Ge_x/Si$ HBT that employs a *pseudomorphic* $Si_{1-x}Ge_x$ layer for the base. Pseudomorphic layers are thin films (typically $< 1000$ Å) that conform to the lattice pattern of the underlying substrate. Such layers must be kept thin because, if grown thicker than a certain critical value, the stressed film relaxes, creating numerous defects and a lattice constant different from that of the substrate. For information about the current status and recent applications of Si/SiGe HBTs, see D. L. Harame et al., "Si/SiGe Epitaxial-Base Transistors," IEEE Trans. on Electron Devices, **42**, 455–482 (March 1995).

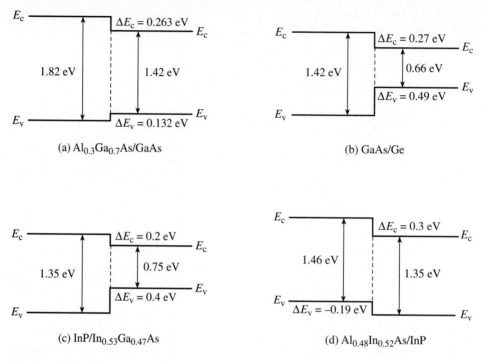

**Figure 11.19**   Heterojunction band alignments for the (a) $Al_{0.3}Ga_{0.7}As/GaAs$, (b) $GaAs/Ge$, (c) $InP/In_{0.53}Ga_{0.47}As$, and (d) $Al_{0.48}In_{0.52}As/InP$ lattice-matched systems. (Data from Tiwari and Frank[15].)

AlGaAs/GaAs is the most extensively researched and technologically advanced of the heterojunction systems. The idealized cross section of an $Al_{0.3}Ga_{0.7}As/GaAs$ *Npn* HBT is shown in Fig. 11.20. (By convention, an uppercase letter, the *N* in *Npn,* is used to identify the wider band gap side of a heterojunction.) The required $n^+$-GaAs subcollector, *n*-GaAs collector, $p^+$-GaAs base, and $N$-$Al_{0.3}Ga_{0.7}As$ emitter films are deposited in succession on a semi-insulating (S.I.) GaAs substrate. The S.I. GaAs substrate and the S.I. sidewalls, the latter formed by deep proton implantation into the deposited layers, dielectrically isolate the HBT from other devices on the wafer. The structure is completed by the appropriate etching of layers and the deposition of metal contacts.

If the E–B heterojunction is assumed to be abrupt, the $Al_{0.3}Ga_{0.7}As/GaAs$ *Npn* HBT can be modeled to first order by the equilibrium energy band diagram in Fig. 11.21(a). Except for the wider emitter band gap and the band discontinuities at the E–B interface, the abrupt-junction HBT diagram is seen to be nearly identical to the energy band diagram of an equivalently doped BJT. Naturally, since $N_E \ll N_B \gg N_C$, the E–B and C–B depletion regions lie almost exclusively in the emitter and collector, respectively. Although modern deposition techniques are capable of growing heterojunctions that are abrupt on an atomic scale, it is very common to enhance device performance by intentionally growing a compositionally graded junction; i.e., the composition is slowly changed from GaAs to

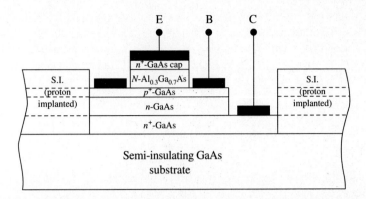

**Figure 11.20**    Idealized cross section of an $Al_{0.3}Ga_{0.7}As/GaAs$ *Npn* HBT.

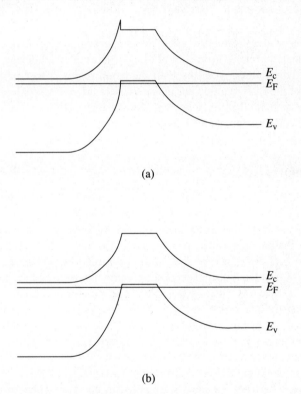

**Figure 11.21**    Equilibrium energy band diagram for the $Al_{0.3}Ga_{0.7}As/GaAs$ *Npn* HBT assuming the E–B heterojunction is (a) abrupt and (b) compositionally graded.

$Al_{0.3}Ga_{0.7}As$ over a distance of a few hundred angstroms. This eliminates the sharp band discontinuities and gives rise to the modified energy band diagram in Fig. 11.21(b).

The basic operational analysis of an HBT with a compositionally graded E–B junction is essentially identical to that of a standard BJT. In fact, the BJT analytical results may be taken over directly if appropriate modifications are made to account for the different $n_i$ values of the HBT emitter and base. Specifically, Eq. (11.31) and later relationships were written in a simplified form by noting $n_{E0}/p_{B0} = N_B/N_E$ in a standard BJT. In an HBT, however, $n_{E0}/p_{B0} = (n_{iE}^2/N_E)/(n_{iB}^2/N_B)$; the $N_B/N_E$ in BJT relationships must be replaced by $(N_B/N_E)(n_{iE}^2/n_{iB}^2)$. Assuming $W \ll L_B$, the emitter efficiency of the graded HBT then becomes

$$\gamma = \frac{1}{1 + \dfrac{D_E}{D_B}\dfrac{W}{L_E}\dfrac{N_B}{N_E}\dfrac{n_{iE}^2}{n_{iB}^2}} \tag{11.62}$$

The $N_E \ll N_B$ in an HBT is offset by the fact that $n_{iB}$ of the narrow gap base is much greater than $n_{iE}$ of the wide gap emitter. Even though $N_B/N_E \sim 100$, for example, the $n_{iE}^2/n_{iB}^2$ ratio in a typical $Al_{0.3}Ga_{0.7}As/GaAs$ HBT is $\sim 10^{-5}$, yielding a theoretical emitter efficiency very close to unity.

As a practical matter it should be noted that emitter efficiencies in actual HBTs tend to be below the value computed from Eq. (11.62). The emitter efficiency is lowered by a significant $I_{R-G}$ contribution from the E–B depletion region, particularly from exposed surface regions of the E–B junction periphery. Proper surface passivation in GaAs devices poses more of a problem than in Si devices. Nevertheless, HBTs are now produced with very respectable gains and unmatched high-frequency performance.

## 11.4 SUMMARY

In this chapter we took on the task of modeling the steady-state response of the bipolar junction transistor. Initially permitting the quasineutral width of the base to be arbitrary, an ideal transistor analysis paralleling the ideal diode analysis was performed to obtain general relationships for the minority carrier distribution in the base (Eq. 11.26), the performance parameters (Eqs. 11.31–11.34), and the terminal currents (Eqs. 11.35 and 11.36). Subsequently the generalized relationships were simplified, making use of the fact that the quasineutral width of the base is typically much less than a minority carrier diffusion length in a well-made transistor. The resulting expression for the minority carrier distribution in the base (Eq. 11.40) was noted to be a linear function of position. The performance parameter relationships appropriate for a well-made transistor were recorded as Eqs. (11.41)–(11.44). Simplified handling of the expressions for the terminal currents was accomplished by introducing the Ebers–Moll model (Fig. 11.3) and equations (Eqs. 11.47).

Several deviations from the ideal were noted in comparing theory and experiment. Bias-dependent variation of the quasineutral width of the base, referred to as base width modulation, causes a number of deviations including the quasilinear upward slant of the common emitter output characteristics. Punch-through (total depletion of the base) or avalanche breakdown of the C–B junction, whichever occurs first, limits the maximum output voltage that can be applied to the transistor under active-mode common-base operation. As far as active-mode common-emitter operation is concerned, carrier multiplication and an internal feedback mechanism can combine to make $V_{CE0} < V_{CB0}$. Other cited phenomena leading to deviations from the ideal include the series resistance external to the "intrinsic" transistor, the voltage drop in the base across the face of the emitter giving rise to current crowding, recombination–generation in the E–B and C–B depletion regions, and the graded doping of the base. Special note was also made of the Gummel and $\beta_{dc}$ versus $I_C$ plots that are often presented as a graphical measure of transistor performance.

Finally, the section on modern BJT structures served the dual purpose of introducing special transistor structures, the polysilicon emitter BJT and the HBT, while simultaneously providing sample practical use of previously established concepts and information.

## PROBLEMS

| CHAPTER 11 PROBLEM INFORMATION TABLE | | | | |
|---|---|---|---|---|
| *Problem* | *Complete After* | *Difficulty Level* | *Suggested Point Weighting* | *Short Description* |
| 11.1 | 11.4 | 1 | 20 (1 each part except b-2, c-3, j-2, l-2) | Quick quiz |
| 11.2 | 11.1.1 | 2 | 10 (2 each part) | Deduce info given *I*-plot |
| 11.3 | 11.1.2 | 3 | 18 (1/2 each equation) | *npn* equation modifications |
| ● 11.4 | 11.1.3 | 3 | 17 (1 each answer) | Current and parameter calc. |
| ● 11.5 | " | 2 | 10 | $\Delta p_B(x)$ plot, $W/L_B$ varied |
| ● 11.6 | " | 2 | 10 | $\Delta p_B(x)$ plots for four modes |
| 11.7 | " | 1 | 12 (3 each part) | Sketch carrier distributions |
| 11.8 | " | 2 | 10 (a-1, b-1, c-2, d-1, e-5) | Deduce info from dist. plots |
| 11.9 | " | 2–3 | 12 ($\gamma$-5, $\alpha_T$-5, $\beta_{dc}$-2) | Parametric effect table |
| ● 11.10 | " | 3 | 12 (3 each plot) | $\beta_{dc}$ vs. $\tau_B$, $N_B$, $\tau_E$, $N_E$ plots |
| 11.11 | " | 3 | 12 (a-4, b::e-2) | Shallow emitter |
| 11.12 | 11.1.4 | 3–4 | 14/"diode" (a-6, b::e-2) | "Diodes" |
| 11.13 | " | 3 | 10 (5 each part) | Apply Ebers–Moll circuit |
| 11.14 | " | 3–4 | 25 (a-5, b-10, c-10) | Derive computational eqs. |
| 11.15 | 11.2.2 | 3–4 | 15 (a-10, b-5) | Emitter input *W*-sensitivity |
| 11.16 | 11.2.3 | 2 | 8 | Punch-through calculation |
| 11.17 | 11.2.4 | 2 | 8 (a-3, b-3, c-2) | Compare two transistors |
| 11.18 | " | 2 | 10 | Determine Early voltage |
| ● 11.19 | " | 3–4 | 20/"diode" (a-10, b-10) | Plot "diode" $I$–$V_A$ |
| ● 11.20 | 11.2.8 | 2 | 12 (a-10, b-2) | Gummel and gain plots |
| 11.21 | " | 3 | 10 | $V_{EC}$ dependence of gain plot |

**11.1** Quick Quiz

(a) In the ideal transistor analysis, what assumptions are made about the regional (E, B, C) doping concentrations and quasineutral widths?

(b) Roughly sketch the minority carrier distribution in the base of a *pnp* BJT under active mode biasing when $W/L_B \ll 1$ and $W/L_B \gg 1$.

(c) A *pnp* BJT with $W/L_B \ll 1$ is active mode biased. Given $\Delta p_{Bmax} = 10 p_{B0}$, make sketches of $\Delta p_B(x)$, $\Delta p_B(x)/p_{B0}$, and $p_B(x)$ in the base.

(d) What are typical sizes for $\gamma$, $\alpha_T$, $\alpha_{dc}$, and $\beta_{dc}$?

(e) Establish a general expression for $I_B = I_B(V_{EB}, V_{CB})$ in terms of the Ebers–Moll parameters.

(f) $\alpha_F = 0.9944$, $\alpha_R = 0.4286$, $I_{F0} = 4.749 \times 10^{-15}$ A. Determine $I_{R0}$.

(g) Referring to Fig. 11.5(d), what is the cause of the quasilinear slope in the characteristics at the lower $V_{EC}$ values? What is the cause of the sharp upward curvature in the characteristics as $V_{EC} \rightarrow V_{CE0}$?

(h) It is established that $V_{CB0}$ of a BJT is limited by punch-through. Nevertheless, a student claims $V_{CE0} < V_{CB0}$ in the given device. Is this possible? Explain.

(i) What is the cause of "current crowding"?

(j) Cite the beneficial effects associated with a graded base.

(k) What is a Gummel plot?

(l) What is the cause of the decrease in $\beta_{dc}$ at low and high $I_C$ values as exemplified by Fig. 11.16?

(m) Would there be any advantage to constructing a Si BJT where the entire emitter is made of polysilicon? Explain.

(n) Define "heterojunction."

(o) What are the differences between a BJT and an HBT?

**11.2** The electron and hole currents inside a *pnp* BJT biased in the active mode are plotted in Fig. P11.2. All the currents are referenced to $I_1$, the hole current injected into the base. Determine:

(a) The emitter efficiency ($\gamma$);

(b) The base transport factor ($\alpha_T$);

(c) The common emitter d.c. current gain ($\beta_{dc}$);

(d) The base current ($I_B$).

(e) For the given transistor, is the recombination–generation current arising from the depletion regions negligible as assumed in the ideal transistor analysis? Explain.

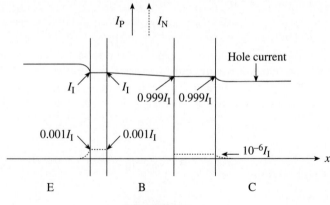

**Figure P11.2**

**11.3** Indicate the required revisions to the equations in Subsections 11.1.1 and 11.1.2 if the device under analysis is taken to be an *npn* BJT.

• **11.4** Sample computations (preferably computer-aided) are to be performed to gauge the magnitude of the BJT current components and the size of the performance parameters. Assume room temperature operation, a *pnp* BJT, active mode biasing with $V_{EB} = 0.7$ V and $V_{CB} = -5$ V, $W \cong W_B$, $A = 10^{-4}$ cm$^2$, and material parameters as listed in Table 11.1 of Subsection 11.1.3. Employing the relationships developed in the ideal transistor analysis (Subsections 11.1.1 and 11.1.2), calculate:

(a)  $\cosh(W/L_B)$, $\sinh(W/L_B)$, and $W/L_B$ (all to five significant figures);

(b)  $I_{Ep}$, $I_{En}$, $I_{Cp}$, and $I_{Cn}$;

(c)  $I_E$, $I_C$, and $I_B$;

(d)  $I_{B1}$, $I_{B2}$, and $I_{B3}$ as defined in Subsection 10.4;

(e)  $\gamma$, $\alpha_T$, $\alpha_{dc}$, and $\beta_{dc}$.

• **11.5** Examine how the minority carrier distribution in the base of a *pnp* BJT varies with the $W/L_B$ ratio. Taking $V_{CB} = 0$ so that $\Delta p_B(W) = 0$, construct a multicurve plot of $\Delta p_B(x)/\Delta p_B(0)$ versus $x/W$ corresponding to $W/L_B = 10, 5, 1, 0.5$, and 0.1. Note that $0 \le \Delta p_B(x)/\Delta p_B(0) \le 1$ and $0 \le x/W \le 1$. Comment on the resultant plot.

• **11.6** Graphically illustrate how the minority carrier distribution in the base of a BJT varies with the biasing mode. Construct four individual "demonstration" plots (subplots in MATLAB) of $\Delta p_B(x)/p_{B0}$ versus $x/W$ corresponding respectively to each of the four biasing modes. For the purposes of the illustration, assume $W/L_B \ll 1$ and restrict $\Delta p_B(0)/p_{B0}$ and $\Delta p_B(W)/p_{B0}$ to be $\le 10$. For axis scaling employ $x_{min} = 0$, $x_{max} = 1$, $y_{min} = -2$, and $y_{max} = 10$. Appropriately label the four plots.

**11.7** The equilibrium majority and minority carrier concentrations in the quasineutral regions of a BJT are shown as dashed lines in Figs. P11.7(a) and P11.7(b), respectively. These figures are intended to be *linear* plots, with a break in the x-axis depletion regions to accommodate the different depletion widths associated with the various biasing modes. Note that the carrier concentrations in the three transistor regions are not drawn to the proper relative scale, but only qualitatively reflect the fact that $N_E \gg N_B > N_C$. Employing solid lines and remembering the figures are intended to be linear plots, sketch the majority and minority carrier distributions in the respective quasineutral regions of the $W \ll L_B$ transistor under

(a)  active mode biasing;

(b)  inverted mode biasing;

(c)  saturation biasing;

(d)  cutoff biasing.

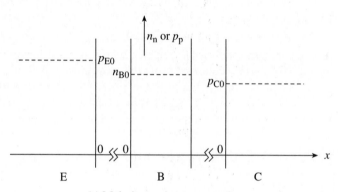

(a) Majority carrier concentrations

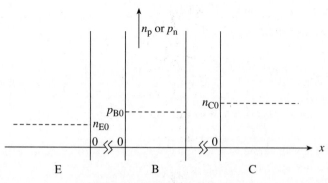

(b) Minority carrier concentrations

**Figure P11.7**

**11.8** The $\Delta n_E/n_{E0}$, $\Delta p_B/p_{B0}$, and $\Delta n_C/n_{C0}$ distributions in the quasineutral regions of a *pnp* BJT are as sketched in Fig. P11.8. Determine:

(a) The polarity of $V_{EB}$;

(b) The polarity of $V_{CB}$;

(c) The magnitude of $V_{CB}$;

(d) The biasing mode;

(e) Repeat parts (a)–(d) if the device is an *npn* BJT.

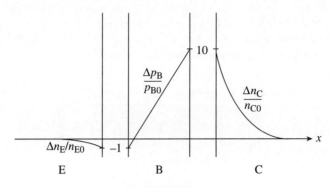

**Figure P11.8**

**11.9** Complete the table below by indicating whether the noted change in a BJT device parameter *increases, decreases,* or has *no effect* on the listed performance parameters.

| Change | Effect on $\gamma$ | Effect on $\alpha_T$ | Effect on $\beta_{dc}$ |
|---|---|---|---|
| Increase $W_B$ | | | |
| Increase $\tau_B$ | | | |
| Increase $N_B$ | | | |
| Increase $\tau_E$ | | | |
| Increase $N_E$ | | | |

● **11.10** Let the lifetimes and dopings specified in Table 11.1 be the reference values $\tau_{B0}$, $N_{B0}$, and so on. Making use of the mobility fit relationships quoted in Exercise 3.1, construct four individual log-log plots (subplots in MATLAB) of $\beta_{dc}$ versus $\tau_B/\tau_{B0}$, $N_B/N_{B0}$, $\tau_E/\tau_{E0}$, and $N_E/N_{E0}$. In each case vary the independent variable from 0.1 to 10. Assume $T = 300$ K and $W = W_B = 2$ $\mu$m in all computations. Neglect the fact that the emitter is degenerately doped over a portion of the specified range. Do your results here agree with the answers to Problem 11.9?

**11.11** The emitter in modern BJTs is often made very thin to achieve high operating speeds. In this problem we examine the effect of employing a "shallow" emitter on the performance parameters. Consider the *pnp* BJT pictured in Fig. P11.11 where, like the base, the emitter is of finite width. Let $W_E$ be the quasineutral width of the emitter and assume $\Delta n_E = 0$ at the metallic emitter contact ($x'' = W_E$).

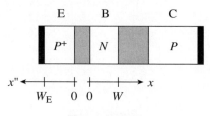

**Figure P11.11**

(a) Paralleling the ideal transistor analysis, obtain revised expressions for $\Delta n_E$ and $I_{En}$.

(b) Establish revised expressions for the performance parameters analogous to Eqs. (11.31)–(11.34).

(c) Establish the revised expressions for the performance parameters when $W/L_B \ll 1$ and $W_E/L_E \ll 1$.

(d) Referring to the part (b) and (c) answers, how are $\gamma$ and $\beta_{dc}$ affected as $W_E$ is systematically decreased?

(e) Make a sketch similar to Fig. 11.2 showing the minority carrier distribution in a shallow ($W_E/L_E \ll 1$) emitter under active mode biasing.

**11.12** When one of the transistor terminals is left floating or two of the terminals are connected together, the transistor becomes a diode-like two-terminal device. The six possible "diode" connections are pictured in Fig. P11.12. Answer the questions that follow after choosing a specific connection for analysis. It is suggested that at least one open-circuit connection (a, c, or e) and one short-circuit connection (b, d, or f) be analyzed.

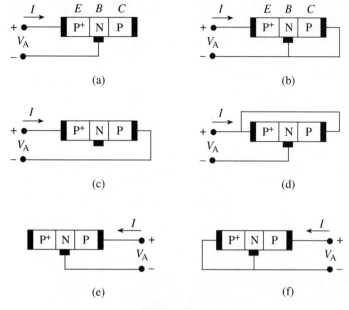

**Figure P11.12**

(a) Derive the $I$–$V_A$ relationship for the "diode" by appropriately manipulating and combining the Ebers–Moll equations. $I$ should be expressed only in terms of $V_A$ and the Ebers–Moll parameters.

(b) Develop expressions for $\Delta p_B(0)/p_{B0}$ and $\Delta p_B(W)/p_{B0}$ in terms of $V_A$ and the Ebers–Moll parameters.

(c) Simplify the part (b) results by setting $\alpha_F = \alpha_R = \alpha$ and $I_{F0} = I_{R0} = I_0$. (The simplifications here would be valid for a transistor where the material parameters of the emitter and collector are identical.)

(d) Utilizing the part (c) relationships and assuming $W \ll L_B$, sketch the minority carrier distribution in the base of the transistor if $V_A \gg kT/q$.

(e) Repeat part (d) for a reverse bias where $-V_A \gg kT/q$.

**11.13** The common emitter output characteristics of a *pnp* BJT for small values of $V_{EC}$ are pictured in Fig. P11.13. The d.c. operating point of the transistor lies *at the boundary between active mode and saturation mode biasing* as illustrated in the figure.

(a) Referring to Fig. 11.3 in Subsection 11.1.4, draw the simplified large signal equivalent circuit for the transistor at the given operating point.

(b) Employing the simplified equivalent circuit of part (a), or working directly with the Ebers–Moll equations, obtain an expression for $V_{EC}$ at the specified operating point. Your answer should be in terms of $I_B$ and the Ebers–Moll parameters.

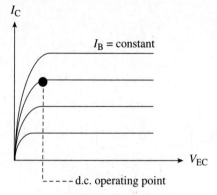

**Figure P11.13**

**11.14** The computational equations for the BJT characteristics quoted in Exercise 11.7 were established by appropriately combining and/or rearranging the Ebers–Moll equations. Perform the algebraic manipulations necessary to derive the quoted equation for

(a) The common base output $[I_C = I_C(V_{CB}, I_E)]$;

(b) The common emitter input $[I_B = I_B(V_{EB}, V_{EC})]$;

(c) The common emitter output $[I_C = I_C(V_{EC}, I_B)]$.

**11.15** In modern high-performance transistors, the gain is usually limited by the emitter efficiency, which implies $D_E N_B W / D_B N_E L_E \gg (1/2)(W/L_B)^2$.

(a) Show that the common emitter input characteristics of such devices are expected to be relatively insensitive to base width modulation.

(b) Sketch the expected form of the common emitter input characteristics if the transistor gain is *not* limited by the emitter efficiency; i.e., $D_E N_B W / D_B N_E L_E \sim (1/2)(W/L_B)^2$.

**11.16** How is the Exercise 11.8 result modified if $V_{EB} = 0.7$ V?

**11.17** Two *pnp* BJTs are identical except that the emitter and collector region doping are interchanged as illustrated in Fig. P11.17.

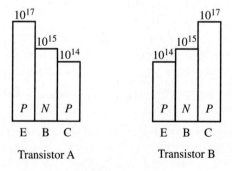

**Figure P11.17**

(a) Which transistor is expected to have the greater emitter efficiency? Explain.

(b) Which transistor will exhibit the greater sensitivity to base width modulation under active mode biasing? Explain.

(c) If limited by avalanche breakdown of the C–B junction, which transistor will exhibit the larger $V_{CB0}$? Explain.

**11.18** For $V_{EC} \gtrsim 20$ V the characteristics displayed in the middle figure of Exercise 11.10 can be closely modeled by a linear relationship of the form $I_C = B(V_{EC} - V_\varepsilon)I_B$, where $B$ and $V_\varepsilon$ are constants. Working with the middle figure (the "base width modulation included" figure), or preferably working with numerical values deduced from the BJTplus program itself, determine the values of $B$ and $V_\varepsilon$ required to fit each of the four $I_B > 0$ characteristics. Note that $V_\varepsilon$, referred to as the "Early voltage," is the voltage intercept that would result if the linear region of a characteristic is extrapolated into the $V_{EC}$ axis. Comment on your numerical results.

● **11.19** One or more $I$–$V_A$ relationships for diode-like operation of the transistor were established in answering Problem 11.12(a). Adapt the BJTplus and subsidiary programs of Exercise 11.10 to generate plots of the $I$–$V_A$ characteristics. The BJT0 program provided for Exercise 11.10 may be used without modification to input the device parameters for sample computations. If necessary for clarity, display the forward and reverse bias portions of the characteristics on separate plots.

(a) Initially construct the $I$–$V_A$ program(s) assuming $W = W_B$ and neglecting both base width modulation and carrier multiplication. Make a hardcopy of the resulting characteristics.

(b) Subsequently include both base width modulation and carrier multiplication. Be sure to take into account carrier multiplication in the E–B depletion region when a large reverse bias is applied across the E–B junction. Make a hardcopy of sample computations; compare the part (a) and (b) results.

● **11.20** Gummel and Gain Plots.
(a) Construct a theoretical Gummel plot similar to Fig. 11.15 employing the device parameters specified in the BJT0 subprogram of Exercise 11.10. (The parameters in the BJT0 subprogram provide a reasonable match to the 2N2605 *pnp* BJT characteristics displayed in Figs. 11.4 and 11.5.) Assume active mode biasing and a $V_{CB} < 0$ such that the $V_{CB}$ terms in Eqs. (11.47) can be neglected. Let the $I_C$ and $I_B = I_E - I_C$ versus $V_{EB}$ characteristics computed from Eqs. (11.47) be known as the ideal device characteristics. Approximately account for the recombination–generation current arising from the E–B depletion region by adding

$$I_{R\text{-}G} = I_{02}(e^{qV_{EB}/n_2 kT} - 1)$$

to the ideal-device base current. Use $I_{02} = 10^{-14}$ A and $n_2 = 1.5$ in calculating the corrected base current. To approximately account for high-level injection and other collector current nonidealities, multiply the ideal-device collector current by the correction factor

$$\Gamma = \frac{1}{1 + e^{q(V_{EB} - 0.75)/2kT}}$$

i.e., calculate the corrected $I_C$ employing $I_C = \Gamma I_C$(ideal). Show both the ideal and corrected characteristics on your Gummel plot making appropriate use of dashed and solid lines. Limit the plot output to $10^{-9}$ A $\leq I \leq 10^{-1}$ A and 0.3 V $\leq V_{EB} \leq 0.8$ V.

(b) Expand the part (a) program to additionally display a $\beta_{dc}$ versus $I_C$ or gain plot similar to Fig. 11.16. Use the MATLAB pause function to interrupt execution of the program between plots. Set the axes limits to obtain a plot covering the range $10^{-1} \leq \beta_{dc} \leq 10^3$ and $10^{-9}$ A $\leq I_C \leq 10^{-1}$ A.

**11.21** Increasing the $V_{EC}$ value employed in acquiring $\beta_{dc}$ versus $I_C$ data invariably leads to a small but perceptible increase in $\beta_{dc}$ over most of the $I_C$ range. Sample experimental data are presented in Fig. P11.21. Formulate and support an explanation as to the cause of the observed $V_{EC}$ dependence.

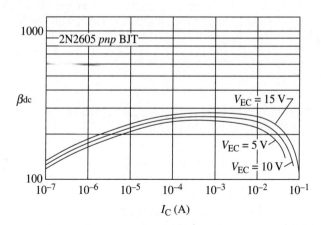

**Figure P11.21**

# 12 BJT Dynamic Response Modeling

Major applications of the BJT include use as a small-signal amplifier in wide-band or tuned circuits and use as a switch in digital logic circuits. In this chapter we examine the first-order modeling of the small-signal response and the transient (switching) response of the transistor. The small-signal response is routinely described or modeled in terms of a small-signal equivalent circuit. Small-signal equivalent circuits are employed, for example, to calculate amplifier signal gain, input impedances, and output impedances. In modern applications the desire for high gain and large bandwidth leads to the almost exclusive use of the BJT in the common emitter configuration. Consequently, the coverage herein emphasizes the development of equivalent circuits appropriate for the noted configuration. Relative to the transient response, digital electronics and other switching applications require the BJT to be switched rapidly between the cutoff "off"-state and the saturation "on"-state. An examination of the switching response and a simplified analysis of the observed time delays are presented in the second section of the chapter.

## 12.1 SMALL-SIGNAL EQUIVALENT CIRCUITS

### 12.1.1 Generalized Two-Port Model

Consider Fig. 12.1(a) where the BJT is assumed to be connected in the common emitter configuration and is modeled as a "black box" with two ports. Note that for the present analysis (and contrary to the *pnp* BJT analyses in Chapter 11), all currents are defined to be positive when flowing *into* a device terminal. Likewise, all voltages are referenced to the terminal common to the input and output. With the BJT assumed to be connected in the common emitter configuration, the d.c. input current, the d.c. output current, the d.c. input voltage, and the d.c. output voltage are respectively $I_B$, $I_C$, $V_{BE}$, and $V_{CE}$. The corresponding a.c. quantities are $i_b$, $i_c$, $v_{be}$, and $v_{ce}$. In the preceding chapter, it was established that $I_B$ could be written as an exclusive function of the applied d.c. voltages; that is $I_B = I_B(V_{BE}, V_{CE})$. When a.c. potentials $v_{be}$ and $v_{ce}$ are superimposed on the d.c. terminal voltages, the input current quite generally becomes $i_B(V_{BE} + v_{be}, V_{CE} + v_{ce}) = I_B(V_{BE}, V_{CE}) + i_b$. Similarly, it is readily established that the d.c. output current can also be expressed as an exclusive function of the d.c. input and output voltages; that is $I_C = I_C(V_{BE}, V_{CE})$. Thus, when $v_{be}$ and $v_{ce}$ are respectively added to $V_{BE}$ and $V_{CE}$, the total output current analogously becomes $i_C(V_{BE} + v_{be}, V_{CE} + v_{ce}) = I_C(V_{BE}, V_{CE}) + i_c$.

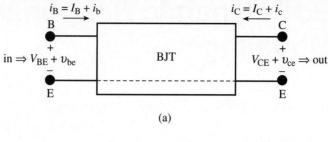

(a)

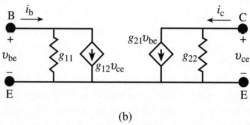

(b)

**Figure 12.1** (a) The BJT viewed as a two-port network and connected in the common emitter configuration. (b) Low-frequency small-signal equivalent circuit characterizing the a.c. response of the BJT.

The foregoing was primarily a statement of fact. We now make a critical assumption; namely, *we assume the transistor can follow the changes in potential quasistatically* so that $i_B(V_{BE} + v_{be}, V_{CE} + v_{ce}) = I_B(V_{BE} + v_{be}, V_{CE} + v_{ce})$ and $i_C(V_{BE} + v_{be}, V_{CE} + v_{ce}) = I_C(V_{BE} + v_{be}, V_{CE} + v_{ce})$. This allows us to write

$$I_B(V_{BE} + v_{be}, V_{CE} + v_{ce}) = I_B(V_{BE}, V_{CE}) + i_b \tag{12.1a}$$

$$I_C(V_{BE} + v_{be}, V_{CE} + v_{ce}) = I_C(V_{BE}, V_{CE}) + i_c \tag{12.1b}$$

or

$$i_b = I_B(V_{BE} + v_{be}, V_{CE} + v_{ce}) - I_B(V_{BE}, V_{CE}) \tag{12.2a}$$

$$i_c = I_C(V_{BE} + v_{be}, V_{CE} + v_{ce}) - I_C(V_{BE}, V_{CE}) \tag{12.2b}$$

Next, expanding the first term on the right-hand side of Eqs. (12.2) in a Taylor series about the d.c. operating point, and keeping only first-order terms in the expansion (higher-order

$v_{be}$ and $v_{ce}$ terms are assumed negligible), one obtains

$$I_B(V_{BE}+v_{be}, V_{CE}+v_{ce}) = I_B(V_{BE}, V_{CE}) + \left.\frac{\partial I_B}{\partial V_{BE}}\right|_{V_{CE}} v_{be} + \left.\frac{\partial I_B}{\partial V_{CE}}\right|_{V_{BE}} v_{ce} \quad (12.3a)$$

$$I_C(V_{BE}+v_{be}, V_{CE}+v_{ce}) = I_C(V_{BE}, V_{CE}) + \left.\frac{\partial I_C}{\partial V_{BE}}\right|_{V_{CE}} v_{be} + \left.\frac{\partial I_C}{\partial V_{CE}}\right|_{V_{BE}} v_{ce} \quad (12.3b)$$

which when substituted into Eqs. (12.2) gives

$$i_b = \left.\frac{\partial I_B}{\partial V_{BE}}\right|_{V_{CE}} v_{be} + \left.\frac{\partial I_B}{\partial V_{CE}}\right|_{V_{BE}} v_{ce} \quad (12.4a)$$

$$i_c = \left.\frac{\partial I_C}{\partial V_{BE}}\right|_{V_{CE}} v_{be} + \left.\frac{\partial I_C}{\partial V_{CE}}\right|_{V_{BE}} v_{ce} \quad (12.4b)$$

Dimensionally the partial derivatives in Eqs. (12.4) are conductances. Introducing

$$g_{11} \equiv \left.\frac{\partial I_B}{\partial V_{BE}}\right|_{V_{CE}} = \left.\frac{\partial I_B}{\partial V_{EB}}\right|_{V_{EC}} \quad ; \quad g_{12} \equiv \left.\frac{\partial I_B}{\partial V_{CE}}\right|_{V_{BE}} = \left.\frac{\partial I_B}{\partial V_{EC}}\right|_{V_{EB}}$$

$$\Uparrow \qquad\qquad \Uparrow \qquad\qquad\qquad \Uparrow \qquad\qquad \Uparrow$$

$$npn \qquad pnp \qquad\qquad\quad npn \qquad pnp$$

$$+I_B \text{ in} \quad +I_B \text{ out} \qquad\quad +I_B \text{ in} \quad +I_B \text{ out} \qquad (12.5a,b)$$

$$g_{21} \equiv \left.\frac{\partial I_C}{\partial V_{BE}}\right|_{V_{CE}} = \left.\frac{\partial I_C}{\partial V_{EB}}\right|_{V_{EC}} \quad ; \quad g_{22} \equiv \left.\frac{\partial I_C}{\partial V_{CE}}\right|_{V_{BE}} = \left.\frac{\partial I_C}{\partial V_{EC}}\right|_{V_{EB}}$$

$$\Uparrow \qquad\qquad \Uparrow \qquad\qquad\qquad \Uparrow \qquad\qquad \Uparrow$$

$$npn \qquad pnp \qquad\qquad\quad npn \qquad pnp$$

$$+I_C \text{ in} \quad +I_C \text{ out} \qquad\quad +I_C \text{ in} \quad +I_C \text{ out} \qquad (12.5c,d)$$

we can then write

$$i_b = g_{11} v_{be} + g_{12} v_{ce} \quad (12.6a)$$

$$i_c = g_{21} v_{be} + g_{22} v_{ce} \quad (12.6b)$$

Equations (12.6a) and (12.6b) may be viewed as the a.c. current node equations for the base and collector terminals, respectively. These equations indicate the $i_b$ and $i_c$ currents flowing into the terminals divide into two components. One component ($g_{11} v_{be}$ at the input and $g_{22}v_{ce}$ at the output) is just the current through a conductance connected across the given port. The second current component ($g_{12}v_{ce}$ at the input and $g_{21}v_{be}$ at the output) is controlled by the voltage at the opposite port and is logically associated with a current generator. We therefore infer the small-signal circuit consistent with the derived equations is as pictured in Fig. 12.1(b). [The reader should verify that the node equations for the B and C terminals in Fig. 12.1(b) are indeed Eqs. (12.6a) and (12.6b), respectively.]

Several comments are in order concerning the Fig. 12.1(b) result. First, the small-signal equivalent circuit is understood to be applicable at low operational frequencies where capacitive effects associated with the transistor junctions can be neglected. Second, although a viable equivalent circuit, an alternative form of the circuit to be established in the next subsection is normally preferred in practical computations. Lastly, it should be noted that the development leading to Fig. 12.1(b) was of a purely mathematical nature. The Fig. 12.1(b) result could be readily modified to apply to other BJT configurations or even to other three-terminal devices by simply changing the identity of the two-port terminals. The only tie to a specific device or configuration is through the conductance definitions. If desired, Eqs. (12.5) and the d.c. current relationships established in Chapter 11 could be employed to obtain explicit relationships for the small-signal conductances.

## 12.1.2 Hybrid-Pi Models

In performing small-signal analyses, the Hybrid-Pi equivalent circuit is by far and away the most frequently used model. The model has several advantages that make it particularly attractive to circuit design engineers. Notably, the model parameters are readily related to the d.c. operating point variables and the temperature variation of the parameters is easily deduced. Simplified and complete versions of the basic Hybrid-Pi model are shown respectively in Figs. 12.2(a) and (b). The cited low-frequency equivalent circuits apply to the "intrinsic transistor" under active-mode operation. The Hybrid-Pi model gets its name from the Fig. 12.2(b) $\pi$-arrangement of circuit elements with "hybrid" (a combination of conductance and resistance) units. $g_m$ is the *transconductance,* a measure of the forward voltage gain, $r_o$ the *output resistance,* $r_\pi$ the *input resistance,* and $r_\mu$ the *feedthrough resistance.*

The simplified model of Fig. 12.2(a), which finds extensive use in first-order analyses, can be derived directly from the generalized two-port model. Under active-mode biasing the Eqs. (11.47) Ebers–Moll relationships for a *pnp* BJT simplify to

$$I_E \cong I_{F0}e^{qV_{EB}/kT} \qquad \dots V_{EB} > 0,\ V_{CB} < 0 \qquad (12.7a)$$

$$I_C \cong \alpha_F I_{F0}e^{qV_{EB}/kT} \qquad \dots V_{EB} > 0,\ V_{CB} < 0 \qquad (12.7b)$$

and

$$I_B = I_E - I_C \cong (1 - \alpha_F)I_{F0}e^{qV_{EB}/kT} \qquad (12.7c)$$

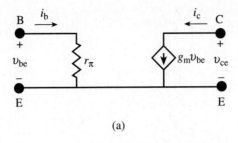

(a)

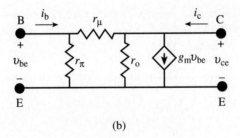

(b)

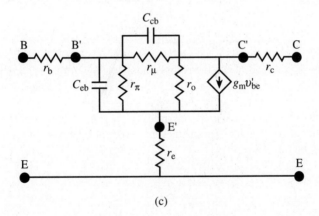

(c)

**Figure 12.2**  Hybrid-Pi equivalent circuits. (a) Simplified and (b) complete low-frequency Hybrid-Pi equivalent circuits that model the intrinsic transistor under active-mode common-emitter operation. (c) High-frequency equivalent circuit including the parasitic series resistances.

If base-width modulation is assumed to be negligible, $\alpha_F$ and $I_{F0}$ are constants independent of bias and

$$g_{12} \equiv \left. \frac{\partial I_B}{\partial V_{EC}} \right|_{V_{EB}} \cong 0 \tag{12.8a}$$

$$g_{22} \equiv \left. \frac{\partial I_C}{\partial V_{EC}} \right|_{V_{EB}} \cong 0 \tag{12.8b}$$

With $g_{12} = g_{22} = 0$ we obtain the desired result; Fig. 12.1(b) reduces to precisely the same form as Fig. 12.2(a). We also conclude $g_m \cong g_{21}$ and $r_\pi \cong 1/g_{11}$. Explicitly evaluating $g_m$ and $r_\pi$ using Eqs. (12.7) yields

$$g_m \cong \left. \frac{\partial I_C}{\partial V_{EB}} \right|_{V_{EC}} \cong \frac{q I_C}{kT} \tag{12.9a}$$

$$r_\pi \cong \frac{1}{(\partial I_B/\partial V_{EB})|_{V_{EC}}} \cong \frac{1}{q I_B/kT} \cong \frac{I_C}{I_B} \frac{1}{g_m} = \frac{\beta_{dc}}{g_m} \tag{12.9b}$$

Note that $g_m$ and $r_\pi$ are readily evaluated given a knowledge of the d.c. operating point.

The four-element low-frequency model of Fig. 12.2(b) is used when base-width modulation cannot be ignored or a more precise analysis is to be performed. Although somewhat more involved than the simplified model derivation, the Fig. 12.2(b) circuit parameters can be related to the two-port model parameters in a straightforward fashion. It is left as an exercise to show that in general

$$g_m = g_{21} - g_{12} \tag{12.10a}$$

$$r_o = 1/(g_{22} + g_{12}) \tag{12.10b}$$

$$r_\pi = 1/(g_{11} + g_{12}) \tag{12.10c}$$

$$r_\mu = -1/g_{12} \tag{12.10d}$$

Typically $g_{12}$ is several orders of magnitude smaller than the other two-port model conductances and may be neglected in computing $g_m$, $r_o$, and $r_\pi$. Equations (12.9) may of course be used to evaluate $g_m$ and $r_\pi$ if $g_{12}$ is much less than $g_{21}$ and $g_{11}$.

Finally, the model shown in Fig. 12.2(c) is appropriate for use at higher frequencies. Included in the model are the parasitic series resistances ($r_b$, $r_c$, and $r_e$) that connect the intrinsic transistor to the external device terminals (see Subsection 11.2.5). $C_{cb}$ and $C_{eb}$

respectively model the collector-base and emitter-base *pn* junction capacitances. Under active-mode biasing the $C_{cb}$ associated with the reverse-biased C–B junction is primarily a junction capacitance, while $C_{eb}$ associated with the forward-biased E–B junction includes both junction and diffusion components. The appropriate junction capacitance relationships from Chapter 7 may be used without modification in evaluating $C_{cb}$ and $C_{eb}$. However, the diffusion component of $C_{eb}$ must be modified as outlined in Problem 12.7. An additional capacitance between the C and B terminals and a capacitance from the C′ terminal to a substrate terminal may be necessary to properly model transistors incorporated in ICs. It should be noted that the Fig. 12.2(c) model is typically valid at frequencies up to roughly 500 MHz. The model becomes increasingly inaccurate above 500 MHz because it does not take into account the delays in signal propagation across the various device regions.

## 12.2 TRANSIENT (SWITCHING) RESPONSE

The BJT finds extensive utilization as an electronic switch in applications ranging from simple discrete-device circuits to complex IC logic circuits. When compared to a *pn* junction diode, the BJT offers intrinsically faster switching speeds and the advantage of a third lead, which greatly facilitates the switching process. Like the *pn* junction diode, a time delay in progressing between the "on" and "off" states can be attributed for the most part to the build-up or removal of excess minority carrier charge. Since switching speed is a major concern in many applications, it is reasonable to examine the switching process in some detail.

### 12.2.1 Qualitative Observations

For the purposes of the discussion, we assume a switching circuit of the form pictured in Fig. 12.3(a). $V_{CC}$ supplies the emitter-to-collector d.c. bias, and $R_L$ is the output load resistor. Switching is accomplished by pulsing the input power supply between positive and negative voltages. Under steady-state conditions with $v_s < 0$, the BJT is biased into the cutoff mode, making $I_C \cong 0$. In the described state the transistor is "off" and its operating point lies near the bottom of the load line shown superimposed on the characteristics in Fig. 12.3(b). Conversely, under steady-state conditions with $v_s > 0$, the applied voltages are typically such that both junctions are forward biased. Operating in the saturation mode the transistor is in the "on" state with $I_C \cong V_{CC}/R_L$.

Referring to Fig. 12.4, let us now examine the transient response itself. We begin with the turn-on transient that takes place after $v_s$ is pulsed from $-\xi V_S$ to $+V_S$ at $t = 0$. With $V_S$ assumed to be much greater than $v_{EB}$, the base current as shown in Fig. 12.4(b) jumps up to $i_B \cong V_S/R_S$ and remains constant at this value as long as $v_s > 0$. The corresponding idealized $i_C$ response is sketched in Fig. 12.4(c). Prior to $t = 0$ the transistor is cutoff biased and the minority carrier distribution in the quasineutral base is characterized by plot (i) of Fig. 12.4(d). Following the $t = 0$ turn-on pulse, minority carrier holes begin piling up in the base [plot (ii) of Fig. 12.4(d)]. $i_C$ in turn increases roughly in proportion to the slope of

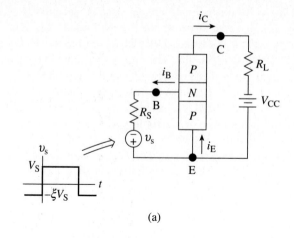

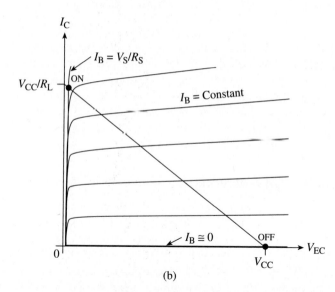

**Figure 12.3** BJT switching fundamentals. (a) Idealized switching circuit. (b) Load line with the "on" and "off" steady-state endpoints superimposed on the common emitter output characteristics.

$\Delta p_B(x, t)$ evaluated at $x = W$. The increase in $i_C$ tracks the build-up of minority carriers in the quasineutral base as long as the transistor remains active mode biased. Eventually, however, $\Delta p_B(0, t) = (n_i^2/N_B) \exp(q v_{EB}/kT)$ reaches the point where $v_{EB} = v_{EC}$ or $v_{CB} = 0$, and the transistor subsequently becomes saturation biased. Once saturation biased, the minority carrier build-up is completed as pictured in plot (iii) of Fig. 12.4(d). Consistent with the cited plot, $i_C$ increases only slightly during this portion of the transient to finally attain its on-state value.

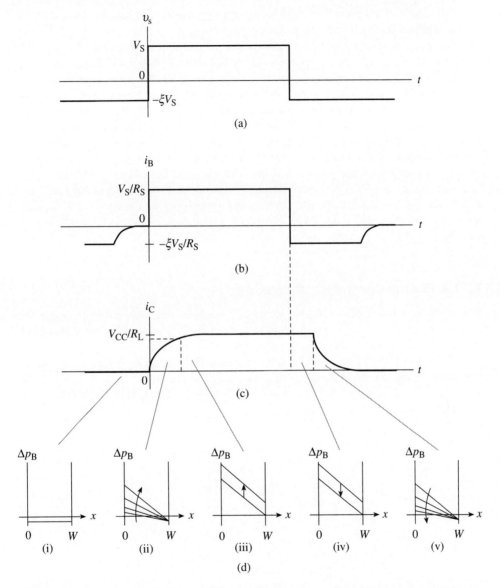

**Figure 12.4** BJT switching characteristics. (a) Applied input voltage. (b) Input current ($i_B$) as a function of time. (c) Output current ($i_C$) as a function of time. (d) Approximate minority carrier distribution in the quasineutral base at various times during the transient response.

The $i_C$ turn-off transient also exhibits two distinct phases. The situation inside the base of the transistor during the two phases is approximately modeled by plots (iv) and (v) of Fig. 12.4(d), respectively. During the initial phase, occurring immediately after $v_s$ is pulsed from $+V_S$ to $-\xi V_S$, $i_C$ remains approximately constant while the stored charge associated with saturation biasing is removed from the quasineutral base. The time delay here is obviously similar in nature to the storage delay time encountered in the *pn* junction analysis. The second turn-off phase begins when sufficient stored charge has been removed so that the transistor becomes active mode biased. Once active mode biased, $i_C$ monotonically decreases in proportion to the stored charge remaining in the quasineutral base. Ultimately the total removal of the minority carrier charge leads to cutoff biasing and the off-state. Although of secondary importance, it is interesting to note from Fig. 12.4(b) that the $i_B$ response during turn-off is essentially identical to the *i-t* turn-off transient of the *pn* junction diode. The base current abruptly reverses and remains constant at $i_B = -\xi V_S/R_S$ until the BJT enters the cutoff mode. With increasing reverse bias across the E–B junction, $i_B$ subsequently decreases to approximately zero.

## 12.2.2  Charge Control Relationships

In performing a quantitative analysis of the BJT transient response, we will employ the charge control approach. Before launching into the analysis it is necessary to first establish the charge control relationships specifically applicable to the BJT.

The *pn* junction charge control results established in Subsection 6.3.1 can be applied directly to the quasineutral base of the BJT with only minor modifications. Working with a *pnp* BJT, making required symbol changes, and accounting for the finite width ($0 \leq x \leq W$) of the quasineutral base, we infer

$$\frac{dQ_B}{dt} = -A \int_{J_P(0)}^{J_P(W)} dJ_P - \frac{Q_B}{\tau_B} \qquad (12.11)$$

where

$$Q_B = \left( \begin{array}{c} \text{Excess hole charge} \\ \text{in quasineutral base} \end{array} \right) = qA \int_0^W \Delta p_B(x,t)\, dx \qquad (12.12)$$

If $i_E \cong AJ_P(0)$ and $i_C \cong AJ_P(W)$, which is equivalent to assuming the emitter and collector currents are dominated by the hole diffusion currents respectively flowing across the E–B and C–B junctions, we conclude

$$-A \int_{J_P(0)}^{J_P(W)} dJ_P = A[J_P(0) - J_P(W)] \cong i_E - i_C = i_B \qquad (12.13)$$

and

$$\boxed{\frac{dQ_B}{dt} = i_B - \frac{Q_B}{\tau_B}}$$

(12.14)

Note as an aside that if Eq. (12.14) is applied under steady-state conditions, $dQ_B/dt = 0$ and $i_B \rightarrow I_B = Q_B/\tau_B$. This result may be interpreted as follows: There are a total of $Q_B/q$ excess minority carriers in the quasineutral base. The average lifetime of an excess minority carrier is $\tau_B$. Thus every $\tau_B$ seconds the $Q_B$ excess store of charge is eliminated by recombination. Under steady-state conditions a current equal to the eliminated charge per second, $I_B = Q_B/\tau_B$, must flow into the base to replace the eliminated carriers. The noted result is valid provided the current associated with recombination in the quasineutral base is the dominant component of $I_B$; that is, carrier injection from the base into the emitter and other base current components are negligible as implicitly assumed in the derivation of Eq. (12.14).

Continuing the development of charge control relationships, let us next examine the collector current. Consistent with previous approximations, we can write

$$i_C \cong AJ_P(W) = -qAD_B \left.\frac{\partial \Delta p_B(x,t)}{\partial x}\right|_{x=W}$$

(12.15)

If we assume $\Delta p_B(x, t)$ has the same functional form as $\Delta p_B(x)$ under steady-state conditions, then in the $W \ll L_B$ limit

$$\Delta p_B(x,t) = \Delta p_B(0,t) + [\Delta p_B(W,t) - \Delta p_B(0,t)]\frac{x}{W}$$

(12.16)

which under active mode biasing simplifies to

$$\Delta p_B(x,t) \cong \Delta p_B(0,t)\left(1 - \frac{x}{W}\right)$$

(12.17)

Thus, taking the BJT to be active mode biased,

$$\left.\frac{\partial \Delta p_B(x,t)}{\partial x}\right|_{x=W} \cong -\frac{\Delta p_B(0,t)}{W}$$

(12.18)

$$Q_B = qA\int_0^W \Delta p_B(x,t)\,dx \cong \frac{qAW}{2}\Delta p_B(0,t)$$

(12.19)

and

$$i_C \cong -qAD_B \left.\frac{\partial \Delta p_B}{\partial x}\right|_{x=W} \cong \frac{qAD_B}{W}\Delta p_B(0,t) = \frac{Q_B}{(W^2/2D_B)} \tag{12.20}$$

or

$$\boxed{i_C \cong \frac{Q_B}{\tau_t}} \qquad \text{(active mode)} \tag{12.21}$$

where

$$\boxed{\tau_t \equiv \frac{W^2}{2D_B}} \tag{12.22}$$

Assuming $Q_B(t)$ can be determined by solving Eq. (12.14), Eq. (12.21) then provides a simple means of deducing the instantaneous collector current. The $\tau_t$ parameter finds extensive use in approximate analyses and is known as the *base transit time*. It is interpreted as the average time taken by minority carriers to diffuse across the quasineutral base. According to Eq. (12.21), one may view the collector current to be the result of the stored charge in the quasineutral base dropping into the collector every $\tau_t$ seconds. Also note that $\beta_{dc} \equiv I_C/I_B = \tau_B/\tau_t$.

### 12.2.3  Quantitative Analysis

**Turn-on Transient**

During the turn-on transient $i_B \cong V_S/R_S \equiv I_{BB} = $ constant and

$$\frac{dQ_B}{dt} = I_{BB} - \frac{Q_B}{\tau_B} \tag{12.23}$$

The general solution of the (12.23) differential equation is

$$Q_B(t) = I_{BB}\tau_B + Ae^{-t/\tau_B} \tag{12.24}$$

where $A$ is an arbitrary constant. To a very good approximation, $Q_B(0) = 0$. This is true since the transistor is cutoff biased at the beginning of the transient. Substituting $Q_B(0) = 0$ into Eq. (12.24) when $t = 0$, and solving for $A$, yields $A = -I_{BB}\tau_B$. Thus

$$Q_B(t) = I_{BB}\tau_B(1 - e^{-t/\tau_B}) \qquad \ldots t \geq 0 \tag{12.25}$$

and

$$
i_C(t) = \begin{cases} \dfrac{Q_B(t)}{\tau_t} = \dfrac{I_{BB}\tau_B}{\tau_t}\,(1 - e^{-t/\tau_B}) & \ldots\, 0 \le t \le t_r & (12.26a) \\[4mm] \dfrac{V_{CC}}{R_L} \equiv I_{CC} & \ldots\, t \ge t_r & (12.26b) \end{cases}
$$

The rise time, $t_r$, introduced above is understood to be the time period over which the BJT is active mode biased.

The turn-on solution is illustrated graphically on the left-hand side of Fig. 12.5. Exhibiting an exponential dependence, $Q_B(t)$ increases smoothly from zero to its maximum value of $Q_B(\infty) = I_{BB}\tau_B$. $i_C(t)$ tracks $Q_B(t)$ until the BJT becomes saturation biased at $t = t_r$. Consistent with the qualitative observations, $i_C(t)$ is subsequently assumed to remain constant at $i_C(t_r) \cong V_{CC}/R_L \equiv I_{CC}$. Setting $i_C(t_r) = I_{CC}$ in Eq. (12.26a), one can solve for $t_r$ to obtain

$$
t_r = \tau_B \ln\left[\frac{1}{1 - \dfrac{I_{CC}\tau_t}{I_{BB}\tau_B}}\right] \tag{12.27}
$$

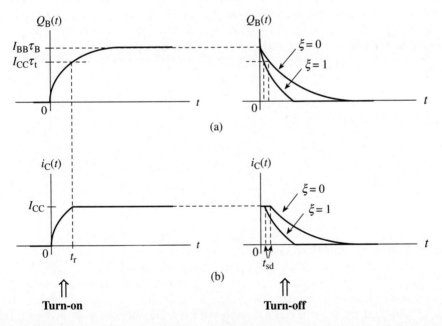

**Figure 12.5**  Transient response solution. (a) Stored charge in the quasineutral base and (b) collector current as a function of time. The turn-on transient is shown on the left and the turn-off transient on the right. The $\xi = 0$ and $\xi = 1$ labels identify turn-off solutions corresponding respectively to $i_B \cong 0$ and $i_B = -I_{BB}$ during the turn-off transient.

In agreement with intuition, $t_r$ can be decreased by increasing $I_{BB}$ (faster stored-charge build-up) or decreasing $I_{CC}$ (less active-mode charge storage). We also note that substitution of the Eq. (12.27) expression for $t_r$ into Eq. (12.25) yields $Q_B(t_r) = I_{CC}\tau_t$. This value for the stored charge at the active-mode/saturation-mode transition point was appropriately positioned along the $Q_B(t)$ axis in Fig. 12.5(a) and will be required in completing the turn-off analysis.

## Turn-off Transient

During the saturation-mode and active-mode portions of the turn-off transient, $i_B \cong -\xi V_S/R_S = -\xi I_{BB} = $ constant. The charge stored in the quasineutral base therefore obeys the charge control relationship

$$\frac{dQ_B}{dt} = -\xi I_{BB} - \frac{Q_B}{\tau_B} \tag{12.28}$$

which has the general solution

$$Q_B(t) = -\xi I_{BB}\tau_B + Ae^{-t/\tau_B} \tag{12.29}$$

$A$ is an arbitrary constant and $t = 0$ has been reset to the beginning of the turn-off transient. Since the total stored charge at the beginning of the turn-off transient is just equal to the stored charge at the end of the turn-on transient, $Q_B(0)|_{\text{turn-off}} = Q_B(\infty)|_{\text{turn-on}} = I_{BB}\tau_B$. Substituting $Q_B(0) = I_{BB}\tau_B$ into Eq. (12.29) when $t = 0$, and solving for $A$, yields $A = (1 + \xi)I_{BB}\tau_B$. We therefore conclude

$$Q_B(t) = I_{BB}\tau_B[(1 + \xi)e^{-t/\tau_B} - \xi] \quad \ldots t \geq 0 \tag{12.30}$$

and

$$i_C(t) = \begin{cases} I_{CC} & \ldots 0 \leq t \leq t_{sd} \quad (12.31a) \\[2mm] \dfrac{Q_B(t)}{\tau_t} = \dfrac{I_{BB}\tau_B}{\tau_t}[(1 + \xi)e^{-t/\tau_B} - \xi] & \ldots t \geq t_{sd} \quad (12.31b) \end{cases}$$

The storage delay time, $t_{sd}$, introduced above is understood to be the time period over which the BJT is saturation-mode biased.

The turn-off solution is illustrated graphically on the right-hand side of Fig. 12.5 for the assumed values of $\xi = 0$ and $\xi = 1$. $\xi = 0$ corresponds to $v_s = 0$ and $i_B \cong 0$ during the turn-off transient; $v_s = -V_S$ and $i_B \cong -I_{BB}$ if $\xi = 1$. It is immediately obvious from Fig. 12.5 that both the constant (saturation-mode) and falling (active-mode) portions of the $i_C$ transient are decreased when $\xi > 0$. This is to be expected since an $i_B < 0$ aids the withdrawal of stored charge from the quasineutral base. At the saturation-mode/active-mode transition point $t = t_{sd}$ and $Q_B(t_{sd})\big|_{turn\text{-}off} = Q_B(t_r)\big|_{turn\text{-}on} = I_{CC}\tau_t$. Substituting the transition point values into Eq. (12.31b) and solving for $t_{sd}$, we obtain

$$t_{sd} = \tau_B \ln\left[\frac{1 + \xi}{\dfrac{I_{CC}\tau_t}{I_{BB}\tau_B} + \xi}\right] \tag{12.32}$$

Since $I_{CC}\tau_t/I_{BB}\tau_B < 1$, Eq. (12.32) confirms that $t_{sd}$ decreases when $\xi > 0$. Generally speaking, decreasing $\tau_B$ (faster recombination), decreasing $I_{BB}$ (less total stored charge), and increasing $I_{CC}$ (relatively less saturation-mode charge) also decreases $t_{sd}$.

### 12.2.4 Practical Considerations

The foregoing analysis was highly simplified to provide physical insight and to achieve a solution of the generally correct form without becoming bogged down in an excessive amount of mathematics. Readers interested in a more precise solution are encouraged to consult the BJT supplemental reference cited at the end of Part II. A more realistic sketch of the $i_C$ transient response is presented in Fig. 12.6. Other than a "rounding" near transition points, the only major difference between Fig. 12.6 and earlier $i_C$ plots is an added time delay, $t_d$, at the beginning of the transient. The added time delay arises from the charging of the E–B junction capacitance in going from cutoff to the active mode. Figure 12.6 also illustrates the measurement-based definitions of the rise time ($t_r$), the storage delay time ($t_{sd}$), and the fall time ($t_f$), parameters often found in device specification sheets.

It should come as no surprise that various methods have been employed to speed up the turn-on and turn-off transients. A common procedure to speed up the turn-on transient

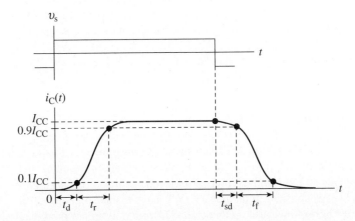

**Figure 12.6** Actual form of the collector-current switching transient and the measurement-based definitions of $t_d$, $t_r$, $t_{sd}$, and $t_f$. The input voltage waveform is included for reference purposes.

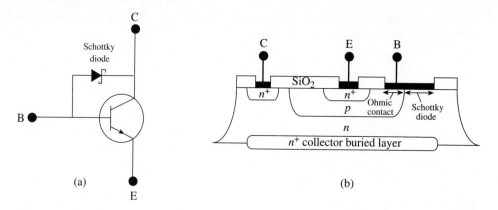

**Figure 12.7** Schottky diode clamped *npn* transistor. (a) Schematic circuit representation. (b) Physical implementation.

is to add a capacitor to the input circuitry, such as placing a capacitor in parallel with $R_S$ in Fig. 12.3(a). The capacitive discharge after $v_s \rightarrow V_S$ provides a momentary pulse of current that accelerates the build-up of stored charge and thereby reduces the time to saturation.

A speed-up of the turn-off transient can and has been achieved by adding recombination–generation centers to the BJT base. The addition of R–G centers, such as gold in Si, reduces $\tau_B$ and therefore increases the rate of carrier removal from the base. In another method to speed up the turn-off transient, a Schottky diode is connected between the collector and base as shown in Fig. 12.7. This arrangement is known as a *Schottky diode clamp*. The Schottky diode, a device formed from a rectifying metal-semiconductor contact (see Chapter 14), conducts at a much lower forward-bias voltage than a *pn* junction. Consequently, when the transistor enters the saturation mode, the Schottky diode begins to conduct and "clamps" the C–B junction voltage at a relatively low positive value. This greatly reduces the saturation-mode build-up of stored charge. Moreover, there is very little minority-carrier charge storage in a Schottky diode. With less charge to be removed from the BJT and very little charge stored in the Schottky diode, the turn-off time is dramatically reduced.

## 12.3 SUMMARY

In this chapter we first examined the small signal response and then the transient or switching response of the bipolar junction transistor. A small-signal equivalent circuit for the BJT connected in the common emitter configuration, a generalized two-port model valid at low frequencies, was established employing a purely mathematical argument. The development was of interest in itself because it can be applied with only minor modifications to essentially any three-terminal device. Subsequently, the popular Hybrid-Pi equivalent circuit was introduced and related to the two-port model. Simplified and high-frequency versions of the Hybrid-Pi circuit were also reviewed.

Because of the importance of switching speed in many applications, the transient response was examined in some detail. The transient response was first described qualitatively, concentrating on the build-up and removal of excess minority carriers in the quasineutral base. Similar to the *pn* junction diode, stored charge must be added or removed in progressing between the off-state (cutoff) and the on-state (saturation). Charge control relationships specifically applicable to the BJT were next established assuming for simplicity that recombination in the quasineutral base and minority carrier diffusion at the quasineutral base boundaries totally dominated the observed terminal currents. The concept of a base transit time, the average time required for diffusing minority carriers to cross the quasineutral base, was introduced in the course of the development. Quantitative expressions for the stored charge, the $i_C$ transient, and characteristic time periods were derived for both the turn-on and turn-off transients. A concluding subsection cited practical considerations including methods that have been employed to speed up the transient response.

## PROBLEMS

| CHAPTER 12   PROBLEM INFORMATION TABLE | | | | |
|---|---|---|---|---|
| Problem | Complete After | Difficulty Level | Suggested Point Weighting | Short Description |
| 12.1 | 12.3 | 1 | 10 (1 each part) | Quick quiz |
| 12.2 | 12.1.1 | 2–3 | 8 | Derive common base circuit |
| 12.3 | 12.1.2 | 1 | 5 | Determine $g_m$, $r_\pi$ |
| 12.4 | " | 3 | 8 | Confirm Eqs. (12.10) |
| ● 12.5 | " | 4 | 20 | Calculate H-Pi parameters |
| * 12.6 | " | 4 | 20 (a-14, b-6) | Derive/compute $f_T$ |
| 12.7 | 12.2.2 | 3–4 | 10 | E–B diffusion capacitance |
| 12.8 | 12.2.3 | 2 | 5 | Estimate $I_{CC}\tau_t / I_{BB}\tau_B$ |
| ● 12.9 | " | 2 | 8 (4 each part) | $t_r$ and $t_{sd}$ plots |
| * 12.10 | 12.2.4 | 3 | 10 (a-6, b-4) | $t_f$ derivation and plot |

**12.1** Quick Quiz

Answer the following questions as concisely as possible.

(a) In words, what is the quasistatic assumption? How is the assumption utilized in the derivation of the generalized two-port model?

(b) Why are there separate definitions for the *npn* and *pnp* $g_{ij}$ conductances?

(c) What is the origin of the "Hybrid-Pi" name?

(d) State the names of the following Hybrid-Pi parameters: $g_m$, $r_o$, $r_\pi$, and $r_\mu$.

(e) What is the justification for adding capacitors to the low-frequency Hybrid-Pi model to obtain the high-frequency model?

(f) Describe what happens to $i_B$, $i_C$, and the minority-carrier concentration in the quasi-neutral base once the transistor enters the saturation mode during the turn-on transient.

(g) Give both the word and mathematical definitions of the "base transit time."

(h) How are $\tau_B$ and $\tau_t$ related to $\beta_{dc}$?

(i) Why does an $i_B < 0$ ($\xi > 0$) during the turn-off transient give rise to a reduction in both the storage-delay and fall times?

(j) What is a Schottky diode clamp, and what purpose does it serve?

**12.2** Establish a small-signal equivalent circuit analogous to Fig. 12.1(b) appropriate for the *common base* configuration. Provide a summary of the parameter definitions similar to Eqs. (12.5).

**12.3** The common emitter output characteristics of a 2N2605 *pnp* BJT were displayed in Fig. 11.5(d). The simplified Hybrid-Pi equivalent circuit is to be used in modeling the low-frequency a.c. response of the transistor. Suppose $I_B = 5\ \mu A$ and $V_{EC} = 10$ V at the d.c. operating point. Determine the values of $g_m$ and $r_\pi$ to be employed in the Hybrid-Pi model.

**12.4** Perform the circuit analysis and mathematical manipulations required to confirm Eqs. (12.10).

● **12.5** The MATLAB program BJTplus, provided as a solution to Exercise 11.10, computes $I_B = I_B(V_{EB}, V_{EC})$ and $I_C = I_C(V_{EC}, I_B)$ with and without taking into account base-width modulation. Starting from scratch or borrowing from the BJTplus and subsidiary program code, write a MATLAB (computer) program that automatically calculates the values of the Hybrid-Pi [Fig. 12.2(b)] parameters. Provide for alternative computations with and without taking into account base-width modulation. For simplicity, let the input variables be $V_{EB}$ and $V_{EC}$. Employ the same device and material parameters as those found in the BJT0 subprogram. (The MATLAB scripts for both BJT0 and BJTplus are listed in Appendix M.)
Run your program with $V_{EC} = 10$ V and $V_{EB}$ chosen so that $I_C \cong 1$ mA. Present computational results for $g_m$, $r_o$, $r_\pi$, and $r_\mu$ both with and without accounting for base-width modulation. Also compute $g_m$ and $r_\pi$ using Eqs. (12.9). Discuss your results.

**12.6** The unity beta frequency, $f_T$, is often quoted as a figure of merit in assessing the high-frequency response of a BJT. By definition, $f_T$ is the signal frequency at which $|i_c/i_b| = 1$ when the BJT common emitter output is a.c. short-circuited ($v_{ce} = 0$). Silicon BJTs intended for use in RF and microwave applications are available with an $f_T = 10$ GHz. "General purpose" Si transistors typically have an $f_T$ in the 100's of MHz.

(a) Making use of the high-frequency Hybrid-Pi model, derive a general expression for $i_c/i_b$ when $v_{ce} = 0$.

● (b) The 2N3906 is a general-purpose Si transistor. Determine the $f_T$ of a 2N3906 *pnp* BJT when d.c. biased such that $V_{EB} = 0.68$ V, $V_{EC} = 10$ V, $I_B = 5.57$ $\mu$A, and $I_C = 1.00$ mA. The high-frequency Hybrid-Pi parameters for the 2N3906 at the given d.c. operating point are tabulated below.

| Parameter | Value |
|:---:|:---:|
| $g_m$ | $3.86 \times 10^{-2}$ S |
| $r_o$ | $2.00 \times 10^4$ $\Omega$ |
| $r_\pi$ | $4.65 \times 10^3$ $\Omega$ |
| $r_\mu$ | $3.59 \times 10^6$ $\Omega$ |
| $C_{eb}$ | 23.6 pF |
| $C_{cb}$ | 2.32 pF |
| $r_b$ | 10 $\Omega$ |
| $r_c$ | 2.8 $\Omega$ |
| $r_e$ | 0 |

**12.7** Show that the diffusion component of $C_{eb}$ for a *pnp* BJT under active mode biasing is given by

$$C_{eb|diff} = \frac{2}{3}\left(\frac{\tau_t}{kT/q}\right)\left(qA\,\frac{D_B}{W}\,\frac{n_i^2}{N_B}\right)e^{qV_{EB}/kT} \cong \frac{2}{3}\left(\frac{\tau_t}{kT/q}\right)I_E \cong \frac{2}{3}g_m\tau_t$$

HINT: $C_{eb|diff}$ is the same as the diffusion capacitance, $C_D$, to be expected from a *narrow-base* diode. Noting the $I_{DIFF}$ flowing in a narrow-base diode is given by Eqs. (6.68)/(6.69), revise the diffusion admittance derivation in Subsection 7.3.2 to obtain the $C_D$ for a narrow-base diode. The revision requires use of the expansion, $\text{ctnh}(\xi) = \cosh(\xi)/\sinh(\xi) \cong (1/\xi)(1 + \xi^2/3)$ if $\xi \ll 1$. Make appropriate symbol identifications in transferring the result to the BJT.

**12.8** Refer to the Fig. 12.3(b) characteristic. Suppose $V_{CC}/R_L = 5$ mA, $V_S/R_S = 30$ $\mu$A, and $I_B$ was stepped in 5 $\mu$A increments to obtain the $I_B$ = constant curves. *Estimate* $I_{CC}\tau_t/I_{BB}\tau_B$ for switching between the "off" and "on" points shown in the figure.

● **12.9** Rise and Storage-Delay Times

(a) Let $x = I_{CC}\tau_t/I_{BB}\tau_B$. Employing Eq. (12.27) and noting $0 \leq x \leq 1$, construct a plot of $t_r/\tau_B$ versus $x$.

(b) Employing Eq. (12.32), construct a plot of $t_{sd}/\tau_B$ versus $x$ simultaneously displaying curves corresponding to $\xi = 0$ and $\xi = 1$.

**12.10** Fall Time

(a) Derive an expression for the fall time, $t_f$, using the measurements-based definition presented in Fig. 12.6 and the Eq. (12.31b) $i_C$ versus $t$ relationship.

● (b) Plot $t_f/\tau_B$ versus $x \equiv I_{CC}\tau_t/I_{BB}\tau_B$ when $\xi = 0$ and $\xi = 1$. Note that $0 \leq x \leq 1$. Discuss your results.

# 13 PNPN Devices

In earlier chapters we progressed through the analyses of the two-region, one-junction diode and the three-region, two-junction transistor. Containing four alternately doped regions and three interacting *pn* junctions, PNPN devices, or *thyristors,* embody the next (and final) increment in junction complexity. Thyristors, often large area devices packaged in metal studs for optimum heat dissipation, find extensive use as high-power rectifiers and electronic switches. Although typical ratings are considerably lower, thyristors are available with current ratings in excess of 5 kA and voltage ratings exceeding 10 kV. For the most part, we concentrate herein on the best-known and leading member of the PNPN device family, the silicon controlled rectifier (SCR). One section is devoted to describing the physical nature of the device and the rather interesting "negative resistance" or bistable switching characteristics. A second investigates the internal operation of the device giving rise to the switching characteristics. Practical details related to turn-on and turn-off are considered in a third. The concluding section briefly surveys other members of the PNPN device family.

## 13.1 SILICON CONTROLLED RECTIFIER (SCR)

The silicon controlled rectifier is pictured schematically in Fig. 13.1(a). The SCR contains four alternately doped regions labeled P1 through N4 in the figure. $J_{12}, J_{23}$, and $J_{34}$ identify the P1–N2, N2–P3, and P3–N4 junctions, respectively. The outer P-region and the contact to the region is called the anode (A); the outer N-region and the contact to the region is called the cathode (K). The inner N- and P-regions are referred to as bases. The third device terminal, the gate (G), is connected to the P3 base. We should interject that a similar two-terminal PNPN device constructed without a gate lead is known as the PNPN diode or Shockley diode. Note that $V_{AK}$ is the voltage applied between the anode and cathode, $I_{AK}$ the current flowing into the anode, and $I_G$ the current flowing into the gate.

The SCR is typically fabricated by starting with a properly thinned, lowly doped *n*-type Si wafer ($N_D \sim 5 \times 10^{13}/\text{cm}^3$), the interior of which eventually becomes the N2 base. Next, *p*-type dopant is diffused into the two sides of the wafer, yielding almost identical $J_{12}$ and $J_{23}$ junctions. Finally, *n*-type dopant is diffused into one side of the wafer forming the heavily doped N4-cathode. The resulting doping profile is sketched in Fig. 13.1(b). As might be inferred from the fabrication description, and as emphasized in Fig. 13.1(b), the N2 and P3 base regions are of moderate electrical width. Unlike the wide-base diode where the widths of the P- and N-regions are assumed to be much greater than a minority-carrier diffusion length, or the BJT where the width of the base is routinely much less than a

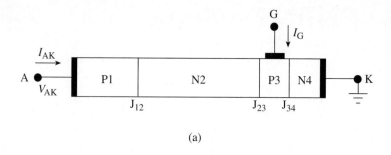

(a)

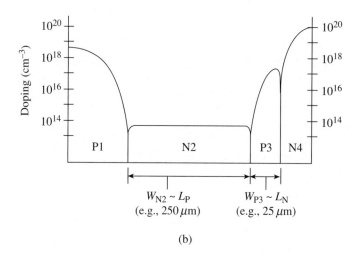

(b)

**Figure 13.1** The silicon controlled rectifier (SCR). (a) Schematic representation showing the doping regions, junctions, device terminals, and current plus voltage variables. (b) Typical doping profile.

minority-carrier diffusion length, the widths of the SCR base regions must be roughly comparable to a diffusion length. This, as we will see, is critical to the proper operation of the SCR.

The general form of the characteristics derived from an SCR is sketched in Fig. 13.2. The $V_{AK} < 0$ portion of the characteristics is $I_G$ independent and very similar to that of a reverse-biased *pn* junction diode. The device conducts little current until breaking down at $V_{AK} = -V_{BR}$. $V_{BR}$ in SCR work is referred to as the reverse-bias holding or *blocking voltage*. The forward-bias characteristics, on the other hand, are quite distinctive. If $I_G = 0$, systematically increasing $V_{AK}$ from zero causes little current to flow through the device until $V_{AK}$ exceeds $V_{BF}$, the maximum forward-bias blocking voltage. Once $V_{BF}$ is exceeded, however, the device switches from the high-impedance blocking mode to the low-impedance conducting mode. Operation then lies along the forward-bias diode-like char-

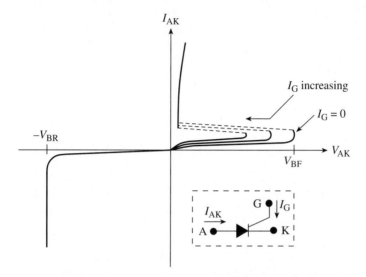

**Figure 13.2**  SCR characteristics and circuit symbol.

acteristic near $V_{AK} = 0$. Applying an $I_G > 0$ lowers the forward-bias blocking voltage and permits the device to enter the conducting mode at a lower applied $V_{AK}$. $V_{BR}$ and $V_{BF}$, it should be noted, can be in the hundreds or even thousands of volts, while the voltage drop in the conducting mode is routinely on the order of a volt.

## 13.2  SCR OPERATIONAL THEORY

We wish to explain the observed SCR characteristics. To simplify the initial considerations, let $I_G = 0$. With $I_G = 0$, the operation of the SCR is equivalent to that of the PNPN diode.

The question arises as to how to begin the analysis. A reasonable approach would be to start with the simplest possible model and subsequently consider more sophisticated models as required to obtain agreement with the observed characteristics. The simplest possible model for the SCR would be three *pn* junction diodes in series as envisioned in Fig. 13.3(a).

Working with the diode model, we note that if $V_{AK} < 0$, the middle diode becomes forward biased and may be replaced with a short circuit. This leads to the simplified model of Fig. 13.3(b). Further observe from Fig. 13.1(b) that the doping on the lightly doped side of the $J_{34}$ junction is $\sim 10^{17}/cm^3$, while the doping on the lightly doped side of the $J_{12}$ junction is $\lesssim 10^{14}/cm^3$. The $J_{34}$ junction therefore breaks down at a relatively small $V_{AK} < 0$, and the SCR reverse-bias characteristic is predicted to be basically the same as the reverse-bias characteristic of the $J_{12}$ *pn* junction. In a similar manner, if $V_{AK} > 0$, the $J_{12}$ and $J_{34}$ diodes become forward biased and may be replaced with short circuits, leading to the simplified model of Fig. 13.3(c). Consequently, the forward-biased SCR characteristic is

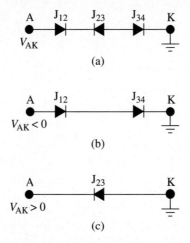

**Figure 13.3**   The diode model. (a) Arbitrary $V_{AK}$; (b) $V_{AK} < 0$; (c) $V_{AK} > 0$.

predicted to be basically the same as the reverse-bias characteristic of the $J_{23}$ diode. The $J_{12}$ diode dominating the $V_{AK} < 0$ characteristic and the $J_{23}$ diode dominating the $V_{AK} > 0$ characteristic are in turn expected to have similar characteristics because of the almost identical doping profiles of the two junctions. In other words, the SCR characteristic based on the diode model is concluded to be nearly symmetrical about $V_{AK} = 0$ and of the form pictured in Fig. 13.4.

Although providing insight and yielding the correct general form of both the reverse-bias and forward-bias blocking characteristics, the diode model fails to predict the forward-bias switching characteristics. Moreover, according to the diode model, $V_{BF}$ is approximately equal to $V_{BR}$, which in turn is approximately equal to the breakdown voltage of the $J_{12}$ (or $J_{23}$) junction. The $J_{12}$ breakdown voltage is understood to be the smaller of the

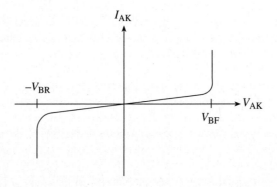

**Figure 13.4**   Predicted SCR characteristics based on the diode model.

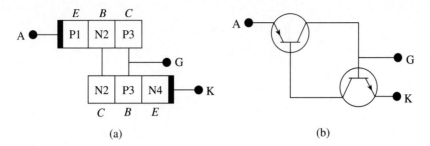

**Figure 13.5** The two-transistor model. (a) Diagrammatic and (b) equivalent circuit representations.

avalanche breakdown voltage and the punch-through voltage required to totally deplete the lowly doped N2 region. As it turns out, the blocking voltages predicted by the diode model are upper limits; the observed blocking voltages are invariably somewhat smaller.

The critical defect of the diode model is that it does not take into account the interaction between the three junctions. An interaction must exist because the widths of the N2 and P3 base regions are comparable to the respective minority carrier diffusion lengths. Correcting the defect, the interaction between junctions is taken into account in the *two-transistor model*. In the two-transistor model illustrated in Fig. 13.5, the SCR is conceptually subdivided into two interconnected transistors. The P1–N2–P3 regions form a *pnp* transistor and the N2–P3–N4 regions form an *npn* transistor. The P1 region is identified as the emitter of the *pnp* transistor, and the N4 region is taken to be the emitter of the *npn* transistor. Shorting wires connect the regions that are common to the two transistors. Please note for future reference that the BJTs in the model are intrinsically low-gain transistors. Because the base widths are only comparable to the diffusion lengths, the transport factors and therefore the gains will be considerably smaller than unity under normal operating conditions.

Let us now use the two-transistor model to explain how the SCR manages to switch from the high-impedance blocking mode to the low-impedance conducting mode. Figure 13.6 is included to assist in the discussion. Assume the SCR is in the blocking mode with $I_G = 0$ and $V_{AK} > 0$. With $V_{AK} > 0$ both transistors are active-mode biased and majority carriers from the P1 and N4 emitters are injected into the adjacent transistor bases ([1] in Fig. 13.6). A fraction of the injected carriers successfully diffuse across the base region [2] and enter the base of the other transistor [3]. Forming an excess of majority carriers that cannot be shunted into an external terminal, the carriers coming from the other transistor lead to an increase in the base-to-emitter injection, which in turn gives rise to added emitter-to-base injection [4]. These added carriers follow the same path as the originally injected carriers leading to even more injection. Clearly the process feeds back on itself or is regenerative.[†] At low applied $V_{AK}$ the process is stable and there is negligible

---

[†]Except for the source of the majority carriers entering the base, the process described here is identical to the regenerative process introduced in Subsection 11.2.4 to explain the reduction in $V_{CE0}$ of a BJT compared to $V_{CB0}$.

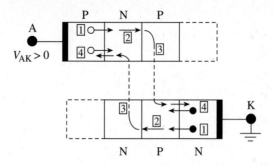

**Figure 13.6**  Use of the two-transistor model to describe the regenerative process that leads to switching. $\boxed{1}$ Initial carrier injection. $\boxed{2}$ Diffusion across the quasineutral base. $\boxed{3}$ Injected carriers enter the base of the other transistor. $\boxed{4}$ Additional injection induced by the majority carrier excess in the base.

carrier build-up about the $J_{23}$ junction. However, with increasing $V_{AK}$ a significant carrier build-up reduces the reverse bias and ultimately forward biases the $J_{23}$ junction. (Remember, a *pn* junction is forward biased if a minority carrier concentration in excess of the equilibrium value exists at the edges of the depletion region.) In terms of the two-transistor model, the transistors become saturation biased. The SCR enters the low-impedance conducting mode when all three junctions become forward biased.

The foregoing qualitative arguments outlined the internal process whereby the SCR switches, but it was rather vague about determining the critical switching voltage. Actually, quantitative considerations employing the two-transistor model are necessary to pin down the precise requirement for switching. Let us again assume the SCR is in the blocking mode with $I_G = 0$ and $V_{AK} > 0$. Being active-mode biased under the assumed conditions, the transistors in the model can be approximately represented by the large signal equivalent circuits shown in Figs. 13.7(a) and (b). These circuits follow from the Ebers–Moll equivalent circuit of Fig. 11.3. When active mode biased, the $\alpha_R I_R$ current generator in the Ebers–Moll circuit is negligible, $I_F \cong I_E$, and $I_R \cong -I_{R0}$. If the Fig. 13.7(a) and (b) equivalent circuits are connected according to the arrangement in Fig. 13.5, the result is Fig. 13.7(c). In Fig. 13.7(c) subscript 1 is associated with the *pnp* transistor and subscript 2 with the *npn* transistor. Equating the current flowing into and out of the E2 node in Fig. 13.7(c) yields

$$I_{AK} = \alpha_1 I_{AK} + I_{R01} + \alpha_2 I_{AK} + I_{R02} \tag{13.1}$$

or

$$I_{AK} = \frac{I_{R01} + I_{R02}}{1 - (\alpha_1 + \alpha_2)} \tag{13.2}$$

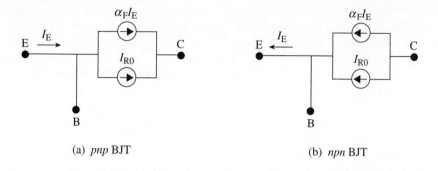

(a) *pnp* BJT

(b) *npn* BJT

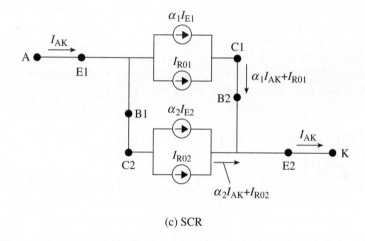

(c) SCR

**Figure 13.7** Simplified large-signal equivalent circuits for (a) an active-mode biased *pnp* transistor, (b) an active-mode biased *npn* transistor, and (c) an SCR in the blocking mode with $I_G = 0$ and $V_{AK} > 0$.

From Eq. (13.2) one infers that the critical switching voltage is reached when $\boxed{\alpha_1 + \alpha_2 \Rightarrow 1}$. As stressed earlier, BJTs in the model are intrinsically low-gain transistors; that is, the SCR is constructed so that $\alpha_1 + \alpha_2 < 1$ at small $V_{AK} > 0$. Due to a variety of mechanisms to be detailed shortly, $\alpha_1 + \alpha_2$ progressively increases with increasing $V_{AK}$. Thus eventually $\alpha_1 + \alpha_2$ approaches unity and the SCR switches into the conducting mode.

The mechanisms that cause $\alpha_1 + \alpha_2$ to increase with increasing $V_{AK}$ were all described in the BJT discussion. They include base-width modulation, an increase in the emitter efficiency, and carrier multiplication in the C–B junctions. In general, one can write

$$\alpha_{dc} = M\gamma\alpha_T \tag{13.3}$$

where, reviewing previous definitions, $\alpha_T$ is the base transport factor, $\gamma$ the emitter efficiency, and $M$ the carrier multiplication factor. When $V_{AK} > 0$ and the SCR is in the blocking mode, the $J_{23}$ junction is reverse biased. Most of the depletion region about the $J_{23}$ junction extends into the lightly doped N2 base of the *pnp* transistor. Thus increasing $V_{AK}$ causes considerable base-width modulation and a corresponding increase in $\alpha_T$. The discussion on BJT "figures of merit" in Subsection 11.2.8 pointed out that recombination–generation current reduces the BJT emitter efficiency at low injection levels. Increasing the injection level, increasing $V_{AK}$ in the SCR, is therefore also expected to increase the injection efficiency. Finally, if the increase in $\alpha_T$ and $\gamma$ are insufficient to cause $\alpha_1 + \alpha_2$ to approach unity, carrier multiplication in the reverse-biased $J_{23}$ junction will ultimately trigger the SCR. Obviously, as previously asserted, the breakdown voltage of the $J_{23}$ junction marks the upper limit of the forward-bias blocking voltage.

The only remaining experimental observation to be explained is the decrease in the forward-bias blocking voltage with increasing $I_G > 0$. The reason for the decrease may be obvious to the reader in light of the previous discussion. Very simply, a positive gate current into the P3 region causes majority carriers to be added to the base of the N2–P3–N4 transistor. Injection from the N4 emitter is thereby enhanced and the regenerative triggering process becomes unstable at a lower $V_{AK}$. Typically, only a few milliamps of gate current are necessary to turn on a multi-ampere anode current.

## 13.3  PRACTICAL TURN-ON/TURN-OFF CONSIDERATIONS

### 13.3.1 Circuit Operation

To provide a complete picture of SCR operation, it is useful to examine turn-on and turn-off from an external or circuit point of view. Consider the device biasing configuration and $V_{AK} > 0$ characteristics with superimposed load line shown in Fig. 13.8. We assume that

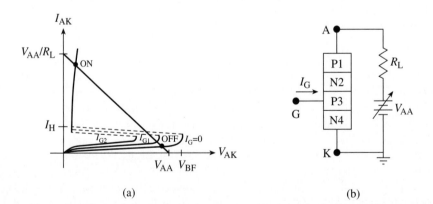

(a)                                                                (b)

**Figure 13.8**  (a) $V_{AK} > 0$ SCR characteristics and superimposed load line. (b) Assumed device biasing configuration.

the SCR is initially biased in the blocking mode at the "off" point in part (a) of the figure. One way to switch the device "on" is to increase the applied voltage $V_{AA}$. This causes a parallel displacement of the load line in the direction of increasing $V_{AK}$. When the load line no longer intersects the blocking-mode characteristic the SCR turns on. Alternatively, with $V_{AA}$ as shown, $I_G$ can be increased from zero until the high-impedance portion of the $I_G \neq 0$ curve [corresponding, for example, to the $I_{G2}$ curve in Fig. 13.8(a)] no longer intersects the load line. With no high-impedance intersection point, the device operation then switches to the pictured "on" point. It should be noted that once the SCR "latches on", the gate current can be turned off and the SCR will remain in the conducting mode.

If being operated at the "on" point shown in Fig. 13.8(a), the SCR can be turned off by decreasing $V_{AA}$. Once $I_{AK}$ is decreased below $I_H$, the low-impedance conducting mode can no longer be maintained and the device reverts to the blocking mode. Theoretically, an $I_G < 0$ or extracting gate current can also be used to turn off the SCR. However, the charge removal rate from the gated P3 region must be faster than the charge entry rate into the region. This necessitates a special device design to handle the relatively high gate currents. An $I_G \approx -10$ A, for example, would typically be required to turn off an $I_{AK} = 100$ A. Devices specially designed to provide $I_G$ turn-off are called GTO (gate turn-off) SCRs.

### 13.3.2  Additional Triggering Mechanisms

Thus far we have described triggering initiated by the electrical injection of majority carriers into the base regions of the SCR. The required carrier injection can also be achieved by shining a sufficiently intense pulse of $h\nu > E_G$ light on the device. Provision must be made of course to allow light penetration into a base region, usually the P3 region, of the device. The greatest advantage of photo-triggering is that it permits complete electrical isolation of the device input circuit. A specially fabricated SCR designed to accept a photo-input is called a light-activated SCR (LASCR) or light-activated switch (LAS).

SCRs can also be turned on by increasing the device operating temperature. The forward blocking voltage decreases with increasing temperature. Raising the operating temperature therefore has the same effect as applying an $I_G > 0$. The temperature dependence is quite strong, limiting the maximum operating temperature of SCRs to roughly 125–150°C.

Because of its extremely high gain [$I_{AK}(\text{on})/I_G(\text{turn-on}) \sim 10^4 - 10^6$], the SCR is susceptible to inadvertent switching due to noise pulses in the gate circuitry. This can be remedied at the expense of triggering sensitivity by using the shorted-cathode configuration described in the next subsection. An unexpected rise in the device temperature can likewise lead to inadvertent switching. Packaging of high-power devices in metal studs optimizes heat dissipation and minimizes temperature fluctuations.

### 13.3.3  Shorted-Cathode Configuration

If constructed as described in Section 13.1, $\alpha_1 + \alpha_2$, and hence $V_{BF}$ and the forward-bias characteristics in general, are extremely sensitive to the dopings and widths of the base regions. Only a slight variation in these parameters leads to a significant variation in the

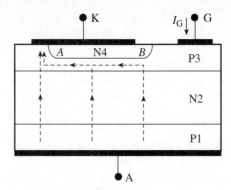

**Figure 13.9** The SCR shorted-cathode configuration.

device characteristics. Historically, the first commercial venture seeking to fabricate and sell PNPN diodes was a financial failure because the turn-on voltage could not be controlled within acceptable tolerances. The now-standard shorted-cathode configuration pictured in Fig. 13.9 was subsequently introduced to obtain reproducible switching characteristics.

The key feature of the shorted-cathode configuration is the extension of the cathode metallization over the P3 region. The extension adds an ohmic contact to the P3 region at a point remote from the gate. At low $I_{AK}$ current levels, the P3/N4 short essentially eliminates injection from the N4 region, and the gain of the N2–P3–N4 transistor is effectively zero. However, at higher current levels the lateral flow of current beneath the N4 region creates a forward-bias P3-to-N4 voltage drop that increases from point A to point B in Fig. 13.9. When the potential at the point B reaches a critical value, ~0.7 V or so, significant N4 to P3 injection takes place near the point B. In rapid succession, the $I_{AK}$ current then increases, the injecting portion of N4 widens from the point B toward the point A, and the gain of the N2-P3-N4 transistor rises abruptly. Shortly thereafter, injection is uniform across the face of the N4 region, $\alpha_1 + \alpha_2 \to 1$, and the SCR switches into the conducting mode. Note that the shorted-cathode arrangement makes it possible to obtain reproducible switching characteristics by simply controlling the resistance of the P3 region beneath the cathode.

### 13.3.4  *di/dt* and *dv/dt* Effects

A large, short-duration $I_G$ pulse is sometimes used to minimize the time required to turn on an SCR. However, if the gate current increases too rapidly in a shorted-cathode SCR, the injection current across the P3–N4 junction may rise to excessive levels near the gate before the injection becomes uniform across the face of the cathode. The concentration of current near the gate edge of the junction causes localized heating, which in turn can lead to thermal device failure. The described failure mechanism is known as *di/dt burnout*. Obviously, care must be taken to limit the *di/dt* of the turn-on pulse.

The application of a pulsed anode-to-cathode voltage also gives rise to a significant transient effect. Specifically, the SCR conducts prematurely; the forward-bias blocking voltage is less if the anode voltage is applied at a rapid rate. In fact, it is observed that the reduction in the rise time of an applied voltage pulse leads to a systematic decrease in the voltage required to turn on an SCR. Premature conduction likewise occurs under high-frequency a.c. operation or if the device is exposed to a high-voltage noise spike. This phenomenon is referred to as the *dv/dt effect*.

The *dv/dt* effect is directly related to the charging of the $J_{23}$ junction capacitance. The depletion width about the $J_{23}$ junction widens as the anode-to-cathode voltage is increased positively. The associated majority carrier current induces additional injection from the P1 and N4 emitters and thereby enhances the gains of the component transistors. The increased gain accounts for the premature conduction. The greater *dv/dt,* the greater the enhanced gain, the greater the reduction in the blocking voltage.

## 13.3.5  Triggering Time

For a PNPN device to switch from the blocking mode to the conducting mode, it is first necessary for carriers injected from the anode and cathode to cross the adjacent base. The average time taken by minority carriers to diffuse across a quasineutral base region was noted to be $W^2/2D_B$ in Chapter 12. If $W_2$ and $W_3$ are the respective quasineutral widths of the N2 and P3 base regions, then the corresponding transit times across the regions are $t_1 = W_2^2/2D_P$ and $t_2 = W_3^2/2D_N$. As a first-order approximation, the triggering time is taken to be the geometrical mean of the two transit times:

$$t_{ON} \approx \sqrt{t_1 t_2} = \frac{W_2 W_3}{2\sqrt{D_P D_N}} \tag{13.4}$$

The important point to be derived from the $t_{ON}$ result is that the base widths should be small for rapid switching. However, the smaller the base widths, the larger the component transistor gains and the smaller the punch-through voltages. Smaller base widths are therefore expected to lower the blocking voltages and decrease the power-handling capability of the device. Clearly, there is a trade-off between high-power handling capability and rapid switching.

## 13.3.6  Switching Advantages/Disadvantages

Both the SCR and BJT can function as electronic switches. Both the SCR and high-power versions of the BJT are used in power-control applications. Derived in large part from the preceding discussion and following Navon[16], the comparative advantages and disadvantages of the SCR are summarized below.

*Advantages*

(1) The SCR requires very little gate current to turn on very large anode-to-cathode currents.

(2) The SCR can block *both* polarities of an a.c. signal.

(3) The SCR has a very high blocking voltage capability combined with a low voltage drop in the conducting mode.

(4) Unlike the BJT, the SCR is not subject to current crowding when operating in the conducting mode under steady-state conditions.

### Disadvantages

(1) Whereas elimination of the base current turns off the BJT, SCRs cannot be turned off by setting $I_G = 0$.

(2) SCRs cannot operate at high frequencies.

(3) SCRs are prone to turn on by noise voltage spikes.

(4) SCRs have a limited temperature range of operation.

## 13.4  OTHER PNPN DEVICES

There are a number of other PNPN devices sold commercially in addition to those already mentioned. We briefly survey four of the more common structures.

A schematic representation of the dual-gate SCR or silicon controlled switch (SCS) is shown in Fig. 13.10. The SCS can be triggered by either gate and offers greater flexibility to the circuit designer. Devices that exhibit a symmetrical response to positive and negative applied voltages are said to be "bilateral." The idealized cross section and a sketch of the characteristics exhibited by a bilateral diode or DIAC are presented in Fig. 13.11. Figure 13.12 displays the cross section and characteristics of the bilateral SCR or TRIAC.

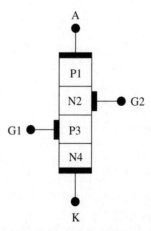

**Figure 13.10**  Schematic representation of the dual-gate SCR or silicon-controlled switch (SCS).

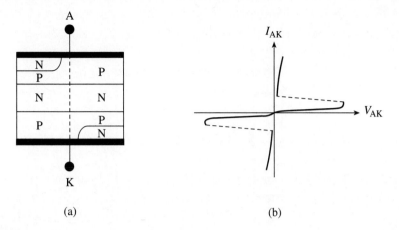

**Figure 13.11**  The DIAC (Diode AC). (a) Idealized cross section and (b) general form of the device characteristics.

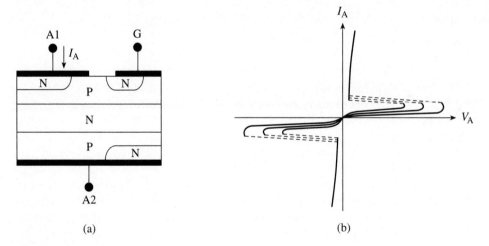

**Figure 13.12**  The TRIAC (Triode AC). (a) Idealized cross section and (b) general form of the device characteristics.

As is obvious from its cross section, the DIAC is just an integrated version of two oppositely oriented PNPN diodes connected in parallel. Similarly, the TRIAC functions as two oppositely oriented SCRs connected in parallel. The bilateral devices are especially suited for use in a.c. power-control applications.

Lastly, there is the so-called programmable unijunction transistor (PUT). Contrary to its name, the PUT is not a unijunction transistor! A unijunction transistor (UJT) is a three-

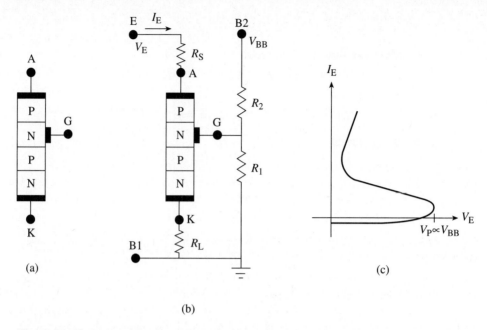

**Figure 13.13**   PUT (programmable unijunction transistor). (a) Structure; (b) circuit connection to simulate UJT characteristics; (c) *I–V* characteristics of a PUT.

lead device with two ohmic contacts at the ends of a bar-like resistive region. A third lead is connected to one side of a *pn* junction formed along the bar. The UJT exhibits switching characteristics grossly similar to those of the SCR. As illustrated in Fig. 13.13(a), the programmable unijunction transistor is actually an anode-gated PNPN device. Differing from a conventional SCR, the gate contact in a PUT is made to the N2 base instead of the P3 base. When connected as shown in Fig. 13.13(b), the PUT is functionally equivalent to a UJT, yielding characteristics like those sketched in Fig. 13.13(c). The characteristics can be modified by simply changing the external $R_1$ and $R_2$ resistors, thereby giving rise to the "programmability" of the structure.

# 14 MS Contacts and Schottky Diodes

The metal-semiconductor (MS) contact plays a very important role of one kind or another in all solid-state devices. When in the form of a non-rectifying or ohmic contact, it is the critical link between the semiconductor and the outside world. The rectifying MS contact, referred to as the Schottky diode or the MS diode, is found in a number of device structures and is an important device in its own right. Physically and functionally, there are close similarities between the MS diode and the asymmetrical ($p^+$-$n$ or $n^+$-$p$) step junction diode. Indeed, a large portion of the $pn$ diode analysis can be applied directly to the MS diode with only minor modifications.

We initiate the MS analysis by establishing the equilibrium energy band diagram for an ideal contact. With the aid of the diagram, one can readily distinguish between rectifying and ohmic contacts. The next section treats the electrostatics, $I$–$V$ characteristics, a.c. response, and transient response of the Schottky diode. The chapter concludes with a presentation of select practical information about MS contacts.

## 14.1 IDEAL MS CONTACTS

An ideal MS contact has the following properties: (1) The metal and semiconductor are assumed to be in intimate contact on an atomic scale, with no layers of any type (such as an oxide) between the components. (2) There is no interdiffusion or intermixing of the metal and semiconductor. (3) There are no adsorbed impurities or surface charges at the MS interface.

The initial task at hand is to construct the energy band diagram appropriate for an ideal MS contact under equilibrium conditions. The surface-included energy band diagrams for the individual, electrically isolated metal and semiconductor components are pictured in Fig. 14.1. Flat band (zero field) conditions are assumed to exist throughout the semiconductor. In both diagrams the vertical line where the energy bands are abruptly terminated is meant to represent a surface. The cross-hatching on the diagrams identifies allowed states that are nearly completely filled with electrons.

Several key energies and energy differences are readily introduced with the aid of Fig. 14.1. The ledge at the top of the vertical line denotes the minimum energy an electron must possess to completely free itself from the material and is called the *vacuum level, $E_0$.* The energy difference between the vacuum level and the Fermi energy is known as the

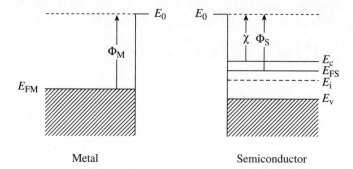

**Figure 14.1**  Surface-included energy band diagrams for a metal (left) and $n$-type semiconductor (right).

*workfunction* ($\Phi$) of the material. The metal workfunction, $\Phi_M$, is an invariant fundamental property of the specified metal. The value of $\Phi_M$ ranges from 3.66 eV for magnesium to 5.15 eV for nickel. The semiconductor workfunction, $\Phi_S$, is composed of two distinct parts; that is

$$\Phi_S = \chi + (E_c - E_F)_{FB} \tag{14.1}$$

The *electron affinity*, $\chi \equiv (E_0 - E_c)|_{\text{surface}}$, is an invariant fundamental property of the specified semiconductor. $\chi = 4.0$ eV, 4.03 eV, and 4.07 eV for Ge, Si, and GaAs, respectively. Conversely, $(E_c - E_F)_{FB}$, the energy difference between $E_c$ and $E_F$ under flat band or zero field conditions, is a computable function of the semiconductor doping.

Suppose now the $\Phi_M > \Phi_S$ metal and $n$-type semiconductor of Fig. 14.1 are brought together to form an ideal MS contact. Let us assume the contact formation is accomplished almost instantaneously so that there is negligible electron transfer between the components during the contacting process. If this be the case, then an instant after formation the energy band diagram for the contact will be as pictured in Fig. 14.2(a). In this figure the isolated energy band diagrams are vertically aligned to the common $E_0$ reference level and simply abutted at the mutual interface. It should be emphasized that $\Phi_M$ and $\chi$ are material constants and remain unaffected by the contacting process.

Since $E_{FS} \neq E_{FM}$, the MS contact characterized by Fig. 14.2(a) is obviously not in equilibrium. Under equilibrium conditions the Fermi level in a material or a group of materials in intimate contact must be invariant with position (see Subsection 3.2.4). Consequently, a short time after the conceptual contact formation, electrons will begin transferring from the semiconductor to the metal given the situation pictured in Fig. 14.2(a). The net loss of electrons from the semiconductor creates a surface depletion region and a growing barrier to electron transfer from the semiconductor to the metal. This will continue until the transfer rate across the interface is the same in both directions and $E_F$ is the same

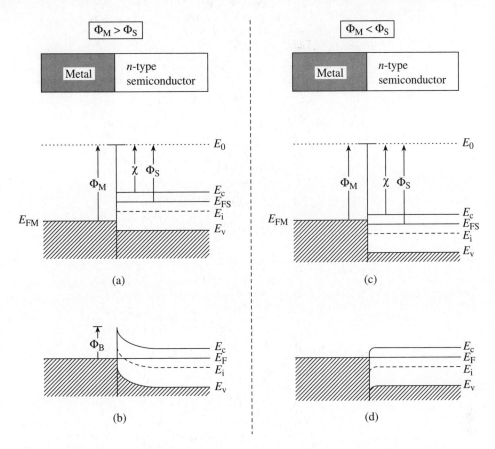

**Figure 14.2**  Energy band diagrams for ideal MS contacts between a metal and an $n$-type semiconductor: $\Phi_M > \Phi_S$ system (a) an instant after contact formation and (b) under equilibrium conditions;
$\Phi_M < \Phi_S$ system (c) an instant after contact formation and (d) under equilibrium conditions.

throughout the structure. The net result, the equilibrium energy band diagram for an ideal
$\Phi_M > \Phi_S$ metal to $n$-type semiconductor contact, is shown in Fig. 14.2(b). In drawing this
figure, extraneous lines such as the $E_0$ reference level and the portion of the vertical surface
line above $E_c$ have been removed. Also note that

$$\Phi_B = \Phi_M - \chi \quad \ldots \text{ ideal MS}(n\text{-type}) \text{ contact} \quad (14.2)$$

where $\Phi_B$ is the surface potential-energy barrier encountered by electrons with $E = E_F$ in
the metal. Finally, if the entire argument is repeated for a metal and $n$-type semiconductor
where $\Phi_M < \Phi_S$, one obtains the equilibrium energy band diagram shown in Fig. 14.2(d).

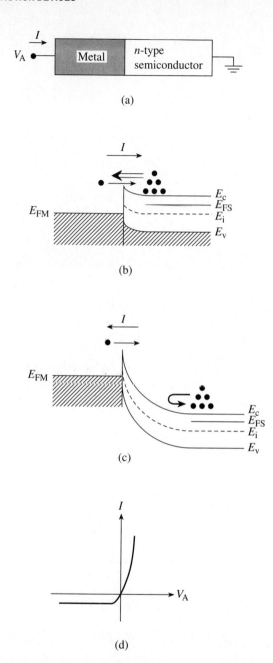

**Figure 14.3**  Response of the $\Phi_M > \Phi_S$ ($n$-type) MS contact to an applied d.c. bias. (a) Definition of current and voltage polarities. (b) Energy band diagram and carrier activity when $V_A > 0$. (c) Energy band diagram and carrier activity when $V_A < 0$. (d) Deduced general form of the $I$–$V$ characteristics.

Let us next qualitatively examine the effect of biasing the two MS structures in Fig. 14.2. As specified in Fig. 14.3(a), the semiconductor is taken to be grounded and $V_A$ applied to the metal. The current $I$ is defined to be positive when current flows from the metal to the semiconductor.

Consider first the $\Phi_M > \Phi_S$ MS contact. Applying a $V_A > 0$ as in Fig. 14.3(b) lowers $E_{FM}$ below $E_{FS}$, reduces the barrier seen by electrons in the semiconductor, and therefore permits a net flow of electrons from the semiconductor to the metal. Increasing $V_A$ leads to a rapidly rising forward bias current, since an exponentially increasing number of electrons from the semiconductor are able to surmount the surface barrier. On the other hand, applying a $V_A < 0$ raises $E_{FM}$ above $E_{FS}$ as pictured in Fig. 14.3(c). This all but blocks the flow of electrons from the semiconductor to the metal. Some electrons in the metal will be able to surmount the $\Phi_B$ barrier, but the associated reverse-bias current should be relatively small. Moreover, since $\Phi_B$ is ideally the same for all reverse biases, the reverse current is expected to remain constant after the reverse bias exceeds a few $kT/q$ volts. Clearly, we have just described rectifying characteristics similar to that displayed by a *pn* junction diode. The ideal *n*-type semiconductor to metal contact where $\Phi_M > \Phi_S$ is identified as an MS diode.

The response to an applied bias is considerably different for the $\Phi_M < \Phi_S$ MS contact. There is no barrier of any kind in the Fig. 14.2(d) structure for electron flow from the semiconductor to the metal. Thus even a small $V_A > 0$ gives rise to a large forward bias current. Under reverse biasing there is a small barrier for electron flow from the metal to the semiconductor, but the barrier essentially vanishes if the reverse bias exceeds a few tenths of a volt. Large reverse currents are expected at relatively small reverse biases, and the reverse current definitely does not saturate. The behavior here is obviously non-rectifying or ohmic-like.

The overall conclusion is that an ideal MS contact formed from a metal and an *n*-type semiconductor will be a rectifying contact if $\Phi_M > \Phi_S$ and an ohmic-like contact if $\Phi_M < \Phi_S$. Parallel arguments applied to an ideal MS contact formed from a metal and a *p*-type semiconductor lead to the conclusion that the contact will be rectifying if $\Phi_M < \Phi_S$ and ohmic-like if $\Phi_M > \Phi_S$. These conclusions are summarized in Table 14.1. It should be re-emphasized that all results and conclusions in this section are contingent upon the MS contact being ideal. Required modifications imposed by the often nonideal nature of real MS contacts are discussed in Section 14.3.

**Table 14.1**   Electrical Nature of Ideal MS Contacts.

|  | *n-type* Semiconductor | *p-type* Semiconductor |
|---|---|---|
| $\Phi_M > \Phi_S$ | Rectifying | Ohmic |
| $\Phi_M < \Phi_S$ | Ohmic | Rectifying |

**Exercise 14.1**

**P:** (a) Construct the equilibrium energy band diagram appropriate for an ideal $p$-type semiconductor to metal contact where $\Phi_M < \Phi_S$.

(b) Repeat part (a) when $\Phi_M > \Phi_S$.

(c) Verify that an ideal MS contact formed from a metal and a $p$-type semiconductor will be rectifying if $\Phi_M < \Phi_S$ and ohmic-like if $\Phi_M > \Phi_S$.

(d) Establish an expression for the barrier height, $\Phi_B \equiv E_{FM} - E_{v|\text{interface}}$, of the rectifying $p$-type contact.

**S:** (a)/(b) The "prescription" for drawing the equilibrium energy band diagram established in the text can be summarized as follows: (i) Draw the surface-included energy band diagrams for the individual components. (ii) Vertically align the diagrams to the common $E_0$ reference level, and join the diagrams at the mutual interface. (iii) Without changing the interfacial positioning of the semiconductor bands, move the field-free semiconductor bulk (the region far from the interface) up or down until $E_F$ is constant everywhere. (iv) Appropriately connect up the $E_c$, $E_i$, and $E_v$ at the interface with the field-free positioning of the bands in the semiconductor bulk. (v) Eliminate extraneous lines.

Following the cited prescription one obtains the equilibrium energy band diagrams shown below.

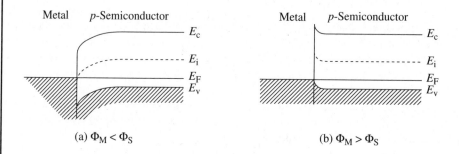

(a) $\Phi_M < \Phi_S$          (b) $\Phi_M > \Phi_S$

(c) Hole flow under bias must be examined to determine whether the given MS contacts are rectifying or ohmic. Empty electronic states in the metal, which decrease exponentially with energy below the Fermi level, can be thought of as holes for the purpose of the discussion. For the $\Phi_M < \Phi_S$ contact, there is clearly a barrier to hole flow in both directions under equilibrium conditions. Moving $E_{FM}$ upward relative to $E_{FS}$ reduces the barrier to hole flow from the semiconductor to the metal. The resulting S $\rightarrow$ M hole current is expected to increase exponentially with increased separation between $E_{FM}$ and $E_{FS}$. Reversing the bias blocks hole flow from the semiconductor to the metal, leaving only a saturating hole current from the metal to the semiconductor. The $\Phi_M < \Phi_S$ contact is obviously rectifying. For the $\Phi_M > \Phi_S$ contact, there is no barrier to hole flow from the semiconductor to the metal. Moreover, the small barrier to hole flow from the metal to the semiconductor vanishes if $E_{FM}$ is moved only slightly downward relative to $E_{FS}$. The $\Phi_M > \Phi_S$ contact is concluded to be ohmic-like, thereby completing the required verification.

(d) Since

$$E_{c|\text{interface}} - E_{FM} = \Phi_M - \chi$$

it follows that

$$\Phi_B = E_{FM} - E_{v|\text{interface}} = (E_c - E_v) - (E_{c|\text{interface}} - E_{FM})$$

or

$$\Phi_B = E_G + \chi - \Phi_M \qquad \ldots \text{ ideal MS}(p\text{-type}) \text{ contact}$$

## 14.2 SCHOTTKY DIODE

Having established the basic nature of the rectifying MS contact, we undertake here a more quantitative analysis of the Schottky (MS) diode. Following the usual outline, the analysis includes a survey of the d.c., a.c., and transient characteristics preceded by an examination of the device electrostatics. Strong parallels with the $pn$ junction diode permit a relatively condensed presentation. Throughout the discussion we take the semiconductor to be $n$-type and uniformly doped. The assumed current and applied voltage polarities are as specified in Fig. 14.3(a).

### 14.2.1 Electrostatics

#### Built-in Voltage

Like in the $pn$ junction diode, there is a voltage drop or built-in voltage across the MS diode under equilibrium conditions. Referring to Fig. 14.4(a), the built-in voltage ($V_{bi}$) is readily deduced to be

$$V_{bi} = \frac{1}{q} \left[ \Phi_B - (E_c - E_F)_{FB} \right] \tag{14.3}$$

where as previously noted $\Phi_B = \Phi_M - \chi$ for an ideal MS($n$-type) contact.

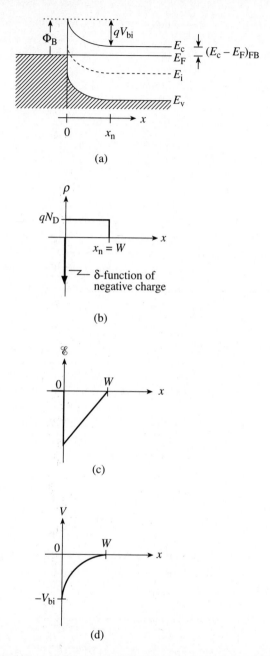

**Figure 14.4** Electrostatic variables in an MS (*n*-type) diode under equilibrium conditions. (a) Equilibrium energy band diagram. (b)–(d) Charge density, electric field, and electrostatic potential as a function of position.

## $\rho$, $\mathscr{E}$, and $V$

A further examination of Fig. 14.4(a) reveals there is a depletion of electrons in the semi-conductor adjacent to the MS interface. Completely analogous to the *pn* junction diode, the depletion region can be viewed as extending a distance $x_n$ into the *n*-type semiconductor, with a net positive charge arising from the donors in the depleted region. Unlike the *pn* junction, however, there is no negative *p*-side acceptor charge to balance the positive *n*-side donor charge. Rather, an essentially $\delta$-function of negative charge (extra electrons) piles up in the metal immediately adjacent to the interface. The charge density inside the structure is therefore concluded to be roughly as pictured in Fig. 14.4(b). Formally invoking the depletion approximation, we can write

$$\rho \cong \begin{cases} qN_D & \ldots 0 \le x \le W \\ \\ 0 & \ldots x > W \end{cases}$$

(14.4a)

(14.4b)

Note that $x_n$ can be equated to the depletion width $W$, since the depletion region is totally contained inside the semiconductor.

Given the $\delta$-function nature of the charge on the metal side of the interface, it automatically follows that $\mathscr{E} = 0$ and $V = $ constant inside the metal. Consequently, no further consideration need be given to the metal.

On the semiconductor side of the MS contact, the electric field and charge density are related through Poisson's equation (Eq. 5.2); namely,

$$\frac{d\mathscr{E}}{dx} = \frac{\rho}{K_S \varepsilon_0} \cong \frac{qN_D}{K_S \varepsilon_0} \quad \ldots 0 \le x \le W$$

(14.5)

Separating variables and integrating Eq. (14.5) from an arbitrary point $x$ in the depletion region to $x = W$ where $\mathscr{E} = 0$, we obtain

$$\int_{\mathscr{E}(x)}^{0} d\mathscr{E}' = \int_{x}^{W} \frac{qN_D}{K_S \varepsilon_0} dx'$$

(14.6)

or

$$\mathscr{E}(x) = -\frac{qN_D}{K_S \varepsilon_0} (W - x) \quad \ldots 0 \le x \le W$$

(14.7)

Plotted in Fig. 14.4(c), the Eq. (14.7) result is identical to the Eq. (5.21) solution for the $\mathscr{E}$-field on the *n*-side of a $p^+$-*n* step junction if $x_n$ is equated to $W$.

Seeking the solution for the electrostatic potential in the semiconductor, we next note

$$\frac{dV}{dx} = -\mathscr{E} = \frac{qN_D}{K_S\varepsilon_0}(W - x) \qquad \ldots 0 \leq x \leq W \tag{14.8}$$

Once again separating variables and integrating from an arbitrary point $x$ in the depletion region to $x = W$ where the potential is arbitrarily set equal to zero, we obtain

$$\int_{V(x)}^{0} dV' = \int_{x}^{W} \frac{qN_D}{K_S\varepsilon_0}(W - x')dx' \tag{14.9}$$

or

$$\boxed{V(x) = -\frac{qN_D}{2K_S\varepsilon_0}(W - x)^2 \qquad \ldots 0 \leq x \leq W} \tag{14.10}$$

At first glance the solution for $V(x)$ would appear to differ from that in the *pn* junction analysis. Actually, the solutions are totally equivalent. In treating the *pn* junction, the potential was set equal to zero on the far *p*-side of the junction; an MS diode solution of precisely the same form would result if the metal were the $V = 0$ reference. With the analysis focusing on a single semiconductor region as in the MS diode, however, the surface-side edge of the semiconductor bulk is the preferred reference point.

Under equilibrium conditions the potential drop across the depletion region is $V_{bi}$, $V = -V_{bi}$ at $x = 0$, and the $V(x)$ versus $x$ dependence is as sketched in Fig. 14.4(d). If $V_A \neq 0$, then $V_{bi} \rightarrow V_{bi} - V_A$ and $V = -(V_{bi} - V_A)$ at $x = 0$. The simple $V_{bi} \rightarrow V_{bi} - V_A$ replacement assumes of course that the back contact to the diode is ohmic and that the $IR$ potential drop across the semiconductor bulk is negligible.

## Depletion Width

Since $V(0) = -(V_{bi} - V_A)$, evaluating Eq. (14.10) at $x = 0$ gives

$$-(V_{bi} - V_A) = -\frac{qN_D}{2K_S\varepsilon_0}W^2 \tag{14.11}$$

Thus, identical to the situation in a $p^+$-$n$ step junction,

$$\boxed{W = \left[\frac{2K_S\varepsilon_0}{qN_D}(V_{bi} - V_A)\right]^{1/2}} \tag{14.12}$$

---

**Exercise 14.2**

**P:** Copper is deposited on a carefully prepared $n$-type silicon substrate to form an ideal Schottky diode. $\Phi_M \cong 4.65$ eV, $\chi = 4.03$ eV, $N_D = 10^{16}/cm^3$, and $T = 300$ K. Determine

(a) $\Phi_B$,

(b) $V_{bi}$,

(c) $W$ if $V_A = 0$, and

(d) $|\mathcal{E}|_{max}$ if $V_A = 0$.

**S:** (a) $\Phi_B = \Phi_M - \chi = \mathbf{0.62\ eV}$

(b) $(E_c - E_F)_{FB} \cong \dfrac{E_G}{2} - kT \ln\left(\dfrac{N_D}{n_i}\right) = 0.56 - (0.0259) \ln\left(\dfrac{10^{16}}{10^{10}}\right) \cong 0.20$ eV

$V_{bi} = \dfrac{1}{q} [\Phi_B - (E_c - E_F)_{FB}] = \mathbf{0.42\ V}$

(c) $W = \left[\dfrac{2K_S \varepsilon_0}{qN_D} (V_{bi} - V_A)\right]^{1/2} = \left[\dfrac{(2)(11.8)(8.85 \times 10^{-14})}{(1.6 \times 10^{-19})(10^{16})} (0.42)\right]^{1/2}$

$= \mathbf{0.234\ \mu m}$

(d) $|\mathcal{E}|_{max} = |\mathcal{E}|_{x=0}| = \dfrac{qN_D}{K_S \varepsilon_0} W = \dfrac{(1.6 \times 10^{-19})(10^{16})(2.34 \times 10^{-5})}{(11.8)(8.85 \times 10^{-14})}$

$= \mathbf{3.59 \times 10^4\ V/cm}$

---

### 14.2.2 I–V Characteristics

Whereas the MS diode electrostatics and the general shape of the MS diode $I$–$V$ characteristics are very similar to those of a $pn$ junction diode, the details of the d.c. current flow are decidedly different. In a $p^+$-$n$ diode, as reviewed in Fig. 14.5(a), the dominant components of the current typically arise from recombination in the depletion region under small forward biases and hole injection from the $p^+$ to the $n$-side of the diode under larger forward biases. The electron injection from the lighter doped $n$ to the $p^+$-side is always negligible. In an MS($n$-type) diode, as pictured in Fig. 14.5(b), the recombination and hole-injection currents still exist. However, because of the relatively low potential barrier seen by electrons in the semiconductor, electron injection from the semiconductor into the metal routinely dominates the observed current. Stated another way, the electron injection leads to a very large forward bias current before the recombination and diffusion (hole-injection) currents become important. The situation under reverse bias is similar. Electron flow from the metal to the semiconductor as previously pictured in Fig. 14.3(c) totally dominates the

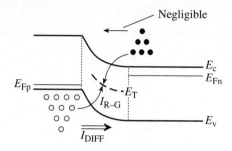

(a) $p^+$–$n$ junction diode

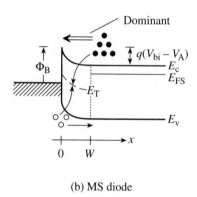

(b) MS diode

**Figure 14.5**  Negligible and dominant current components in a forward-biased (a) $p^+$-$n$ junction diode and (b) MS diode.

observed current. The reverse-bias hole diffusion current and the R–G current associated with carrier generation in the depletion region are typically negligible. Reflecting the fact that minority carriers in the semiconductor normally play an insignificant role in determining the $I$–$V$ and other characteristics, the MS diode is often said to be a "majority carrier device."

The current resulting from majority carrier electron or hole injection over the potential barrier in an MS diode is referred to as the *thermionic emission current.* To establish a quantitative expression for the thermionic emission current, we treat an $n$-type device and focus initially on electron injection from the semiconductor into the metal. The $x$-coordinate is assumed to be normal to the MS interface and directed into the semiconductor as shown in Fig. 14.5(b).

Consider an electron entering the depletion region from the semiconductor bulk. The electron is capable of surmounting the surface barrier and crossing into the metal if it has a $v_x$ velocity directed toward the interface such that

$$\text{KE}_x = \frac{1}{2} m_n^* v_x^2 \geq q(V_{bi} - V_A) \tag{14.13}$$

or

$$|v_x| \geq v_{min} \equiv \left[ \frac{2q}{m_n^*} (V_{bi} - V_A) \right]^{1/2} \tag{14.14}$$

Suppose there are $n(v_x)$ electrons/cm$^3$ in the semiconductor bulk with a negatively directed $v_x$ sufficient to surmount the barrier. Paralleling the drift current derivation in Subsection 3.1.2, the current associated with this set of electrons will be

$$I_{S\bullet \to M, v_x} = -qA v_x n(v_x) \tag{14.15}$$

Summing over all electrons in the conduction band with $v_x$ velocities capable of surmounting the barrier then gives

$$I_{S\bullet \to M} = -qA \int_{-\infty}^{-v_{min}} v_x n(v_x)\, dv_x \tag{14.16}$$

For a nondegenerate semiconductor it can be shown that[17]

$$n(v_x) = \left( \frac{4\pi kTm_n^{*2}}{h^3} \right) e^{(E_F - E_c)/kT} e^{-(m_n^*/2kT)v_x^2} \tag{14.17}$$

Substituting Eq. (14.17) into Eq. (14.16), integrating, and simplifying the result, one obtains

$$I_{S\bullet \to M} = A\mathscr{A}^* T^2 e^{-\Phi_B/kT} e^{qV_A/kT} \tag{14.18}$$

where

$$\mathscr{A}^* \equiv \left( \frac{m_n^*}{m_0} \right) \mathscr{A} \tag{14.19}$$

and

$$\mathscr{A} \equiv \frac{4\pi q m_0 k^2}{h^3} = 120 \text{ amps}/(\text{cm}^2 - \text{K}^2) \tag{14.20}$$

The constant $\mathscr{A}$ was introduced in a related analysis of electron emission from metals and has since become known as *Richardson's constant*.

Electrons crossing the interface in the opposite direction from the metal into the semiconductor always see the same potential barrier, $\Phi_B$. Consequently,

$$I_{M\bullet \to S}(V_A) = I_{M\bullet \to S}(V_A = 0) \tag{14.21}$$

Moreover, under equilibrium conditions the $M\bullet \to S$ and $S\bullet \to M$ currents across the barrier must precisely balance, or

$$I_{M\bullet \to S}(V_A = 0) = -I_{S\bullet \to M}(V_A = 0) = -A\mathscr{A}^*T^2 e^{-\Phi_B/kT} \tag{14.22}$$

The total current at an arbitrary $V_A$ is of course given by

$$I = I_{S\bullet \to M} + I_{M\bullet \to S} = I_{S\bullet \to M} + I_{M\bullet \to S}(V_A = 0) \tag{14.23}$$

Combining Eqs. (14.18) and (14.22), we therefore conclude

$$I = I_s(e^{qV_A/kT} - 1) \tag{14.24}$$

$$I_s \equiv A\mathscr{A}^*T^2 e^{-\Phi_B/kT} \tag{14.25}$$

The Eq. (14.24)/(14.25) result is clearly the MS diode version of the ideal diode equation. For forward biases greater than a few $kT/q$ volts, the exponential term in Eq. (14.24) dominates and $I \to I_s \exp(qV_A/kT)$. For reverse biases greater than a few $kT/q$ volts, the exponential term becomes negligible and the current is predicted to saturate at $I = -I_s$. As formulated, the theory implies $I$ would remain constant at $-I_s$ for reverse biases of unlimited magnitude.

Representative experimental $I$–$V$ characteristics derived from a MBR040 MS diode are displayed in Fig. 14.6. For forward biases of $V_A \leq 0.35$ V, experiment (the solid line curve in Fig. 14.6a) and theory are in almost perfect agreement. Over the range 0.1 V $\leq V_A \leq 0.35$ V the slope of the forward bias semilog plot is very close to $q/kT$. Like in the *pn* junction diode, the decrease in slope at larger forward biases is typically caused by an appreciable voltage drop across the bulk series resistance.

Turning to the reverse-bias characteristic in Fig. 14.6(b), we note two significant deviations from the ideal. First, as should have been anticipated, and again paralleling the *pn* junction diode, a breakdown phenomenon limits the maximum magnitude of the reverse bias voltage. In the absence of edge effects, the expected $V_{BR}$ due to avalanching in an MS diode is essentially identical to that of an equivalently doped $p^+$-$n$ or $n^+$-$p$ diode. Second, the reverse-bias current does not saturate. Differing from the *pn* junction diode, the observed behavior here cannot be attributed to the recombination–generation current. Rather, the systematic increase in the magnitude of the reverse-bias current is primarily the result of a phenomenon known as *Schottky barrier lowering*. $\Phi_B$ is not a bias-independent constant as assumed in the ideal theory, but decreases slightly with reverse biases of increasing magnitude. Specifically,

$$\Phi_B = \Phi_{B0} - \Delta\Phi_B \tag{14.26}$$

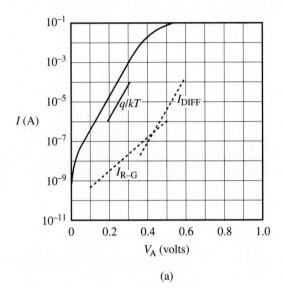

(a)

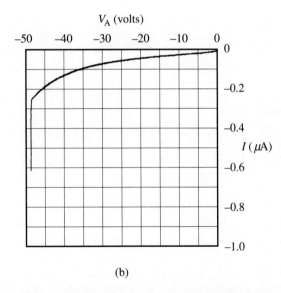

(b)

**Figure 14.6**  Measured $I–V$ characteristics derived from a MBR040 MS diode: (a) forward bias; (b) reverse bias. The dashed lines in (a) are theoretical estimates of the diffusion ($I_{DIFF}$) and recombination–generation ($I_{R–G}$) currents flowing in the diode. The experimental data were obtained employing an HP4145B Semiconductor Parameter Analyzer.

where $\Phi_{B0}$ is the barrier height when $\mathscr{E} = 0$ at the MS interface and

$$\Delta\Phi_B = q\left[\frac{q|\mathscr{E}_S|}{4\pi K_S \varepsilon_0}\right]^{1/2} \tag{14.27}$$

$\mathscr{E}_S$, the electric field at the semiconductor surface, is computed using the previously established electrostatic relationships. Since $I_s$ varies as $\exp(-\Phi_B/kT)$, even a small decrease in $\Phi_B$ gives rise to a readily noticeable increase in the magnitude of the reverse-bias current. We should point out that, at reverse biases approaching $-V_{BR}$, avalanche multiplication may play a role in enhancing the observed current.

Finally, a comment is in order concerning the dashed lines in Fig. 14.6(a) labeled $I_{DIFF}$ and $I_{R-G}$. These are theoretical estimates of the diffusion current and recombination current flowing in the given diode. As previously indicated, the $I_{DIFF}$ and $I_{R-G}$ components are seen to be totally negligible compared to the observed thermionic emission current.

---

## (C) Exercise 14.3

**P:** Assume the $0.1 \text{ V} \leq V_A \leq 0.35 \text{ V}$ portion of the forward-bias characteristic in Fig. 14.6(a) can be modeled by an equation of the form

$$I = I_s e^{qV_A/n_1 kT}$$

where $I_s$ and $n_1$ are constants. According to Eq. (14.24), the "ideality factor," $n_1$, should be equal to unity but is usually determined to be slightly greater than unity. Employing the point-by-point data provided in the table below, perform a least squares fit to determine the values of $I_s$ and $n_1$ that yield an optimum match to the given MS diode characteristics.

| $V_A$ (volts) | $I$ (amps) |
|:---:|:---:|
| 0.10 | $4.047 \times 10^{-7}$ |
| 0.15 | $2.792 \times 10^{-6}$ |
| 0.20 | $1.890 \times 10^{-5}$ |
| 0.25 | $1.263 \times 10^{-4}$ |
| 0.30 | $8.084 \times 10^{-4}$ |
| 0.35 | $4.487 \times 10^{-3}$ |

**S:** In logarithmic form the fit equation becomes

$$\ln(I) = \ln(I_s) + \frac{q}{n_1 kT} V_A$$

Associating $V_A$ with $x$ and $\ln(I)$ with $y$, the MATLAB `polyfit` function can be used to perform the required least squares fit to the experimental data. The best fit values derived from the program listed below are $\boxed{n_1 = 1.03}$ and $\boxed{I_s = 1.02 \times 10^{-8} \text{ amps}}$.

MATLAB program script...

```
%Least-Squares fit to MS diode I-V data
%ln(I)=ln(Is)+qVA/n1kT
I=[4.047e-7 2.792e-6 1.890e-5 1.263e-4 8.084e-4 4.487e-3];
VA=[0.1 0.15 0.2 0.25 0.3 0.35];
y=log(I);
c=polyfit(VA,y,1);      %least squares fit function; c(1)=slope, c(2)=ln(Is)
slope=c(1);
format compact
n1=1/(0.0259*slope)      %kT/q=0.0259V
Is=exp(c(2))
```

## 14.2.3 a.c. Response

A small a.c. signal superimposed on a d.c. reverse bias gives rise to a charge fluctuation inside the diode as pictured in Fig. 14.7. A variation in the depletion width about its equilibrium value and the associated change in charge inside the semiconductor balance fluctuations in the $\delta$-function of charge at the MS interface. Majority carriers move rapidly in and out of the semiconductor to facilitate the variation in the depletion width. With the charge fluctuation taking place along two planes separated by a distance $W$, the described a.c. situation is physically identical to what takes place inside a parallel plate capacitor. One can therefore write by analogy

$$C = \frac{K_S \varepsilon_0 A}{W} \tag{14.28}$$

or, making use of Eq. (14.12),

$$C = \frac{K_S \varepsilon_0 A}{\left[ \dfrac{2 K_S \varepsilon_0}{q N_D} (V_{bi} - V_A) \right]^{1/2}} \tag{14.29}$$

for a uniformly doped $n$-type semiconductor. Also note for future reference that if both sides of Eq. (14.29) are inverted and then squared, one obtains

$$\frac{1}{C^2} = \frac{2}{q N_D K_S \varepsilon_0 A^2} (V_{bi} - V_A) \tag{14.30}$$

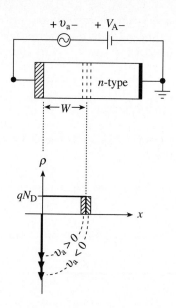

**Figure 14.7**  Charge fluctuations inside an MS ($n$-type) diode in response to an applied a.c. signal. $|v_\mathrm{a}| \ll V_\mathrm{bi} - V_\mathrm{A}$.

Reverse bias $C\text{–}V$ data derived from a commercial MBR040 MS diode are presented in Fig. 14.8(a). The data were obtained using the measurement set-up previously described in Subsection 7.2.3 and schematically pictured in Fig. 7.6. Variation of the capacitance with the applied d.c. bias is generally as expected. Consistent with the predicted $1/\sqrt{V_\mathrm{bi} - V_\mathrm{A}}$ dependence, the capacitance systematically decreases at a slower and slower rate as the reverse bias is increased in magnitude. Figure 14.8(b) provides a more detailed verification of the theory. The $1/C^2$ versus $V_\mathrm{A}$ plot of the experimental data exhibits a nearly straight line dependence in agreement with Eq. (14.30). Per Eq. (14.30) the semiconductor doping could be deduced from the slope of the straight line fitted to the plot points and $V_\mathrm{bi}$ from the extrapolated $1/C^2 = 0$ intercept. In addition, once both the semiconductor doping and $V_\mathrm{bi}$ are known, $\Phi_\mathrm{B}$ can be computed employing Eq. (14.3).

Although the reverse bias behavior is essentially identical to that of an asymmetrically doped $pn$ junction diode, the forward bias a.c. response of the MS diode is significantly different. In the MS diode the diffusion component of the current is typically negligible. Thus there is very little minority carrier injection and storage within the semiconductor. Since it is stored minority carriers that give rise to the diffusion admittance, *the MS diode does not exhibit a diffusion capacitance or diffusion conductance.* There is of course a forward-bias depletion-region capacitance, a potentially large parallel conductance $G = dI/dV_\mathrm{A}$, and a bulk series resistance ($R_\mathrm{S}$) that must be included under certain circumstances. However, even at a.c. frequencies routinely approaching or into the GHz range, $C$ and $G$ remain frequency independent.

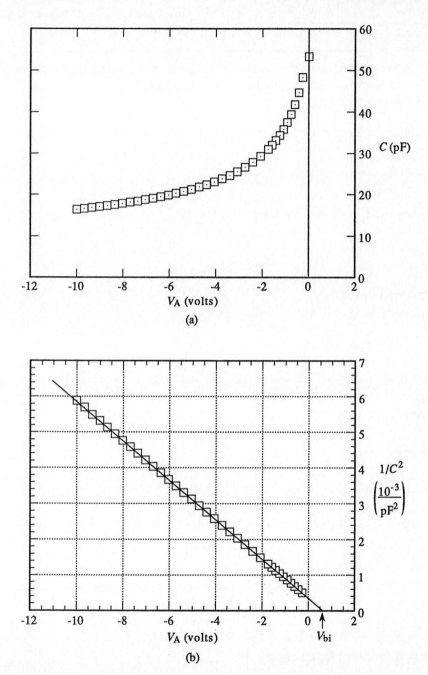

**Figure 14.8** (a) Sample $C-V$ data derived from a MBR040 MS diode. (b) $1/C^2$ versus $V_A$ plot constructed from the experimental $C-V$ data. [*Note:* To correct for the encapsulation-related stray capacitance shunting the MS diode, 3.4 pF was subtracted from all measured capacitance values before constructing the part (b) plot.]

**Exercise 14.4**

**P:** Working directly with the plotted data in Fig. 14.8(b), estimate the barrier height ($\Phi_B$) inside the MS diode. Assume an $n$-type device, $A = 1.5 \times 10^{-3}$ cm$^2$, and room temperature operation.

**S:** From the extrapolated $1/C^2$ intercept of the straight line drawn through the plot points, we roughly estimate

$$V_{bi} \cong 0.6 \text{ V}$$

The slope of the straight line is approximately

$$\text{slope} = -\frac{6 \times 10^{-3}/\text{pF}^2}{10.3 + 0.6} = -5.5 \times 10^{20}/\text{F}^2\text{-V}$$

and therefore

$$N_D = \frac{2}{qK_S\varepsilon_0 A^2|\text{slope}|}$$

$$\cong \frac{2}{(1.6 \times 10^{-19})(11.8)(8.85 \times 10^{-14})(1.5 \times 10^{-3})^2(5.5 \times 10^{20})}$$

$$= 9.7 \times 10^{15}/\text{cm}^3$$

Noting

$$(E_c - E_F)_{FB} \cong \frac{E_G}{2} - kT \ln\left(\frac{N_D}{n_i}\right) = 0.56 - 0.0259 \ln\left(\frac{9.7 \times 10^{15}}{10^{10}}\right)$$

$$= 0.20 \text{ eV}$$

we conclude, making use of Eq. (14.3),

$$\Phi_B = qV_{bi} + (E_c - E_F)_{FB} \cong 0.6 + 0.2 = \mathbf{0.8 \text{ eV}}$$

## 14.2.4 Transient Response

A very rapid transient response is the most distinctive characteristic of the MS diode. In *pn* junction devices the excess minority carriers stored in the quasineutral regions of the semiconductor must be removed before the device can be switched from the forward-bias on-

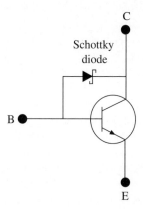

**Figure 14.9**   Schematic circuit representation of a Schottky-diode-clamped *npn* BJT.

state to the reverse-bias off-state. In an MS diode there is very little minority carrier injection and storage within the semiconductor because the diffusion component of the current is typically negligible. The reverse recovery time of commercial MS diodes is routinely only a few nanoseconds. The MS diode response time can in fact be limited, not by the stored charge, but by the internal *RC* time constant associated with the junction capacitance and the bulk series resistance. Small area devices, sometimes called *hot carrier diodes,* are commercially produced with a maximum $C \lesssim 1$ pF and sub-nanosecond response times.

As first noted in the Chapter 12 discussion of the BJT transient response, the MS diode has been used to speed up the BJT turn-off transient. The arrangement initially pictured in Fig. 12.7(a) and shown again in Fig. 14.9, an arrangement where an MS (Schottky) diode is connected between the base and collector of a BJT, is referred to as a *Schottky diode clamp.* Basically, when the transistor enters the saturation mode during the turn-on transient ($v_{BE} > 0$ and $v_{BC} > 0$), the MS diode begins to conduct and "clamps" the C–B junction at a relatively low forward bias. In other words, use is made of the fact that the MS diode conducts at a lower forward bias than a *pn* junction. Since the C–B junction is held at a relatively low voltage, there is minimal charge storage in the BJT. With less charge to be removed from the BJT and very little charge stored in the Schottky diode, the turn-off time is thereby significantly reduced.

## 14.3  PRACTICAL CONTACT CONSIDERATIONS

### 14.3.1 Rectifying Contacts

At the beginning of the chapter, an MS contact was defined to be ideal if the metal and semiconductor were in intimate contact on an atomic scale, there was no intermixing of components, and there were no adsorbed impurities or surface charges at the MS interface. Unfortunately, even though great progress has been made in recent years toward achieving the ideal MS contact, real MS structures invariably turn out to be nonideal. Silicon devices, for example, are likely to contain a thin (5 Å–25 Å) oxide layer between the metal and semiconductor. A native oxide layer forms almost instantly when Si is exposed to the atmosphere. Arsenic precipitates are believed to form on GaAs surfaces. Allowed electronic states that charge and discharge are known to exist on essentially all semiconductor surfaces. In extreme cases, the cited nonidealities can lead to an ill-functioning or non-functioning MS diode. The minimized nonidealities in modern-day structures, however, primarily affect the $\Phi_B$ barrier height characterizing the contact.

Contrary to Eq. (14.2), $\Phi_B \neq \Phi_M - \chi$ in most real $n$-type diodes. A similar statement can be made for $p$-type diodes. In Si, GaAs, and the majority of other semiconductors, surface charges tend to fix or "pin" the equilibrium Fermi level at a specific energy within the surface band gap. Because of this pinning effect, the observed barrier height normally varies only slightly with the metal used to fabricate the diode. Regardless of the metal employed in forming a GaAs MS diode, for example, $\Phi_B \cong 2E_G/3 = 0.95$ eV for $n$-type devices and $\Phi_B \cong E_G/3 = 0.47$ eV for $p$-type devices. Since $\Phi_B$ cannot be predicted from a prior knowledge of $\Phi_M$ and $\chi$, measurements must be performed to accurately determine $\Phi_B$ for a given MS system and fabrication procedure. An $C–V$ based approach similar to that outlined in Exercise 14.4 is one of the more popular measurement techniques. It should be emphasized that the formalism developed in the previous section still applies provided one employs the measured $\Phi_B$ in all relevant expressions.

### 14.3.2 Ohmic Contacts

Metal-semiconductor contacts that have a low impedance regardless of the biasing polarity are an essential part of just about every modern device structure. Although metal-semiconductor combinations where $\Phi_M < \Phi_S$ ($n$-type) and $\Phi_M > \Phi_S$ ($p$-type) ideally yield ohmic contacts, it follows from the barrier height comments in the preceding subsection that the noted metal-semiconductor combinations could produce rectifying contacts in practice. As a case in point, because the surface Fermi level tends to be pinned at $E_c - 2E_G/3$ in GaAs, the deposition of *any* metal on $n$-type GaAs forms a barrier-type contact. It is reasonable then to ask, "How are ohmic contacts achieved in practice?"

Ohmic contacts are usually produced in practice by heavily doping the surface region of the semiconductor immediately beneath the contact. In Si processing, for example, an $n^+$ on $n$ region as shown in Fig. 14.10(a) would be created prior to deposition of the metal. The reason this procedure leads to a low-impedance contact can be explained with the aid of Fig. 14.10(b). The equilibrium barrier height is to first order unaffected by an increase in the semiconductor doping. However, the depletion width, and hence the width of the barrier, systematically decrease with increased semiconductor doping. When the semiconductor doping exceeds $\sim 10^{17}$/cm$^3$, significant tunneling can take place through the thin upper portion of the barrier. For dopings exceeding $\sim 10^{19}$/cm$^3$, the entire barrier becomes so narrow that even low energy majority carriers can readily transfer between the semiconductor and metal via the tunneling process. In other words, although the barrier exists, it effectively becomes transparent to carrier flow when the contact is formed on a heavily doped semiconductor.

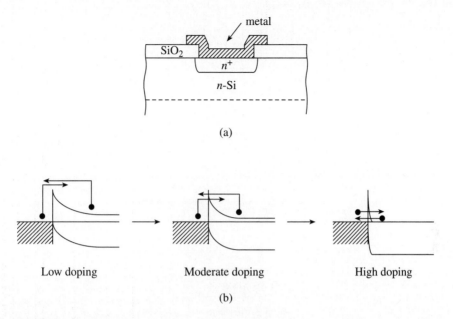

(a)

Low doping          Moderate doping          High doping

(b)

**Figure 14.10**   Ohmic contact formation. (a) Heavy doping of the semiconductor beneath the MS contact to facilitate ohmic contact formation. (b) Emission currents across a barrier-type contact as a function of doping. The emission is shown varying from solely thermionic emission at low semiconductor dopings to predominantly field emission (tunneling through the barrier) at high semiconductor dopings.

Whereas heavy doping of the underlying semiconductor is a key step in ohmic contact formation, it is seldom the entire story. Annealing or heating the device structure is routinely necessary to minimize the contact resistance. Aluminum is the most widely used contacting metal in the fabrication of discrete Si devices. Heating to approximately $475°\,C$ for a few minutes in a nitrogen atmosphere allows the aluminum to penetrate the native oxide layer on the Si surface and facilitates a certain amount of beneficial Al-Si interdiffusion. In contacts formed over shallow $p^+$-$n$ or $n^+$-$p$ junctions, however, nonuniform interdiffusion over the contact area can lead to penetration of the aluminum through the junction and junction shorting. Pictured schematically in Fig. 14.11(a), the nonuniform penetration of Al into the Si is called *spiking*. A small percentage of Si is sometimes added to the deposited Al film to inhibit Si diffusion from the substrate and the resultant spiking.

Another consideration is the stability of the ohmic contact during subsequent processing. Aluminum is not an acceptable contact material in complex integrated circuits where additional processing at temperatures in excess of $500°\,C$ must be performed after contact formation. Rather, high temperature stability is achieved by employing the metal-like silicon compounds (silicides) formed with members of the refractory metal family (Mo, Ta, Ti, W). Titanium silicide, $TiSi_2$, is presently in widespread use as a contact material. A Ti film deposited over the Si contact area is converted to $TiSi_2$ by heating in an inert atmosphere. Since Si is consumed in the process, the silicide-silicon interface moves a short

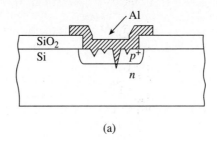

(a)

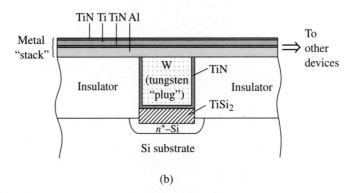

(b)

**Figure 14.11** (a) Visualization of spiking beneath an Al-Si contact. (b) Illustration of contact and interconnect metallization representative of that found in modern ICs.

distance into the Si, thereby eliminating surface defects and minimizing contamination. The net result is a clean, planar, temperature-stable, ohmic contact. Figure 14.11(b) illustrates the $TiSi_2$ contact and interconnect metallization representative of that found in modern ICs.

## 14.4 SUMMARY

The physical properties of an ideal MS contact were specified and the methodology was presented for constructing the energy band diagram characterizing a given metal-semiconductor system. Energy band diagrams corresponding to different combinations of the $\Phi_M$ and $\Phi_S$ workfunctions were employed to probe the basic nature of the contacts. Ideally, the combinations $\Phi_S(p\text{-type}) > \Phi_M > \Phi_S(n\text{-type})$ and $\Phi_S(p\text{-type}) < \Phi_M < \Phi_S(n\text{-type})$ were concluded to yield rectifying and ohmic contacts, respectively. Operation of the rectifying contact, called a Schottky diode or MS diode when utilized as a device, was next examined in greater detail. The device electrostatics and reverse-bias a.c. response of the MS diode were found to be all but identical to that of a comparably doped $p^+$-$n$ or $n^+$-$p$ junction diode. Differing from the *pn* junction diode, the thermionic emission current associated with majority carrier injection over the surface barrier was identified as the dominant d.c. current component. Because the diffusion current is typically small by comparison, the related minority carrier injection and storage are likewise small. As a consequence, the MS diode does not exhibit a significant diffusion admittance and switches between the on- and off-states at an extremely rapid rate. Although the basic description of device operation is unaffected, the $\Phi_B$ barrier height in actual Schottky diodes is seldom equal to the ideal device value. Moreover, the ohmic or rectifying nature of an MS contact may be different from that expected from ideal-contact considerations. An ohmic contact is usually produced in practice by heavily doping the surface region of the semiconductor immediately beneath the contact.

# PROBLEMS

| CHAPTER 14 | PROBLEM INFORMATION TABLE | | | |
|---|---|---|---|---|
| *Problem* | *Complete After* | *Difficulty Level* | *Suggested Point Weighting* | *Short Description* |
| 14.1 | 14.4 | 1 | 10 (1 each part) | Quick quiz |
| 14.2 | 14.1 | 2 | 10 (5 each part)/comb. | Draw/use band diagram |
| 14.3 | 14.2.1 | 2 | 8 (2 each part) | Parameter computation |
| ● 14.4 | " | 2 | 8 | $W$ versus $N_D$ plot |
| ● 14.5 | " | 4 | 30 (a-18, b-8, c-4) | Autodraw band diagram |
| 14.6 | 14.2.2 | 2 | 8 | Confirm Eq. (14.18) |
| 14.7 | " | 2–3 | 15 (a-4, b-5, c-4, d-2) | MS photodiode |
| ● 14.8 | " | 1 | 8 | Effect of $R_S$ on $I$–$V$ |
| 14.9 | " | 2 | 8 | Injection ratio |
| ● 14.10 | " | 3 | 15 (a-10, b-4, c-1) | Schottky barrier lowering |
| ● 14.11 | 14.2.3 | 2 | 12 | $C$–$V$ data analysis |
| 14.12 | " | 4 | 20 (a-12, b-6, c-2) | Linearly graded MS diode |

**14.1** Quick Quiz

(a) What are the differences between an MS diode, a Schottky diode, and a hot carrier diode?

(b) An ideal MS contact is formed between a metal and a semiconductor with $\Phi_M = \chi$. Under what conditions will the contact be ohmic-like, and under what conditions will the contact be rectifying?

(c) The solutions for the electrostatic variables $\rho$, $\mathscr{E}$, and $V$ on the lightly doped side of a $p^+$-$n$ or $n^+$-$p$ junction diode and inside the semiconductor of a comparably doped MS diode are all but identical. Are there any significant differences in the electrostatic formulations?

(d) Name the dominant current component in MS diodes.

(e) Somehow the $I_{M \bullet \to S}$ component of the current in an MS($n$-type) diode was determined without a detailed mathematical analysis of electron injection from the metal into the semiconductor. How was this accomplished?

(f) Explain why MS diodes do not exhibit a diffusion capacitance or diffusion conductance.

(g) Explain why MS diodes switch very rapidly from the forward-bias on-state to the reverse-bias off-state.

(h) What is a "Schottky diode clamp"?

(i) Describe the usual procedure followed in forming practical ohmic contacts.

(j) What is "spiking"?

**14.2** A number of ideal MS contacts are formed on Ge, Si, and GaAs substrates maintained at room temperature. For one or more of the MS parameter combinations listed below,

(a) Draw the equilibrium energy band diagram characterizing the ideal MS contact.

(b) Present an argument similar to the one summarized in Fig. 14.3 to confirm the ohmic or rectifying nature of the contact.

combination A: $\Phi_M = 4.75$ eV, $\chi(\text{Ge}) = 4.00$ eV, $N_D = 10^{16}/\text{cm}^3$.
combination B: $\Phi_M = 4.75$ eV, $\chi(\text{Ge}) = 4.00$ eV, $N_A = 10^{15}/\text{cm}^3$.
combination C: $\Phi_M = 4.00$ eV, $\chi(\text{Si}) = 4.03$ eV, $N_D = 10^{15}/\text{cm}^3$.
combination D: $\Phi_M = 4.25$ eV, $\chi(\text{Si}) = 4.03$ eV, $N_A = 10^{16}/\text{cm}^3$.
combination E: $\Phi_M = 4.75$ eV, $\chi(\text{GaAs}) = 4.07$ eV, $N_D = 10^{16}/\text{cm}^3$.
combination F: $\Phi_M = 4.75$ eV, $\chi(\text{GaAs}) = 4.07$ eV, $N_A = 10^{17}/\text{cm}^3$.

**14.3** An ideal rectifying contact is formed by depositing gold ($\Phi_M = 5.10$ eV) on an $N_D = 10^{15}/\text{cm}^3$ doped silicon substrate maintained at room temperature. Calculate

(a) $\Phi_B$,

(b) $V_{bi}$,

(c) $W$ under equilibrium conditions,

(d) $|\mathscr{E}|_{max}$ in the semiconductor under equilibrium conditions.

● **14.4** Construct a plot of the equilibrium depletion width versus the $N_D$ doping concentration in silicon MS diodes maintained at $T = 300$ K. Vary $N_D$ over the range $10^{14}/\text{cm}^3 \le N_D \le 10^{17}/\text{cm}^3$ and include curves corresponding to $\Phi_B = 0.5$ eV, 0.6 eV, and 0.7 eV.

● **14.5** (a) Paralleling Exercise 5.4, construct a MATLAB (computer) program that draws the equilibrium energy band diagram for a nondegenerately $N_D$-doped silicon MS diode maintained at room temperature. Let $\Phi_B$ and the $N_D$ doping of the semiconductor be input parameters. Be careful to exclude parameter combinations that would yield a nonrectifying contact.

(b) Generalize the part (a) program so that it also draws the equilibrium energy band diagram for $N_A$ doped silicon diodes.

(c) Revise the (a)/(b) programs so they can be used to generate the equilibrium energy band diagrams for GaAs MS diodes.

**14.6** Fill in the missing steps in the derivation of Eq. (14.18). Accepting Eqs. (14.16) and (14.17) to be accurate, perform and record the mathematical manipulations leading to Eq. (14.18).

**14.7** MS Photodiode

An $N_D$-doped silicon MS diode is illuminated, thereby generating electron-hole pairs inside the semiconductor.

(a) If the diode terminals are taken to be *short-circuited,* draw the energy band diagram for the device and picture what happens to the photogenerated carriers created in the semiconductor near the MS interface. With the current and voltage polarities as specified in Fig. 14.3(a), what is the polarity of the short-circuit photocurrent?

(b) If the diode terminals are taken to be *open-circuited,* draw the energy band diagram characterizing the illuminated device. Remember the total current must be identically zero under open-circuit conditions. Explain how you arrived at the form of your diagram.

(c) Assume the light is uniformly absorbed throughout the semiconductor producing a photogeneration rate of $G_L$ electron-hole pairs per cm$^3$-sec and giving rise to low-level injection. For an arbitrary applied bias, and following the simplified procedure outlined in Subsection 9.2.1, derive an expression for the photocurrent ($I_L$) flowing in the MS photodiode.

(d) Sketch the expected general form of the $I = I_{dark} + I_L$ versus $V_A$ characteristic exhibited by an illuminated MS photodiode. Is your sketch consistent with the answers to parts (a) and (b)?

● **14.8** The quasineutral portion of the semiconductor and the ohmic back contact of Schottky diodes introduce a resistance $R_S$ in series with the current flowing across the rectifying MS junction. Graphically illustrate the effect of the series resistance on the diode $I$–$V$ characteristic. Taking $I_s = 10^{-8}$ A, construct a semilog plot of the forward-bias $I$–$V$ characteristics when $R_S = 0$, $0.1\ \Omega$, $1.0\ \Omega$, and $10\ \Omega$. Limit your plotted output to $0 \le V_A \le 0.6$ V and $10^{-9}$ A $\le I \le 10^{-1}$ A.

**14.9** The *minority-carrier injection ratio* is often cited as a quantity of interest in characterizing MS diodes. By definition, it is the number of minority carriers injected into the semiconductor per majority carrier injected from the semiconductor into the metal when the device is forward biased. Mathematically, the ratio is just $I_{DIFF}/I_{TE}$. $I_{DIFF}$ and $I_{TE}$ are respectively the diffusion and thermionic emission currents flowing in the MS diode. Estimate the minority carrier injection ratio in a silicon MS diode where $\mathscr{A}^* = 140$ amps/cm$^2$-K$^2$, $\Phi_B = 0.72$ eV, $N_D = 10^{16}$/cm$^3$, $\tau_p = 10^{-6}$ sec, and $T = 300$ K. $I_{DIFF}$ in the given diode may be equated to the $I_{DIFF}$ flowing in a $p^+$-$n$ step junction diode with equivalent device parameters.

● **14.10** In this problem we wish to examine the magnitude and effect of Schottky barrier lowering. Parameters will be employed similar to those of the MS diode yielding the Fig. 14.6 characteristics.

(a) Given a silicon MS diode with $N_D = 10^{16}$/cm$^3$ operating at room temperature, compute and plot $\Delta\Phi_B$ versus $V_A$ for $-50$ V $\leq V_A \leq 0$.

(b) Compute and plot $I_s(V_A)/I_s(V_A = 0)$ versus $V_A$ for $-50$ V $\leq V_A \leq 0$.

(c) Comment on your results.

● **14.11** A subset of the $1/C^2$ versus $V_A$ data plotted in Fig. 14.8(b) is reproduced in Table P14.11. Analyze the data employing a least squares fit to determine $V_{bi}$, $N_D$, and $\Phi_B$. Assume $A = 1.5 \times 10^{-3}$ cm$^2$. Compare your results with the approximate values obtained in Exercise 14.4.

**Table P14.11**

| $-V_A$ (volts) | $1/C^2 (10^{21}$/farad$^2)$ |
|:---:|:---:|
| 1.09 | 0.953 |
| 2.08 | 1.494 |
| 3.07 | 2.035 |
| 4.06 | 2.579 |
| 5.05 | 3.125 |
| 6.04 | 3.673 |
| 7.03 | 4.217 |
| 8.02 | 4.763 |
| 9.01 | 5.320 |
| 10.00 | 5.890 |

**14.12** The doping profile inside the semiconductor component of an MS diode is linearly graded; i.e., $N_D(x) = ax$.

(a) Derive solutions for $\rho$, $\mathscr{E}$, $V$, and $W$ inside the semiconductor.

(b) Indicate how $V_{bi}$ is to be determined and computed.

(c) Establish an expression for the junction (depletion region) capacitance.

# R2 Part II
## SUPPLEMENT AND REVIEW

### ALTERNATIVE / SUPPLEMENTAL READING LIST

| Author(s) | Type (A–Alt., S–Supp.) | Level | Relevant Chapters |
|---|---|---|---|
| *General References* | | | |
| Streetman | A | Undergraduate | 5–7, 11 |
| Neamen | A | Undergraduate | 7–10 |
| Tyagi | A/S | Advanced Undergrad Introductory Grad | 6–10, 12–14, 18 |
| *For Selected Topics* | | | |
| Navon | A | Undergraduate | 12 (PNPN) |
| Sah | S | Undergraduate to Graduate (variable) | 7 (BJT, HBT and PNPN) |
| Yang | A/S | Undergraduate | 3 (*pn* electrostatics) 6 (BJT technology) 7 (MS diodes) |

(1) D. H. Navon, *Semiconductor Microdevices and Materials,* Holt, Rinehart and Winston, New York, © 1986.

(2) D. A. Neamen, *Semiconductor Physics and Devices, Basic Principles,* Irwin, Homewood, IL, © 1992.

(3) C. T. Sah, *Fundamentals of Solid-State Electronics,* World Scientific, Singapore, © 1991.

(4) B. G. Streetman, *Solid State Electronic Devices,* 4th edition, Prentice Hall, Englewood Cliffs, NJ, © 1995.

R2

(5) M. S. Tyagi, *Introduction to Semiconductor Materials and Devices,* John Wiley & Sons, New York, © 1991.

(6) E. S. Yang, *Microelectronic Devices,* McGraw Hill, New York, © 1988.

## FIGURE SOURCES / CITED REFERENCES

(1) S. M. Sze, *Physics of Semiconductor Devices,* 2nd edition, John Wiley & Sons, New York, © 1981.

(2) C. T. Sah, R. H. Noyce, and W. Shockley, "Carrier Generation and Recombination in P-N Junctions and P-N Junction Characteristics," Proceedings IRE, **45,** 1228 (Sept. 1957).

(3) D. K. Schroder, *Semiconductor Material and Device Characterization,* John Wiley and Sons, New York, © 1990.

(4) R. H. Kingston, "Switching Times in Junction Diodes and Junction Transistors," Proceedings IRE, **42,** 829–834 (1954).

(5) R. H. Dean and C. J. Neuse, "A Refined Step-Recovery Technique for Measuring Minority Carrier Lifetimes and Related Parameters in Asymmetric *p-n* Junction Diodes," IEEE Transactions on Electron Devices, **ED-18,** 151–158 (March 1971).

(6) E. S. Yang, *Microelectronic Devices,* McGraw Hill, New York, © 1988; page 383.

(7) M. A. Green, A. W. Blakers, J. Zhao, A. M. Milne, A. Wang, and X. Dai, "Characterization of 23-Percent Efficient Silicon Solar Cells," IEEE Transactions on Electron Devices, **37,** 331–336 (Feb. 1990).

(8) Photovoltaics: Program Overview Fiscal Year 1992, report produced by the National Renewable Energy Laboratory, DOE/CH10093-190 DE93000055, March 1993.

(9) M. G. Craford, "LEDs Challenge the Incandescents," IEEE Circuits and Devices, **8,** 24–29 (Sept. 1992).

(10) M. G. Craford, "Recent Developments in Light-Emitting-Diode Technology," IEEE Transactions on Electron Devices, **ED-24,** 935–943 (July 1977).

(11) S. S. Ahmed et al., "A Triple Diffused Approach for High Performance 0.8 $\mu$m BiCMOS Technology," Solid State Technology, **35,** 33 (Oct. 1992). Also see the other articles in the special series on BiCMOS Processing, edited by D. Gupta and M. T. Bohr, started in Solid State Technology, June 1992.

**R2**

(12) M. Lundstrom, "III-V Heterojunction Bipolar Transistors," Chapter 1 in *Heterojunction Transistors and Small Size Effects in Devices,* edited by M. Willander, Studentlitteratur and Chartwell Bratt, © 1992.

(13) I. R. C. Post, P. Ashburn, and G. R. Wolstenholme, "Polysilicon Emitters for Bipolar Transistors: A Review and Re-Evaluation of Theory and Experiment," IEEE Transactions on Electron Devices, **39,** 1717 (July 1992).

(14) For information about MBE, consult the device fabrication references listed in the R1 mini-chapter.

(15) S. Tiwari and D. J. Frank, "Empirical Fit to Band Discontinuities and Barrier Heights in III-V Alloy Systems," Applied Physics Letters, **60,** 630 (Feb. 3, 1992).

(16) D. H. Navon, *Semiconductor Microdevices and Materials,* Holt, Rinehart and Winston, New York, © 1986; pp. 410, 411.

(17) See, for example, S. M. Sze, *Physics of Semiconductor Devices,* 2nd edition, John Wiley and Sons, New York, © 1981; pp. 255, 256.

## REVIEW LIST OF TERMS

**R2**

Defining the following terms using your own words provides a rapid review of the Part II material.

(1) wide-base diode
(2) narrow-base diode
(3) ideal diode
(4) law of the junction
(5) quasineutral region
(6) breakdown
(7) impact ionization
(8) avalanching
(9) multiplication factor
(10) Zener process
(11) tunneling
(12) conductivity modulation
(13) punch-through (in narrow-base diode)
(14) high-level injection
(15) quasistatic
(16) hyperabrupt
(17) varactor
(18) abrupt

(19) profiling
(20) junction capacitance
(21) diffusion admittance
(22) turn-off transient
(23) storage delay time
(24) reverse recovery time
(25) reverse injection
(26) step recovery diode
(27) photodetector
(28) solar cell
(29) LED
(30) *p-i-n* diode
(31) avalanche photodiode
(32) fill factor
(33) shadowing (in solar cells)
(34) texturing (in solar cells)
(35) concentrator solar cells
(36) isoelectric trap

(37)  total internal reflection
(38)  BJT
(39)  emitter, base, collector
(40)  common base
(41)  common emitter
(42)  active mode
(43)  saturation mode
(44)  cutoff mode
(45)  inverted mode
(46)  buried layer
(47)  quasineutral base width
(48)  emitter efficiency
(49)  base transport factor
(50)  common base d.c. gain
(51)  common emitter d.c. gain
(52)  performance parameters
(53)  Ebers–Moll equations, model
(54)  forward gain
(55)  base-width modulation
(56)  Early effect
(57)  punch-through (in BJT)
(58)  regenerative
(59)  phototransistor
(60)  intrinsic transistor
(61)  base series resistance
(62)  current crowding
(63)  graded base (in BJT)
(64)  Gummel plot
(65)  BiCMOS
(66)  shallow emitter
(67)  polysilicon emitter
(68)  HBT

(69)  heterojunction
(70)  band alignment
(71)  lattice-matched
(72)  graded junction (in HBT)
(73)  Hybrid-Pi equivalent circuit
(74)  transconductance
(75)  base transit time
(76)  rise, storage delay, and fall times
(77)  Schottky diode clamp
(78)  thyristors
(79)  SCR
(80)  anode, cathode, gate
(81)  blocking voltages
(82)  two-transistor model
(83)  GTO SCR
(84)  LAS
(85)  $di/dt$ burnout
(86)  $dv/dt$ effect
(87)  DIAC, TRIAC
(88)  PUT
(89)  Schottky diode
(90)  vacuum level
(91)  metal workfunction
(92)  electron affinity
(93)  thermionic emission
(94)  Richardson's constant
(95)  Schottky barrier lowering
(96)  hot carrier diode
(97)  Fermi-level pinning
(98)  field emission
(99)  Al spiking
(100)  silicide

## PART II—REVIEW PROBLEM SETS AND ANSWERS

The following problem sets were designed assuming a knowledge—at times an integrated knowledge—of the subject matter in Chapters 5–11 of Part II. The sets could serve as a review or as a means of evaluating the reader's mastery of the subject. Problem Set A is adapted from a one-hour "open-book" examination; Problem Sets B and C are combinations of select problems from "closed-book" examinations. The answers to Problem Sets A and B are included at the end of this section. Answers are not provided for Problem Set C so that it can be used as a homework set with integrated-knowledge-type questions on the *pn* junction diode.

## Problem Set A

### Problem A1

A *pn* junction has the doping profile sketched below. Throughout this problem, assume the carrier concentrations may be neglected ($n = 0$, $p = 0$) in the $0 \le x \le x_i$ region of the diode.

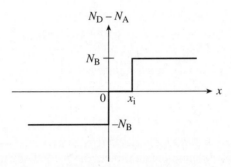

(a) What is the built-in voltage across the junction? Justify your answer.

(b) Invoking the depletion approximation, make a sketch of the charge density inside the diode. Label significant $\rho$ and $x$ values.

(c) Obtain an analytical solution for the electric field, $\mathcal{E}(x)$, at all points inside the depletion region ($-x_p \le x \le x_n$). Show all work and make a sketch of the deduced $\mathcal{E}(x)$ versus $x$.

(d) In a standard *pn* step junction $N_A x_p = N_D x_n$. How are $x_n$ and $x_p$ related here?

(e) Draw the energy band diagram for the diode under equilibrium conditions. Clearly identify the points $x = 0$ and $x = x_i$ on your diagram. Also indicate how your diagram differs from that of a simple *pn* step junction where $N_A = N_D = N_B$.

### Problem A2

Two silicon $p^+$-$n$ step junction diodes maintained at 300 K are physically identical except for the *n*-side doping. In diode #1, $N_D = 10^{14}/cm^3$; in diode #2, $N_D = 10^{16}/cm^3$. Compare the operation of the two diodes by answering the questions that follow.

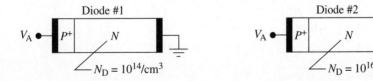

(a) Which diode will exhibit the larger built-in voltage ($V_{bi}$)? Explain.

(b) Which diode will exhibit the larger breakdown voltage ($V_{BR}$)? Explain.

(c) Which diode will exhibit the larger junction capacitance ($C_J$) at a given reverse bias voltage when $|V_A| \gg V_{bi}$? Explain.

(d) If the diodes are assumed to be ideal and reverse biased at a $|V_A| >$ few volts, which diode will support the larger $|I|$? Explain.

(e) If the diodes are *not* assumed to be ideal, which diode will support the larger $|I|$ when reverse biased at a $|V_A| >$ few volts? Explain.

(f) Which diode will exhibit the larger diffusion capacitance ($C_D$) at a given applied forward bias and frequency? Assume operation in the forward bias "ideal" region. Explain.

(g) Which diode will exhibit the larger storage delay time ($t_s$) under transient conditions if the $I_F/I_R$ ratio is the same? Explain.

### Problem A3

(a) The diagram below pictures the major carrier activity in a *pnp* BJT under active mode biasing. Construct a similar diagram picturing the major carrier activity in the *same pnp* BJT under *inverted* mode biasing. (Include a few words of explanation as necessary to forestall a misinterpretation of your diagram.)

(b) The pictured BJT is accidentally connected up backward so that the collector functions as the emitter and the emitter functions as the collector. In the backward connection with $V_{CB} > 0$ and $V_{EB} < 0$ the device exhibits a lower gain and is more sensitive to base-width modulation. (i) Explain why the backward connection leads to a lower gain. (ii) Explain why the backward connection leads to a greater sensitivity to base-width modulation.

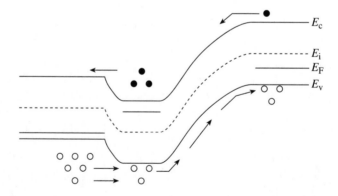

### Problem Set B

### Problem B1

A *pn* junction diode has the doping profile shown in the following sketch. Mathematically, $N_D - N_A = N_0 [\exp(\alpha x) - 1]$, where $N_0$ and $\alpha$ are constants.

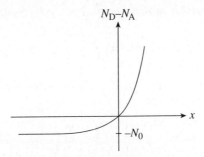

(a) Give a concise statement of the depletion approximation.

(b) Invoking the depletion approximation, make a sketch of the charge density inside the diode.

(c) Establish an expression for the electric field, $\mathcal{E}(x)$, inside the depletion region.

(d) *Indicate* how you would complete the electrostatic development to eventually obtain an expression for the depletion width, $W$. Be as specific as possible about the equations to be solved and the boundary conditions to be employed. Don't waste time actually performing the mathematical manipulations. Organize your answer into steps—step 1, step 2, etc.

## Problem B2

The energy band diagram given below characterizes a Si step junction diode maintained at room temperature. Note that $E_v(-\infty) = E_c(+\infty)$. Also, $x_n + x_p = 2 \times 10^{-4}$ cm, $A = 10^{-3}$ cm$^2$, $\tau_n = \tau_p = \tau_0 = 10^{-6}$ sec, $\mu_n(p\text{-side}) = 1352$ cm$^2$/V-sec, $\mu_p(n\text{-side}) = 459$ cm$^2$/V-sec, $K_S = 11.8$, and $\varepsilon_0 = 8.85 \times 10^{-14}$ F/cm.

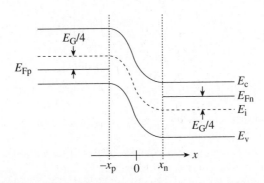

(a) What is the magnitude of the reverse-bias voltage $(V_A)$ being applied to the diode? Explain how you arrived at your answer.

(b) Determine $V_{bi}$, the built-in voltage.

(c) Compute the recombination–generation current flowing through the diode at the pictured bias point.

(d) Compute the diffusion current flowing through the diode at the pictured bias point.

(e) What will be the junction capacitance ($C_J$) exhibited by the diode at the pictured bias point?

(f) On a linear scale, sketch the minority carrier concentrations versus position for the $x \le -x_p$ and $x \ge x_n$ portions of the diode.

(g) Would you expect the device to exhibit a significant diffusion capacitance ($C_D$) at the applied bias point? Explain.

(h) If the diode were pulsed from the pictured bias point to a larger reverse bias at $t = 0$, would you expect to observe a current transient characterized by a storage delay time $t_s$? Explain.

## Problem B3

Two Si *pnp* transistors, BJT #1 and BJT #2, are identical except $W_{B1} > W_{B2}$. $N_E \gg N_B > N_C$ and $W \ll L_B$ in both transistors. Under the same active mode biasing conditions, which transistor will exhibit

(a) the larger emitter efficiency? Explain

(b) the larger base transport factor? Explain.

(c) the larger $\beta_{dc}$? Explain.

(d) the greater sensitivity to base-width modulation? Explain.

(e) the larger punch-through voltage? Explain.

If the maximum output current is assumed to be limited by carrier multiplication and avalanching, which transistor will exhibit

(f) the larger $V_{CB0}$? Explain.

(g) the larger $V_{CE0}$? Explain.

## Problem Set C

### Problem C1

(a) Which of the following assumptions is *not* invoked in deriving the ideal diode equation?
  (i) No recombination–generation in the depletion region.
  (ii) Low-level injection.
  (iii) Narrow-base diode; i.e., the $n$ and $p$ quasineutral widths are much less than the respective minority carrier diffusion lengths.
  (iv) No "other" process; i.e., no photogeneration, avalanching, tunneling, etc.

(b) Under reverse biasing and small forward biasing, the dominant current component in most Si *pn* junction diodes maintained at room temperature is which of the following?
  (i) The diffusion current.
  (ii) The R–G current.
  (iii) The ideal-diode current.
  (iv) The drift current.

(c) Which of the following statements is *incorrect?*

   (i) *pn* junction breakdown is a reversible process.

   (ii) For the Zener process to occur in a *pn* junction diode, the depletion width must be very narrow ($\leq 10^{-6}$ cm).

   (iii) The avalanche breakdown voltage varies roughly as the inverse of the doping concentration on the lightly doped side of $p^+$-$n$ and $n^+$-$p$ junctions.

   (iv) In Si diodes maintained at room temperature, avalanching is the dominant process causing breakdown if $V_{BR} \lesssim 4.5$ V.

(d) Which of the following statements about the junction capacitance ($C_J$) is *correct?*

   (i) $C_J$ always varies as $1/\sqrt{V_{bi} - V_A}$.

   (ii) The minimum observable $C_J$ will occur at $V_{BR}$.

   (iii) $C_J$ vanishes under forward biasing.

   (iv) $C_J$ is associated physically with fluctuations in the minority carrier concentrations at the edges of the depletion region.

## Problem C2

Minority carrier concentration versus position plots and sketches are often used to describe the situation inside semiconductor devices. A linear plot of the minority carrier concentration on the *n*-side of two *ideal $p^+$-$n$ diodes* maintained at room temperature is pictured below. The *n*-side doping ($N_D$) and the cross-sectional area ($A$) are the same in both diodes. Assume low-level injection conditions prevail.

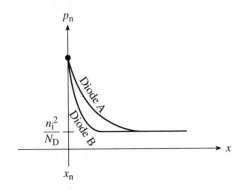

(a) The diodes are [(i) forward biased, (ii) zero biased, (iii) reverse biased].

(b) The magnitude of the bias applied to Diode B is [(i) larger than, (ii) the same as, (iii) smaller than] the magnitude of the bias applied to Diode A.

(c) The magnitude of the d.c. current, $|I|$, flowing through Diode B is [(i) significantly larger than, (ii) roughly the same as, (iii) significantly smaller than] the magnitude of the d.c. current flowing through Diode A.

(d) The breakdown voltage ($V_{BR}$) of Diode B is [(i) significantly larger than, (ii) roughly the same as, (iii) significantly less than] the breakdown voltage of Diode A.

(e) Diodes A and B are tested in the same switching circuit. $I_F/I_R$ is the same for both diodes. Which diode will exhibit the larger storage delay time (larger $t_s$)? [(i) Diode A; (ii) Diode B; (iii) $t_s$ will be essentially the same for both diodes.]

## Problem C3

The steady-state carrier concentrations inside a *pn* junction diode maintained at room temperature are as pictured below.

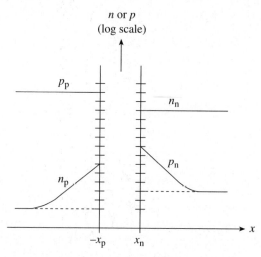

(a) Is the diode forward or reverse biased? Explain how you arrived at your answer.

(b) Do low-level injection conditions prevail inside the diode? Explain how you arrived at your answer.

(c) *Qualitatively,* what is the physical relationship between the pile-up or store of minority carriers near the depletion region edges and the diffusion capacitance ($C_D$)?

(d) *Qualitatively,* what is the physical relationship between the pile-up or store of minority carriers near the depletion region edges and the storage delay time ($t_s$) observed during the turn-off transient?

## Problem C4

The reverse-bias current–voltage ($I$–$V_A$), junction capacitance ($C_J$–$V_A$), and turn-off transient ($i$–$t$) characteristics derived from a $p^+$-$n$ Si step junction diode maintained at room temperature are sketched to the right. After reproducing the figures, answer the following questions by adding a *dashed line* to the appropriate characteristic. Note that an answer of *no effect* (a dashed line the same as the given characteristic) is possible. In such cases write *no effect.*

(a–c) *Roughly* indicate how the $I$–$V_A$, $C_J$–$V_A$, and $i$–$t$ characteristics are modified if the $n$-side doping ($N_D$) is *increased* by a factor of 2. All other parameters remain the same.

(d–f) *Roughly* indicate how the $I$–$V_A$, $C_J$–$V_A$, and $i$–$t$ characteristics are modified if the minority carrier lifetime on the $n$-side ($\tau_p$) and the effective depletion-region generation lifetime ($\tau_0$) are *increased* by a factor of 2. All other parameters remain the same.

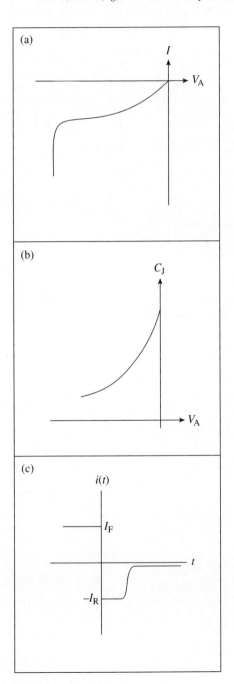

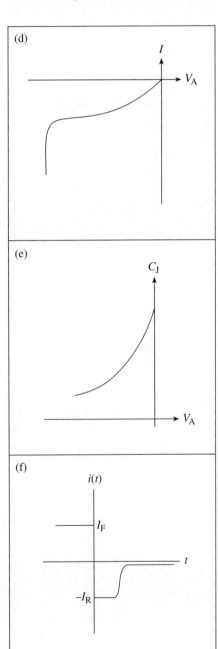

## Answers—Set A

### Problem A1

(a) Paralleling the derivation in Subsection 5.1.4, one can write $V_{bi} = (kT/q)\ln[n(x_n)/n(-x_p)]$. Here $n(x_n) = n(\infty) = N_B$ and $n(-x_p) = n(-\infty) = n_i^2/N_B$. Thus

$$\boxed{V_{bi} = \frac{kT}{q}\ln\left(\frac{N_B^2}{n_i^2}\right) = \frac{2kT}{q}\ln\left(\frac{N_B}{n_i}\right)}$$

(b)

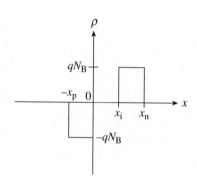

(c)

$$\frac{d\mathcal{E}}{dx} = \frac{\rho}{K_S\varepsilon_0} \cong \begin{cases} \dfrac{-qN_B}{K_S\varepsilon_0} & \cdots \; -x_p \le x \le 0 \\[2mm] 0 & \cdots \; 0 \le x \le x_i \\[2mm] \dfrac{qN_B}{K_S\varepsilon_0} & \cdots \; x_i \le x \le x_n \end{cases}$$

$$\int_0^{\mathcal{E}(x)} d\mathcal{E}' = -\int_{-x_p}^x \frac{qN_B}{K_S\varepsilon_0}\,dx' \Rightarrow \mathcal{E}(x) = -\frac{qN_B}{K_S\varepsilon_0}(x + x_p) \quad \cdots \; -x_p \le x \le 0$$

$$\mathcal{E}(x) = \text{constant} = \mathcal{E}(0) = -\frac{qN_B}{K_S\varepsilon_0}x_p \quad \cdots \; 0 \le x \le x_i$$

$$\int_{\mathcal{E}(x)}^{\mathcal{E}(x_n)=0} d\mathcal{E}' = \int_x^{x_n} \frac{qN_B}{K_S\varepsilon_0}\,dx' \Rightarrow \mathcal{E}(x) = -\frac{qN_B}{K_S\varepsilon_0}(x_n - x) \quad \cdots \; x_i \le x \le x_n$$

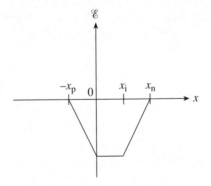

(d) The $(+)$ and $(-)$ charge areas on the part (b) $\rho$-plot must be equal, or

$$(-qN_B)(-x_p) = qN_B(x_n - x_i) \quad \Rightarrow \quad \boxed{x_p = x_n - x_i}$$

Alternatively, the $\mathcal{E}$-field must be continuous at $x = x_i$, giving

$$-\frac{qN_B}{K_S\varepsilon_0}\, x_p = -\frac{qN_B}{K_S\varepsilon_0}\,(x_n - x_i) \quad \Rightarrow \quad x_p = x_n - x_i$$

(e)

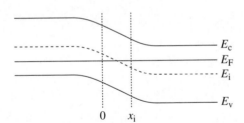

A constant energy band slope or straight lines in the $0 \le x \le x_i$ region is the only difference. Note that $(E_i - E_F)|_{p\text{-side}} = (E_F - E_i)|_{n\text{-side}}$ since $N_A = N_D = N_B$.

## Problem A2

(a) $\boxed{\text{Diode \#2}}$
    $V_{bi} = (1/q)\,[(E_i - E_F)|_{p\text{-side}} + (E_F - E_i)|_{n\text{-side}}].\ (E_i - E_F)|_{p\text{-side}}$ is the same for both diodes. $(E_F - E_i)|_{n\text{-side}} = kT\ln(N_D/n_i)$ is larger for the heavier doped diode. Thus $V_{bi2} > V_{bi1}$. Note that $p^+$ implies degenerate or very heavy doping. Thus the standard relationship, $V_{bi} = (kT/q)\ln(N_A N_D/n_i^2)$ cannot be used to compute $V_{bi}$.

(b) [Diode #1]

$V_{BR}$ is roughly proportional to $1/N_D$ in a $p^+$-$n$ diode. Also, Fig. 6.11 shows $V_{BR} > 1000$ V if $N_D = 10^{14}/\text{cm}^3$, while $V_{BR} \cong 60$ V if $N_D = 10^{16}/\text{cm}^3$.

(c) [Diode #2]

$$C_J = \frac{K_S \varepsilon_0 A}{W}$$

For a $p^+$-$n$ step junction

$$W = \left[\frac{2K_S \varepsilon_0}{qN_D} (V_{bi} - V_A)\right]^{1/2} \cong \left[\frac{2K_S \varepsilon_0}{qN_D} (-V_A)\right]^{1/2}$$

where the latter form of the $W$ equation follows from the fact that $|V_A| \gg V_{bi}$. Thus, $W_1 > W_2$ and $C_{J2} > C_{J1}$.

(d) [Diode #1]

For an ideal $p^+$-$n$ diode reverse biased to $|V_A| >$ few volts, we can write

$$I \cong -I_0 \cong -qA \frac{D_P}{L_P} \frac{n_i^2}{N_D}$$

Now $1/N_{D1} > 1/N_{D2}$. Also

$$\frac{D_P}{L_P} = \frac{D_P}{\sqrt{D_P \tau_p}} = \sqrt{\frac{D_P}{\tau_p}} = \sqrt{\frac{kT}{q} \frac{\mu_p}{\tau_p}}$$

Since $\mu_p$ decreases with increasing $N_D$, $D_{P1}/L_{P1} > D_{P2}/L_{P2}$. Both of the cited factors in the $I$ expression are such as to make $I_{01} > I_{02}$.

(e) [Diode #1]

In a real Si diode maintained at 300 K the reverse current is typically dominated by the recombination–generation current component.

$$I_{R-G} = -qA \frac{n_i}{2\tau_0} W \quad \ldots \text{ given } -V_A > \text{ few volts}$$

From the Eq. (6.44) definition of $\tau_0$ we infer $\tau_0 \propto 1/N_T$ (both $\tau_n$ and $\tau_p$ are proportional to $1/N_T$). Only the doping is specified as being different in the two diodes, not $N_T$. Thus $\tau_0$ is the same for both diodes. However, $W_1 > W_2$ from part (c). Consequently, $|I_{R-G1}| > |I_{R-G2}|$.

(f) Diode #1

Assuming the diodes are forward biased into the ideal region of operation, $C_D \propto G_0 \propto (I + I_0) \propto I_0$. $G_0$ is the diode low-frequency conductance. However, $I_{01} > I_{02}$ from part (d). Thus $C_{D1} > C_{D2}$.

(g) Same for both diodes

Examining either the approximate solution [Eq. (8.8)] or the more accurate solution [Eq. (8.9)] for $t_s$, one concludes $t_{s1} = t_{s2}$ if the $I_F/I_R$ ratio and $\tau_p \propto 1/N_T$ are the same for the two diodes. The $I_F/I_R$ ratio is noted to be the same in the problem statement. $N_T$ and therefore $\tau_p$ are also inferred to be the same in the two diodes, since only the $N_D$ doping is specified as being different.

## Problem A3

(a)

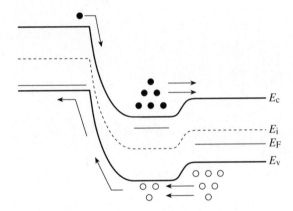

(b) (i) The efficiency of the injecting collector is far below that of the emitter and $\alpha_{dc} = \gamma a_T$ is thereby degraded. The E–B junction has $N_E \gg N_B$, making hole injection into the base much greater than electron injection from the base into the emitter under active mode biasing. On the other hand, $N_C < N_B$, giving rise to significant electron injection from the base into the collector under "backward" (inverted) operation.

(ii) When used as an amplifier, the larger depletion region is associated with the reverse-biased junction. With the C–B junction reverse biased under active mode operation, most of the depletion region extends into the collector because $N_B > N_C$. However, when the E–B junction is reverse biased, as is the case under "backward" operation, most of the E–B depletion region extends into the *base* because $N_E \gg N_B$. Much larger variations of $W$ for equivalent biasing conditions are expected under backward operation.

## Answers—Set B

### Problem B1

(a) In the depletion approximation one makes the following simplifying assumptions: (i) $p$ and $n$ are much less than $|N_D - N_A|$ in a $-x_p \leq x \leq x_n$ region about the metallurgical junction. (ii) $\rho = 0$ elsewhere.

(b)

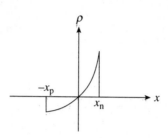

(c)
$$\frac{d\mathscr{E}}{dx} = \frac{\rho}{K_S \varepsilon_0} \cong \frac{qN_0}{K_S \varepsilon_0}(e^{\alpha x} - 1) \qquad \ldots \; -x_p \leq x \leq x_n$$

$$\int_0^{\mathscr{E}(x)} d\mathscr{E}' = \frac{qN_0}{K_S \varepsilon_0} \int_{-x_p}^{x} (e^{\alpha x'} - 1)\, dx' = \frac{qN_0}{K_S \varepsilon_0}[e^{\alpha x'}/\alpha - x']|_{-x_p}^{x}$$

$$\boxed{\mathscr{E}(x) = \frac{qN_0}{K_S \varepsilon_0}\left[\frac{1}{\alpha}(e^{\alpha x} - e^{-\alpha x_p}) - (x + x_p)\right]}$$

(d) Step 1: Solve for $V(x)$ in the $-x_p \leq x \leq x_n$ depletion region.
$$\frac{dV}{dx} = -\mathscr{E}(x); \qquad V = 0 \text{ at } x = -x_p$$

Step 2: Two equations with $x_n$, $x_p$, and $V_{bi}$ as unknowns are obtained by imposing the boundary conditions, $\mathscr{E}(x_n) = 0$ and $V(x_n) = V_{bi} - V_A$. Once $V_{bi}$ is known, these two equations can be solved for the $x_n$, $x_p$, and $W = x_n + x_p$ resulting from a given $V_A$. In this particular problem, the solution will have to be obtained numerically.

Step 3: $V_{bi}$ is obtained by a procedure paralleling the linearly graded analysis in Subsection 5.2.5.

### Problem B2

(a) $V_A = -\dfrac{1}{q}(E_{Fp} - E_{Fn}) = -\dfrac{E_G}{2q} = -0.56$ V

(b) $V_{\text{bi}} = \dfrac{1}{q}[(E_{\text{i}} - E_{\text{F}})|_{p\text{-side}} + (E_{\text{F}} - E_{\text{i}})|_{n\text{-side}}] = \dfrac{E_{\text{G}}}{2q} = \mathbf{0.56\ V}$

(c) $I_{\text{R-G}} = -qA\dfrac{n_{\text{i}}}{2\tau_0} W = -(1.6 \times 10^{-19})(10^{-3})\left(\dfrac{10^{10}}{2 \times 10^{-6}}\right)(2 \times 10^{-4})$

$\qquad = \mathbf{-1.6 \times 10^{-10}\ A}$

(d) $D_{\text{N}} = \dfrac{kT}{q}\mu_{\text{n}} = (0.0259)(1352) = 35.02\ \text{cm}^2/\text{sec}$

$\quad L_{\text{N}} = \sqrt{D_{\text{N}}\tau_{\text{n}}} = [(35.02)(10^{-6})]^{1/2} = 5.92 \times 10^{-3}\ \text{cm}$

$\quad D_{\text{P}} = \dfrac{kT}{q}\mu_{\text{p}} = (0.0259)(459) = 11.89\ \text{cm}^2/\text{sec}$

$\quad L_{\text{P}} = \sqrt{D_{\text{P}}\tau_{\text{p}}} = [(11.89)(10^{-6})]^{1/2} = 3.45 \times 10^{-3}\ \text{cm}$

$\quad N_{\text{A}} = N_{\text{D}} = n_{\text{i}}e^{E_{\text{G}}/4kT} = (10^{10})(e^{1.12/[4(0.0259)]}) = 4.96 \times 10^{14}/\text{cm}^3$

$\quad I_{\text{Diff}} \cong -I_0 = -qA\left(\dfrac{D_{\text{N}}}{L_{\text{N}}}\dfrac{n_{\text{i}}^2}{N_{\text{A}}} + \dfrac{D_{\text{P}}}{L_{\text{P}}}\dfrac{n_{\text{i}}^2}{N_{\text{D}}}\right)$

$\qquad = -(1.6 \times 10^{-19})(10^{-3})\left[\left(\dfrac{35.02}{5.92 \times 10^{-3}}\right)\left(\dfrac{10^{20}}{4.96 \times 10^{14}}\right)\right.$

$\qquad\qquad \left. + \left(\dfrac{11.89}{3.45 \times 10^{-3}}\right)\left(\dfrac{10^{20}}{4.96 \times 10^{14}}\right)\right]$

$\qquad = \mathbf{-3.02 \times 10^{-13}\ A}$

(e) $C_{\text{J}} = \dfrac{K_{\text{S}}\varepsilon_0 A}{W} = \dfrac{K_{\text{S}}\varepsilon_0 A}{x_{\text{n}} + x_{\text{p}}} = \dfrac{(11.8)(8.85 \times 10^{-14})(10^{-3})}{2 \times 10^{-4}} = \mathbf{5.22\ pF}$

(f)

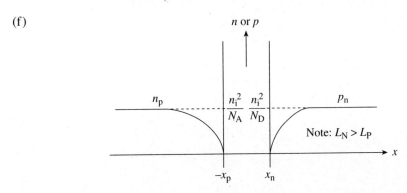

(g) No. There is no store of minority carriers adjacent to the depletion region.

(h) No. Again, there is no store of minority carriers adjacent to the depletion region. (The device is already reverse biased prior to applying the voltage pulse.)

## Problem B3

(a) $\boxed{\text{BJT \#2}}$  If $W_{B1} > W_{B2}$, then $W_1 > W_2$ under the same biasing conditions and

$$\gamma_1 = \frac{1}{1 + \dfrac{D_E}{D_B} \dfrac{N_B}{N_E} \dfrac{W_1}{L_E}} < \gamma_2 = \frac{1}{1 + \dfrac{D_E}{D_B} \dfrac{N_B}{N_E} \dfrac{W_2}{L_E}}$$

(b) $\boxed{\text{BJT \#2}}$ $\qquad \alpha_{T1} = \dfrac{1}{1 + \dfrac{1}{2}\left(\dfrac{W_1}{L_B}\right)^2} < \alpha_{T2} = \dfrac{1}{1 + \dfrac{1}{2}\left(\dfrac{W_2}{L_B}\right)^2}$

Physically, fewer carriers are lost via recombination in crossing the narrower base.

(c) $\boxed{\text{BJT \#2}}$  With $\gamma_1 < \gamma_2$ and $\alpha_{T1} < \alpha_{T2}$, $\alpha_{dc1} < \alpha_{dc2}$ since $\alpha_{dc} = \gamma\alpha_T$. Noting that $\beta_{dc} = \alpha_{dc}/(1 - \alpha_{dc})$, we conclude $\beta_{dc2} > \beta_{dc1}$.

(d) $\boxed{\text{BJT \#2}}$  Changes in bias give rise to the same $\Delta W$ in the two transistors. However, with $W_1 > W_2$, $\Delta W/W_2 > \Delta W/W_1$, making BJT #2 more sensitive to base width modulation.

(e) $\boxed{\text{BJT \#1}}$  Because $W_{B1} > W_{B2}$ and changes in bias give rise to the same $\Delta W$, it will clearly take a larger applied $V_{CB}$ to completely deplete the wider BJT #1 base under active mode biasing.

(f) $\boxed{\text{Same for \#1 and \#2}}$  $V_{CB0}$ is equal to the $V_{BR}$ of the C–B junction. The avalanche breakdown voltage of the C–B junction should be the same for both transistors since the junction dopings are the same.

(g) $\boxed{\text{BJT \#1}}$  According to Eq. (11.54), $V_{CE0} = V_{CB0}/(\beta_{dc} + 1)^{1/m}$, where $3 \leq m \leq 6$. Now, as noted in answering part (c), $\beta_{dc1} < \beta_{dc2}$. Thus, $V_{CE01} > V_{CE02}$.

# PART III

# FIELD EFFECT DEVICES

# 15 Field Effect Introduction— The J-FET and MESFET

## 15.1 GENERAL INTRODUCTION

Historically, the field effect phenomenon was the basis for the first type of solid-state transistor ever proposed. Field effect transistors predate the bipolar junction transistor by approximately 20 years. As recorded in a series of patents filed in the 1920s and 1930s, J. E. Lilienfeld in the United States and O. Heil working in Germany independently conceived a transistor structure of the form shown in Fig. 15.1. The device worked on the principle that a voltage applied to the metallic plate modulated the conductance of the underlying semiconductor, which in turn modulated the current flowing between ohmic contacts A and B. This phenomenon, where the conductivity of a semiconductor is modulated by an electric field applied normal to the surface of the semiconductor, has been named the *field effect*.

The early field effect transistor proposals were of course somewhat ahead of their time. Modern-day semiconductor materials were just not available and technological immaturity, in general, retarded the development of field effect structures for many years. A practical implementation had to await the successful development of other solid-state devices, notably the bipolar junction transistor, in the late 1940s and early 1950s. The first modern-day field effect device, the junction field effect transistor (J-FET), was proposed and analyzed by W. Shockley[1] in 1952. Operational J-FETs were subsequently built by Dacey

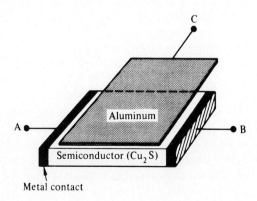

**Figure 15.1** Idealization of the Lilienfeld transistor.

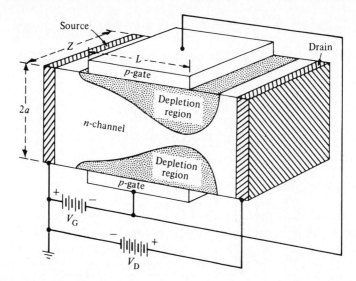

**Figure 15.2**  Schematic diagram of the junction field effect transistor. (From Dacey and Ross[3], © 1955 AT&T. Reprinted with permission.)

and Ross[2] in 1953. As pictured schematically in Fig. 15.2 (which was adapted from an early Dacey and Ross publication[3]), *pn* junctions replaced the metallic plate in the Lilienfeld structure, the A and B contacts became known as the source and drain, and the field-effect electrode was named the gate. In the J-FET it is the depletion regions associated with the *pn* junctions that directly modulate the semiconductor conductivity between the source and drain contacts.

A key component of present-day microelectronics, the metal-oxide-semiconductor field effect transistor (MOSFET), achieved practical status during the early 1960s. The now-familiar planar version of the structure pictured in Fig. 15.3, a structure with thermally grown $SiO_2$ functioning as the gate insulator, a surface-inversion channel, and islands doped opposite to the substrate acting as the source and drain, was first reported by Kahng and Atalla[4] in 1960. Following intense development efforts, commercial MOSFETs produced by Fairchild Semiconductor and by RCA became available in late 1964. Figure 15.3 originally appeared on the cover page of the 1964 applications bulletin describing the early Fairchild MOSFET[5].

A major development occurring in the latter half of the 1960s was the invention by Dennard[6] of the one-transistor dynamic memory cell used in the random access memory (DRAM). The DRAM cell (see Fig. 15.4) is an integrated combination of a charge storage element (a capacitor or *pn* junction) and a MOSFET utilized as a switch. A quote from a recent review article[8] best underscores the significance of the invention and its impact. "The 1-T DRAM cell is now the most abundant man-made object on this planet earth."

The early 1970s saw the introduction of still another significant field effect structure the charge-coupled device (CCD). Physically, a CCD might be viewed as a MOSFET with

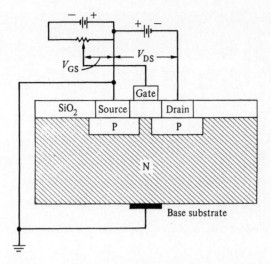

**Figure 15.3**   Cross section of a MOSFET with correct biasing polarity. (From MacDougall[5].)

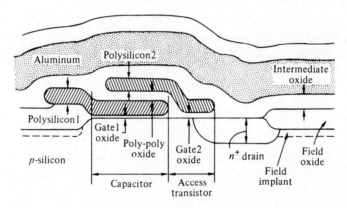

**Figure 15.4**   Schematic cross section of a DRAM cell. The planar, single-capacitor cell of the type shown was employed in 16 K–256 K memories. (From Wyns and Anderson[7], © 1989 IEEE.)

a segmented gate. As pictured in Fig. 15.5, the proper application of biases to the CCD gates induces the systematic movement or transfer of stored charge along the surface channel and into the device output. The image sensing element in the camcorder, a combination home TV camera and video recorder, is a CCD containing a two-dimensional array of gates. Stored charge is produced in proportion to the optical image incident on the two-dimensional gate array. The electrical analog of the optical image is subsequently transferred to the output and converted into an electrical signal.

The invention, development, and evolution of field effect devices continues to the pres-

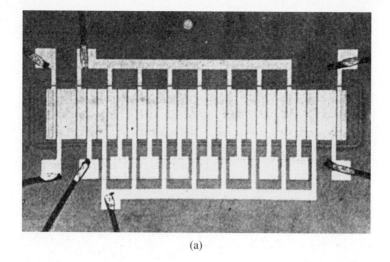

(a)

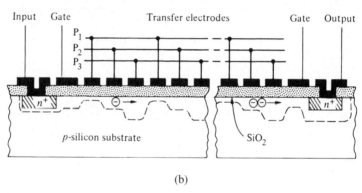

(b)

**Figure 15.5** First charge-coupled device comprising eight three-phase elements and input-output gates and diodes, shown (a) in plan view and (b) schematically in its cross-sectional view. (From Tompsett et al.[9] Reprinted with permission.)

ent. Notably, in progressing through the 1980s, complementary MOS or CMOS became the integrated circuit technology of choice. Figure 15.6 shows a cut-away view of a fundamental building block in CMOS circuitry, the CMOS inverter. The term "complementary" denotes the fact that, unlike other MOS circuit technologies, both *n*-channel (electron current carrying) MOSFETs and *p*-channel (hole current carrying) MOSFETs are fabricated on the same chip. Although CMOS came to the forefront in the 1980s because of its lower power dissipation and other circuit-related advantages, the change was evolutionary— CMOS is not a new circuit technology. It was initially conceived by Wanlass[10] in the early 1960s and was even mass produced as part of LED watch circuitry in 1972.

Whereas the use of CMOS exemplifies continuing evolutionary changes, there is also continuing field effect device invention and development. The modulation doped field effect transistor (MODFET) pictured in Fig. 15.7 is an excellent example. Although the elec-

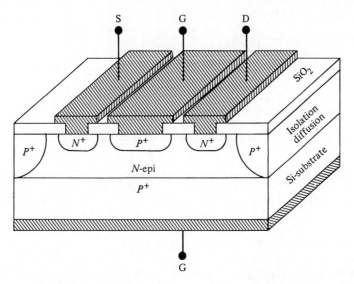

**Figure 15.9**   Perspective view of a modern epilayer J-FET.

semiconductor surface. Changes in the depletion widths, in turn, modulate the conductance between the ohmic source and drain contacts.

The J-FET was initially named the unipolar transistor to distinguish it from the bipolar junction transistor and to emphasize that only one type of carrier was involved in the operation of the new device. Specifically, for the structure pictured in Fig. 15.8, normal operation of the transistor can be described totally in terms of the electrons flowing in the $n$-region from the source to the drain. The source (S) terminal gets its name from the fact that the carriers contributing to the current move from the external circuit into the semiconductor at this electrode. The carriers leave the semiconductor, or are "drained" from the semiconductor, at the drain (D) electrode. The gate is so named because of its control or gating action. The modern version of the J-FET shown in Fig. 15.9, although somewhat different in physical appearance, is functionally equivalent to the original Shockley structure.

## 15.2.2  Qualitative Theory of Operation

To establish the basic principles of J-FET operation, we will assume standard biasing conditions and treat the symmetrical, somewhat idealized, Shockley structure of Fig. 15.8. Given an $n$-type region between the source and drain, standard operational conditions prevail in the J-FET when the top and bottom gates are tied together, $V_G \leq 0$, and $V_D \geq 0$, as illustrated in Fig. 15.10. Note that with $V_G \leq 0$, the $pn$ junctions are always zero or reverse biased. Also, $V_D \geq 0$ ensures that the electrons in the $n$-region move from the source to the drain (in agreement with the naming of the S and D terminals). Our approach here will be to systematically change the terminal voltages and examine what is happening inside the device.

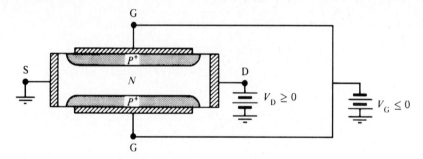

**Figure 15.10**  Specification of the device structure and biasing conditions assumed in the qualitative analysis.

First suppose that the gate terminal is grounded, $V_G = 0$, and the drain voltage is increased in small steps starting from $V_D = 0$. At $V_D = 0$ (remember $V_G$ is also zero) the device is in thermal equilibrium and about all one sees inside the structure are small depletion regions about the top and bottom $p^+$-$n$ junctions (see Fig. 15.11a). The depletion regions extend, of course, primarily into the lightly doped, central $n$-region of the device. Stepping $V_D$ to small positive voltages yields the situation pictured in Fig. 15.11(b). A current, $I_D$, begins to flow into the drain and through the nondepleted $n$-region sandwiched between the two $p^+$-$n$ junctions. The nondepleted, current-carrying region, we might note, is referred to as the *channel*. For small $V_D$, the channel looks and acts like a simple resistor, and the resulting variation of $I_D$ with $V_D$ is linear (see Fig. 15.12a).

When $V_D$ is increased above a few tenths of a volt, the device typically enters a new phase of operation. To gain insight into the revised situation, refer to Fig. 15.11(c), where an arbitrarily chosen potential of 5 V is assumed to exist at the drain terminal. Since the source is grounded, it naturally follows that somewhere in the channel the potential takes on the values of 1, 2, 3, and 4 volts, with the potential increasing as one progresses from the source to the drain. The $p^+$ sides of the $p^+$-$n$ junctions, however, are being held at zero bias. Consequently, the applied drain bias leads indirectly to a reverse biasing of the gate junctions and an increase in the junction depletion widths. Moreover, the top and bottom depletion regions progressively widen in going down the channel from the source to the drain (see Fig. 15.11d). Still thinking of the channel region (the nondepleted $n$-region) as a resistor, but no longer a simple resistor, one would expect the loss of conductive volume to increase the source-to-drain resistance and reduce the $\Delta I_D$ resulting from a given change in drain voltage. This is precisely the situation pictured in Fig. 15.12(b). The slope of the $I_D$-$V_D$ characteristic decreases at larger drain biases because of the channel-narrowing effect.

Continuing to increase the drain voltage obviously causes the channel to narrow more and more, especially near the drain, until eventually the top and bottom depletion regions touch in the near vicinity of the drain, as pictured in Fig. 15.11(e). The complete depletion of the channel, touching of the top and bottom depletion regions, is an important special condition and is referred to as "*pinch-off*." When the channel pinches off inside the device, the slope of the $I_D$-$V_D$ characteristic becomes approximately zero (see Fig. 15.12c), and

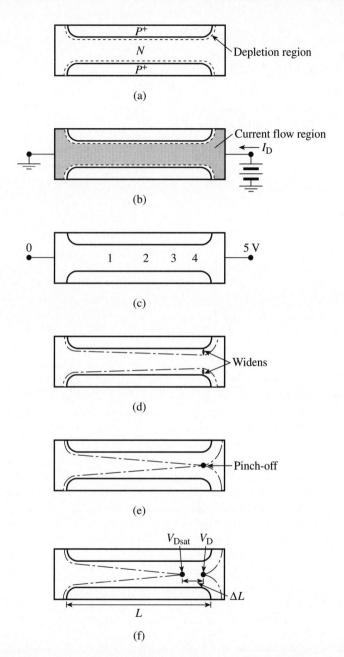

**Figure 15.11**  Visualization of various phases of $V_G = 0$ J-FET operation. (a) Equilibrium ($V_D = 0$, $V_G = 0$); (b) small $V_D$ biasing; (c) voltage drop down the channel for an arbitrarily assumed $V_D = 5$ V; (d) channel narrowing under moderate $V_D$ biasing; (e) pinch-off; (f) postpinch-off ($V_D > V_{Dsat}$).

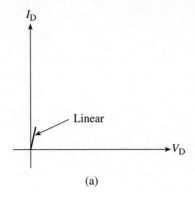

(a)

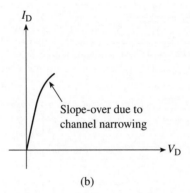

(b)

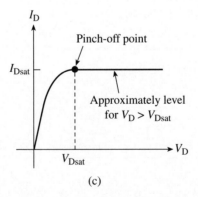

(c)

**Figure 15.12**   General form of the $I_D$–$V_D$ characteristics. (a) Linear, simple resistor, variation for very small drain voltages. (b) Slope-over at moderate drain biases due to channel narrowing. (c) Pinch-off and saturation for drain voltages in excess of $V_{Dsat}$.

the drain bias at the pinch-off point is given the special designation $V_{Dsat}$. For drain biases in excess of $V_{Dsat}$, the $I_D$–$V_D$ characteristic saturates, that is, remains approximately constant at the $I_{Dsat}$ value.

The statements presented without explanation in the preceding paragraph are totally factual. The $I_D$–$V_D$ characteristic does level off or saturate when the channel pinches off. At first glance, however, the facts appear to run contrary to physical intuition. Should not pinch-off totally eliminate any current flow in the channel? How can one account for the fact that $V_D$ voltages in excess of $V_{Dsat}$ have essentially no effect on the drain current?

In answer to the first question, let us suppose $I_D$ at pinch-off was identically zero. If $I_D$ were zero, there would be no current in the channel at any point and the voltage down the channel would be the same as at $V_D = 0$, namely, zero everywhere. If the channel potential is zero everywhere, the *pn* junctions would be zero biased and the channel in turn would be completely open from the source to the drain, clearly contradicting the initial assumption of a pinched-off channel. In other words, a current must flow in the J-FET to induce and maintain the pinched-off condition. Perhaps the conceptual difficulty often encountered with pinch-off arises from the need for a large current to flow through a depletion region. Remember, depletion regions are not totally devoid of carriers. Rather, the carrier numbers are just small compared to the background doping concentration ($N_D$ or $N_A$) and may still approach densities $\sim 10^{12}/\text{cm}^3$ or greater. Moreover, the passage of large currents through a depletion region is not unusual in solid-state devices. For example, a large current flows through the depletion region in a forward-biased diode and through both depletion regions in a bipolar junction transistor.

With regard to the saturation of $I_D$ for drain biases in excess of $V_{Dsat}$, there is a very simple physical explanation. When the drain bias is increased above $V_{Dsat}$, the pinched-off portion of the channel widens from just a point into a depleted channel section $\Delta L$ in extent. As shown in Fig. 15.11(f), the voltage on the drain side of the $\Delta L$ section is $V_D$, while the voltage on the source side of the section is $V_{Dsat}$. In other words, the applied drain voltage in excess of $V_{Dsat}$, $V_D - V_{Dsat}$, is dropped across the depleted section of the channel. Now, assuming $\Delta L \ll L$, the usual case, the source-to-pinch-off region of the device will be essentially identical in shape and will have the same endpoint voltages (zero and $V_{Dsat}$) as were present at the start of saturation. If the shape of a conducting region and the potential applied across the region do not change, then the current through the region must also remain invariant. This explains the approximate constancy of the drain current for post-pinch-off biasing. [Naturally, if $\Delta L$ is comparable to $L$, then the same voltage drop ($V_{Dsat}$) will appear across a shorter channel section ($L - \Delta L$) and the postpinch-off $I_D$ will increase perceptibly with increasing $V_D > V_{Dsat}$. This effect is especially noticeable in short channel (small $L$) devices.]

Another approach to explaining the saturation of the $I_D$–$V_D$ characteristics makes use of an analogous situation in everyday life, namely, a waterfall. The water-flow rate over a falls is controlled not by the height of the falls, but by the flow rate down the rapids leading to the falls. Thus, assuming an identical rapids region, the water-flow rate at the bottom of the two falls pictured in Fig. 15.13 is precisely the same, even though the heights of the falls are different. The rapids region is analogous to the source side of the channel in the J-FET, the falls proper corresponds to the pinched-off $\Delta L$ section at the drain end of

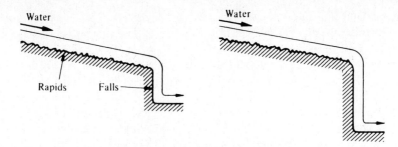

**Figure 15.13** The waterfalls analogy.

the channel, and the height of the falls corresponds to the $V_D - V_{Dsat}$ potential drop across the $\Delta L$ section.

Thus far we have established the expected variation of $I_D$ with $V_D$ when $V_G = 0$. To complete the discussion we need to investigate the operation of the J-FET when $V_G < 0$. It turns out that $V_G < 0$ operation is very similar to $V_G = 0$ operation with three minor modifications. First, if $V_G < 0$, the top and bottom $p^+$-$n$ junctions are reverse biased even when $V_D = 0$. A reverse bias on the junctions increases the width of the depletion regions and shrinks the $V_D = 0$ width of the channel. Consequently, the resistance of the channel increases at a given $V_D$ value and the linear portion of the $I_D$–$V_D$ characteristic exhibits a smaller slope when $V_G < 0$ (see Fig. 15.14a). Second, because the channel is narrower at $V_D = 0$, the channel also becomes pinched off at a smaller drain bias. Therefore, as pictured in Fig. 15.14(b), $V_{Dsat}$ and $I_{Dsat}$ when $V_G < 0$ are smaller than $V_{Dsat}$ and $I_{Dsat}$ when $V_G = 0$. Finally, note that for sufficiently negative $V_G$ biases it is possible to deplete the entire channel even with $V_D = 0$ (see Fig. 15.14c). The gate bias, $V_G = V_P$, where the gate voltage first totally depletes the entire channel with $V_D$ set equal to zero, is referred to as the pinch-off gate voltage. For $V_G \le V_P$ the drain current is identically zero for all drain biases.[†]

### 15.2.3 Quantitative $I_D$– $V_D$ Relationships

*Wanted:* a quantitative expression for the drain current as a function of the terminal voltages; that is, $I_D = I_D(V_D, V_G)$.

*Device Specification.* The precise device structure, dimensions, and assumed coordinate orientations are as specified in Fig. 15.15. The $y$-axis is directed down the channel from the source to the drain while the $x$-coordinate is oriented normal to the $p^+$-$n$ metallurgical junctions, $L$ is the channel length, $Z$ is the $p^+$-$n$ junction lateral width, and $2a$ is the distance between the top and bottom metallurgical junctions. Note that $y = 0$ and $y = L$ are

---

[†] If the drain bias is made very large, the $p^+$-$n$ junctions in the vicinity of the drain will eventually break down, leading to a very rapid increase in $I_D$ with $V_D$ for any gate bias. This breakdown has been omitted from all of the theoretical characteristics sketched herein.

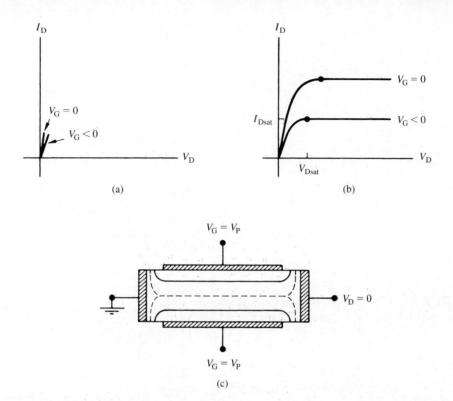

**Figure 15.14** Modification of the $I_D$–$V_D$ characteristics when $V_G < 0$. (a) Decrease in the linear slope of the characteristics for small drain voltages. (b) Decrease in the saturation current and saturation drain voltage. (c) Gate pinch-off.

slightly removed from the source and drain contacts, respectively. $V(y)$ is the potential and $W(y) = W_{top}(y) = W_{bottom}(y)$ is the junction depletion width at an arbitrary point $y$ in the channel. $W(y)$ lies almost totally in the $n$-region because of the $p^+$-$n$ nature of the junctions.

*Basic Assumptions.* (1) The junctions are $p^+$-$n$ *step* junctions and the $n$-region is uniformly doped with a donor concentration equal to $N_D$. (2) The device is structurally symmetrical about the $x = a$ plane as shown in Fig. 15.15, and the symmetry is maintained by operating the device with the same $V_G$ bias applied to the top and bottom gates. (3) Current flow is confined to the nondepleted portions of the $n$-region and is directed exclusively in the $y$-direction. (4) $W(y)$ can be increased to at least a width $a$ without inducing breakdown in the $p^+$-$n$ junctions. (We implicitly assumed this to be the case in the qualitative discussion.) (5) Voltage drops from the source to $y = 0$ and from $y = L$ to the drain are negligible. (6) $L \gg a$.

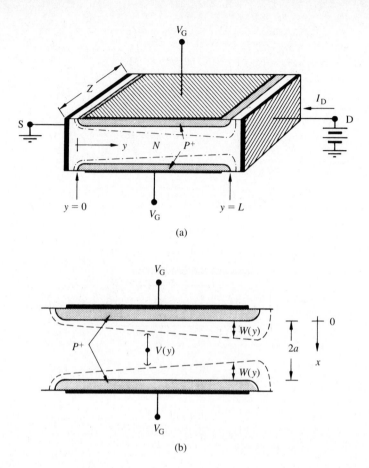

**Figure 15.15**   Device structure, dimensions, and coordinate orientations assumed in the quantitative analysis. (a) Overall diagram. (b) Expanded view of the channel region.

For drain and gate voltages below pinch-off, $0 \leq V_D \leq V_{Dsat}$ and $0 \geq V_G \geq V_P$, the derivation of the desired $I_D$–$V_D$ relationship proceeds as follows: In general one can write

$$\mathbf{J_N} = q\mu_n n \boldsymbol{\mathscr{E}} + qD_N\nabla n \tag{15.1}$$

Within the conducting channel $n \simeq N_D$ and the current is flowing almost exclusively in the $y$-direction. Moreover, with $n \simeq N_D$, the diffusion component of the current $(qD_N\nabla n)$ should be relatively small. Under the cited conditions Eq. (15.1) reduces to

$$J_N = J_{Ny} = q\mu_n N_D\mathscr{E}_y = -q\mu_n N_D \frac{dV}{dy} \qquad \text{(in the conducting channel)} \tag{15.2}$$

Since there are no carrier sinks or sources in the device, the current flowing through any cross-sectional plane within the channel must be equal to $I_D$. Thus, integrating the current density over the cross-sectional area of the conducting channel at an arbitrary point $y$ yields

$$I_D = - \int \int J_{Ny} \, dx \, dz = -Z \int_{W(y)}^{2a-W(y)} J_{Ny} \, dx = 2Z \int_{W(y)}^{a} q\mu_n N_D \frac{dV}{dy} \, dx \qquad (15.3a)$$

$$= 2qZ\mu_n N_D a \frac{dV}{dy} \left(1 - \frac{W}{a}\right) \qquad (15.3b)$$

The minus sign appears in the general formula for $I_D$ because $I_D$ is defined to be positive in the $-y$ direction. Use was also made of the fact that the structure is symmetrical about the $x = a$ plane.

Remembering that $I_D$ is independent of $y$, one can recast Eq. (15.3b) into a more useful form by integrating $I_D$ over the length of the channel. Specifically,

$$\int_0^L I_D \, dy = I_D L = 2qZ\mu_n N_D a \int_{V(0)\simeq0}^{V(L)\simeq V_D} \left[1 - \frac{W(V)}{a}\right] dV \qquad (15.4)$$

or

$$I_D = \frac{2qZ\mu_n N_D a}{L} \int_0^{V_D} \left[1 - \frac{W(V)}{a}\right] dV \qquad (15.5)$$

To proceed any further, we need an analytical expression for $W$ as a function of $V$. It should be recognized that the electrostatic problem inside the J-FET is really two-dimensional in nature. To obtain an exact expression for $W$ as a function of $V$ would necessitate the solution of Poisson's equation taking into account both the $x$ and $y$ variation of the electrostatic variables. Fortunately, with $L \gg a$ (basic assumption 6), the $y$-direction dependence of the electrostatic variables (potential, electric field, and charge density) in the J-FET depletion regions is expected to be a much more gradual function of position than the $x$-direction dependence of the same variables. This allows us to invoke the *gradual channel approximation,* an approximation that is encountered frequently in the analysis of field-effect devices. Specifically, $d\mathscr{E}_y/dy$ is neglected in solving Poisson's equation for the electrostatic variables in the FET depletion regions. In essence, a one-dimensional $x$-direction solution is obtained for each channel $y$-value.

For the problem under consideration, invoking the gradual channel approximation simply means that we can approximate $W$ at every point $y$ using Eq. (5.38), the one-dimensional expression established in Section 5.2. Thus, for the given $p^+$-$n$ step junctions,

$$W(V) \cong \left[\frac{2K_S\varepsilon_0}{qN_D}(V_{bi} - V_A)\right]^{1/2} = \left[\frac{2K_S\varepsilon_0}{qN_D}(V_{bi} + V - V_G)\right]^{1/2} \qquad (15.6)$$

where, as is evident from Fig. 15.15(b), $V_A = V_G - V(y)$ is the applied potential drop across the junction at a given point $y$. It is next convenient to note that $W \rightarrow a$ when $V_D = 0$ ($V = 0$) and $V_G = V_P$. Thus substituting into Eq. (15.6) yields

$$a = \left[ \frac{2K_S \varepsilon_0}{q N_D} (V_{bi} - V_P) \right]^{1/2} \tag{15.7}$$

and

$$\frac{W(V)}{a} = \left( \frac{V_{bi} + V - V_G}{V_{bi} - V_P} \right)^{1/2} \tag{15.8}$$

Finally, substituting $W(V)/a$ from Eq. (15.8) into Eq. (15.5) and performing the indicated integration, one obtains

$$I_D = \frac{2qZ\mu_n N_D a}{L} \left\{ V_D - \frac{2}{3}(V_{bi} - V_P) \left[ \left( \frac{V_D + V_{bi} - V_G}{V_{bi} - V_P} \right)^{3/2} - \left( \frac{V_{bi} - V_G}{V_{bi} - V_P} \right)^{3/2} \right] \right\}$$
$$\text{for } 0 \leq V_D \leq V_{Dsat}; \qquad V_P \leq V_G \leq 0$$

$$\tag{15.9}$$

Equation (15.9) could be simplified by introducing $G_0 \equiv 2qZ\mu_n N_D a/L$; physically, $G_0$ is the channel conductance one would observe if there were no depletion regions. We have retained $2qZ\mu_n N_D a/L$ in the text expressions so that the major parametic dependencies are immediately obvious.

It should be reemphasized that the foregoing development, and Eq. (15.9) in particular, apply only below pinch-off. In fact, the computed $I_D$ versus $V_D$ for a given $V_G$ actually begins to decrease if $V_D$ values in excess of $V_{Dsat}$ are inadvertently substituted into Eq. (15.9). As pointed out in the qualitative discussion, however, $I_D$ is approximately constant if $V_D$ exceeds $V_{Dsat}$. To first order, then, the postpinch-off portion of the characteristics can be modeled by simply setting

$$I_D|_{V_D > V_{Dsat}} = I_D|_{V_D = V_{Dsat}} \equiv I_{Dsat} \tag{15.10a}$$

or

$$I_{Dsat} = \frac{2qZ\mu_n N_D a}{L} \left\{ V_{Dsat} - \frac{2}{3}(V_{bi} - V_P) \left[ \left( \frac{V_{Dsat} + V_{bi} - V_G}{V_{bi} - V_P} \right)^{3/2} - \left( \frac{V_{bi} - V_G}{V_{bi} - V_P} \right)^{3/2} \right] \right\}$$

$$\tag{15.10b}$$

The $I_{Dsat}$ relationship can be simplified somewhat by noting that pinch-off at the drain end of the channel implies $W \to a$ when $V(L) = V_{Dsat}$. Therefore, from Eq. (15.6),

$$a = \left[ \frac{2K_S \varepsilon_0}{qN_D} (V_{bi} + V_{Dsat} - V_G) \right]^{1/2} \tag{15.11}$$

Comparing Eqs. (15.7) and (15.11), one concludes

$$\boxed{V_{Dsat} = V_G - V_P} \tag{15.12}$$

and

$$\boxed{I_{Dsat} = \frac{2qZ\mu_n N_D a}{L} \left\{ V_G - V_P - \frac{2}{3} (V_{bi} - V_P) \left[ 1 - \left( \frac{V_{bi} - V_G}{V_{bi} - V_P} \right)^{3/2} \right] \right\}} \tag{15.13}$$

Theoretical $I_D$–$V_D$ characteristics computed using Eqs. (15.9) and (15.13) are presented in Fig. 15.16. For comparison purposes a sample set of experimental characteristics are displayed in Fig. 15.17. Generally speaking, the theory does a reasonably adequate job of modeling the experimental observations. A somewhat improved agreement between experiment and theory can be achieved by lifting the assumption of negligible voltage drops in the regions of the device between the active channel and the source/drain contacts (see Fig. 15.18). Exercise 15.3 examines the required revisions to the theory when the source and drain resistances are taken into account. Finally, it should be pointed out that in saturation most J-FET characteristics can be closely modeled by the simple relationship

$$\boxed{I_{Dsat} = I_{D0}(1 - V_G/V_P)^2} \qquad \text{where } I_{D0} = I_{Dsat|V_G=0} \tag{15.14}$$

Although appearing totally different than Eq. (15.13), the semi-empirical "square-law" relationship of Eq. (15.14) yields similar numerical results and is much easier to use in performing first-order circuit calculations where the J-FET is viewed as a "black-box." Equation (15.13), on the other hand, is indispensable if one wishes to investigate the dependence of the J-FET characteristics on temperature, channel doping, or some other fundamental device parameter.

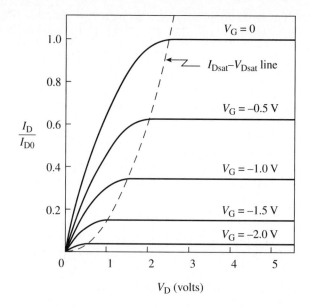

**Figure 15.16**   Normalized theoretical $I_D$–$V_D$ characteristics assuming $V_{bi} = 1$ V and $V_P = -2.5$ V. $I_{D0} = I_{Dsat|V_G=0}$.

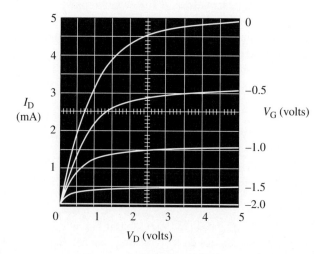

**Figure 15.17**   Sample experimental $I_D$–$V_D$ characteristics. (Characteristics were derived from a TI 2N3823 $n$-channel J-FET.)

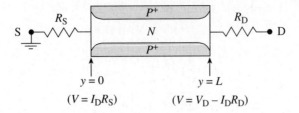

**Figure 15.18**  Modified model for the J-FET including the resistance of the semiconductor regions between the ends of the active channel and the source/drain terminals.

---

### Exercise 15.1

**P:** An $n$-channel Si J-FET is to be constructed with a channel doping of $N_D = 10^{16}/$ cm$^3$. Assuming room temperature operation, determine the maximum junction-to-junction half-width ($a$) that can be employed in constructing the J-FET.

**S:** Let $W_{BR}$ be the depletion width when the $pn$ junctions in the J-FET are biased to the breakdown voltage. For the J-FET to exhibit saturating characteristics, we must have $a \leq W_{BR}$. As read from Fig. 6.11, $V_{BR} \cong 55$ V for the given $N_D = 10^{16}/$cm$^3$ $p^+$-$n$ junctions. Also, $V_{bi} = 0.92$ V, as inferred from the figure associated with Exercise 5.1 in Subsection 5.1.4. Thus

$$a_{max} = W_{BR} = \left[\frac{2K_S\varepsilon_0}{qN_D}(V_{bi} + V_{BR})\right]^{1/2} = \left[\frac{(2)(11.8)(8.85 \times 10^{-14})}{(1.6 \times 10^{-19})(10^{16})}(0.92 + 55)\right]^{1/2}$$

or

$$\boxed{a_{max} = 2.70 \; \mu\text{m}}$$

---

### (C) Exercise 15.2

**P:** Construct a plot comparing the $I_{Dsat}/I_{D0}$ versus $V_G/V_P$ computed using the phenomenological relationship (Eq. 15.13) and the square-law relationship (Eq. 15.14). Assume $V_{bi} = 1$ V and $V_P = -2.5$ V or, equivalently, set $V_{bi}/V_P = -0.4$. How does the comparison vary with the $V_{bi}/V_P$ ratio?

**S:** Setting $V_G = 0$ in Eq. (15.13) yields

$$I_{D0} = \frac{2qZ\mu_n N_D a}{L}\left\{-V_P - \frac{2}{3}(V_{bi} - V_P)\left[1 - \left(\frac{V_{bi}}{V_{bi} - V_P}\right)^{3/2}\right]\right\}$$

Thus in the phenomenological theory (after some rearrangement),

$$\frac{I_{Dsat}}{I_{D0}} = \frac{V_G/V_P - 1 - \frac{2}{3}(V_{bi}/V_P - 1)\left[1 - \left(\frac{V_{bi}/V_P - V_G/V_P}{V_{bi}/V_P - 1}\right)^{3/2}\right]}{-1 - \frac{2}{3}(V_{bi}/V_P - 1)\left[1 - \left(\frac{V_{bi}/V_P}{V_{bi}/V_P - 1}\right)^{3/2}\right]}$$

According to the square-law relationship,

$$\frac{I_{Dsat}}{I_{D0}} = \left(1 - \frac{V_G}{V_P}\right)^2$$

A comparative plot of the two $I_{Dsat}/I_{D0}$ relationships and the script of the MATLAB program used to generate the plot is reproduced below. Given the difference in functional forms, the agreement is quite good. By changing the $z \equiv V_{bi}/V_P$ value in the computer program, the agreement is found to improve with increasing $|V_{bi}/V_P|$.

MATLAB program script...

%Exercise 15.2...Comparison of IDsat Relationships

%Computational parameters
clear
z=-0.4;            %z=Vbi/VP
x=linspace(0,1);   %x=VG/VP

%P-Theory (y=IDsat/ID0)
Num=x-1-(2/3)*(z-1)*(1-((z-x)./(z-1)).^(1.5));
Den=-1-(2/3)*(z-1)*(1-(z./(z-1)).^(1.5));
yP=Num./Den;

%Square-law Theory
yS=(1-x).^2;

%Plotting result
close
plot(x,yP,x,yS,'--');  grid
xlabel('VG/VP');  ylabel('IDsat/ID0');
text(0.38,0.4,'Square-Law')
text(0.27,0.3,'Eq.(15.13)')
text(0.8,0.83,'Vbi/VP=-0.4')

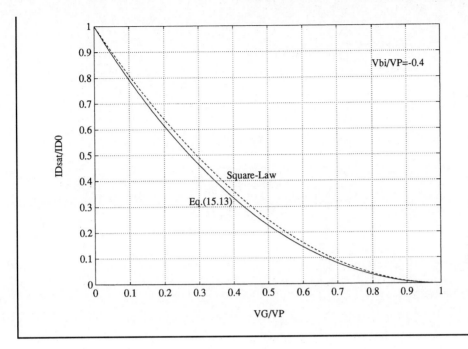

---

## Exercise 15.3

**P:** Show that the source and drain resistances at the ends of the active channel ($R_S$ and $R_D$ in Fig. 15.18) are appropriately taken into account by replacing $V_D$ with $V_D - I_D(R_S + R_D)$ and $V_G$ with $V_G - I_D R_S$ in Eqs. (15.9), (15.12), and (15.13).

**S:** With the voltage drops at the channel ends taken into account, the channel voltages at $y = 0$ and $y = L$ are $V(0) = I_D R_S$ and $V(L) = V_D - I_D R_D$, respectively. Inserting the revised voltage limits into Eq. (15.4), and likewise modifying Eq. (15.5), we find

$$I_D = \frac{2qZ\mu_n N_D a}{L} \int_{I_D R_S}^{V_D - I_D R_D} \left[ 1 - \frac{W(V)}{a} \right] dV \qquad (15.5')$$

Since the Eq. (15.8) expression for $W(V)/a$ remains unchanged, the integral in Eq. (15.5') is readily evaluated. The integration yields

$$I_D = \frac{2qZ\mu_n N_D a}{L} \left\{ V_D - I_D(R_S + R_D) \right.$$
$$- \frac{2}{3}(V_{bi} - V_P)\left[\left(\frac{V_D - I_D R_D + V_{bi} - V_G}{V_{bi} - V_P}\right)^{3/2} \right.$$
$$\left. \left. - \left(\frac{I_D R_S + V_{bi} - V_G}{V_{bi} - V_P}\right)^{3/2}\right]\right\} \qquad (15.9')$$

Turning next to the modification of Eq. (15.12), we note that when $V_D = V_{Dsat}$, $W \to a$, $V(L) = V_{Dsat} - I_{Dsat}R_D$, and from Eq. (15.6),

$$a = \left[\frac{2K_S\varepsilon_0}{qN_D}(V_{bi} + V_{Dsat} - I_{Dsat}R_D - V_G)\right]^{1/2} \qquad (15.11')$$

However,

$$a = \left[\frac{2K_S\varepsilon_0}{qN_D}(V_{bi} - V_P)\right]^{1/2} \qquad (15.7)$$

and clearly,

$$V_{Dsat} - I_{Dsat}R_D = V_G - V_P \qquad (15.12')$$

Finally, setting $V_D = V_{Dsat}$ and $I_D = I_{Dsat}$ in Eq. (15.9'), and simplifying the result using Eq. (15.12'), one obtains

$$I_{Dsat} = \frac{2qZ\mu_n N_D a}{L}\left\{V_G - V_P - I_{Dsat}R_S - \frac{2}{3}(V_{bi} - V_P) \times \right. \qquad (15.13')$$
$$\left. \left[1 - \left(\frac{I_{Dsat}R_S + V_{bi} - V_G}{V_{bi} - V_P}\right)^{3/2}\right]\right\}$$

Note that replacing $V_D$ by $V_D - I_D(R_S + R_D)$ and $V_G$ by $V_G - I_D R_S$ in Eqs. (15.9), (15.12), and (15.13) does indeed yield Eqs. (15.9'), (15.12'), and (15.13'), respectively.

A few words are in order concerning the foregoing results. First, it is obvious that the primed equations cannot be solved for $I_D$ or $I_{Dsat}$ as a closed-form function of $V_D$ and $V_G$. Nevertheless, numerical iteration techniques can be used to readily calculate the revised current–voltage characteristics. Second, this problem could have been rapidly completed by (i) changing the integration variable in Eq. (15.5′) to $V' = V - I_D R_S$ and (ii) requiring the original and revised versions of Eq. (15.5) to have the same general form. The change-of-variable approach, however, is not very informative.

### 15.2.4 a.c. Response

The a.c. response of the J-FET, routinely expressed in terms of the J-FET small-signal equivalent circuit, is most conveniently established by considering the two-port network shown in Fig. 15.19(a). Initially we restrict our considerations to low operational frequencies where capacitive effects may be neglected.

Let us begin by examining the device input. Under standard d.c. biasing conditions the input port between the gate and source is connected across a reverse-biased diode on the inside of the structure. A reverse-biased diode, however, behaves (to first order) like an open circuit at low frequencies. It is standard practice, therefore, to model the input to the J-FET by an open circuit.

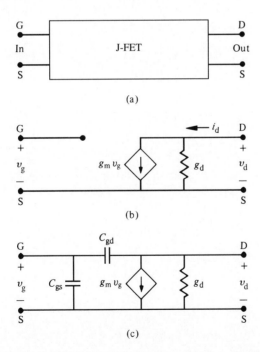

**Figure 15.19** (a) The J-FET viewed as a two-port network. (b) Low-frequency and (c) high-frequency small-signal equivalent circuits characterizing the a.c. response of the J-FET.

At the output port the d.c. drain current has already been established to be a function of $V_D$ and $V_G$; that is, $I_D = I_D(V_D, V_G)$. When the a.c. drain and gate potentials, $v_d$ and $v_g$, are respectively added to the d.c. drain and gate terminal voltages, $V_D$ and $V_G$, the drain current through the structure is modified to $I_D(V_D, V_G) + i_d$, where $i_d$ is the a.c. component of the drain current. Provided the device can follow the a.c. changes in potential quasistatically, which is assumed to be the case at low operational frequencies, one can state

$$i_d + I_D(V_D, V_G) = I_D(V_D + v_d, V_G + v_g) \tag{15.15a}$$

and

$$i_d = I_D(V_D + v_d, V_G + v_g) - I_D(V_D, V_G) \tag{15.15b}$$

Expanding the first term on the right-hand side of Eq. (15.15b) in a Taylor series about the d.c. operating point, and keeping only first-order terms in the expansion (higher-order terms are negligible), one obtains

$$I_D(V_D + v_d, V_G + v_g) = I_D(V_D, V_G) + \left.\frac{\partial I_D}{\partial V_D}\right|_{V_G} v_d + \left.\frac{\partial I_D}{\partial V_G}\right|_{V_D} v_g \tag{15.16}$$

which when substituted into Eq. (15.15b) gives

$$i_d = \left.\frac{\partial I_D}{\partial V_D}\right|_{V_G} v_d + \left.\frac{\partial I_D}{\partial V_G}\right|_{V_D} v_g \tag{15.17}$$

Dimensionally the partial derivatives in Eq. (15.17) are conductances. Introducing

$$g_d \equiv \left.\frac{\partial I_D}{\partial V_D}\right|_{V_G = \text{constant}} \quad \dots \text{the drain or channel conductance} \tag{15.18a}$$

$$g_m \equiv \left.\frac{\partial I_D}{\partial V_G}\right|_{V_D = \text{constant}} \quad \dots \text{transconductance or mutual conductance} \tag{15.18b}$$

we can then write

$$i_d = g_d v_d + g_m v_g \tag{15.19}$$

Eq. (15.19) may be viewed as the a.c.-current node equation for the drain terminal and, by inspection, leads to the output portion of the circuit displayed in Fig. 15.19(b). Since, as

**Table 15.1**  J-FET Small-Signal Parameters. Entries in the table were obtained by direct differentiation of Eqs. (15.9) and (15.13). $G_0 \equiv 2qZ\mu_n N_D a/L$.

| *Below pinch-off* $(V_D \leq V_{Dsat})$ | *Above pinch-off* $(V_D \geq V_{Dsat})$ |
|---|---|
| $$g_d = G_0\left[1 - \left(\frac{V_D + V_{bi} - V_G}{V_{bi} - V_P}\right)^{1/2}\right]$$ | $$g_d = 0$$ |
| $$g_m = G_0\left[\left(\frac{V_D + V_{bi} - V_G}{V_{bi} - V_P}\right)^{1/2} - \left(\frac{V_{bi} - V_G}{V_{bi} - V_P}\right)^{1/2}\right]$$ | $$g_m = G_0\left[1 - \left(\frac{V_{bi} - V_G}{V_{bi} - V_P}\right)^{1/2}\right]$$ |

concluded earlier, the gate-to-source or input portion of the device is simply an open circuit, Fig. 15.19(b) is the desired small-signal equivalent circuit characterizing the low-frequency a.c. response of the J-FET.

The $g_m$ parameter in the modeling of field effect transistors plays a role analogous to the gain parameters ($\alpha$ and $\beta$) previously encountered in the BJT analysis. As its name indicates, $g_d$ may be viewed as either the device output admittance or the a.c. conductance of the channel between the source and drain. Explicit $g_d$ and $g_m$ relationships obtained by direct differentiation of Eqs. (15.9) and (15.13), using the Eq. (15.18) definitions, are catalogued in Table 15.1. Note that the $g_d = 0$ result in Table 15.1 for operation of the device under saturation conditions is consistent with the theoretically zero slope of the $I_D$–$V_D$ characteristics when $V_D \geq V_{Dsat}$.

At higher operating frequencies often encountered in practical applications, the Fig. 15.19(b) circuit must be modified to take into account the capacitive coupling between the gate and the drain/source. Being *pn* junctions, the J-FET gate junctions are in general represented by a small-signal equivalent circuit similar to that presented in Fig. 7.2. However, since the J-FET gate junctions are normally reverse biased, the admittance of the junctions is adequately represented by the depletion region capacitance. This capacitance must be partially connected between the gate and source and partially connected between the gate and drain. The resulting high-frequency equivalent circuit is pictured in Fig. 15.19(c).

---

### Exercise 15.4

**P:** Using the low-frequency equivalent circuit of Fig. 15.19(b), and assuming the J-FET is saturation biased, show that the source and drain resistances at the ends of the active channel give rise to a reduced effective transconductance

$$g'_m = \frac{g_m}{1 + g_m R_S}$$

**S:** Under saturation conditions $g_d \to 0$. Open-circuiting $g_d$ and adding $R_S$ and $R_D$ to the proper nodes in Fig. 15.19(b) yields the working equivalent circuit displayed in Fig. E15.4. As deduced from Fig. E15.4,

$$i_d = g_m v'_g = g_m(v_g - i_d R_S)$$

Thus

$$i_d = \frac{g_m}{1 + g_m R_S} v_g \equiv g'_m v_g$$

where

$$g'_m = \frac{g_m}{1 + g_m R_S}$$

It is interesting to note that $R_D$ does not affect the transconductance if $g_d = 0$. $R_S$, on the other hand, causes a decrease in the effective transconductance of the J-FET if $g_m R_S$ is comparable to unity.

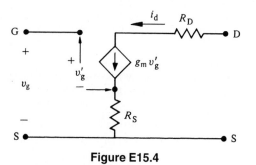

**Figure E15.4**

## 15.3 MESFET

### 15.3.1 General Information

The similarity between MS (Schottky) diodes and *pn* junction diodes was noted in Chapter 14. In particular, a depletion region, which can be modulated by the applied voltage, also forms beneath a rectifying MS contact. It should therefore come as no surprise that field-effect transistors can be built with rectifying metal-semiconductor gates. A simplified perspective view of a GaAs MESFET and an idealized cross section of the structure are pictured in Fig. 15.20.

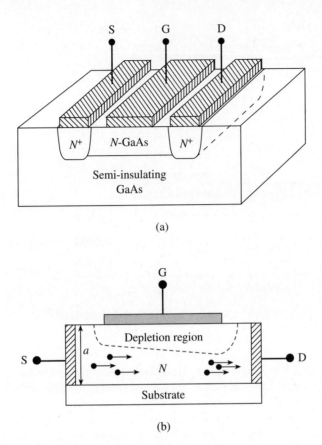

(a)

(b)

**Figure 15.20**  The MESFET. (a) Simplified perspective view of a GaAs MESFET. (b) Idealized MESFET cross section. (Note that the total $n$-region thickness in a single-gate structure is taken to be $a$.)

As might be inferred from Fig. 15.20, the MESFETs in commercial production are GaAs-based $n$-channel structures. The MESFETs are primarily used in high-frequency applications where GaAs is preferred over Si because of its superior electron-transport properties. Historically, difficulties were encountered in fabricating metal-insulator-GaAs FETs with acceptable characteristics. MESFET fabrication was developed as an alternative and is presently the most mature of GaAs fabrication technologies. GaAs MESFETs are the heart of monolithic microwave integrated circuits (MMICs) and are also sold as discrete devices for use in amplifiers and oscillators that must operate at frequencies in excess of 5 GHz.

There are two basic types of MESFETs—the depletion-mode or D-MESFET and the enhancement-mode or E-MESFET. Cross sections of the two types of MESFETs are shown

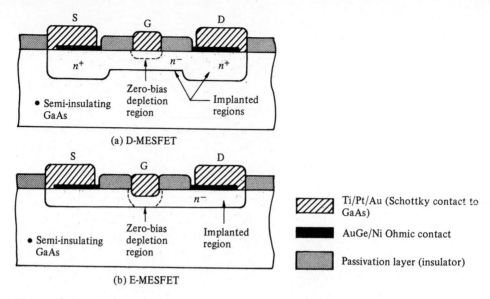

**Figure 15.21**   GaAs MESFET structure types. (a) Depletion-mode or D-MESFET; (b) enhancement-mode or E-MESFET.

in Fig. 15.21. Similar to the previously described J-FET, a bias applied to the gate of a D-MESFET further depletes the subgate region and reduces the channel conductance. The E-MESFET, on the other hand, is fabricated so that the built-in voltage associated with the metal-semiconductor contact is sufficient to totally deplete the channel. A forward bias must be applied to the E-MESFET gate to reduce the depletion width and obtain a channel current. The preponderance of applications including MMICs involve analog circuitry and make exclusive use of D-MESFETs. Digital logic circuits incorporate both D- and E-MESFETs.

It should be noted that high operational frequencies necessitate the fabrication of devices with extremely short channel lengths. The gate or active channel length in commercial devices is routinely $\lesssim 1$ $\mu$m. The discrete GaAs MESFETs listed in the 1993 Hewlett-Packard catalogue of communications components, for example, are quoted to have a nominal gate length of 0.25 $\mu$m.

## 15.3.2  Short-Channel Considerations

Given the structural similarities between the MESFET and J-FET, the $I_D$–$V_D$ theory developed for the J-FET could be applied with only minor modifications to the MESFET provided all derivational assumptions are satisfied. Unfortunately, some of the derivational assumptions are at best questionable for the typical short-channel MESFET. For one, the $L \gg a$ assumption and the associated gradual channel approximation are clearly suspect. Second, the electric field in the FET channel is implicitly assumed to be sufficiently low so that the carrier drift velocity is given by $v_d = \mu_0 \mathscr{E}_y$, where $|\mu_0|$ is the usual low-field mobility. However, if a drain voltage of $V_D = 1$ V is applied across a channel length of $L = 1$ $\mu$m, the *average* magnitude of the electric field in the channel will be $|\overline{\mathscr{E}_y}| = V_D/L = 10^4$ V/cm. Since $|\mathscr{E}_y|$ increases as one progresses down the channel toward the drain, the magnitude of the electric field near the drain will be even larger than the cited average. Referring to Fig. 3.4 in Part I, we find the electron drift velocity in Si is significantly different from the extrapolated linear dependence at an electric field of $10^4$ V/cm. Failure of the low-field assumption occurs at an even smaller $|\mathscr{E}_y|$-value in GaAs.

In the following discussion, we examine three modifications to the long channel theory that have been proposed in the device literature. Each of the approximate short channel models has its validity limits and utility under certain conditions.

## Variable Mobility Model

The nonlinear variation of $v_d$ with $\mathscr{E}_y$ as the carriers progress down the MESFET channel can be approximately taken into account using Eq. (3.3) from Part I. Specifically, assuming Eq. (3.3) with $\beta = 1$ adequately models the $v_d$ versus $\mathscr{E}$ dependence, one can write

$$v_d = \frac{\mu_0 \mathscr{E}}{1 + \dfrac{\mu_0 \mathscr{E}}{v_{sat}}} \tag{15.20}$$

and

$$\mu(\mathscr{E}) \equiv \left|\frac{v_d}{\mathscr{E}}\right| = \frac{|\mu_0|}{1 + \dfrac{\mu_0 \mathscr{E}}{v_{sat}}} \tag{15.21}$$

where $\mu(\mathscr{E})$ is the field-dependent mobility and $v_{sat}$ is the saturation drift velocity. If $\mu(\mathscr{E})$ with $\mu_0 = -\mu_n$ and $\mathscr{E} \to \mathscr{E}_y = -dV/dy$ is now substituted for $\mu_n$ in Eq. (15.1) of the long-channel J-FET derivation, and subsequent equations are appropriately modified, one ultimately obtains

$$I_D = \frac{I_D(\text{long-channel})}{1 + \dfrac{\mu_n V_D}{v_{sat} L}} \quad \ldots 0 \le V_D \le V_{Dsat} \tag{15.22}$$

$I_D(\text{long-channel})$ is the $I_D$ computed using Eq. (15.9). Above pinch-off, $I_D(\text{long-channel}) \to I_{Dsat}(\text{long-channel})$ and $V_D \to V_G - V_P$. In addition, $2a$ must be replaced by $a$ in the long-channel current equations if the MESFET is a single-gate structure as pictured in Fig. 15.20.

Examining Eq. (15.22), we note that the drain current is always reduced relative to the long-channel case. This is to be expected since $\mu(\mathscr{E}) \le \mu_n$. Also, $V_D/L$ in the denominator of Eq. (15.22) is just $|\overline{\mathcal{E}_y}|$. It is therefore reasonable to identify $\mu_n V_D/L = \mu_n|\overline{\mathcal{E}_y}|$ as the corresponding "average" drift velocity, $\overline{v_d}$. Equation (15.22) obviously reduces to the long-channel result if $\overline{v_d} \ll v_{sat}$.

Overall, the variable mobility model provides an acceptable description of the non-linear $v_d$ versus $\mathcal{E}_y$ dependence in the FET channel and is adequate as a first-order correction to the long-channel theory when treating FETs of moderate channel length (typically $L \gtrsim 10~\mu m$). However, the model does not account for the failure of the gradual channel approximation if $L \sim a$. As a consequence, additional considerations are necessary to properly describe the observed characteristics of short-channel MESFETs.

## Saturated Velocity Model

Current saturation in long-channel devices is always caused by a pinching-off or constriction of the channel near the drain. An alternative mechanism can give rise to current saturation in short-channel devices where the gradual channel approximation is no longer valid.

Suppose $|\mathcal{E}_y|$ in a short $n$-channel device is sufficiently large so that $v_d \to v_{sat}$ at a point $y_1 < L$ in the FET channel. With $v_d = v_{sat}$, the current flowing at the point $y_1$ will be

$$I(y_1) = qv_{sat}N_D Z[a - W(y_1)] \tag{15.23}$$

The continuity of the channel current requires $I(y)$ to be the same at all points in the channel and, in particular, $I(y > y_1) = I(y_1)$. Since $v_d = v_{sat}$ at $y = y_1$, there can be no further increase in $v_d$ at points $y > y_1$. It therefore follows from Eq. (15.23) that, to satisfy the $I(y > y_1) = I(y_1)$ requirement, $W(y)$ must be constant for all $y_1 \leq y \leq L$ as envisioned in Fig. 15.22. In total contrast to the assumption made in the gradual channel approximation, the described behavior can occur only if there is an appreciable $\mathcal{E}_y$ field in the subgate depletion region, with some of the field lines terminating on charges external to the gated region as illustrated in Fig. 15.22. Moreover, even though the channel is only partially constricted, the drain current through the device pictured in Fig. 15.22 will have saturated. The application of a larger drain bias will merely cause the $y_1$ point in the chan-

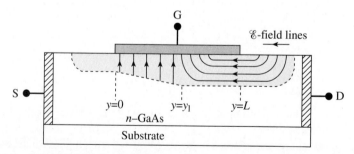

**Figure 15.22**  Approximate representation of the depletion region (shaded area) and electric field in a short channel MESFET where the drift velocity has reached its maximum value ($v_d = v_{sat}$) at the point $y_1$ in the channel.

nel to move closer to the source. In general, drain current saturation ($I_D = I_{Dsat}$) occurs when the channel pinches off *or* when $v_d \to v_{sat}$ at $y = L$.

If velocity saturation occurs in a MESFET with a channel length of only a few microns or less, then the depletion width in the $y_1 \leq y \leq L$ region of the device will differ by only a small amount from the depletion width at $y = 0$. For the envisioned special situation, $W(y_1)$ can be replaced by $W(0)$ in Eq. (15.23). This leads to the approximation

$$I_{Dsat} = I(y_1) \cong q v_{sat} N_D Z [a - W(0)] \tag{15.24}$$

with

$$W(0) = \left[ \frac{2 K_S \varepsilon_0}{q N_D} (V_{bi} - V_G) \right]^{1/2} = a \left( \frac{V_{bi} - V_G}{V_{bi} - V_P} \right)^{1/2} \tag{15.25}$$

Note that $W(0)$, the depletion width at $y = 0$, is computed assuming the gradual channel approximation can be applied to the source end of the channel. The saturation drain currents computed using Eqs. (15.24) and (15.25) are found to be in fairly good agreement with experimental results derived from short-channel ($L \sim 1 \ \mu m$) GaAs MESFETs.

## Two-Region Model

Although useful, the saturated velocity model provides only an expression for $I_{Dsat}$. The two-region model, on the other hand, provides an entire characteristic consistent with the saturated velocity model.

In the two-region model the analysis is broken into two parts corresponding to the two spatial or drift velocity regions pictured in Fig. 15.22. For $0 \leq y \leq y_1$ the gradual channel approximation and long-channel theory are assumed to hold *with* $v_d = \mu_0 \mathscr{E}_y$ *throughout the region*. The $v_d = v_{sat}$ model is applied to the $y_1 \leq y \leq L$ portion of the channel. The transition point ($y_1$) is taken to occur at the $y$-value where $\mu_0 \mathscr{E}_y = v_{sat}$. Naturally, the long-channel theory is applied throughout the channel if $\mu_0 \mathscr{E}_{y|y=L} < v_{sat}$.

Paralleling the long-channel theory, the $I_D$–$V_D$ characteristics are computed using Eq. (15.9) or the single-gate equivalent when $V_D \leq V_{Dsat}$. Likewise, one sets $I_{D|V_D>V_{sat}} = I_{D|V_D=V_{sat}} \equiv I_{Dsat}$. However, in general $V_{Dsat} \neq V_G - V_P$ and Eq. (15.13) is *not* used to compute $I_{Dsat}$. In the two-region model, drain current saturation first occurs when $\mu_0 \mathscr{E}_y = v_{sat}$ or $\mathscr{E}_y = v_{sat}/\mu_0$ at the drain end of the channel. If the long-channel relationships are solved for $V(y)$ in the channel (see Problem 15.3), the resulting expression differentiated with respect to $y$ to obtain $\mathscr{E}_y$, and $V(L)$ equated to $V_{Dsat}$ when $\mathscr{E}_{y|y=L} = v_{sat}/\mu_0 \equiv \mathscr{E}_{sat}$, one obtains

$$\mathscr{E}_{sat} L = \frac{V_{Dsat} - \frac{2}{3} (V_{bi} - V_P) \left[ \left( \frac{V_{Dsat} + V_{bi} - V_G}{V_{bi} - V_P} \right)^{3/2} - \left( \frac{V_{bi} - V_G}{V_{bi} - V_P} \right)^{3/2} \right]}{\left( \frac{V_{Dsat} + V_{bi} - V_G}{V_{bi} - V_P} \right)^{1/2} - 1} \tag{15.26}$$

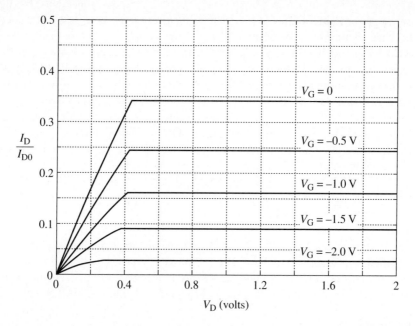

**Figure 15.23**   Normalized theoretical $I_D$–$V_D$ characteristics of a short $n$-channel MESFET based on the two-region model. $V_{bi} = 1$ V, $V_P = -2.5$ V, $\mathscr{E}_{sat} = -5 \times 10^3$ V/cm, and $L = 1$ $\mu$m. The drain current is normalized to the $V_G = 0$ saturation current ($I_{D0}$) of an equivalent long-channel FET.

$V_{Dsat}$ for a given $V_G$ and set of device parameters is determined from a numerical solution of Eq. (15.26).

A sample set of theoretical $I_D$–$V_D$ characteristics based on the two-region model is displayed in Fig. 15.23. An $\mathscr{E}_{sat} = -5 \times 10^3$ V/cm appropriate for GaAs $n$-channel MESFETs and $L = 1$ $\mu$m was assumed in establishing the characteristics. For comparison with long-channel results, the characteristics were normalized to $I_{D0} = I_{Dsat|V_G=0}$ computed using the single-gate version of Eq. (15.13).

Examining Fig. 15.23 we note that $V_{Dsat}$ for most $V_G$ values occurs at a significantly lower voltage and $I_D/I_{D0}$ is reduced relative to the comparable long-channel characteristics presented in Fig. 15.16. Also, $I_{Dsat}$ exhibits a $V_G$ dependence approximately described by Eq. (15.24). Although somewhat crude in appearance with a noticeable discontinuity in slope at $V_D = V_{Dsat}$, the characteristics resulting from the two-region model afford a reasonable first-order match to the observed short-channel characteristics. It should be acknowledged, however, that an accurate modeling of the short-channel characteristics inherently requires the numerical two-dimensional solution of the electrostatic and current equations.

## 15.4 SUMMARY

The modulation of the conductivity of a semiconductor by the application of an electric field normal to the surface of the semiconductor is known as the field effect. The chapter began with a general historical introduction to the field effect family of devices. The remainder of the chapter, devoted to the J-FET and MESFET, provided a specific introduction to the special terminology and typical analytical procedures associated with FETs. Steady-state J-FET operation was examined both qualitatively and quantitatively, with the quantitative analysis including the classic long-channel $I_D$–$V_D$ derivation based on the gradual channel approximation. The channel conductance ($g_d$) and the transconductance ($g_m$) were noted to be key parameters in modeling the a.c. response of the J-FET. The MESFET discussion emphasized short-channel considerations and the required modifications to the steady-state long-channel J-FET analysis.

## PROBLEMS

| CHAPTER 15 | PROBLEM INFORMATION TABLE | | | |
|---|---|---|---|---|
| Problem | Complete After | Difficulty Level | Suggested Point Weighting | Short Description |
| 15.1 | 15.4 | 1 | 10 (1 each part) | Quick quiz |
| 15.2 | 15.2.3 | 2 | 9 (3 each part) | Compositristor |
| 15.3 | " | 3 | 10 (a-7, b-3) | Relationship for $V(y)$ |
| 15.4 | " | 2 | 5 | $V_{Dsat}$ from $I_D$ maximum |
| 15.5 | " | 3 | 25 (a-3, b-5, c-3, d-5, e-2, f-7) | J-FET operation with $V_{GB} = 0$. |
| 15.6 | " | 3 | 12 (a-3, b-9) | Linearly graded channel |
| ● 15.7 | " | 4 | 25 | $R_S$ and $R_D$ included $I_D$–$V_D$ |
| 15.8 | 15.2.4 | 2 | 9 (a-6, b-3) | Pinch resistor |
| 15.9 | " | 3 | 12 (a-3, b-6, c-3) | $f_{max}$ |
| * 15.10 | " | 3 | 18 (a-12, b-6) | $g_m$ $T$-dependence |
| 15.11 | 15.3.2 | 3 | 8 | Back-gated MESFET |
| 15.12 | " | 2–3 | 8 | Derive Eq. (15.22) |
| 15.13 | " | 2 | 5 | Derive Eq. (15.26) |
| ● 15.14 | " | 3 | 18 (a-14, b-2, c-2) | Two region $I_D$–$V_D$ |

**15.1** Quick Quiz.

Answer the following questions as concisely as possible.

(a) Define "field effect."

(b) Precisely what is the "channel" in J-FET terminology?

(c) For a $p$-channel J-FET (a J-FET with $n^+$-$p$ gating junctions and a $p$-region between the source and drain), does the drain current flow into or out of the drain contact under normal operating conditions? Explain.

(d) What is the "gradual channel approximation"?

(e) What is meant by the term "pinch-off"?

(f) What is the mathematical definition of the drain conductance? of the transconductance?

(g) Draw the small-signal equivalent circuit characterizing the low-frequency a.c. response of a J-FET under *saturation* conditions. (Assume the d.c. characteristics of the device are similar to those shown in Fig. 15.16.)

(h) What do MESFET, D-MESFET, and E-MESFET stand for?

(i) Why is the magnitude of the electric field in the channel of concern in modeling short-channel MESFETs?

(j) Stated concisely, what is the primary difference between the long-channel and the two-region short-channel $I_D$–$V_D$ theories?

**15.2** In this problem you will be asked various questions about the device shown in Fig. P15.2. The device, which might be called a compositristor (composite transistor) is formed from a uniformly doped $n$-type bar. Ohmic contacts are made to the top and bottom of the bar and are connected to the outside world through leads D and B, respectively. $p^+$-$n$ step junctions are formed on the two sides of the bar and are connected to the outside world through contacts E and C. As shown in the figure, $d$ is the separation between the two $p^+$ regions and $L$ is the lateral length of the $p^+$ regions. *To receive full credit, you must indicate your reasoning* in addition to answering each of the following.

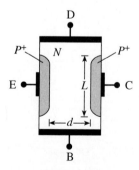

**Figure P15.2**

(a) *Given:* D–B tied together; $d \ll L_P$, where $L_P$ is the minority carrier diffusion length in the $n$-region; $V_{EB} > 0$; $V_{CB} < 0$. *Problem:* Sketch the current ($I_C$) flowing *out* of the C contact as a function of $V_{CB}$ if $I_E$ is held constant at various different values.

(b) *Given:* E–C tied together; D–B tied together; $d \gg L_P$. *Problem:* Sketch the current flowing into the E–C leads as a function of the voltage applied from E–C to D–B.

(c) *Given:* E–C tied together; $d < 2W_{BR}$, where $W_{BR}$ is the depletion width when the *pn* junctions in the compositristor are biased to the breakdown voltage; $L$ is almost equal to the total length of the bar between D and B; $V_{DB} > 0$; $V_{EB} < 0$. *Problem:* Sketch the current ($I_D$) flowing into the contact D as a function of the $V_{DB}$ voltage if $V_{EB}$ is held constant at various different values.

**15.3** As shown in Fig. 15.11(c), the variation in voltage down the length of the J-FET channel is typically a nonlinear function of position.

(a) Derive an expression that can be used to compute the point in the channel (that is, $y/L$) where a given channel voltage $0 \leq V(y) \leq V_D$ is expected to occur. HINT: Let $L \to y$ and $V_D \to V(y)$ in Eq. (15.4), solve for $y$, evaluate the remaining integral, and then form the ratio $y/L$ from your result.

(b) Assuming $V_G = 0$, $V_D = 5$ V, $V_{bi} = 1$ V, and $V_P = -8$ V, calculate $y/L$ for $V(y) = 1$, 2, 3, and 4 V. How do the computed $y/L$ values compare with the positioning of the voltages shown in Fig. 15.11(c)?

**15.4** If Eq. (15.9) is used to compute $I_D$ as a function of $V_D$ for a given $V_G$, and if $V_D$ is allowed to increase above $V_{Dsat}$, one finds $I_D$ to be a peaked function of $V_D$ maximizing at $V_{Dsat}$. The foregoing suggests a second way to establish the Eq. (15.12) relationship for $V_{Dsat}$. Specifically, show that the standard mathematical procedure for determining extrema points of a function can be used to derive Eq. (15.12) directly from Eq. (15.9).

**15.5** Suppose, as shown in Fig. P15.5, the bottom gate lead of a long-channel J-FET is tied to the source and grounded.

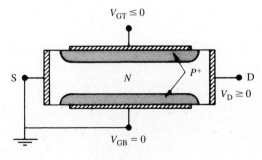

**Figure P15.5**

(a) Sketch an outline of the depletion regions inside the $V_{GB} = 0$ device when $V_{GT}$ is made sufficiently large to pinch off the channel with $V_D = 0$.

(b) If $V_P$ (the pinch-off gate voltage) $= -8$ V when the two gates are tied together, $V_{bi} = 1$ V, and assuming $p^+$-$n$ step junctions, determine $V_{PT}$ (the top gate pinch-off voltage) when $V_{GB} = 0$ and $V_D = 0$. Is your answer here consistent with the sketch in part (a)? Explain.

(c) Assuming $V_{PT} < V_{GT} < 0$, sketch an outline of the depletion regions inside the device when the *drain* voltage is increased to the pinch-off point.

(d) Derive an expression that specifies $V_{Dsat}$ in terms of $V_{PT}$, $V_{bi}$, and $V_{GT}$ for $V_{GB} = 0$ operation. (Your answer should contain only voltages. Make no attempt to actually solve for $V_{Dsat}$.)

(e) In light of your answers to parts (c) and (d), will $V_{Dsat}$ for $V_{GB} = 0$ operation be greater than or less than the $V_{Dsat}$ for $V_{GB} = V_{GT}$ operation? Explain.

(f) Derive an expression for $I_D$ as a function of $V_D$ and $V_{GT}$ analogous to Eq. (15.9).

**15.6** A J-FET is constructed with the *gate-to-gate* doping profile shown in Fig. P15.6. Specifically assume the $p^+$-region doping is much greater than the maximum $n$-region doping and make other obvious assumptions as required.

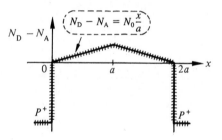

**Figure P15.6**

(a) Establish an expression for the depletion width ($W$) associated with one of the linearly graded junctions. Let $V_A$ be the applied voltage drop across the junction. (It may be helpful to refer to Subsection 7.2.2 or to review the linearly graded junction analysis in Subsection 5.2.5.)

(b) Neglecting the $\mu_n$ doping dependence and assuming that the left-hand and right-hand gates are tied together, appropriately modify the text J-FET analysis to obtain the below-pinch-off $I_D$-$V_D$ relationship for this linearly graded junction. (*Caution:* More than the $W(V)/a$ expression must be modified.)

● **15.7** Graphically illustrate the effect of the $R_S$ and $R_D$ resistances on the $I_D$-$V_D$ characteristics. Referring to Exercise 15.3, write a MATLAB (computer) program that can be used to calculate and plot the J-FET $I_D$-$V_D$ characteristics for arbitrary values of $R_S$ and $R_D$. Normalize all $I_D$ values to $I_{D0} = I_{Dsat|V_G=0}$ *as computed from Eq. (15.13).* (The normalized current relationships should involve only $I_D/I_{D0}$ or $I_{Dsat}/I_{D0}$, $V_D$, $V_G$, and the parameters $V_{bi}$, $V_P$, $G_0R_S$, and $G_0R_D$, where $G_0 \equiv 2qZ\mu_n N_D a/L$.) Run your program assuming $V_{bi} = 1$ V and $V_P = -2.5$ V; successively set $G_0R_S = G_0R_D = 0, 0.1,$ and $0.5.$ Compare your results with Fig. 15.16. Comment on the comparison.

**15.8** Bipolar integrated circuits occasionally make use of *pinch resistors.* The two-terminal pinch resistor is essentially a J-FET with the *gate(s) internally shorted to the source.* Exhibiting a voltage-dependent nonlinear resistance, the device finds use in applications that require large values of resistance, but where the precise values are not critical.

(a) Assuming the text development can be used without modification, establish general expressions for the d.c. conductance ($G = 1/\text{Resistance} = I/V$) and the a.c. conductance ($g = dI/dV$) of the pinch resistor.

(b) Compute the value of the d.c. and a.c. resistances ($R = 1/G$ and $r = 1/g$) at an applied voltage of $V_{Dsat}/2.$ Employ $Z/L = 1$, $a = 0.5$ $\mu$m, $N_D = 10^{16}/\text{cm}^3$, $V_{bi} = 1$ V, and $V_P = -2$ V.

**15.9** The maximum frequency of operation or cutoff frequency of the J-FET is given by

$$f_{max} = \frac{g_m}{2\pi C_G}$$

where $C_G$ is the capacitance of the *pn* junction gates.

(a) Derive the above expression for $f_{max}$ by presenting an argument analogous to that found in Subsection 17.3.2.

(b) Show that for a J-FET one can write

$$f_{max} = \frac{g_m}{2\pi C_G} \leq \frac{q\mu_n N_D a^2}{2\pi K_S \varepsilon_0 L^2}$$

(c) Given a silicon J-FET with $N_D = 10^{16}/\text{cm}^3$, $a = 0.5$ $\mu$m, and $L = 5$ $\mu$m, compute the limiting value of the cutoff frequency.

* **15.10** In this problem we wish to explore the temperature dependence of the J-FET transconductance.

(a) Making use of parametric temperature information or relationships from earlier chapters, compute and construct a log-log plot of $g_m(T)$ normalized to $g_m(300 \text{ K})$ versus $T(\text{K})$ for $225 \text{ K} \leq T \leq 475 \text{ K}.$ At a minimum, include data points at 50 K intervals. Set $V_G = 0$ and take the device to be saturation biased ($V_D \geq V_{Dsat}$). Assume an *n*-channel silicon device with $N_D = 10^{16}/\text{cm}^3$, $N_A$ (of the $p^+$ regions) $= 5 \times 10^{17}/\text{cm}^3$, and $a = 0.6$ $\mu$m. [NOTE: Neglecting the very small change in device dimensions with temperature, we conclude from Eq. (15.7) that $V_{bi} - V_P$ must be temperature indepen-dent. Separately, however, $V_{bi}$ and $V_P$ *are* temperature dependent.]

(b) Assuming $g_m(T)/g_m(300 \text{ K}) \propto T(\text{K})^{-n}$, determine **n**. Also, superimposed on the part (a) plot and spanning the same temperature range, plot $\mu_n(T)/\mu_n(300 \text{ K})$ versus $T$. Briefly discuss your results.

**15.11** Some $n$-channel MESFETs are built on $p^+$ substrates so they can be "back-gated"; i.e., a top MS diode gate is paired with a bottom $p^+$-$n$ gate. Let $V_{biT}$ be the built-in voltage of the top (MS) gate, $V_{biB}$ the built-in voltage of the bottom ($pn$) gate, $V_{GT}$ the top gate voltage, $V_{GB}$ the bottom gate voltage, $V_P$ the top gate voltage required to pinch off the channel when $V_{GB} = V_D = 0$, $2a$ the top-to-bottom junction width, and $L$ the length of the channel as defined by the shorter MS gate. Invoking the two-region model, derive an ex-pression for the below-saturation $I_D$ as a function of $V_D$.

**15.12** Derive Eq. (15.22).

**15.13** Utilizing the solution to Problem 15.3(a) and following the procedure outlined in the text, derive Eq. (15.26).

● **15.14** (a) Construct a MATLAB (computer) program to calculate and plot the $n$-channel FET $I_D$–$V_D$ characteristics predicted by the two-region model. Normalize $I_D$ to the $V_G = 0$ saturation current ($I_{D0}$) of an equivalent long-channel FET. Take $V_{bi}$, $V_P$, $\mathscr{E}_{sat}$, and $L$ to be input parameters. Run and check your program employing the parameters used to generate Fig. 15.23.

(b) Run your program employing $V_{bi} = 1$ V, $V_P = -2.5$ V, $\mathscr{E}_{sat} = -10^4$ V/cm (charac-teristic of silicon devices), and $L = 100$ $\mu$m. Compare your program output with Fig. 15.16.

(c) For $V_{bi} = 1$ V and $V_P = -2.5$ V, at what value of $\mathscr{E}_{sat}L$ does the long-channel theory begin to fail? Define "fail" as a reduction of the computed $I_{Dsat}/I_{D0}$ to less than 0.95 when $V_G = 0$.

# 16 MOS Fundamentals

The metal–oxide ($SiO_2$)–semiconductor (Si) or MOS structure is, without a doubt, the core structure in modern-day microelectronics. Even ostensibly *pn* junction type devices incorporate the MOS structure in some functional and/or physical manner. A quasi-MOS device, as noted in the Section 15.1 General Introduction, was first proposed in the 1920s. The dawn of modern history, however, is generally attributed to D. Kahng and M. M. Atalla who filed for patents on the Si–$SiO_2$ based field effect transistor in 1960. The MOS designation, it should be noted, is reserved for the technologically dominant metal–$SiO_2$–Si system. The more general designation, metal–insulator–semiconductor (MIS), is used to identify similar device structures composed of an insulator other than $SiO_2$ or a semiconductor other than Si.

This chapter is intended to serve as an introduction to MOS structural and device fundamentals. The two-terminal MOS-capacitor or MOS-C is both the simplest of MOS devices and the structural heart of all MOS devices. We begin with a precise specification of the "ideal" MOS-C structure. Energy band and block-charge diagrams are next constructed and utilized to qualitatively visualize the charge, electric field, and band bending inside the MOS-C under static biasing conditions. Quantitative relationships for the electrostatic variables inside the semiconductor are then developed and subsequently related to the voltage applied to the metallic gate. Capacitance is of course the primary electrical observable exhibited by an MOS-capacitor. The MOS-C capacitance–voltage ($C$–$V$) characteristics are important not only from a fundamental but also a practical viewpoint. In the final section of the chapter, our knowledge of the internal workings of the MOS structure is used to explain and analyze the normally observed form of the MOS-C $C$–$V$ characteristics. The chapter concludes with an examination of computed ideal-structure characteristics, comments about measurement procedures, and other relevant $C$–$V$ considerations.

## 16.1 IDEAL STRUCTURE DEFINITION

As pictured in Fig. 16.1, the MOS-capacitor is a simple two-terminal device composed of a thin ($0.01\ \mu m$–$1.0\ \mu m$) $SiO_2$ layer sandwiched between a silicon substrate and a metallic field plate. The most common field plate materials are aluminum and heavily doped polycrystalline silicon.[†] A second metallic layer present along the back or bottom side of the

---

[†] Heavily doped Si is metallic in nature. Polysilicon gates, used extensively in complex MOS device structures, are deposited by a chemical-vapor process and then heavily *n*- or *p*-doped by either diffusion or ion implantation.

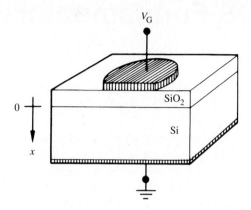

**Figure 16.1**    The metal–oxide–semiconductor capacitor.

semiconductor provides an electrical contact to the silicon substrate. The terminal connected to the field plate and the field plate itself are referred to as the gate; the silicon terminal, which is normally grounded, is simply called the back or substrate contact.

The ideal MOS structure has the following explicit properties: (1) the metallic gate is sufficiently thick so that it can be considered an equipotential region under a.c. as well as d.c. biasing conditions; (2) the oxide is a *perfect insulator* with *zero current* flowing through the oxide layer under *all* static biasing conditions; (3) there are no charge centers located in the oxide or at the oxide–semiconductor interface; (4) the semiconductor is uniformly doped; (5) the semiconductor is sufficiently thick so that, regardless of the applied gate potential, a field-free region (the so-called Si "bulk") is encountered before reaching the back contact; (6) an *ohmic* contact has been established between the semiconductor and the metal on the back side of the device; (7) the MOS-C is a one-dimensional structure with all variables taken to be a function only of the $x$-coordinate (see Fig. 16.1); and (8) $\Phi_M = \Phi_S = \chi + (E_c - E_F)_{FB}$. The material parameters appearing in idealization 8 were previously introduced in Section 14.1 and will be reviewed in the next section.

All of the listed idealizations can be approached in practice and the ideal MOS structure is fairly realistic. For example, the resistivity of $SiO_2$ can be as high as $10^{18}$ ohm-cm, and the d.c. leakage current through the layer is indeed negligible for typical oxide thicknesses and applied voltages. Moreover, even very thin gates can be considered equipotential regions and ohmic back contacts are quite easy to achieve in practice. Similar statements can be made concerning most of the other idealizations. Special note, however, should be made of idealization 8. The $\Phi_M = \Phi_S$ requirement could be omitted and will in fact be eliminated in Chapter 18. The requirement has only been included at this point to avoid unnecessary complications in the initial description of the static behavior.

## 16.2 ELECTROSTATICS—MOSTLY QUALITATIVE

### 16.2.1 Visualization Aids

#### Energy Band Diagram

The energy band diagram is an indispensable aid in visualizing the internal status of the MOS structure under static biasing conditions. The task at hand is to construct the diagram appropriate for the ideal MOS structure under equilibrium (zero-bias) conditions.

Figure 16.2 shows the surface-included energy band diagrams for the individual components of the MOS structure. In each case the abrupt termination of the diagram in a vertical line designates a surface. The ledge at the top of the vertical line, known as the vacuum level, denotes the minimum energy ($E_0$) an electron must possess to completely free itself from the material. The energy difference between the vacuum level and the Fermi energy in a metal is known as the metal workfunction, $\Phi_M$. In the semiconductor the height of the surface energy barrier is specified in terms of the electron affinity, $\chi$, the energy difference between the vacuum level and the conduction band edge at the surface. $\chi$ is used instead of $E_0 - E_F$ because the latter quantity is not a constant in semiconductors, but varies as a function of doping and band bending near the surface. Note that $(E_c - E_F)_{FB}$ is the energy difference between $E_c$ and $E_F$ in the flat band (FB) or field-free portion of the semiconductor. The remaining component, the insulator, is in essence modeled as an intrinsic wide-gap semiconductor where the surface barrier is again specified in terms of the electron affinity.

The conceptual formation of the MOS zero-bias band diagram from the individual components involves a two-step process. First the metal and semiconductor are brought together until they are a distance $x_o$ apart and the two-component system is allowed to equilibrate. Once the system is in equilibrium the metal and semiconductor Fermi levels

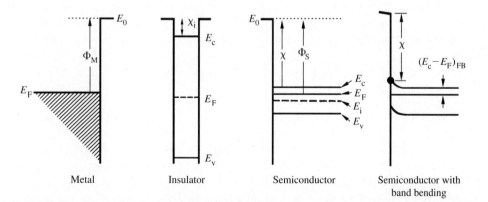

**Figure 16.2** Individual energy band diagrams for the metal, insulator, and semiconductor components of the MOS structure. The diagram labeled "semiconductor with band bending" defines $(E_c - E_F)_{FB}$ and shows $\chi$ to be invariant with band bending. The value of $\chi$, it should be emphasized, is measured relative to $E_c$ at the semiconductor surface.

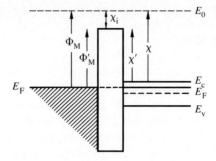

**Figure 16.3**  Equilibrium energy band diagram for an ideal MOS structure.

must be at the same energy (see Subsection 3.2.4). Moreover, the vacuum levels of the M and S components must also be in alignment because we have specified $\Phi_M = \Phi_S$. The foregoing implies that there are no charges or electric fields anywhere in the metal–gap–semiconductor system. Next the insulator of thickness $x_o$ is inserted into the empty space between the metal and semiconductor components. Given the zero electric field in the $x_o$ gap, the only effect of inserting the insulator is to slightly lower the barrier between the M and S components. Thus the equilibrium energy band diagram for the ideal MOS structure is concluded to be of the form pictured in Fig. 16.3.

### Block Charge Diagrams

Complementary in nature to the energy band diagram, block charge diagrams provide information about the approximate charge distribution inside the MOS structure. As just noted in the energy band diagram discussion, there are no charges anywhere inside the ideal MOS structure under equilibrium conditions. However, when a bias is applied to the MOS-C, charge appears within the metal and semiconductor near the metal–oxide and oxide–semiconductor interfaces. A sample block charge diagram is shown in Fig. 16.4.

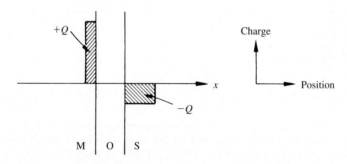

**Figure 16.4**  Sample block charge diagram.

Note that no attempt is made to represent the exact charge distributions inside the structure. Rather, a squared-off or block approximation is employed and hence the resulting figure is called a block charge diagram. Block charge diagrams are intended to be qualitative in nature; the magnitude and spatial extent of the charges should be interpreted with this fact in mind. Nevertheless, because the electric field is zero in the interior of both the metal and the semiconductor (see idealization 5), the charges within the structure must sum to zero according to Gauss's law. Consequently, in constructing block charge diagrams, the area representing positive charges is always drawn equal to the area representing negative charges.

## 16.2.2 Effect of an Applied Bias

### General Observations

Before examining specific-case situations, it is useful to establish general ground rules as to how one modifies the MOS energy band diagram in response to an applied bias. Assume normal operating conditions where the back side of the MOS-C is grounded and let $V_G$ be the d.c. bias applied to the gate.

With $V_G \neq 0$ we note first of all that the *semiconductor Fermi energy is unaffected by the bias and remains invariant (level on the diagram) as a function of position.* This is a direct consequence of the assumed zero current flow through the structure under all static biasing conditions. In essence, the semiconductor always remains in equilibrium independent of the bias applied to the MOS-C gate. Second, as in a *pn* junction, the applied bias separates the Fermi energies at the two ends of the structure by an amount equal to $qV_G$; that is,

$$E_F(\text{metal}) \ - \ E_F(\text{semiconductor}) \ = \ -qV_G \qquad (16.1)$$

Conceptually, the metal and semiconductor Fermi levels may be thought of as "handles" connected to the outside world. In applying a bias, one grabs onto the handles and rearranges the relative up-and-down positioning of the Fermi levels. The back contact is grounded and the semiconductor-side handle therefore remains fixed in position. The metal-side handle, on the other hand, is moved downward if $V_G > 0$ and upward if $V_G < 0$.

Since the barrier heights are fixed quantities, the movement of the metal Fermi level obviously leads in turn to a distortion in other features of the band diagram. The situation is akin to bending a rubber doll out of shape. Viewed another way, $V_G \neq 0$ causes potential drops and $E_c$ ($E_v$) band bending interior to the structure. No band bending occurs, of course, in the metal because it is an equipotential region. In the oxide and semiconductor, however, the energy bands must exhibit an upward slope (increasing $E$ going from the gate toward the back contact) when $V_G > 0$ and a downward slope when $V_G < 0$. Moreover, the application of Poisson's equation to the oxide, taken to be an ideal insulator with no carriers or charge centers, yields $d\mathscr{E}_{\text{oxide}}/dx = 0$ and therefore $\mathscr{E}_{\text{oxide}} = $ constant. Hence, the slope of the energy bands in the oxide is a constant—$E_c$ and $E_v$ are linear functions of position. Naturally, band bending in the semiconductor is expected to be somewhat more complex in its functional form, but per idealization 5, must always vanish ($\mathscr{E} \to 0$) before reaching the back contact.

## Specific Biasing Regions

Given the general principles just discussed, it is now a relatively simple matter to describe the internal status of the ideal MOS structure under various static biasing conditions. Taking the Si substrate to be $n$-type, consider first the application of a positive bias. The application of $V_G > 0$ lowers $E_F$ in the metal relative to $E_F$ in the semiconductor and causes a positive sloping of the energy bands in both the insulator and semiconductor. The resulting energy band diagram is shown in Fig. 16.5(a). The major conclusion to be derived from Fig. 16.5(a) is that the electron concentration inside the semiconductor, $n = n_i \exp[(E_F - E_i)/kT]$, increases as one approaches the oxide–semiconductor interface. This particular situation, where the majority carrier concentration is greater near the oxide–semiconductor interface than in the bulk of the semiconductor, is known as *accumulation*.

When viewed from a charge standpoint, the application of $V_G > 0$ places positive charges on the MOS-C gate. To maintain a balance of charge, negatively charged electrons must be drawn toward the semiconductor–insulator interface—the same conclusion established previously by using the energy band diagram. Thus the charge inside the device as a function of position can be approximated as shown in Fig. 16.5(b).

Consider next the application of a *small* negative potential to the MOS-C gate. The application of a small $V_G < 0$ slightly raises $E_F$ in the metal relative to $E_F$ in the semiconductor and causes a small negative sloping of the energy bands in both the insulator and semiconductor, as displayed in Fig. 16.5(c). From the diagram it is clear that the concentration of majority carrier electrons has been decreased, depleted, in the vicinity of the oxide–semiconductor interface. A similar conclusion results from charge considerations. Setting $V_G < 0$ places a minus charge on the gate, which in turn repels electrons from the oxide–semiconductor interface and exposes the positively charged donor sites. The approximate charge distribution is therefore as shown in Fig. 16.5(d). This situation, where the electron and hole concentrations at the oxide–semiconductor interface are less than the background doping concentration ($N_A$ or $N_D$), is known for obvious reasons as *depletion*.

Finally, suppose a larger and larger negative bias is applied to the MOS-C gate. As $V_G$ is increased negatively from the situation pictured in Fig. 16.5(c), the bands at the semiconductor surface will bend up more and more. The hole concentration at the surface ($p_s$) will likewise increase systematically from less than $n_i$ when $E_i$(surface) $< E_F$, to $n_i$ when $E_i$(surface) $= E_F$, to greater than $n_i$ when $E_i$(surface) exceeds $E_F$. Eventually, the hole concentration increases to the point shown in Fig. 16.5(e) and (f), where

$$E_i(\text{surface}) - E_i(\text{bulk}) = 2[E_F - E_i(\text{bulk})] \qquad (16.2)$$

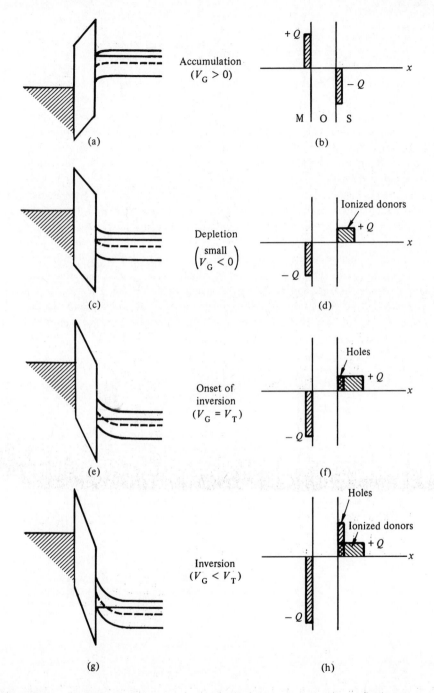

**Figure 16.5** Energy band diagrams and corresponding block charge diagrams describing the static state in an ideal $n$-type MOS-capacitor.

and

$$p_s = n_i e^{[E_i(\text{surface}) - E_F]/kT} = n_i e^{[E_F - E_i(\text{bulk})]/kT} = n_{\text{bulk}} = N_D \qquad (16.3)$$

Clearly, when $p_s = N_D$ for the special applied bias $V_G = V_T$ the surface is no longer depleted. Moreover, for further increases in negative bias ($V_G < V_T$), $p_s$ exceeds $n_{\text{bulk}} = N_D$ and the surface region appears to change in character from $n$-type to $p$-type. In accordance with the change in character observation, the $V_G < V_T$ situation where the minority carrier concentration at the surface exceeds the bulk majority carrier concentration is referred to as *inversion*. Energy band and block charge diagrams depicting the inversion condition are displayed in Fig. 16.5(g) and (h).

If analogous biasing considerations are performed for an ideal $p$-type device, the results will be as shown in Fig. 16.6. It is important to note from this figure that biasing regions in a $p$-type device are reversed in polarity relative to the voltage regions in an $n$-type device; that is, accumulation in a $p$-type device occurs when $V_G < 0$, and so forth.

In summary, then, one can distinguish three physically distinct biasing regions—accumulation, depletion, and inversion. For an ideal $n$-type device, accumulation occurs when $V_G > 0$, depletion when $V_T < V_G < 0$, and inversion when $V_G < V_T$. The cited voltage polarities are simply reversed for an ideal $p$-type device. No band bending in the semiconductor or *flat band* at $V_G = 0$ marks the dividing line between accumulation and depletion. The dividing line at $V_G = V_T$ is simply called the depletion–inversion transition

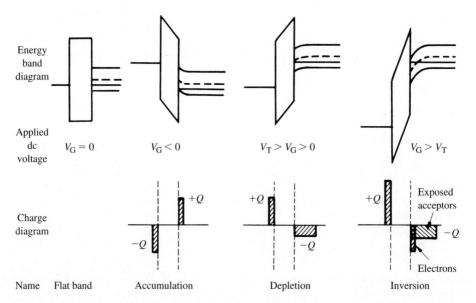

**Figure 16.6**  Energy band and block charge diagrams for a $p$-type device under flat band, accumulation, depletion, and inversion conditions.

point, with Eq. (16.2) quantitatively specifying the onset of inversion for both *n*- and *p*-type devices.

---

**Exercise 16.1**

**P:** Construct line plots (with $V_G$ plotted along the *x*-axis) that visually identify the voltage ranges corresponding to accumulation, depletion, and inversion in ideal *n*- and *p*-type MOS devices.

**S:** The "plots" shown below are in essence a graphical representation of the word summary given at the end of the preceding section. Note that *acc, depl,* and *inv* are standard abbreviations for accumulation, depletion, and inversion, respectively.

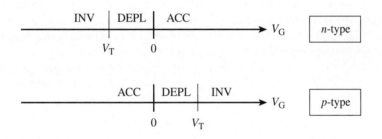

---

# 16.3  ELECTROSTATICS—QUANTITATIVE FORMULATION

## 16.3.1 Semiconductor Electrostatics

### Preparatory Considerations

The purpose of this section is to establish analytical relationships for the charge density ($\rho$), the electric field ($\mathscr{E}$), and the electrostatic potential existing inside an ideal MOS-C under static biasing conditions. The task is simplified by noting that the metal is an equipotential region. Charge appearing near the metal–oxide interface resides only a few Ångstroms (1 Ångstrom = $10^{-8}$ cm) into the metal and, to a high degree of precision, may be modeled as a δ-function of charge at the M–O interface. Since by assumption there is no charge in the oxide (idealization 3), the magnitude of the charge in the metal is simply equal to the sum of the charges inside the semiconductor. Also, as noted previously, with no charges in the oxide, it follows that the electric field is constant in the oxide and the potential is a linear function of position. In other words, solving for the electrostatic variables inside an ideal MOS-C essentially reduces to solving for the electrostatic variables inside the semiconductor component of the MOS-C.

The mathematical description of the electrostatics inside the semiconductor is established in a relatively straightforward manner beginning with Poisson's equation. In the following analysis we invoke the depletion approximation to obtain a first-order closed-form solution. The development closely parallels the presentation in the *pn* junction analysis of Chapter 5. It should be pointed out, however, that an *exact* solution is possible in the MOS-C case. The exact solution stems from simplifications associated with the fact that the semiconductor in an ideal MOS-C is *always* in equilibrium regardless of the applied d.c. bias. For reference purposes the exact solution is presented in Appendix B.

In performing the analysis, $\phi(x)$ is taken to be the potential inside the semiconductor at a given point $x$; $x$ is understood to be the depth into the semiconductor as measured from the oxide–semiconductor interface [see Fig. 16.7(a)]. The symbol $\phi$, instead of $V$, is used in MOS-C work to avoid possible confusion with externally applied potentials. In accordance with idealization 5, the electric field ($\mathscr{E} = -d\phi/dx$) is assumed to vanish as one proceeds into the semiconductor substrate. Following standard convention, the potential is chosen to be zero in the field-free region of the substrate referred to as the semiconductor bulk. $\phi$ evaluated at the oxide–semiconductor interface (at $x = 0$) is given the special symbol, $\phi_S$, and is known as the surface potential.

Figure 16.7(b) indicates how $\phi(x)$ is related to band bending on the energy band diagram. As shown (and consistent with Eq. 3.12),

$$\phi(x) = \frac{1}{q} [E_i(\text{bulk}) - E_i(x)] \tag{16.4}$$

and

$$\phi_S = \frac{1}{q} [E_i(\text{bulk}) - E_i(\text{surface})] \tag{16.5}$$

Figure 16.7(b) also introduces an important material parameter; namely,

$$\phi_F = \frac{1}{q} [E_i(\text{bulk}) - E_F] \tag{16.6}$$

$\phi_F$ is clearly related to the semiconductor doping. For one, the sign of $\phi_F$ indicates the doping type; directly from the definition one concludes $\phi_F > 0$ if the semiconductor is *p*-type and $\phi_F < 0$ if the semiconductor is *n*-type. More importantly, the magnitude of $\phi_F$ is functionally related to the doping concentration. Given a nondegenerate Si substrate maintained at or near room temperature, we know from Chapter 2 that

$$p_{\text{bulk}} = n_i e^{[E_i(\text{bulk}) - E_F]/kT} = N_A \quad \dots \text{ if } N_A \gg N_D \tag{16.7a}$$

$$n_{\text{bulk}} = n_i e^{[E_F - E_i(\text{bulk})]/kT} = N_D \quad \dots \text{ if } N_D \gg N_A \tag{16.7b}$$

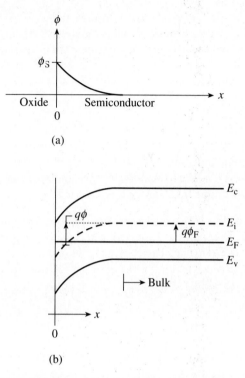

**Figure 16.7** Electrostatic parameters: (a) Graphical definition of $\phi$ and $\phi_S$. (b) Relationship between $\phi(x)$ and band bending; graphical definition of $\phi_F$.

Thus, combining Eqs. (16.6) and (16.7) yields

$$\phi_F = \begin{cases} \dfrac{kT}{q}\ln(N_A/n_i) & \ldots \; p\text{-type semiconductor} \qquad (16.8a) \\[2ex] -\dfrac{kT}{q}\ln(N_D/n_i) & \ldots \; n\text{-type semiconductor} \qquad (16.8b) \end{cases}$$

Extensive use will be made of the $\phi_S$ and $\phi_F$ parameters throughout the MOS discussion. Our immediate interest in these parameters involves their use in quantitatively specifying the biasing state inside the semiconductor. Clearly, under flat band conditions $\phi_S = 0$. Moreover, substituting Eqs. (16.5) and (16.6) into Eq. (16.2), one concludes

$$\phi_S = 2\phi_F \qquad \text{at the depletion–inversion transition point} \qquad (16.9)$$

With $\phi_F > 0$ in a $p$-type semiconductor, it follows that $\phi_S < 0$ if the semiconductor is accumulated, $0 < \phi_S < 2\phi_F$ if the semiconductor is depleted, and $\phi_S > 2\phi_F$ if the semiconductor is inverted. For an $n$-type semiconductor the inequalities are merely reversed.

---

### Exercise 16.2

**P:** (a) Construct line plots (with $\phi_S$ plotted along the $x$-axis) that visually identify the surface potential ranges corresponding to accumulation, depletion, and inversion in ideal $n$-type and $p$-type MOS devices.

(b) For each of the $\phi_F$, $\phi_S$ parameter sets listed below, indicate the doping type and the specified biasing condition. Also draw the corresponding energy band diagram and block charge diagram that characterize the static state of the ideal MOS system.

(i)  $\dfrac{\phi_F}{kT/q} = 12, \dfrac{\phi_S}{kT/q} = 12$

(ii)  $\dfrac{\phi_F}{kT/q} = -9, \dfrac{\phi_S}{kT/q} = 3$

(iii)  $\dfrac{\phi_F}{kT/q} = -9, \dfrac{\phi_S}{kT/q} = -18$

(iv)  $\dfrac{\phi_F}{kT/q} = 15, \dfrac{\phi_S}{kT/q} = 36$

(v)  $\dfrac{\phi_F}{kT/q} = -15, \dfrac{\phi_S}{kT/q} = 0$

**S:** (a) Converting the discussion at the end of the preceding subsection into a graphical representation yields

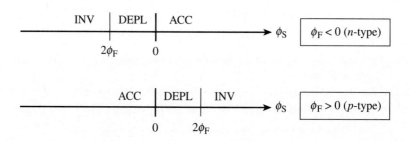

| (b) Set | Doping | Biasing Condition | Energy Band Diagram | Block Charge Diagram |
|---------|--------|-------------------|---------------------|----------------------|
| (i) | $p$ | Depletion | $E_F$ | $+Q$   $-Q$ |
| (ii) | $n$ | Accumulation | $E_F$ | |
| (iii) | $n$ | Depl / Inv Transition | $E_F$ | Holes |
| (iv) | $p$ | Inversion | $E_F$ | Electrons |
| (v) | $n$ | Flat Band | $E_F$ | M   O   S |

### Delta-Depletion Solution

The approximate closed-form solution for the electrostatic variables based in part on the depletion approximation is conveniently divided into three segments corresponding to the three biasing regions of accumulation, depletion, and inversion.

Let us first consider accumulation. Figure 16.8 displays charge density and potential plots constructed using the exact solution found in Appendix B. After verifying the general correlation between the Fig. 16.8 plots and the $p$-bulk semiconductor portion of the dia-

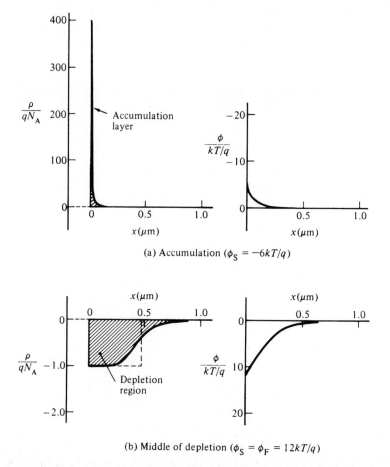

(a) Accumulation ($\phi_S = -6kT/q$)

(b) Middle of depletion ($\phi_S = \phi_F = 12kT/q$)

**Figure 16.8**  Exact solution for the charge density and potential inside the semiconductor component of an MOS-C assuming $\phi_F = 12kT/q$ and $T = 300$ K ($kT/q = 0.0259$ V). (a) Accumulation ($\phi_S = -6kT/q$), (b) middle of depletion ($\phi_S = \phi_F = 12kT/q$), (c) onset of inversion ($\phi_S = 2\phi_F = 24kT/q$), and (d) heavily inverted ($\phi_S = 2\phi_F + 6kT/q = 30kT/q$). The $p$-diagrams were drawn on a linear scale and the $+\phi$ axes oriented downward to enhance the correlation with the diagrams sketched in Fig. 16.6. The dashed lines on the part (b) through (d) $\rho$-plots outline the depletion approximation version of the charge distribution.

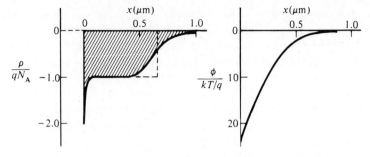

(c) Onset of inversion ($\phi_S = 2\phi_F = 24kT/q$)

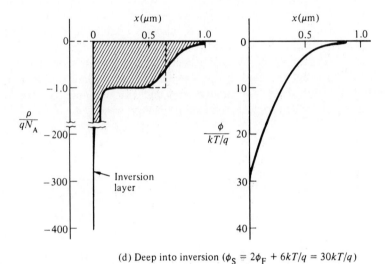

(d) Deep into inversion ($\phi_S = 2\phi_F + 6kT/q = 30kT/q$)

**Figure 16.8**  *Continued.*

grams sketched in Fig. 16.6, specifically note from Fig. 16.8(a) that *the charge associated with majority carrier accumulation resides in an extremely narrow portion of the semiconductor immediately adjacent to the oxide–semiconductor interface.* By comparison, the depleted portion of the semiconductor under moderate depletion biasing shown in Fig. 16.8(b) extends much deeper into the semiconductor. Given the narrow extent of the accumulation layer, it would appear reasonable as a first order approximation to replace the accumulation charge with a δ-function of equal charge positioned at the oxide–semiconductor interface. Indeed, we have just described the delta-depletion solution for accumulation. Because of the assumed δ-function of charge at $x = 0$, it automatically follows that the electric field and electrostatic potential are identically zero for all $x > 0$ under accumulation biasing in the delta-depletion solution. This is clearly somewhat inaccurate, but acceptable as a first-order approximation.

Turning next to inversion, we note from Fig. 16.8(d) that, like the accumulation layer charge, *the charge associated with minority carrier inversion resides in an extremely narrow portion of the semiconductor immediately adjacent to the oxide–semiconductor interface.* Moreover, in comparing the depleted semiconductor regions when $\phi_S = \phi_F$ (middle of depletion), $\phi_S = 2\phi_F$ (onset of inversion), and $\phi_S = 2\phi_F + 6kT/q$ (inversion), we find the depletion width increases substantially with increased depletion biasing, *but increases only slightly once the semiconductor inverts.* Based on the first of the foregoing observations, the actual inversion layer charge is approximately modeled in the delta-depletion solution by a δ-function of equal charge positioned at the oxide–semiconductor interface. To account for the second observation, it is additionally assumed the δ-function of charge added in inversion *precisely* balances the charge added to the MOS-C gate. As a consequence, in the delta-depletion solution for inversion biases, the depletion region charge, the $x > 0$ electric field, and the $x > 0$ electrostatic potential remain fixed at their $\phi_S = 2\phi_F$ values. In other words, the inversion bias solution is established by merely adding a δ-function of surface charge to the solution existing at the end of depletion.

The remaining biasing region to be considered is depletion. In the standard depletion approximation the actual depletion charge is replaced with a squared-off distribution terminated abruptly a distance $x = W$ into the semiconductor. Assuming a $p$-type semiconductor and invoking the depletion approximation, one can write

$$\rho = q(p - n + N_D - N_A) \cong -qN_A \qquad (0 \le x \le W) \qquad (16.10)$$

Poisson's equation then reduces to

$$\frac{d\mathscr{E}}{dx} = \frac{\rho}{K_S \varepsilon_0} \cong -\frac{qN_A}{K_S \varepsilon_0} \qquad (0 \le x \le W) \qquad (16.11)$$

The straightforward integration of Eq. (16.11) employing the boundary condition $\mathscr{E} = 0$ at $x = W$ next yields

$$\mathscr{E}(x) = -\frac{d\phi}{dx} = \frac{qN_A}{K_S \varepsilon_0}(W - x) \qquad (0 \le x \le W) \qquad (16.12)$$

A second integration with $\phi = 0$ at $x = W$ gives

$$\phi(x) = \frac{qN_A}{2K_S \varepsilon_0}(W - x)^2 \qquad (0 \le x \le W) \qquad (16.13)$$

The final unknown in the electrostatic relationships, the depletion width $W$, is determined from Eq. (16.13) by applying the boundary condition $\phi = \phi_S$ at $x = 0$. We obtain

$$\phi_S = \frac{qN_A}{2K_S \varepsilon_0} W^2 \qquad (16.14)$$

and therefore

$$W = \left[ \frac{2K_S \varepsilon_0}{qN_A} \phi_S \right]^{1/2} \tag{16.15}$$

Taken together, Eqs. (16.10), (16.12), (16.13), and (16.15) constitute the desired depletion bias solution. For an $n$-bulk device $N_A$ in the preceding equations is replaced by $-N_D$.

Before concluding, special note should be made of the depletion width, $W_T$, existing at the depletion–inversion transition point. In the delta-depletion formulation, $W_T$ is of course the maximum attainable equilibrium depletion width. Since $W = W_T$ when $\phi_S = 2\phi_F$, simple substitution into Eq. (16.15) yields

$$W_T = \left[ \frac{2K_S \varepsilon_0}{qN_A} (2\phi_F) \right]^{1/2} \tag{16.16}$$

A plot of $W_T$ versus doping covering the typical range of MOS doping concentrations is displayed in Fig. 16.9.

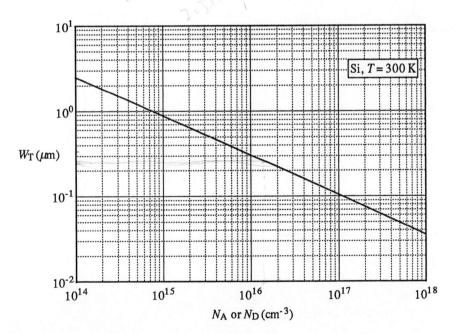

**Figure 16.9** Doping dependence of the maximum equilibrium depletion width inside silicon devices maintained at 300 K.

## 16.3.2 Gate Voltage Relationship

In Subsection 16.3.1 the biasing state was described in terms of the semiconductor surface potential, $\phi_S$. Results formulated in this manner are dependent only on the properties of the semiconductor. $\phi_S$, however, is an *internal* system constraint or boundary condition. It is the *externally* applied gate potential, $V_G$, that is subject to direct control. Thus, if the results of Subsection 16.3.1 are to be utilized in practical problems, an expression relating $V_G$ and $\phi_S$ must be established. This subsection is devoted to deriving the required relationship.

We begin by noting that $V_G$ in the ideal structure is dropped partly across the oxide and partly across the semiconductor, or symbolically,

$$V_G = \Delta\phi_{\text{semi}} + \Delta\phi_{\text{ox}} \tag{16.17}$$

Because $\phi = 0$ in the semiconductor bulk, however, the voltage drop across the semiconductor is simply

$$\Delta\phi_{\text{semi}} = \phi(x = 0) = \phi_S \tag{16.18}$$

The task of developing a relationship between $V_G$ and $\phi_S$ is therefore reduced to the problem of expressing $\Delta\phi_{\text{ox}}$ in terms of $\phi_S$.

As stated previously (Subsection 16.2.2), in an ideal insulator with no carriers or charge centers

$$\frac{d\mathscr{E}_{\text{ox}}}{dx} = 0 \tag{16.19}$$

and

$$\mathscr{E}_{\text{ox}} = -\frac{d\phi_{\text{ox}}}{dx} = \text{constant} \tag{16.20}$$

Therefore

$$\Delta\phi_{\text{ox}} = \int_{-x_0}^{0} \mathscr{E}_{\text{ox}} \, dx = x_0 \mathscr{E}_{\text{ox}} \tag{16.21}$$

where $x_0$ is the oxide thickness. The next step is to relate $\mathscr{E}_{\text{ox}}$ to the electric field in the semiconductor. The well-known boundary condition on the fields normal to an interface between two dissimilar materials requires

$$(D_{\text{semi}} - D_{\text{ox}})|_{\text{O-S interface}} = Q_{\text{O-S}} \tag{16.22}$$

where $D = \varepsilon\mathscr{E}$ is the dielectric displacement and $Q_{O-S}$ is the charge/unit area located at the interface. Since $Q_{O-S} = 0$ in the idealized structure (idealization 3),[†]

$$D_{ox} = D_{semi}|_{x=0} \tag{16.23}$$

$$\mathscr{E}_{ox} = \frac{K_S}{K_O}\mathscr{E}_S \tag{16.24}$$

and

$$\Delta\phi_{ox} = \frac{K_S}{K_O} x_o \mathscr{E}_S \tag{16.25}$$

$K_S$ is the semiconductor dielectric constant; $K_O$, the oxide dielectric constant; and $\mathscr{E}_S$, the electric field in the semiconductor at the oxide–semiconductor interface.[‡] Finally, substituting Eqs. (16.18) and (16.25) into Eq. (16.17), and recognizing that $\mathscr{E}_S$ is a known or readily determined function of $\phi_S$, we obtain

$$V_G = \phi_S + \frac{K_S}{K_O} x_o \mathscr{E}_S \tag{16.26}$$

If the results of the delta-depletion solution are employed, a combination of Eqs. (16.12) and (16.15) gives

$$\mathscr{E}_S = \left[\frac{2qN_A}{K_S\varepsilon_0}\phi_S\right]^{1/2} \tag{16.27}$$

and

$$V_G = \phi_S + \frac{K_S}{K_O} x_o \sqrt{\frac{2qN_A}{K_S\varepsilon_0}\phi_S} \qquad (0 \le \phi_S \le 2\phi_F) \tag{16.28}$$

---

[†] If the delta-depletion formulation is invoked, the δ-function layers of carrier charge at the O–S interface would technically contribute a $Q_{O-S}$ under accumulation and inversion conditions. However, $\phi_S = 0$ for all accumulation biases and $\phi_S = 2\phi_F$ for all inversion biases in the delta-depletion solution. In the cited formulation, therefore, the $V_G - \phi_S$ relationship we are deriving would only be used in performing depletion-bias calculations.

[‡] Since $K_S = 11.8$ for silicon and $K_O = 3.9$ for $SiO_2$, we conclude from Eq. (16.24) that $\mathscr{E}_{ox} \cong 3\mathscr{E}_S$ in an MOS system with no charge at the Si–SiO$_2$ interface. All energy band diagrams in this chapter should be consistent with this observation.

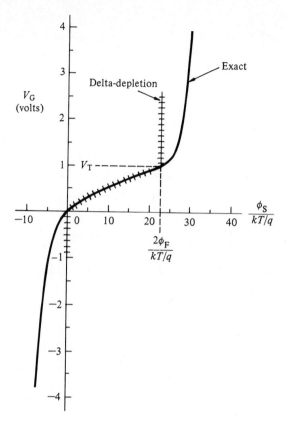

**Figure 16.10** Typical interrelationship between the applied gate voltage and the semiconductor surface potential; ┼┼┼┼┼┼ delta-depletion solution, ───── exact solution ($x_o = 0.1$ $\mu$m, $N_A = 10^{15}/cm^3$, $T = 300$ K).

The $V_G - \phi_S$ dependence calculated from Eq. (16.28) employing a typical set of device parameters is displayed in Fig. 16.10. Also shown is the corresponding exact dependence. The figure nicely illustrates certain important features of the gate voltage relationship. For one, $\phi_S$ is a rather rapidly varying function of $V_G$ when the device is depletion biased. However, when the semiconductor is accumulated ($\phi_S < 0$) or inverted ($\phi_S > 2\phi_F$), it takes a large change in gate voltage to produce a small change in $\phi_S$. This implies the gate voltage divides proportionally between the oxide and the semiconductor under depletion biasing. Under accumulation and inversion biasing, on the other hand, changes in the applied potential are dropped almost totally across the oxide. Also note that the depletion bias region is only slightly greater than 1 volt in extent. Since the character of the semiconductor changes drastically in progressing from one side of the depletion bias region to the other, we are led to anticipate a significant variation in the electrical characteristics over a rather narrow range of voltages.

### Exercise 16.3

**P:** An MOS-C is maintained at $T = 300$ K, $x_o = 0.1$ $\mu$m, and the silicon doping is $N_A = 10^{15}/cm^3$. Compute:

(a)  $\phi_F$ in $kT/q$ units and in volts

(b)  $W$ when $\phi_S = \phi_F$

(c)  $\mathcal{E}_S$ when $\phi_S = \phi_F$

(d)  $V_G$ when $\phi_S = \phi_F$

**S:** (a)

$$\frac{\phi_F}{kT/q} = \ln(N_A/n_i) = \ln\left(\frac{10^{15}}{10^{10}}\right) = \textbf{11.51}$$

$$\phi_F = 11.51 \ (kT/q) = (11.51)(0.0259) = \textbf{0.298 V}$$

(b)  Using Eq. (16.15),

$$W = \left[\frac{2K_S\varepsilon_0}{qN_A}\phi_F\right]^{1/2} = \left[\frac{2(11.8)(8.85 \times 10^{-14})(0.298)}{(1.6 \times 10^{-19})(10^{15})}\right]^{1/2} = \textbf{0.624 } \boldsymbol{\mu}\textbf{m}$$

(c)  Evaluating Eq. (16.12) at $x = 0$ yields $\mathcal{E}_S$. Thus

$$\mathcal{E}_S = \frac{qN_A}{K_S\varepsilon_0}W = \frac{(1.6 \times 10^{-19})(10^{15})(6.24 \times 10^{-5})}{(11.8)(8.85 \times 10^{-14})} = \textbf{9.56} \times \textbf{10}^3 \textbf{ V/cm}$$

(d)  Substituting into Eq. (16.26) gives

$$V_G = \phi_F + \frac{K_S}{K_O}x_0\mathcal{E}_S \quad \dots \mathcal{E}_S \text{ evaluated at } \phi_F$$

$$= 0.298 + \frac{(11.8)(10^{-5})(9.56 \times 10^3)}{3.9} = \textbf{0.587 V}$$

*Comment:* The manipulations and results in this exercise are fairly representative. In $kT/q$ units, $|\phi_F|$ typically ranges between 9 and 18 at $T = 300$ K. For the non-degenerate dopings used in MOS devices, one also expects $|\phi_F|$ to be less than one-half a Si band gap expressed in volts ($< 0.56$ V at $T = 300$ K). As required, the calculated $W$ is less than $W_T$ associated with the given doping (see Fig. 16.9). The only possible surprise is the size of the surface electric field; $\mathcal{E}_S \sim 10^4$ V/cm. Finally, the device parameters assumed in this exercise are identical to those used in constructing Fig. 16.10. As must be the case, the computed $V_G$ agrees with the value read from the Fig. 16.10 plot.

## 16.4 CAPACITANCE–VOLTAGE CHARACTERISTICS

With d.c. current flow blocked by the oxide, the major observable exhibited by MOS-Cs is capacitance. The capacitance varies as a function of the applied gate voltage and the measured capacitance–voltage ($C$–$V$) characteristic is of considerable practical importance. To the device specialist, the MOS-C $C$–$V$ characteristic is like a picture window, a window revealing the internal nature of the structure. The characteristic serves as a powerful diagnostic tool for identifying deviations from the ideal in both the oxide and the semiconductor. For the reasons cited, MOS-C $C$–$V$ characteristics are routinely monitored during MOS device fabrication.

In most laboratories and fabrication facilities, the $C$–$V$ measurements are performed with automated equipment such as that previously described in Subsection 7.2.3 and schematically pictured in Fig. 7.6. The MOS-C is positioned on a probing station, normally housed in a light-tight box to exclude room light, and is connected by shielded cables to a $C$–$V$ meter. The meter superimposes a small a.c. signal on top of a preselected d.c. voltage and detects the resulting a.c. current flowing through the test structure. The a.c. signal is typically 15 mV rms or less, and a common signal frequency is 1 MHz. Built-in provisions are made for slowly changing the d.c. voltage to obtain a continuous (or quasi-continuous) capacitance versus voltage characteristic. Output from the meter is usually displayed on a computer monitor and a hardcopy of the data is produced employing a printer or plotter. Somewhat more sophisticated commercial equipment combining a high-frequency (1 MHz) $C$–$V$ meter with low-frequency (quasistatic) measurement capabilities is pictured in Fig. 16.11(a).

This section is primarily devoted to modeling the observed form of the MOS-C $C$–$V$ characteristic in the so-called low-frequency and high-frequency limits. These limiting-case designations originally referred to the frequency of the a.c. signal used in the capacitance measurement. Although the theoretical treatment in this chapter is restricted to the ideal structure, practical measurement considerations are included at the end of the section.

### 16.4.1 Theory and Analysis

#### Qualitative Theory

High- and low-frequency $C$–$V$ data derived from a representative MOS-capacitor are displayed in Fig. 16.11(b). To explain the observed form of the $C$–$V$ characteristics, let us consider how the charge inside an $n$-type MOS-C responds to the applied a.c. signal as the d.c. bias is systematically changed from accumulation, through depletion, to inversion. We

(a)

(b)

**Figure 16.11** (a) The Keithley simultaneous high- and low-frequency $C$–$V$ system. (b) Sample MOS-C high- and low-frequency capacitance–voltage characteristics. The device was fabricated on $N_D = 9.1 \times 10^{14}/cm^3$ (100) Si; $x_o = 0.119$ $\mu$m. [Part (a) appears in the Keithley 1993–1994 *Test & Measurement Catalog*. Photograph courtesy of Keithley Instruments, Inc.]

begin with accumulation. In accumulation the d.c. state is characterized by the pileup of majority carriers right at the oxide–semiconductor interface. Furthermore, under accumulation conditions the state of the system can be changed very rapidly. For typical semiconductor dopings, the majority carriers, the only carriers involved in the operation of the accumulated device, can equilibrate with a time constant on the order of $10^{-10}$ to $10^{-13}$ sec. Consequently, at standard probing frequencies of 1 MHz or less it is reasonable to assume the device can follow the applied a.c. signal quasistatically, with the small a.c. signal adding or subtracting a small $\Delta Q$ on the two sides of the oxide as shown in Fig. 16.12(a). Since the a.c. signal merely adds or subtracts a charge close to the edges of an insulator, the charge configuration inside the accumulated MOS-C is essentially that of an ordinary parallel-plate capacitor. For either low or high probing frequencies we therefore

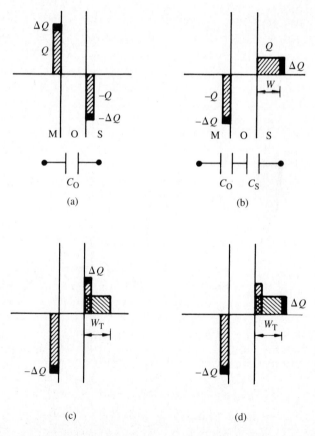

(a)

(b)

(c)

(d)

**Figure 16.12**   a.c. charge fluctuations inside an $n$-type MOS-capacitor under d.c. biasing conditions corresponding to (a) accumulation, (b) depletion, (c) inversion when $\omega \to 0$, and (d) inversion when $\omega \to \infty$. Equivalent circuit models appropriate for accumulation and depletion biasing are also shown beneath the block charge diagrams in parts (a) and (b), respectively.

conclude

$$C(\text{acc}) \simeq C_O = \frac{K_O \varepsilon_0 A_G}{x_o} \tag{16.29}$$

where $A_G$ is the area of the MOS-C gate.

Under depletion biasing the d.c. state of an $n$-type MOS structure is characterized by a $-Q$ charge on the gate and a $+Q$ depletion layer charge in the semiconductor. The depletion layer charge is directly related, of course, to the withdrawal of majority carriers from an effective width $W$ adjacent to the oxide–semiconductor interface. Thus, once again, only majority carriers are involved in the operation of the device and the charge state inside the system can be changed very rapidly. As pictured in Fig. 16.12(b), when the a.c. signal places an increased negative charge on the MOS-C gate, the depletion layer inside the semiconductor widens almost instantaneously; that is, the depletion width quasistatically fluctuates about its d.c. value in response to the applied a.c. signal. If the stationary d.c. charge in Fig. 16.12(b) is conceptually eliminated, all that remains is a small fluctuating charge on the two sides of a double-layer insulator. For all probing frequencies this situation is clearly analogous to two parallel plate capacitors ($C_O$ and $C_S$) in series, where

$$C_O = \frac{K_O \varepsilon_0 A_G}{x_o} \qquad \text{(oxide capacitance)} \tag{16.30a}$$

$$C_S = \frac{K_S \varepsilon_0 A_G}{W} \qquad \text{(semiconductor capacitance)} \tag{16.30b}$$

and

$$C(\text{depl}) = \frac{C_O C_S}{C_O + C_S} = \frac{C_O}{1 + \dfrac{K_O W}{K_S x_o}} \tag{16.31}$$

Note from Eq. (16.31) that, because $W$ increases with increased depletion biasing, $C(\text{depl})$ correspondingly decreases as the d.c. bias is changed from flat band to the onset of inversion.

Once inversion is achieved we know that an appreciable number of minority carriers pile up near the oxide–semiconductor interface in response to the applied d.c. bias. Also, the d.c. width of the depletion layer tends to maximize at $W_T$. The a.c. charge response, however, is not immediately obvious. The inversion layer charge might conceivably fluctuate in response to the a.c. signal as illustrated in Fig. 16.12(c). Alternatively, the semiconductor charge required to balance $\Delta Q$ changes in the gate charge might result from small variations in the depletion width as pictured in Fig. 16.12(d). Even a combination of the two extremes is a logical possibility. The problem is to ascertain which alternative describes the actual a.c. charge fluctuation inside an MOS-C. As it turns out, the observed charge fluctuation depends on the frequency of the a.c. signal used in the capacitance measurement.

First of all, if the measurement frequency is very low ($\omega \to 0$), minority carriers can be generated or annihilated in response to the applied a.c. signal and the time-varying a.c. state is essentially a succession of d.c. states. Just as in accumulation, one has a situation (Fig. 16.12c) where charge is being added or subtracted close to the edges of a single-layer insulator. We therefore conclude

$$C(\text{inv}) \simeq C_O \qquad \text{for } \omega \to 0 \tag{16.32}$$

If, on the other hand, the measurement frequency is very high ($\omega \to \infty$), the relatively sluggish generation–recombination process will not be able to supply or eliminate minority carriers in response to the applied a.c. signal. The number of minority carriers in the inversion layer therefore remains fixed at its d.c. value and the depletion width simply fluctuates about the $W_T$ d.c. value. Similar to depletion biasing, this situation (Fig. 16.12d) is equivalent to two parallel-plate capacitors in series and

$$C(\text{inv}) = \frac{C_O C_S}{C_O + C_S} = \frac{C_O}{1 + \dfrac{K_O W_T}{K_S x_o}} \qquad \text{for } \omega \to \infty \tag{16.33}$$

$W_T$ being a constant independent of the d.c. inversion bias makes $C(\text{inv})_{\omega \to \infty} = C(\text{depl})_{\text{minimum}} = $ constant for all inversion biases. Finally, if the measurement frequency is such that a *portion* of the inversion layer charge can be created/annihilated in response to the a.c. signal, an inversion capacitance intermediate between the high- and low-frequency limits will be observed.

An overall theory can now be constructed by combining the results of the foregoing accumulation, depletion, and inversion considerations. Specifically, we expect the MOS-C capacitance to be approximately constant at $C_O$ under accumulation biases, to decrease as the d.c. bias progresses through depletion, and to be approximately constant again under inversion biases at a value equal to $\sim C_O$ if $\omega \to 0$ or $C(\text{depl})_{\text{min}}$ if $\omega \to \infty$. Moreover, for an $n$-type device, accumulating gate voltages (where $C \simeq C_O$) are positive, inverting gate voltages are negative, and the decreasing-capacitance, depletion bias region is on the order of a volt or so in width. Quite obviously, this theory for the capacitance–voltage characteristics is in good agreement with the experimental MOS-C $C-V_G$ characteristics presented in Fig. 16.11(b).

---

### Exercise 16.4

**P:** Complete the following table making use of the ideal-structure $C-V$ characteristic and the block charge diagrams included in Figure E16.4. For each of the biasing conditions named in the table, employ letters (a–g) to identify the corresponding bias point or points on the ideal MOS-C $C-V$ characteristic. Likewise, use a number (1–5) to identify the block charge diagram associated with each of the biasing conditions.

| Bias Condition | Capacitance (a–g) | Charge Diagram (1–5) |
|---|---|---|
| Accumulation | | |
| Depletion | | |
| Inversion | | |
| Flat band | | |
| Depl/inv transition | | |

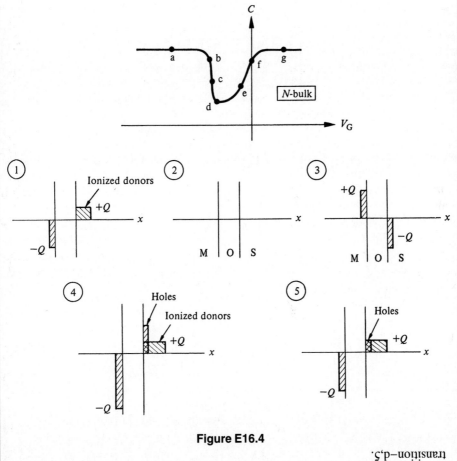

**Figure E16.4**

transition–d,5.

**S:** Accumulation–g,3; depletion–e,1; inversion–a/b/c,4; flat band–f,2; depl/inv

## Delta-Depletion Analysis

Building on the development in the previous subsection, it is relatively easy to establish a first-order quantitative theory based on the delta-depletion formulation. Specifically, in the delta-depletion formulation the charge blocks representing accumulation and inversion layers in Fig. 16.12 are formally replaced by δ-functions of charge *right at* the oxide–semiconductor interface. Consequently, $C$ in the delta-depletion solution is *precisely* equal to $C_O$ for accumulation biases and for inversion biases in the low-frequency limit. On the other hand, the depletion relationship and the high-frequency inversion relationship (Eqs. 16.31 and 16.33, respectively) can be used without modification. The block charge modeling of the depletion regions in Fig. 16.12 conforms exactly with the simplified charge distributions assumed in the depletion approximation. Within the framework of the delta-depletion formulation, therefore,

$$
C = \begin{cases}
C_O & \text{acc} & (16.34a) \\[2ex]
\dfrac{C_O}{1 + \dfrac{K_O W}{K_S x_o}} & \text{depl} & (16.34b) \\[4ex]
C_O & \text{inv } (\omega \to 0) & (16.34c) \\[2ex]
\dfrac{C_O}{1 + \dfrac{K_O W_T}{K_S x_o}} & \text{inv } (\omega \to \infty) & (16.34d)
\end{cases}
$$

Given a set of device parameters, one can compute $C_O$ and $W_T$ from previous relationships. For the analytical solution to be complete, however, the depletion-bias $W$ in Eq. (16.34b) must be expressed as a function of $V_G$. Inverting Eq. (16.28) to obtain $\phi_S$ (or more precisely, $\sqrt{\phi_S}$) as a function of $V_G$, and then substituting the result into Eq. (16.15), yields the required expression. We find

$$
W = \frac{K_S}{K_O} x_o \left[ \sqrt{1 + \frac{V_G}{V_\delta}} - 1 \right] \tag{16.35}
$$

where

$$
V_\delta \equiv \frac{q}{2} \frac{K_S x_o^2}{K_O^2 \varepsilon_0} N_A \qquad
\begin{array}{l} \cdots \; p\text{-bulk device} \\ (\text{for } n\text{-bulk } N_A \to -N_D) \end{array} \tag{16.36}
$$

Note that if Eq. (16.35) is substituted into Eq. (16.34b), one obtains the very simple result

$$C = \frac{C_O}{\sqrt{1 + \dfrac{V_G}{V_\delta}}} \qquad \text{(depletion biases)} \qquad (16.37)$$

A sample set of low- and high-frequency $C–V$ characteristics constructed using the results of the delta-depletion analysis is displayed in Fig. 16.13.

## 16.4.2 Computations and Observations

### Exact Computations

The delta-depletion characteristics, as typified by Fig. 16.13, are a rather crude representation of reality. The first-order theory does a credible job for gate voltages comfortably within a given biasing region, but fails badly in the neighborhood of the transition points going from accumulation to depletion and from depletion to inversion. A more accurate modeling of the observed characteristics is often required in practical applications and is established by working with the exact-charge distribution inside the MOS-capacitor. The results of the exact-charge analysis are presented in Appendix C. Although the derivation

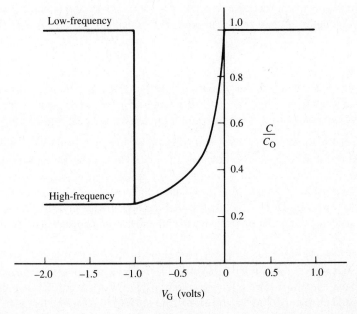

**Figure 16.13** Sample set of low- and high-frequency $C–V$ characteristics established using the delta-depletion theory ($x_o = 0.1~\mu m$, $N_D = 10^{15}/cm^3$, $T = 300$ K).

of the exact-charge relationships is beyond the scope of this text, the results themselves are quite tractable. Highly accurate ideal-structure $C–V$ characteristics can be readily constructed.

A number of sample $C–V$ characteristics calculated using the exact-charge relationships are displayed in Figs. 16.14 to 16.16. These figures, respectively, exhibit the general effect of varying the doping concentration (Fig. 16.14), the oxide thickness (Fig. 16.15), and the device temperature (Fig. 16.16). Note in particular from Fig. 16.14 the significant increase in the high-frequency inversion capacitance and the substantial widening of the depletion bias region with increased doping. In fact, at very high dopings (not shown) the capacitance approaches a constant independent of bias. This should not be an unexpected result, for with increased doping the semiconductor begins to look more and more like a metal and the MOS-C should be expected to react more and more like a standard capacitor. As illustrated in Fig. 16.15, an increase in the oxide thickness also widens the depletion bias region and affects the high-frequency inversion capacitance. The increased width of the depletion bias region with increased $x_o$ is simply a consequence of a proportionate increase in the voltage drop across the oxide component of the structure. Finally, Fig. 16.16 nicely displays the moderate sensitivity of the inversion-bias capacitance and the near insensitivity of the depletion-bias capacitance to changes in temperature.

---

### (C) Exercise 16.5

**P:** (a) Employing the exact-charge relationships found in Appendix C, write a MATLAB (computer) program that can be used to calculate and plot *low-frequency* $C/C_O$ versus $V_G$ characteristics. The program is to calculate $C/C_O$ and the corresponding $V_G$ for $U_S = \phi_S/(kT/q)$ stepped in one-unit or less increments from $U_S = U_F - 21$ to $U_F + 21$. $U_F = \phi_F/(kT/q)$. Employ $K_S = 11.8$ and $K_O = 3.9$; let $T = 300$ K. Only $N_A$ and $x_o$ are to be considered input variables. Setting $N_A = 10^{15}/cm^3$, use your program to plot out the low-frequency $C/C_O$ versus $V_G$ curves for $x_o = 0.1$ $\mu$m, $0.2$ $\mu$m, and $0.3$ $\mu$m. Compare your program results with Fig. 16.15.

(b) Again employing the exact-charge relationships found in Appendix C, follow the same general computational procedures noted in part (a) to generate *high-frequency* $C/C_O$ versus $V_G$ characteristics. Setting $x_o = 0.1$ $\mu$m, output high-frequency $C/C_O$ versus $V_G$ curves for $N_A = 10^{14}/cm^3$, $10^{15}/cm^3$, and $10^{16}/cm^3$. Compare your program results with Fig. 16.14. (Be advised that the high-frequency calculation is far more involved and computation intensive than the low-frequency calculation.)

**S:** (a)/(b) The construction of a program or programs to generate exact-charge $C–V$ characteristics is a very informative exercise. The reader is urged to at least complete part (a) of the exercise before referring to the MOS_CV MATLAB file included on disk and listed in Appendix M. In any event, characteristics generated by the reader's program or the MOS_CV program should compare favorably with the $C–V$ curves plotted in Figs. 16.14 and 16.15.

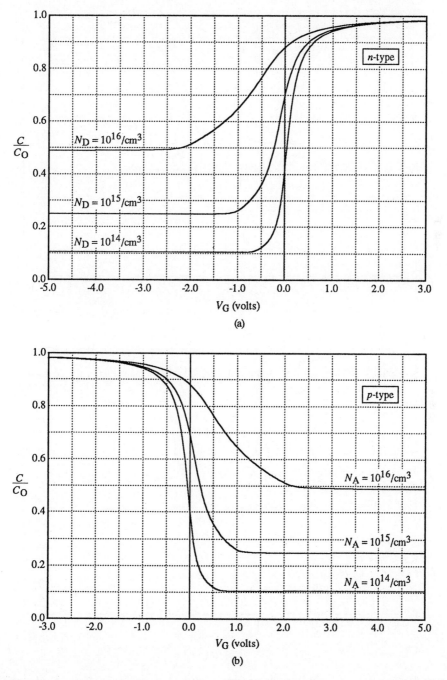

**Figure 16.14** Doping dependence of the MOS-C high-frequency $C$–$V_G$ characteristics. Sample (a) $n$-type and (b) $p$-type characteristics computed employing the exact-charge theory ($x_o = 0.1\ \mu m$, $T = 300$ K).

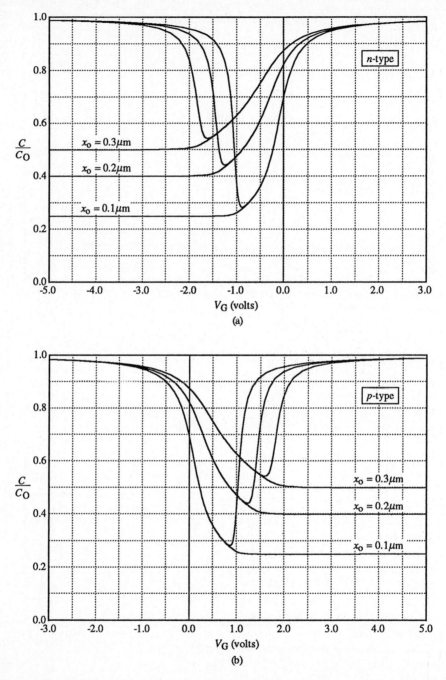

**Figure 16.15**   Oxide thickness dependence of the low- and high-frequency MOS-C $C$–$V_G$ characteristics. Sample (a) $n$-type and (b) $p$-type characteristics computed employing the exact-charge theory ($N_A$ or $N_D = 10^{15}/\mathrm{cm}^3$, $T = 300$ K).

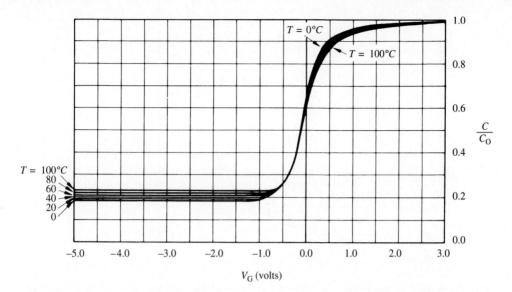

**Figure 16.16** Temperature dependence of the high-frequency $C$–$V_G$ characteristics. (Exact-charge theory, $x_o = 0.1 \ \mu m$, $N_D = 5 \times 10^{14}/cm^3$.)

## Practical Observations

In the discussion to this point we have more or less sidestepped any clarification of precisely what was meant by "low frequency" and "high frequency" in terms of actual measurement frequencies. One might wonder, will a 100 Hz a.c. signal typically yield low-frequency $C$–$V$ characteristics? Perhaps surprisingly, the answer to the question is *no*. Given modern-day MOS-Cs with their long carrier lifetimes and low carrier generation rates, even probing frequencies as low as 10 Hz, the practical limit in bridge-type measurements, will yield high-frequency type characteristics. If an MOS-C low-frequency characteristic is required, indirect means such as the quasistatic technique[13] must be employed to construct the characteristic. In the quasistatic technique a slow (typically 10–100 mV/sec) linear voltage ramp is applied to the MOS-C gate and the *current* into the gate is monitored as a function of the gate voltage. As is readily confirmed, the quasistatic displacement current flowing through the device is directly proportional to the low-frequency capacitance; properly calibrated, the measured current versus voltage data replicates the desired low-frequency $C$–$V$ characteristic.

On the high-frequency side, one cannot actually let $\omega \rightarrow \infty$ and expect to observe a high-frequency characteristic. Measurement frequencies, in fact, seldom exceed 1 MHz. At higher frequencies the resistance of the semiconductor bulk comes into play and lowers the observed capacitance. At even higher frequencies ($\gtrsim 1$ GHz) one must begin to worry about the response time of the majority carriers.

Normally, it is the high-frequency characteristic that is routinely recorded and the standard, almost universal measurement frequency is 1 MHz. This is not to say the high-

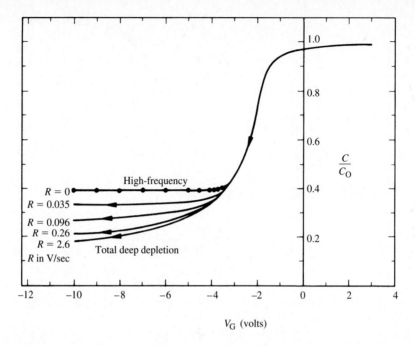

**Figure 16.17**  Measured $C-V$ characteristics as a function of the ramp rate ($R$). In inversion the high-frequency capacitance was obtained by stopping the ramp and allowing the device to equilibrate.

frequency characteristics can be recorded at 1 MHz without exercising a certain amount of caution. Suppose, for example, the $C-V$ measurement is performed as described earlier in this chapter, with the d.c. voltage being ramped from accumulation into inversion to obtain a continuous capacitance-versus-voltage characteristic. Figure 16.17 illustrates the usual results of such a measurement performed at various ramp rates. Note that at even the slowest ramp rate one does not properly plot out the inversion portion of the high-frequency characteristic. One must stop the ramp in inversion and allow the device to equilibrate, or slowly sweep the device backward from inversion toward accumulation, to accurately record the high-frequency inversion capacitance.

The discussion in the preceding paragraph really serves two purposes, the second being an entry into the important topic of deep depletion. Let us examine the ramped measurement (Fig. 16.17) in greater detail. When the ramp voltage is in accumulation or depletion, only majority carriers are involved in the operation of the device, and the d.c. charge configuration inside the structure rapidly reacts to the changing gate bias. As the ramp progresses from depletion into the inversion bias region, however, a significant number of minority carriers are required to attain an equilibrium charge distribution within the MOS-C. The minority carriers were not present prior to entering the inversion bias region, cannot enter the semiconductor from the remote back contact or across the oxide, and therefore must be created in the near-surface region of the semiconductor. The generation process,

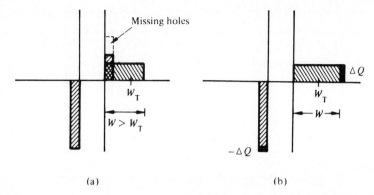

**Figure 16.18** (a) Nonequilibrium charge configuration inside an *n*-type MOS-capacitor under deep-depletion conditions. (b) a.c. charge fluctuations inside the MOS-C when the semiconductor is totally deep depleted.

as we have noted several times, is rather sluggish and has difficulty supplying the minority carriers needed for the structure to equilibrate. Thus, as pictured in Fig. 16.18(a), the semi-conductor is driven into a *nonequilibrium* condition where, in balancing the charge added to the MOS-C gate, the depletion width becomes greater than $W_T$ to offset the missing minority carriers. The described condition, the nonequilibrium condition where there is a deficit of minority carriers and a depletion width in excess of the equilibrium value, is referred to as *deep depletion.*

The existence of a $W > W_T$ explains the reduced values of $C$ observed in the ramp measurement. Moreover, the decrease in capacitance with increased ramp rate indicates a greater deficit of minority carriers and a wider depletion width. This is logical, since the greater the ramp rate, the fewer the number of minority carriers generated prior to arriving at a given inversion bias.

The limiting case as far as deep-depletion is concerned occurs when the semiconductor is totally devoid of minority carriers—totally deep depleted. Except for a wider depletion width, the total deep depletion condition shown in Fig. 16.18(b) is precisely the same as the simple depletion condition pictured in Fig. 16.12(b). Consequently, by analogy, and based on the delta-depletion formulation, the limiting-case capacitance exhibited by the structure under deep depletion conditions should be

$$C = \frac{C_O}{\sqrt{1 + \dfrac{V_G}{V_\delta}}} \qquad \begin{array}{l} \text{total deep depletion} \\ (V_G > V_T \ p\text{-type}; \ V_G < V_T \ n\text{-type}) \end{array} \tag{16.38}$$

Equation (16.38) is in excellent agreement with experimental observations and is essentially identical to the result obtained from an exact-charge analysis. The 2.6 V/sec ramp rate curve shown in Fig. 16.17 is an example of a total-deep-depletion characteristic.

The deeply depleted condition, we should note, is central to the operation of the dynamic random access memory (DRAM) and the charge-coupled devices (CCDs) mentioned in the Section 15.1 General Introduction. DRAMs use deeply depleted MOS-capacitors as storage elements. In CCD imagers, carrier charge is optically produced and temporarily stored in partially deep depleted potential wells beneath an array of MOS-C gates.

---

### Exercise 16.6

**P:** The experimental $C-V$ characteristic shown in Fig. E16.6 was observed under the following conditions: The d.c. bias was changed very slowly from point (1) to point (2). At point (2) the $V_G$ sweep rate was increased substantially. Upon arriving at point (3) the sweep was stopped and the capacitance decayed to point (4). Qualitatively explain the observed characteristic.

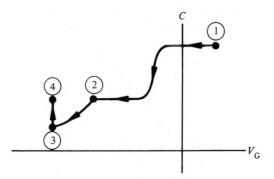

**Figure E16.6**

**S:** In going from point (1) to point (2) the semiconductor clearly equilibrates at each d.c. bias point because of the very slow speed of the ramp. One therefore observes the standard high-frequency $C-V$ characteristic. With the ramp rate increased at point (2), the semiconductor is no longer able to equilibrate and is consequently driven into deep depletion. $W$ becomes larger than $W_T$ and $C$ decreases below the minimum high-frequency equilibrium value. When the sweep is stopped and the capacitance increases from point (3) to point (4), equilibrium is systematically restored inside the device through the generation of minority carrier holes in the near-surface region. As the holes are generated they add to the inversion layer, the depletion width correspondingly decreases, and $C$ increases back to the equilibrium high-frequency value.

## 16.5 SUMMARY AND CONCLUDING COMMENTS

This chapter was devoted to introducing basic MOS terminology, concepts, visualization aids, analytical procedures, and the like. We began by describing the MOS-capacitor and clearly defining what was envisioned as the ideal MOS structure. The ideal MOS-C serves not only as a convenient tool for the introduction of MOS fundamentals, but also provides a point of reference for understanding and analyzing the more complex behavior of real MOS structures. The reference nature of the ideal structure will become more apparent in Chapter 18 where some of the idealizations will be removed and the ensuing perturbations on the device characteristics will be carefully examined.

The internal state of the MOS-C under static-biasing conditions was qualitatively described using energy band and block charge diagrams. With the aid of the diagrams the terms accumulation, flat band, depletion, and inversion were given a physical interpretation. Accumulation corresponds to the pileup of majority carriers at the oxide–semiconductor interface; flat band, to no bending of the semiconductor bands, or equivalently, to no charge in the semiconductor; depletion, to the repulsion of majority carriers from the interface leaving behind an uncompensated impurity-ion charge; and inversion, to the pileup of minority carriers at the oxide–semiconductor interface.

As a point of information, it should be mentioned that, in certain analyses, it is convenient and reasonable to divide the depletion bias region as defined herein into two subregions. In some MOS publications one therefore finds the term *depletion* is only used in referring to band bendings between $\phi_S = 0$ and $\phi_S = \phi_F$. *Weak inversion* is used to describe band bendings from $\phi_S = \phi_F$ to $\phi_S = 2\phi_F$. In addition, *strong inversion* (implying more inversion than weak inversion) replaces inversion as defined herein.

Quantitative expressions for the charge density, electric field, and electrostatic potential inside an MOS-C are established by solving Poisson's equation. A first-order solution based on the depletion approximation and δ-function modeling of the carrier charges was presented in the chapter proper. An exact solution for the electrostatic variables can be found in Appendix B.

The qualitative and quantitative formalism that had been developed was next applied to modeling the MOS-C $C$–$V$ characteristics in the low- and high-frequency limits. Sample ideal-structure characteristics were presented and examined for parametric dependencies. Finally, the practical meaning of "low-frequency" and "high-frequency" was clarified, and deep depletion was noted to occur when the MOS-C is not allowed to equilibrate under inversion biasing. Deep depletion is a nonequilibrium condition where there is a deficit of minority carriers and a depletion width in excess of the equilibrium value.

## PROBLEMS

| CHAPTER 16   PROBLEM INFORMATION TABLE | | | | |
|---|---|---|---|---|
| Problem | Complete After | Difficulty Level | Suggested Point Weighting | Short Description |
| 16.1 | 16.3.1 | 2 | 10 (one each diagram) | $\phi_F$, $\phi_S$ sets/infer diagrams |
| 16.2 | " | 2 | 8 (a-2, b-3, c-3) | Examine Fig. 16.8(c) |
| ● 16.3 | " | 2 | 8 | GaAs $W_T$ vs. doping plot |
| * 16.4 | 16.3.2 | 2–3 | 16 (a::d-2, e-8) | Calculate $\phi_F$, $W$, $\mathscr{E}_S$, $V_G$ |
| ● 16.5 | " | 3–4 | 15 (a-6, b-9) | Exact $V_G$–$\phi_S$ calculation |
| 16.6 | 16.4.1 | 2–3 | 10 (a-4, b-6) | Derive/use Eq. (16.35) |
| 16.7 | " | 3 | 20 (2 each part) | Interpret $E$-band diagram |
| 16.8 | " | 2 | 5 (1/2 each entry) | $C$ and $E$-band vs. bias table |
| 16.9 | " | 3 | 16 (2 each part) | Given $E$-band, deduce info |
| 16.10 | " | 3 | 8 (a-2, b-2, c-4) | Intrinsic MOS-C |
| 16.11 | 16.4.2 | 3 | 15 (5 each part) | SOS-C |
| 16.12 | " | 2 | 8 (4 each part) | Given $C$–$V$, answer questions |
| 16.13 | " | 2–3 | 10 (2 each part) | Given $C$–$V$, deduce info |
| 16.14 | " | 2–3 | 16 (2 each part) | Given MOS-C, ans. questions |
| 16.15 | " | 2–3 | 12 (2 each part) | Given Q-diag., deduce info |
| ● 16.16 | " | 3 | 10 | Deep depletion $C$–$V$ |
| 16.17 | " | 3 | 10 (a-2, b-3, c-5) | $x_o$ and doping determination |

In all computations involving an MOS (Si–SiO$_2$) capacitor maintained at $T = 300$ K employ $kT/q = 0.0259$ V, $n_i = 10^{10}/cm^3$, $K_S = 11.8$, and $K_O = 3.9$.

**16.1** For the $\phi_F$, $\phi_S$ parameter sets listed below first indicate the specified biasing condition and then draw the energy band diagram and block charge diagram that characterize the static state of the system. Assume the MOS structure to be ideal.

(a) $\dfrac{\phi_F}{kT/q} = 18$, $\dfrac{\phi_S}{kT/q} = 9$

(b) $\dfrac{\phi_F}{kT/q} = -12$, $\dfrac{\phi_S}{kT/q} = 0$

(c) $\dfrac{\phi_F}{kT/q} = 12$, $\dfrac{\phi_S}{kT/q} = 24$

(d) $\dfrac{\phi_F}{kT/q} = -15$, $\dfrac{\phi_S}{kT/q} = 3$

(e) $\dfrac{\phi_F}{kT/q} = 9$, $\dfrac{\phi_S}{kT/q} = 21$

**16.2** Let us examine Fig. 16.8, particularly Fig. 16.8(c), more closely.

(a) Draw the block charge diagram describing the charge situation inside an ideal $p$-bulk MOS-C biased at the onset of inversion.

(b) Is your part (a) diagram in agreement with the plot of $\rho/qN_A$ versus $x$ in Fig. 16.8(c)? Explain why the $\rho/qN_A$ plot has a spike-like nature near $x = 0$ and shows a value of $\rho/qN_A = -2$ at $x = 0$.

(c) Noting that $\phi_F/(kT/q) = 12$ and $T = 300$ K was assumed in constructing Fig. 16.8, determine $W_T$. Is the deduced $W_T$ consistent with the approximate charge distribution shown in Fig. 16.8(c)?

● **16.3** Construct a $W_T$ versus doping plot similar to Fig. 16.9 that is appropriate for GaAs. Assume $T = 300$ K; $K_S = 12.85$.

**16.4** An MOS-C is maintained at $T = 300$ K, $x_o = 0.1$ $\mu$m, and the Si doping is $N_D = 10^{15}/\text{cm}^3$. Compute:

(a) $\phi_F$ in $kT/q$ units and in volts;

(b) $W$ when $\phi_S = 2\phi_F$;

(c) $\mathscr{E}_S$ when $\phi_S = 2\phi_F$;

(d) $V_G = V_T$ when $\phi_S = 2\phi_F$. (How is this result related to Fig. 16.10?)

● (e) With the Si doping ($N_A$ or $N_D$), $x_o$, and the depletion-bias $\phi_S$ value of the MOS-C maintained at 300 K as input variables, write a computer program that automatically calculates $\phi_F$, $W$, $\mathscr{E}_S$, and $V_G$. Check the output of your program against the results recorded in Exercise 16.3. Use the program to verify the manually generated answers to parts (a)–(d) of this problem.

● **16.5** (a) Making use of Appendix B, show that the exact-solution equivalent of Eq. (16.28) is

$$V_G = \frac{kT}{q}\left[ U_S + \hat{U}_S \frac{K_S x_o}{K_O L_D} F(U_S, U_F) \right]$$

where $U_S \equiv \phi_S/(kT/q)$ and $U_F \equiv \phi_F/(kT/q)$. See Eqs. (B.5), (B.17), and (B.18) in Appendix B for the definitions of $L_D$, $F(U_S, U_F)$, and $\hat{U}_S$, respectively.

(b) Construct a computer program that calculates $V_G$ as a function of $U_S$ using the part (a) relationship. Only the semiconductor doping and $x_o$ are to be considered input variables; let $T = 300$ K and step $U_S$ in one-unit increments from $U_S = U_F - 21$ to $U_S = U_F + 21$. Run the program assuming $x_o = 0.1$ $\mu$m and $N_D = 10^{15}/\text{cm}^3$. Compare your numerical results with the exact solution curve in Fig. 16.10.

**16.6** (a) Following the approach suggested in the text, derive Eq. (16.35).

(b) Assuming $x_o = 0.1$ $\mu$m, $N_D = 10^{15}$/cm$^3$, and $T = 300$ K, compute:
  (i) $W_T$;
  (ii) $C/C_O$ inversion ($\omega \to \infty$);
  (iii) $V_T$ (delta-depletion theory).
  (iv) Comment on the comparison of your $C$ and $V$ results with Fig. 16.13.

**16.7** The energy band diagram for an ideal $x_o = 0.2$ $\mu$m MOS-C operated at $T = 300$ K is sketched in Fig. P16.7. Note that the applied gate voltage causes band bending in the semiconductor such that $E_F = E_i$ at the Si–SiO$_2$ interface. Invoke the delta-depletion approximation as required in answering the questions that follow.

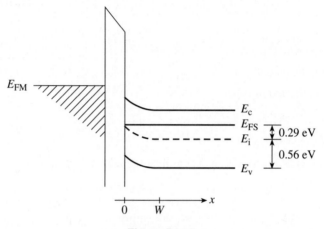

**Figure P16.7**

(a) Sketch the electrostatic potential ($\phi$) inside the semiconductor as a function of position.

(b) Roughly sketch the electric field ($\mathcal{E}$) inside the oxide and semiconductor as a function of position.

(c) Do equilibrium conditions prevail *inside the semiconductor*? Explain.

(d) Roughly sketch the electron concentration versus position inside the semiconductor.

(e) What is the electron concentration at the Si–SiO$_2$ interface?

(f) $N_D = ?$

(g) $\phi_S = ?$

(h) $V_G = ?$

(i) What is the voltage drop ($\Delta\phi_{ox}$) across the oxide?

(j) What is the normalized small-signal capacitance, $C/C_O$, of the MOS-C at the pictured bias point?

**16.8** Complete the following table making use of the ideal-structure $C-V$ characteristic and energy band diagrams in Figure P16.8. For each of the biasing conditions named in the table, employ letters (a–e) to identify the corresponding bias point on the ideal MOS-C $C-V$ characteristic. Likewise, use a number (1–5) to identify the band diagram associated with each of the biasing conditions.

| Bias Condition | Capacitance (a–e) | Band Diagram (1–5) |
|---|---|---|
| Inversion | | |
| Depletion | | |
| Flat band | | |
| $V_G = V_T$ | | |
| Accumulation | | |

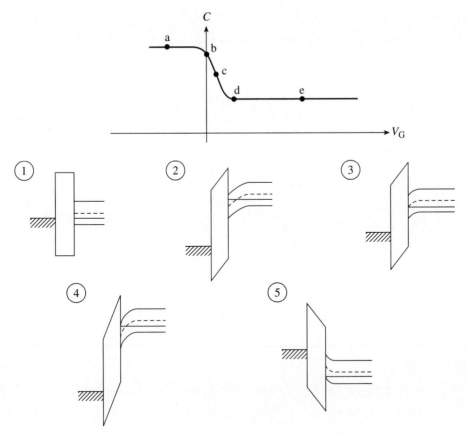

**Figure P16.8**

**16.9** Figure P16.9 is a dimensioned energy band diagram for an ideal MOS-C operated at $T = 300$ K with $V_G \neq 0$. Note that $E_F = E_i$ at the Si–SiO$_2$ interface.

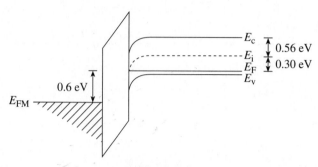

**Figure P16.9**

(a) Do equilibrium conditions prevail *inside the semiconductor*?

(b) $\phi_F = ?$

(c) $\phi_S = ?$

(d) $V_G = ?$

(e) $x_o = ?$

(f) Draw the block charge diagram corresponding to the state pictured in the energy band diagram. For reference purposes, include the maximum equilibrium depletion width, $W_T$, on your diagram.

(g) Sketch the general shape of the *low-frequency* C–V characteristic to be expected from the given MOS-C. Place an $\times$ on the C–V characteristic at a point that roughly corresponds to the state pictured in the energy band diagram.

(h) For the specific bias point pictured in the energy band diagram, which of the following is the correct expression for the capacitance exhibited by the structure? Explain.

$$\text{(i)} \quad C = \frac{C_O}{1 + \dfrac{K_O W_T}{K_S x_o}}, \qquad \text{(ii)} \quad C = \frac{C_O}{\sqrt{1 + \dfrac{V_T}{2V_\delta}}},$$

$$\text{(iii)} \quad C = \frac{C_O}{1 + \dfrac{K_O W_T}{\sqrt{2}\, K_S x_o}}, \qquad \text{(iv)} \quad C = \frac{C_O}{\sqrt{1 + \dfrac{V_T}{V_\delta}}}$$

**16.10** An ideal MOS capacitor is constructed on a substrate of *intrinsic* silicon.

(a) Sketch the energy band diagram for the capacitor under flat band conditions. Include the diagram for all three components of the MOS-C, properly position the Fermi level in the metal and semiconductor, and label the energy levels.

(b) Construct block charge diagrams representing the capacitor under the conditions of positive and negative gate bias.

(c) Invoking the delta-depletion approximation, sketch the low-frequency $C$–$V$ characteristic for the given MOS-C. Justify the form of your sketch in each region of operation.

**16.11** With modern-day processing it is possible to produce semiconductor–oxide–semiconductor (SOS) capacitors in which a semiconductor replaces the metallic gate in a standard MOS-C. Answer the questions posed below assuming an SOS-C composed of two *identical n-type* nondegenerate silicon electrodes, an *ideal structure,* and a biasing arrangement as defined by Fig. P16.11. Include any comments which may help to forestall a misinterpretation of the requested pictorial answers.

(a) Draw the energy band diagram for the structure when (i) $V_G = 0$, (ii) $V_G > 0$ but small, (iii) $V_G > 0$ and very large, (iv) $V_G < 0$ but small, and (v) $V_G < 0$ and very large.

(b) Draw the block charge diagrams corresponding to the five biasing conditions considered in part (a).

(c) Sketch the expected shape of the high-frequency $C$–$V$ characteristic for the SOS-C described in this problem. For reference purposes, also sketch on the same plot the high-frequency $C$–$V_G$ characteristic of an MOS-C assumed to have the same semiconductor doping and oxide thickness as the SOS-C.

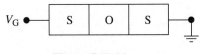

**Figure P16.11**

**16.12** (a) Consider the $C$–$V$ curves shown in Fig. P16.12(a). For which curve or curves is an *equilibrium* inversion layer present when $V_G > V_T$? Explain.

(b) $C$–$V$ curves derived from two MOS-Cs with the same gate area ($A_G$) are compared in Fig. P16.12(b). The MOS-C exhibiting curve $b$ has (choose one: a thinner, the same, a thicker) oxide and (choose one: a lower, the same, a higher) doping than the MOS-C exhibiting curve $a$. Briefly explain how you arrived at your chosen answers.

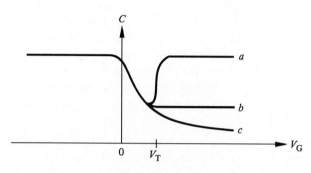

**Figure P16.12(a)**

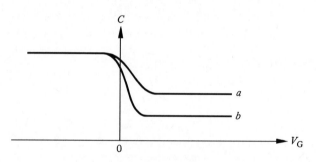

**Figure P16.12(b)**

**16.13** The $C–V$ characteristic exhibited by an MOS-C (assumed to be ideal) is displayed in Figure P16.13.

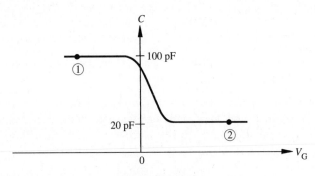

**Figure P16.13**

(a) Is the semiconductor component of the MOS-C doped $n$-type or $p$-type? Indicate how you arrived at your answer.

(b) Draw the MOS-C energy band diagram corresponding to point (2) on the $C–V$ characteristic. (Be sure to include the diagrams for all three components of the MOS-C, show the proper band bending in both the oxide and semiconductor, and properly position the Fermi level in the metal and semiconductor.)

(c) Draw the block charge diagram corresponding to point (1) on the $C–V$ characteristic.

(d) If the area of the MOS-C is $3 \times 10^{-3}$ cm$^2$, what is the oxide thickness $(x_o)$?

(e) Invoking the delta-depletion approximation, determine $W_T$ and the associated semiconductor doping concentration for the given MOS-C.

**16.14** An ideal MOS-C is operated at $T = 300$ K. $x_o = 0.1$ $\mu$m, $N_D = 2 \times 10^{15}$/cm$^3$, and $A_G = 10^{-3}$ cm$^2$.

(a) Sketch the *general shape* of the *high-frequency* $C–V$ characteristic to be expected from the given device.

(b) Defining $C_{MAX}$ to be the maximum high-frequency capacitance, determine $C_{MAX}$.

(c) Defining $C_{MIN}$ to be the minimum high-frequency capacitance, estimate $C_{MIN}$ employing the delta-depletion approximation.

(d) If $V_G = V_T$, determine $\phi_S$. (Give both a symbolic and a numerical answer.)

(e) Compute $V_T$.

(f) Suppose the gate bias is such that $\phi_S = 3\phi_F/2$. Draw the MOS-C energy band diagram corresponding to the specified gate bias. (Be sure to include the diagrams for all three components of the MOS-C, show the proper band bending in both the oxide and semiconductor, and properly position the Fermi level in the metal and semiconductor.)

(g) Suppose the gate bias is such the $\phi_S = 5\phi_F/2$. Draw the block charge diagram corresponding to the specified gate bias.

(h) The $C$–$V$ characteristic of the device is measured as the d.c. bias is *rapidly* swept from accumulation into inversion. Using a dashed line, sketch the expected form of the resulting $C$–$V$ characteristic on the same set of coordinates as the part (a) answer.

**16.15** The d.c. state of an ideal MOS-capacitor is characterized by the block charge diagram shown in Fig. P16.15.

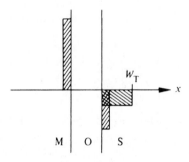

**Figure P16.15**

(a) Is the semiconductor $n$- or $p$-type? Explain.

(b) Is the device accumulation, depletion, or inversion biased? Explain.

(c) Draw the energy band diagram corresponding to the charge state pictured in the block charge diagram.

(d) By appropriately modifying the block charge diagram, characterize the charge state inside the MOS-C when a *high-frequency* a.c. signal is applied to the device.

(e) Sketch the general shape of the high-frequency $C$–$V$ characteristic to be expected from the structure. Place an $\times$ on the $C$–$V$ characteristic at the point which roughly corresponds to the charge state pictured in the Fig. P16.15 block charge diagram.

(f) While biased at the same gate voltage giving rise to the Fig. P16.15 diagram, the MOS-C is somehow *totally deep depleted*. Draw the block charge diagram describing the new state of the system.

● **16.16** Write a MATLAB (computer) program that can be used to calculate and plot total-deep-depletion $C/C_O$ versus $V_G$ characteristics. To simplify the programming, take the MOS-C to be $p$-type. Let $V_I$ be the gate voltage when $\phi_S = \phi_F$. Following the procedure outlined in Exercise 16.5, use the low-frequency relationships found in Appendix C to compute the $V_G \leq V_I$ ($\phi_S \leq \phi_F$) portion of the characteristic. Calculate $C/C_O$ versus $V_G$ for $V_G > V_I$ employing Eq. (16.38); stop the computation at $V_G = 5V_T$. Combine the calculated $C/C_O$ values from the two voltage ranges into a single plot. Setting $x_o = 0.2\ \mu m$ and $N_A = 7.8 \times 10^{14}/cm^3$, run your program and compare the plotted output with the total deep depletion curve in Fig. 16.17. (It will be necessary to mirror your computed $p$-type characteristic about a vertical line through $V_G = 0$ to obtain the desired $n$-type characteristic.)

**16.17** The oxide thickness ($x_o$) and doping concentration ($N_A$ or $N_D$) required in constructing $C$–$V$ characteristics and in modeling MOS devices are often deduced directly from the measured high-frequency MOS-C $C$–$V$ data. Let us explore the determination procedure employing the high-frequency $C$–$V$ data recorded in Fig. 16.17.

(a) The measured capacitance with the MOS-C *heavily accumulated* is used to determine the oxide thickness. The MOS-C yielding the characteristic shown in Fig. 16.17 exhibited a maximum capacitance ($C_O$) of 82 pF and had a gate area of $A_G = 4.75 \times 10^{-3}\ cm^2$. Determine $x_o$ from the given data.

(b) The measured high-frequency capacitance with the MOS-C *heavily inverted* is used to determine the semiconductor doping concentration. Referring to Fig. 16.17, record the minimum high-frequency value of $C/C_O$ observed when the device is biased far into inversion ($V_G < -4$ V). Per Eq. (C.1), found in Appendix C, associate a $W_{eff}$ with the observed $C/C_O$. Calculate $W_{eff}(inv)$ using the $x_o$ determined in part (a). Divide $W_{eff}(inv)$ by $L_D = 2.91 \times 10^{-3}$ cm to obtain the experimental value of $W_{eff}(inv)/L_D$. (The intrinsic Debye length, $L_D$, is defined in Appendices B and C. The value quoted here is appropriate for $T = 300$ K.)

(c) To an accuracy better than 0.05% over the range $9 \leq |U_F| \leq 18$, it has been established that[14]

$$\frac{W_{eff}(inv)}{L_D} \cong 2e^{-|U_F|/2} \{2|U_F| - 1 + \ln[1.15(|U_F| - 1)]\}^{1/2}$$

where $U_F \equiv \phi_F/(kT/q)$. Using manual or computer-based iterative techniques, determine the value of $U_F$ needed to match the part (b) experimental value of $W_{eff}(inv)/L_D$. Determine $U_F$ to four significant figures and compute the corresponding $N_D$ assuming $T = 300$ K. [NOTE: The $N_D$ value deduced from Fig. 16.9 corresponding to $W_T = W_{eff}(inv)$ provides an excellent first guess for $N_D$, which in turn can be used to compute a first-guess value for $U_F$.]

# 17 MOSFETs—*The Essentials*

MOSFET-based integrated circuits have become the dominant technology in the semiconductor industry. There are literally hundreds of MOS-transistor circuits in production today, ranging from rather simple logic gates used in digital-signal processing to custom designs with both logic and memory functions on the same silicon chip. MOS products are found in a mind-boggling number of electronic systems, including the now commonplace personal computer. Initially the MOS-transistor was identified by several competing acronyms, namely, metal–oxide–semiconductor transistor (MOST), insulated gate field effect transistor (IGFET), and metal–oxide–semiconductor field effect transistor (MOSFET). (PIGFET and MISFET were even suggested with a smile at one time or another.) With the passage of time, however, the transistor structure has commonly come to be known as the MOSFET. In this chapter we are concerned with describing the operation of the MOSFET and modeling the device characteristics. We continue to assume the MOS structure to be ideal. Moreover, the development focuses on the basic transistor configuration, the long-channel (or large-dimension) enhancement-mode MOSFET. An examination of small-dimension effects and structural variations is undertaken in Chapter 19. We begin here with a qualitative discussion of MOSFET operation and d.c. current flow inside the structure, progress through a quantitative analysis of the d.c. ($I_D$–$V_D$) characteristics, and conclude with an examination of the a.c. response.

## 17.1 QUALITATIVE THEORY OF OPERATION

Figure 17.1(a) shows a cut-away view, and Fig. 17.1(b) a simplified cross-sectional view, of the basic MOS-transistor configuration. Physically, the MOSFET is essentially nothing more than an MOS-capacitor with two *pn* junctions placed immediately adjacent to the region of the semiconductor controlled by the MOS-gate. The Si substrate can be either *p*-type (as pictured) or *n*-type; $p^+$ junction islands are of course required in *n*-bulk devices. Also shown in Fig. 17.1(b) are the standard terminal and d.c. voltage designations. The drain current ($I_D$), which flows in response to the applied terminal voltages, is the primary d.c. observable. Consistent with the naming of the device leads, the current flow is always such that carriers (electrons in the present case) enter the structure through the source (S), leave through the drain (D), and are subject to the control or gating action of the gate (G). The voltage applied to the gate relative to ground is $V_G$, while the drain voltage relative to ground is $V_D$. Unless stated otherwise, we will assume that the source and back are grounded. Please note that under normal operational conditions the drain bias is always

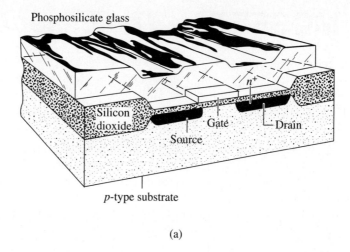

(a)

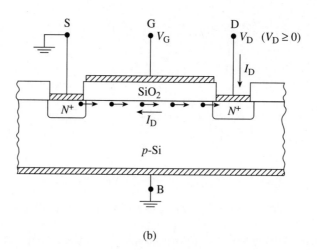

(b)

**Figure 17.1** The basic MOSFET structure. (a) Idealized cut-away view of a *p*-bulk (*n*-channel) MOSFET. (b) Simplified cross-sectional view of the *p*-bulk (*n*-channel) structure showing the terminal designations, carrier and current flow directions, and standard biasing conditions. [(a) From Beadle, Tsai, and Plummer[15]. Reprinted with permission of AT&T.]

such as to reverse bias the drain *pn* junction ($V_D \geq 0$ for the Fig. 17.1(b) device). Finally, reflecting the normal flow of electrons within the structure, the drain current for the *p*-bulk device is taken to be positive when flowing from the external circuit into the drain terminal.

To determine how the drain current is expected to vary as a function of the applied terminal voltages, let us first conceptually set $V_D = 0$ and examine the situation inside the structure as a function of the imposed gate voltage. When $V_G$ is accumulation or depletion biased ($V_G \leq V_T$, where $V_T$ is the depletion–inversion transition-point voltage), the gated region between the source and drain islands contains either an excess or deficit of holes and very few electrons. Thus, looking along the surface between the $n^+$ islands under the cited conditions, one effectively sees an open circuit. When $V_G$ is inversion biased ($V_G > V_T$), however, an inversion layer containing mobile electrons is formed adjacent to the Si surface. Now looking along the surface between the $n^+$ islands one sees, as pictured in Fig. 17.2(a), an induced "*n*-type" region (the inversion layer) or conducting *channel* connecting the source and drain islands. Naturally, the greater the inversion bias, the greater the pileup of electrons at the Si surface and the greater the conductance of the inversion layer. An inverting gate bias, therefore, creates or induces a source-to-drain channel and determines the maximum conductance of the channel.

Turning next to the action of the drain bias, suppose an inversion bias $V_G > V_T$ is applied to the gate and the drain voltage is increased in small steps starting from $V_D = 0$. At $V_D = 0$ the situation inside the device is as pictured in Fig. 17.2(a), thermal equilibrium obviously prevails, and the drain current is identically zero. With $V_D$ stepped to small positive voltages, the surface channel merely acts like a simple resistor and a drain current proportional to $V_D$ begins to flow into the drain terminal. The portion of the $I_D$–$V_D$ relationship corresponding to small $V_D$ biases is shown as the line from the origin to point A in Fig. 17.3. Any $V_D > 0$ bias, it should be interjected, simultaneously reverse biases the drain *pn* junction, and the resulting reverse bias junction current flowing into the Si substrate does contribute to $I_D$. In well-made devices, however, the junction current is totally negligible compared to the channel current, provided $V_D$ is less than the junction breakdown voltage.

Once $V_D$ is increased above a few tenths of a volt, the device enters a new phase of operation. Specifically, the voltage drop from the drain to the source associated with the flow of current in the channel starts to negate the inverting effect of the gate. As pictured in Fig. 17.2(b), the depletion region widens in going down the channel from the source to the drain and the number of inversion layer carriers correspondingly decreases. The reduced number of carriers decreases the channel conductance, which in turn is reflected as a decrease in the slope of the observed $I_D$–$V_D$ characteristic. Continuing to increase the drain voltage causes a progressive reduction in the channel carrier concentration and the systematic slope-over in the $I_D$–$V_D$ characteristic noted in Fig. 17.3. The greatest decrease in channel carriers occurs near the drain, and eventually the inversion layer completely vanishes ($n_{|\text{surface}}$ drops below $N_A$) in the near vicinity of the drain (see Fig. 17.2c). The onset of surface depletion at the drain end of the channel, the special situation where the channel carrier concentration at the Si–SiO$_2$ interface immediately adjacent to the drain becomes equal to the bulk doping concentration, is referred to as *pinch-off*. When the

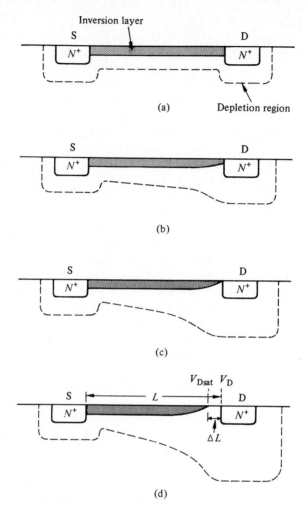

**Figure 17.2** Visualization of various phases of $V_G > V_T$ MOSFET operation. (a) $V_D = 0$; (b) channel (inversion layer) narrowing under moderate $V_D$ biasing; (c) pinch-off; and (d) postpinch-off ($V_D > V_{Dsat}$) operation. (Note that the inversion layer widths, depletion widths, etc. are not drawn to scale.)

channel pinches off inside the device, the point B is reached on the Fig. 17.3 characteristic; that is, the slope of the $I_D-V_D$ characteristic becomes approximately zero.

For drain voltages in excess of the pinch-off voltage, $V_{Dsat}$, the pinched-off portion of the channel widens from just a point into a depleted channel section $\Delta L$ in extent (see Fig. 17.2d). Being a region with few carriers, and hence low conductance, the pinched-off $\Delta L$ section absorbs most of the voltage drop in excess of $V_{Dsat}$. Given a long-channel

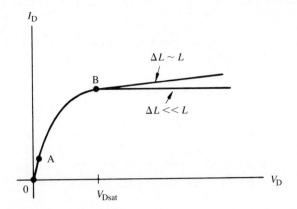

**Figure 17.3**   General variation of $I_D$ with $V_D$ for a given $V_G > V_T$.

device where $\Delta L \ll L$, the source to pinch-off region of the MOSFET will be essentially identical in shape and will have the same endpoint voltages for all $V_D \geq V_{Dsat}$. When the shape of a conductive region and the potential applied across the region do not change, the current through the region must also remain invariant. Thus, $I_D$ remains approximately constant for drain voltages in excess of $V_{Dsat}$ provided $\Delta L \ll L$. If $\Delta L$ is comparable to $L$, the same voltage drop ($V_{Dsat}$) will appear across a shorter channel ($L - \Delta L$) and, as noted in Fig. 17.3, the post-pinch-off $I_D$ in such devices will increase somewhat with increasing $V_D > V_{Dsat}$.

Thus far we have examined the response of the MOSFET to the separate manipulation of the gate and drain biases. To establish a complete set of $I_D$–$V_D$ characteristics, it is necessary to combine the results derived from the separate considerations. Clearly, for $V_G \leq V_T$, the gate bias does not create a surface channel and $I_D \simeq 0$ for all drain biases below the junction breakdown voltage. For all $V_G > V_T$ a characteristic of the form shown in Fig. 17.3 will be observed. Since the conductance of the channel increases with increasing $V_G$, it follows that the initial slope of the $I_D$–$V_D$ characteristic will likewise increase with increasing $V_G$. Moreover, the greater the number of inversion layer carriers present when $V_D = 0$, the larger the drain voltage required to achieve pinch-off. Thus $V_{Dsat}$ must increase with increasing $V_G$. From the foregoing arguments one concludes that the variation of $I_D$ with $V_D$ and $V_G$ must be of the form displayed in Fig. 17.4.

The $I_D$–$V_D$ characteristics just established confirm the "transistor" nature of the MOSFET structure. Transistor action results, of course, when $I_D$ flowing in an output circuit is modulated by an input voltage applied to the gate. Relative to terminology, we should note that the portion of the characteristics where $V_D > V_{Dsat}$ for a given $V_G$ is referred to as the *saturation* region of operation; the portion of the characteristics where $V_D < V_{Dsat}$ is called the *linear* (or sometimes *triode*) region of operation. Also, the MOSFET is known as an *n-channel* device when the channel carriers are electrons; when the channel carriers are holes the MOSFET is designated a *p-channel* device.

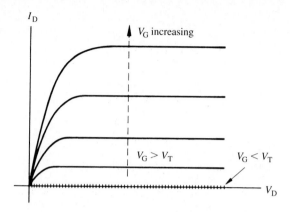

**Figure 17.4** General form of the $I_D$–$V_D$ characteristics expected from a long channel ($\Delta L \ll L$) MOSFET.

---

**Exercise 17.1**

**P:** Suppose the preceding section is to be rewritten with the illustrative MOSFET changed from an *n*-channel to a *p*-channel device. Indicate how the figures in the section must be modified if the MOSFET used for illustrative purposes is a *p*-channel device.

**S:** The required modifications can be summarized as follows:

*Figure 17.1(a)* . . . *p*-type substrate $\rightarrow$ *n*-type substrate; $n^+$ drain/source $\rightarrow p^+$ drain/source.

*Figure 17.1(b)* . . . holes replace electrons as the channel carriers; *p*-Si bulk (substrate) $\rightarrow$ *n*-Si bulk; $V_D \geq 0 \rightarrow V_D \leq 0$ in the polarity comment near the drain terminal; the arrows are reversed to show positive $I_D$ current flow from the source to the drain and out of the drain contact. (IEEE convention always defines the current flowing into a device terminal to be positive. Strictly adhering to the convention would make $I_D \leq 0$ in *p*-channel MOSFETs under normal operating conditions. Herein we prefer to deal with positive drain currents.)

*Figure 17.2* . . . change the source/drain dopings to $p^+$.

*Figure 17.3* . . . $V_D \rightarrow -V_D$ and $V_{Dsat} \rightarrow -V_{Dsat}$ along the *x*-axis of the plot. Alternatively, the existing $V_D$ labeling could be retained and the characteristics appropriately redrawn along the negative portion of the *x*-axis. (Adhering to IEEE convention, the characteristics are sometimes drawn in an upside-down fashion in the third plot quadrant where both $I_D \leq 0$ and $V_D \leq 0$.)

*Figure 17.4*... replace "$V_G$ increasing" with "$-V_G$ increasing" and $V_D$ with $-V_D$. Since $V_T < 0$ in a *p*-channel device, no current flows if $V_G > V_T$ and current is observed if $V_G < V_T$.

In addition to the noted changes to the figures proper, the inequality signs must be reversed in the Fig. 17.2 and 17.3 captions.

## 17.2 QUANTITATIVE $I_D$–$V_D$ RELATIONSHIPS

During the course of MOSFET development there has evolved a hierarchy of long-channel $I_D$–$V_D$ formulations that provide progressively increased accuracy at the expense of increased complexity. We will examine two of the formulations: the "square-law" theory and the "bulk-charge" theory. The former provides very simple relationships; the latter is a much more accurate representation of reality. Interestingly, all but the final derivational steps in the two theories are identical. Comments relative to more exacting long-channel theories can be found at the end of the section.

### 17.2.1 Preliminary Considerations

#### Threshold Voltage

From the qualitative description of MOSFET operation it should be obvious that the parameter $V_T$ plays a prominent role in determining the precise nature of the device characteristics. In MOSFET analyses $V_T$ is commonly called the *threshold* or *turn-on* voltage. The transistor starts to carry current (turns on) at the onset of inversion. A computational expression for the important $V_T$ parameter is readily established using the results of Subsection 16.3.2 and the fact that $V_G = V_T$ when $\phi_S = 2\phi_F$. Specifically, given an ideal *n*-channel (or *p*-bulk) device, simple substitution into Eq. (16.28) yields

$$V_T = 2\phi_F + \frac{K_S x_o}{K_O}\sqrt{\frac{4qN_A}{K_S\varepsilon_0}\phi_F} \qquad \text{... ideal } n\text{-channel} \atop (p\text{-bulk) devices} \qquad (17.1a)$$

Analogously,

$$V_T = 2\phi_F - \frac{K_S x_o}{K_O}\sqrt{\frac{4qN_D}{K_S\varepsilon_0}(-\phi_F)} \qquad \text{... ideal } p\text{-channel} \atop (n\text{-bulk) devices} \qquad (17.1b)$$

## Effective Mobility

In deriving quantitative expressions for the MOSFET d.c. characteristics one encounters a new parameter known as the "effective mobility." The carrier mobilities, $\mu_n$ and $\mu_p$, were first described in Section 3.1 and were noted to be a measure of the ease of carrier motion within a semiconductor crystal. In the semiconductor bulk, that is, at a point far removed from the semiconductor surface, the carrier mobilities are typically determined by the amount of lattice scattering and ionized impurity scattering taking place inside the material. For a given temperature and semiconductor doping, these bulk mobilities ($\mu_n$ and $\mu_p$) are well-defined and well-documented material constants. Carrier motion in a MOSFET, however, takes place in a surface inversion layer where the gate-induced electric field acts so as to accelerate the carriers toward the surface. The inversion layer carriers therefore experience motion impeding collisions with the Si surface (see Fig. 17.5) in addition to lattice and ionized impurity scattering. The additional surface scattering mechanism lowers the mobility of the carriers, with the carriers constrained nearest the Si surface experiencing the greatest reduction in mobility. The resulting average mobility of the inversion layer carriers is called the *effective mobility* and is given the symbol $\bar{\mu}_n$ or $\bar{\mu}_p$.

Seeking to establish a formal mathematical expression for the effective mobility, let us consider an $n$-channel device with the structure and dimensions specified in Fig. 17.6. Let $x$ be the depth into the semiconductor measured from the oxide–semiconductor interface, $y$ the distance along the channel measured from the source, $x_c(y)$ the channel depth, $n(x, y)$ the electron concentration at a point $(x, y)$ in the channel, and $\mu_n(x, y)$ the mobility of carriers at the $(x, y)$ point in the channel. Invoking the standard averaging procedure, the effective mobility of carriers an arbitrary distance $y$ from the source is given by

$$\bar{\mu}_n = \frac{\displaystyle\int_0^{x_c(y)} \mu_n(x, y)n(x, y)\, dx}{\displaystyle\int_0^{x_c(y)} n(x, y)\, dx} \tag{17.2}$$

For future reference it is useful to note that the electronic charge/cm$^2$ in the channel at an arbitrary point $y$ is

$$Q_N(y) = -q \int_0^{x_c(y)} n(x, y)\, dx \tag{17.3}$$

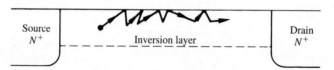

**Figure 17.5** Visualization of surface scattering at the Si–SiO$_2$ interface.

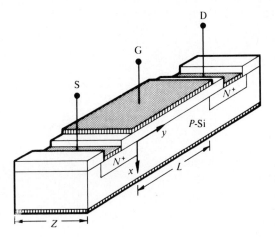

**Figure 17.6**   Device structure, dimensions, and coordinate orientations assumed in the quantitative analysis.

leading to the alternative expression

$$\bar{\mu}_n = -\frac{q}{Q_N(y)} \int_0^{x_c(y)} \mu_n(x, y) n(x, y) \, dx \qquad (17.4)$$

If the drain voltage is small, the channel depth and carrier charge will be more or less uniform from source to drain and the effective mobility will be essentially the same for all $y$-values. When the drain voltage becomes large, on the other hand, $x_c$ and $Q_N$ vary with position, and it is reasonable to expect that $\bar{\mu}_n$ likewise varies somewhat in going down the channel from the source to the drain. Fortunately, in long-channel devices the cited $y$-dependence can typically be neglected without introducing a significant error. *Thus, in this chapter we will subsequently consider $\bar{\mu}_n$ to be a device parameter that is independent of $y$ and the applied drain voltage $V_D$.*

Relative to the dependence of $\bar{\mu}_n$ on the applied *gate* voltage, increased inversion biasing increases the $x$-direction electric field acting on the carriers and confines the carriers closer to the oxide–semiconductor interface. Surface scattering is enhanced and $\bar{\mu}_n$ therefore decreases with increased inversion biasing—a dependence that cannot be ignored. The exact $\bar{\mu}_n$ versus $V_G$ dependence varies from device to device but generally follows the form displayed in Fig. 17.7. To first order the dependence can be modeled by the empirical relationship

$$\bar{\mu}_n = \frac{\mu_0}{1 + \theta(V_G - V_T)} \qquad (17.5)$$

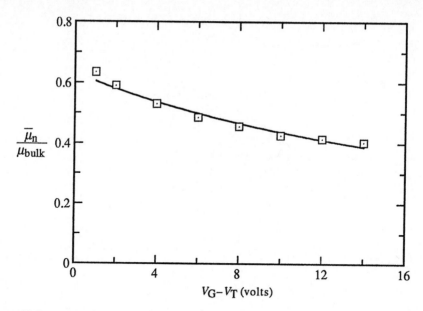

**Figure 17.7**  Sample variation of $\bar{\mu}_n$ with the applied gate voltage ($V_D \cong 0$). Data (▫) from Sun and Plummer[16]. Equation (17.5) was used to construct the solid-line curve. ($\mu_{bulk} = 1340$ cm²/V-sec, $\mu_0 = 847$ cm²/V-sec, and $\theta = 0.0446$/V.)

where $\mu_0$ and $\theta$ are constants. Equation (17.5) was used to construct the solid line curve in Fig. 17.7 after determining $\mu_0$ and $\theta$ from a least-squares fit to the experimental data. Finally, note from Fig. 17.7 that the surface scattering phenomenon can be rather significant, giving rise to effective mobilities considerably below the bulk $\mu_n$.

## 17.2.2  Square-Law Theory

The MOSFET under analysis is taken to be a long-channel device with the structure and dimensions as specified in Fig. 17.6. The figure also indicates the assumed coordinate orientations.

For gate voltages above turn-on ($V_G \geq V_T$), and drain voltages below pinch-off ($0 \leq V_D \leq V_{Dsat}$), the derivation of the square-law $I_D - V_D$ relationship proceeds as follows: In general one can write

$$\mathbf{J}_N = q\mu_n n \mathscr{E} + qD_N \nabla n \tag{17.6}$$

Within the conducting channel the current flow is almost exclusively in the $y$-direction. Moreover, the diffusion component of the current is often found to be negligible when dealing with the more numerous carrier at a given point inside a semiconductor. Thus, based on precedent established in similar problems, it is reasonable to neglect the diffusion component of the current $(qD_N\nabla n)$ in Eq. (17.6). Implementing the suggested simplifications yields

$$J_N \cong J_{Ny} \cong q\mu_n n\mathscr{E}_y = -q\mu_n n\frac{d\phi}{dy} \qquad \text{(in the conducting channel)} \qquad (17.7)$$

All of the quantities in Eq. (17.7)—$\mu_n$, $n$, and $J_{Ny}$—are, of course, $x$- and $y$-position dependent. $J_{Ny}$, like $n$, is expected to be quite large at $x = 0^+$ and to drop off rapidly as one moves into the semiconductor bulk.

Since current flow is restricted to the surface channel, the current passing through any cross-sectional plane within the channel must be equal to $I_D$. That is,[†]

$$I_D = -\iint J_{Ny}\, dx\, dz = -Z\int_0^{x_c(y)} J_{Ny}\, dx \qquad (17.8a)$$

$$= \left(-Z\frac{d\phi}{dy}\right)\left(-q\int_0^{x_c(y)} \mu_n(x,y)n(x,y)\, dx\right) \qquad (17.8b)$$

Upon examining Eq. (17.8b), note that the second bracket on the right-hand side of the equation is just $\bar{\mu}_n Q_N$ (see Eq. 17.4). Eq. (17.8b) thus simplifies to

$$I_D = -Z\bar{\mu}_n Q_N\frac{d\phi}{dy} \qquad (17.9)$$

Next, realizing that $I_D$ is independent of $y$, we can recast Eq. (17.9) into a more useful form by integrating $I_D$ over the length of the channel. Specifically,

$$\int_0^L I_D\, dy = I_D L = -Z\int_0^{V_D} \bar{\mu}_n Q_N\, d\phi \qquad (17.10)$$

or, with $\bar{\mu}_n$ being position independent,

$$I_D = -\frac{Z\bar{\mu}_n}{L}\int_0^{V_D} Q_N\, d\phi \qquad (17.11)$$

An analytical expression relating $Q_N$ to the channel potential $\phi$ at an arbitrary point $y$ is obviously required to complete the derivation. Working to establish the required

---

[†] (a) The minus sign appears in the general formula for $I_D$ because $I_D$ is defined to be positive in the $-y$-direction. (b) Generally speaking, $\phi$ and $d\phi/dy$ are functions of $x$. The very narrow extent of the inversion layer dictates, however, that $\phi$ is a weak function of $x$ ($\phi \approx \phi_s$) in the channel region. Therefore, in writing down the final form of Eq. (17.8), $d\phi/dy$ was taken to be constant over the $x$-width of the channel.

expression, we recall that the equilibrium inversion-layer charge in an MOS-C almost precisely balances the charge added to the MOS-C gate when $V_G$ exceeds $V_T$. In other words

$$\Delta Q_{\text{gate}}\left(\frac{\text{charge}}{\text{cm}^2}\right) = -\Delta Q_{\text{semi}}\left(\frac{\text{charge}}{\text{cm}^2}\right) \cong -Q_N \quad \ldots V_G \geq V_T \quad (17.12)$$

Because the charges are added immediately adjacent to the edges of the oxide, we can also assert

$$\Delta Q_{\text{gate}}\left(\frac{\text{charge}}{\text{cm}^2}\right) \cong C_o \Delta V_G = C_o(V_G - V_T) \quad \ldots V_G \geq V_T \quad (17.13)$$

and therefore

$$Q_N \cong -C_o(V_G - V_T) \quad \ldots V_G \geq V_T \quad (17.14)$$

where

$$C_o \equiv \frac{C_O}{A_G} = \frac{K_O \varepsilon_0}{x_o} \quad (17.15)$$

is the oxide capacitance per unit area of the gate.

Whereas the entire back side in an MOS-C is grounded, the bottom-side "plate" potential in a MOSFET varies from zero at the source to $V_D$ at the drain. As envisioned in Fig. 17.8, the MOSFET can be likened to a resistive-plate capacitor where the plate-to-plate potential difference is $V_G$ at the source, $V_G - V_D$ at the drain, and $V_G - \phi$ at an arbitrary point $y$. Clearly, the potential drop $V_G - \phi$ at an arbitrary point $y$ in the MOSFET functionally replaces the uniform $V_G$ potential drop in an MOS-C. Utilizing Eq. (17.14) we therefore conclude

$$Q_N(y) \cong -C_o(V_G - V_T - \phi) \quad (17.16)$$

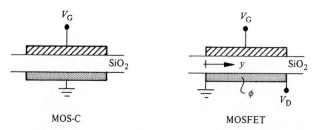

MOS-C                    MOSFET

**Figure 17.8**  Capacitor-like model for determining the charge in the MOSFET channel.

An explicit $I_D - V_D$ relationship can now be established by simply substituting the Eq. (17.16) expression for $Q_N$ into Eq. (17.11) and integrating. The result is

$$I_D = \frac{Z\bar{\mu}_n C_o}{L}\left[(V_G - V_T)V_D - \frac{V_D^2}{2}\right] \quad \left(\begin{matrix} 0 \le V_D \le V_{Dsat} \\ V_G \ge V_T \end{matrix}\right) \tag{17.17}$$

It should be reemphasized that the foregoing development and Eq. (17.17), in particular, apply only below pinch-off. In fact, the computed $I_D$ versus $V_D$ for a given $V_G$ actually begins to decrease if $V_D$ values in excess of $V_{Dsat}$ are inadvertently substituted into Eq. (17.17). As pointed out in the qualitative discussion, however, $I_D$ is approximately constant if $V_D$ exceeds $V_{Dsat}$. To first order, then, the post pinch-off portion of the characteristics can be modeled by simply setting

$$I_{D|V_D>V_{Dsat}} = I_{D|V_D=V_{Dsat}} \equiv I_{Dsat} \tag{17.18}$$

or

$$I_{Dsat} = \frac{Z\bar{\mu}_n C_o}{L}\left[(V_G - V_T)V_{Dsat} - \frac{V_{Dsat}^2}{2}\right] \tag{17.19}$$

The $I_{Dsat}$ relationship can be simplified somewhat by noting that pinch-off at the drain end of the channel implies $Q_N(L) \to 0$ when $\phi(L) = V_D \to V_{Dsat}$. Thus from Eq. (17.16)

$$Q_N(L) = -C_o(V_G - V_T - V_{Dsat}) = 0 \tag{17.20}$$

or

$$V_{Dsat} = V_G - V_T \tag{17.21}$$

and

$$I_{Dsat} = \frac{Z\bar{\mu}_n C_o}{2L}(V_G - V_T)^2 \tag{17.22}$$

Neglecting $\bar{\mu}_n$'s dependence on $V_G$, Eq. (17.22) predicts a saturation drain current that varies as the square of the gate voltage above turn-on, the so-called "square-law" dependence.

## (C) Exercise 17.2

**P:** If the square-law $I_D$ divided by $Z\bar{\mu}_n C_o/L$ is plotted versus $V_D$ for select values of $V_G - V_T$, the resulting normalized characteristics are device independent—the same characteristics are obtained for any combination of $Z$, $L$, $\bar{\mu}_n$, $x_o$, and $N_A$. Construct such a "universal" plot specifically showing the characteristics corresponding to $V_G - V_T = 1$, 2, 3, and 4 V.

**S:** The requested plot is presented in Fig. E17.2. The MATLAB program listed below was used to generate the plot.

MATLAB program script . . .

```
% "Universal" ID-VD Characteristics /// Square-Law Theory

%Initialization
close
clear

%Let VGT = VG - VT;
for VGT=4:-1:1,

    %Primary Computation
    VD=linspace(0,VGT);
    ID=VGT.*VD-VD.*VD./2;
    IDsat=VGT*VGT/2;
    VD=[VD,9];
    ID=[ID,IDsat];

    %Plotting and Labeling
    if VGT==4,
    plot(VD,ID);  grid;
    axis([0 10 0 10]);
    xlabel('VD (volts)');  ylabel('ID/(ZµCo/L)');
    text(8,IDsat+0.2,'VG-VT=4V');
    hold on
    else,
    plot(VD,ID);
    %The following 'if' labels VG-VT curves < 4
    if VGT==3,
    text(8,IDsat+0.2,'VG-VT=3V');
    elseif VGT==2,
    text(8,IDsat+0.2,'VG-VT=2V');
    else,
    text(8,IDsat+0.2,'VG-VT=1V');
```

end
    end
   end
end
hold off

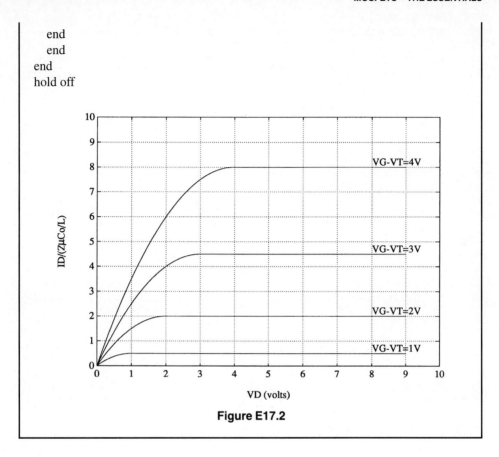

VD (volts)

**Figure E17.2**

### 17.2.3 Bulk-Charge Theory

Although appearing very reasonable and sound on first inspection, close scrutiny reveals that the square-law theory contains a major flaw. The capacitor-like model used in the square-law analysis assumed changes in gate charge going down the MOSFET channel were balanced solely by changes in $Q_N$. This is equivalent to implicitly assuming the depletion width at all channel points from the source to the drain remains fixed at $W_T$ even under $V_D \neq 0$ biasing. In reality, as pictured in Figs. 17.2(b) to (d), the depletion width widens in progressing from the source to the drain when $V_D \neq 0$. This point-to-point variation in the depletion layer or "bulk" charge must be included in any charge balance relationship.

With changes in the depletion width, $W(y)$, taken into account, one more accurately deduces

$$Q_N(y) = -C_o(V_G - V_T - \phi) + qN_A[W(y) - W_T] \qquad (17.23)$$

where, making use of the delta-depletion results in Chapter 16,

$$W(y) = \left[ \frac{2K_S \varepsilon_0}{qN_A} (2\phi_F + \phi) \right]^{1/2} \tag{17.24}$$

$$W_T = \left[ \frac{2K_S \varepsilon_0}{qN_A} (2\phi_F) \right]^{1/2} \tag{17.25}$$

Thus, combining Eqs. (17.23) to (17.25) and introducing

$$V_W \equiv \frac{qN_A W_T}{C_o} \tag{17.26}$$

one obtains the bulk-charge theory analogue of Eq. (17.16), namely,

$$Q_N(y) = -C_o \left[ V_G - V_T - \phi - V_W \left( \sqrt{1 + \frac{\phi}{2\phi_F}} - 1 \right) \right] \tag{17.27}$$

The predicted $I_D - V_D$ relationship based on the bulk-charge formulation is next readily obtained by substituting Eq. (17.27) into Eq. (17.11) and integrating. The end result is

$$I_D = \frac{Z\bar{\mu}_n C_o}{L} \left\{ (V_G - V_T)V_D - \frac{V_D^2}{2} - \frac{4}{3} V_W \phi_F \left[ \left( 1 + \frac{V_D}{2\phi_F} \right)^{3/2} - \left( 1 + \frac{3V_D}{4\phi_F} \right) \right] \right\}$$

$$\text{for } 0 \le V_D \le V_{Dsat}$$
$$\text{and } V_G \ge V_T \tag{17.28}$$

As in the square-law analysis, the post pinch-off portion of the characteristics are approximately modeled by setting $I_D$ evaluated at $V_D > V_{Dsat}$ equal to $I_D$ at $V_D = V_{Dsat}$. Likewise, an expression for $V_{Dsat}$ can be obtained by noting $Q_N(y)|_{y=L} \to 0$ in Eq. (17.27) when $\phi(L) = V_D \to V_{Dsat}$. One finds

$$V_{Dsat} = V_G - V_T - V_W \left\{ \left[ \frac{V_G - V_T}{2\phi_F} + \left( 1 + \frac{V_W}{4\phi_F} \right)^2 \right]^{1/2} - \left( 1 + \frac{V_W}{4\phi_F} \right) \right\} \tag{17.29}$$

Having concluded the mathematical development, let us examine the results and make appropriate comments. First of all, it should be recognized that the primary asset of the square-law theory is its simplicity. General trends, basic interrelationships, and the like can

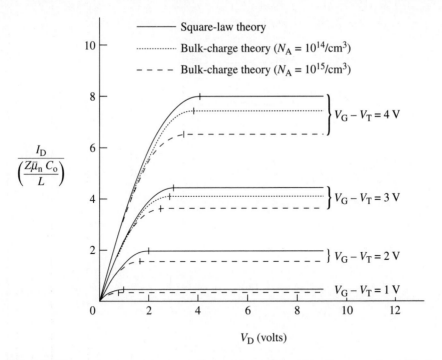

**Figure 17.9** Comparison of the $I_D$–$V_D$ characteristics derived from the square-law and bulk-charge theories. The bulk-charge curves were computed assuming $x_o = 0.1\ \mu m$ and $T = 300$ K.

be established using the square-law formulation without an excessive amount of mathematical entanglement. The bulk-charge theory, on the other hand, is in good agreement with the experimental characteristics derived from long-channel MOSFETs. It should also be noted that, although Eqs. (17.28) and (17.29) are decidedly more complex than their square-law analogues, the added terms, the terms not appearing in Eqs. (17.17) and (17.21), respectively, are always negative and act primarily to reduce $I_D$ and $V_{Dsat}$ for a given set of operational conditions. Figure 17.9, which compares the two theories, confirms the foregoing observation and also illustrates another well-known property—the accuracy of the square-law theory improves as the substrate doping is decreased. In fact, the bulk-charge theory mathematically reduces to the square-law theory as $N_A$ (or $N_D$) $\rightarrow 0$ and $x_o \rightarrow 0$.

---

**Exercise 17.3**

**P:** Suppose the gate and drain of an ideal $n$-channel MOSFET are tied together as pictured on the following page. $x_o = 500$ Å, $N_A = 10^{16}/cm^3$, $Z/L = 10$, $\bar{\mu}_n = 625\ cm^2/V\text{-sec}$, and $T = 300$ K. Employing the bulk-charge theory, determine $I_D$ when

(a) $V_G = V_D = 1$ V;
(b) $V_G = V_D = 3$ V.

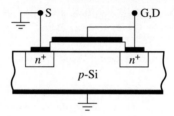

**S:** (a) This is somewhat of a trick question. We note

$$\phi_F = \frac{kT}{q} \ln(N_A/n_i) = 0.0259 \ln(10^{16}/10^{10}) = 0.358 \text{ V}$$

and

$$V_T = 2\phi_F + \frac{K_S}{K_O} x_o \sqrt{\frac{4qN_A}{K_S \varepsilon_0} \phi_F}$$

$$= 0.716 + \frac{(11.8)(5 \times 10^{-6})}{(3.9)} \left[ \frac{(4)(1.6 \times 10^{-19})(10^{16})}{(11.8)(8.85 \times 10^{-14})} (0.358) \right]^{1/2} = 1.42 \text{ V}$$

With $V_G = 1$ V, $V_G < V_T$ and therefore the transistor is turned off; $\boxed{I_D = 0}$.

(b) Computing $I_D$ by simply substituting $V_G = V_D = 3$ V into Eq. (17.28) gives an incorrect result. As noted in the text and as illustrated in Fig. 17.9, $V_{Dsat}$ in the bulk-charge theory occurs at a lower voltage than the square-law $V_{Dsat} = V_G - V_T$. Since $V_D = V_G$, it follows that $V_D = V_G > V_G - V_T > V_{Dsat}$. *With the gate and drain tied together, the MOSFET is saturated biased for all $V_D = V_G > V_T$.*
    Respectively employing Eq. (17.15), (17.25), and (17.26), we find $C_o = 6.90 \times 10^{-8}$ F/cm², $W_T = 3.06 \times 10^{-5}$ cm, and $V_W = 0.71$ V. Substituting into Eq. (17.29) then yields $V_{Dsat} = 1.15$ V. Finally, computing $I_D$ using Eq. (17.28) with $V_D = V_{Dsat} = 1.15$ V gives $\boxed{I_D = 0.382 \text{ mA}}$.

## 17.2.4 Charge-Sheet and Exact-Charge Theories

Both the square-law and bulk-charge theories suffer from two severe inherent limitations. For one, the charge in the MOSFET channel ($Q_N$ in the preceding analyses) was assumed to be identically zero for gate voltages at and below the threshold voltage. In an actual device, the channel charge becomes small, but does not completely vanish. As a consequence, a residual drain current flows between the source and drain at and below the threshold voltage. This residual drain current is called the *subthreshold current* and its precise value is often of interest. Second, the square-law and bulk-charge relationships do not self-saturate—it is necessary to artificially construct the above pinch-off portion of the characteristics.

The noted failings are removed in the *charge-sheet* and *exact-charge* formulations; either theory can be used to compute the subthreshold current, and both theories are self-saturating. The $I_D$–$V_D$ computational relationships resulting from the charge-sheet and exact-charge models are reproduced in Appendix D. The exact-charge results are established by working with the exact-charge distribution inside the MOSFET. Although not overly complex, the exact-charge result does involve integrals. The charge-sheet model may be viewed as a simplified version of the exact formulation. Of the theories discussed, the charge-sheet formulation provides the best trade-off between accuracy and complexity. Sample $I_D$–$V_D$ characteristics resulting from the cited more-exacting theories are shown in Fig. 17.10. Computed and observed subthreshold characteristics of $I_D$ versus $V_G$ with $V_D$ held constant are presented in Fig. 17.11.

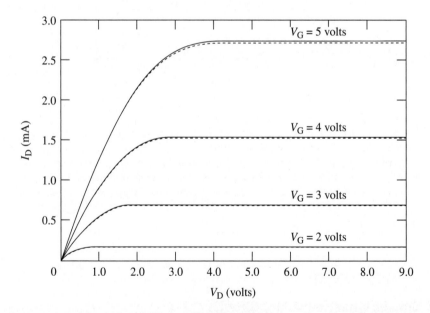

**Figure 17.10**   Theoretical current-voltage characteristics of an *n*-channel MOSFET with $x_o = 0.05\ \mu$m, $N_A = 10^{15}$/cm$^3$, $\bar{\mu}_n = 550$ cm$^2$/V-sec, $L = 7\ \mu$m, $Z = 70\ \mu$m and $T = 23°$C. The solid-line curves were derived from the exact-charge result while the dashed-line curves were computed using the charge-sheet theory. (Reprinted from Pierret and Shields[17], © 1983, with kind permission from Elsevier Science Ltd.)

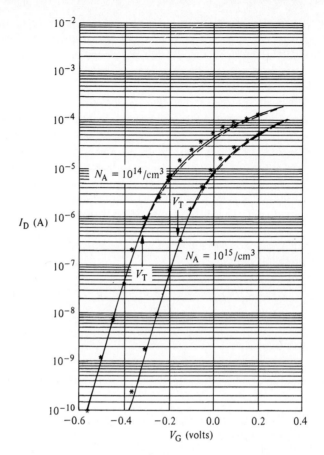

**Figure 17.11**    Subthreshold transfer characteristics of *n*-channel MOSFETs having the same parameters as the Fig. 17.10 device except $N_A = 10^{14}$/cm³ or $N_A = 10^{15}$/cm³ and $x_o = 0.013$ $\mu$m. The (*) are experimental data. Setting $V_D = 1$ V, the solid- and dashed-line curves were computed respectively from the exact-charge and charge-sheet relationships found in Appendix D. The ideal-device curves were shifted along the voltage axis to enhance the comparison with the experimental data. (Reprinted from Pierret and Shields[17], © 1983, with kind permission from Elsevier Science Ltd.)

## 17.3  a.c. RESPONSE

### 17.3.1 Small-Signal Equivalent Circuits

The a.c. response of the MOSFET, routinely expressed in terms of a small-signal equivalent circuit, is most conveniently established by considering the two-port network shown in Fig. 17.12(a). Initially we restrict our considerations to low operational frequencies where capacitive effects may be neglected. It should be noted that the following development and results are very similar to those of the J-FET presentation in Subsection 15.2.4.

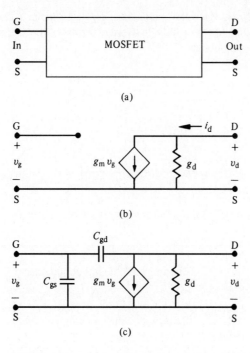

**Figure 17.12** (a) The MOSFET viewed as a two-port network. (b) Low-frequency and (c) high-frequency small-signal equivalent circuits characterizing the a.c. response of the MOSFET.

We begin by examining the device input. Looking into the input port between the gate and grounded source/substrate, one sees a capacitor. A capacitor, however, behaves (to first order) like an open circuit at low frequencies. It is standard practice, therefore, to model the low-frequency input to the MOSFET by an open circuit.

At the output port the d.c. drain current has been established to be a function of $V_D$ and $V_G$; that is, $I_D = I_D(V_D, V_G)$. When a.c. drain and gate potentials, $v_d$ and $v_g$, are respectively added to the d.c. drain and gate terminal voltages, $V_D$ and $V_G$, the drain current through the structure is modified to $I_D(V_D, V_G) + i_d$, where $i_d$ is the a.c. component of the drain current. Provided the device can follow the a.c. changes in potential, which is certainly the case at low operational frequencies, one can state

$$i_d + I_D(V_D, V_G) = I_D(V_D + v_d, V_G + v_g) \qquad (17.30a)$$

and

$$i_d = I_D(V_D + v_d, V_G + v_g) - I_D(V_D, V_G) \qquad (17.30b)$$

Expanding the first term on the right-hand side of Eq. (17.30b) in a Taylor series about the d.c. operating point, and keeping only first-order terms in the expansion (higher-order terms are negligible), one obtains

$$I_D(V_D+v_d, V_G+v_g) = I_D(V_D, V_G) + \left. \frac{\partial I_D}{\partial V_D}\right|_{V_G} v_d + \left. \frac{\partial I_D}{\partial V_G}\right|_{V_D} v_g \qquad (17.31)$$

which when substituted into Eq. (17.30b) gives

$$i_d = \left. \frac{\partial I_D}{\partial V_D}\right|_{V_G} v_d + \left. \frac{\partial I_D}{\partial V_G}\right|_{V_D} v_g \qquad (17.32)$$

Dimensionally the partial derivatives in Eq. (17.32) are conductances. Introducing

$$g_d \equiv \left. \frac{\partial I_D}{\partial V_D}\right|_{V_G=\text{constant}} \qquad \ldots \text{ the drain or channel conductance} \qquad (17.33a)$$

$$g_m \equiv \left. \frac{\partial I_D}{\partial V_G}\right|_{V_D=\text{constant}} \qquad \ldots \text{ transconductance or mutual conductance} \qquad (17.33b)$$

we can then write

$$i_d = g_d v_d + g_m v_g \qquad (17.34)$$

Equation (17.34) may be viewed as the a.c.-current node equation for the drain terminal and, by inspection, leads to the output portion of the circuit displayed in Fig. 17.12(b). Since, as concluded earlier, the gate-to-source or input portion of the device is simply an open circuit, Fig. 17.12(b) then is the desired small-signal equivalent circuit characterizing the low-frequency a.c. response of the MOSFET.

For field-effect transistors the $g_m$ parameter plays a role analogous to the $\alpha$ and $\beta$ in the modeling of bipolar junction transistors. As its name indicates, $g_d$ may be viewed as either the device output admittance or the a.c. conductance of the channel between the source and drain. Explicit $g_d$ and $g_m$ relationships obtained by direct differentiation of Eqs. (17.17), (17.22), and (17.28) using the Eq. (17.33) definitions are catalogued in Table 17.1.

At the higher operational frequencies often encountered in practical applications, the Fig. 17.12(b) circuit must be modified to take into account capacitive coupling between the device terminals. The required modification is shown in Fig. 17.12(c). A capacitor between the drain and source terminals at the output has been omitted in Fig. 17.12(c) because the drain-to-source capacitance is typically negligible. $C_{gd}$, which provides undesirable feedback between the input and output, is associated in large part with the so-called overlap capacitance—the capacitance resulting from the portion of the gate that overlaps the drain

**Table 17.1**    MOSFET Small-Signal Parameters.[†]

|  | *Below pinch-off* ($V_D \leq V_{Dsat}$) | *Above pinch-off* ($V_D > V_{Dsat}$) |
|---|---|---|
| Square law | $g_d = \dfrac{Z\bar{\mu}_n C_o}{L}(V_G - V_T - V_D)$ | $g_d = 0$ |
| Bulk charge | $g_d = \dfrac{Z\bar{\mu}_n C_o}{L}[V_G - V_T - V_D - V_W(\sqrt{1 + V_D/2\phi_F} - 1)]$ | $g_d = 0$ |
| Square law | $g_m = \dfrac{Z\bar{\mu}_n C_o}{L}V_D$ | $g_m = \dfrac{Z\bar{\mu}_n C_o}{L}(V_G - V_T)$ |
| Bulk charge | $g_m = \dfrac{Z\bar{\mu}_n C_o}{L}V_D$ | $g_m = \dfrac{Z\bar{\mu}_n C_o}{L}V_{Dsat}$ with $V_{Dsat}$ per Eq. (17.29) |

[†]Entries in the table were obtained by direct differentiation of Eqs. (17.17), (17.22), and (17.28). The variation of $\bar{\mu}_n$ with $V_G$ was neglected in establishing the $g_m$ expressions.

island. The overlap capacitance is minimized by forming a thicker oxide in the overlap region or preferably through the use of self-aligned gate procedures. In the self-aligned gate fabrication process a MOSFET gate material that can withstand high-temperature processing, usually polysilicon, is deposited first. After the gate is defined, the source and drain islands are subsequently formed abutting the gate by diffusion or ion implantation. The remaining capacitor shown in Fig. 17.12(c), $C_{gs}$, is associated primarily with the capacitance of the MOS gate.

## 17.3.2 Cutoff Frequency

Given the small-signal equivalent circuit of Fig. 17.12(c), it is possible to estimate the maximum operating frequency or cutoff frequency of an MOS transistor. Let $f_{max}$ be defined as the frequency where the MOSFET is no longer amplifying the input signal under optimum conditions—that is, the frequency where the absolute value of the output current to input current ratio is unity when the output of the transistor is short-circuited. By inspection, the input current with the output short-circuited is

$$i_{in} = j\omega(C_{gs} + C_{gd})v_g \simeq j(2\pi f)C_O v_g \quad (j = \sqrt{-1}) \tag{17.35}$$

where $C_{gd}$ is taken to be small and $C_{gs} \simeq C_O$. Likewise, the output current is

$$i_{out} \simeq g_m v_g \tag{17.36}$$

Thus, setting $|i_{out}/i_{in}| = 1$ and solving for $f = f_{max}$, one obtains

$$f_{max} = \frac{g_m}{2\pi C_O} = \frac{\bar{\mu}_n V_D}{2\pi L^2} \qquad \text{if } V_D \leq V_{Dsat} \qquad (17.37)$$

The latter form of Eq. (17.37) was established using the below pinch-off $g_m$ entry in Table 17.1. The important point to note is that the channel length $L$ is the key parameter in determining $f_{max}$; increased MOSFET operating frequencies are achieved by decreasing the channel length.

### 17.3.3 Small-Signal Characteristics

Representative plots of selected small-signal characteristics that have received special attention in the device literature are shown in Fig. 17.13. $g_d$ versus $V_G$ with $V_D = 0$ has been used to obtain a reasonably accurate estimate of $V_T$. This is accomplished by extrapolating the linear portion of the $g_d$–$V_G$ characteristics into the $V_G$ axis and equating the voltage intercept to $V_T$. The basis for this procedure can be understood by referring to the below pinch-off $g_d$ entries in Table 17.1. With $V_D = 0$ the drain conductance in both the square-law and bulk-charge theories reduces to

$$g_d = \frac{Z\bar{\mu}_n C_o}{L}(V_G - V_T) \qquad (V_D = 0) \qquad (17.38)$$

To first order, then, $g_d$ is predicted to be a linear function of $V_G$, going to zero when $V_G = V_T$. The experimental characteristic does not completely vanish at $V_G = V_T$ because there is a small minority-carrier concentration in the surface channel at the depletion–inversion transition point. This residual concentration is neglected in both the square-law and bulk-charge theories. The $g_d$ versus $V_G$ characteristic with $V_D = 0$ has also been used to deduce the effective mobility. Since $g_d$ is directly proportional to $\bar{\mu}_n$ according to Eq. (17.38), $\bar{\mu}_n$ versus $V_G$ can be computed readily from the $g_d$–$V_G$ data. This mobility measurement method is accurate provided the device contains a low density of interfacial traps (see Subsection 18.2.4). A moderate-to-large density of interfacial traps would spread out the $g_d$–$V_G$ characteristic and yield a fallaciously low value for $\bar{\mu}_n$.

The second characteristic in Fig. 17.13 typifies the gate capacitance versus $V_G$ dependence derived from the MOSFET when the drain is grounded. The MOSFET $C_G$–$V_G$ ($V_D = 0$) characteristic has been used for diagnostic purposes in much the same manner as the MOS-C $C$–$V_G$ characteristic. The MOSFET characteristic can, in fact, be modeled to first order by the low-frequency MOS-C $C$–$V_G$ theory. Unlike the MOS-C, however, a low-frequency type characteristic is observed even when the MOSFET is probed at frequencies exceeding 1 MHz. A low-frequency characteristic is obtained because the source and drain islands supply the minority carriers required for the structure to follow the a.c. fluctuations

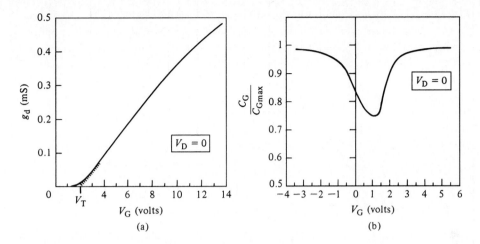

**Figure 17.13**  MOSFET small-signal characteristics. (a) $g_d$ versus $V_G$ with $V_D = 0$; (b) $C_G$ versus $V_G$ with $V_D = 0$.

in the gate potential when the device is inversion biased. Minority carriers merely use the surface channel to flow laterally into and out of the MOS gate area in response to the applied a.c. signal.

---

**Exercise 17.4**

**P:** (a) The transconductance ($g_m$) plotted as a function of $V_G$ with $V_D$ held constant is another small-signal characteristic encountered quite often in the MOSFET literature. Deduce the expected general form of the $g_m$–$V_G$ characteristic. Specifically, making use of the square-law entries in Table 17.1 and assuming $V_T = 2$ V, plot $g_m/(Z\bar{\mu}_n C_o/L)$ versus $V_G$ ($0 \le V_G \le 10$ V) for $V_D = 2$, 4, and 6 V. Neglect the variation of $\bar{\mu}_n$ with $V_G$.

(b) Compare similar experimental $g_m$–$V_G$ characteristics with the part (a) theoretical characteristics. Speculate as to the origin of any differences in the shape of the characteristics.

**S:** (a) Let $\xi = Z\bar{\mu}_n C_o/L$. As deduced from the square-law entries in Table 17.1,

$$\frac{g_m}{\xi} = \begin{cases} V_G - V_T & \ldots \, 0 \le V_G - V_T \le V_D \\ V_D & \ldots \, V_G - V_T \ge V_D \end{cases}$$

For all $V_D$, $g_m/\xi = 0$ if $V_G < V_T = 2$ V. For $V_G > V_T$ but $V_G - V_T < V_D$, the device is saturation biased and $g_m/\xi$ increases linearly with $V_G$. Once $V_G$ reaches the voltage where $V_G - V_T = V_D$, the device drops out of saturation and $g_m/\xi = V_D =$ constant. The $g_m$–$V_G$ characteristics are therefore concluded to be of the form shown below

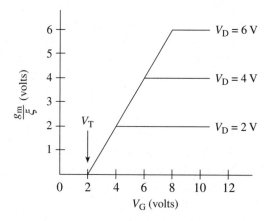

(b) Experimental $g_m$–$V_G$ characteristics derived from an $n$-channel MOSFET are presented below. The data are from Part III reference [18]. The most noticeable difference between theory and experiment is the decrease in the measured $g_m$ at $V_G > V_D + V_T$. This falloff in $g_m$ is readily explained—it is caused by the decrease in $\bar{\mu}_n$ with $V_G$ that was neglected in drawing the theoretical characteristics.

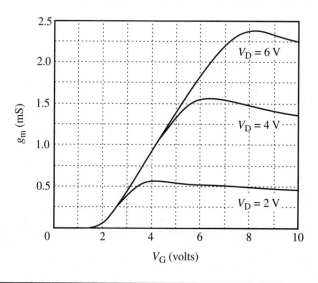

## 17.4 SUMMARY

This chapter was intended to provide an introduction to MOSFET terminology, operation, and analysis. The MOS structure was assumed to be ideal and considerations were limited to the basic transistor configuration. We began with a qualitative discussion of MOSFET operation and d.c. current flow inside the structure. When biased into inversion, the induced surface inversion layer forms a conducting channel between the source and drain contacts. The greater the applied gate voltage in excess of turn-on, the larger the conductance of the internal channel at a given drain voltage. A nonzero drain voltage in turn initiates current flow between the source and drain. The current flow is proportional to $V_D$ at low drain voltages, slopes over due to channel narrowing as $V_D$ is increased, and eventually saturates once the internal channel vanishes or pinches off near the drain.

The quantitative analysis of the MOSFET d.c. characteristics, considered next, was subject to two notable complications. First of all, carriers in a surface channel experience motion-impeding collisions with the Si surface, which lower the mobility of the carriers and necessitate the introduction of an effective carrier mobility. Second, the carrier concentration and therefore the current density in the surface channel are strong functions of position, dropping off rapidly as one proceeds into the semiconductor bulk. Nonetheless, the first-order results for the MOSFET current–voltage relationship are surprisingly simple. The results of the first-order theory, referred to herein as the square-law theory, are contained in Eqs. (17.17), (17.21), and (17.22). The bulk-charge theory, a second formulation culminating in Eqs. (17.28) and (17.29), provides a more accurate representation of reality at the expense of complexity. Even more-exacting formulations were discussed briefly, with the reader referred to Appendix D for computational details.

The last section of the chapter was devoted to the a.c. response of the MOSFET. The equivalent circuits of Figs. 17.12(b) and (c), respectively, specify the small-signal response at low and high frequencies. With the aid of the Fig. 17.12(c) circuit it was established that a short channel length is necessary to achieve high-frequency, high-speed operation. It was also pointed out that useful information can often be extracted from the small-signal parameters ($g_d$, $g_m$, $C_G$) monitored as a function of the d.c. terminal voltages.

Although designed to be self-contained, the MOSFET development in the present chapter does closely parallel the J-FET presentation in Chapter 15. The reader may find it a useful exercise to note similarities and differences in the operation and analysis of the two devices. In Chapter 18 we examine the impact of nonidealities on MOSFET operation. As noted previously, an examination of small-dimension effects and structural variations is undertaken in Chapter 19.

## PROBLEMS

| CHAPTER 17 | PROBLEM INFORMATION TABLE | | | |
|---|---|---|---|---|
| Problem | Complete After | Difficulty Level | Suggested Point Weighting | Short Description |
| 17.1 | 17.4 | 1 | 10 (1 each part) | Quick quiz |
| 17.2 | " | 2 | 10 (a, c, f-2; b, d, e, g-1) | Sample computations |
| 17.3 | 17.1 | 2 | 10 (a-3, b-2, c-5) | Sketch diagrams |
| ● 17.4 | 17.2.1 | 2 | 10 | Plot $V_T$ vs. $N_A$ for select $x_o$ |
| ● 17.5 | " | 3 | 10 | $V_T$ temperature dependence |
| 17.6 | 17.2.2 | 2 | 5 | $V_{Dsat}$ from $I_D$ maximum |
| 17.7 | " | 2–3 | 10 | $n \rightarrow p$ in Subsection 17.2.2 |
| ● 17.8 | " | 1 | 5 | Plot $I_D$–$V_D$, $\bar{\mu}_n$ per Eq. (17.5) |
| 17.9 | " | 3 | 10 (5 each part) | $V_D = V_G - V_B$ |
| 17.10 | " | 3–4 | 10 | $I_D$ versus $V_G - V_T$ |
| 17.11 | " | 3–4 | 10 (a-8, b-2) | Circular MOSFET |
| ● 17.12 | " | 3 | 12 (a-2, b-10 discuss-2) | $I_D$ temperature dependence |
| 17.13 | " | 3–4 | 12 | Include $R_S$, $R_D$ in theory |
| 17.14 | 17.2.3 | 3–4 | 10 | Derive Eq. (17.29) |
| ● 17.15 | " | 3 | 10 | Plot bulk-charge $I_D$–$V_D$ |
| ● 17.16 | 17.2.4 | 4 | 25 (a-10, b-15) | Plot sheet, exact $I_D$–$V_D$ |
| 17.17 | 17.3.1 | 2 | 5 | Verify Table 17.1 entries |
| 17.18 | " | 2 | 5 (a-3, b-2) | MOSFET $Q_N$, $g_d$ |
| 17.19 | 17.3.2 | 3 | 10 | Deduce $Z$, $L$ for match |
| 17.20 | " | 2–3 | 15 (a-3, b::g-2) | General MOSFET review |
| 17.21 | 17.3.3 | 3 | 12 (4 each part) | Small signal characteristics |
| 17.22 | 17.4 | 2 | 12 | Compare FETs |

**17.1** Quick Quiz

Answer the following questions as concisely as possible.

(a) Why are the current-carrying contacts in the MOSFET referred to as the *source* and *drain*?

(b) Precisely what is the *channel* in MOSFET terminology?

(c) When referring to the MOSFET $I_D$–$V_D$ characteristics, what exactly is the *saturation* region of operation?

(d) What is the relationship between the depletion-inversion transition point voltage introduced in the MOS-C discussion and the threshold (turn-on) voltage introduced in the MOSFET discussion?

(e) Why is the mobility in the surface channel of a MOSFET different from the carrier mobility in the semiconductor bulk?

(f) Why is the $I_D$–$V_D$ theoretical formulation of Subsection 17.2.2 referred to as the *square-law* theory?

(g) Why is the $I_D$–$V_D$ theoretical formulation of Subsection 17.2.3 referred to as the *bulk-charge* theory?

(h) What variables are plotted in displaying the "subthreshold transfer characteristics"?

(i) What is the mathematical definition of the drain conductance? the transconductance?

(j) Why is the observed MOSFET $C_G$–$V_G(V_D = 0)$ curve typically a low-frequency characteristic even at a measurement frequency of 1 MHz?

**17.2** Sample Calculations

Simple numerical computations help to establish the expected size of device variables. An ideal *n*-channel MOSFET maintained at $T = 300$ K is characterized by the following parameters: $Z = 50\ \mu m$, $L = 5\ \mu m$, $x_o = 0.05\ \mu m$, $N_A = 10^{15}/cm^3$, and $\bar{\mu}_n = 800\ cm^2/$V-sec (assumed independent of $V_G$). Determine:

(a) $V_T$;

(b) $I_{Dsat}$ (square-law theory) if $V_G = 2$ V;

(c) $I_{Dsat}$ (bulk-charge theory) if $V_G = 2$ V;

(d) $g_d$ if $V_G = 2$ V and $V_D = 0$;

(e) $g_m$ (square-law theory) if $V_G = 2$ V and $V_D = 2$ V;

(f) $g_m$ (bulk-charge theory) if $V_G = 2$ V and $V_D = 2$ V;

(g) $f_{max}$ if $V_G = 2$ V and $V_D = 1$ V.

**17.3** Given an ideal *p*-channel MOSFET maintained at room temperature:

(a) Assuming $V_D = 0$, sketch the MOS energy band diagram for the gate region of the given transistor at threshold.

(b) Assuming $V_D = 0$, sketch the MOS block charge diagram for the gate region of the given transistor at threshold.

(c) Sketch the inversion layer and depletion region inside the MOSFET at pinch-off. Show and label all parts of the transistor.

● **17.4** Construct a plot of $V_T$ versus $N_A$ for ideal *n*-channel MOSFETs operated at room temperature. Superimpose on the same plot the curves corresponding to $x_o = 0.01, 0.02$,

0.05, and 0.1 $\mu$m. Let the semiconductor doping vary over the range $10^{14}$/cm$^3 \leq N_A \leq 10^{18}$/cm$^3$; limit the plotted $V_T$ to $0 \leq V_T \leq 3$ V.

● **17.5** Explore the temperature dependence of the MOSFET threshold voltage. Considering an ideal $n$-channel MOSFET, taking $x_o$ and $N_A$ to be input parameters, and perhaps referring to Exercise 2.4, construct a computer program that calculates and plots $V_T$ versus $T$ over the range $200$ K $\leq T \leq 400$ K. Record the output of the program when $x_o = 0.1$ $\mu$m and $N_A = 10^{16}$/cm$^3$. Describe the general nature of the $V_T$ versus $T$ dependence.

**17.6** If Eq. (17.17) is used to compute $I_D$ as a function of $V_D$ for a given $V_G$, and if $V_D$ is allowed to increase above $V_{Dsat}$, one finds $I_D$ to be a peaked function of $V_D$ maximizing at $V_{Dsat}$. The foregoing suggests a second way to establish the Eq. (17.21) relationship for $V_{Dsat}$. Specifically, show that the standard mathematical procedure for determining extrema points of a function can be used to derive Eq. (17.21) directly from Eq. (17.17).

**17.7** Suppose Subsection 17.2.2 is to be rewritten with the illustrative MOSFET changed from an $n$-channel to a $p$-channel device. Indicate how the equations in the subsection must be modified if the MOSFET used for illustrative purposes is a $p$-channel device.

● **17.8** The $\bar{\mu}_n$ in relationships throughout the chapter can be replaced by the Eq. (17.5) expression to approximately account for the dependence of $\bar{\mu}_n$ on the applied gate voltage. Modify the computational program in Exercise 17.2 to illustrate the effect of incorporating the $\bar{\mu}_n$–$V_G$ dependence. Introducing the new normalizing factor, $Z\mu_0 C_o/L$, superimpose plots of the $I_D/(Z\mu_0 C_o/L)$ versus $V_D$ characteristics corresponding to $\theta = 0$ and $\theta = 0.05/$V. Note that $\theta = 0$ should yield characteristics identical to Fig. E17.2, while $\theta = 0.05/$V is roughly the value used to fit the $\bar{\mu}_n$–$V_G$ data of Fig. 17.7.

**17.9** Suppose a battery $V_B \geq 0$ is connected between the gate and drain of an ideal $n$-channel MOSFET as pictured in Fig. P17.9. Using the square-law results,

(a) Sketch $I_D$ versus $V_D$ ($V_D \geq 0$) if $V_B = V_T/2$;

(b) Sketch $I_D$ versus $V_D$ ($V_D \geq 0$) if $V_B = 2V_T$.

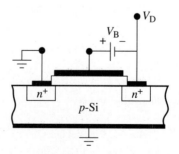

**Figure P17.9**

**17.10** The most widely encountered MOSFET characteristics are a plot of $I_D$ versus $V_D$ with $V_G$ or $V_G-V_T$ held constant at select values. An alternative plot of $I_D$ versus $V_G$ or $V_G-V_T$ with $V_D$ held constant at select values is sometimes useful. Sketch the shape of the $I_D$ versus $V_G-V_T$ characteristics to be expected from an ideal $n$-channel MOSFET. Specifically show the characteristics corresponding to $V_D = 1, 2, 3,$ and 4 V. Explain how you arrived at your sketch.

**17.11** A linear device geometry and a rectangular gate of length $L$ by width $Z$ were explicitly assumed in the text derivation of $I_D-V_D$ relationships. However, MOSFETs have been built with circular geometry as pictured (top view) in Fig. P17.11.

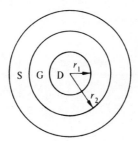

**Figure P17.11**

(a) If $r_1$ and $r_2$ are the inside and outside diameters of the gated area, show that in the square-law formulation one obtains

$$I_D = \frac{2\pi}{\ln(r_2/r_1)} \bar{\mu}_n C_o \left[ (V_G - V_T)V_D - \frac{V_D^2}{2} \right]$$

for below pinch-off operation of a MOSFET with circular geometry. To derive the above result, use cylindrical coordinates $(r, \theta, z)$ and appropriately modify Eqs. (17.7) through (17.17).

(b) Setting $r_2 = r_1 + L$ and $Z = 2\pi r_1$, show that the part (a) result reduces to the linear geometry result, Eq. (17.17), in the limit where $L/r_1 \ll 1$.

● **17.12** As a follow-up to Problem 17.5, we wish to explore the temperature dependence of the MOSFET saturation current. Consider an ideal $n$-channel MOSFET with $x_o = 0.1$ $\mu$m and $N_A = 10^{16}/\text{cm}^3$. Assume $\bar{\mu}_n$ in the MOSFET channel has the same temperature dependence as $\mu_n$ in the semiconductor bulk.

(a) Establish an expression for $I_{Dsat}(T)/I_{Dsat}(300 \text{ K})$ utilizing the square-law theory.

(b) Setting $V_G = 3$ V, compute and plot $I_{Dsat}(T)/I_{Dsat}(300 \text{ K})$ versus $T$ over the range $200 \text{ K} \leq T \leq 400 \text{ K}$. Superimpose a plot of $\mu_n(T)/\mu_n(300 \text{ K})$ on the same set of coordinates. Repeat the computation and plotting taking $V_G = 10$ V. Discuss your results.

**17.13** As pictured in Fig. P17.13, resistances $R_S$ and $R_D$ exist between the source/drain terminals and the channel proper. These resistances arise from a combination of the metal–Si contact resistance and the bulk resistance of the source/drain islands. Typically $R_S$ and $R_D$ are negligible in long-channel MOSFETs. However, as the dimensions of MOSFETs are reduced to achieve higher operating frequencies and higher packing densities, $R_S$ and $R_D$ have become increasingly important. Working with the square-law theory, show that the source and drain resistances are appropriately taken into account by replacing $V_D$ with $V_D-I_D(R_S + R_D)$ and $V_G$ with $V_G-I_DR_S$ in Eqs. (17.17), (17.21), and (17.22).

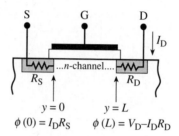

**Figure P17.13**

**17.14** Derive Eq. (17.29).

● **17.15** Write a computer program that can be used to calculate and plot $I_D/(Z\bar{\mu}_n C_o/L)$ versus $V_D$ characteristics based on the bulk-charge theory. Use your program to verify the accuracy of the bulk-charge characteristics pictured in Fig. 17.9.

● **17.16** Utilizing the relationships in Appendix D, construct computer programs to calculate and plot the $I_D$–$V_D$ characteristics based on (a) the charge-sheet theory and (b) the exact-charge theory. As a check, run your programs to obtain results that can be compared with the characteristics shown in Fig. 17.10 and/or Fig. 17.11.

**17.17** Perform the mathematical manipulations to verify the bulk-charge $g_d$ and $g_m$ entries in Table 17.1.

**17.18** Biases $V_G = 3$ V and $V_D = 0$ are applied to an ideal $n$-channel MOSFET with $Z = 70 \ \mu$m, $L = 7 \ \mu$m, $\bar{\mu}_n = 550$ cm²/V-sec, $x_o = 0.05 \ \mu$m, and $V_T = 1$ V. Making use of the square-law theory,

(a) determine the inversion layer charge/cm² at the midpoint ($y = L/2$) of the channel.

(b) determine the drain conductance ($g_d$) at the specified bias point.

**17.19** A complementary pair of ideal $n$-channel and $p$-channel MOSFETs are to be designed so that the devices exhibit the same $g_m$ and $f_{max}$ when equivalently biased and operated at $T = 300$ K. The structural parameters of the $n$-channel device are $Z = 50$ $\mu$m, $L = 5$ $\mu$m, $x_o = 0.05$ $\mu$m, and $N_A = 10^{15}$/cm$^3$. The $p$-channel device has the same oxide thickness and doping concentration, but, because of the lower hole mobility, must have different gate dimensions. Determine the required $Z$ and $L$ of the $p$-channel device. Assume the effective mobility of carriers in both devices is one-half the bulk mobility.

**17.20** General MOSFET Review

An $I_D$–$V_D$ characteristic derived from an ideal MOSFET is pictured in Fig. P17.20. Note that $I_{Dsat} = 10^{-3}$ A and $V_{Dsat} = 5$ V for the given characteristic. Answer the questions that follow making use of the square-law theory and the information conveyed in the figure.

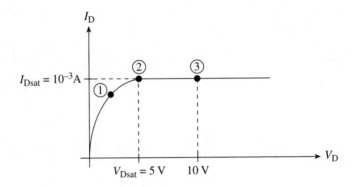

**Figure P17.20**

(a) Carefully sketch the inversion layer and depletion region inside the MOSFET corresponding to point (1) on the pictured characteristic. Show and label all parts of the transistor.

(b) Given a turn-on voltage of $V_T = 1$ V, what is the gate voltage one must apply to the MOSFET gate to obtain the pictured characteristic?

(c) If $x_o = 0.1$ $\mu$m, what is the inversion-layer charge/cm$^2$ at the drain end of the channel when the MOSFET is biased at point (2) on the characteristic?

(d) Suppose the gate voltage is readjusted so that $V_G - V_T = 3$ V. For the new condition, determine $I_D$ if $V_D = 4$ V.

(e) Determine $g_d$ if the quiescent operating point of the MOSFET is point (3) on the pictured characteristic.

(f) Determine $g_m$ if the quiescent operating point of the MOSFET is point (3) on the pictured characteristic.

(g) If $V_D = 0$ (i.e., the drain is shorted to the source and back), sketch the general shape of the $C_G$ (gate capacitance) versus $V_G$ characteristic to be expected from the MOSFET.

**17.21** Making free use of the square-law entries in Table 17.1, ignoring the variation of $\bar{\mu}_n$ with $V_G$, and *employing only one set of coordinates per each part of the problem,* draw

(a)  $g_d/(Z\bar{\mu}_n C_o/L)$ versus $V_G$ ($0 \le V_G \le 5$ V) if $V_T = 1$ V and $V_D = 0, 1,$ and 2 V.

(b)  $g_d/(Z\bar{\mu}_n C_o/L)$ versus $V_D$ ($0 \le V_D \le 5$ V) when $V_G - V_T = 1, 2,$ and 3 V.

(c)  $g_m/(Z\bar{\mu}_n C_o/L)$ versus $V_D$ ($0 \le V_D \le 5$ V) when $V_G - V_T = 1, 2,$ and 3 V.

**17.22** Compare the MOSFET and J-FET; concisely describe similarities and differences in structure, operation, and analysis.

# 18 Nonideal MOS

The ideal structure provides a convenient vehicle for establishing the basic principles of MOS theory in a clear and uncomplicated fashion. Real MOS device structures, however, are never perfectly ideal. In this chapter we examine well-documented deviations from the ideal that have been encountered in MOS device structures. The effect of a nonideality on device characteristics, its identified or suspected physical origin, and methods implemented to minimize the nonideality are noted. Because of ease of fabrication and functional simplicity, the MOS-capacitor has long been the test structure of choice for probing nonidealities. It is understandable, therefore, that the vast majority of nonideal effects are illustrated using MOS-C $C$–$V$ data. The description herein likewise relies heavily on the comparison of real and ideal MOS-C $C$–$V$ characteristics. Nevertheless, any deviation from the ideal has a comparable impact on the MOS transistor. To underscore this fact, the chapter concludes with a section dealing exclusively with the MOSFET. We discuss how nonidealities can affect the MOSFET threshold voltage, practical ramifications, and in-use methods for adjusting the threshold voltage.

## 18.1 METAL–SEMICONDUCTOR WORKFUNCTION DIFFERENCE

The energy band diagrams for the isolated components of an $Al$–$SiO_2$–($p$-type) Si system are drawn roughly to scale in Fig. 18.1(a). Upon examining this figure, we see that in a real device the energy difference between the Fermi energy and the vacuum level is unlikely to be the same in the isolated metal and semiconductor components of the system; that is, in contrast to the ideal structure, $\Phi_M \neq \Phi_S = \chi + (E_c - E_F)_{FB}$. To correctly describe real systems, the ideal theory must be modified to account for this metal–semiconductor workfunction difference.

In working toward the required modification, let us first construct the equilibrium ($V_G = 0$) energy band diagram appropriate for the sample system of Fig. 18.1(a). We begin by conceptually connecting a wire between the outer ends of the metal and semiconductor. The two materials are then brought together in a vacuum until they are a distance $x_o$ apart. The connecting wire facilitates the transfer of charge between the metal and semiconductor and helps maintain the system in an equilibrium state where the respective Fermi levels "line up" as the materials are brought together. With the metal $E_F$ and semiconductor $E_F$ at the same energy, and $\Phi_M \neq \chi + (E_c - E_F)_{FB}$, the vacuum levels in the two materials must be at different energies. Thus an electric field, $\mathscr{E}_{vac}$, develops between the components, with the Si vacuum level above the Al vacuum level given the situation pictured in

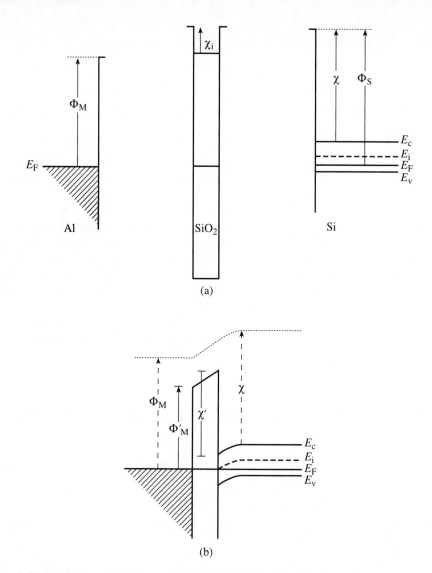

**Figure 18.1**  (a) To scale energy band diagrams for the isolated components of the Al–SiO$_2$–Si system. (b) Equilibrium ($V_G = 0$) energy band diagram typical of real MOS structures.

Fig. 18.1(a). Moreover, band bending occurs inside the semiconductor—$K_S \mathscr{E}_S$ must equal $\mathscr{E}_{vac}$. $\mathscr{E}_{vac}$ and the semiconductor band bending increase of course as the components are brought closer and closer together. Once the metal and semiconductor are positioned a distance $x_o$ apart, the insulator is next inserted into the empty space between the other two components. The addition of the insulator simply lowers the effective surface barriers

($\Phi_M \to \Phi_M - \chi_i = \Phi'_M$ and $\chi \to \chi - \chi_i = \chi'$) and reduces the electric field in the $x_o$ region ($K_O > 1$). The resulting equilibrium energy band diagram typical of real MOS systems is shown in Fig. 18.1(b).

The point to be derived from the preceding argument and Fig. 18.1(b) is that the work-function difference modifies the relationship between the semiconductor surface potential and the applied gate voltage. Specifically, setting $V_G = 0$ does not give rise to flat band conditions inside the semiconductor. Like in a *pn* junction or MS diode, there is a built-in potential. The precise value of the built-in potential, $V_{bi}$, can be determined by equating the energies from the Fermi level to the top of the band diagram as viewed from the two sides of the insulator in Fig. 18.1(b). One obtains

$$\underbrace{\Phi'_M + q\Delta\phi_{ox}}_{\text{metal side}} = \underbrace{(E_c - E_F)_{FB} - q\phi_S + \chi'}_{\text{semiconductor side}} \tag{18.1}$$

Thus, taking the metal to be the zero-potential reference point (the usual procedure in defining built-in potentials), we find

$$V_{bi} = -(\phi_S + \Delta\phi_{ox}) = \phi_{MS} \tag{18.2}$$

where

$$\boxed{\phi_{MS} \equiv \frac{1}{q}(\Phi_M - \Phi_S) = \frac{1}{q}[\Phi'_M - \chi' - (E_c - E_F)_{FB}]} \tag{18.3}$$

Perhaps the result here should have been intuitively obvious: The built-in potential inside a $\Phi_M \neq \Phi_S$, but otherwise ideal, MOS structure is just the metal–semiconductor work-function difference expressed in volts.

In dealing with any nonideality, a major concern is the effect of the nonideality on device characteristics. Generally speaking, one would like to know how the given non-ideality perturbs the ideal-device characteristics. To illustrate the general determination procedure and to specifically ascertain the effect of a $\phi_{MS} \neq 0$, let us suppose Fig. 18.1(b) is the energy band diagram for an MOS-C. Also let the broken-line curve in Fig. 18.2 be the expected form of the high-frequency $C$–$V$ characteristic exhibited by an ideal version of this $p$-bulk MOS-C. Flat band for the ideal device occurs, of course, at a gate bias of zero volts. On the other hand, from a cursory inspection of Fig. 18.1(b) one infers that a negative bias must be applied to the nonideal device to achieve flat band conditions. In fact, a gate voltage $V_G = \phi_{MS}$ (where $\phi_{MS} < 0$ for the given device) must be applied to offset the built-in voltage and achieve a $\phi_S = 0$. Since both devices will exhibit the same capacitance under flat-band conditions, we conclude the flat-band point for the real device will be displaced laterally $\phi_{MS}$ volts along the voltage axis.

As it turns out, we can begin the argument just presented at any point along the ideal-device characteristic. There is a one-to-one correspondence between the degree of band

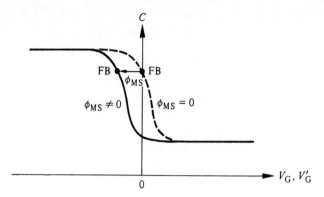

**Figure 18.2**   Effect of a $\phi_{MS} \neq 0$ on the MOS-C high-frequency $C$–$V$ characteristic.

bending or $\phi_S$ and the observed capacitance. Thus, regardless of the reference point along the ideal-device $C$–$V$ characteristic, one must always apply an added $\phi_{MS}$ volts to the gate of the real device to achieve the same degree of band bending and hence observe the same capacitance. In other words, as pictured in Fig. 18.2, the entire real-device $C$–$V$ characteristic will be shifted $\phi_{MS}$ volts along the voltage axis relative to the ideal-device characteristic.

In the preceding discussion the effect of the $\phi_{MS} \neq 0$ nonideality was described in graphical terms. Alternatively, one can generate a mathematical expression for the voltage shift, $\Delta V_G$, between the ideal and real $C$–$V$ curves. If $V_G'$ is the voltage applied to the gate of the ideal device to achieve a given capacitance, and $V_G$ the real-device gate voltage required to achieve the same capacitance, then, simply converting the $C$–$V$ curve discussion into mathematical terms,

$$\Delta V_G = (V_G - V_G')\big|^{\text{same } \phi_S}_{\text{(or same } C)} = \phi_{MS} \tag{18.4}$$

It should be interjected that it is common practice to use $V_G$ for the gate voltage when referring to the ideal structure. For presentation clarity, we have herein limited the use of $V_G'$ to this chapter where simultaneous reference is made to real and ideal devices.

The actual $\Delta V_G = \phi_{MS}$ value for a given MOS structure is routinely computed from Eq. (18.3) using the $\Phi_M' - \chi'$ appropriate for the system and the $(E_c - E_F)_{FB}$ deduced from a knowledge of the doping concentration inside the semiconductor. The $\phi_{MS}$ ($T = 300$ K) for the commercially important $n^+$ poly-Si-gate and Al-gate systems are graphed as a function of doping in Fig. 18.3. The experimentally determined $\Phi_M' - \chi'$ values for a number of other metal–silicon combinations are listed in Table 18.1. Note from Fig. 18.3 and the $\Phi_M' - \chi'$ values listed in Table 18.1 that $\phi_{MS}$ is more often than not a negative quantity, especially for $p$-type devices, and is typically quite small—on the order of one volt or less.

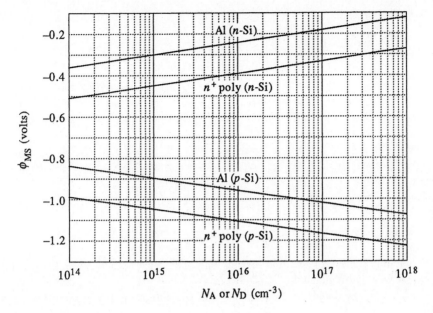

**Figure 18.3**  Workfunction difference as a function of the $n$- and $p$-type dopant concentration in $n^+$ poly-Si-gate and Al-gate $SiO_2$–Si structures. ($T = 300$ K. $\Phi_M' - \chi' = -0.18$ eV for the $n^+$ poly-Si-gate structure; $\Phi_M' - \chi' = -0.03$ eV for the Al-gate structure.)

**Table 18.1**  Barrier Height Differences in Selected Metal–$SiO_2$–Si Structures. (Data from Kar[19].)

| Metal Gate Material | $\Phi_M - \chi = \Phi_M' - \chi'$ (eV) |
| --- | --- |
| Ag | 0.73 |
| Au | 0.82 |
| Cr | −0.06 |
| Cu | 0.63 |
| Mg | −1.05 |
| Sn | −0.83 |

**Exercise 18.1**

**P:** It is possible to build an MOS-C with a $\phi_{MS} = 0$ through the proper choice of gate material and Si doping concentration. Restricting the Si doping to be in the range $10^{14}/\text{cm}^3 \leq N_A$ or $N_D \leq 10^{18}/\text{cm}^3$, and assuming operation at $T = 300$ K, identify the gate-material/doping-concentration combination(s) that gives rise to a $\phi_{MS} = 0$. Employ the $\Phi_M' - \chi'$ values given in Table 18.1.

**S:** Noting that

$$(E_c - E_F)_{FB} = E_c - E_i + (E_i - E_F)_{FB}$$
$$\cong E_G/2 - kT \ln(N_D/n_i) \quad \ldots n\text{-type Si}$$
$$\cong E_G/2 + kT \ln(N_A/n_i) \quad \ldots p\text{-type Si}$$

and employing $kT = 0.0259$ eV, $E_G = 1.12$ eV, and $n_i = 10^{10}/\text{cm}^3$, one calculates

$$0.08 \text{ eV} \leq (E_c - E_F)_{FB} \leq 0.32 \text{ eV} \quad \ldots \text{ if } 10^{14}/\text{cm}^3 \leq N_D \leq 10^{18}/\text{cm}^3$$
$$0.80 \text{ eV} \leq (E_c - E_F)_{FB} \leq 1.04 \text{ eV} \quad \ldots \text{ if } 10^{14}/\text{cm}^3 \leq N_A \leq 10^{18}/\text{cm}^3$$

Since $\phi_{MS} = (1/q)[\Phi'_M - \chi' - (E_c - E_F)_{FB}]$, to achieve a $\phi_{MS} = 0$ clearly requires

$$0.08 \text{ eV} \leq \Phi'_M - \chi' \leq 0.32 \text{ eV} \quad \ldots \text{ or} \ldots \quad 0.80 \text{ eV} \leq \Phi'_M - \chi' \leq 1.04 \text{ eV}$$

Examining Table 18.1, we find that the only gate material that meets the general requirement is Au with a $\Phi'_M - \chi' = 0.82$ eV.

The specific doping of the Au $p$-Si MOS-C exhibiting a $\phi_{MS} = 0$ must be such that

$$(E_c - E_F)_{FB} = \Phi'_M - \chi' = 0.82 \text{ eV}$$

or

$$(E_i - E_F)_{FB} = 0.26 \text{ eV}$$

and

$$N_A = n_i e^{(E_i - E_F)_{FB}/kT} = 10^{10} e^{0.26/0.0259} = 2.29 \times 10^{14}/\text{cm}^3$$

Thus

---

Au gate; $N_A = 2.29 \times 10^{14}/\text{cm}^3 \Rightarrow$ MOS-C with $\phi_{MS} = 0$

---

## 18.2 OXIDE CHARGES

### 18.2.1 General Information

As might be inferred from the comments at the end of Section 18.1, $\phi_{MS} \neq 0$ is a relatively minor nonideality. The voltage shift associated with $\phi_{MS} \neq 0$ is small, totally predictable, and incapable of causing device instabilities. Oxide charge, on the other hand, can give rise

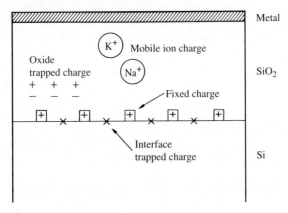

**Figure 18.4** Nature and location of charge centers in thermally grown SiO$_2$–Si structures. (Adapted from Deal[20], © 1980 IEEE.)

to far more significant effects, including large voltage shifts and instabilities. Through extensive research a number of distinct charge centers have been identified actually within the oxide or at the Si–SiO$_2$ interface. The nature and position of the oxide charges are summarized in Fig. 18.4.

To establish the general effect of oxide charges, let us postulate the existence of a charge distribution, $\rho_{ox}(x)$, that varies in an arbitrary manner across the width of the oxide layer. Note from the Fig. 18.5 visualization of the charge distribution that, for convenience in this particular analysis, *the origin of the x-coordinate has been relocated at the metal–oxide interface*. With the addition of the charge centers, a portion of the $V_G$–$\phi_S$ derivation presented in Subsection 16.3.2 is no longer valid and must be revised. Specifically, in place of Eqs. (16.19) to (16.21), one has, respectively,

$$\frac{d\mathscr{E}_{ox}}{dx} = \frac{\rho_{ox}(x)}{K_O \varepsilon_0} \tag{18.5}$$

$$\mathscr{E}_{ox}(x) = -\frac{d\phi_{ox}}{dx} = \mathscr{E}_{ox}(x_o) - \frac{1}{K_O \varepsilon_0} \int_x^{x_o} \rho_{ox}(x')dx' \tag{18.6}$$

and

$$\Delta\phi_{ox} = x_o \mathscr{E}_{ox}(x_o) - \frac{1}{K_O \varepsilon_0} \int_0^{x_o} \int_x^{x_o} \rho_{ox}(x')dx'\ dx \tag{18.7}$$

The double integral in Eq. (18.7) can be reduced to a single integral employing integration by parts. Moreover, $\mathscr{E}_{ox}(x_o) = K_S \mathscr{E}_S / K_O$ if a plane of charge [other than one possibly

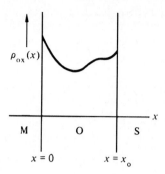

**Figure 18.5**   Arbitrary distribution of oxide charges.

included in $\rho_{\mathrm{ox}}(x_0)$] is excluded from the oxide–semiconductor interface. Performing the indicated modifications yields

$$\Delta\phi_{\mathrm{ox}} = \frac{K_S}{K_O} x_0 \mathscr{E}_S - \frac{1}{K_O \varepsilon_0} \int_0^{x_0} x \rho_{\mathrm{ox}}(x)\, dx \tag{18.8}$$

Since $V_G = \phi_S + \Delta\phi_{\mathrm{ox}}$ for a structure that has charge centers in the oxide but is otherwise ideal, one obtains

$$V_G = \phi_S + \frac{K_S}{K_O} x_0 \mathscr{E}_S - \frac{1}{K_O \varepsilon_0} \int_0^{x_0} x \rho_{\mathrm{ox}}(x)\, dx \tag{18.9}$$

However, for an ideal device

$$V_G' = \phi_S + \frac{K_S}{K_O} x_0 \mathscr{E}_S \tag{18.10}$$

Thus

$$\boxed{\Delta V_G \left( \begin{array}{c} \text{oxide} \\ \text{charges} \end{array} \right) = (V_G - V_G')\big|_{\text{same } \phi_S} = -\frac{1}{K_O \varepsilon_0} \int_0^{x_0} x \rho_{\mathrm{ox}}(x)\, dx} \tag{18.11}$$

As emphasized in the development, the voltage translation specified by Eq. (18.11) is valid for an arbitrary charge distribution and is added to the Eq. (18.4) voltage translation due to $\phi_{\mathrm{MS}}$. In the following subsections we systematically review known information about the various types of charge centers and examine their specific effect on MOS device characteristics.

## 18.2.2 Mobile Ions

The most perplexing and serious problem encountered in the development of MOS devices can be described as follows: First, the as-fabricated early (c. 1960) devices exhibited $C-V$ characteristics that were sometimes shifted negatively by *tens* of volts with respect to the theoretical characteristics. Second, when subjected to bias-temperature (BT) stressing, a common reliability-testing procedure where a device is heated under bias to accelerate device-degrading processes, the MOS structures displayed a severe instability. The negative shift in the characteristics was increased additional tens of volts after the device was biased positively and heated up to 150°C or so. Negative bias-temperature stressing had the reverse effect: The $C-V$ curve measured at room temperature after stressing shifted positively or toward the theoretical curve. In extreme cases the instability could even be observed by simply biasing the device at room temperature. One might sweep the $C-V$ characteristics for a given device, go out to lunch leaving the device positively biased, and return to repeat the $C-V$ measurement only to find the characteristics had shifted a volt or so toward negative biases. Note that the characteristics were always shifted in the direction opposite to the applied gate polarity and that the observed curves were always to the negative side of the theoretical curves. The nature and extent of the problem is nicely summarized in Fig. 18.6.

From a practical standpoint, the nonideality causing the as-fabricated translation and instability of the MOS device characteristics had to be identified and eliminated. A device whose effective operating point uncontrollably changes as a function of time is fairly useless. It is now well established that the large as-fabricated shifting and the related instability can be traced to mobile ions inside the oxide, principally $Na^+$.

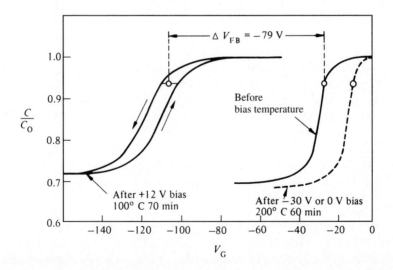

**Figure 18.6** Illustration of the large as-fabricated $C-V$ curve shifting and bias-temperature instability observed with early MOS devices. All $C-V$ curves were taken at room temperature; $x_o = 0.68 \mu m$. The arrows adjacent to the after $+BT$ curves indicate the direction of the voltage sweep. (From Kerr et al.[21], © 1964 by IBM Corporation. Reprinted with permission.)

If $\rho_{ion}(x)$ is taken to be the ionic charge distribution with

$$Q_M \equiv \int_0^{x_o} \rho_{ion}(x)\,dx \tag{18.12}$$

being the total ionic charge within the oxide per unit area of the gate, then it clearly follows from Eq. (18.11) that

$$\boxed{\Delta V_G \left( \begin{matrix} \text{mobile} \\ \text{ions} \end{matrix} \right) = -\frac{1}{K_O \varepsilon_0} \int_0^{x_o} x\rho_{ion}(x)\,dx} \tag{18.13}$$

Note from Eq. (18.13) that positive ions in the oxide would give rise to a negative shift in the $C$–$V$ characteristics as observed experimentally, while negative ions would give rise to a positive shift in disagreement with experimental observations. Furthermore, because the integrand in Eq. (18.13) varies as $x\rho_{ion}(x)$, $\Delta V_G$ is *sensitive to the exact position of the ions in the oxide.* If, for example, the same $Q_M$ charge per unit gate area is positioned (a) near the metal and (b) near the semiconductor as shown in Fig. 18.7(a), one computes $\Delta V_G(a) = -(0.05)Q_M/C_o$ and $\Delta V_G(b) = -(0.95)Q_M/C_o$, where $C_o = K_O \varepsilon_0 / x_o$. For the cited example, the shift is predicted to be some 19 times larger when the ions are located near the oxide–semiconductor interface! Indeed, based on the preceding observations, it is reasonable to speculate that a large as-fabricated negative shift in the measured $C$–$V$ characteristics and the attendant instability is caused by *positive* ions in the oxide that *move around or redistribute* under bias-temperature stressing. The required ion movement, away from the metal for $+$BT stressing and toward the metal for $-$BT stressing, is in fact consistent with the direction of ion motion grossly expected from the repulsive/attractive action of other charges within the structure (see Fig. 18.7b).

Actual verification of the mobile ion model and identification of the culprit (the ionic species) rivals some of the best courtroom dramas. The suspects were first indicted because of their past history and their accessibility to the scene of the crime. Long before the fabrication of the first MOS device, as far back as 1888, researchers had demonstrated that $Na^+$, $Li^+$, and $K^+$ ions could move through quartz, crystalline $SiO_2$, at temperatures below 250°C. Furthermore, alkali ions, especially sodium ions, were abundant in chemical reagents, in glass apparatus, on the hands of laboratory personnel, and in the tungsten evaporation boats used in forming the metallic gate. With the suspect identified, great care was taken to avoid alkali ion contamination in the formation of the MOS structure. The net result was devices that showed essentially no change in their $C$–$V$ characteristics after they were subjected to either positive or negative biases for many hours at temperatures up to 200°C. Next, other carefully processed devices were purposely contaminated by rinsing the oxidized Si wafers in a dilute solution of NaCl (or LiCl) prior to metallization. As expected, the purposely contaminated devices exhibited severe instabilities under bias-temperature stressing. Finally, sodium was positively identified in the oxides of normally

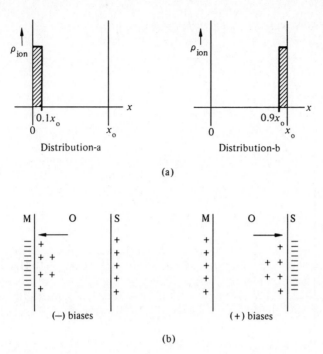

**Figure 18.7**  (a) Two hypothetical ionic charge distributions involving the same total number of ions situated near the metal (distribution-a) and near the semiconductor (distribution-b). (b) Expected motion of positive mobile ions within the oxide under ($+$) and ($-$) bias-temperature stressing.

fabricated devices (no intentional contamination) through the use of the neutron activation technique; that is, the oxides were bombarded with a sufficient number of neutrons to create a radioactive species of sodium. The analysis of the resultant radioactivity directly confirmed the presence of sodium within the oxide.

Although care to eliminate alkali-ion contamination throughout the fabrication process did lead to stable MOS devices, MOSFET manufacturers encountered difficulties in attaining and maintaining the required degree of quality control in production-line facilities. Thus, in addition to alkali-ion reduction efforts, special fabrication procedures were developed and implemented to minimize the effects of residual alkali-ion contamination. Two different procedures found widespread usage: phosphorus stabilization and chlorine neutralization.

In phosphorus stabilization the oxidized Si wafer is simply placed in a phosphorus diffusion furnace for a short period of time. During the diffusion, as illustrated in Fig. 18.8(a), phosphorus enters the outer portion of the $SiO_2$ film and becomes incorporated into the bonding structure, thereby forming a new thin layer referred to as a phosphosilicate glass. At the diffusion temperature the sodium ions are extremely mobile and invariably wander into the phosphorus-laden region of the oxide. Once in the phosphosilicate glass the ions

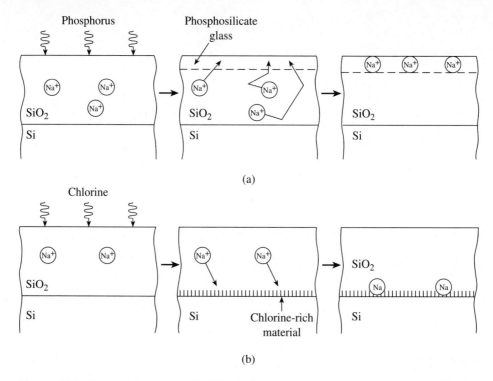

**Figure 18.8** Pictorial description of MOS stabilization procedures: (a) phosphorus stabilization; (b) chlorine neutralization.

become trapped and stay trapped when the system is cooled to room temperature. In this way the alkali ions are "gettered" or drawn out of the major portion of the oxide, are positioned near the outer interface where they give rise to the least amount of as-fabricated $C-V$ curve shifting, and are held firmly in place during normal operating conditions. The phosphosilicate glass layer, it should be noted, also blocks any subsequent contamination associated with the gate metallization or other poststabilization processing steps.

Chlorine neutralization involves a totally different approach. A small amount of chlorine in the form of a chlorine-containing compound is introduced into the furnace ambient during the growth of the $SiO_2$ layer. As pictured in Fig. 18.8(b), the chlorine enters the oxide and reacts to form a new material, believed to be a chlorosiloxane, located at the oxide–silicon interface. Stabilization occurs when the ionic sodium migrates into the vicinity of the oxide–silicon interface, becomes trapped, and is *neutralized*. Once neutralized, the sodium has no effect on the MOS device characteristics.

The shrinking size of MOS device dimensions, calling for gate-oxide thicknesses on the order of 100 Å, limits the use of the phosphorus stabilization procedure in present-day structures. Because of a potential polarization problem, the phosphosilicate glass can be only a small fraction of the overall oxide thickness. With $x_o \sim 100$ Å, the gettering volume becomes difficult to control. Chlorine neutralization continues to be employed, with careful control of the chlorine concentration to avoid an oxide thickness variation that can accompany the process. Neutralization coupled with improvements in the purity of fabrication materials (chemicals, gases, etc.) and upgraded processing procedures now permit the routine fabrication of stable MOS devices. It is common practice, nevertheless, to closely monitor furnace tubes and the processing in general to detect the onset of ionic contamination. As an extension of the chlorine neutralization procedure, chlorine has come to be widely employed in the pre-oxidation cleaning of furnace tubes. Also, a phosphosilicate glass layer is invariably deposited via chemical vapor techniques to form a protective coating on ICs. The layer helps to minimize ionic contamination subsequent to device fabrication.

---

### Exercise 18.2

**P:** Positive bias-temperature ($+$BT) and negative bias-temperature ($-$BT) stressing performed for a sufficient amount of time to respectively cause the mobile ions to pile up at the O–S and M–O interfaces, and the voltage displacement between the corresponding $C$–$V$ curves, are routinely used to deduce the total mobile-ion charge/cm$^2$ ($Q_M$) inside MOS-Cs. Suppose an MOS-C with $x_o = 0.1$ $\mu$m exhibits the post-stressing $C$–$V$ characteristics pictured below. Assuming the mobile ions are all piled up in a $\delta$-function distribution adjacent to the O–S interface after $+$BT stressing, and all piled up in a $\delta$-function distribution adjacent to the M–O interface after $-$BT stressing, determine $Q_M/q$.

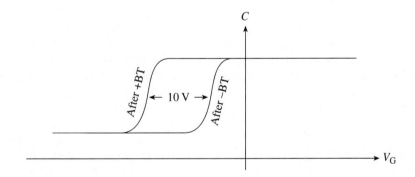

**S:** After $+$BT stressing, $\rho_{\text{ion}} = Q_M \delta(x_o)$. Substituting into Eq. (18.13) therefore yields

$$\Delta V_G \left( \begin{array}{c} \text{mobile ions} \\ \text{after } +\text{BT} \end{array} \right) = -\frac{1}{K_O \varepsilon_0} \int_0^{x_o} x Q_M \delta(x_o)\, dx = -\frac{x_o}{K_O \varepsilon_0} Q_M = -\frac{Q_M}{C_o}$$

After $-$BT stressing, $\rho_{ion} = Q_M \delta(0)$ and

$$\Delta V_G \left( \begin{array}{c} \text{mobile ions} \\ \text{after } -\text{BT} \end{array} \right) = -\frac{1}{K_O \varepsilon_0} \int_0^{x_o} x Q_M \delta(0)\, dx = 0$$

Assuming all other nonidealities are unaffected by the bias-temperature stressing, the displacement between the two $C$–$V$ curves is merely due to the difference between the above $\Delta V_G(+\text{BT})$ and $\Delta V_G(-\text{BT})$; that is

$$\Delta V_G(+\text{BT}) - \Delta V_G(-\text{BT}) = -10 \text{ V} = -Q_M / C_o$$

Thus

$$\frac{Q_M}{q} = -\frac{C_o}{q}[\Delta V_G(+\text{BT}) - \Delta V_G(-\text{BT})]$$

$$= \frac{(3.9)(8.85 \times 10^{-14})}{(1.6 \times 10^{-19})(10^{-5})}(10) = \mathbf{2.16 \times 10^{12} / cm^2}$$

## 18.2.3 The Fixed Charge

The gross perturbation associated with mobile ions in the oxide tended to obscure or cover up the effects of other deviations from the ideal. Indeed, with the successful elimination of the mobile-ion problem, it became possible to perform a more exacting examination of the device characteristics. The results were rather intriguing. Even in structures free of mobile ions, and after correcting for $\phi_{MS} \neq 0$, the observed $C$–$V$ characteristics were still translated up to a few volts toward *negative* biases relative to the theoretical characteristics. The possibility of mobile-ion contamination had been eliminated because the structures were stable under bias-temperature stressing. Moreover, for a given set of fabrication conditions the observed $\Delta V_G$ was *completely reproducible*. Confirming data were obtained from devices independently fabricated by a number of workers at different locations. Subsequent testing (by etching the oxide away in small steps and through photo-measurements) revealed the unexplained $\Delta V_G$ shift was caused by a charge residing within the oxide very close to the oxide–semiconductor interface. Because this quasi-interfacial charge was reproducibly fabricated into the structure and was fixed in position under bias-temperature stressing, the nonideality became known as the "built-in" or "fixed" oxide charge.

In modeling the quantitative effect of the fixed charge on the $C$–$V$ characteristics, it is typically assumed the charge is located immediately adjacent to the oxide–semiconductor interface. Under this assumption one can write

$$\rho_{ox}(x) = Q_F \delta(x_o) \tag{18.14}$$

where $\delta(x_o)$ is a delta-function positioned at the oxide–semiconductor interface and $Q_F$ is the fixed oxide charge per unit area of the gate. Substituting the Eq. (18.14) charge "distribution" into Eq. (18.11) and simplifying yields

$$\Delta V_G \left( \frac{\text{fixed}}{\text{charge}} \right) = -\frac{Q_F}{C_o} \tag{18.15}$$

From Eq. (18.15) it is obvious that, like the mobile-ion charge, the fixed oxide charge must be *positive* to account for the negative $\Delta V_G$'s observed experimentally. Other relevant information about the fixed oxide charge can be summarized as follows:

(1) The fixed charge is independent of the oxide thickness, the semiconductor doping concentration, and the semiconductor doping type (*n* or *p*).

(2) The fixed charge varies as a function of the Si surface orientation; $Q_F$ is largest on {111} surfaces, smallest on {100} surfaces, and the ratio of the fixed charge on the two surfaces is approximately 3:1.

(3) $Q_F$ is a strong function of the oxidation conditions such as the oxidizing ambient and furnace temperature. As displayed in Fig. 18.9, the fixed charge decreases more or less linearly with increasing oxidation temperatures. It should be emphasized, however, that only the *terminal* oxidation conditions are important. If, for example, a Si wafer is first oxidized in water vapor at 1000°C for 1 h, and then exposed to a dry $O_2$ ambient at 1200°C for a sufficiently long time to achieve a steady-state condition (~5 min), the $Q_F$ value will reflect only the dry oxidation process at 1200°C.

(4) Annealing (that is, heating) of an oxidized Si wafer in an Ar or $N_2$ atmosphere for a time sufficient to achieve a steady-state condition reduces $Q_F$ to the value observed for dry oxidations at 1200°C. In other words, regardless of the oxidation conditions, the fixed charge can always be reduced to a minimum by annealing in an inert atmosphere.

The preceding experimental facts all provide clues to the physical origin of the fixed oxide charge. For one, although doping impurities from the semiconductor diffuse into the oxide during the high-temperature oxidation process, the fixed charge was noted to be independent of the semiconductor doping concentration and doping type. The existence of ionized doping impurities within the oxide can therefore be eliminated as a possible source of $Q_F$. Second, the combination of the interfacial positioning of the fixed charge, the Si-surface orientation dependence, and the sensitivity of $Q_F$ to the terminal oxidation conditions suggests that the fixed charge is intimately related to the oxidizing reaction at the Si–$SiO_2$ interface. In this regard, it should be understood that, during the thermal formation of the $SiO_2$ layer, the oxidizing species diffuses through the oxide and reacts at the Si–$SiO_2$ interface to form more $SiO_2$. Thus, the last oxide formed, the portion of the oxide controlled by the terminal oxidation conditions, lies closest to the Si–$SiO_2$ interface and contains the fixed oxide charge. From considerations such as these, it has been postulated that

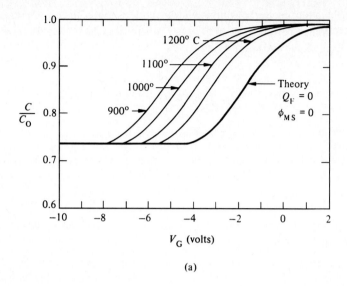

(a)

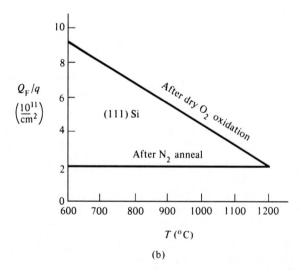

(b)

**Figure 18.9** Effect of the oxidation temperature and annealing on the fixed charge in MOS structures. (a) Measured $C–V$ characteristics after dry $O_2$ oxidations at various temperatures [$x_o = 0.2\ \mu$m, $N_D = 1.4 \times 10^{16}/\text{cm}^3$, (111) Si surface orientation]. (b) Fixed charge concentrations—the so-called *oxidation triangle* specifying the expected $Q_F/q$ after dry $O_2$ oxidation and after inert ambient annealing. [(a) From Deal et al.[22] Reprinted by permission of the publisher, The Electrochemical Society, Inc.]

the fixed oxide charge is due to *excess ionic silicon* that has broken away from the silicon proper and is waiting to react in the vicinity of the Si–SiO$_2$ interface when the oxidation process is abruptly terminated. The monolayer of oxide adjacent to the Si surface has in fact been experimentally determined to be $x < 2$ SiO$_x$, which is consistent with the excess-Si hypothesis. Annealing in an inert atmosphere, a standard procedure for minimizing the fixed oxide charge, apparently reduces the excess reaction components and thereby lowers $Q_F$.

---

### Exercise 18.3

**P:** An MOS-C is characterized by the energy band diagram shown in Fig. E18.3.

(a) Roughly sketch the electric field ($\mathscr{E}$) inside the oxide and semiconductor as a function of position.

(b) Is there an ionic charge ($Q_M$) distributed throughout the SiO$_2$? Explain.

(c) Is a fixed charge likely to exist at the Si–SiO$_2$ interface? Explain.

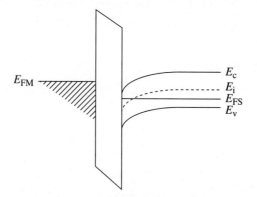

**Figure E18.3**

**S:** (a) The electric field is directly proportional to the slope of the energy bands. The deduced $\mathscr{E}$ versus $x$ dependence is sketched below.

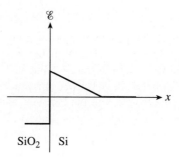

(b) If $\rho_{ox} = 0$, $\mathscr{E}_{ox} =$ constant and the oxide energy bands are a linear-function of position. If $\rho_{ox} \neq 0$, if there is charge distributed throughout the oxide, $\mathscr{E}_{ox}$ becomes a function of position and the oxide energy bands will in turn exhibit curvature. Since the oxide energy bands are a linear function of position in Fig. E18.3 and $\mathscr{E}_{ox} =$ constant, we conclude $\boxed{Q_M \cong 0}$.

(c) The normal component of the $D$-field, where $D = K\varepsilon_0 \mathscr{E}$, must be continuous if there is no plane of charge at an interface between two dissimilar materials (see Subsection 16.3.2). When a plane of charge does exist, there is a discontinuity in the $D$-field equal to the charge/cm² along the interface. Clearly, with $D = K_O \varepsilon_0 \mathscr{E}_{ox} < 0$ on the oxide side of the interface and $D = K_S \varepsilon_0 \mathscr{E}_S > 0$ on the semiconductor side of the interface, there must be a plane of charge at the Si–SiO$_2$ interface in the device characterized by Fig. E18.3. Moreover, since $Q_{interface} = D_S - D_{ox} = K_S \varepsilon_0 \mathscr{E}_S - K_O \varepsilon_0 \mathscr{E}_{ox}$, the interfacial charge must be *positive*. The fixed charge closely approximates a plane of positive charge at the Si–SiO$_2$ interface and we suspect $\boxed{Q_F \neq 0}$. (In general a $D$-field discontinuity at the Si–SiO$_2$ interface can arise from other sources of interface charge. These include the mobile ion charge drifted to the O–S interface during +BT stressing and the interfacial trap charge discussed in the next subsection.)

### 18.2.4  Interfacial Traps

Judged in terms of their wide-ranging and degrading effect on the operational behavior of MIS devices, insulator–semiconductor interfacial traps must be considered the most important nonideality encountered in MIS structures. A common manifestation of a significant interfacial trap concentration within an MOS-C is the distorted or spread-out nature of the $C–V$ characteristics. This is nicely illustrated in Fig. 18.10, which displays two $C–V$ curves derived from the same device before and after minimizing the number of Si–SiO$_2$ interfacial traps inside the structure.

From prior chapters the reader is familiar with donors, acceptors, and recombination–generation (R–G) centers, which introduce localized electronic states in the bulk of a semiconductor. Interfacial traps (also referred to as surface states or interface states) are allowed energy states in which electrons are localized in the vicinity of a material's *surface*. All of the bulk centers are found to add levels to the energy band diagram within the forbidden band gap. Donors, acceptors, and R–G centers respectively introduce bulk levels near $E_c$, $E_v$, and $E_i$. Analogously, as modeled in Fig. 18.11, interfacial traps introduce energy levels in the forbidden band gap at the Si–SiO$_2$ interface. Note, however, that interface states can, and normally do, introduce levels distributed throughout the band gap. Interface levels can also occur at energies greater than $E_c$ or less than $E_v$, but such levels are usually obscured by the much larger density of conduction or valence band states.

Figure 18.12 provides some insight into the behavior and significance of the levels. When an $n$-bulk MOS-C is biased into inversion as shown in Fig. 18.12(a), the Fermi level at the surface lies close to $E_v$. For the given situation essentially all of the interfacial traps

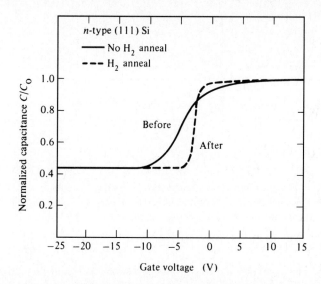

**Figure 18.10** *C–V* characteristics derived from the same MOS-C before (——) and after (----) minimizing the number of Si–SiO$_2$ interfacial traps inside the structure. (From Razouk and Deal[23]. Reprinted by permission of the publisher, The Electrochemical Society, Inc.)

will be empty because, to a first-order approximation, all energy levels above $E_F$ are empty and all energy levels below $E_F$ are filled. Moreover, if the states are assumed to be donor-like in nature (that is, positively charged when empty and neutral when filled with an electron), the net charge per unit area associated with the interfacial traps, $Q_{IT}$, will be positive. Changing the gate bias to achieve depletion conditions (Fig. 18.12b) positions the Fermi level somewhere near the middle of the band gap at the surface. Since the interface levels always remain fixed in energy relative to $E_c$ and $E_v$ at the surface, depletion biasing

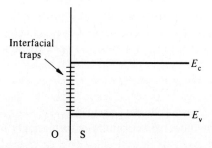

**Figure 18.11** Electrical modeling of interfacial traps as allowed electronic levels localized in space at the oxide–semiconductor interface.

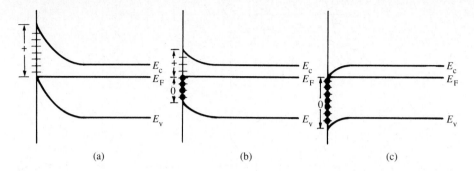

**Figure 18.12** Filling of the interface levels under (a) inversion, (b) depletion, and (c) accumulation biasing in an *n*-type device. The charge state exhibited by donor-like interfacial traps ["+" (plus) or "0" (neutral)] is noted to the left of the respective diagrams.

obviously draws electrons into the lower interface state levels and $Q_{IT}$ reflects the added negative charge: $Q_{IT}$ (depletion) $<$ $Q_{IT}$ (inversion). Finally, with the MOS-C accumulation biased (Fig. 18.12c), electrons fill most of the interfacial traps and $Q_{IT}$ approaches its minimum value. The point is that the interfacial traps charge and discharge as a function of bias, thereby affecting the charge distribution inside the device, the $V_G$–$\phi_S$ relationship, and the device characteristics in an understandable but somewhat complex manner.

The gross effect of interfacial traps on the $V_G$–$\phi_S$ relationship is actually quite easy to establish. Since $Q_{IT}$, like $Q_F$, is located right at the Si–SiO$_2$ interface, we can write by analogy with the fixed charge result

$$\Delta V_G \left( \begin{array}{c} \text{interfacial} \\ \text{traps} \end{array} \right) = -\frac{Q_{IT}(\phi_S)}{C_o} \tag{18.16}$$

As emphasized in Eq. (18.16), the result here differs from the fixed-charge result in that $Q_{IT}$ varies with $\phi_S$, while $Q_F$ is a constant independent of $\phi_S$.

Combined with the earlier considerations on the filling of interface levels, Eq. (18.16) helps to explain the form of the $C$–$V$ characteristics presented in Fig. 18.10. Assuming donor-like interfacial traps, $Q_{IT}$ takes on its largest positive value under inversion conditions and gives rise to a moderately large negative shift in the $C$–$V$ characteristics. In progressing through depletion toward accumulation, $Q_{IT}$ decreases, and the translation in the $C$–$V$ curve likewise decreases as observed experimentally. Once in accumulation $\Delta V_G$ should continue to decrease and still remain negative according to the Fig. 18.12 model. The Fig. 18.10 data, on the other hand, exhibits an increasingly *positive* shift in the characteristics with increased accumulation biasing. This discrepancy can be traced to the donor-like assumption. In actual MOS devices the interfacial traps in the upper half of the band gap are believed to be acceptor-like in nature (that is, neutral when empty and negative when filled with an electron). Thus, upon reaching flat band (or roughly flat band), $Q_{IT}$ passes through zero and becomes increasingly negative as more and more upper band gap states are filled with electrons. Qualitatively, then, we can explain the observed characteristics. A complete quantitative description would require a detailed knowledge of the interfacial trap concentration versus energy and additional theoretical considerations to establish an explicit expression for $Q_{IT}$ as a function of $\phi_S$.

Although models that detail the electrical behavior of the interfacial traps exist, the *physical origin* of the traps has not been totally clarified. The weight of experimental evidence, however, supports the view that the interfacial traps primarily arise from unsatisfied chemical bonds or so-called "dangling bonds" at the surface of the semiconductor. When the silicon lattice is abruptly terminated along a given plane to form a surface, one of the four surface-atom bonds is left dangling as pictured in Fig. 18.13(a). Logically, the thermal formation of the $SiO_2$ layer ties up some but not all of the Si-surface bonds. It is the remaining dangling bonds that become the interfacial traps (see Fig. 18.13b).

To add support to the foregoing physical model, let us perform a simple feasibility calculation. On a (100) surface there are $6.8 \times 10^{14}$ Si atoms per $cm^2$. If 1/1000 of these form interfacial traps and one electronic charge is associated with each trap, the structure would contain a $Q_{IT}/q = 6.8 \times 10^{11}/cm^2$. Choosing an $x_o = 0.1 \ \mu m$ and substituting into Eq. (18.16), we obtain a $\Delta V_G$(interfacial traps) = 3.15 V. Clearly, it only takes a relatively small number of residual dangling bonds to significantly perturb the device characteristics and readily account for observed interfacial trap concentrations.

The overall interfacial trap concentration and the distribution or density of states as a function of energy in the band gap (given the symbol $D_{IT}$ with units of states/$cm^2$-eV) are extremely sensitive to even minor fabrication details and vary significantly from device to device. Nevertheless, reproducible general trends have been recorded. The interfacial trap density, like the fixed oxide charge, is greatest on {111} Si surfaces, smallest on {100} surfaces, and the ratio of midgap states on the two surfaces is approximately $3:1$. After oxidation in a dry $O_2$ ambient, $D_{IT}$ is relatively high, $\sim 10^{11}$ to $10^{12}$ states/$cm^2$-eV at midgap, with the density increasing for increased oxidation temperatures in a manner also paralleling the fixed oxide charge. Annealing at high temperatures ($\geq 600°C$) in an inert ambient, however, does *not* minimize $D_{IT}$. Rather, as will be described shortly, annealing in the presence of hydrogen at relatively low temperatures ($\leq 500°C$) minimizes $D_{IT}$. $D_{IT}$ at midgap after an ideal interface-state anneal is typically $\leq 10^{10}/cm^2$-eV and the distribution of states as a function of energy is of the form sketched in Fig. 18.14. As shown in this figure, the interfacial trap density is more or less constant over the midgap region and increases rapidly as one approaches the band edges. Lastly, the states near the two band edges are usually about equal in number and opposite in their charging character; that is, states near the conduction and valence bands are believed to be acceptor-like and donor-like in nature, respectively.

The very important annealing of MOS structures to minimize the interfacial trap concentration is routinely accomplished in one of two ways, namely, through postmetallization annealing or hydrogen ($H_2$) ambient annealing. In the postmetallization process, which

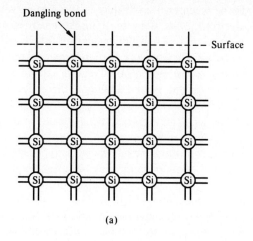

(a)

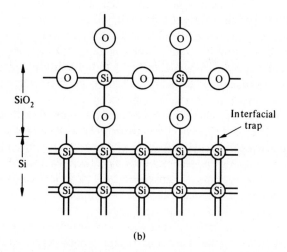

(b)

**Figure 18.13** Physical model for the interfacial traps. (a) "Dangling bonds," which occur when the Si lattice is abruptly terminated along a given plane to form a surface. (b) Postoxidation dangling bonds (relative number greatly exaggerated) that become the interfacial traps. [Part (b) adapted from Deal[24].]

requires a chemically active gate material such as Al or Cr, the metallized structure is simply placed in a nitrogen ambient at ~450°C for 5 to 10 min. During the formation of MOS structures, minute amounts of water vapor inevitably become adsorbed on the SiO$_2$ surface. At the postmetallization annealing temperature the active gate material reacts with the water vapor on the oxide surface to release a hydrogen species thought to be atomic hydrogen. As pictured in Fig. 18.15, the hydrogen species subsequently migrates through

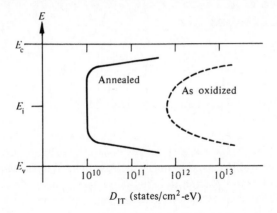

**Figure 18.14** Energy distribution of interface states within the band gap. General form and approximate magnitude of the interfacial trap density observed before and after an interface state anneal.

the $SiO_2$ layer to the $Si-SiO_2$ interface where it attaches itself to a dangling Si bond, thereby making the bond electrically inactive. The hydrogen-ambient process operates on a similar principle, except the hydrogen is supplied directly in the ambient and the structure need not be metallized.

Even though we originally stated that the interfacial trap problem was of paramount importance, it is a challenge to convey the true significance and scope of the problem. Simply stated, if the thermal oxide didn't tie up most of the dangling Si bonds, and if an annealing process were not available for reducing the remaining bonds or interfacial traps to an acceptable level, MOS devices would merely be a laboratory curiosity. Indeed, high interfacial-trap concentrations have severely stunted the development of other insulator–semiconductor systems.

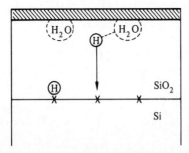

**Figure 18.15** Model for the annihilation of interface states during the postmetallization annealing process. (H)'s represent the active hydrogen species involved in the process; X's represent interface states.

## 18.2.5 Induced Charges

### Radiation Effects

Radiation damage in solid-state devices has been a major concern of space and military experts since the launch of the first Telstar satellite through the van Allen radiation belt in the early 1960s. Radiation temporarily disabled the Telstar satellite and can have debilitating effects on most solid-state devices and systems. Radiation in the form of x-rays, energetic electrons, protons, and heavy ionized particles all have a similar effect on MOS devices. After irradiation, MOS device structures invariably exhibit both an increase in the apparent fixed charge within the oxide and an increase in the interfacial-trap concentration.

The sequence of events leading to radiation-induced damage is summarized in Fig. 18.16. The primary effect directly related to the ionizing radiation is the generation of electron-hole pairs throughout the oxide. A percentage of the generated electrons and holes recombine immediately. The electric field in the oxide operates to separate the surviving carriers, accelerating electrons and holes in opposite directions. Electrons, which have a relatively high mobility in $SiO_2$, are rapidly (nsec) swept out of the oxide. Holes, on the other hand, tend to be trapped near their point of origin. Over a period of time the holes migrate to the Si–$SiO_2$ interface (assuming $\mathcal{E}_{ox}$ is positive as shown in Fig. 18.16) where they either recombine with electrons from the silicon or become trapped at deep-level sites. Once trapped in the near vicinity of the interface, the holes mimic the fixed charge thereby giving rise to an apparent increase in $Q_F$. The process leading to interface-state creation is less well understood. Some of the interfacial traps are created immediately upon exposure

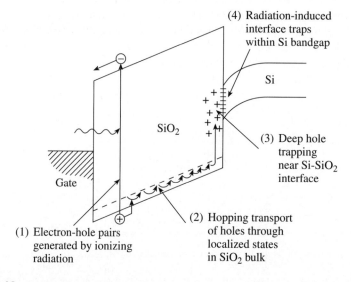

**Figure 18.16**  Response to ionizing radiation and the resulting damage in MOS structures. (From Srour and McGarrity[25], © 1988 IEEE.)

to ionizing radiation, while the remainder are created in proportion to the number of holes reaching the Si–SiO$_2$ interface. It has been proposed that the energy released in the deep-level trapping of holes breaks the Si–H bonds associated with deactivated interfacial traps.

Over a period of days to years the trapped-hole charge tends to be slowly reduced at room temperature by the capture of electrons injected into the oxide from the metal or the silicon. Naturally, removal of the trapped-hole charge is greatly accelerated by thermal annealing; a standard interfacial-trap anneal totally removes both the trapped holes and the induced interfacial traps. It should be noted, however, that once subjected to ionizing radiation and recovered through low-temperature annealing, MOS devices exhibit greater sensitivity to subsequent radiation. This has been attributed to the creation of neutral traps within the oxide in addition to the induced charges. Higher-temperature annealing ($T > 600°C$) has been found to be partially effective in removing the neutral traps.

Thermal annealing can be readily performed to remove radiation damage that occurs during fabrication.[†] It is also feasible to increase the ambient temperature of completed devices to 100°C or so to accelerate the removal of trapped holes caused by ionizing radiation. After-the-fact recovery in completed devices, however, is relatively limited. It is preferable to "harden" the devices. The oxide is hardened (i.e., its sensitivity to radiation reduced) by employing empirically established optimum growth conditions such as oxidation temperatures below 1000°C. Other hardening procedures include Al-shielding, which stops the majority of energetic electrons encountered in space, and increasing the threshold voltage of MOSFETs so they are less sensitive to $\Delta V_G$ changes caused by radiation. (The general topic of MOSFET threshold adjustment is discussed in Section 18.3.) Somewhat fortuitously, the reduction in oxide thickness that accompanies reductions in device dimensions is also leading to harder MOS devices. $\Delta V_G$ for both the fixed charge and interfacial traps is proportional to $1/C_o = x_o/K_O\varepsilon_0$ and therefore automatically decreases with decreasing $x_o$. The improvement may also be due, in part, to the smaller hole trapping cross sections at the higher oxide fields that exist across the thinner oxides. Projections even suggest hole trapping and the associated interfacial-trap production might actually vanish as the oxide thickness drops below 100 Å. It is envisioned that electrons from the metal and semiconductor can tunnel into all parts of a very thin oxide and rapidly annihilate the trapped holes.

### Negative-Bias Instability

Negative-bias instability is a significant perturbation of MOS device characteristics that occurs as a direct result of stressing the structure with a large negative bias at elevated temperatures. Typical stress conditions would be a negative gate bias sufficient to produce a field of $2 \times 10^6$ V/cm in the oxide and $T > 250°C$. The instability is characterized by a large *negative* shift along the voltage axis and a distortion of the MOS-C $C$–$V$ curve.

---

[†]Ion implantation, electron-beam evaporation of metals, deposition of special-purpose thin films over the SiO$_2$ layer in a hostile plasma environment (sputtering), electron-beam and x-ray lithography, and a number of other fabrication processes can lead to varying degrees of radiation damage.

Similar to ionizing radiation, the stress clearly causes an increase in the apparent fixed charge within the oxide and an increase in the interfacial-trap concentration. An added peak in $D_{IT}$ near midgap is considered to be a distinctive signature of the instability. Note that the $C$–$V$ curve shifting related to the negative-bias instability is opposite to that caused by alkali-ion contamination.

The exact mechanism causing the negative-bias instability has not been established. However, because large hole concentrations are adjacent to the oxide under the conditions of the negative stress, it has been proposed that the instability may result from hole injection from the Si into the oxide and the subsequent trapping of the holes at deep-level sites near the Si–SiO$_2$ interface. The radiation-induced and stress-induced damage would then have a related origin. Indeed, it has been found that the negative-bias instability is enhanced if the MOS structure is first exposed to ionizing radiation. Sensitivity to the instability can be minimized by annealing the device structure in hydrogen at 800°C–900°C prior to gate deposition.

## 18.2.6 $\Delta V_G$ Summary

In the first two sections of this chapter we cited and summarized four of the most commonly encountered deviations from the ideal; specifically, the metal–semiconductor workfunction difference, mobile ions in the oxide, fixed oxide charge, and interfacial traps. It was also noted that ionizing radiation and stressing with a large negative bias at elevated temperatures give rise to additional oxide charges. In both cases there is an increase in $Q_{IT}$ and the apparent $Q_F$.

The combined effect of the analyzed nonidealities on the $V_G$–$\phi_S$ relationship is described by

$$\Delta V_G = (V_G - V_G')|_{\text{same} \phi_S} = \phi_{MS} - \frac{Q_F}{C_o} - \frac{Q_M \gamma_M}{C_o} - \frac{Q_{IT}(\phi_S)}{C_o} \qquad (18.17)$$

where

$$\gamma_M \equiv \frac{\displaystyle\int_0^{x_o} x \rho_{\text{ion}}(x)\,dx}{\displaystyle x_o \int_0^{x_o} \rho_{\text{ion}}(x)\,dx} \qquad (18.18)$$

In writing down Eq. (18.17) we recast the mobile ion contribution (Eq. 18.13) to emphasize the similarity between the three terms associated with oxide charges. $\gamma_M$ is a unitless quantity representing the centroid of the mobile ion charge in the oxide normalized to the width of the oxide layer; $\gamma_M = 0$ if the mobile ions are all at the metal–oxide interface, while

$\gamma_M = 1$ if the mobile ions are all piled up at the Si–SiO$_2$ interface. As a general rule, $Q_M$, $Q_F$, and $\phi_{MS}$ all lead to a negative parallel translation of the $C$–$V$ characteristics along the voltage axis relative to the ideal theory. $\Delta V_G$ due to $Q_{IT}$, on the other hand, can be either positive or negative, depends on the applied bias, and tends to distort or spread out the characteristics.

From the discussion of nonidealities, it is safe to conclude that real MOS devices are not intrinsically perfect; the devices are in fact intrinsically imperfect. Through an extensive research effort, however, procedures have been developed for minimizing the net effect of nonidealities in MOS structures. Although constant checks must be run to maintain quality control, manufacturers today routinely fabricate near-ideal devices. We should mention that, whereas the described or similar nonidealities apply to all MIS structures, the specific minimization procedures outlined herein apply only to the thermally grown SiO$_2$-Si system.

---

### Exercise 18.4

**P:** Identify the physical cause of the listed oxide charges. Place the appropriate letter in the box preceding the charge.

| *Oxide Charge* | *Physical Cause* |
|---|---|
| ☐ Fixed charge | (a) Phosphorus ions (P$^+$) |
| ☐ Mobile ion charge | (b) Sodium ions (Na$^+$) |
| ☐ Interfacial traps | (c) Nitrogen ions (N$^+$) |
| ☐ Apparent fixed charge resulting from ionizing radiation | (d) Dangling bonds at the Si surface |
| | (e) Trapped electrons |
| | (f) Ionized Si waiting to be oxidized |
| | (g) Ionized oxygen waiting to form SiO$_2$ |
| | (h) Trapped holes |

**S:** $Q_F$–f, $Q_M$–b, $Q_{IT}$–d, apparent $Q_F$–h.

---

### Exercise 18.5

**P:** An Al–SiO$_2$–Si MOS-C is to be formed on an $N_A = 10^{15}$/cm$^3$ (100)-oriented Si wafer. Thermal oxidation in dry O$_2$ at 1000°C is followed by an $N_2$ anneal sufficient to achieve a steady-state condition. The structure is next phosphorus gettered so that an inadvertent alkali ion concentration, $Q_M/q = 2 \times 10^{11}$/cm$^2$, winds up being distributed in the oxide as shown in Fig. E18.5. The device is postmetallization annealed, but there is a residual interfacial trap density. A constant $D_{IT} = 2 \times 10^{10}/$ cm²-eV of acceptor-like states is found to exist for all band gap energies. Determine the expected flat band voltage of the MOS-C if $T = 300$ K and $x_o = 0.1 \ \mu$m.

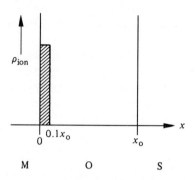

**Figure E18.5**

**S:** Under flat-band conditions, $\phi_S = 0$ and $V'_G = 0$. It therefore follows from Eq. (18.17) that

$$V_{FB} \equiv V_G|_{\phi_S=0} = \phi_{MS} - \frac{Q_F}{C_o} - \frac{Q_M \gamma_M}{C_o} - \frac{Q_{IT}(0)}{C_o}$$

where

$$C_o = \frac{K_O \varepsilon_0}{x_o} = \frac{(3.9)(8.85 \times 10^{-14})}{10^{-5}} = 3.45 \times 10^{-8} \text{F/cm}^2$$

We must systematically evaluate each of the terms in the $V_{FB}$ relationship.

$\phi_{MS} \dots$ Given $N_A = 10^{15}/\text{cm}^3$ and an Al gate, we conclude from Fig. 18.3 that $\phi_{MS} = -0.90$ V.

$Q_F/C_o \dots$ The $N_2$ anneal after the dry oxidation minimizes the fixed oxide charge. Figure 18.9(b) indicates an expected $Q_F/q = 2 \times 10^{11}/\text{cm}^2$ for a (111)-oriented Si Substrate. It was pointed out in the text discussion that $Q_F$ is approximately three times smaller on (100) surfaces. Thus for the given device $Q_F/q \cong 6.67 \times 10^{10}/\text{cm}^2$ and

$$\frac{Q_F}{C_o} = \frac{q(Q_F/q)}{C_o} = \frac{(1.6 \times 10^{-19})(6.67 \times 10^{10})}{(3.45 \times 10^{-8})} = 0.31 \text{ V}$$

$Q_M \gamma_M / C_o$ ... Referring to the specified mobile ion distribution in Fig. E18.5, and making use of Eq. (18.18), we conclude

$$\gamma_M = \frac{\displaystyle\int_0^{0.1x_o} x\rho_{max}\, dx}{\displaystyle x_o \int_0^{0.1x_o} \rho_{max}\, dx} = \frac{\left.\dfrac{x^2}{2}\right|_0^{0.1x_o}}{\left. x_o x \right|_0^{0.1x_o}} = 0.05$$

and

$$\frac{Q_M \gamma_M}{C_o} = \frac{q(Q_M/q)\gamma_M}{C_o} = \frac{(1.6 \times 10^{-19})(2 \times 10^{11})(0.05)}{(3.45 \times 10^{-8})} = 0.046\ \text{V}$$

$Q_{IT}(0)/C_o$ ... Acceptor-like centers are $(-)$ charged if filled with an electron and neutral if empty. Under equilibrium conditions the interfacial traps are mostly filled below the Fermi level and mostly empty above the Fermi level. The filling and charge states under flat-band conditions are therefore as pictured below.

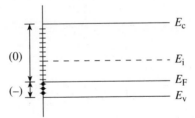

Because $D_{IT} = $ constant, $Q_{IT}(0) = -qD_{IT}\Delta E$, where $\Delta E$ is the range of energies in the surface band gap containing negatively charged interfacial traps. Consequently,

$$\Delta E = E_F - E_v \cong E_G/2 - (E_i - E_F) = E_G/2 - kT \ln(N_A/n_i)$$
$$= 0.56 - 0.0259 \ln(10^{15}/10^{10}) = 0.26\ \text{eV}$$

and

$$\frac{Q_{IT}}{C_o} = -\frac{qD_{IT}\Delta E}{C_o} = -\frac{(1.6 \times 10^{-19})(2 \times 10^{10})(0.26)}{(3.45 \times 10^{-8})} = -0.024\ \text{V}$$

Finally, summing the individual terms in the $V_{FB}$ relationship yields $\boxed{V_{FB} = -1.23\ \text{V}}$.

## 18.3  MOSFET THRESHOLD CONSIDERATIONS

Thus far we have described the effect of nonidealities in terms of MOS-C $C-V$ curve shifting and distortion. The $V_T$ point on the $C-V$ curve, corresponding to the threshold voltage of a MOSFET, is of course directly affected by any $\Delta V_G$ displacement. Figure 18.17(a) graphically illustrates the general effect of changes in the threshold voltage on the MOSFET $I_D-V_D$ characteristics. A hypothetical $p$-channel MOSFET with an ideal device $V'_T = -1$ V was assumed in constructing the figure. The $\Delta V_G$ leading to the perturbed

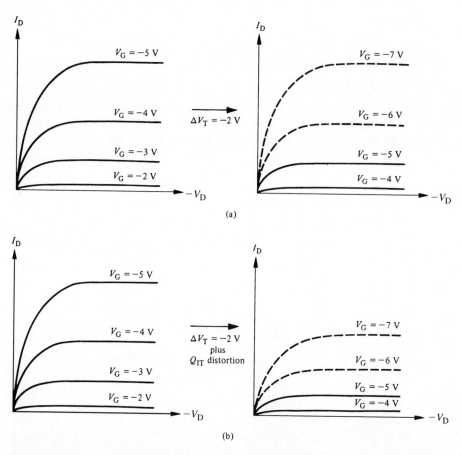

**Figure 18.17**  General effect of nonidealities on the MOSFET current-voltage characteristics. The ideal characteristics of a hypothetical $p$-channel MOSFET with $V'_T = -1$ V are pictured on the left and the perturbed characteristics are shown on the right. (a) Effect of a simple shift in the threshold voltage caused by $\phi_{MS} \neq 0$, the fixed charge, and/or mobile ions. (b) $g_m$ degrading effect of interfacial traps.

characteristics in Fig. 18.17(a) is taken to be caused by $\phi_{MS}$, the fixed charge, and/or alkali ions in the oxide. Note that, analogous to the parallel translation of the MOS-C $C-V$ characteristics, the form of the perturbed MOSFET characteristics is unchanged, but larger $|V_G|$ values are needed to achieve comparable $I_D$ current levels. If present in large densities, the interfacial traps can additionally decrease the change in current resulting from a stepped increase in gate voltage. This effect, equivalent to a reduction in the $g_m$ of the transistor, is pictured in Fig. 18.17(b).

While it is true that both the mobile ion and interfacial trap problems were minimized early in MOSFET development, the remaining nonidealities, primarily through their effect on $V_T$, have had a very large impact on fabrication technology, device design, and modes of operation. In this section we examine the reasons for the impact and methods that have evolved for adjusting the MOSFET threshold.

### 18.3.1  $V_T$ Relationships

For discussion purposes it is desirable to establish an expression for the expected threshold voltage, $V_T$, of a real MOSFET. Let $V'_T$ be the threshold voltage of an ideal version of the given MOSFET. Simply evaluating Eq. (18.17) at $\phi_S = 2\phi_F$ then gives

$$V_T = V'_T + \phi_{MS} - \frac{Q_F}{C_o} - \frac{Q_M\gamma_M}{C_o} - \frac{Q_{IT}(2\phi_F)}{C_o} \qquad (18.19)$$

Although Eq. (18.19) can be used directly to compute $V_T$, it is standard practice to express the threshold shift in terms of the real-device flat-band voltage. Under flat-band conditions $\phi_S = 0$, $V'_T = 0$, and from Eq. (18.17),

$$\boxed{V_{FB} \equiv V_{G|\phi_S} = 0 = \phi_{MS} - \frac{Q_F}{C_o} - \frac{Q_M\gamma_M}{C_o} - \frac{Q_{IT}(0)}{C_o}} \qquad (18.20)$$

If $Q_{IT}$ changes little in going from $\phi_S = 0$ to $\phi_S = 2\phi_F$, a reasonably good approximation in well-made devices, the nonideality-related terms in Eq. (18.19) can be replaced by $V_{FB}$ and one obtains

$$\boxed{V_T = V'_T + V_{FB}} \qquad (18.21)$$

where, repeating Eqs. (17.1),

$$V'_T = 2\phi_F \pm \frac{K_S}{K_O} x_o \sqrt{\frac{4qN_B}{K_S\varepsilon_0}}(\pm\phi_F)$$

(+) for *n*-channel devices
(−) for *p*-channel devices
$N_B = N_A$ or $N_D$ as appropriate

(18.22)

## 18.3.2 Threshold, Terminology, and Technology

As an entry into the discussion, let us perform a simple threshold voltage computation employing relationships developed in the preceding subsection. Suppose the gate material is Al, the Si surface orientation is (111), $T = 300$ K, $x_o = 0.1$ $\mu$m, $N_A = 10^{15}$/cm$^3$, $Q_F/q = 2 \times 10^{11}$/cm$^2$, $Q_M = 0$, and $Q_{IT} = 0$. For the given *n*-channel device, one computes $\phi_{MS} = -0.90$ V, $-Q_F/C_o = -0.93$ V, $V_{FB} = -1.83$ V, $V'_T = 1.00$ V, and $V_T = -0.83$ V. Observe that whereas $V'_T$ is positive, as expected, *nonidealities of a very realistic magnitude cause $V_T$ to be negative*. Since an *n*-channel device turns on for gate voltages $V_G > V_T$, the device in question is already "on" at a gate bias of zero volts. Actually, negative biases must be applied to deplete the surface channel and turn the device off! For a *p*-channel device with identical parameters (except, of course, for an $N_D$ doped substrate), one obtains a $V'_T = -1.00$ V, $V_{FB} = -1.23$ V, and $V_T = -2.23$ V. In the *p*-channel case the considered nonidealities merely increase the negative voltage required to achieve turn-on.

As illustrated in Fig. 18.18, a MOSFET is called an *enhancement mode* device when it is "off" at $V_G = 0$ V. When a MOSFET is "on" at $V_G = 0$ V, it is called a *depletion mode* device. Routinely fabricated *p*-channel MOSFETs constructed in the standard configuration are ideally and practically enhancement mode devices, *n*-channel MOSFETs are also ideally enhancement mode devices. However, because nonidealities tend to shift the threshold voltage toward negative biases in the manner indicated in our sample calculation, early *n*-channel MOS transistors were typically of the depletion mode type. Up until approximately 1977 this difference in behavior led to the total dominance of *PMOS* technology over *NMOS* technology; that is, ICs incorporating *p*-channel MOSFETs dominated the commercial marketplace. Subsequently, as explained under the heading of threshold adjustment, NMOS, which is to be preferred because of the greater mobility of electrons compared to holes, benefited from technological innovations widely implemented in the late 1970s and is now incorporated in the majority of newly designed ICs.

While on the topic of the threshold voltage in practical devices, it is relevant to note that the inversion threshold of regions adjacent to the device proper is also of concern. Consider, for example, the unmetallized region between the two *n*-channel MOS transistors pictured in Fig. 18.19(a). If the potential at the unmetallized outer oxide surface is assumed to be zero (normally a fairly reasonable assumption) and if the threshold voltage for the *n*-channel transistors is negative, then the intermediate region between the two transistors

### n-channel MOSFET

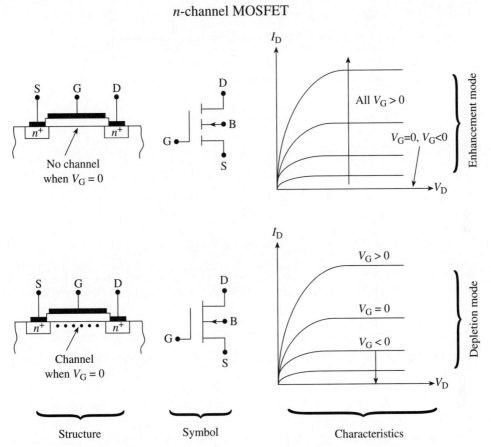

**Figure 18.18** MOSFET operational modes. $V_G = 0$ channel status, circuit symbol, and $I_D$–$V_D$ characteristics of n-channel enhancement-mode and depletion-mode MOSFETs.

will be inverted. In other words, a conducting path, a pseudo-channel, will exist between the transistors. This undesirable condition was another nuisance in early NMOS technology, where, as already noted, nonidealities tended to invert the surface of the semiconductor in the absence of an applied gate bias. Unless special precautions are taken, unwanted pseudo-channels between devices can also arise in both n- and p-channel ICs from the potential applied to the metal overlays supplying the gate and drain biases. To avoid this problem, the oxide in the non-gated portions of the IC, referred to as the *field-oxide,* is typically much thicker than the *gate-oxide* in the active regions of the structure (see Fig. 18.19b). The idea behind the use of the thicker oxide can be understood by referring to Eqs. (18.20) and (18.22). Both $V_{FB}$ and $V'_T$ contain terms that are proportional to $x_o$.

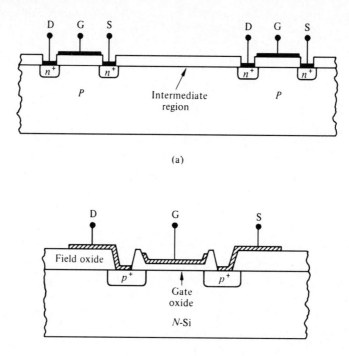

(a)

(b)

**Figure 18.19**   (a) Visualization of the intermediate region between two MOSFETs. (b) Identification of the gate-oxide and field-oxide regions in practical MOSFET structures.

Thus employing an $x_o$ (field-oxide) $\gg x_o$ (gate-oxide) increases $|V_T|$ in the field-oxide areas relative to the gated areas in PMOS (and modern NMOS) structures. Inversion of the field-oxide regions is thereby avoided at potentials normally required for IC operation.

### 18.3.3  Threshold Adjustment

Several physical factors affect the threshold voltage and can therefore be used to vary the $V_T$ actually exhibited by a given MOSFET. We have, in fact, already cited the adjustment of $V_T$ through a variation of the oxide thickness. Obviously, the substrate doping can also be varied to increase or decrease the threshold voltage. However, although strongly influencing the observed $V_T$ value, the gate-oxide thickness and substrate doping are predetermined in large part by other design restraints.

Other factors that play a significant role in determining $V_T$ are the substrate surface orientation and the material used in forming the MOS gate. As first noted in Subsection 18.2.3, the $Q_F$ in MOS devices constructed on (100) surfaces is approximately three times smaller than the $Q_F$ in devices constructed on (111) surfaces. The use of (100) substrates therefore reduces the $\Delta V_G$ associated with the fixed oxide charge. The use of a polysilicon instead of an A1 gate, on the other hand, changes $\phi_{MS}$. Given a polysilicon gate the effective "metal" workfuncton becomes

$$\text{``}\Phi_M\text{''} = \chi_{Si} + (E_c - E_F)_{poly\text{-}Si} \tag{18.23}$$

and

$$\phi_{MS} = \frac{1}{q} [(E_c - E_F)_{poly\text{-}Si} - (E_c - E_F)_{FB,crystalline\text{-}Si}] \tag{18.24}$$

If the calculation in Subsection 18.3.2 is revised assuming a (100) surface orientation $(Q_F/q = 2/3 \times 10^{11}/cm^2)$ and a $p$-type polysilicon gate where $E_F = E_v$, one obtains a $\phi_{MS} = +0.26$ V, $V_{FB} = -0.05$ V, and $V_T = +0.95$ V. Thus positive NMOS thresholds are possible in (100)-oriented structures incorporating $p$-type polysilicon gates.

Although the foregoing calculation shows positive threshold voltages are possible, the $V_T$ in actual structures may be only nominally positive, or the structure impractical—in forming $n$-channel MOSFETs, the polysilicon is conveniently doped $n$-type, the same as the drain and source islands, not $p$-type as assumed above. Also, a larger threshold voltage may be desired, or one may desire to modify the threshold attainable in a PMOS structure, or tailoring of the threshold for both $n$- and $p$-channel devices on the same IC chip may be required. For a number of reasons, it is very desirable to have a flexible threshold adjustment process where $V_T$ can be controlled essentially at will. In modern device processing this is accomplished through the use of *ion implantation.*

The general ion-implantation process was described in Chapter 4. To adjust the threshold voltage, a relatively small, precisely controlled number of either boron or phosphorus ions are implanted into the near-surface region of the semiconductor. When the MOS structure is depletion or inversion biased, the implanted dopant adds to the exposed dopant-ion charge near the oxide–semiconductor interface and thereby translates the $V_T$ exhibited by the structure. The implantation of boron causes a positive shift in the threshold voltage; phosphorus implantation causes a negative voltage shift. For shallow implants the procedure may be viewed to first order as placing an additional "fixed" charge at the oxide–semiconductor interface. If $N_I$ is the number of implanted ions/cm² and $Q_I = \pm qN_I$ is the implant-related donor $(+)$ or acceptor $(-)$ charge/cm² at the oxide–semiconductor interface, then, by analogy with the fixed charge analysis

$$\Delta V_G \left( \begin{matrix} \text{implanted} \\ \text{ions} \end{matrix} \right) = -\frac{Q_I}{C_o} \tag{18.25}$$

Assuming, for example, an $N_I = 5 \times 10^{11}$ boron ions/cm² and $x_o = 0.1$ $\mu$m, one computes a threshold adjustment of $+2.32$ V.

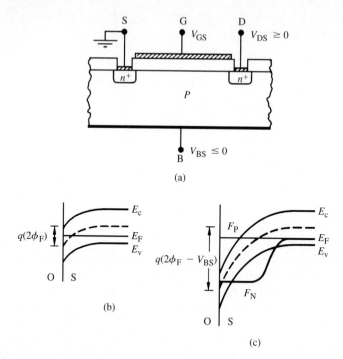

**Figure 18.20**    The back-biased MOSFET. (a) Cross-sectional view indicating the double sub-scripted voltage variables used in the analysis. Also shown are the semiconductor energy band diagrams corresponding to the onset of inversion when (b) $V_{BS} = 0$ and (c) $V_{BS} < 0$.

### 18.3.4 Back Biasing

Reverse biasing the back contact or bulk of a MOS transistor relative to the source is another method that has been employed to adjust the threshold potential. This electrical method of adjustment, which predates ion implantation, makes use of the so-called *body effect* or substrate-bias effect.

To explain the effect, let us consider the *n*-channel MOSFET shown in Fig. 18.20(a). If the back-to-source potential difference ($V_{BS}$) is zero, inversion occurs of course when the voltage drop across the semiconductor ($\phi_S$) equals $2\phi_F$ as pictured in Fig. 18.20(b). If $V_{BS} < 0$, the semiconductor still attempts to invert when $\phi_S$ reaches $2\phi_F$. However, with $V_{BS} < 0$ any inversion-layer carriers that do appear at the semiconductor surface migrate laterally into the source and drain because these regions are at a lower potential. Not until $\phi_S = 2\phi_F - V_{BS}$, as pictured in Fig. 18.20(c), will the surface invert and normal transistor action begin. In essence, back biasing changes the inversion point in the semiconductor from $2\phi_F$ to $2\phi_F - V_{BS}$. The ideal device threshold potential given by Eq. (18.22) is in turn modified to

$$V'_{\text{GB}|\text{at threshold}} = 2\phi_{\text{F}} - V_{\text{BS}} \pm \frac{K_{\text{S}}}{K_{\text{O}}} x_{\text{o}} \sqrt{\frac{2qN_{\text{B}}}{K_{\text{S}}\varepsilon_0} (\pm 2\phi_{\text{F}} \mp V_{\text{BS}})}$$

$$(+), \quad V_{\text{BS}} < 0 \text{ for } n\text{-channel}$$
$$(-), \quad V_{\text{BS}} > 0 \text{ for } p\text{-channel} \qquad (18.26)$$

or, since $V'_{\text{GB}|\text{at threshold}} = V'_{\text{GS}|\text{at threshold}} - V_{\text{BS}}$,

$$V'_{\text{GS}|\text{at threshold}} = 2\phi_{\text{F}} \pm \frac{K_{\text{S}}}{K_{\text{O}}} x_{\text{o}} \sqrt{\frac{2qN_{\text{B}}}{K_{\text{S}}\varepsilon_0} (\pm 2\phi_{\text{F}} \mp V_{\text{BS}})} \qquad (18.27)$$

Finally, introducing $\Delta V'_{\text{T}} \equiv (V'_{\text{GS}|\text{at threshold}} - V'_{\text{T}})$, we obtain

$$\Delta V'_{\text{T}} = (V'_{\text{T}} - 2\phi_{\text{F}}) \left[ \sqrt{1 - \frac{V_{\text{BS}}}{2\phi_{\text{F}}}} - 1 \right] \quad \begin{array}{l} \phi_{\text{F}} > 0, \ V_{\text{BS}} < 0 \text{ for } n\text{-channel} \\ \phi_{\text{F}} < 0, \ V_{\text{BS}} > 0 \text{ for } p\text{-channel} \end{array} \quad (18.28)$$

Having established Eq. (18.28), we make the following observations concerning back biasing or the body effect: (1) Back biasing always increases the magnitude of the ideal device threshold voltage. It therefore makes the $p$-channel threshold of actual devices more negative and the $n$-channel threshold more positive—it cannot be used to reduce the negative threshold of a $p$-channel MOSFET. (2) The current–voltage relationships developed in Chapter 17 are still valid when $V_{\text{BS}} \neq 0$ provided $2\phi_{\text{F}} \rightarrow 2\phi_{\text{F}} - V_{\text{BS}}$, $V_{\text{G}} \rightarrow V_{\text{GS}}$, $V_{\text{D}} \rightarrow V_{\text{DS}}$, and $V_{\text{T}}$ is interpreted as $V_{\text{GS}|\text{at threshold}}$. (3) Care must be exercised in describing back-biased structures to properly identify voltage differences through the use of double-subscripted voltage variables.

### 18.3.5 Threshold Summary

All nonidealities shift the MOSFET threshold voltage; interfacial traps in addition reduce the low-frequency $g_{\text{m}}$ of the transistor. A two-step procedure is followed in computing the expected threshold voltage of a real MOSFET. The flat-band voltage is first deduced from known information about device nonidealities. Subsequently $V_{\text{FB}}$ is added to the threshold voltage of an ideal version of the given MOSFET. A transistor that is normally "off" when $V_{\text{G}} = 0$ is referred to as an enhancement mode device; a depletion mode MOSFET is "on" or conducting when $V_{\text{G}} = 0$. Because of residual nonidealities, real $n$-channel MOSFETs are typically depletion-mode devices. The attainment of $n$-channel enhancement mode MOSFETs, and threshold adjustment in general, is now accomplished through the use of ion implantation. Biasing the back contact relative to the source has also been employed to adjust the threshold voltage.

## Exercise 18.6

**P:** The $C_G-V_G(V_D = 0)$ characteristic derived from an $n$-channel MOSFET is pictured in Fig. E18.6.

(a) What is the threshold voltage, $V_T$, of the transistor? Explain how you arrived at your answer.

(b) Is the MOSFET a depletion mode or an enhancement mode device? Explain.

(c) Sketch the general form of the $I_D-V_D$ characteristics expected from the device, specifically labeling those characteristics corresponding to $V_G = -2, -1, 0, 1,$ and 2 V.

(d) Given $\phi_{MS} = -1$ V, $Q_{IT} = 0$, and the fact that the device is stable under bias-temperature stressing, how do you explain the observed flat band voltage of $-2.2$ V?

(e) If the MOSFET is doped such that $\phi_F = 0.3$ V, what substrate bias ($V_{BS}$) must be applied to attain a $V_{GS|\text{at threshold}} = 1$ V?

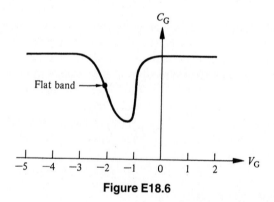

**Figure E18.6**

**S:** (a) $\boxed{V_T = -1\text{ V}}$. In progressing toward positive biases, the $C$–$V$ curve increases toward $C = C_O$ at roughly the inversion–depletion transition point. $V_G$ at the inversion–depletion transition point is the threshold voltage in MOSFETs.

(b) $\boxed{\text{Depletion Mode}}$. With $V_T = -1$ V, the MOSFET is clearly "on" or conducting when $V_G = 0$.

(c) The general form of the $I_D$–$V_D$ characteristics expected from the device is sketched below.

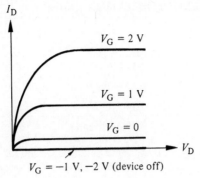

$$V_G = -1 \text{ V}, -2 \text{ V (device off)}$$

(d) Since the device is stable under bias-temperature stressing, one concludes that $Q_M = 0$. With $Q_M = 0$, $Q_{IT} = 0$, and $\phi_{MS} = -1$ V unable to account for the entire flat-band shift of $-2.2$ V, the residual shifting can only be attributed to the $\boxed{\text{fixed charge}}$.

(e) If Eq. (18.28) is solved for $V_{BS}$, one obtains

$$V_{BS} = 2\phi_F \left[ 1 - \left( 1 + \frac{\Delta V_T'}{V_T' - 2\phi_F} \right)^2 \right]$$

Noting

$$V_{GS|\text{at threshold}} - V_T = V_{GS|\text{at threshold}}' - V_T' = \Delta V_T' = 2 \text{ V}$$

and

$$V_T' = V_T - V_{FB} = -1.0 \text{ V} + 2.2 \text{ V} = 1.2 \text{ V}$$

direct substitution into the $V_{BS}$ espression then yields

$$V_{BS} = 0.6 \left[ 1 - \left( 1 + \frac{2}{1.2 - 0.6} \right)^2 \right] = -10.7 \text{ V}$$

## PROBLEMS

| Problem | Complete After | Difficulty Level | Suggested Point Weighting | Short Description |
|---|---|---|---|---|
| 18.1 | 18.3.5 | 1 | 10 (1 each part) | Quick quiz |
| 18.2 | 18.1 | 2 | 6 (2 each part) | Polysilicon-gate MOS-C |
| ● 18.3 | " | 2 | 10 | $\phi_{MS}$ versus doping plot |
| 18.4 | 18.2.2 | 2 | 8 (4 each part) | Mobile ion calculation |
| 18.5 | 18.2.3 | 3 | 10 (a-1, b-6, c-3) | Fixed charge calculation |
| 18.6 | " | 2 | 10 (5 each part) | $\phi_{MS}$, $Q_F$ from $V_{FB}$–$x_o$ plot |
| 18.7 | 18.2.4 | 2 | 8 (4 each part) | Infer $Q$-info from $E$-band |
| 18.8 | " | 3–4 | 14 (a-2, b-6, c-6) | Deduce $Q_F$ from $C$–$V$ data |
| 18.9 | " | 3 | 10 | Sketch pre-annealed $C$–$V$ |
| 18.10 | " | 2–3 | 8 | Dangling bond calculation |
| 18.11 | " | 3 | 10 (a-3, b-2, c-3, d-2) | Single-level interfacial trap |
| 18.12 | 18.2.5 | 3 | 10 | Half-irradiated MOS-C |
| 18.13 | 18.3.1 | 2 | 6 (a-2, b-4) | BT-stressed MOSFET |
| 18.14 | 18.3.3 | 2–3 | 10 (2 each) | Modification to $V_T$ |
| 18.15 | " | 2 | 8 (a-4, b-3, c-1) | Compute MOSFET $V_{FB}$, $V_T$ |
| 18.16 | " | 2 | 8 | Compute required $N_I$ |

**18.1 Quick Quiz**

Answer the following questions as concisely as possible.

(a) What is the difference between $\Phi_M - \chi$ and $\Phi_M' - \chi'$?

(b) What is involved in subjecting a MOS-C or MOSFET to a bias-temperature stress?

(c) What is believed to be the physical origin of the fixed oxide charge in MOS structures?

(d) The interfacial trap density ($D_{IT}$) is observed to depend on the silicon surface orientation. Describe the observed orientation dependence.

(e) What is the net effect of ionizing radiation on MOS structures?

(f) When performing bias-temperature stress experiments, how does one distinguish between the negative-bias instability and the voltage instability arising from alkali ions?

(g) Under what circumstances would $V_T \neq V_T' + V_{FB}$?

(h) Explain what is meant by the term "depletion mode" transistor.

(i) What is the difference between the "field-oxide" and the "gate-oxide" in MOSFETs?

(j) Precisely what is the "body effect"?

**18.2** Consider a polysilicon-gate MOS-C where $E_F - E_c = 0.2$ eV in the heavily doped gate and $E_F - E_c = -0.2$ eV in the nondegenerately doped silicon substrate. Assume the structure to be ideal (other than an obvious $\phi_{MS} \neq 0$) and $\chi'$(polysilicon) $= \chi'$(crystalline-Si).

(a) Sketch the energy band diagram for the polysilicon-gate MOS-C under flat-band conditions.

(b) What is the "metal"–semiconductor workfunction difference for the cited polysilicon-gate MOS-C?

(c) Will the given MOS-C be accumulation, depletion, or inversion biased when $V_G = 0$? Explain.

● **18.3** With the $\Phi_M' - \chi'$ of the MOS system as an input parameter, write a computer program that will output a plot similar to Fig. 18.3. Use your program to confirm the Al and $n^+$ poly plots in Fig. 18.3. Also output a plot appropriate for MOS-Cs with $p^+$ polysilicon gates doped such that $E_F = E_v$.

**18.4** (a) An MOS-C is found to possess a uniform distribution of sodium ions in the oxide; that is, $\rho_{ion}(x) = \rho_0 = $ constant for all $x$ in the oxide. Compute the $\Delta V_G$ shift resulting from this distribution of ions if $\rho_0/q = 10^{18}/\text{cm}^3$ and $x_o = 0.1\ \mu\text{m}$.

(b) After positive bias-temperature (+BT) stressing, the sodium ions of part (a) all pile up immediately adjacent to the oxide–semiconductor interface. Determine the $\Delta V_G$ after +BT stressing.

**18.5** In modeling the quantitative effect of the fixed charge, it is typically assumed the charge is located immediately adjacent to the oxide–semiconductor interface. Suppose the charge is actually distributed a short distance into the oxide from the Si–SiO$_2$ interface.

(a) For reference purposes, write down the standard result for $\Delta V_G$(fixed charge).

(b) Determine the expected $\Delta V_G$ shift caused by an equivalent amount of charge distributed in a linearly increasing fashion from zero at a distance $\Delta x$ from the Si–SiO$_2$ interface to $2Q_F/\Delta x$ right at the Si–SiO$_2$ interface.

(c) Compute $\Delta V_G$(part b)$/\Delta V_G$(part a) assuming $\Delta x = 10$ Å $= 10^{-7}$cm and $x_o = 0.1\ \mu\text{m} = 10^{-5}$cm. Repeat for $x_o = 0.01\ \mu\text{m} = 10^{-6}$cm. Comment on your results.

**18.6** It is possible to determine $\phi_{MS}$ and $Q_F$ for a given fabrication process and metal–SiO$_2$–Si system by fabricating a set of MOS-Cs with different oxide thicknesses. The flat-band voltage ($V_{FB}$) for each device in the set is measured and plotted as a function of $x_o$. The fabrication process is assumed to yield ideal devices other than $\phi_{MS} \neq 0$ and $Q_F \neq 0$.

(a) Indicate how $\phi_{MS}$ and $Q_F$ can be deduced from the $V_{FB}$ versus $x_o$ plot.

(b) Given the $V_{FB}$ versus $x_o$ data tabulated below, determine $\phi_{MS}$ and $Q_F/q$.

| $x_o$ ($\mu$m) | $V_{FB}$ (V) |
|---|---|
| 0.1 | −0.91 |
| 0.15 | −1.04 |
| 0.2 | −1.2 |
| 0.25 | −1.33 |
| 0.3 | −1.52 |

**18.7** An MOS-C is characterized by the energy band diagram in Fig. P18.7. Assume that the interfacial trap density in the structure is negligible.

(a) Is $Q_M = 0$ or $Q_M \neq 0$ in the oxide? Explain.

(b) Is $Q_F = 0$ or $Q_F \neq 0$ at the Si–SiO$_2$ interface? Explain.

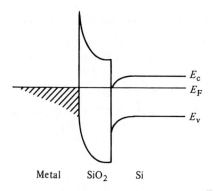

Metal    SiO$_2$    Si

**Figure P18.7**

**18.8** The $C–V$ characteristic exhibited by an Al–SiO$_2$–Si capacitor maintained at $T = 300$ K is reproduced in Fig. P18.8. Note that $C_{MAX} = 200$ pF, $C_{MIN} = 67$ pF, and the flat-band voltage $V_{FB} = -0.71$ V. There are no mobile ions in the oxide ($Q_M = 0$), the interfacial trap density is negligible ($Q_{IT} = 0$), and $A_G = 2.9 \times 10^{-3}$ cm$^2$.

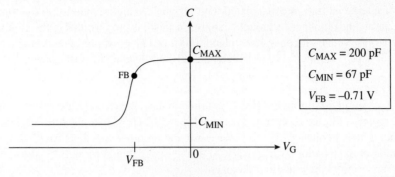

**Figure P18.8**

(a) Adding to a copy of Fig. P18.8 and using a dashed line, sketch the $C-V$ curve one would expect from an *ideal* version of the given MOS-C. Also indicate the flat-band point on the ideal-device characteristic.

(b) Ascertain the Si doping concentration. (An approximate determination using the delta-depletion $C-V$ formulation is acceptable.)

(c) Determine the fixed charge/cm$^2$, $Q_F$, in the given device.

**18.9** The $C-V$ characteristics of an MOS-C are measured before and after performing a postmetallization anneal to eliminate interfacial traps. The AFTER $C-V$ characteristic exhibited by the device is traced in Fig. P18.9. Assuming *acceptor-like* interfacial traps, sketch the $C-V$ characteristic of the device BEFORE annealing on the same set of coordinates. Explain how you arrived at your answer.

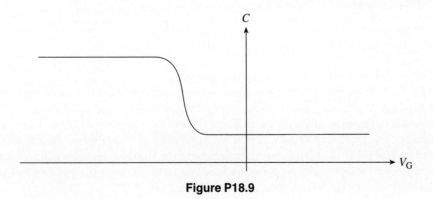

**Figure P18.9**

**18.10** If interfacial traps are associated with residual "dangling bonds" at the Si surface, and assuming the number of residual "dangling bonds" is proportional to the number of Si surface atoms, which silicon surface plane, (100) or (110), would be expected to exhibit the higher density of interfacial traps? Record all work leading to your answer.

**18.11** A rather unusual $n$-bulk MOS-C is found to have interfacial traps at only one band-gap energy, $E_{IT}$, located right at midgap (see Fig. P18.11). Answer the following questions assuming a high-frequency $C$–$V$ measurement and an otherwise ideal MOS-C. Include a few words of explanation, if necessary, to convey your thought process and to prevent a misinterpretation of the requested sketches.

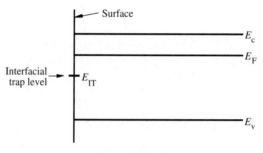

**Figure P18.11**

(a) Sketch the general form of the expected MOS-C $C$–$V$ characteristic if the interface states giving rise to the $E_{IT}$ level are donor-like in nature. (Assume the number of states is sufficiently large to perturb the ideal-device characteristic.)

(b) Repeat part (a) if the interface states are acceptor-like in nature.

(c) Repeat part (a) assuming the interface states are donor-like but the energy level is located very close to the conduction band (say $E_c - E_{IT} = 0.001$ eV).

(d) Repeat part (a) for donor-like interfacial traps that introduce an energy level very close to the valence band (say $E_{IT} - E_v = 0.001$ eV).

**18.12** The as-fabricated $C$–$V$ characteristic derived from an MOS-C is pictured in Fig. P18.12. After fabrication, *precisely one-half* of the gate area is exposed to ionizing radiation, which increases the apparent fixed charge ($Q_F$) in the exposed half of the device. The increase in interfacial trap density caused by the irradiation is negligible. Sketch the expected form of the $C$–$V$ characteristic after irradiation on the same set of coordinates as the pre-irradiation characteristic. Also explain how you arrived at your answer. *HINT*: Think of the two halves of the MOS-C as separate capacitors.

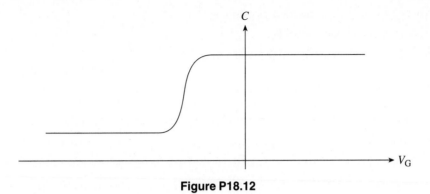

**Figure P18.12**

**18.13** The before bias-temperature stressing $I_D$–$V_D$ characteristics of a MOSFET are sketched in Fig. P18.13(a). The $g_d$–$V_G$ ($V_D = 0$) characteristics of the same device before and after bias-temperature stressing using a positive bias are shown in Fig. P18.13(b).

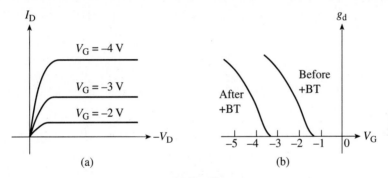

**Figure P18.13**

(a) What was the probable cause of the shift in the $g_d$–$V_G$ characteristic after positive bias-temperature stressing?

(b) Sketch the $I_D$–$V_D$ characteristics *after* bias-temperature stressing for gate voltages of $V_G = -2, -3$, and $-4$ V. Also record your reasoning.

**18.14** An $n$-channel ($p$-bulk) MOSFET is ideal except $\phi_{MS} \neq 0$. Indicate how the following modifications to the structure, taken independently, will affect the threshold voltage $V_T$. Include a few words of explanation in each case.

(a) Ionizing radiation causes an apparent $Q_F \neq 0$.

(b) The gate material is changed from Al to Cu.

(c) There is an increase in the substrate doping.

(d) The oxide thickness is decreased.

(e) Boron ions are implanted into the near surface region of the silicon.

**18.15** A MOSFET is fabricated with $\phi_{MS} = -0.46$ V, $Q_F/q = 2 \times 10^{11}/cm^2$, $Q_M = 0$, $Q_{IT} = 0$, and $Q_I/q = -4 \times 10^{11}/cm^2$, $x_o = 0.05$ $\mu$m, $A_G = 10^{-3}$ cm$^2$, and $N_D = 10^{15}/$cm$^3$. $Q_I$ is the ion charge implanted immediately adjacent to the Si–SiO$_2$ interface.

(a) Determine $V_{FB}$.

(b) Determine $V_T$.

(c) Is the given MOSFET an enhancement mode or depletion mode device? Explain.

**18.16** Given an Al–SiO$_2$–Si MOSFET, $T = 300$ K, a Si-substrate doping of $N_A = 10^{17}/$cm$^3$, $x_o = 100$ Å, $Q_F/q = 10^{11}/cm^2$, no interfacial traps and no mobile ions in the oxide. Determine the boron ions/cm$^2$ ($N_I$) that must be implanted into the structure to achieve a $V_T = 0.5$ V. (Assume the implanted ions create an added negative charge right at the Si–SiO$_2$ interface.)

# 19 Modern FET Structures

To achieve higher operating speeds and increased packing densities, FET device structures have been subjected to greater and greater miniaturization. The decrease in FET device dimensions, in itself, can lead to major modifications in the observed device characteristics. Small dimension effects, also referred to as short-channel effects or small geometry effects, include, for example, shifts in the threshold voltage and an increase in the subthreshold current. The cited modifications in device behavior are of major importance in practical applications. Notably, an accurate prediction of the threshold voltage is needed to determine logic levels, noise margins, speed, and node voltages, while the subthreshold current affects the off-state power dissipation, dynamic logic clock speeds, and memory refresh times. A large portion of this chapter is devoted to the description and discussion of small dimension effects. It should be understood from the outset that small dimension effects are generally undesirable and are minimized or avoided in commercial structures through the proper scaling of device dimensions or modifications in device design. Relative to modifications in device design, the chapter concludes with a brief survey of select implemented and developmental FET structures.

## 19.1 SMALL DIMENSION EFFECTS

### 19.1.1 Introduction

In 1965 the smallest MOSFETs had an $L \sim 1$ mil $= 25$ $\mu$m. By 1990 industry-standard versions of MOS device structures had attained submicron dimensions. The steady progression toward smaller and smaller FETs is pictured schematically in Fig. 19.1. Projections into the near future are summarized in Table 19.1.

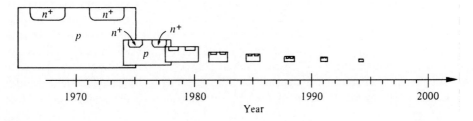

**Figure 19.1** The "shrinking" MOSFET. The relative scaling of the MOSFET channel length in the figure roughly depicts the decrease in minimum feature size of production-line MOS DRAMs.

**Table 19.1**    Silicon IC Projections[26]. (DRAM is an acronym for dynamic random access memory. I/O is an abbreviation for input/output. Wiring levels refer to the number of metallization levels that interconnect devices on an IC.)

| YEAR of first DRAM shipment | | 1995 | 1998 | 2001 | 2004 | 2007 | 2010 |
|---|---|---|---|---|---|---|---|
| Minimum feature size ($\mu$m) | | 0.35 | 0.25 | 0.18 | 0.13 | 0.10 | 0.07 |
| Chip size (mm$^2$) | • Microprocessor | 250 | 300 | 360 | 430 | 520 | 620 |
| | • DRAM | 190 | 280 | 420 | 640 | 960 | 1400 |
| Bits/chip in DRAMs | | 64M | 256M | 1G | 4G | 16G | 64G |
| Si wafer diameter (mm) | | 200 | 200 | 300 | 300 | 400 | 400 |
| Wiring levels (maximum in on-chip logic) | | 4-5 | 5 | 5-6 | 6 | 6-7 | 7-8 |
| I/O pads per chip (*high performance*) | | 900 | 1350 | 2000 | 2600 | 3600 | 4800 |
| Power supply voltage (V) | • Desktop | 3.3 | 2.5 | 1.8 | 1.5 | 1.2 | 0.9 |
| | • Portable | 2.5 | 1.8-2.5 | 0.9-1.8 | 0.9 | 0.9 | 0.9 |
| Chip frequency (MHz) *high performance* | • On-chip | 300 | 450 | 600 | 800 | 1000 | 1100 |
| | • Chip-to-board | 150 | 200 | 250 | 300 | 375 | 475 |

The departure from long-channel behavior, which can accompany the noted decrease in device dimensions, is nicely illustrated using observed $I_D$–$V_D$ characteristics. The *onset* of short-channel effects is heralded by a significant upward slant in the post-pinch-off portions of the $I_D$–$V_D$ curves. *Severe* short-channel effects lead to characteristics of the form reproduced in Fig. 19.2. Not only do the $I_D$–$V_D$ characteristics fail to saturate, but $I_D \propto V_D^2$ curves are observed for gate voltages below threshold. (The Fig. 19.2 device should be "off" for $V_G > 0$ V!) Another clear manifestation of short-channel effects is provided by the subthreshold transfer characteristics. A sample long-channel subthreshold characteristic was presented in Fig. 17.11. In long-channel MOSFETs the subthreshold drain current varies exponentially with $V_G$ and is independent of $V_D$ provided $V_D >$ few $kT/q$ volts. In short-channel devices, on the other hand, the subthreshold drain current is found to increase systematically and significantly with increasing $V_D$. Shifts in the observed threshold voltage constitute the third widely quoted indication of small-dimension effects. As is readily confirmed by referring to the $V_T$ relationship in Subsection 18.3.1, the threshold voltage in a long-channel MOSFET is independent of the gate length and width. In short-channel devices, $V_T$ becomes a function of the gate dimensions and the applied biases (see Fig. 19.3).

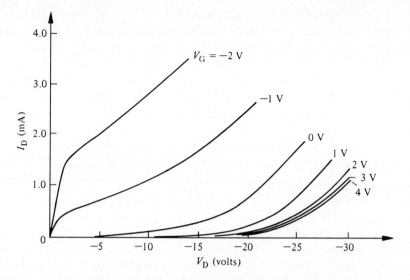

**Figure 19.2** $I_D$–$V_D$ characteristics of a MOSFET exhibiting severe short-channel effects. (Reprinted from Bateman et al.[27], © 1974, with kind permission of Elsevier Science Ltd.)

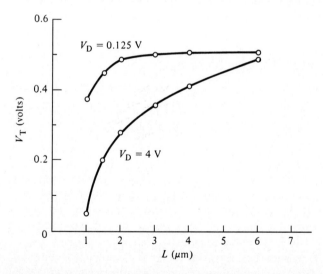

**Figure 19.3** Observed threshold-voltage variation with channel length and applied drain bias in short-channel MOSFETs. $N_A = 8 \times 10^{15}$/cm$^3$, $x_o = 0.028$ $\mu$m, and $r_j = 1$ $\mu$m. (Adapted from Fichtner and Potzl[28].)

The majority of small-dimension effects in MOSFETs are associated with the reduction in the channel length $L$. It is therefore reasonable to introduce and specify a minimum channel length, $L_{min}$, below which significant short-channel effects are expected to occur. Crudely speaking, $L_{min}$ must be greater than the sum of the depletion widths associated with the source and drain junctions. Values in the 0.1 $\mu$m to 1 $\mu$m range are indicated. As suggested by computer simulations and confirmed by experimental observations, a more precise estimate of $L_{min}$ is given by the empirical relationship[29]

$$L_{min} = 0.4[r_j x_o(W_S + W_D)^2]^{1/3} \quad \ldots x_o \text{ in Å}; L_{min}, r_j, \quad (19.1)$$
$$W_S, W_D \text{ in } \mu m$$

$r_j$ is the source/drain junction depth, $x_o$ the oxide thickness, $W_S$ the depletion width at the source junction, and $W_D$ the depletion width at the drain junction. Note from Eq. (19.1) that $L_{min}$ can be made smaller by reducing the depth of the source/drain islands, by reducing the oxide thickness, and/or by increasing the substrate doping, which in turn causes a decrease in $W_S$ and $W_D$. All of the above have in fact been employed to help assure long-channel operation of MOSFETs with increasingly scaled-down dimensions.

The causes underlying departures from long-channel behavior fall into one of three general categories. For one, differences between experiment and long-channel theory may simply arise from a breakdown of assumptions used in the long-channel analysis. Second, a reduction in device dimensions automatically leads to an enhancement of certain effects that are known to occur but are negligible in long-channel devices. Lastly, some departures from long-channel behavior arise from totally new phenomena. All three categories are represented in the following consideration of specific-case effects.

## 19.1.2 Threshold Voltage Modification

### Short Channel

In enhancement-mode short-channel devices $|V_T|$ is found to monotonically decrease with decreasing channel length $L$. Qualitatively, the decrease in threshold voltage can be explained as follows: Before an inversion layer or channel forms beneath the gate, the subgate region must first be depleted ($W \rightarrow W_T$). In a short-channel device the source and drain assist in depleting the region under the gate; that is, a significant portion of the subgate deletion-region charge is balanced by the charge on the other side of the source and drain $pn$ junctions. Thus less gate charge is required to reach the start of inversion and $|V_T|$ decreases. The smaller $L$, the greater the percentage of charge balanced by the source/drain $pn$ junctions, and the greater the reduction in $|V_T|$.

A first-order quantitative expression for the $\Delta V_T$ associated with short-channel effects can be established using straightforward geometric arguments. Although highly simplified, the derivation to be presented is very informative in that it illustrates the general method of analysis. The derivation also indicates how parameters such as the source/drain junction depth enter into the specification of short-channel effects.

As previously established, for an ideal device

$$V_G = \phi_S + \frac{K_S}{K_O} x_o \mathscr{E}_S \tag{19.2}$$

Let $Q_S$ be the total charge/cm$^2$ inside the semiconductor. Applying Gauss' law, it is readily established that

$$Q_S = -K_S \varepsilon_0 \mathscr{E}_S \tag{19.3}$$

Combining Eqs. (19.2) and (19.3), we can therefore write

$$V_G = \phi_S - \frac{x_o Q_S}{K_O \varepsilon_0} = \phi_S - \frac{Q_S}{C_o} \tag{19.4}$$

When $V_G = V_T$, $\phi_S = 2\phi_F$ and $Q_S = Q_B$, where $Q_B$ is the bulk or depletion-region charge per unit area of the gate. Specialized to the threshold point, Eq. (19.4) thus becomes

$$V_T = 2\phi_F - \frac{Q_B}{C_o} \tag{19.5}$$

Next introducing

$$\Delta V_T \equiv V_T(\text{short channel}) - V_T(\text{long channel}) \tag{19.6}$$

taking $Q_{BL}$ and $Q_{BS}$ to be the long-channel and short-channel depletion-region charges/cm$^2$, respectively, and making use of Eq. (19.5), we obtain

$$\Delta V_T = -\frac{1}{C_o}(Q_{BS} - Q_{BL}) = \frac{Q_{BL}}{C_o}\left(1 - \frac{Q_{BS}}{Q_{BL}}\right) \tag{19.7}$$

To complete the derivation, it is necessary to develop expressions for $Q_{BL}$ and $Q_{BS}$ in terms of the device parameters. Working toward this end, consider the short $n$-channel MOSFET pictured in Fig. 19.4. To simplify the analysis, $V_D$ is taken to be small or zero so that $W \cong W_T$ at all points beneath the central portion of the gate. The shaded areas in the figure identify those portions of the subgate region that are assumed to be controlled by the source and drain $pn$ junctions. In a long-channel device, charge in the entire rectangular region of side-length $L$ is balanced by charges on the gate and

$$Q_{BL} = -\frac{qN_A(ZLW_T)}{ZL} = -qN_A W_T \tag{19.8}$$

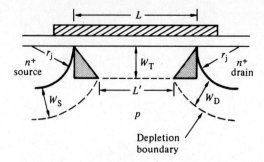

**Figure 19.4**  Cross section of a MOSFET identifying parameters that enter into the short-channel analysis. The shaded portions of the subgate area are assumed to be controlled by the source and drain *pn* junctions. It is also assumed that $V_D \cong 0$.

$ZLW_T$ is the depletion-region volume and $ZL$ the area of the gate. In a short-channel device, the depletion-region charge controlled by the gate is confined to the trapezoidal region of side lengths $L$ and $L'$. Thus

$$
Q_{BS} = -\frac{qN_A\left[\frac{1}{2}(L + L')ZW_T\right]}{ZL} = -qN_A W_T \frac{L + L'}{2L}
\tag{19.9}
$$

Substituting the $Q_{BL}$ and $Q_{BS}$ expressions into Eq. (19.7) then gives

$$
\Delta V_T = -\frac{qN_A W_T}{C_o}\left(1 - \frac{L + L'}{2L}\right)
\tag{19.10}
$$

Next, looking at the source region in Fig. 19.4 and assuming $W_S \cong W_T$, one deduces from geometrical considerations

$$
(r_j + W_T)^2 = \left(r_j + \frac{L - L'}{2}\right)^2 + W_T^2
\tag{19.11}
$$

which can be solved to obtain

$$
L' = L - 2r_j\left[\sqrt{1 + \frac{2W_T}{r_j}} - 1\right]
\tag{19.12}
$$

Finally, eliminating $L'$ in Eq. (19.10) using Eq. (19.12), we conclude

$$\Delta V_T(\text{short channel}) = -\frac{qN_AW_T}{C_o}\frac{r_j}{L}\left(\sqrt{1 + \frac{2W_T}{r_j}} - 1\right) \qquad (19.13)$$

Although a first-order result, the $\Delta V_T$ given by Eq. (19.13) does exhibit the same parametric dependencies initially noted in the $L_{\min}$ discussion. Examining $\Delta V_T/V_T$ (long-channel), which is the relevant quantity in gauging the importance of short-channel effects, one again finds the effects are minimized by reducing $x_o$, reducing $r_j$, and increasing $N_A$.

## Narrow Width

The threshold voltage is also affected when the lateral width $Z$ of a MOSFET becomes comparable to the channel depletion-width $W_T$. In enhancement-mode narrow-width devices, $|V_T|$ is found to monotonically *increase* with decreasing channel width $Z$. Note that the $Z$-dependence of the threshold voltage shift is opposite to the $L$-dependence. The narrow-width effect, however, is explained in much the same manner as the short-channel effect. Referring to the side view of a MOSFET in Fig. 19.5, note that the gate-controlled depletion region extends to the side, lying in part outside the $Z$-width of the gate. In wide-width devices, the gate-controlled charge in the lateral region is totally negligible. In narrow-width devices, on the other hand, the lateral charge becomes comparable to the charge directly beneath the $Z$-width of the gate; that is, there is an increase in the effective charge/cm$^2$ being balanced by the gate charge. Thus, added gate charge is required to reach the start of inversion and $|V_T|$ increases.

Paralleling the short-channel derivation, a quantitative expression for $\Delta V_T$ associated with the narrow-width effect is readily established. If the lateral regions are assumed to be quarter-cylinders of radius $W_T$, the lateral volume controlled by the gate is $(\pi/2)W_T^2L$ and

$$Q_B(\text{narrow width}) = -\frac{qN_A\left(ZLW_T + \dfrac{\pi}{2}W_T^2L\right)}{ZL} = -qN_AW_T\left(1 + \frac{\pi}{2}\frac{W_T}{Z}\right) \qquad (19.14)$$

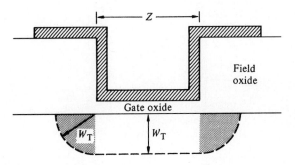

**Figure 19.5** Side view of a MOSFET used to explain and analyze the narrow-width effect.

Replacing $Q_{BS}$ in Eq. (19.7) with the narrow-width $Q_B$, one rapidly concludes

$$\Delta V_T(\text{narrow width}) = \frac{q N_A W_T}{C_o} \frac{\pi W_T}{2Z} \tag{19.15}$$

This result confirms our initial assertion that the narrow-width effect becomes important when $Z$ is comparable to $W_T$.

As a concluding point it should be noted that a combined-effect $\Delta V_T$ must be established for MOSFETs that have both a short channel and a narrow width. The short-channel and narrow-width $\Delta V_T$'s are not simply additive. The combined-effect $\Delta V_T$ and more exacting $\Delta V_T$ expressions for the individual effects can be found in the device literature.[30]

### 19.1.3 Parasitic BJT Action

Containing an oppositely doped region between the source and the drain, the MOSFET bears a striking physical resemblance to a lateral bipolar junction transistor (BJT). Thus, with the distance between the source and drain in a modern MOSFET reduced to a value comparable to the base width in a bipolar transistor, it is not surprising that phenomena have been observed that are normally associated with the operation of BJTs.

One such phenomenon is source to drain *punch-through*. When the source and drain are separated by a few microns or less it becomes possible for the *pn* junction depletion regions around the source and drain to touch or punch-through as pictured in Fig. 19.6. When punch-through occurs, a significant change takes place in the operation of the MOSFET. Notably, the gate loses control of the subgate region except for a small portion of the region immediately adjacent to the Si–SiO$_2$ interface. The source-to-drain current is then no longer constrained to the surface channel, but begins to flow beneath the surface through the touching depletion regions. Analogous to the punch-through current in BJTs, this subsurface "space-charge" current varies as the square of the voltage applied between

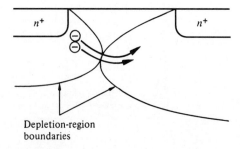

Depletion-region
boundaries

**Figure 19.6**  Punch-through and space-charge current in a short-channel MOSFET.

the source and drain. The $V_G > 0$ characteristics presented in Fig. 19.2 are examples of the $I_D \propto V_D^2$ current that results from source to drain punch-through.

As a practical matter, punch-through in small-dimension MOSFETs is routinely suppressed by increasing the doping of the subgate region and thereby decreasing the source/drain depletion widths. This can be accomplished by increasing the substrate doping. However, increasing the substrate doping has the adverse effect of increasing parasitic capacitances. Consequently, it is common practice to perform a deep-ion implantation to selectively increase the doping of the subgate region.

Parasitic BJT action involving *carrier multiplication and regenerative feedback* leads to a second potentially significant perturbation of the MOSFET characteristics. There is a certain amount of carrier multiplication in the high-field depletion region near the drain in all MOSFETs. In long-channel devices the multiplication is negligible. In short-channel devices, however, carrier multiplication coupled with regenerative feedback can dramatically increase the drain current and place a reduced limit on the maximum operating $V_D$. Catastrophic failure may even occur under extreme conditions.

The multiplication and feedback mechanism operating in small-dimension MOSFETs is very similar to that which lowers $V_{(BR)CEO}$ in a BJT relative to $V_{(BR)CB0}$ (see Subsection 11.2.4). The basic mechanism is best described and explained with the aid of Fig. 19.7. The process is initiated by channel current entering the high-field region in the vicinity of the drain. Upon acceleration in the high-field region, a small percentage of the channel carriers gain a sufficient amount of energy to produce electron–hole pairs through impact ionization. For an *n*-channel device the added electrons drift into the drain, while the added holes are swept into the quasineutral bulk. Because the semiconductor bulk has a finite resistance, the current flow associated with the impact-generated holes gives rise to a potential drop between the depletion region edge and the back contact. This potential drop is

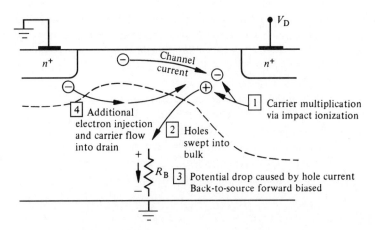

**Figure 19.7**  Visualization of carrier multiplication and regenerative feedback that can give rise to current enhancement in short-channel MOSFETs.

of such a polarity as to foward bias the source *pn* junction. Forward biasing of the source *pn* junction in turn leads to electron injection from the source *pn* junction into the quasineutral bulk, an additional electron flow into the drain, increased carrier multiplication, and so forth. The process is stable as long as the fractional increase in the drain current or the multiplication factor is less than $1/\alpha$, where $\alpha$ is the common-base gain of the parasitic BJT. At high enhanced currents there is the potential for excessive current flow through the device and device failure.

### 19.1.4 Hot-Carrier Effects

#### Oxide Charging

Oxide charging, or charge injection and trapping in the oxide, is a phenomenon that occurs in all MOSFETs. In the vicinity of the drain under operational conditions, channel carriers, and carriers entering the drain depletion region from the substrate, periodically gain a sufficient amount of energy to surmount the $Si–SiO_2$ surface barrier and enter the oxide. Neutral centers in the oxide trap a portion of the injected charge and thereby cause a charge buildup within the oxide. In long-channel devices, oxide charging is the well-known cause of "walk-out"—a progressive increase in the drain breakdown voltage of MOSFETs operated at large $V_D$ biases. Unfortunately, the effects of oxide charging in short-channel MOSFETs are decidedly more serious. This is true because a larger percentage of the gated region is affected in the smaller devices. Specifically, significant changes in $V_T$ and $g_m$ can result from the oxide-charging phenomenon. Moreover, because oxide charging is cumulative over time, the phenomenon tends to limit the useful "life" of a device and must be minimized. A popular approach for minimizing hot-carrier effects, the formation of a lightly doped drain (LDD), is described in Subsection 19.2.1.

#### Velocity Saturation

In the conventional analysis of the long-channel MOSFET, there is no theoretical limitation on the velocity that the carriers can attain in the surface channel. It is implicitly assumed the carrier velocities increase as needed to support the computed current. In reality, the carrier drift velocities inside silicon at $T = 300$ K approach a maximum value of $v_{sat} \cong 10^7$ cm/sec when the accelerating electric field exceeds $\sim 3 \times 10^4$ V/cm for electrons and $\sim 10^5$ V/cm for holes (see Fig. 3.4 in Part I). If, for example, $V_D = 2$ V and $L = 0.5$ $\mu$m, there will obviously be points in the MOSFET surface channel where the accelerating electric field is greater than or equal to $4 \times 10^4$ V/cm. Limitation of the channel current due to velocity saturation is clearly a possibility in short-channel devices.

Velocity saturation has two main effects on the observed characteristics. First, $I_{Dsat}$ is significantly reduced. The modified $I_{Dsat}$ is approximately described by

$$I_{Dsat} \cong ZC_o(V_G - V_T)v_{sat} \tag{19.16}$$

Second, as can be inferred from Eq. (19.16) and as illustrated in Fig. 19.8, the saturation current exhibits an almost linear dependence on $V_G$–$V_T$ as opposed to the conventional square-law dependence.

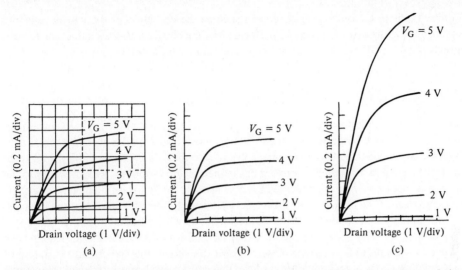

**Figure 19.8**  Illustration of the effects of velocity saturation on the MOSFET $I_D$–$V_D$ characteristics. (a) Experimental characteristics derived from a short-channel MOSFET with $L = 2.7$ $\mu$m, $x_o = 0.05$ $\mu$m, $r_j = 0.4$ $\mu$m, and $N_A$ (substrate) $\cong 10^{15}$/cm$^3$. Comparative theoretical characteristics computed (b) including velocity saturation and (c) ignoring velocity saturation. (From Yamaguchi[31], © 1979 IEEE.)

## Velocity Overshoot / Ballistic Transport

In the Chapter 17 modeling of carrier drift in the MOSFET surface channel, we implicitly assumed that carriers experienced numerous scattering events in traveling between the source and the drain. This is equivalent to assuming the channel length ($L$) is much greater than the average distance ($l$) between scattering events. Clearly, if the MOSFET channel length is reduced to a value comparable to $l$, fundamental revisions will be required in the analytical formalism. More importantly, however, if even smaller-dimension structures could be built with $L < l$, a large percentage of the carriers would then travel from the source to the drain without experiencing a single scattering event. The envisioned projectile-like motion of the carriers is referred to as *ballistic transport*.

Observable ballistic effects are theoretically possible in GaAs structures where $L \lesssim$ 0.3 $\mu$m. Somewhat shorter lengths are required in Si devices. Experimentally, both Si and GaAs FETs with channel lengths ~0.1 $\mu$m have been fabricated in research laboratories. Thus, FET structures with $L \lesssim l$ are attainable with present-day technology. Moreover, referring to Table 19.1, such structures are expected to become commonplace in the near future.

Practically speaking, ballistic transport is of interest because it could lead to super-fast devices. With reduced scattering, the average velocity of carriers transversing the channel can exceed $v_{sat}$. This is referred to as *velocity overshoot*. Average velocities up to 35% larger than the saturation velocity have been observed in an $L = 0.12$ $\mu$m MOSFET[32].

Admittedly there are other considerations, such as the operation of the carrier injecting source, that can limit ballistic device performance. Nonetheless, ballistic effects have been observed and are likely to play a role in the operation of future FETs.

## 19.2  SELECT STRUCTURE SURVEY

In discussing the essentials of MOSFET operation, we made use of the basic enhancement-mode structure. Use of the basic structure allows one to focus on the development of concepts and the understanding of phenomena. Naturally, given the maturity of MOSFET technology, there exists a significant number of distinct device variations. Modifications to the basic structure have been implemented to solve specific problems or to enhance a specific device characteristic. It is also true that closely related FET devices fabricated in GaAs and other compound semiconductors invariably take a somewhat different form than those fabricated in Si. A brief survey of select MOSFET and MOSFET-like structures has been included in this section to provide some feel for the variety that exists and the nature of major modifications. It should be emphasized that the surveyed device structures constitute only a sampling, with a bias toward structures likely to be encountered in the recent FET literature.

### 19.2.1 MOSFET Structures

#### LDD Transistors

As described in the preceding section, reduced dimension devices are more susceptible to hot-carrier effects. The field-aided injection and subsequent trapping of carriers in the gate oxide near the drain can lead to serious device degradation. The degradation effects are further worsened by the common practice of using bias voltages that have not been scaled down in proportion to the device dimensions. The lightly doped drain (LDD) structure shown in Fig. 19.9 helps to minimize hot-carrier effects. The feature of note is the lightly doped drain region between the end of the channel and the drain proper (the $n^-$ region in Fig. 19.9). The reduced doping gradient in going from the channel to the drain proper lowers the $\mathcal{E}$-field in the vicinity of the drain and shifts the position of the peak $\mathcal{E}$-field toward the end of the channel. Carrier injection into the oxide is thereby reduced and oxide charging correspondingly minimized.

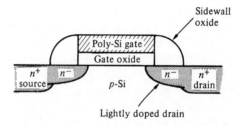

**Figure 19.9**  Cross section of a lightly doped drain (LDD) structure.

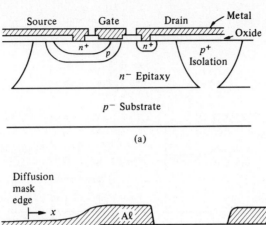

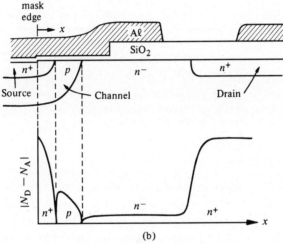

**Figure 19.10** (a) Cross section of a DMOS structure. (b) Magnified cross section of the channel region and lateral doping profile. (From Pocha et al.[33], © 1974 IEEE.)

## DMOS

A double-diffused MOSFET (DMOS) structure is pictured in Fig. 19.10. The structure is distinctive in that the channel region is formed, and the channel length specified, by the difference in the lateral extent of two impurity profiles. A $p$-type dopant (e.g., boron) and an $n$-type dopant (e.g., phosphorus) are admitted and diffused into the Si through the same oxide mask opening. The $p$-type dopant, which is introduced first, diffuses slightly deeper and farther to the side than the $n$-type dopant. The result is the simultaneous formation of the source and channel regions as clearly shown in the magnified cross section of Fig. 19.10(b). The most important physical characteristic of DMOS is a short channel length ($\sim 1$ $\mu$m) that can be established without using small-dimension lithographic masks. The DMOS structure thus boasts high-frequency operation, which is combined with a high drain breakdown voltage. It has been used in high-frequency analog applications and high-voltage/high-power circuits. Although first introduced in the early 1970s, variations of the DMOS structure, notably power-DMOS structures, continue to be developed.

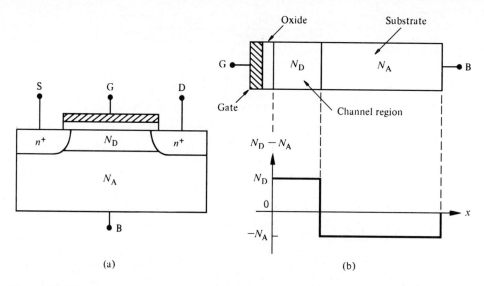

**Figure 19.11** (a) Cross section of a buried-channel MOSFET. (b) Approximate subgate doping profile. (Adapted from Van der Tol and Chamberlain[34], © 1989 IEEE.)

## Buried-Channel MOSFET

Figure 19.11 shows the cross section of a buried-channel MOSFET and the approximate subgate-doping profile inside the transistor. The unique structural feature, a surface layer beneath the gate with the same doping as the source and drain islands, is typically formed by ion implantation. With a *pn* junction bottom gate and an MOS top gate, the buried-channel MOSFET is physically and functionally a hybrid J-FET/MOSFET structure. It can be designed to function as a depletion-mode or enhancement-mode device depending on the thickness and doping of the surface layer. The buried-channel MOSFET gets its name from the fact that channel conduction can be made to take place away from the oxide-semiconductor interface. This leads to inherently higher carrier mobilities, a reduced interfacial trap interaction, and a decreased sensitivity to hot-carrier effects.

## SiGe Devices

With the development of the ultrahigh-vacuum chemical-vapor-deposition (UHV/CVD) process[35], it has become possible to deposit high-quality $Si_{1-x}Ge_x$ alloy films on a production-line basis. The process permits atomic-level control of the film thickness, precise control of the film composition, and minimization of background contamination. Films covering the entire range from 0 to 100% Ge content are readily deposited.

Reflecting the fact that the Ge lattice constant is $a = 5.65$ Å compared to the smaller $a = 5.43$ Å of silicon, SiGe alloys preferably exhibit a lattice constant larger than Si. However, if sufficiently thin (typically $\leq 1000$ Å), SiGe alloy films deposited on a Si substrate are *pseudomorphic* in nature; that is, they conform atom-by-atom to the lattice pattern

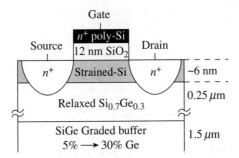

**Figure 19.12**   SiGe enhanced-mobility MOSFET. The *n*-channel MOSFET incorporates a strained-Si surface channel. (From Welser et al.[36], © 1994 IEEE.)

of the underlying Si substrate. Si heterojunction bipolar transistors (HBTs) fabricated with a pseudomorphic SiGe base layer have achieved the highest operating frequency of any Si bipolar device.

Because of the constrained lattice spacing, pseudomorphic films are subject to a considerable amount of internal stress. Growing a SiGe alloy film beyond a critical thickness causes defects to nucleate spontaneously. The defect generation relieves the stress and allows the film to relax to its preferred lattice constant. Relaxed SiGe films are of particular interest in the present context because they have been used in fabricating "enhanced mobility" MOSFETs as illustrated in Fig. 19.12.

In fabricating the Fig. 19.12 device a compositionally graded SiGe buffer layer, a relaxed $Si_{0.7}Ge_{0.3}$ layer, and a thin pseudomorphic Si film are sequentially grown on a Si substrate. With strain-relief defects primarily confined to the buffer layer, the relaxed $Si_{0.7}Ge_{0.3}$ layer effectively functions as a quality substrate for the pseudomorphic layer. Conforming to the larger lattice constant of $Si_{0.7}Ge_{0.3}$ places the pseudomorphic Si film under tension. The strain in the Si layer in turn gives rise to reduced carrier scattering and a lower electron effective mass parallel to the $Si–SiO_2$ interface. These combine to enhance the carrier mobility in the Si surface channel. The pictured device, for example, exhibited a factor of $\sim 2$ enhancement in the observed low-field mobility over a standard MOSFET. The enhanced mobility, translating into increased current drive at higher fields, has the potential to extend the performance limits of existing MOS technology.

## SOI Structures

The term *silicon-on-insulator* (SOI) is used to describe structures where devices are fabricated in single-crystal Si layers formed *over* an insulating film or substrate. The first realization of SOI structures involved crystalline films of Si deposited on properly oriented sapphire substrates (SOS). Subsequently, laser annealing techniques were used to crystallize amorphous Si films deposited on insulators such as $SiO_2$ and $Si_3N_4$. In both of these approaches the Si film quality has impeded their widespread utilization. A more recent approach, epitaxial layer overgrowth (ELO), has solved the film quality problem. In ELO

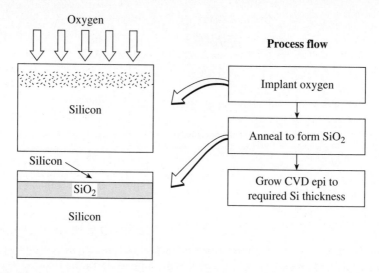

**Figure 19.13**  Basic steps in the production of a SIMOX wafer. (From Hostack et al.[37] Reprinted with permission.)

a hole is opened in the insulator to expose a small section of the underlying crystalline-Si substrate. With the exposed substrate acting as a seed, epitaxial Si is then grown up through the hole and sideways over the insulator. Unfortunately, ELO is not considered to be a production-line technology. Among all the SOI technologies only SIMOX and BESOI are currently considered production-line technologies that are capable of yielding Si films of acceptable quality. Both SIMOX and BESOI wafers are available commercially.

In BESOI (*BondEd* SOI) a $SiO_2$ layer, which is to become the insulating layer, is thermally grown on one Si wafer. A second Si wafer is next bonded to the top of the first wafer and annealed. Finally, the top wafer is ground, polished, and chemically etched until the desired surface-layer thickness is attained. In SIMOX (*Separation by IMplantation of OXygen*) the SOI structure is formed by creating an $SiO_2$ layer just beneath the surface of a bulk-Si wafer as illustrated in Fig. 19.13. The SIMOX process begins with the ion-implantation of oxygen beneath the Si surface. The $O_2$ implant is normally performed employing an accelerating energy of 150–200 keV and involves a total dose of $1-2 \times 10^{18}$ atoms/cm$^2$. Annealing the structure at an elevated temperature (typically 1300°C for roughly 6 hours) next allows the implanted $O_2$ to react with the Si to form a buried oxide (BOX) layer. The anneal also minimizes defects in the surface Si layer. Additional Si is sometimes deposited by standard epitaxial techniques to increase the final thickness of the Si surface layer. Required thicknesses range from 500 Å to 2000 Å for CMOS applications and from 0.3 $\mu$m to 10 $\mu$m for thick-film bipolar applications.

The numerous projected advantages of SOI-based ICs, particularly MOS ICs, help to explain the extensive research effort expended to develop a viable SOI technology. The dielectric isolation of individual circuit elements possible with SOI reduces parasitic capacitances, thereby giving rise to higher operating speeds, and totally eliminates latch-up[†]. SOI-based MOSFETs are also projected to consume less power, operate at higher temperatures, provide improved radiation hardness, and exhibit reduced short-channel effects. From a design standpoint, SOI permits higher packing densities and easier scaling to smaller linewidths. From a fabrication standpoint SOI leads to simpler processing because it permits a reduction in the number of masking (lithographic) steps. Perhaps most importantly, the wafers derived from SIMOX and BESOI are totally compatible with production-line equipment and procedures. The most extensive use of SOI to date is in the production of radiation-hardened SRAMs for military applications. Expanded use is anticipated in the areas of very high speed very large scale CMOS ICs, low-power supply-voltage devices, and DRAMS.[38]

## 19.2.2 MODFET (HEMT)

The last member of the FET family to be surveyed is the modulation doped field effect transistor (MODFET). Although MODFET appears to be the preferred acronym for the device structure, it is often alternatively identified as the high electron mobility transistor (HEMT). When first introduced, the structure was sometimes called the selectively doped heterostructure transistor (SDHT) or the two-dimensional electron-gas field effect transistor (TEGFET). A perspective view of an AlGaAs/GaAs MODFET was presented in Fig. 15.7 as part of the general introduction to field-effect devices; a simplified cross section of the structure is shown in Fig. 19.14(a).

The structural similarities of the MODFET and MOSFET are obvious from Fig. 19.14(a) if one thinks of the AlGaAs layer as an insulator. The MODFET does differ from the MOSFET in that the "insulating" layer, the AlGaAs, is doped. Moreover, the GaAs epilayer is nominally undoped. The term "modulation doping" arises from the modulation of the dopant source to selectively dope only the AlGaAs layer during the sequential deposition of the GaAs and AlGaAs layers. Doping of the AlGaAs helps induce a surface channel adjacent to the AlGaAs/GaAs interface. Because the current-carrying GaAs layer is nominally undoped, there is only minimal scattering from residual (unintentional) dopant impurities in the GaAs surface channel. Very high electron mobilities are observed at room temperature, and an even greater mobility enhancement relative to doped-channel devices like the MESFET is obtained at liquid nitrogen temperatures. This is why the structure is alternatively identified as the "high electron mobility" transistor.[‡]

Greater detail of the gated region and the electrostatic situation inside the MODFET are pictured in Fig. 19.14(b). AlGaAs has a wider band gap than GaAs and, as previously illustrated in Fig. 11.19(a), energy band offsets arise at the AlGaAs/GaAs interface. In the

---

[†] Latch-up is the term for a major problem in digital CMOS circuits where the circuitry gets stuck in a specific logic state. Simply stated, latch-up is caused by an internal feedback mechanism associated with parasitic PNPN-like action.

[‡] Although the HEMT acronym continues to be employed, it is now recognized that the low-field mobility is not the key parameter in determining the characteristics of the device under normal operating conditions. Rather, analogous to MESFET operation described in Section 15.3, it is the high-field velocity that becomes the critical parameter when large electric fields exist along the FET channel.

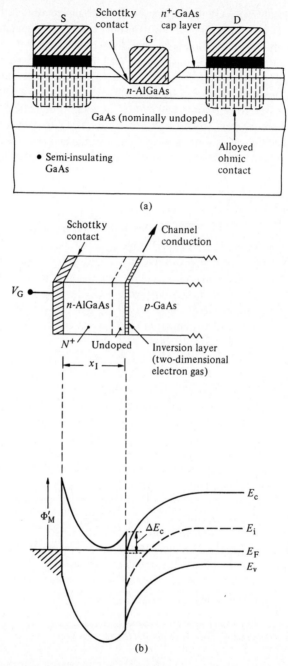

**Figure 19.14**  The basic MODFET or HEMT structure. (a) Simplified cross section. (b) Section through the transistor beneath the gate and the associated energy band diagram. [Part (b) from Pierret and Lundstrom[39], © 1984 IEEE.]

MODFET the conduction band offset ($\Delta E_c$) becomes an electron containment barrier. As already noted, the AlGaAs is doped to induce an inversion or accumulation layer of electrons at the GaAs surface. (The conducting electron layer is usually referred to as a two-dimensional electron gas in the MODFET literature.) The AlGaAs layer is also made sufficiently thin so that it is totally depleted under equilibrium conditions by the built-in potential associated with the Schottky gate contact. The resulting uncompensated dopant ions in the AlGaAs give rise to an energy band diagram very similar to that of a MOSFET with sodium ions distributed throughout the oxide (compare Figs. 19.14b and P18.7). We should note that, because AlGaAs/GaAs is a lattice-matched system, the interface between the two materials is essentially free of fixed charges and interfacial traps, a prerequisite of proper device operation.

Although exhibiting acceptable electrical characteristics similar to those of other FETs at low current levels, a material-related property of the AlGaAs/GaAs MODFET leads to a degradation of device performance at moderate and high current levels. Specifically, above a certain carrier density, the relatively small conduction band discontinuity at the AlGaAs/GaAs interface permits channel charge to spill over into the AlGaAs. Consequently, the carrier density in the channel tends to saturate as a function of the gate potential. Device performance is thereby degraded. Seeking a solution to the problem has led to the development of two second-generation MODFET structures. In one structure a pseudomorphic $In_x Ga_{1-x}As$ layer is positioned between the AlGaAs and GaAs. The addition of In to GaAs reduces the semiconductor band gap and increases the $\Delta E_c$ containment barrier. The larger $\Delta E_c$ in turn permits higher carrier densities to be induced in the InGaAs surface channel. The cross section of a pseudomorphic MODFET or PHEMT sold commercially by Hewlett-Packard is reproduced in Fig. 19.15. The other second-generation MODFET features an alternative lattice-matched system—namely, the $Al_{0.48}In_{0.52}As/In_{0.53}Ga_{0.47}As/$ InP system, where InP is the substrate material. The $\Delta E_c$ at the AlInAs/InGaAs interface is approximately two times the conduction band discontinuity of the AlGaAs/GaAs system. Even though the AlInAs/InGaAs MODFET technology is relatively immature, it already

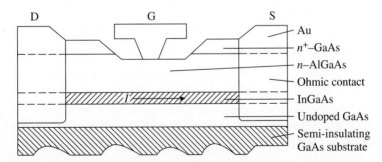

**Figure 19.15**  Cross section of a commercially available pseudomorphic MODFET or PHEMT. (From the 1993 Hewlett-Packard catalogue of communication components.[40] Reproduced courtesy of Hewlett Packard Co., Components Group.)

boasts the highest current-gain cutoff frequency (>250 GHz) yet reported for a three-terminal device.

Relative to high-frequency operation, note the mushroom or "T" shape of the gate in Fig. 19.15. At the routinely submicron gate-lengths found in MODFETs, the effect of parasitic resistances often masks intrinsic device performance. While maintaining a small "footprint," the T shape of the gate metallization increases its current-carrying cross-sectional area and thus reduces its resistance.

At present, the MODFET is gradually replacing GaAs MESFETs in many advanced commercial and military systems that demand ultra-high performance. It is an ideal candidate for applications involving low-noise amplification at microwave and millimeter wave frequencies. Although originally conceived as a ultra-high-speed digital logic device, the compound-semiconductor MODFET is now considered unlikely to replace Si devices in the near future.

## PROBLEMS

| CHAPTER 19 PROBLEM INFORMATION TABLE | | | | |
| --- | --- | --- | --- | --- |
| Problem | Complete After | Difficulty Level | Suggested Point Weighting | Short Description |
| 19.1 | 19.2.2 | 1 | 15 (1 each part) | Quick quiz |
| 19.2 | 19.2.2 | 1 | 5 (1/2 each answer) | Acronyms and abbreviations |
| 19.3 | 19.1.2 | 2–3 | 12 (a-5, b-5, c-2) | $L_{min}$, $\Delta V_T$ calculations |
| 19.4 | 19.2.1 | 1 | 10 (2 each part) | Identify unique features |
| 19.5 | 19.2.2 | 2–3 | 30 (5 each part) | Summarize journal articles |

**19.1** Quick Quiz
Answer the following questions as concisely as possible.

(a) Small-dimension effects are generally undesirable and are minimized or avoided in commercial MOSFET structures through the proper scaling of device dimensions or modifications in device design. True or false?

(b) Name the three most commonly cited indications of short-channel effects.

(c) Relative to threshold-voltage modification, how do the short-channel and narrow-width effects differ? In what ways are they alike?

(d) For a fixed channel length and channel width, name two other MOSFET parameters that can be adjusted to minimize small-dimension effects. Also indicate how the parameters must be adjusted to reduce small-dimension effects.

(e) Name two BJT-like phenomena that have been observed in short-channel MOSFETs.

(f) Describe what happens inside a MOSFET affected by the hot-carrier phenomenon known as *oxide charging*.

(g) Why is oxide charging associated with hot-carrier effects more important in short-channel devices?

(h) How can you tell whether velocity saturation is affecting the $I_D$–$V_D$ characteristics of a MOSFET?

(i) Indicate what is meant by *ballistic transport*.

(j) Indicate what is meant by *velocity overshoot*.

(k) Describe a popular approach for minimizing the oxide charging problem.

(l) What exactly is a *pseudomorphic* film?

(m) Cite two examples of pseudomorphic films in device structures.

(n) Briefly summarize the SIMOX process.

(o) What is the difference between a MODFET and a HEMT?

**19.2** Acronyms and Abbreviations
Indicate what the following acronyms and abbreviations stand for: LDD, DMOS, SOI, SOS, ELO, SIMOX, BESOI, BOX, MODFET, and PHEMT.

**19.3** (a) Making use of the parameters supplied in the figure caption, compute the expected $L_{min}$ for the Fig. 19.3 device when $V_D = 0.125$ V. Assume $n^+$-$p$ drain and source step junctions and that the source and back are grounded. Comment on your computational result.

(b) Utilizing Eq. (19.13), what is the $\Delta V_T$ expected for the Fig. 19.3 device when $L = 1$ $\mu$m and $V_D = 0.125$ V?

(c) Can Eq. (19.13) be applied to compute an expected $\Delta V_T$ for comparison with the $V_D = 4$ V data? Explain.

**19.4** Briefly indicate the unique physical feature of the following transistor structures:

(a) LDD transistors

(b) DMOS

(c) Buried-channel MOSFET

(d) Enhanced-mobility (strained-Si) MOSFET

(e) SOI structures

**19.5** Read the following journal articles and prepare a one- to two-page summary of each article.

(a) J. J. Sanchez, K. K. Hsueh, and T. A. DeMassa, "Drain-Engineered Hot-Electron Resistant Device Structures: A Review," IEEE Trans. on Electron Devices, **36**, 1125 (June 1989).

(b) M. J. Van der Tol and S. G. Chamberlain, "Potential and Electron Distribution Model for the Buried-Channel MOSFET," IEEE Trans. on Electron Devices, **36**, 670 (April 1989).

(c) B. S. Meyerson, "UHV/CVD Growth of Si and Si:Ge Alloys: Chemistry, Physics, and Device Applications," Proc. IEEE, **80**, 1592 (Oct. 1992).

(d) B. S. Meyerson, "Ultrahigh-Vacuum CVD Process Makes SiGe Devices," Solid State Technology, **37**, 53 (Feb. 1994).

(e) L. Peters, "SOI Takes Over Where Silicon Leaves Off," Semiconductor International, **16**, 48 (March 1993).

(f) L. D. Nguyen, L. E. Larson, and U. K. Mishra, "Ultra-High-Speed Modulation-Doped Field-Effect Transistors: A Tutorial Review," Proc. IEEE, **80**, 494 (April 1992).

# R3 Part III
## SUPPLEMENT AND REVIEW

### ALTERNATIVE / SUPPLEMENTAL READING LIST

| Author(s) | Type (A–Alt., S–Supp.) | Level | Relevant Chapters |
|---|---|---|---|
| General References | | | |
| Neamen | A | Undergraduate | 11–13 |
| Singh | A | Undergraduate | 8, 9 |
| Yang | A/S | Undergraduate | 8–13 |
| For Selected Topics | | | |
| Ladbrooke | S | Advanced Undergrad to Professional | 5, 6 (MESFETs) 7 (MODFETs) |
| Pulfrey and Tarr | A/S | Undergraduate | 7 (MOSFETs) |
| Sah | S | Undergraduate to Graduate (variable) | 6 (MOSFETs) |
| Schroder | S | Graduate | 3, 4 (CCDs) 6 (Modern FET) |

(1) P. H. Ladbrooke, *MMIC Design: GaAs FETs and HEMTs,* Artech House, Boston, © 1989.

(2) D. A. Neamen, *Semiconductor Physics and Devices, Basic Principles,* Irwin, Homewood, IL, © 1992.

(3) D. L. Pulfrey and N. G. Tarr, *Introduction to Microelectronic Devices,* Prentice Hall, Englewood Cliffs, NJ, © 1989.

(4) C. T. Sah, *Fundamentals of Solid-State Electronics,* World Scientific, Singapore, © 1991.

**R3**

(5) D. K. Schroder, *Advanced MOS Devices,* Volume VII in the Modular Series on Solid State Devices, edited by G. W. Neudeck and R. F. Pierret, Addison-Wesley, Reading, MA, © 1987.

(6) J. Singh, *Semiconductor Devices, an Introduction,* McGraw-Hill, New York, © 1994.

(7) E. S. Yang, *Microelectronic Devices,* McGraw Hill, New York, © 1988.

## FIGURE SOURCES / CITED REFERENCES

(1) W. Shockley, "A Unipolar Field-Effect Transistor," Proceedings IRE, **40,** 1365 (Nov. 1952).

(2) G. C. Dacey and I. M. Ross, "Unipolar Field-Effect Transistor," Proceedings IRE, **41,** 970 (Aug. 1953).

(3) G. C. Dacey and I. M. Ross, "The Field-Effect Transistor," Bell System Technical Journal, **34,** 1149 (Nov. 1955).

(4) D. Kahng and M. M. Atalla, "Silicon-Silicon Dioxide Field Induced Surface Devices," presented at the IRE-AIEE Solid-State Device Research Conference, Carnegie Institute of Technology, Pittsburgh, PA, 1960.

(5) J. S. MacDougall, "Applications of the Silicon Planar II MOSFET," Application Bulletin, Fairchild Semiconductor, Nov. 1964.

(6) R. H. Dennard, "Field-Effect Transistor Memory," U.S. Patent 3 387 286, application filed July 14, 1967, granted June 4, 1968.

(7) P. Wyns and R. L. Anderson, "Low-Temperature Operation of Silicon Dynamic Random-Access Memories," IEEE Transactions on Electron Devices, **36,** 1423 (August 1989).

(8) C. T. Sah, "Evolution of the MOS Transistor—From Conception to VLSI," Proceedings IEEE, **76,** 1280 (Oct. 1988).

(9) M. F. Tompsett, G. F. Amelio, and G. E. Smith, "Charge Coupled 8-Bit Shift Register," Applied Physics Letters, **17,** 111 (1970). C. H. Sequin and M. F. Tompsett, *Charge Transfer Devices,* Advances in Electronics and Electron Physics, Supplement **8,** Academic Press, New York, © 1975.

(10) F. M. Wanlass and C. T. Sah, "Nanowatt Logic Using Field-Effect Metal-Oxide-Semiconductor Triodes," in Technical Digest of IEEE 1963 Int. Solid-State Circuit Conf., pp. 32–33, Feb. 20, 1963. F. M. Wanlass, "Low Stand-By Power Complementary Field-Effect Circuitry," U.S. Patent 3 356 858 filed June 18, 1963, issued Dec. 5, 1967.

(11) C. G. Fonstad, *Microelectronic Devices and Circuits,* McGraw-Hill, New York, © 1994.

(12) T. J. Drummond, W. T. Masselink, and H. Morkoç, "Modulation-Doped GaAs/(Al,Ga)As Heterojunction Field-Effect Transistors: MODFETs," Proceedings IEEE, **74,** 773 (June 1986).

(13) M. Kuhn, "A Quasi-Static Technique for MOS *C–V* and Surface State Measurements," Solid-State Electronics, **13,** 873 (1970).

(14) A. Berman and D. R. Kerr, "Inversion Charge Redistribution Model of High-Frequency MOS Capacitance," Solid-State Electronics, **17,** 735 (July 1974).

(15) W. E. Beadle, J. C. C. Tsai, and R. D. Plummer, *Quick Reference Manual for Silicon Integrated Circuit Technology,* Wiley, New York, © 1985.

(16) S. C. Sun and J. D. Plummer, "Electron Mobility in Inversion and Accumulation Layers on Thermally Oxidized Silicon Surfaces," IEEE Transactions on Electron Devices, **ED-27,** 1497 (August 1980).

(17) R. F. Pierret and J. A. Shields, "Simplified Long-Channel MOSFET Theory," Solid-State Electronics, **26,** 143 (1983).

(18) "Practical Applications of the 4145A Semiconductor Parameter Analyzer: DC Parameter Analysis of Semiconductor Devices," Hewlett-Packard Application Note 315, March 1982.

(19) S. Kar, "Determination of Si-Metal Work Function Differences by MOS Capacitance Technique," Solid-State Electronics, **18,** 169 (1975).

(20) B. E. Deal, "Standardized Terminology for Oxide Charges Associated with Thermally Oxidized Silicon," IEEE Transactions on Electron Devices, **ED-27,** 606 (March 1980).

(21) D. R. Kerr, J. S. Logan, P. J. Burkhardt, and W. A. Pliskin, "Stabilization of $SiO_2$ Passivation Layers with $P_2O_5$," IBM Journal of Research & Development, **8,** 376 (1964).

(22) B. E. Deal, M. Sklar, A. S. Grove, and E. H. Snow, "Characteristics of the Surface-State Charge ($Q_{SS}$) of Thermally Oxidized Silicon," Journal of the Electrochemical Society, **114,** 266 (1967).

(23) R. R. Razouk and B. E. Deal, "Dependence of Interface State Density on Silicon Thermal Oxidation Process Variables," Journal of the Electrochemical Society, **126,** 1573 (1979).

(24) B. E. Deal, "The Current Understanding of Charges in the Thermally Oxidized Silicon Structure," Journal of the Electrochemical Society, **121,** 198C (1974).

(25) J. R. Srour and J. M. McGarrity, "Radiation Effects on Microelectronics in Space," Proceedings IEEE, **76,** 1443 (1988).

(26) (a) P. Singer, "1995: Looking Down the Road to Quarter-Micron Production," Semiconductor International, **18,** 46 (Jan. 1995). (b) "Processes of the Future: Updated Roadmap Identifies Technical Strategic Challenges," Solid State Technology, **38,** 42 (Feb. 1995).

(27) I. M. Bateman, G. A. Armstrong, and J. A. Magowan, "Drain Voltage Limitations of MOS Transistors," Solid-State Electronics, **17,** 539 (June 1974).

(28) W. Fichtner and H. W. Potzl, "MOS Modelling by Analytical Approximations. I. Subthreshold Current and Threshold Voltage," International Journal of Electronics, **46,** 33 (1979).

(29) J. R. Brews, W. Fichtner, E. H. Nicollian, and S. M. Sze, "Generalized Guide for MOSFET Minimization," IEEE Electron Device Letters, **EDL-1,** 2 (1980).

(30) T. A. DeMassa and H. S. Chien, "Threshold Voltage of Small-Geometry Si MOS-FETs," Solid-State Electronics, **29,** 409 (1986).

(31) K. Yamaguchi, "Field-Dependent Mobility Model for Two-Dimensional Numerical Analysis of MOSFET's," IEEE Transactions on Electron Devices, **ED-26,** 1068 (July 1979).

(32) F. Assaderaghi, P. K. Ko, and C. Hu, "Observation of Velocity Overshoot in Silicon Inversion Layers," IEEE Electron Device Letters, **14,** 484 (Oct. 1993).

(33) M. D. Pocha, A. G. Gonzalez, and R. W. Dutton, "Threshold Voltage Controllability in Double-Diffused-MOS Transistors," IEEE Transactions on Electron Devices, **ED-21,** 778 (1974).

(34) M. J. Van der Tol and S. G. Chamberlain, "Potential and Electron Distribution Model for the Buried-Channel MOSFET," IEEE Transactions on Electron Devices, **36,** 670 (April 1989).

(35) B. S. Meyerson, "UHV/CVD Growth of Si and Si:Ge Alloys: Chemistry, Physics, and Device Applications," Proceedings IEEE, **80,** 1592 (Oct. 1992).

(36) J. Welser, J. L. Hoyt, and J. F. Gibbons, "Electron Mobility Enhancement in Strained-Si N-Type Metal-Oxide-Semiconductor Field-Effect Transistors," IEEE Electron Device Letters, **15,** 100 (March 1994).

(37) H. H. Hosack, T. W. Houston, and G. P. Pollack, "SIMOX Silicon-on-Insulator: Materials and Devices," Solid State Technology, **33,** 61 (Dec. 1990).

(38) L. Peters, "SOI Takes Over Where Silicon Leaves Off," Semiconductor International, **16,** 48 (March 1993).

(39) R. F. Pierret and M. S. Lundstrom, "Correspondence Between MOS and Modulation-Doped Structures," IEEE Transactions on Electron Devices, **ED-31,** 383 (March 1984).

(40) Hewlett-Packard Communications Components, GaAs and Silicon Products Designer's Catalog, © 1993.

**R3**

# REVIEW LIST OF TERMS

Defining the following terms using your own words provides a rapid review of the Part III material.

(1) field effect
(2) J-FET
(3) DRAM
(4) CCD
(5) MOSFET
(6) CMOS
(7) MODFET
(8) MESFET
(9) MOS-C
(10) source
(11) drain
(12) gate
(13) channel
(14) pinch-off
(15) $I_D-V_D$ saturation
(16) gradual channel approximation
(17) square-law relationship
(18) drain conductance
(19) transconductance
(20) MMIC
(21) D-MESFET
(22) E-MESFET
(23) long channel
(24) short channel
(25) MIS
(26) bulk
(27) block charge diagram
(28) accumulation
(29) flat band (MOS)
(30) depletion (MOS)
(31) inversion
(32) weak inversion
(33) strong inversion
(34) surface potential
(35) delta-depletion approximation
(36) oxide capacitance
(37) semiconductor capacitance
(38) quasistatic technique
(39) deep depletion
(40) total deep depletion

(41) $n$-channel, $p$-channel
(42) saturation (of MOSFET $I_D-V_D$)
(43) linear (triode) region
(44) threshold voltage
(45) effective mobility
(46) bulk charge
(47) subthreshold transfer characteristics
(48) overlap capacitance
(49) self-aligned gate
(50) flat-band voltage
(51) mobile ions
(52) fixed charge
(53) interfacial traps
(54) trapped charge
(55) bias-temperature (BT) stressing
(56) phosphorus stabilization
(57) chlorine neutralization
(58) oxidation triangle
(59) donor-like trap
(60) acceptor-like trap
(61) dangling bonds
(62) postmetallization annealing
(63) hydrogen-ambient annealing
(64) "hardened" oxides
(65) negative-bias instability
(66) enhancement mode MOSFET
(67) depletion mode MOSFET
(68) gate-oxide
(69) field-oxide
(70) body effect (substrate-bias effect)
(71) oxide charging
(72) ballistic transport
(73) velocity overshoot
(74) LDD
(75) DMOS
(76) pseudomorphic
(77) SOI
(78) SIMOX
(79) modulation-doped
(80) PHEMT

R3

## PART III—REVIEW PROBLEM SETS AND ANSWERS

The following problem sets were designed assuming a knowledge—at times an integrated knowledge—of the subject matter in Chapters 16–18 of Part III. The sets could serve as a review or as a means of evaluating the reader's mastery of the subject. Problem Set A was adapted from a one-hour "open-book" examination; Problem Set B was adapted from a one-hour "closed-book" examination. An answer key is included at the end of the problem sets.

### Problem Set A

### Problem A1

The energy band diagram for a $p$-Si/SiO$_2$/$n$-Si (SOS) capacitor under *flat-band* conditions is given below. To achieve the pictured state, there must of course be a non-zero voltage applied to the SOS-C gate. The SOS-C is ideal except for a non-zero workfunction difference. $T = 300$ K, $N_A$($p$-side) $= 10^{15}$/cm$^3$, $N_D$($n$-side) $= 10^{15}$/cm$^3$, $n_i = 10^{10}$/cm$^3$, $x_o = 5 \times 10^{-6}$ cm, and $A_G = 10^{-3}$ cm$^2$.

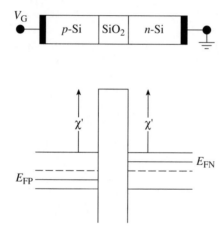

(a) What is the voltage being applied to the $p$-Si/SiO$_2$/$n$-Si SOS-C to achieve the pictured flat-band condition? Give both the polarity and magnitude of $V_G$.

(b) Sketch the energy band diagram and the associated block charge diagram for the SOS-C when a *large positive* gate voltage (say $V_G > 5$ V) is applied to the device. Add descriptive words to your sketches as necessary to forestall a misinterpretation of your answer.

(c) Sketch the energy band diagram and the associated block charge diagram for the SOS-C when a *large negative* gate voltage is applied to the device.

(d) Make a sketch of the *high-frequency* $C$–$V_G$ characteristic to be expected from the SOS-C described in this problem. Explain how you arrived at your $C$–$V_G$ sketch.

(e) Invoking the delta-depletion approximation, determine the *minimum* capacitance exhibited by the device? Give both a symbolic and numerical answer.

## Problem A2

Refer to Fig. 17.10 in Subsection 17.2.4. For the purposes of this problem, consider the characteristics plotted in the figure to be *experimental* characteristics of a MOSFET under test.

(a) Completing the figure below, carefully sketch the inversion layer and depletion region inside the given MOSFET corresponding to the biasing point of $V_G = V_D = 5$ V. Also insert the doping type of the source and drain islands.

(b) Invoking the square-law relationships, and making use only of the plotted characteristics (ignore the figure caption information), roughly determine the MOSFET threshold voltage ($V_T$). Explain how you arrived at your answer.

(c) Given only the parametric values in the figure caption (ignoring the characteristics themselves), what is the expected MOSFET threshold voltage ($V_T$)?

(d) Making use only of the plotted characteristics, estimate $g_d$ if the quiescent operating point of the MOSFET is $V_G = 5$ V, $V_D = 0$. Indicate how you arrived at your answer.

(e) Determine $g_m$ if the quiescent operating point of the MOSFET is $V_G = V_D = 5$ V. Indicate how you arrived at your answer.

## Problem A3

The three parts of this problem are similar. A pair of MOS-Cs are taken to be identical except for one physical difference. The physical difference causes a voltage displacement of the $C$–$V$ characteristics derived from the two devices as illustrated below. Defining $\delta V_G = V_{FB1} - V_{FB2}$, where $V_{FB1}$ and $V_{FB2}$ are the flat-band gate voltages of devices #1 and #2 respectively, your job in each case will be to determine $\delta V_G$.

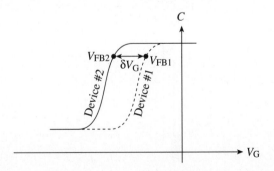

(a) Two MOS-Cs are identical in every way except device #1 has an aluminum gate while device #2 has a gate made of gold. Determine the expected value of $\delta V_G$. Record your work.

(b) Two MOS-Cs are identical in every way except device #1 is fabricated on a (100)-oriented Si surface and device #2 is fabricated on a (111)-oriented Si surface. In both devices the interfacial trap charge is negligible ($Q_{IT} = 0$). Also, both devices were annealed in $N_2$ after a dry $O_2$ oxidation to minimize the fixed charge. Assuming $C_o = 3 \times 10^{-8}$ farads/cm², determine (approximately) the expected value of $\delta V_G$. Record your work.

(c) Two MOS-Cs are identical in every way except the $Na^+$ ions in device #1 are all piled up right at the *metal–oxide* interface, while in device #2 the $Na^+$ ions are all piled up right at the *oxide–semiconductor* interface. Taking $C_o = 3 \times 10^{-8}$ farads/cm² and $Q_M/q = 5 \times 10^{11}$/cm², determine the expected value of $\delta V_G$. Record your work. NOTE: You are to assume δ-function type ionic distributions in working this problem.

## Problem Set B

### I.  MOS Fundamentals

A totally dimensioned energy band diagram for an M"O"S-C recently fabricated in a research laboratory is shown below. (The "O" is actually ZnSe and the semiconductor is GaAs.) The device is maintained at $T = 300$ K, $kT/q = 0.0259$ V, $n_i = 2.25 \times 10^6$/cm³, $K_S = 12.85$, $K_O = 9.0$, and $x_o = 0.1$ μm. It has also been established that $Q_M = 0$, $Q_F = 0$, and $Q_{IT} = 0$. Use the cited energy band diagram and the given information in answering Problems 1–10.

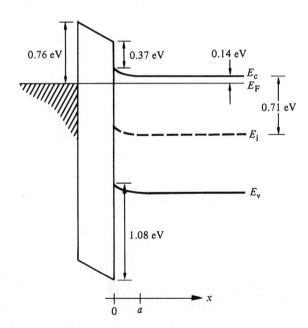

(1) Sketch the electrostatic potential ($\phi$) inside the semiconductor as a function of position. (Let $\phi = 0$ in the semiconductor bulk.)

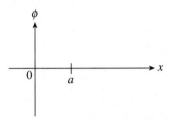

(2) Roughly sketch the electric field ($\mathscr{E}$) inside the semiconductor as a function of position.

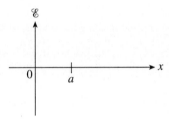

(3) Do equilibrium conditions prevail *inside the semiconductor*?

   (a) Yes

   (b) No

   (c) Can't be determined

(4) $N_D = ?$

   (a) $4.03 \times 10^{20}/\text{cm}^3$

   (b) $8.13 \times 10^{15}/\text{cm}^3$

   (c) $1.00 \times 10^{15}/\text{cm}^3$

   (d) $5.01 \times 10^{8}/\text{cm}^3$

(5) $V_G = ?$

   (a) $-0.57$ V

   (b) $-0.39$ V

   (c) $0$ V

   (d) $0.39$ V

   (e) $0.57$ V

R3

(6) For the pictured condition the M"O"S-C is

(a) Accumulated.

(b) Depleted.

(c) Inverted.

(d) Biased at the depletion–inversion transition point.

(7) What is the metal–semiconductor workfunction difference ($\phi_{MS}$)?

(a) −0.39 V

(b) −0.25 V

(c) 0 V

(d) 0.25 V

(e) 0.39 V

(8) What voltage must be applied to the gate to achieve flat-band conditions?

(a) −0.39 V

(b) −0.25 V

(c) 0 V

(d) 0.25 V

(e) 0.39 V

(9) Invoking the delta-depletion approximation, determine the normalized low-frequency small-signal capacitance, $C/C_O$, at the pictured bias point.

(a) 0.25

(b) 0.41

(c) 0.56

(d) 0.83

(10) As noted in the energy band figure, $a$ is the distance from the "oxide"–semiconductor interface to the quasineutral semiconductor bulk. Determine the length of $a$ at the pictured bias point.

(a) 0.112 $\mu$m

(b) 0.205 $\mu$m

(c) 0.428 $\mu$m

(d) 0.813 $\mu$m

R3

## II.  MOSFET

A standard MOSFET is fabricated with $\phi_{MS} = -0.89$ V, $Q_M = 0$, $Q_{IT} = 0$, $Q_F/q = 5 \times 10^{10}/cm^2$, $x_o = 500$ Å, $A_G = 10^{-3}$ cm², and $N_A = 10^{15}/cm^3$. Assume $T = 300$ K.

(11)  Determine the flat-band gate voltage, $V_{FB}$.

  (a)  $-2.05$ V

  (b)  $-1.01$ V

  (c)  $-0.89$ V

  (d)  0 V

(12)  Determine the gate voltage at the onset of inversion, $V_T$.

  (a)  $-1.01$ V

  (b)  $-0.21$ V

  (c)  0.80 V

  (d)  1.81 V

(13)  The given MOSFET is

  (a)  An enhancement-mode MOSFET.

  (b)  A depletion-mode MOSFET.

  (c)  A built-in channel MOSFET.

(14)  If the internal condition inside the MOSFET is as shown below to the left, identify the corresponding operational point on the $I_D$–$V_D$ characteristic at the right.

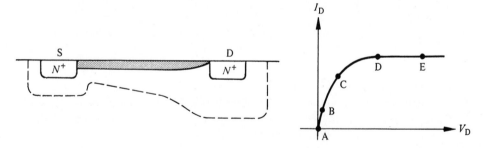

(15)  At $V_G - V_T = 3$ V and $V_D = 1$ V the MOSFET exhibits a drain current of $I_D = 2.5 \times 10^{-4}$ amp. Using the square-law formulation, determine the drain current if $V_G - V_T = 3$ V and $V_D = 4$ V.

  (a)  $3.5 \times 10^{-4}$ amp

  (b)  $4.0 \times 10^{-4}$ amp

  (c)  $4.5 \times 10^{-4}$ amp

  (d)  $1.0 \times 10^{-3}$ amp

**R3**

## III. True or False

(16) The "field effect" is the phenomenon where carriers are accelerated by an electric field impressed parallel to the surface of the semiconductor.

(a) True

(b) False

(17) The electron affinity ($\chi$) of a semiconductor is the difference in energy between the vacuum level and $E_c$ at the surface of the semiconductor.

(a) True

(b) False

(18) The "quasistatic technique" is employed in measuring the low-frequency MOS-C $C$–$V$ characteristics.

(a) True

(b) False

(19) The nonequilibrium condition where there is a deficit of minority carriers and a depletion width in excess of the equilibrium value is referred to as "deep inversion."

(a) True

(b) False

(20) The voltage shift due to mobile ions in the oxide is at a minimum when the ions are located midway between the gate and semiconductor.

(a) True

(b) False

(21) The interfacial trap charge ($Q_{IT}$) is typically a function of the applied gate voltage.

(a) True

(b) False

(22) The "bulk-charge" theory for the d.c. characteristics of a MOSFET derives its name from the fact that, in this theory, one properly accounts for changes in the "bulk" or depletion-region charge beneath the MOSFET channel.

(a) True

(b) False

(23) Let $g_d$ be the drain or channel conductance of a MOSFET. By definition, at low frequencies $g_d = \partial I_D / \partial V_G |_{V_D = \text{constant}}$.

(a) True

(b) False

R3

(24) The mobility of carriers in surface inversion layers or channels is typically lower than the bulk mobility of the same carriers because of the added scattering associated with the depletion region charge.

(a) True

(b) False

(25) In modern-day MOS structures the "M" in MOS is often heavily doped polycrystalline Si.

(a) True

(b) False

## Answers—Set A

### Problem A1

(a) $V_G = \dfrac{1}{q}(E_{FN} - E_{FP}) = \dfrac{1}{q}[(E_{FN} - E_i) + (E_i - E_{FP})] = \dfrac{kT}{q}[\ln(N_D/n_i) + \ln(N_A/n_i)]$

$= 2(0.0259)\ \ln(10^{15}/10^{10}) = \mathbf{0.596\ V}$

(b)

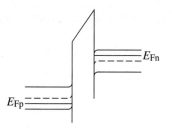
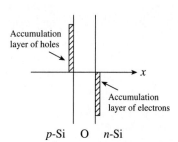

(c)

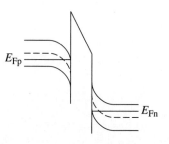
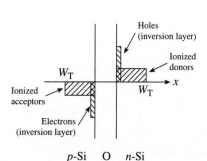

(d) When $V_G > 0$, both semiconductor components are accumulated. Thus $C$ approaches $C_O$ at large positive gate biases. When $V_G < 0$, the two semiconductors components first deplete and then invert. Inversion occurs at the same bias voltage for the two sides of the SOS-C because $N_A(p\text{-side}) = N_D(n\text{-side})$. The high-frequency capacitance is therefore expected to smoothly decrease to $C_{min}$ at large negative biases. As sketched below, the deduced characteristic should look very similar to a standard $n$-bulk high-frequency MOS-C $C\text{–}V$ curve.

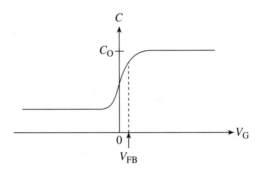

(e) Reflecting on the answers to previous parts of the problem, particularly part (c), we are led to model the SOS-C as three capacitors in series. In the equivalent circuit for the SOS-C shown below, $C_{Sp}$ and $C_{Sn}$ are respectively the $p$- and $n$-side semiconductor capacitance.

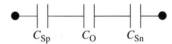

Because $N_A(p\text{-side}) = N_D(n\text{-side})$, $C_{Sp} = C_{Sn} = C_S$ and

$$\frac{1}{C} = \frac{1}{C_O} + \frac{1}{C_{Sp}} + \frac{1}{C_{Sn}} = \frac{1}{C_O} + \frac{2}{C_S}$$

or

$$C = \frac{C_O C_S}{C_S + 2C_O} = \frac{C_O}{1 + \dfrac{2C_O}{C_S}} = \frac{C_O}{1 + \dfrac{2K_O W}{K_S x_o}}$$

The minimum capacitance occurs under inversion biasing where $W = W_T$. Performing the indicated computations gives

$$\phi_F = \frac{kT}{q} \ln(N_A/n_i) = (0.0259) \ln(10^{15}/10^{10}) = 0.298 \text{ V}$$

$$W_T = \left[\frac{2K_S\varepsilon_0}{qN_A}(2\phi_F)\right]^{1/2} = \left[\frac{2(11.8)(8.85 \times 10^{-14})(0.596)}{(1.6 \times 10^{-19})(10^{15})}\right]^{1/2} = 8.82 \times 10^{-5} \text{ cm}$$

$$C_O = \frac{K_O\varepsilon_0A_G}{x_o} = \frac{(3.9)(8.85 \times 10^{-14})(10^{-3})}{(5 \times 10^{-6})} = 69.0 \text{ pF}$$

and

$$C_{min} = \frac{C_O}{1 + \dfrac{2K_OW_T}{K_Sx_o}} = \frac{69.0}{1 + \dfrac{2(3.9)(8.82 \times 10^{-5})}{(11.8)(5 \times 10^{-6})}} = \mathbf{5.45 \text{ pF}}$$

## Problem A2

(a) When $V_G = V_D = 5$ V the MOSFET is biased above pinch-off.

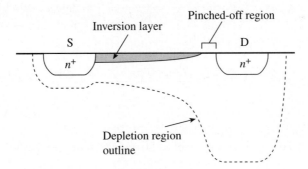

(b) In the square-law theory, $V_{Dsat} = V_G - V_T$ or $V_T = V_G - V_{Dsat}$. From an inspection of the $V_G = 5$ V characteristic in Fig. 17.10, we estimate $V_{Dsat} \cong 4$ V. Thus $V_T = 5 - 4 \cong \mathbf{1 \text{ V}}$.

(c) Since $V_T = V_T'$ for the given device (the characteristics were generated assuming an ideal device),

$$\phi_F = \frac{kT}{q} \ln(N_A/n_i) = (0.0259) \ln(10^{15}/10^{10}) = 0.298 \text{ V}$$

and

$$V_T = 2\phi_F + \frac{K_S x_o}{K_O} \sqrt{\frac{4qN_A}{K_S \varepsilon_0}} \phi_F$$

$$= 2(0.298) + \frac{(11.8)(5 \times 10^{-6})}{(3.9)} \sqrt{\frac{(4)(1.6 \times 10^{-19})(10^{15})}{(11.8)(8.85 \times 10^{-14})}} (0.298) = \mathbf{0.80 \ V}$$

(d) Since $g_d \equiv \partial I_D / \partial V_D |_{V_G}$, it follows that $g_d$ can be deduced from the slope of the appropriate $V_G = $ constant $I_D$–$V_D$ characteristic evaluated at the $V_D$ operating point. Specifically, the slope of the $V_G = 5$ V characteristic in Fig. 17.10 evaluated at $V_D = 0$ yields

$$g_d \cong (3.0 \times 10^{-3} \ A)/(2.3 \ V) = \mathbf{1.30 \times 10^{-3} \ S}$$

(e) Inspecting the device characteristics, we find the MOSFET is saturation biased when $V_G = V_D = 5$ V. When biased above pinch-off, $g_m = (Z\bar{\mu}_n C_o/L)V_{Dsat}$ using either the square-law or bulk-charge entry in Table 17.1. $V_{Dsat} \cong 4$ V when $V_G = 5$ V as noted in answering part (b). Also

$$C_o = \frac{K_O \varepsilon_0}{x_o} = \frac{(3.9)(8.85 \times 10^{-14})}{5 \times 10^{-6}} = 6.90 \times 10^{-8} \ F/cm^2$$

Thus

$$g_m = \frac{Z\bar{\mu}_n C_o}{L} V_{Dsat} \cong \frac{(70)(550)(6.9 \times 10^{-8})(4)}{7} = \mathbf{1.52 \times 10^{-3} \ S}$$

## Problem A3

As can be inferred from Eq. (18.17), or as Eq. (18.20) states explicitly,

$$V_{FB} = \phi_{MS} - \frac{Q_F}{C_o} - \frac{Q_M \gamma_M}{C_o} - \frac{Q_{IT}(0)}{C_o}$$

This basic relationship is employed in all three parts of the problem.

(a) The material used to form the gate only affects $\phi_{MS}$. Consequently,

$$\delta V_G = V_{FB1} - V_{FB2} = \phi_{MS}|_{Al} - \phi_{MS}|_{Au}$$

or, since $(E_c - E_F)_{FB}$ is the same for both devices,

$$\delta V_G = \frac{1}{q} [(\Phi'_M - \chi')_{Al} - (\Phi'_M - \chi')_{Au}]$$
$$= -0.03 \ V - 0.82 \ V = \mathbf{-0.85 \ V}$$

The $\Phi_M' - \chi'$ for Al was noted in the Fig. 18.3 caption, while the $\Phi_M' - \chi'$ for Au is listed in Table 18.1.

(b) In general the silicon surface orientation affects both $Q_F$ and $Q_{IT}$. However, the problem statement indicates $Q_{IT} = 0$. Thus here

$$\delta V_G = V_{FB1} - V_{FB2} = -\frac{Q_F}{C_o}\Big|_{(100)} + \frac{Q_F}{C_o}\Big|_{(111)}$$

As read from Fig. 18.9(b), there is a residual $Q_F/q = 2 \times 10^{11}/\text{cm}^2$ in the MOS-C fabricated on a (111)-oriented Si surface. Moreover, it was pointed out in the text discussion that $Q_F$ is approximately three times smaller on (100) surfaces. (Similar observations were made in Exercise 18.5.) We can therefore write

$$\delta V_G \cong \frac{2}{3}\frac{Q_F}{C_o}\Big|_{(111)} = \frac{(2)(1.6 \times 10^{-19})(2 \times 10^{11})}{(3)(3 \times 10^{-8})} = \textbf{0.71 V}$$

(c) Being identical except for the sodium ion distribution, the flat-band voltages of the two MOS-Cs will exhibit the voltage displacement

$$\delta V_G = V_{FB1} - V_{FB2} = -\frac{Q_M \gamma_M}{C_o}\Big|_{\#1} + \frac{Q_M \gamma_M}{C_o}\Big|_{\#2}$$

As discussed in Subsection 18.2.6, $\gamma_M = 0$ if the ions are piled up adjacent to the metal–oxide interface, whereas $\gamma_M = 1$ if the ions are piled up adjacent to the oxide–semiconductor interface. With $\gamma_{M1} = 0$ and $\gamma_{M2} = 1$,

$$\delta V_G = \frac{Q_M}{C_o} = \frac{(1.6 \times 10^{-19})(5 \times 10^{11})}{(3 \times 10^{-8})} = \textbf{2.67 V}$$

## Answers—Set B

(1) ... $\phi$ has the same shape as the "upside down" of the bands.

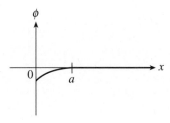

(2) ... $\mathcal{E}$ is proportional to the slope of the bands.

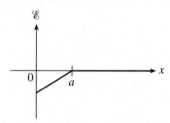

(3) a ... The device is zero biased; thus the semiconductor must be in equilibrium. $E_F$ is indeed shown energy independent inside the semiconductor.

(4) b ... $N_D = n_i e^{(E_F - E_i)/kT} = (2.25 \times 10^6)e^{0.57/0.0259} = 8.13 \times 10^{15}/\text{cm}^3$.

(5) c ... Per Eq. (16.1), $V_G = -(1/q)[E_F(\text{metal}) - E_F(\text{semi})] = 0$.

(6) b

(7) d ... $\phi_{MS} = (1/q)[\Phi'_M - \chi' - (E_c - E_F)_{FB}] = 0.76 - 0.37 - 0.14 = 0.25$ V

(8) d ... Since $Q_M = Q_F = Q_{IT} = 0$, $\Delta V_G = (V_G - V'_G)_{|\text{same } \phi_S} = \phi_{MS}$.
Under flat-band conditions $V'_G = 0$ and $V_G = V_{FB} = \phi_{MS} = 0.25$ V.

(9) c ... The delta-depletion analysis gave (Eq. 16.37),

$$\frac{C}{C_O} = \frac{1}{\sqrt{1 + V'_G/V_\delta}}$$

It should be pointed out that $V_G$, not $V'_G$, was actually used in text Eq. (16.37). However, an ideal structure was assumed throughout Chapter 16 (all the $V_G$ in Chapter 16 are in reality $V'_G$). As noted in the answer to Problem 8, $\Delta V_G = (V_G - V'_G)_{|\text{same } \phi_S} = \phi_{MS}$. Since $V_G = 0$ for the pictured bias point, the corresponding $V'_G = -\phi_{MS}$.

$$V_\delta = -\frac{q}{2} \frac{K_S x_0^2}{K_O^2 \varepsilon_0} N_D = -\frac{(1.6 \times 10^{-19})(12.85)(10^{-5})^2(8.13 \times 10^{15})}{(2)(9)^2(8.85 \times 10^{-14})} = -0.117 \text{ V}$$

$$\frac{C}{C_O} = \frac{1}{\sqrt{1 + \dfrac{0.25}{0.117}}} = 0.56$$

(10) a ... The quantity $a$ is of course just the depletion width $W$. Since $C/C_O$ was determined in Problem 9, $W$ can be computed using Eq. (16.34b).

$$\frac{C}{C_O} = \frac{1}{1 + \left(\dfrac{K_O W}{K_S x_o}\right)}$$

$$W = \frac{K_S}{K_O} x_o \left(\frac{C_O}{C} - 1\right) = \frac{(12.85)(10^{-5})}{(9)} \left(\frac{1}{0.56} - 1\right) = 0.112 \ \mu\text{m}$$

(11) b . . . $V_{FB} = \phi_{MS} - \dfrac{Q_F}{C_o} = \phi_{MS} - q\dfrac{x_o}{K_O \varepsilon_0}\dfrac{Q_F}{q}$

$$= -0.89 - \frac{(1.6 \times 10^{-19})(5 \times 10^{-6})(5 \times 10^{10})}{(3.9)(8.85 \times 10^{-14})}$$

$$= -1.01 \ \text{V}$$

(12) b . . . $V_T = V_T' + V_{FB}$

$$V_T' = 2\phi_F + \frac{K_S}{K_O} x_o \sqrt{\frac{4qN_A}{K_S \varepsilon_0}\phi_F}$$

$$\phi_F = \frac{kT}{q} \ln\left(\frac{N_A}{n_i}\right) = 0.0259 \ln\left(\frac{10^{15}}{10^{10}}\right) = 0.298 \ \text{V}$$

$$V_T' = 2(0.298) + \frac{(11.8)(5 \times 10^{-6})}{(3.9)} \left[\frac{(4)(1.6 \times 10^{-19})(10^{15})(0.298)}{(11.8)(8.85 \times 10^{-14})}\right]^{1/2}$$

$$= 0.80 \ \text{V}$$

$$V_T = 0.80 - 1.01 = -0.21 \ \text{V}$$

(13) b . . . Since the MOSFET conducts for $V_G > V_T$, the device is "on" at $V_G = 0$ and is therefore a depletion-mode MOSFET.

(14) D . . . The MOSFET channel is shown just being pinched off. This corresponds to the start of saturation, point D.

(15) c . . . If $V_G - V_T = 3$ V and $V_D = 1$ V, the MOSFET is biased below pinch-off and, in the square-law formulation,

$$I_D = \frac{Z\bar{\mu}_n C_o}{L}\left[(V_G - V_T)V_D - \frac{V_D^2}{2}\right]$$

or

$$\frac{Z\bar{\mu}_n C_o}{L} = \frac{I_D}{(V_G - V_T)V_D - V_D^2/2} = \frac{2.5 \times 10^{-4}}{3 - 0.5} = 10^{-4} \ \text{amps/volt}^2$$

**R3**

When $V_G - V_T = 3$ V and $V_D = 4$ V, the device is saturation biased, and

$$I_D = I_{Dsat} = \frac{Z\bar{\mu}_n C_o}{2L}(V_G - V_T)^2 = \frac{(10^{-4})(3)^2}{2}$$

$$= 4.5 \times 10^{-4} \text{ amps}$$

(16) b          (17) a          (18) a          (19) b          (20) b

(21) a          (22) a          (23) b          (24) b          (25) a

# Appendix A
# ELEMENTS OF QUANTUM MECHANICS

Before progressing to the modeling of carriers in a crystal, one first must be able to describe the electronic situation inside an isolated semiconductor atom. Unfortunately, the "everyday" descriptive formalism known as classical (Newtonian) mechanics yields inaccurate results when applied to the electrons in semiconductor atoms or, more generally, when applied to any system with atomic dimensions. The mathematical formalism known as *Quantum Mechanics* must be employed in treating atomic dimension systems. Quantum mechanics is a more precise description of nature that reduces to classical mechanics in the limit where the masses and energies of the particles involved are large.

The first section of this appendix contains a discussion of key observations and associated analyses leading to the development of quantum mechanics. This is followed by a brief survey of the basic quantum mechanical formalism. The final section contains a summary of the quantum mechanical solution for the electronic states in atoms—the information needed for the eventual modeling of carriers in a crystal.

## A.1  THE QUANTIZATION CONCEPT

### A.1.1 Blackbody Radiation

It is a well-known fact that a solid object will glow or give off light if it is heated to a sufficiently high temperature. Actually, solid bodies in equilibrium with their surroundings emit a spectrum of radiation at all times. When the temperature of the body is at or below room temperature, however, the radiation is almost exclusively in the infrared and therefore not detectable by the human eye. For an ideal radiator, called a blackbody, the spectrum or wavelength dependence of the emitted radiation is as graphed in Fig. A.1.

Various attempts to explain the observed blackbody spectrum were made in the latter half of the nineteenth century. The most successful of the arguments, all of which were based on classical mechanics, was proposed by Rayleigh and Jeans. Heat energy absorbed by a material was known to cause a vibration of the atoms within the solid. The vibrating atoms were modeled as harmonic oscillators with a spectrum of normal mode frequencies, $\nu = \omega/2\pi$, and a *continuum of allowed energies* distributed in accordance with statistical considerations. The emitted radiation was in essence equated to a sampling of the energy distribution inside the solid. The Rayleigh–Jeans "law" resulting from this analysis is

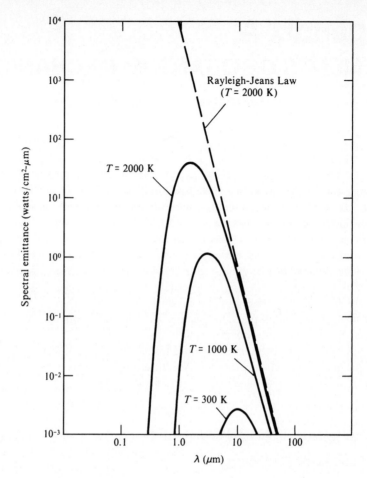

**Figure A.1** Wavelength dependence of the radiation emitted by a blackbody heated to 300 K, 1000 K, and 2000 K. Note that the visible portion of the spectrum is confined to wavelengths 0.4 $\mu$m $\lesssim \lambda \lesssim$ 0.7 $\mu$m. The dashed line is the predicted dependence for $T = 2000$ K based on classical considerations.

shown as a dashed line in Fig. A.1. As is evident from Fig. A.1, the classical theory was in reasonably good agreement with experimental observations at the longer wavelengths. Over the short-wavelength portion of the spectrum, however, there was total divergence between experiment and theory. This came to be known as the "ultraviolet catastrophe," since integration over all wavelengths theoretically predicted an infinite amount of radiated energy.

In 1901 Max Planck provided a detailed theoretical fit to the observed blackbody spectrum. The explanation was based on the then-startling hypothesis that the vibrating atoms in a material could only radiate or absorb energy in discrete packets. Specifically, for a given atomic oscillator vibrating at a frequency $\nu$, Planck postulated that the energy of the oscillator was restricted to the *quantized* values

$$E_\mathbf{n} = \mathbf{n}h\nu = \mathbf{n}\hbar\omega \qquad \mathbf{n} = 0, 1, 2, \cdots \qquad (A.1)$$

An $h$ value of $6.63 \times 10^{-34}$ joule-sec ($\hbar = h/2\pi$) was obtained by matching theory to experiment and has subsequently come to be known as Planck's constant.

The point to be learned from the blackbody discussion is that, for atomic dimension systems, the classical view, which always allows a continuum of energies, is demonstrably incorrect. Extremely small discrete steps in energy, or energy quantization, can occur and is a central feature of quantum mechanics.

## A.1.2 The Bohr Atom

Another experimental observation that puzzled scientists of the nineteenth century was the sharp, discrete spectral lines emitted by heated gases. The first step toward unraveling this puzzle was provided by Rutherford, who advanced the nuclear model for the atom in 1910. Atoms were viewed as being composed of electrons with a small rest mass $m_0$ and charge $-q$ orbiting a massive nucleus with charge $+Zq$, where $Z$ was an integer equal to the number of orbiting electrons. Light emission from heated atoms could then be associated with the energy lost by electrons in going from a higher-energy to a lower-energy orbit. Classically, however, the electrons could assume a continuum of energies and the output spectrum should likewise be continuous—not sharp, discrete spectral lines. The nuclear model itself posed somewhat of a dilemma. According to classical theory, whenever a charged particle is accelerated, the particle will radiate energy. Thus, based on classical arguments, the angularly accelerated electrons in an atom should continuously lose energy and spiral into the nucleus in a relatively short period of time.

In 1913 Niels Bohr proposed a model that both resolved the Rutherford atom dilemma and explained the discrete nature of the spectra emitted by heated gases. Building on Planck's hypothesis, Bohr suggested that the electrons in an atom were restricted to certain well-defined orbits, or, equivalently, assumed that the orbiting electrons could take on only certain (quantized) values of angular momentum $L$.

For the simple hydrogen atom with $Z = 1$ and a circular electron orbit, the Bohr postulate can be expressed mathematically in the following manner:

$$L_\mathbf{n} = m_0 v r_\mathbf{n} = \mathbf{n}\hbar \qquad \mathbf{n} = 1, 2, 3, \cdots \qquad (A.2)$$

where $m_0$ is the electron rest mass, $v$ is the linear electron velocity, and $r_\mathbf{n}$ is the radius of the orbit for a given value of $\mathbf{n}$. Since the electron orbits are assumed to be stable, the centripedal force on the electron ($m_0 v^2/r_\mathbf{n}$) must precisely balance the coulombic attraction

$(q^2/4\pi\varepsilon_0 r_n{}^2$ in rationalized MKS units) between the nucleus and the orbiting electron. Therefore, one can also write

$$\frac{m_0 v^2}{r_n} = \frac{q^2}{4\pi\varepsilon_0 r_n{}^2} \tag{A.3}$$

where $\varepsilon_0$ is the permittivity of free space. Combining Eqs. (A.2) and (A.3), one obtains

$$r_n = \frac{4\pi\varepsilon_0 (n\hbar)^2}{m_0 q^2} \tag{A.4}$$

Next, by examining the kinetic energy (K.E.) and potential energy (P.E.) components of the total electron energy ($E_n$) in the various orbits, we find

$$\text{K.E.} = \frac{1}{2} m_0 v^2 = \frac{1}{2} (q^2/4\pi\varepsilon_0 r_n) \tag{A.5a}$$

and

$$\text{P.E.} = -q^2/4\pi\varepsilon_0 r_n \qquad (\text{P.E. set} = 0 \text{ at } r = \infty) \tag{A.5b}$$

Thus

$$E_n = \text{K.E.} + \text{P.E.} = -\frac{1}{2} (q^2/4\pi\varepsilon_0 r_n) \tag{A.6}$$

or, making use of Eq. (A.4),

$$E_n = -\frac{m_0 q^4}{2(4\pi\varepsilon_0 n\hbar)^2} = -\frac{13.6}{n^2} \text{ eV} \tag{A.7}$$

The *electron volt* (eV) introduced in Eq. (A.7) is a non-MKS unit of energy equal to $1.6 \times 10^{-19}$ joules.

With the electron energies in the hydrogen atom restricted to the values specified by Eq. (A.7), the light energies that can be emitted by the atom upon heating are now discrete in nature and equal to $E_{n'} - E_n$, $n' > n$. As summarized in Fig. A.2, the allowed energy transitions are found to be in excellent agreement with the observed photo-energies.

Although the Bohr model was immensely successful in explaining the hydrogen spectra, numerous attempts to extend the "semi-classical" Bohr analysis to more complex atoms such as helium proved to be futile. Success along these lines had to await further development of the quantum mechanical formalism. Nevertheless, the Bohr analysis reinforced the concept of energy quantization and the attendant failure of classical mechanics

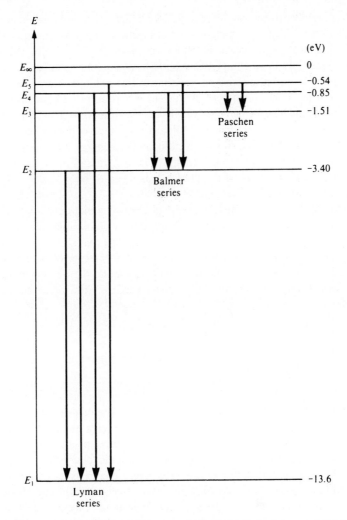

**Figure A.2**  Hydrogen atom energy levels as predicted by the Bohr theory and the transitions corresponding to prominent, experimentally observed, spectral lines.

in dealing with systems on an atomic scale. Moreover, the quantization of angular momentum in the Bohr model clearly extended the quantum concept, seemingly suggesting a general quantization of atomic-scale observables.

## A.1.3 Wave-Particle Duality

An interplay between light and matter was clearly evident in the blackbody and Bohr atom discussions. Those topics can be treated, however, without disturbing the classical viewpoint that electromagnetic radiation (light, x-rays, etc.) is totally wave-like in nature and matter (an atom, an electron) is totally particle-like in nature. A different situation arises in treating the photoelectric effect—the emission of electrons from the illuminated surface of a material. To explain the photoelectric effect, as argued by Einstein in 1905, one must view the impinging light to be composed of particle-like quanta (photons) with an energy $E = h\nu$. The particle-like properties of electromagnetic radiation were later solidified in the explanation of the Compton effect. The deflected portion of an x-ray beam directed at solids was found to undergo a change in frequency. The observed change in frequency was precisely what one would expect from a "billiard ball" type collision between the x-ray quanta and electrons in the solid. In such a collision both energy and momentum must be conserved. Noting that $E = h\nu = mc^2$, where $m$ is the "mass" of the photon and $c$ the velocity of light, the momentum of the photon was taken to be $p = mc = h\nu/c = h/\lambda$, $\lambda$ being the wavelength of the electromagnetic radiation.

By the mid-1920s the wave-particle duality of electromagnetic radiation was an established fact. Noting this fact and the general reciprocity of physical laws, Louis de Broglie in 1925 made a rather interesting conjecture. He suggested that since electromagnetic radiation exhibited particle-like properties, particles should be expected to exhibit wave-like properties. De Broglie further hypothesized that, paralleling the photon momentum calculation, the *wavelength* characteristic of a given particle with momentum $p$ could be computed from[†]

$$\boxed{p = h/\lambda} \qquad \ldots \text{ de Broglie hypothesis} \qquad (A.8)$$

Although pure conjecture at the time, the de Broglie hypothesis was quickly substantiated. Evidence of the wave-like properties of matter was first obtained by Davisson and Germer from an experiment performed in 1927. In their experiment, a low-energy beam of electrons was directed perpendicularly at the surface of a nickel crystal. The energy of the electrons was chosen such that the wavelength of the electrons as computed from the de Broglie relationship was comparable to the nearest-neighbor distance between nickel atoms. If the electrons behaved as simple particles, one would expect the electrons to scatter more or less randomly in all directions from the surface of the nickel crystal (assumed to be rough on an atomic scale). The angular distribution actually observed was quite similar to the interference pattern produced by light diffracted from a grating. In fact, the angular positions of maxima and minima of electron intensity could be predicted accurately using the de Broglie wavelength and assuming wave-like reflection from atomic planes inside the nickel crystal. Later experiments performed by other researchers likewise confirmed the inherent wave-like properties of heavier particles such as protons and neutrons.

In summary, then, based on experimental evidence—a portion of which has been dis-

---

[†]Use of the symbol $p$ to represent the momentum of a particle is confined to this appendix. Throughout the remainder of the text $p$ is defined as the hole concentration (initially introduced in Subsection 2.3.3).

cussed herein under the headings of blackbody radiation, the Bohr atom, and the wave-particle duality—one is led to conclude that classical mechanics does not accurately describe the action of particles on an atomic scale. Experiments point to a quantization of observables (energy, angular momentum, etc.) and to the inherent wave-like nature of all matter.

## A.2  BASIC FORMALISM

The accumulation of experimental data and physical explanations in the early twentieth century that were at odds with the classical laws of physics emphasized the need for a revised formulation of mechanics. In 1926 Schrödinger not only provided the required revision, but established a unified scheme valid for describing both the microscopic and macroscopic universes. The formulation, called *wave mechanics,* incorporated the physical notions of quantization first advanced by Planck and the wave-like nature of matter hypothesized by de Broglie. It should be mentioned that at almost the same time an alternative formulation called *matrix mechanics* was advanced by Heisenberg. Although very different in their mathematical orientations, the two formulations were later shown to be precisely equivalent and were merged under the general heading of *quantum mechanics.* Herein we will restrict ourselves to the Schrödinger wave mechanical description, which is somewhat simpler mathematically and more readily related to the physics of a particular problem. Our general approach will be to present the five basic postulates of wave mechanics and to subsequently discuss the postulates to provide some insight into the formulation.

For a single-particle system, the five basic postulates of wave mechanics are as follows:

(1) There exists a wavefunction, $\Psi = \Psi(x, y, z, t)$, from which one can ascertain the dynamic behavior of the system and all desired system variables. $\Psi$ might be called the "describing function" for the system. Mathematically, $\Psi$ is permitted to be a complex quantity (with real and imaginary parts) and will, in general, be a function of the space coordinates $(x, y, z)$ and time $t$.

(2) The $\Psi$ for a given system and specified system constraints is determined by solving the equation,

$$-\frac{\hbar^2}{2m} \nabla^2\Psi + U(x, y, z)\Psi = -\frac{\hbar}{i} \frac{\partial\Psi}{\partial t} \qquad (A.9)$$

where $m$ is the mass of the particle, $U$ is the potential energy of the system, and $i = \sqrt{-1}$. Eq. (A.9) is referred to as the time-dependent Schrödinger equation, or simply, the wave equation.

(3) $\Psi$ and $\nabla\Psi$ must be finite, continuous, and single-valued for all values of $x$, $y$, $z$ and $t$.

(4) If $\Psi^*$ is the complex conjugate of $\Psi$, $\Psi^*\Psi \, d\mathcal{V} = |\Psi|^2 \, d\mathcal{V}$ is to be identified as the probability that the particle will be found in the spatial volume element $d\mathcal{V}$. Hence, by implication,

$$\int_{\mathcal{V}} \Psi^*\Psi \, d\mathcal{V} = 1 \tag{A.10}$$

where $\int_{\mathcal{V}}$ indicates an integration over all space.

(5) One can associate a unique mathematical operator with each dynamic system variable such as position or momentum. The value—or, more precisely, the expectation value—of a given system variable is in turn obtained by "operating" on the wavefunction. Specifically, taking $\alpha$ to be the system variable of interest and $\alpha_{op}$ the associated mathematical operator, the desired expectation value, $\langle \alpha \rangle$, is computed from

$$\langle \alpha \rangle = \int_{\mathcal{V}} \Psi^* \alpha_{op} \Psi \, d\mathcal{V} \tag{A.11}$$

The unique mathematical operator associated with a given system variable has been established by requiring the wave mechanical expectation value to approach the corresponding value derived from classical mechanics in the large-mass/high-energy limit. An abbreviated listing of dynamic variables and associated operators is presented in Table A.1.

The solution of problems using wave mechanics is in principle quite straightforward. Subject to the constraints (boundary conditions) inherent in a problem and the additional

**Table A.1**  Dynamic Variable/Operator Correspondence.

| Dynamic Variable ($\alpha$) | | Mathematical Operator ($\alpha_{op}$) | | Expectation Value—$\langle \alpha \rangle$ |
|---|---|---|---|---|
| $x, y, z$ | $\leftrightarrow$ | $x, y, z$ | $\cdots$ | $\langle x \rangle = \int_{\mathcal{V}} \Psi^* x \Psi \, d\mathcal{V}$ |
| $f(x, y, z)$ | $\leftrightarrow$ | $f(x, y, z)$ | | |
| $p_x, p_y, p_z$ | $\leftrightarrow$ | $\dfrac{\hbar}{i}\dfrac{\partial}{\partial x}, \dfrac{\hbar}{i}\dfrac{\partial}{\partial y}, \dfrac{\hbar}{i}\dfrac{\partial}{\partial z}$ | $\cdots$ | $\langle p_x \rangle = \int_{\mathcal{V}} \Psi^* \dfrac{\hbar}{i}\dfrac{\partial \Psi}{\partial x} \, d\mathcal{V}$ |
| $E$ | $\leftrightarrow$ | $-\dfrac{\hbar}{i}\dfrac{\partial}{\partial t}$ | | |

constraints imposed by postulates 3 and 4, one solves Schrödinger's equation for the system wavefunction $\Psi$. Once $\Psi$ is known, system variables of interest can be deduced from Eq. (A.11) per the postulate 5 recipe. The straightforward approach, however, is often difficult to implement. Except for simple problems of an idealized nature and a very select number of practical problems, it is usually impossible to obtain a closed-form solution to Schrödinger's equation. Nevertheless, in many problems the constraints imposed on the solution can be used to deduce information about the system variables, notably the allowed system energies, without actually solving for the system wavefunction. Another common approach is to use expansions, trial (approximate) wavefunctions, or limiting-case solutions to deduce information of interest.

Finally, a comment is in order concerning the "derivation" of Schrödinger's equation and the origin of the other basic postulates. Although excellent theoretical arguments can be presented to justify the form of the equation, Schrödinger's equation is essentially an empirical relationship. Like Newton's laws, Schrödinger's equation and the other basic postulates of quantum mechanics constitute a generalized mathematical description of the physical world extrapolated from specific empirical observations. Relative to the validity of the formulation, it can only be stated that, whenever subject to test by experiment, the predictions of the quantum mechanical formulation have been found to be in agreement with observations to within the limit of experimental uncertainty, which in many cases has been extremely small.

# A.3  ELECTRONIC STATES IN ATOMS

We examine here the application of the quantum mechanical formalism to the hydrogen atom and the solution results for atoms in general. It should be reiterated that the overall goal of the appendix is to provide information about the electronic states in isolated semiconductor atoms as a prelude to the eventual modeling of carriers in a semiconductor crystal. The hydrogen atom is the logical place to begin the quantum mechanical analysis because it is the simplest of atoms and because results can be compared with the semiclassical Bohr solution. Although the hydrogen atom analysis yields a complete closed-form solution, the treatment and solution are hardly trivial. We will only indicate the solution procedure and review key results. Information about the electronic states in multielectron atoms is extrapolated from the hydrogen atom results.

## A.3.1 The Hydrogen Atom

The hydrogen atom consists of a relatively massive nucleus with charge $+q$ surrounded by an electron with charge $-q$. With little error, the nucleus can be considered fixed in space and the problem reduced to a single particle system (the electron) that is assumed to have a fixed total energy $E$. In other words, the hydrogen atom is taken to be isolated in space and not subject to any perturbations that could lead to a change in the total energy.

For any single-particle system with a fixed total energy $E$, the position and time coordinates can be separated yielding a general solution of the form

$$\Psi(x,\ y,\ z,\ t) = \psi(x,\ y,\ z)e^{-iEt/\hbar} \tag{A.12}$$

Direct substitution of Eq. (A.12) into Eq. (A.9), and the subsequent simplification and rearrangement of the result, gives

$$\nabla^2\psi + \frac{2m}{\hbar^2}[E - U(x,\ y,\ z)]\psi = 0 \tag{A.13}$$

Eq. (A.13), which must be solved to obtain $\psi(x,\ y,\ z)$, is known as the time-independent Schrödinger equation.

In the hydrogen atom $m = m_0$ and the $-q$ electron is electrostatically attracted to the $+q$ nucleus at the origin of coordinates. As noted in the Bohr analysis, the potential energy associated with the electrostatic attraction is

$$U = -\frac{q^2}{4\pi\varepsilon_0 r} \tag{A.14}$$

where $r = \sqrt{x^2 + y^2 + z^2}$ is the distance from the nucleus. Thus the equation to be solved takes on the specific form

$$\nabla^2\psi + \frac{2m_0}{\hbar^2}\left(E + \frac{q^2}{4\pi\varepsilon_0 r}\right)\psi = 0 \tag{A.15}$$

In principle one could seek a solution to Eq. (A.15) employing Cartesian $(x,\ y,\ z)$ coordinates. However, given the spherically symmetric nature of the potential energy, it is more convenient to employ spherical $(r,\ \theta,\ \phi)$ coordinates. In spherical coordinates the desired wavefunction solution becomes $\psi(r,\ \theta,\ \phi)$ and

$$\nabla^2\psi = \frac{1}{r^2}\frac{\partial}{\partial r}\left(r^2\frac{\partial\psi}{\partial r}\right) + \frac{1}{r^2\sin\theta}\frac{\partial}{\partial\theta}\left(\sin\theta\frac{\partial\psi}{\partial\theta}\right) + \frac{1}{r^2\sin^2\theta}\frac{\partial^2\psi}{\partial\phi^2} \tag{A.16}$$

Equation (A.15) can be solved using the separation-of-variables technique where one assumes the wavefunction can be written as the product of three functions separately dependent on $r$, $\theta$, and $\phi$. The procedure yields an ordered set of bound-state $(E < 0)$ wavefunction solutions. Arising from the separation constants, and associated with each solution, there is a unique group of three *quantum numbers*. The standard symbols, allowed values, and full names of the three parameters are as follows:

$$n = 1, 2, 3 \cdots \qquad \ldots \text{principal quantum number}$$
$$l = 0, 1, 2, \cdots n - 1 \qquad \ldots \text{azimuthal quantum number}$$
$$m = -l \text{ to } l \qquad \ldots \text{magnetic orbital quantum number}$$

The $\psi_{n,l,m}(r, \theta, \phi)$ solutions corresponding to $n = 1$ and $n = 2$ are presented in Table A.2 for illustrative and reference purposes. The $a_0$ appearing in the solutions is the *Bohr radius* and is numerically equal to the ground state Bohr orbit; that is, $a_0 = 4\pi\varepsilon_0\hbar^2/m_0q^2$ as deduced from Eq. (A.4).

Let us examine and comment on the results. Suppose first of all that the $\psi_{1,0,0}$ solution is substituted into Eq. (A.15) and the resulting expression solved for $E$. One obtains

$$E_{1,0,0} = -\frac{\hbar^2}{2m_0}\frac{1}{a_0^2} = -\frac{m_0q^4}{2(4\pi\varepsilon_0\hbar)^2} \tag{A.17}$$

Note that $E_{1,0,0}$ is identical to $E_1$ of the Bohr analysis. Similarly, if the $n = 2$ wavefunctions in Table A.2 are substituted into Eq. (A.15) and the resulting expressions solved for $E$, one obtains

$$E_{2,0,0} = E_{2,1,-1} = E_{2,1,0} = E_{2,1,1} = -\frac{1}{4}\left[\frac{m_0q^4}{2(4\pi\varepsilon_0\hbar)^2}\right] \tag{A.18}$$

**Table A.2**   The Hydrogen Atom $\psi_{n,l,m}$ Solutions Corresponding to $n = 1$ and $n = 2$. $a_0 = 4\pi\varepsilon_0\hbar^2/m_0q^2 =$ Bohr radius. (J. L. Powell and B. Crasemann, *Quantum Mechanics,* Addison-Wesley Publishing Co., Reading, MA, © 1961.)

$$\psi_{1,0,0} = \frac{1}{\sqrt{\pi}\,a_0^{3/2}}\,e^{-r/a_0}$$

$$\psi_{2,0,0} = \frac{1 - r/2a_0}{2\sqrt{2\pi}\,a_0^{3/2}}\,e^{-r/2a_0}$$

$$\psi_{2,1,-1} = \frac{r/2a_0}{4\sqrt{\pi}\,a_0^{3/2}}\,e^{-r/2a_0}\,e^{-i\phi}\,\sin\theta$$

$$\psi_{2,1,0} = \frac{r/2a_0}{2\sqrt{2\pi}\,a_0^{3/2}}\,e^{-r/2a_0}\,\cos\theta$$

$$\psi_{2,1,1} = -\frac{r/2a_0}{4\sqrt{\pi}\,a_0^{3/2}}\,e^{-r/2a_0}\,e^{i\phi}\,\sin\theta$$

The $\mathbf{n} = 2$ states are all associated with the same total energy, and the energy is identical to $E_2$ of the Bohr analysis. The general point to be made is that the quantum analysis yields the same predicted energy levels as the Bohr analysis. Moreover, knowledge of the principal quantum number, $\mathbf{n}$, completely specifies the total energy of an electron in a particular state. Clearly, $\psi_{1,0,0}$ corresponds to the ground state while wavefunctions associated with larger $\mathbf{n}$-values correspond to excited states.

When there is more than one allowed state at a given energy, the states are said to be *degenerate*. The $l$ and $m$ of degenerate states come into play if, for example, the hydrogen atom were perturbed by a magnetic field. Because of the different spatial distribution of the wavefunctions, the interaction with the magnetic field would cause a splitting of the energy levels and thereby remove the degeneracy.

While on the topic of degenerate states, it is convenient to point out that a fourth quantum number is actually required to completely specify a quantum state. More precise analyses indicate electrons and other subatomic particles exhibit a property call *spin*, which becomes important in particle-particle interactions. The electron is visualized as spinning about an axis through its center in either a clockwise or counterclockwise direction. This gives rise to two spin states often referred to as spin-up and spin-down. The associated spin quantum number, $\mathbf{s}$, can take on the values of $\mathbf{s} = +\frac{1}{2}$ and $\mathbf{s} = -\frac{1}{2}$. Spin causes a two-fold degeneracy to be associated with each of the states in Table A.2.

A comment is also in order concerning the spatial distribution of the allowed states. As noted in the section on basic formalism, $\Psi^*\Psi d\mathcal{V}$ represents the probability that a particle will be found in a spatial volume element $d\mathcal{V}$. To provide a specific example, the probability of finding an electron in the ground state at a distance between $r$ and $r + dr$ from the nucleus is equal to $4\pi r^2|\psi_{1,0,0}|^2 dr$. A plot of $4\pi r^2|\psi_{1,0,0}|^2$ versus $r/a_0$ is displayed in Fig. A.3(a). Whereas the probability of finding the ground-state electron increases to a maximum at the Bohr radius, and the peak probability progressively moves to larger $r$ as $\mathbf{n}$ is increased, there is significant probability of finding the electron over a range of distances from the nucleus. This is in total contrast to the Bohr model where the electron is assumed to be in an orbit at an $r$ = constant distance from the nucleus. In fact, the electron is sometimes conceived as a charge "cloud" distributed in proportion to the $|\psi|^2 d\mathcal{V}$ probability as illustrated in Fig. A.3(b). The $\psi_{1,0,0}$ wavefunction used in constructing Fig. A.3(b) is of course spherically symmetric. Wavefunctions with $l \neq 0$ would exhibit charge clouds with an angular dependence.

## A.3.2 Multi-Electron Atoms

The wavefunction solutions, energy levels, and probability distributions established for the hydrogen atom are specific to the hydrogen atom and cannot be applied without modification to more complex atoms. However, the allowed electronic states in multi-electron atoms are uniquely characterized employing the same set of four quantum numbers ($\mathbf{n}$, $l$, $\mathbf{m}$, and $\mathbf{s}$) introduced in the hydrogen atom analysis. The same general energy order also applies; $\mathbf{n} = 1$ is associated with the lowest energy state, $\mathbf{n} = 2$ with the next lowest energy state, and so on. The foregoing, coupled with restrictions placed on multi-electron systems, per-

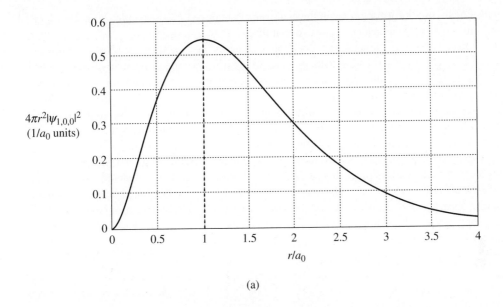

(a)

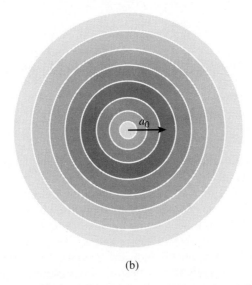

(b)

**Figure A.3**  For a hydrogen atom in the $\psi_{1,0,0}$ ground state: (a) probability of finding the electron at a distance $r$ from the nucleus; (b) cloud-like representation of the electronic charge.

mits one to infer information about the electronic structure of more complex atoms without actually solving for the electronic wavefunctions.

One of the restrictions referred to above goes by the name of the *Pauli Exclusion Principle*. The Pauli Exclusion Principle dictates that no two electrons in a system can be characterized by the same set of quantum numbers. For example, one and only one electron in a multi-electron atom can have $\mathbf{n} = 1$, $l = 0$, $\mathbf{m} = 0$, and $\mathbf{s} = \frac{1}{2}$. A second implicit restriction is that the electronic configuration be such as to minimize the system energy when a multi-electron atom is in its ground state. As a general rule this means electrons populate states with the lowest possible $\mathbf{n}$-values.

Electronic information pertinent to the first fourteen elements (up to Si) in the Periodic Table of the Elements is presented in Table A.3. The top portion of Table A.3 lists the sets of four quantum numbers corresponding to the lowest energy states. The bottom-line entry is the spectroscopic designation for the state specified by the quantum number set. The number in the bottom-line entry gives the $\mathbf{n}$-value while the letter identifies the $l$-value according to the scheme

$$l = 0, \quad 1, \quad 2, \quad 3 \quad (4, 5 \cdots)$$
$$\uparrow \quad \uparrow \quad \uparrow \quad \uparrow$$
$$s \quad p \quad d \quad f \quad (g, h \cdots)$$

The rather odd lettering of the first four $l$-values stems from early spectroscopic work where the transitions between states were associated with spectral lines named *s*harp, *p*rincipal, *d*iffuse, and *f*undamental. Generally speaking, in multi-electron atoms the *s*-states have a slightly lower energy than *p*-states and therefore appear first in the listing of states. *p*-levels corresponding to a given $\mathbf{n}$-value have the same energy. The bottom portion of Table A.3 shows the ground-state electronic configuration in elements up to Si. The reader should verify that the arrangements are consistent with previously cited facts and restrictions. The spectroscopic shorthand notation for the electron configuration is given in the far-right column. The superscript on a letter in the shorthand notation indicates the number of electrons with the same $\mathbf{n}l$ combination.

Table A.3 is very useful for inferring major features of the electronic configuration in an isolated Si atom. Si is of particular interest of course because it is presently the preeminent semiconductor material. From the table we see that the complete filling of allowed states with a given $\mathbf{n}$-value leads to an extremely stable, tightly bound, electronic configuration; namely, it leads to the inert gases Helium and Neon. Understandably, as envisioned in Fig. A.4, the two $\mathbf{n} = 1$ and eight $\mathbf{n} = 2$ electrons in Si likewise populate deep-lying energy levels tightly bound to the nucleus of the atom. The binding is so strong, in fact, that these ten electrons remain essentially unperturbed during chemical reactions or normal atom-atom interactions, with the ten-electron-plus-nucleus combination often being referred to as the *core* of the atom. The remaining four electrons are an "add-on" to the stable Neon configuration and are expected to be rather weakly bound. They are collectively called *valence electrons* because of their strong participation in chemical reactions

**Table A.3** Energy States and the Electronic Configuration in Elements 1–14. Atoms are assumed to be in the ground state.

| Quantum Numbers | | | | | | | | | | | | | | | | | | |
|---|---|---|---|---|---|---|---|---|---|---|---|---|---|---|---|---|---|---|
| **n** | 1 | 1 | 2 | 2 | 2 | 2 | 2 | 2 | 2 | 2 | 3 | 3 | 3 | 3 | 3 | 3 | 3 | 3 |
| **l** | 0 | 0 | 0 | 0 | 1 | 1 | 1 | 1 | 1 | 1 | 0 | 0 | 1 | 1 | 1 | 1 | 1 | 1 |
| **m** | 0 | 0 | 0 | 0 | -1 | -1 | 0 | 0 | 1 | 1 | 0 | 0 | -1 | -1 | 0 | 0 | 1 | 1 |
| **s** | $\frac{1}{2}$ | $-\frac{1}{2}$ | $\frac{1}{2}$ | $-\frac{1}{2}$ | $\frac{1}{2}$ | $-\frac{1}{2}$ | $\frac{1}{2}$ | $-\frac{1}{2}$ | $\frac{1}{2}$ | $-\frac{1}{2}$ | $\frac{1}{2}$ | $-\frac{1}{2}$ | $\frac{1}{2}$ | $-\frac{1}{2}$ | $\frac{1}{2}$ | $-\frac{1}{2}$ | $\frac{1}{2}$ | $-\frac{1}{2}$ |
| *State* | $1s$ | $1s$ | $2s$ | $2s$ | $2p$ | $2p$ | $2p$ | $2p$ | $2p$ | $2p$ | $3s$ | $3s$ | $3p$ | $3p$ | $3p$ | $3p$ | $3p$ | $3p$ |

| Atomic Number | Element | \(1s\) | \(1s\) | \(2s\) | \(2s\) | \(2p\) | \(2p\) | \(2p\) | \(2p\) | \(2p\) | \(2p\) | \(3s\) | \(3s\) | \(3p\) | \(3p\) | Electronic Configuration |
|---|---|---|---|---|---|---|---|---|---|---|---|---|---|---|---|---|
| 1 | H | ↕ | | | | | | | | | | | | | | $1s$ |
| 2 | He | ↑ | ↓ | | | | | | | | | | | | | $1s^2$ |
| 3 | Li | ↑ | ↓ | ↕ | | | | | | | | | | | | $1s^2 2s$ |
| 4 | Be | ↑ | ↓ | ↑ | ↓ | | | | | | | | | | | $1s^2 2s^2$ |
| 5 | B | ↑ | ↓ | ↑ | ↓ | ↑ | | | | | | | | | | $1s^2 2s^2 2p$ |
| 6 | C | ↑ | ↓ | ↑ | ↓ | ↑ | ↓ | | | | | | | | | $1s^2 2s^2 2p^2$ |
| 7 | N | ↑ | ↓ | ↑ | ↓ | ↑ | ↓ | ↑ | | | | | | | | $1s^2 2s^2 2p^3$ |
| 8 | O | ↑ | ↓ | ↑ | ↓ | ↑ | ↓ | ↑ | ↓ | | | | | | | $1s^2 2s^2 2p^4$ |
| 9 | F | ↑ | ↓ | ↑ | ↓ | ↑ | ↓ | ↑ | ↓ | ↑ | | | | | | $1s^2 2s^2 2p^5$ |
| 10 | Ne | ↑ | ↓ | ↑ | ↓ | ↑ | ↓ | ↑ | ↓ | ↑ | ↓ | | | | | $1s^2 2s^2 2p^6$ |
| 11 | Na | ↑ | ↓ | ↑ | ↓ | ↑ | ↓ | ↑ | ↓ | ↑ | ↓ | ↑ | | | | $1s^2 2s^2 2p^6 3s$ |
| 12 | Mg | ↑ | ↓ | ↑ | ↓ | ↑ | ↓ | ↑ | ↓ | ↑ | ↓ | ↑ | ↓ | | | $1s^2 2s^2 2p^6 3s^2$ |
| 13 | Al | ↑ | ↓ | ↑ | ↓ | ↑ | ↓ | ↑ | ↓ | ↑ | ↓ | ↑ | ↓ | ↑ | | $1s^2 2s^2 2p^6 3s^2 3p$ |
| 14 | Si | ↑ | ↓ | ↑ | ↓ | ↑ | ↓ | ↑ | ↓ | ↑ | ↓ | ↑ | ↓ | ↑ | ↓ | $1s^2 2s^2 2p^6 3s^2 3p^2$ |

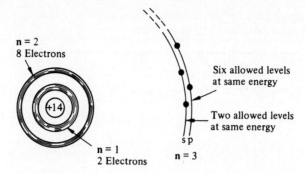

**Figure A.4** Schematic representation of the electronic configuration in an isolated, unperturbed Si atom.

and atom-atom interactions. Reflecting the information in Table A.3, and as emphasized pictorially in Fig. A.4, the valence electrons occupy the two $3s$ states and two of the six available $3p$ states. Finally, we should mention that the electronic configuration in the 32 electron Ge-atom (germanium being the other elemental semiconductor) is essentially identical to the Si-atom configuration except the Ge-core contains 28 electrons.

# Appendix B
## MOS SEMICONDUCTOR ELECTROSTATICS—EXACT SOLUTION

### Definition of Parameters

To streamline the mathematical presentation, it is customary in the exact formulation to introduce the normalized potentials

$$U(x) = \frac{\phi(x)}{kT/q} = \frac{E_i(\text{bulk}) - E_i(x)}{kT} \tag{B.1}$$

$$U_S = \frac{\phi_S}{kT/q} = \frac{E_i(\text{bulk}) - E_i(\text{surface})}{kT} \tag{B.2}$$

and

$$U_F = \frac{\phi_F}{kT/q} = \frac{E_i(\text{bulk}) - E_F}{kT} \tag{B.3}$$

$\phi(x)$, $\phi_S$, and $\phi_F$ were formally defined in Chapter 16 (also see Fig. 16.7). $U(x)$ is clearly the electrostatic potential normalized to $kT/q$ and is usually referred to as "the potential" if no ambiguity exists. Similarly, $U_S = U(x = 0)$ is known as the "surface potential." $U_F$ is simply called the doping parameter. $x$ is of course the depth into the semiconductor as measured from the oxide–semiconductor interface. Because the electric field is assumed to vanish in the semiconductor bulk (idealization 5, Section 16.1), it is permissible to treat the semiconductor as if it extended from $x = 0$ to $x = \infty$. Note that $U(x \to \infty) = 0$ in agreement with the choice of $\phi = 0$ in the semiconductor bulk.

In addition to the normalized potentials, quantitative expressions for the band bending inside of a semiconductor are normally formulated in terms of a special length parameter known as the *intrinsic Debye length*. The Debye length is a characteristic length that was originally introduced in the study of plasmas. (A plasma is a highly ionized gas containing an equal number of positive gas ions and negative electrons.) Whenever a plasma is perturbed by placing a charge in or near it, the mobile species always rearrange so as to shield the plasma proper from the perturbing charge. The Debye length is the shielding distance, or roughly the distance where the electric field emanating from the perturbing charge falls off by a factor of $1/e$. In the bulk, or everywhere under flat-band conditions, the semiconductor can be viewed as a type of plasma with its equal number of ionized impurity sites and mobile electrons or holes. The placement of charge near the semiconductor, on the MOS-C gate for example, then causes the mobile species inside the semiconductor to rearrange so as to shield the semiconductor proper from the perturbing charge. The shielding distance or band-bending region is again on the order of a Debye length, the bulk or extrinsic Debye length $L_B$, where

$$L_B = \left[ \frac{K_S \varepsilon_0 kT}{q^2(n_{bulk} + p_{bulk})} \right]^{1/2} \tag{B.4}$$

Although the bulk Debye length characterization applies only to small deviations from flat band, it is convenient to employ the Debye length appropriate for an *intrinsic* material as a normalizing factor in theoretical expressions. The *intrinsic* Debye length, $L_D$, is obtained from the more general $L_B$ relationship by setting $n_{bulk} = p_{bulk} = n_i$; that is,

$$L_D = \left[ \frac{K_S \varepsilon_0 kT}{2q^2 n_i} \right]^{1/2} \tag{B.5}$$

## Exact Solution

Expressions for the charge density, electric field, and potential as a function of position inside the semiconductor are obtained by solving Poisson's equation. Since the MOS-C is assumed to be a one-dimensional structure (idealization 7, Section 16.1), Poisson's equation simplifies to

$$\frac{d\mathscr{E}}{dx} = \frac{\rho}{K_S \varepsilon_0} = \frac{q}{K_S \varepsilon_0} (p - n + N_D - N_A) \tag{B.6}$$

Maneuvering to recast the equation in a form more amenable to solution, we note

$$\mathscr{E} = \frac{1}{q} \frac{dE_i(x)}{dx} = -\frac{kT}{q} \frac{dU}{dx} \tag{B.7}$$

The first equality in Eq. (B.7) is a restatement of Eq. (3.15) in Part I. The second equality follows from the Eq. (B.1) definition of $U$ and the fact that $dE_i(\text{bulk})/dx = 0$. In a similar vein we can write

$$p = n_i e^{[E_i(x) - E_F]/kT} = n_i e^{U_F - U(x)} \tag{B.8a}$$

$$n = n_i e^{[E_F - E_i(x)]/kT} = n_i e^{U(x) - U_F} \tag{B.8b}$$

Moreover, since $\rho = 0$ and $U = 0$ in the semiconductor bulk,

$$0 = p_{\text{bulk}} - n_{\text{bulk}} + N_{\text{D}} - N_{\text{A}} = n_i e^{U_{\text{F}}} - n_i e^{-U_{\text{F}}} + N_{\text{D}} - N_{\text{A}} \tag{B.9}$$

or

$$N_{\text{D}} - N_{\text{A}} = n_i(e^{-U_{\text{F}}} - e^{U_{\text{F}}}) \tag{B.10}$$

Substituting the foregoing $\mathscr{E}$, $p$, $n$, and $N_{\text{D}} - N_{\text{A}}$ expressions into Eq. (B.6) yields

$$\boxed{\rho = qn_i(e^{U_{\text{F}} - U} - e^{U - U_{\text{F}}} + e^{-U_{\text{F}}} - e^{U_{\text{F}}})} \tag{B.11}$$

and

$$\frac{d^2 U}{dx^2} = \left(\frac{q^2 n_i}{K_{\text{S}}\varepsilon_0 kT}\right)(e^{U - U_{\text{F}}} - e^{U_{\text{F}} - U} + e^{U_{\text{F}}} - e^{-U_{\text{F}}}) \tag{B.12}$$

or, in terms of the intrinsic Debye length,

$$\frac{d^2 U}{dx^2} = \frac{1}{2L_{\text{D}}^2}(e^{U - U_{\text{F}}} - e^{U_{\text{F}} - U} + e^{U_{\text{F}}} - e^{-U_{\text{F}}}) \tag{B.13}$$

We turn next to the main task at hand. Poisson's equation, Eq. (B.13), is to be solved subject to the boundary conditions:

$$\mathscr{E} = 0 \quad\text{or}\quad \frac{dU}{dx} = 0 \quad\text{at } x = \infty \tag{B.14a}$$

and

$$U = U_{\text{S}} \quad\text{at } x = 0 \tag{B.14b}$$

Multiplying both sides of Eq. (B.13) by $dU/dx$, integrating from $x = \infty$ to an arbitrary point $x$, and making use of the Eq. (B.14a) boundary condition, we quickly obtain

$$\mathscr{E}^2 = \left(\frac{kT/q}{L_{\text{D}}}\right)^2 [e^{U_{\text{F}}}(e^{-U} + U - 1) + e^{-U_{\text{F}}}(e^{U} - U - 1)] \tag{B.15}$$

Equation (B.15) is of the form $y^2 = a^2$, which has two roots, $y = a$ and $y = -a$. As can be deduced by inspection using the energy band diagram, we must have $\mathscr{E} > 0$ when $U > 0$ and $\mathscr{E} < 0$ when $U < 0$. Since the right-hand side of Eq. (B.15) is always positive ($a \geq 0$), the proper polarity for the electric field is obtained by choosing the positive root when $U > 0$ and the negative root when $U < 0$. We can therefore write

$$\mathscr{E} = -\frac{kT}{q}\frac{dU}{dx} = \hat{U}_\mathrm{S}\frac{kT}{q}\frac{F(U, U_\mathrm{F})}{L_\mathrm{D}} \tag{B.16}$$

where

$$F(U, U_\mathrm{F}) \equiv [e^{U_\mathrm{F}}(e^{-U} + U - 1) + e^{-U_\mathrm{F}}(e^U - U - 1)]^{1/2} \tag{B.17}$$

and

$$\hat{U}_\mathrm{S} = \begin{cases} +1 & \text{if } U_\mathrm{S} > 0 \\ -1 & \text{if } U_\mathrm{S} < 0 \end{cases} \tag{B.18}$$

To complete the solution, one separates the $U$ and $x$ variables in Eq. (B.16) and, making use of the Eq. (B.14b) boundary condition, integrates from $x = 0$ to an arbitrary point $x$. The end result is Eq. (B.19),

$$\hat{U}_\mathrm{S} \int_U^{U_\mathrm{S}} \frac{dU'}{F(U', U_\mathrm{F})} = \frac{x}{L_\mathrm{D}} \tag{B.19}$$

Although not in a totally explicit form, Eqs. (B.11), (B.16), and (B.19) collectively constitute an exact solution for the electrostatic variables. For a given $U_\mathrm{S}$, numerical techniques can be used to compute $U$ as a function of $x$ from Eq. (B.19). Once $U$ as a function of $x$ is established, direct substitution into Eqs. (B.11) and (B.16) yields $\rho$ and $\mathscr{E}$ as a function of $x$. The Fig. 16.8 plots of $U = \phi/(kT/q)$ versus $x$ and $\rho$ versus $x$ were constructed following the cited procedure.

# Appendix C
## MOS *C–V* SUPPLEMENT

An analysis based on the exact charge distribution inside an ideal MOS-C yields the following capacitance–voltage relationships:[†]

$$C = \frac{C_O}{1 + \left(\dfrac{K_O W_{eff}}{K_S x_o}\right)} \tag{C.1}$$

$$W_{eff} = \begin{cases} \hat{U}_S L_D \left[ \dfrac{2F(U_S, U_F)}{e^{U_F}(1 - e^{-U_S}) + e^{-U_F}(e^{U_S} - 1)} \right] & \ldots \text{ acc} & \text{(C.2a)} \\[4mm] \dfrac{\sqrt{2}L_D}{(e^{U_F} + e^{-U_F})^{1/2}} & \ldots \text{ flat band} & \text{(C.2b)} \\[4mm] \hat{U}_S L_D \left[ \dfrac{2F(U_S, U_F)}{e^{U_F}(1 - e^{-U_S}) + e^{-U_F}(e^{U_S} - 1)/(1 + \Delta)} \right] & \ldots \text{ depl/inv} & \text{(C.2c)} \end{cases}$$

where

$$\Delta = \begin{cases} 0 & \ldots \text{ low frequency limit} & \text{(C.3a)} \\[4mm] \dfrac{(e^{U_S} - U_S - 1)/F(U_S, U_F)}{\displaystyle\int_{0^+}^{U_S} \dfrac{e^{U_F}(1 - e^{-U})(e^U - U - 1)}{2F^3(U, U_F)} dU} & \begin{array}{l} \ldots \text{ high frequency limit} \\ (p\text{-type MOS-C}) \end{array} & \text{(C.3b)} \end{cases}$$

$$F(U, U_F) = [e^{U_F}(e^{-U} + U - 1) + e^{-U_F}(e^U - U - 1)]^{1/2} \tag{C.4}$$

---

[†] Except for Eqs. (C.2c) and (C.3b), the relationships are valid for either *n*- or *p*-type devices. The required modification of Eqs. (C.2c) and (C.3b) for *n*-type devices is noted in the text. For a derivation of the low-frequency relationship see A. S. Grove, B. E. Deal, E. H. Snow, and C. T. Sah, "Investigation of Thermally Oxidised Silicon Surfaces Using Metal-Oxide-Semiconductor Structures," *Solid-State Electronics*, **8**, 145 (1965). The high-frequency result, which includes the so-called rearrangement capacitance or capacitance contribution from the movement of inversion-layer carriers, is adapted from J. R. Brews, "An Improved High-Frequency MOS Capacitance Formula," *J. Appl. Phys.*, **45**, 1276 (1974).

$$F(U_S, U_F) = F(U=U_S, U_F) \tag{C.5}$$

$$L_D = \left[\frac{K_S \varepsilon_0 kT}{2q^2 n_i}\right]^{1/2} \tag{C.6}$$

$$U_F = \frac{\phi_F}{kT/q} \tag{C.7}$$

$$U_S = \frac{\phi_S}{kT/q} \tag{C.8}$$

$$\hat{U}_S = \begin{cases} +1 & \text{if } U_S > 0 \\ -1 & \text{if } U_S < 0 \end{cases} \tag{C.9}$$

and

$$V_G = \frac{kT}{q}\left[U_S + \hat{U}_S \frac{K_S x_0}{K_O L_D} F(U_S, U_F)\right] \tag{C.10}$$

It should be noted that $F(U, U_F)$, $L_D$, $U_F$, $U_S$, and $\hat{U}_S$ come from the exact solution for the semiconductor electrostatics. For additional information about the cited quantities, see Appendix B.

Unlike the delta-depletion result, $C$ cannot be expressed explicitly as a function of $V_G$ in the exact charge formulation. Both variables, however, have been related to $U_S$ and the capacitance expected from the structure for a given applied gate voltage can be computed numerically. The low-frequency computation is simple enough that it can be performed on a hand calculator. The usual and most efficient computational procedure is to calculate $C$ and the corresponding $V_G$ for a set of assumed $U_S$ values. Typically, a sufficient set of $(C, V_G)$ points to construct the $C$–$V_G$ characteristic will be generated if $U_S$ is stepped by whole-number units $(-5, -4, \ldots)$ over the normal operating range of $U_S$ values $(U_F - 21 \le U_S \le U_F + 21$ at room temperature). It should be noted that care must be exercised if $U_S = 0$ is included as one of the computational points. At $U_S = 0$ the Eq. (C.2b) expression for $W_{eff}$ must be employed; the accumulation and depletion/inversion relationships become indeterminate (0/0) if $U_S$ is set equal to zero. Also, Eqs. (C.2c) and (C.3b) are only valid for $p$-type devices. For $n$-type devices $\exp(U_F)[1 - \exp(-U_S)] \rightarrow \exp(-U_F)[1 - \exp(U_S)]$ and $\exp(-U_F)[\exp(U_S) - 1] \rightarrow \exp(U_F)[\exp(-U_S) - 1]$ in Eq. (C.2c), while $[\exp(U_S) - U_S - 1] \rightarrow [\exp(-U_S) + U_S - 1]$ and $\exp(U_F)\{[1 - \exp(-U)] \cdot [\exp(U) - U - 1]\} \rightarrow \exp(-U_F)\{[\exp(U) - 1][\exp(-U) + U - 1]\}$ in Eq. (C.3b). Alternatively, it is possible to obtain an $n$-type characteristic by simply running the calculations for an equivalently doped $p$-type device and then changing the sign of all computed $V_G$ values. The latter procedure works because of the voltage symmetry between ideal $n$- and $p$-type devices.

# Appendix D
## MOS *I–V* SUPPLEMENT

An analysis based on the exact charge distribution inside an ideal *n-channel* (*p*-bulk) MOSFET yields the following current-voltage relationship:

$$I_D\left(\begin{array}{c}\text{exact}\\\text{charge}\end{array}\right) = \frac{Z\bar{\mu}_n C_o}{L}\left[V_G(V_{SL} - V_{SO}) - \frac{1}{2}(V_{SL}^2 - V_{SO}^2)\right]$$

$$+ \frac{Z\bar{\mu}_n C_o}{L}\frac{K_S x_o}{K_O L_D}\left(\frac{kT}{q}\right)^2\left[\int_0^{U_{SO}} F(U, U_F, 0)\,dU - \int_0^{U_{SL}} F(U, U_F, U_D)\,dU\right]$$

$$\text{(D.1)}$$

where

$$F(U, U_F, \xi) \equiv [e^{U_F}(e^{-U} + U - 1) + e^{-U_F}(e^{U-\xi} - U - e^{-\xi})]^{1/2} \quad \text{(D.2)}$$

The corresponding charge-sheet relationship is

$$I_D\left(\begin{array}{c}\text{charge}\\\text{sheet}\end{array}\right) = \frac{Z\bar{\mu}_n C_o}{L}\left\{\left(V_G + \frac{kT}{q}\right)(V_{SL} - V_{SO}) - \frac{1}{2}(V_{SL}^2 - V_{SO}^2)\right.$$

$$\left. + V_B^2\left[\sqrt{U_{SL} - 1} - \sqrt{U_{SO} - 1} - \frac{2}{3}(U_{SL} - 1)^{3/2} + \frac{2}{3}(U_{SO} - 1)^{3/2}\right]\right\}$$

$$\text{(D.3)}$$

where

$$V_B^2 \equiv \left(\frac{kT}{q}\right)^2\frac{K_S x_o}{K_O L_D}\sqrt{\frac{N_A}{n_i}} \quad \text{(D.4)}$$

---

*Note:* The form of the relationships quoted herein are from R. F. Pierret and J. A. Shields, "Simplified Long-Channel MOSFET Theory," *Solid-State Electronics*, **26**, 143 (1983). See H. C. Pao and C. T. Sah, *Solid-State Electronics*, **9**, 927 (1966) for the original exact-charge analysis, and J. R. Brews, *Solid-State Electronics*, **21**, 345 (1978) for the original charge-sheet analysis.

In both theories

$$\phi_F = \frac{kT}{q} U_F \qquad (D.5)$$

$$V_{S0} = \frac{kT}{q} U_{S0} \qquad (D.6)$$

$$V_{SL} = \frac{kT}{q} U_{SL} \qquad (D.7)$$

and

$$V_D = \frac{kT}{q} U_D \qquad (D.8)$$

Finally, the normalized surface potentials at the source ($U_{S0}$) and drain ($U_{SL}$) are respectively computed from

$$V_G = \frac{kT}{q} \left[ U_{S0} + \frac{K_S x_o}{K_O L_D} F(U_{S0}, U_F, 0) \right] \quad \dots (U_{S0} > 0) \qquad (D.9a)$$

and

$$V_G = \frac{kT}{q} \left[ U_{SL} + \frac{K_S x_o}{K_O L_D} F(U_{SL}, U_F, U_D) \right] \quad \dots (U_{SL} > 0) \qquad (D.9b)$$

To generate a set of $I_D$–$V_D$ characteristics, $V_G$ and $V_D$ are systematically stepped over the desired range of operation. For each $V_G$ and $V_D$ combination, Eqs. (D.9a) and (D.9b) are iterated to determine $U_{S0}$ and $U_{SL}$ at the specified operating point. Once $U_{S0}$ and $U_{SL}$ are known, $I_D$ can then be computed using either Eq. (D.1) or Eq. (D.3). The process is repeated for each $V_G$–$V_D$ combination. The characteristics for a $p$-channel device can be established by running the calculations for an equivalently doped and biased $n$-channel device. Naturally, the biasing-polarities must be reversed in plotting the $p$-channel characteristics. For additional information about the exact-charge formalism, the reader is referred to Appendixes B and C.

# Appendix E
## LIST OF SYMBOLS

| | |
|---|---|
| A | anode |
| $A$ | area; arbitrary constant |
| $a$ | lattice constant; grading constant; half-width of the channel region in a J-FET; width of the channel region in a MESFET |
| $a_0$ | Bohr radius |
| $\mathscr{A}$ | Richardson's constant (120 amps/cm$^2$-K$^2$) |
| $\mathscr{A}^*$ | modified Richardson's constant (see Eq. 14.19) |
| $A_G$ | gate area |
| B | base |
| $C$ | capacitance |
| C | collector |
| $c$ | speed of light |
| $C_{cb}$ | collector-to-base capacitance in the high-frequency Hybrid-Pi model |
| $C_D$ | diffusion capacitance |
| $C_{eb}$ | emitter-to-base capacitance in the high-frequency Hybrid-Pi model |
| $C_G$ | MOSFET gate capacitance |
| $C_{gd}$ | gate-to-drain capacitance in the high-frequency, small-signal equivalent circuit for the J-FET and MOSFET |
| $C_{gs}$ | gate-to-source capacitance in the high-frequency, small-signal equivalent circuit for the J-FET and MOSFET |
| $C_J$ | junction or depletion region capacitance |
| $c_n$ | electron capture coefficient |
| $C_O$ | oxide capacitance (pF) |

| | |
|---|---|
| $C_o$ | oxide capacitance per unit area (pF/cm$^2$) |
| $c_p$ | hole capture coefficient |
| $C_S$ | semiconductor capacitance |
| D | drain |
| $D_B$ | minority-carrier diffusion coefficient in the BJT base |
| $D_C$ | minority-carrier diffusion coefficient in the BJT collector |
| $D_E$ | minority-carrier diffusion coefficient in the BJT emitter |
| $D_{IT}$ | density of interfacial traps (states/cm$^2$-eV) |
| $D_N$ | electron diffusion coefficient (cm$^2$/sec) |
| $D_{ox}$ | dielectric displacement in the oxide |
| $D_P$ | hole diffusion coefficient (cm$^2$/sec) |
| $D_{semi}$ | dielectric displacement in the semiconductor |
| E | emitter |
| $E$ | energy |
| $\mathscr{E}, \mathscr{E}$ | electric field |
| $\mathscr{E}_{ox}$ | electric field in the oxide |
| $\mathscr{E}_S$ | surface electric field, electric field in the semiconductor at the oxide–semiconductor interface |
| $\mathscr{E}_y$ | $y$-direction component of the electric field |
| $E_0$ | vacuum level, minimum energy an electron must possess to completely free itself from a material |
| $E_A$ | acceptor energy level |
| $E_B$ | binding energy at dopant (donor, acceptor) sites |
| $E_c$ | minimum conduction band energy |
| $E_D$ | donor energy level |
| $E_F$ | Fermi energy or Fermi level |
| $E_{FM}$ | Fermi level in the metal |
| $E_{Fn}$ | Fermi level on the $n$-side of a $pn$ junction |
| $E_{Fp}$ | Fermi level on the $p$-side of a $pn$ junction |
| $E_{FS}$ | Fermi level in the semiconductor |

| | |
|---|---|
| $E_G$ | band gap or forbidden gap energy |
| $E_H$ | electron binding energy within the hydrogen atom |
| $E_i$ | intrinsic Fermi level |
| $E_n$ | energy corresponding to the **n** quantum number |
| $E_{ph}$ | photon energy ($h\nu$) |
| $E_T$ | trap or R–G center energy level |
| $E_v$ | maximum valence band energy |
| **F** | force |
| $f$ | frequency (Hz) |
| $f(E)$ | Fermi function |
| $F(U, U_F)$ | field function (see Eq. B.17) |
| $F_{1/2}$ | Fermi-Dirac integral of order 1/2 |
| $FF$ | fill factor |
| $f_{max}$ | maximum operational frequency of a J-FET or MOSFET, cutoff frequency |
| $F_N$ | quasi-Fermi level (or energy) for electrons |
| $F_P$ | quasi-Fermi level (or energy) for holes |
| $f_T$ | unity beta frequency of a BJT |
| $G$ | conductance |
| G | gate |
| $G_0$ | low frequency conductance of a *pn* junction diode; channel conductance in a J-FET if there were no depletion regions |
| $g_c(E)$ | density of conduction band states |
| $G_D$ | diffusion conductance |
| $g_d$ | drain or channel conductance |
| $G_L$ | photogeneration rate, number of electron-hole pairs created per $cm^3$-sec |
| $g_m$ | transconductance |
| $g_v(E)$ | density of valence band states |
| $h$ | Planck's constant |
| $\hbar$ | $h/2\pi$ |
| $I$ | current; light intensity |

| | |
|---|---|
| $i$ | a.c. current; $\sqrt{-1}$ in Appendix A |
| $I_0$ | saturation current in an ideal diode; light intensity at $x = 0$ |
| $I_{AK}$ | anode to cathode current |
| $I_B$ | d.c. base current |
| $i_B$ | total (a.c. + d.c.) base current |
| $i_b$ | a.c. base current |
| $I_C$ | d.c. collector current |
| $i_C$ | total (a.c. + d.c.) collector current |
| $i_c$ | a.c. collector current |
| $I_{CB0}$ | collector to base current when $I_E = 0$ |
| $I_{CE0}$ | collector to emitter current when $I_B = 0$ |
| $I_{Cn}$ | d.c. collector current due to electrons |
| $I_{Cp}$ | d.c. collector current due to holes |
| $I_D$ | d.c. drain current in a field-effect transistor |
| $i_d$ | small-signal drain current |
| $I_{D0}$ | $V_G = 0$ saturation drain current in a J-FET |
| $I_{dark}$ | dark current |
| $I_{DIFF}$ | diffusion current (same as the ideal diode current) |
| $i_{diff}$ | a.c. component of the diffusion current |
| $I_{Dsat}$ | saturation drain current |
| $I_E$ | d.c. emitter current |
| $I_{En}$ | d.c. emitter current due to electrons |
| $I_{Ep}$ | d.c. emitter current due to joles |
| $I_F$ | steady-state forward-bias current |
| $I_{F0}$ | effective diode forward saturation current (Ebers–Moll model) |
| $I_G$ | gate current in an SCR |
| $I_L$ | current due to light |
| $I_{M\bullet \rightarrow S}$ | current due to electrons drifting from the metal to the semiconductor in an MS ($n$-type) diode |
| $I_P$ | hole current |

| $I_{P|drift}$ | hole current due to drift |
|---|---|
| $I_R$ | steady-state reverse-bias current |
| $I_{R-G}$ | recombination–generation current |
| $I_{R0}$ | effective diode reverse saturation current (Ebers–Moll model) |
| $I_s$ | reverse-bias saturation current in an MS diode |
| $I_{sc}$ | short-circuit current in a solar cell |
| $I_{S\bullet\to M}$ | current due to electrons drifting from the semiconductor to the metal in an MS ($n$-type) diode |
| $j$ | $\sqrt{-1}$ |
| $\mathbf{J}, J$ | current density (amps/cm$^2$) |
| $\mathbf{J}_{drift}$ | total current density due to drift |
| $\mathbf{J}_N, J_N$ | electron current density |
| $J_{Nx}, J_{Ny}, J_{Nz}$ | $x$, $y$, and $z$ direction components of the electron current density |
| $\mathbf{J}_{N|diff}$ | electron current density due to diffusion |
| $\mathbf{J}_{N|drift}$ | electron current density due to drift |
| $\mathbf{J}_P, J_P$ | hole current density |
| $J_{Px}, J_{Py}, J_{Pz}$ | $x$, $y$, and $z$ direction components of the hole current density |
| $\mathbf{J}_{P|diff}$ | hole current density due to diffusion |
| $\mathbf{J}_{P|drift}$ | hole current density due to drift |
| K | cathode |
| $k$ | Boltzmann constant ($8.617 \times 10^{-5}$ eV/K) |
| $\mathbf{k}$ | wavenumber (parameter proportional to the electron crystal momentum) |
| K.E. | kinetic energy |
| $K_O$ | oxide dielectric constant |
| $K_S$ | semiconductor (usually Si) dielectric constant |
| $L$ | length of the J-FET or MOSFET channel |
| $l$ | azimuthal quantum number |
| $L'$ | reduced channel length defined in Fig. 19.4 |
| $L_B$ | minority carrier diffusion length in the BJT base; extrinsic Debye length |
| $L_C$ | minority carrier diffusion length in the BJT collector |

| | |
|---|---|
| $L_D$ | intrinsic Debye length |
| $L_E$ | minority carrier diffusion length in the BJT emitter |
| $L_{min}$ | minimum MOSFET channel length yielding long-channel behavior |
| $L_N$ | electron minority carrier diffusion length |
| $L_\mathbf{n}$ | angular momentum corresponding to the $\mathbf{n}$ quantum number |
| $L_P$ | hole minority-carrier diffusion length |
| $M$ | carrier multiplication factor |
| $m$ | particle mass |
| $\mathbf{m}$ | magnetic orbital quantum number |
| $m_0$ | electron rest mass |
| $m_n^*$ | electron effective mass |
| $m_p^*$ | hole effective mass |
| $\mathbf{n}$ | energy quantum number |
| $n$ | electron carrier concentration (number of electrons/cm$^3$) |
| $n^+$ | heavily doped $n$-type material |
| $n_0$ | equilibrium electron concentration |
| $n_1$ | defined electron concentration (see Eq. 3.36a) |
| $N_A$ | total number of acceptor atoms/cm$^3$ |
| $N_A^-$ | number of ionized (negatively charged) acceptors/cm$^3$ |
| $N_B$ | bulk semiconductor doping ($N_A$ or $N_D$ as appropriate); doping concentration in the BJT base |
| $n_{bulk}$ | electron concentration in the semiconductor bulk |
| $N_C$ | effective density of conduction band states; doping concentration in the BJT collector |
| $n_{C0}$ | equilibrium electron concentration in the collector of a $pnp$ BJT |
| $N_D$ | total number of donor atoms/cm$^3$ |
| $N_D^+$ | number of ionized (positively charged) donors/cm$^3$ |
| $N_E$ | doping concentration in the BJT emitter |
| $n_{E0}$ | equilibrium electron concentration in the emitter of a $pnp$ BJT |
| $N_I$ | number of implanted ions/cm$^2$ |

| $n_i$ | intrinsic carrier concentration |
|---|---|
| $N_T$ | number of R–G centers/cm$^3$ |
| $N_V$ | effective density of valence band states |
| $p$ | hole concentration (number of holes/cm$^3$); momentum in Appendix A |
| $p^+$ | heavily doped $p$-type material |
| P.E. | potential energy |
| $p_0$ | equilibrium hole concentration |
| $p_1$ | defined hole concentration (see Eq. 3.36b) |
| $p_{B0}$ | equilibrium hole concentration in the base of a $pnp$ BJT |
| $p_{bulk}$ | hole concentration in the semiconductor bulk |
| $p_s$ | hole concentration at the semiconductor surface (number/cm$^3$) |
| $Q$ | general designation for a charge |
| $q$ | magnitude of the electronic charge ($1.60 \times 10^{-19}$ coul) |
| $Q_B$ | excess minority carrier charge in the quasineutral base; bulk or depletion-region charge per unit area of the MOSFET gate |
| $Q_{BL}$ | $Q_B$ in a long-channel MOSFET |
| $Q_{BS}$ | $Q_B$ in a short-channel MOSFET |
| $Q_F$ | fixed oxide charge per unit area at the oxide–semiconductor interface |
| $Q_I$ | implant-related charge/cm$^2$ located at the oxide–semiconductor interface |
| $Q_{IT}$ | net charge per unit area associated with the interfacial traps |
| $Q_M$ | total mobile ion charge within the oxide per unit area of the MOS gate |
| $Q_N$ | total electronic charge/cm$^2$ in the MOSFET channel ($n$-channel device) |
| $Q_{O\text{-}S}$ | charge per unit area located at the oxide–semiconductor interface |
| $Q_P$ | excess hole charge |
| $Q_S$ | total charge in the semiconductor per unit area of the gate |
| $R$ | ramp rate (see Fig. 16.17) |
| $r_B, r_b$ | base resistance |
| $r_C, r_c$ | collector resistance |
| $R_D$ | channel-to-drain resistance in a J-FET or MOSFET |
| $r_E, r_e$ | emitter resistance |

| | |
|---|---|
| $r_j$ | depth of the source and drain islands in a MOSFET |
| $R_L$ | load resistor |
| $r_n$ | radius of the Bohr orbit corresponding to the **n** quantum number |
| $r_o$ | output resistance in the BJT Hybrid-Pi model |
| $R_P$ | projected range in ion implantation |
| $R_S$ | series resistance; sample resistance; source-to-channel resistance in a J-FET or MOSFET |
| $r_\mu$ | feedthrough resistance in the BJT Hybrid-Pi model |
| $r_\pi$ | input resistance in the BJT Hybrid-Pi model |
| S | source |
| $s$ | probe-to-probe spacing in a four-point probe |
| **s** | spin quantum number |
| $T$ | temperature |
| $t$ | time |
| $t_{ON}$ | triggering time in an SCR |
| $TR$ | tuning ratio |
| $t_r$ | recovery time (*pn* diode); rise time (BJT) |
| $t_{rr}$ | reverse recovery time (*pn* diode) |
| $t_s$ | storage delay time (*pn* diode) |
| $t_{sd}$ | storage delay time (BJT) |
| $U$ | potential energy in Appendix A; electrostatic potential normalized to $kT/q$ in Appendices B–D |
| $U_D$ | drain voltage normalized to $kT/q$ |
| $U_F$ | semiconductor doping parameter |
| $U_S$ | normalized surface potential, $U$ evaluated at the oxide–semiconductor interface |
| $\hat{U}_S$ | sign ($\pm$) of $U_S$ |
| $U_{S0}$ | normalized surface potential at $x = 0$ in a MOSFET |
| $U_{SL}$ | normalized surface potential at $x = L$ in a MOSFET |
| $V$ | voltage, electrostatic potential |

| | |
|---|---|
| $\mathcal{V}$ | volume |
| $\boldsymbol{v}$ | velocity |
| $V_A$ | applied d.c. voltage |
| $v_a$ | applied a.c. voltage |
| $V_{AK}$ | anode-to-cathode voltage |
| $V_B$ | defined voltage (see Eq. D.4) |
| $v_{be}$ | a.c. base-to-emitter voltage |
| $V_{BF}$ | forward-bias blocking voltage in PNPN devices |
| $V_{bi}$ | "built-in" junction voltage |
| $V_{BR}$ | reverse-bias $pn$ junction breakdown voltage; reverse-bias blocking voltage in PNPN devices |
| $V_{BS}$ | back-to-source voltage |
| $V_{CB}$ | d.c. collector-to-base voltage |
| $V_{CB0}$ | collector-to-base breakdown voltage when $I_E = 0$ |
| $v_{ce}$ | a.c. collector-to-emitter voltage |
| $V_{CE0}$ | collector-to-emitter breakdown voltage when $I_B = 0$ |
| $V_D$ | d.c. drain voltage |
| $\boldsymbol{v}_d$ | drift velocity vector |
| $v_d$ | drift velocity; a.c. drain voltage |
| $V_{DS}$ | drain-to-source voltage |
| $V_{Dsat}$ | saturation drain voltage |
| $v_{sat}$ | saturation drift velocity |
| $V_{EB}$ | d.c. emitter-to-base voltage |
| $V_{EC}$ | d.c. emitter-to-collector voltage |
| $V_{FB}$ | flat-band voltage |
| $V_G$ | d.c. gate voltage |
| $v_g$ | a.c. gate voltage |
| $V_G'$ | d.c. gate voltage applied to an ideal device |
| $V_{GB}'$ | gate-to-back voltage being applied to an ideal device |
| $V_{GS}$ | gate-to-source voltage |

| | |
|---|---|
| $V'_{GS}$ | gate-to-source voltage being applied to an ideal device |
| $V_J$ | junction voltage |
| $V_{oc}$ | open circuit voltage of a solar cell |
| $V_P$ | pinch-off gate voltage in a J-FET |
| $V_S$ | d.c. source voltage |
| $v_s$ | pulsed source voltage |
| $V_{S0}$ | surface potential at $x = 0$ in a MOSFET |
| $V_{SL}$ | surface potential at $x = L$ in a MOSFET |
| $V_T$ | inversion-depletion transition point gate voltage, MOSFET threshold or turn-on voltage |
| $V'_T$ | ideal device inversion–depletion transition point gate voltage |
| $V_W$ | defined voltage (see Eq. 17.24) |
| $V_\delta$ | defined voltage (see Eq. 16.36) |
| $W$ | depletion width; quasineutral width of the BJT base |
| $W_2$ | quasineutral width of the N2 base in an SCR |
| $W_3$ | quasineutral width of the P3 base in an SCR |
| $W_B$ | total width of the base in a BJT |
| $W_D$ | drain $pn$ junction depletion width in a MOSFET |
| $W_{eff}$ | effective depletion width in a MOSFET (see Appendix C) |
| $W_{N2}$ | width of the N2 base in an SCR |
| $W_{P3}$ | width of the P3 base in an SCR |
| $W_S$ | source $pn$ junction depletion width in a MOSFET |
| $W_T$ | MOS depletion width when the semiconductor is biased at the inversion–depletion transition point |
| $x_c$ | width of the base in a narrow-base $pn$ junction diode; depth of the MOSFET channel |
| $x_n$ | $n$-side width of the $pn$ junction depletion region |
| $x_o$ | oxide thickness |
| $x_p$ | $p$-side width of the $pn$ junction depletion region |
| $Y$ | admittance |

| | |
|---|---|
| $Y_D$ | diffusion admittance |
| $Z$ | width of the J-FET or MOSFET channel |
| $\alpha$ | absorption coefficient |
| $\alpha_{dc}$ | common base d.c. current gain |
| $\alpha_F$ | $\alpha_F = \alpha_{dc}$, forward gain (Ebers–Moll model) |
| $\alpha_R$ | reverse gain (Ebers–Moll model) |
| $\alpha_T$ | base transport factor |
| $\beta_{dc}$ | common emitter d.c. current gain |
| $\chi$ | semiconductor electron affinity |
| $\chi'$ | $\chi' = \chi - \chi_i$, effective semiconductor electron affinity in an MOS structure |
| $\chi_i$ | insulator (oxide) electron affinity |
| $\chi_{Si}$ | silicon electron affinity |
| $\Delta$ | frequency parameter in the exact-charge $C$–$V$ theory (see Eqs. C.3) |
| $\Delta E_c$ | conduction band offset energy in a heterojunction |
| $\Delta E_v$ | valence band offset energy in a heterojunction |
| $\Delta\phi_{ox}$ | voltage drop across the oxide |
| $\Delta\phi_{semi}$ | voltage drop across the semiconductor |
| $\Delta L$ | decrease in the channel length under above-pinch-off conditions |
| $\Delta n$ | $\Delta n = n - n_0$, deviation in the electron concentration from its equilibrium value |
| $\Delta n_C$ | excess electron concentration in the collector of a *pnp* BJT |
| $\Delta n_E$ | excess electron concentration in the emitter of a *pnp* BJT |
| $\Delta n_p$ | $\Delta n$ in *p*-type material |
| $\Delta p$ | $\Delta p = p - p_0$, deviation in the hole concentration from its equilibrium value |
| $\Delta p_B$ | excess hole concentration in the base of a *pnp* BJT |
| $\Delta p_n$ | $\Delta p$ in *n*-type material |
| $\Delta Q$ | general designation for a change in charge |
| $\Delta Q_{gate}$ | change in the gate charge/cm$^2$ |
| $\Delta Q_{semi}$ | change in the charge/cm$^2$ inside the semiconductor |

| | |
|---|---|
| $\Delta R_p$ | straggle in ion implantation |
| $\Delta V_G$ | difference between the actual device and ideal device gate voltage required to achieve a given semiconductor surface potential |
| $\Delta V_T$ | change in the threshold voltage due to small-dimension effects |
| $\Delta V_T'$ | change in the threshold voltage due to back biasing (specifically applied to ideal devices) |
| $\Delta \Phi_B$ | change in the MS barrier height caused by Schottky barrier lowering |
| $\varepsilon$ | permittivity |
| $\varepsilon_0$ | permittivity of free space ($8.85 \times 10^{-14}$ farad/cm) |
| $\phi$ | electrostatic potential inside the semiconductor component of an MOS device |
| $\Phi_B$ | surface potential-energy barrier height in an MS diode |
| $\Phi_{B0}$ | $\Phi_B$ barrier height when $\mathscr{E} = 0$ at the MS interface |
| $\phi_F$ | reference voltage related to the semiconductor doping concentration |
| $\Phi_M$ | metal workfunction |
| $\Phi_M'$ | $\Phi_M' = \Phi_M - \chi_i$, effective metal workfunction in an MOS structure |
| $\phi_{MS}$ | metal–semiconductor workfunction difference expressed in volts |
| $\phi_{ox}$ | voltage inside the oxide |
| $\Phi_S$ | semiconductor workfunction |
| $\phi_S$ | semiconductor surface potential |
| $\Gamma$ | four-point probe correction factor |
| $\gamma$ | emitter efficiency |
| $\gamma_M$ | normalized centroid of mobile ion charge in the oxide |
| $\eta$ | $\eta = (E - E_c)/kT$; power conversion efficiency of a solar cell; external efficiency of an LED |
| $\eta_c$ | $\eta_c = (E_F - E_c)/kT$ |
| $\eta_v$ | $\eta_v = (E_v - E_F)/kT$ |
| $\lambda$ | wavelength of light |
| $\lambda_G$ | wavelength of light corresponding to the semiconductor band gap |
| $\mu_0$ | low-field electron or hole mobility; mobility fit parameter |
| $\mu_{bulk}$ | carrier mobility in the bulk of a semiconductor |

| | |
|---|---|
| $\mu_n$ | electron mobility |
| $\mu_p$ | hole mobility |
| $\bar{\mu}_n$ | effective electron mobility |
| $\bar{\mu}_p$ | effective hole mobility |
| $\nu$ | frequency of light |
| $\rho$ | resistivity (ohm-cm); charge density (coul/cm$^3$) |
| $\rho_{ion}$ | ionic charge density inside the oxide |
| $\rho_{ox}$ | charge density in the oxide |
| $\sigma$ | conductivity |
| $\tau_0$ | defined carrier lifetime (see Eq. 6.44) |
| $\tau_B$ | minority-carrier lifetime in the BJT base |
| $\tau_C$ | minority-carrier lifetime in the BJT collector |
| $\tau_E$ | minority-carrier lifetime in the BJT emitter |
| $\tau_n$ | electron minority-carrier lifetime |
| $\tau_p$ | hole minority-carrier lifetime |
| $\tau_t$ | base transit time |
| $\Psi$ | time-dependent wavefunction |
| $\psi$ | time-independent wavefunction |
| $\omega$ | angular frequency (radians) |

# Appendix M
## MATLAB Program Script

---

**Exercise 10.2 (BJT _ Eband)**

% BJT Equilibrium Energy Band Diagram Generator
% This program plots out the BJT equilibrium energy band diagram

% Original version authored by Aaron Luft as a course project for Prof. Gerry Neudeck
% Major revisions by R. F. Pierret

DOPING=[1e18 -1e16 1e15]; % E, B, and C type and doping concentrations (- = n-type)
WB=1.0e-4; %Total base width in cm; 1.0e-4cm=1micrometer
close

%Constants
T=300;          % Temperature in Kelvin
k=8.617e-5;     % Boltzmann constant eV/K
e0=8.85e-14;    % permittivity of free space (f/cm)
q=1.602e-19;    % charge on an electron (coul)
KS=11.8;        % Dielectric constant of Si at 300K
ni=1.0e10;      % intrinsic conc. of Silicon at 300K
EG=1.12;        % Silicon band gap (eV)
%end constants

%General Computations and Manipulations
NE = DOPING (1);          % Emitter doping and type
NB = DOPING (2);          % Base doping and type
NC = DOPING (3);          % Collector doping and type

sE = sign (NE);
sB = sign (NB);
sC = sign (NC);

NE = abs(NE);             % Emitter doping
NB = abs(NB);             % Base doping
NC = abs(NC);             % Collector doping

Ei_emitter = [ (sE * k * T * log (NE / ni) ) ...
          (-sB * k * T * log ( NB / ni ) ) ];

```
Ei_collector = [ ( (sB * k * T * log ( NB / ni ) ) ...
                ( -sC * k * T * log ( NC / ni ) ) ];

Vbi  = [ (sum (Ei_emitter)) (sum (Ei_collector)) ];
svbi = sign (Vbi);
Vbi  = abs  (Vbi);

                            % Depletion width on emitter side of EB junction
xE = sqrt(2*KS*e0/q*NB*Vbi(1)/(NE*(NB+NE)));
                            % Depletion width on base side of EB junction
xBeb = sqrt(2*KS*e0/q*NE*Vbi(1)/(NB*(NE+NB)));

                            % Depletion width on base side of CB junction
xBcb = sqrt(2*KS*e0/q*NC*Vbi(2)/(NB*(NC+NB)));
                            % Depletion width on collector side of EB junction
xC = sqrt(2*KS*e0/q*NB*Vbi(2)/(NC*(NB+NC)));

W = WB-xBeb-xBcb;

if W < 0
error('For the given DOPING and WB, the base is totally depleted.')
end

if ( xC > xE )          % Adjust the x-axis for optimum looking plot
HIGH_X = 1.5;
LOW_X = xC/xE;
else
HIGH_X = xE/xC;
LOW_X = 1.5;
end

VMAX = 3;                       % Maximum Plot Voltage
plot ( [-LOW_X*xE HIGH_X*xC+WB ] , [ 0 VMAX ] , 'i');
hold on;

% EB JUNCTION
xlft = -LOW_X*xE;               % Leftmost x position
xrght = xBeb + W/2;             % Rightmost x position

x = linspace(xlft, xrght, 200);
sVx = -svbi(1) * sE * sB;

Vx1=sVx * (Vbi(1)-q*NB.*(xBeb-x).^2/(2*KS*e0).*(x<=xBeb)).*(x>=0);
Vx2=sVx * 0.5*q*NE.*(xE+x).^2/(KS*e0).*( x>=-xE & x<0 );
Vx=Vx1+Vx2;                                         % V as a function of x

EF=Vx(1)+VMAX/2-sE*k*T*log(NE/ni);                  % Fermi level

Ec = -Vx+EG/2+VMAX/2;
Ev = -Vx-EG/2+VMAX/2;
Ei = -Vx+VMAX/2;
```

```
LEc = Ec (1);
LEv = Ev (1);
LEi = Ei (1);

% Plot V vs x
plot ( x, Ec );                    % Ec
plot ( x, Ev );                    % Ev
plot ( x, Ei, 'w:');               % Ei
plot ( [xlft 0], [ EF EF ], 'w' );                 % EF on left
plot ( [ 0 0 ], [ 0.15 VMAX-0.15 ], 'w--' );       % Junction center

% CB JUNCTION
xlft = -xBcb-W/2;                  % Leftmost x position
xrght = HIGH_X*xC;                 % Rightmost x position

x = linspace(xlft, xrght, 200);
sVx = -svbi(2) * sC * sB;

Vx1=sVx * ( Vbi(2)-q*NC.*(xC-x).^2/(2*KS*e0).*(x<=xC)).*(x>=0);
Vx2=sVx * 0.5*q*NB.*(xBcb+x).^2/(KS*e0).*( x>=-xBcb & x<0 );
Vx=Vx1+Vx2;    % V as a function of x

OFFSET = (Ec(200))-(-Vx(1)+EG/2+VMAX/2);
Ec = (-Vx+EG/2+VMAX/2) + OFFSET;
Ev = (-Vx-EG/2+VMAX/2) + OFFSET;
Ei = (-Vx+VMAX/2) + OFFSET;

x = x + WB;

% Plot V vs x
plot ( x, Ec );                    % Ec
plot ( x, Ev );                    % Ev
plot ( x, Ei, 'w:');               % Ei
plot ([0 xrght+ WB], [EF EF], 'w');                % EF on right
plot ( [ WB WB ], [ 0.15 VMAX-0.15 ], 'w--' );     % Junction center

if ( sC == -1 )
    RIGHT = 'N';
else
    RIGHT = 'P';
end
if ( sB == -1 )
    MIDL = 'N';
else
    MIDL = 'P';
end
if ( sE == -1 )
    LEFT = 'N';
```

```
else
    LEFT = 'P';
end

A = -LOW_X*xE/2;
B = WB + ((HIGH_X*xC+WB)-WB)/2;

text ( A, 2.5, LEFT );
text ( WB/2, 2.5, MIDL );
text ( B, 2.5, RIGHT);

text ( x(200), Ec (200), 'Ec' );
text ( x(200), Ei (200), 'Ei' );
text ( x(200), Ev (200), 'Ev' );
text ( x(200), EF, 'EF' );

REG = [ LEFT(1) MIDL(1) RIGHT(1) ];
TITLE = [ ('Energy band diagram for the ') (REG) (' device') ];
title (TITLE);
```

---

### Exercise 11.7 (BJT) and Exercise 11.10 (BJT plus)

NOTE: The italicized lines in the BJT/BJTplus listing are added to the BJT program to form the BJTplus program. Subprogram BJT0 is a run-time requirement of both BJT and BJTplus; subprogram BJTmod must also be present when BJTplus is executed. Constants, material parameters, and the $W = W_B$ Ebers–Moll parameters are specified or computed in the BJT0 subprogram. Computations related to base-width modulation are performed in the BJTmod subprogram.

**BJT/BJTplus**
%BJT Common Base/Emitter Input/Output Characteristics
*%Modified version of BJT including Base-Width Modulation and*
    *%Carrier Multiplication*

%Input Ebers-Moll Parameters
BJT0

%Limiting Voltages used in Calculation
VbiE=kT*log(NE*NB/ni^2);
VbiC=kT*log(NC*NB/ni^2);
VCB0=50; VCE0=50;
*VCB0= 60*(NC/1.0e16)^(-3/4);*
*m=6; VCE0=VCB0*(1-aF)^(1/m);*

%Choice of Characteristic *and Special Calculations*
format compact
echo on

```
%THIS PROGRAM COMPUTES BJT INPUT AND OUTPUT CHARACTERISTICS
% Subprograms BJT0 and BJTmod are run-time requirements.
% Modify entries in BJT0 to change device/material parameters.
% Modify axis commands to change plot min/max values.
echo off
close
c=menu('Specify the desired characteristic','Common Base Input',...
   'Common Base Output','Common Emitter Input','Common Emitter Output');
j=input('Specify number of curves per plot...');
if c~=2,
   bw=input('Include base-width modulation? 1-Yes, 2-No...');
   else
   end
ii=2;
if c==4 & bw==1,
   ii=input('Include impact ionization? 1-Yes, 2-No...');
   else
   end

%Calculation Proper
for i=1:j,

   %Common-Base Input Characteristics
   if c==1,
   VCB=-(i-1)*10;
   VEB=0:0.005:VbiE;
   jj=length(VEB);
   if bw==1,
   BJTmod    %Base-Width Modulation subprogram
   else
   end
IE=(IF0.*(exp(VEB/kT)-1) - aR.*IR0.*(exp(VCB/kT)-1))*1.0e3;
%1.0e3 in the preceeding equation changes IE units to mA
if i==1,
plot(VEB,IE);  axis ([0.35 0.85 0 5]);
grid; xlabel('VEB(volts)'); ylabel('IE(mA)');
else plot(VEB,IE);
end

%Common-Base Output Characteristics
elseif c==2,
IE=(j-i)*1.0e-3;
VCB1=2:-0.01:0;
VCB2=0:-VCB0/200:-VCB0;
VCB=[VCB1,VCB2];
```

```
jj=length(VCB);
IC=(aF*IE-(1-aF*aR)*IR0*(exp(VCB/kT)-1))*(1.0e3);
if i==1,
plot(-VCB,IC); axis([-VCB0/10 VCB0 0 1.3e3*IE]);
grid; xlabel('-VCB(volts)'); ylabel('IC(mA)');
text(5,1.1e3*IE,'IEstep=1mA');
else plot(-VCB,IC);
end

else
end

%Common-Emitter Input Characteristics
if c==3,
VEC=(i-1)*5;
VEB=0:0.005:VbiE;
jj=length(VEB);
   if bw==1,
   VCB=VEB-VEC;
   BJTmod
   else
   end
IB0=(1-aF).*IF0+(1-aR).*IR0;
IB1=(1-aF).*IF0+(1-aR).*IR0.*exp(-VEC/kT);
IB=(IB1.*exp(VEB/kT)-IB0)*(1.0e6);
if i==1,
plot(VEB,IB); axis([.35 .85 -5 20]);
grid; xlabel('VEB(volts)'); ylabel ('IB(µA)');
else plot(VEB,IB);
end

%Common-Emitter Output Characteristics
elseif c==4,
IB=(j-i)*2.5e-6;
VECA=0:0.01:VCE0/50;
VECB=VCE0/50:VCE0/200:VCE0;
VEC=[VECA,VECB];
jj=length(VEC);
    if bw==1,
    VEB=0; %Neglect xnEB variation with bias
    VCB=VEB-VEC;
    BJTmod
    else
    end
    if ii==1,
```

```
    M=1.0./(1-(-VCB/VCB0).^m);
    aF=M.*aF;
    else
    end
IB0=(1-aF).*IF0+(1-aR).*IR0;
IB1=(1-aF).*IF0+(1-aR).*IR0.*exp(-VEC/kT);
IC=((aF.*IF0-IR0.*exp(-VEC/kT)).*(IB+IB0)./IB1+IR0-aF.*IR0)*(1.0e3);
if i==1,
jA=length(VECA);
plot(VEC,IC); axis([0 VCE0 0 2.5*IC(jA)]);
grid; xlabel('VEC(volts)'); ylabel('IC(mA)');
text(5,2*IC(jA),'IBstep=2.5μA');
else plot(VEC,IC);
end

    else
    end

hold on
end
hold off
```

**BJT0**

```
%BJT Constants and Ebers-Moll Parameters (subprogram BJT0)

%Universal Constants
q=1.602e-19;
k=8.617e-5;
e0=8.85e-14;

%Device/Miscellaneous Parameters
A=1.0e-4;      %A in cm2
WB=2.5e-4;   %WB in cm
T=300; kT=k*T;

%Material Parameters
ni=1.0e10;
KS=11.8;
NE=1.0e18;
NB=1.5e16;
NC=1.5e15;
   %Mobility Fit Parameters
   NDref=1.3e17; NAref=2.35e17;
   μnmin=92; μpmin=54.3;
   μn0=1268; μp0=406.9;
   an=0.91; ap=0.88;
```

μE=μnmin+μn0./(1+(NE/NDref).^an);
μB=μpmin+μp0./(1+(NB/NAref).^ap);
μC=μnmin+μn0./(1+(NC/NDref).^an);
TauE=1.0e-7;
TauB=1.0e-6;
TauC=1.0e-6;
DE=kT*μE;
DB=kT*μB;
DC=kT*μC;
LE=sqrt(DE*TauE);
LB=sqrt(DB*TauB);
LC=sqrt(DC*TauC);
nE0=ni^2/NE;
pB0=ni^2/NB;
nC0=ni^2/NC;

%Ebers-Moll Parameter Computation (W = WB)
W=WB;
fB=(DB/LB)*pB0*(cosh(W/LB)/sinh(W/LB));
IF0=q*A*((DE/LE)*nE0+fB);
IR0=q*A*((DC/LC)*nC0+fB);
aF=q*A*(DB/LB)*(pB0/sinh(W/LB))/IF0;
aR=q*A*(DB/LB)*(pB0/sinh(W/LB))/IR0;

## BJTmod
%Base-width modulation-included calculation of Ebers-Moll parameters
%Subprogram BJTmod

xnEB=sqrt((2*KS*e0/q)*(NE/(NB*(NE+NB)))*(VbiE-VEB));
xnCB=sqrt((2*KS*e0/q)*(NC/(NB*(NC+NB)))*(VbiC-VCB));
W=WB-xnEB-xnCB;
fB=(DB/LB)*pB0*(cosh(W/LB)./sinh(W/LB));
IF0=q*A.*((DE/LE)*nE0+fB);
IR0=q*A.*((DC/LC)*nC0+fB);
aF=q*A*(DB/LB)*(pB0./sinh(W/LB))./IF0;
aR=q*A*(DB/LB)*(pB0./sinh(W/LB))./IR0;

---

### Exercise 16.5 (MOS_CV)

%LOW and/or HIGH-frequency p-type MOS-C C-V CHARACTERISTICS
%Subprogram CVintgrd is a run-time requirement.

%Initialization and Input
format compact
close
clear

```
s=menu('Choose the desired plot','Low-f C-V','High-f C-V','Both');
NA=input('Please input the bulk doping in /cm3, NA=');
xo=input('Please input the oxide thickness in cm, xo=');
xmin=input('Specify VGmin(volts), VGmin=');
xmax=input('Specify VGmax(volts), VGmax=');
global UF

%Constants and Parameters
e0=8.85e-14;
q=1.6e-19;
k=8.617e-5;
KS=11.8;
KO=3.9;
ni=1.0e10;
T=300;
kT=k*T;

%Computed Constants
UF=log(NA/ni);
LD=sqrt((kT*KS*e0)/(2*q*ni));

%Gate Voltage Computation
US=UF-21:0.5:UF+21;
F=sqrt(exp(UF).*(exp(-US)+US-1)+exp(-UF).*(exp(US)-US-1));
VG=kT*(US+(US./abs(US)).*(KS*xo)/(KO*LD).*F);

%Low-frequency Capacitance Computation
DENOML=exp(UF).*(1-exp(-US))+exp(-UF).*(exp(US)-1);
WL=(US./abs(US)).*LD.*(2*F)./DENOML;
cL=1.0./(1+(KO*WL)./(KS*xo));

%High-frequency Capacitance Computation
if s~=1,
  jj=length(US);
  nn=0;
  for ii=1:jj,
    if US(ii) < 3,
      elseif nn==0,
      INTG=QUAD('CVintgrd',3,US(ii),0.001);
      nn=1;
      else
      INTG=INTG+QUAD('CVintgrd',US(ii-1),US(ii),0.001);
      end
    if US(ii) < 3,
      cH(ii)=cL(ii);
      else
      d=(exp(US(ii))-US(ii)-1)./(F(ii).*exp(UF).*INTG);
```

```
        DENOMH=exp(UF).*(1-exp(-US(ii)))+exp(-UF).*((exp(US(ii))-1)./(1+d));
        WH=LD.*(2*F(ii))./DENOMH;
        cH(ii)=1.0./(1+(KO*WH)./(KS*xo));
        end
    end
else
end

%Plotting the Result
if s==1,
plot(VG,cL);
elseif s==2,
plot(VG,cH);
else
plot(VG,cL,'--',VG,cH);
text(0.8*xmin,.17,'---Low-f','color',[1,1,0]);
text(0.8*xmin,.12,'__High-f','color',[1,0,1]);
end
axis([xmin,xmax,0,1]);
text(0.8*xmin,.27,['NA=',num2str(NA),'/cm3']);
text(0.8*xmin,.22,['xo=',num2str(xo),'cm']);
xlabel('VG (volts)'); ylabel('C/CO'); grid
```

**CVintgrd**

```
function [y] = cvintegrand(U)
global UF
F=sqrt(exp(UF).*(exp(-U)+U-1)+exp(-UF).*(exp(U)-U-1));
y=(1-exp(-U)).*(exp(U)-U-1)./(2*F.^3);
```

# INDEX